SALAMANDERS
OF THE
UNITED STATES AND CANADA

SALAMANDERS
OF THE
UNITED STATES AND CANADA

James W. Petranka

QL
668
C2
P36
1998

Smithsonian Institution Press
Washington and London

© 1998 by the Smithsonian Institution
All rights reserved

Copy editor: Peter Strupp/Princeton Editorial Associates
Production editor: Ruth Spiegel
Designer: Linda McKnight

Library of Congress Cataloging-in-Publication Data
Petranka, James W.
    Salamanders of the United States and Canada / James W. Petranka.
        p. cm.
    Includes bibliographical references (p.) and index.
    ISBN 1-56098-828-2 (alk. paper)
    1. Salamanders—United States. 2. Salamanders—Canada. 3. Salamanders—United States—Identification. 4. Salamanders—Canada—Identification. I. Title.
    QL668.C2P36 1998
    597.8'5'0973—dc21   97-38094

British Library Cataloguing-in-Publication Data available

Color plates were printed in Singapore, not at government expense
Manufactured in the United States of America
05 04 03 02 01 00 99 98 5 4 3 2 1

♻ The recycled paper used in this publication meets the minimum requirements of the American National Standard for Information Sciences—Permanence of Paper for Printed Library Materials ANSI Z39.48-1984.

For permission to reproduce illustrations appearing in this book, please correspond directly with the owners of the works, as identified in the captions and in the list, at the back of the book, of collection localities and photographic credits for the color plates. The Smithsonian Institution Press does not retain reproduction rights for these illustrations individually or maintain a file of addresses for photo sources.

To my wife, Becky

Contents

Preface   xi

Introduction   1

Salamander Identification and Plan of the Book   10

Conservation Biology of Amphibians   15

Key to Adult Salamanders of the United States and Canada   19

Key to Larval Salamanders of the United States and Canada   29

Family Ambystomatidae (Mole Salamanders)   35
- *Ambystoma annulatum* (ringed salamander)   37
- *Ambystoma barbouri* (streamside salamander)   40
- *Ambystoma californiense* (California tiger salamander)   47
- *Ambystoma cingulatum* (flatwoods salamander)   50
- *Ambystoma gracile* (northwestern salamander)   53
- *Ambystoma jeffersonianum* (Jefferson salamander)   58
- *Ambystoma laterale* (blue-spotted salamander)   63
- *Ambystoma mabeei* (Mabee's salamander)   68
- *Ambystoma macrodactylum* (long-toed salamander)   70
- *Ambystoma maculatum* (spotted salamander)   76
- *Ambystoma opacum* (marbled salamander)   88
- *Ambystoma talpoideum* (mole salamander)   96
- *Ambystoma texanum* (small-mouthed salamander)   103
- *Ambystoma tigrinum* (tiger salamander)   108
- Unisexual *Ambystoma* Biotypes   122

Family Amphiumidae (Amphiumas)   131
- *Amphiuma means* (two-toed amphiuma)   132
- *Amphiuma pholeter* (one-toed amphiuma)   134
- *Amphiuma tridactylum* (three-toed amphiuma)   136

Family Cryptobranchidae (Hellbender and Giant Salamanders)   139
- *Cryptobranchus alleganiensis* (hellbender)   140

Family Dicamptodontidae (Pacific Giant Salamanders)   145
    Dicamptodon aterrimus (Idaho giant salamander)   146
    Dicamptodon copei (Cope's giant salamander)   147
    Dicamptodon ensatus (California giant salamander)   150
    Dicamptodon tenebrosus (Pacific giant salamander)   152

Family Plethodontidae (Lungless Salamanders)   157
  SUBFAMILY DESMOGNATHINAE   159
    Desmognathus aeneus (seepage salamander)   159
    Desmognathus apalachicolae (Apalachicola dusky salamander)   162
    Desmognathus auriculatus (southern dusky salamander)   164
    Desmognathus brimleyorum (Ouachita dusky salamander)   167
    Desmognathus carolinensis (Carolina mountain dusky salamander)   169
    Desmognathus fuscus (dusky salamander)   173
    Desmognathus imitator (imitator salamander)   182
    Desmognathus marmoratus (shovel-nosed salamander)   184
    Desmognathus monticola (seal salamander)   187
    Desmognathus ochrophaeus (Allegheny mountain dusky salamander)   192
    Desmognathus ocoee (Ocoee salamander)   196
    Desmognathus orestes (Blue Ridge dusky salamander)   202
    Desmognathus quadramaculatus (black-bellied salamander)   206
    Desmognathus welteri (Black Mountain dusky salamander)   211
    Desmognathus wrighti (pygmy salamander)   213
    Phaeognathus hubrichti (Red Hills salamander)   216

  SUBFAMILY PLETHODONTINAE   219
  Tribe Bolitoglossini   219
    Batrachoseps aridus (desert slender salamander)   219
    Batrachoseps attenuatus (California slender salamander)   220
    Batrachoseps campi (Inyo Mountains salamander)   224
    Batrachoseps gabrieli (San Gabriel Mountain salamander)   225
    Batrachoseps nigriventris (black-bellied slender salamander)   226
    Batrachoseps pacificus (Pacific slender salamander)   228
    Batrachoseps simatus (Kern Canyon slender salamander)   231
    Batrachoseps stebbinsi (Tehachapi slender salamander)   232
    Batrachoseps wrighti (Oregon slender salamander)   234
    Hydromantes brunus (limestone salamander)   236
    Hydromantes platycephalus (Mt. Lyell salamander)   237
    Hydromantes shastae (Shasta salamander)   239

  Tribe Hemidactyliini   241
    Eurycea bislineata (two-lined salamander)   241
    Eurycea guttolineata (three-lined salamander)   249
    Eurycea junaluska (Junaluska salamander)   251
    Eurycea longicauda (long-tailed salamander)   254

Eurycea lucifuga (cave salamander)   258
Eurycea multiplicata (many-ribbed salamander)   262
Eurycea nana (San Marcos salamander)   264
Eurycea neotenes (Texas salamander)   266
Eurycea quadridigitata (dwarf salamander)   269
Eurycea rathbuni (Texas blind salamander)   272
Eurycea robusta (Blanco blind salamander)   275
Eurycea sosorum (Barton Springs salamander)   276
Eurycea tridentifera (Comal blind salamander)   277
Eurycea tynerensis (Oklahoma salamander)   278
Gyrinophilus palleucus (Tennessee cave salamander)   280
Gyrinophilus porphyriticus (spring salamander)   282
Gyrinophilus subterraneus (West Virginia spring salamander)   287
Haideotriton wallacei (Georgia blind salamander)   289
Hemidactylium scutatum (four-toed salamander)   290
Pseudotriton montanus (mud salamander)   295
Pseudotriton ruber (red salamander)   299
Stereochilus marginatus (many-lined salamander)   304
Typhlotriton spelaeus (grotto salamander)   307

Tribe Plethodontini   310
Aneides aeneus (green salamander)   310
Aneides ferreus (clouded salamander)   314
Aneides flavipunctatus (black salamander)   318
Aneides hardii (Sacramento Mountain salamander)   320
Aneides lugubris (arboreal salamander)   322
Ensatina eschscholtzii (ensatina)   325
Plethodon aureolus (Tellico salamander)   332
Plethodon caddoensis (Caddo Mountain salamander)   333
Plethodon cinereus (red-backed salamander)   335
Plethodon dorsalis (zigzag salamander)   346
Plethodon dunni (Dunn's salamander)   349
Plethodon elongatus (Del Norte salamander)   352
Plethodon glutinosus (slimy salamander)   354
Plethodon hoffmani (Valley and Ridge salamander)   361
Plethodon hubrichti (Peaks of Otter salamander)   363
Plethodon idahoensis (Coeur d'Alene salamander)   365
Plethodon jordani (Jordan's salamander)   367
Plethodon kentucki (Cumberland Plateau salamander)   374
Plethodon larselli (Larch Mountain salamander)   377
Plethodon neomexicanus (Jemez Mountains salamander)   380
Plethodon nettingi (Cheat Mountain salamander)   381
Plethodon oconaluftee (southern Appalachian salamander)   383
Plethodon ouachitae (Rich Mountain salamander)   386
Plethodon petraeus (Pigeon Mountain salamander)   389

Plethodon punctatus (white-spotted salamander)   390
Plethodon richmondi (ravine salamander)   392
Plethodon serratus (southern red-backed salamander)   395
Plethodon shenandoah (Shenandoah salamander)   397
Plethodon stormi (Siskiyou Mountains salamander)   399
Plethodon vandykei (Van Dyke's salamander)   401
Plethodon vehiculum (western red-backed salamander)   403
Plethodon websteri (southern zigzag salamander)   407
Plethodon wehrlei (Wehrle's salamander)   409
Plethodon welleri (Weller's salamander)   412
Plethodon yonahlossee (Yonahlossee salamander)   414

Family Proteidae (Waterdogs and Mudpuppy)   417
Necturus alabamensis (Alabama waterdog)   418
Necturus beyeri (Gulf Coast waterdog)   419
Necturus lewisi (Neuse River waterdog)   422
Necturus maculosus (mudpuppy)   425
Necturus punctatus (dwarf waterdog)   429

Family Rhyacotritonidae (Torrent Salamanders)   433
Rhyacotriton cascadae (Cascade torrent salamander)   434
Rhyacotriton kezeri (Columbia torrent salamander)   437
Rhyacotriton olympicus (Olympic torrent salamander)   439
Rhyacotriton variegatus (southern torrent salamander)   441

Family Salamandridae (Newts)   445
Notophthalmus meridionalis (black-spotted newt)   446
Notophthalmus perstriatus (striped newt)   448
Notophthalmus viridescens (eastern newt)   451
Taricha granulosa (rough-skinned newt)   462
Taricha rivularis (red-bellied newt)   469
Taricha torosa (California newt)   473

Family Sirenidae (Sirens)   479
Pseudobranchus axanthus (southern dwarf siren)   480
Pseudobranchus striatus (northern dwarf siren)   482
Siren intermedia (lesser siren)   484
Siren lacertina (greater siren)   489

Glossary   493

Literature Cited   499

Collection Localities and Photographic Credits for Color Plates   579

Taxonomic Index   583

Preface

In 1943 Sherman C. Bishop published his classic treatise, *A Handbook of Salamanders,* which served for many years as the primary reference source on North American salamanders. From 1943 to 1997 researchers published more than 1500 refereed scientific papers on the systematics, ecology, and natural history of North American salamanders. In addition, scientists described more than 30 new species of salamanders from the United States and Canada. Although several state and regional works with accounts of salamander life histories appeared during this period, they did not fulfill the need for a comprehensive summary of the published literature on North American salamanders.

The primary purpose of this work is to provide a reference source for basic researchers, science teachers, naturalists, conservation biologists, foresters, environmental planners, and others who need detailed information on North America's diverse salamander fauna. The focus is on ecology, evolution, biodiversity, behavior, and natural history. For this reason I have excluded most of the voluminous literature on salamander physiology, development, morphology, genetics, anatomy, and infectious disease agents, including parasites. Because of limitations on book length, I have omitted synonymies and highly technical descriptions of species. Instead, I describe key external characters that are useful in determining the sex and species identity of live or freshly preserved specimens.

Rather than present stereotypic descriptions of species and life histories, as is common in many state and regional accounts, I provide summaries of major patterns of geographic variation within species, as well as the general localities where studies were conducted. This approach emphasizes intraspecific differences between local and regional populations and provides a more realistic view of the life history diversity that occurs within each species. In addition, it provides information that can be applied with greater confidence to specific geographic regions that are of interest to resource managers, conservation biologists, and researchers.

One of the most challenging aspects of this work was deciding how many species to recognize. Much of the disagreement among taxonomists regarding current nomenclature applied to salamanders reflects the fact that scientists do not agree on what constitutes a species. Traditionally, the biological species concept has been the primary model for recognizing vertebrate species. This concept uses reproductive isolation as the primary criterion for recognizing species. It is built upon the premise that members of the same species can (or potentially can) freely interbreed with each other in nature, whereas members of different species cannot. The biological species concept can be easily tested when two groups coexist together locally, and in most instances species can be defined based on whether individuals freely interbreed. In some cases this simple test cannot be applied and scientists must use less direct evidence. Such instances include situations in which species reproduce asexually, populations are allopatric (geographically isolated), genetic divergence of groups is not accompanied by morphological divergence, and populations freely interbreed in certain areas of their range, but not in others.

The biological species concept has been challenged by some researchers because of operational problems as well as on philosophical grounds regarding the goals of systematics and taxonomy. Although many alternative definitions of species have been proposed under the general headings of "evolutionary" and "phylogenetic" species, all pose potential problems in terms of their practical use (for literature most pertinent to salamanders, see Cole 1990; Echelle 1990; Frost and Hillis 1990; Highton 1990, 1995; Larson and Chippindale 1993). The biological species concept recognizes polytypic species consisting of two or more subspecies, but many evolutionary and phylogenetic species concepts would treat currently recognized subspecies as separate species.

Evolutionary species concepts are gaining in popularity. Although several versions have been proposed, these generally view a species as a group of populations or an evolutionary lineage that is evolving along a separate evolutionary pathway from other such groups. Thus, evolutionary species have been characterized as having their own "unitary roles and tendencies" (Simpson 1961) or having their own "evolutionary tendencies and historical fates" (Wiley 1978). Specialists using evolutionary species concepts often attempt to delineate the largest evolutionary units that appear to be on separate phylogenetic trajectories. Phylogenetic species concepts also seek to recognize lineages that are on separate evolutionary trajectories, but these differ in trying to delineate the smallest (rather than the largest) evolving units that appear to be on separate evolutionary pathways. Thus, researchers employing phylogenetic species concepts tend to be extreme "splitters." A group consisting of many weakly differentiated allopatric subgroups might be treated by a researcher using the biological species concept as a single polytypic species consisting of several subspecies. In contrast, a researcher using a phylogenetic species concept might consider each allopatric form to be a separate species, whereas a researcher using an evolutionary species concept might recognize an intermediate number of species.

Evolutionary and phylogenetic species concepts have been criticized because they lack objective criteria for determining whether groups are on different evolutionary trajectories—except, of course, for species that coexist locally without interbreeding. Given the large number of genes and known mutation rates for vertebrates, one could argue from a purely statistical standpoint that most allopatric groups within a species' range contain unique mutations and are following different evolutionary trajectories of sorts. Although recognizing each allopatric form as a separate species would be impractical for highly fragmented species, scientists who employ evolutionary and phylogenetic species concepts must ultimately make subjective decisions on what constitutes a "different" evolutionary trajectory or separately evolving lineage. Unfortunately, there is little agreement among experts about where to draw species boundaries using this or other species concepts. A recent move to raise allopatric subspecies of North American salamanders and other vertebrates to the full species level by evoking phylogenetic species arguments (Collins 1991) was met with much criticism and debate (Collins 1992; Dowling 1993; Frost et al. 1992; Montanucci 1992; Van Devender et al. 1992). This exchange illustrates the divergent viewpoints on taxonomic philosophies that abound in today's herpetological community.

The problems associated with defining species reflect the fact that our current hierarchical classification system uses discrete taxonomic categories to stereotype the complex genetic and evolutionary patterns that exist in nature. Salamanders have low dispersal rates relative to other vertebrates and are often constrained by physiology and design to cool, moist microhabitats. Biogeographic and molecular data indicate that many species have undergone one or more cycles of geographic expansion, followed by contraction and fragmentation into isolated populations or regional groups (e.g., Highton 1995). Studies of protein variants, mitochondrial DNA, and other molecular information indicate that regional populations are often genetically differentiated from one another. In many cases, however, genetic differentiation between subgroups is not paralleled by conspicuous differences in external morphology, coloration, color patterning, or behavior (Highton 1995; Larson 1984; Wake 1993).

Many salamander species consist of geographic subgroups that evolutionary biologists employing the biological species concept have traditionally treated as semispecies. Such species contain geographic groups that have evolved beyond the level of geographic races or subspecies but have not yet reached the level of full biological species characterized by complete reproductive closure. Where these forms come into geographic contact, they may hybridize freely or nearly so, but the hybrid zones are often narrower than those seen in other vertebrate groups.

Semispecies complexes with narrow hybrid zones are the source of many of the current controversies in salamander taxonomy, and none of the existing species concepts provides a rigorous set of criteria that will result in consistent interpretations by all members of the scientific community. Some researchers have elected to split semispecies complexes into separate species. The most common rationale is that these are well-differentiated genetic groups with narrow hybrid zones. Narrow hybrid zones have often been used as evidence of genetic incompatibility between forms, and to argue that the parental groups that form hybrid zones are on separate evolutionary trajectories. However, the presence of narrow hybrid zones does not necessarily reflect genetic incompatibility. The width of hybrid zones tends to be positively correlated with the innate dispersal abilities of different vertebrate groups. Salamanders have very low dispersal rates relative to almost all other vertebrates, and narrow hybrid zones of only a few kilometers could be produced by a variety of mechanisms regardless of whether the hybrids have lower, equal, or higher fitness than the homozygous parental types from which they are derived (e.g., Hewitt 1989). In addition, hybrid zones often act as semipermeable genetic filters—genes that exhibit negative heterosis (or genes that are epistatically linked to these) may form the hybrid zone, whereas genes that are advantageous may pass readily through the hybrid zone and be incorporated into opposing genomes. Researchers all too often have documented hybrid zones in salamanders without delineating the mechanisms that have produced them. This information is often critical for determining levels of gene flow and for making taxonomic decisions about members of semispecies complexes.

Some scientists have used molecular data to justify naming regionally differentiated groups as new species—if, in their opinion, genetic divergence (which is indirectly estimated from molecular similarity) is at or above levels that typically occur between closely related species. Unfortunately, molecular similarity is not always a reliable predictor of the degree of reproductive isolation between groups. For that reason, much debate has surfaced as to whether regional groups should be recognized as separate species when molecular similarity is the only criterion used to make such assessments. Scientists have thus far been unable to agree on the validity of this approach. Given the same set of data, certain scientists would split regionally differentiated groups into separate species, whereas others would treat them as members of a single, geographically variable species.

Many changes in North American salamander taxonomy during the last two decades have been controversial. Fewer than 110 to as many as 150 or so species could be recognized in this treatment depending on how a particular scientist interprets taxonomically difficult groups. Here I recognize 127 species (excluding unisexual *Ambystoma*) and rely primarily on the biological species concept for species recognition.

NOMENCLATURE

At this juncture in the evolution of salamander nomenclature, I feel that it is more important to seek nomenclatural stability in the published literature and in field guides than to radically change the existing nomenclature for the sake of being a philosophical purist. In a work of this breadth, it is also important to be as consistent as possible in the treatment of all species. Thus, I have used the following guidelines for interpreting taxonomically difficult groups:

1. In cases in which closely related groups occur allopatrically and show moderate to high levels of genetic divergence, I treat each as a separate species. Examples include *Plethodon idahoensis, Dicamptodon aterrimus,* and *Rhyacotriton olympicus.*
2. In cases in which groups appear to interbreed freely in all areas of geographic contact, I treat them as being conspecific. Examples of described species that are not recognized in this treatment include *Desmognathus santeetlah* and *Plethodon fourchensis.*
3. In cases in which groups form narrow hybrid zones and there is evidence of restricted gene flow or partial reproductive isolation between forms, I treat each form as a separate species. Examples of these species pairs include *Dicamptodon tenebrosus–D. ensatus* and *D. orestes–D. carolinensis.*
4. In cases in which groups extensively interbreed in many areas of a species' range, but are reproductively isolated in others, I treat the complex as a single species. An example is *Ensatina eschscholtzii.* One exception is *Plethodon jordani* and *P. oconaluftee,* for there is evidence that hybridization between these species was triggered by recent anthropogenic disturbance.
5. In cases in which broadly sympatric groups show very limited hybridization, I treat each group as a separate species. Examples of such pairs that occasionally hybridize include *D. ochrophaeus–D. fuscus* and *P. kentucki–P. glutinosus.*
6. In cases involving unisexual *Ambystoma* of hybrid origin, none of the biotypes is formally recognized, and all are referred to by their genomic complements.

7. In cases in which closely related groups that exhibit moderate genetic differentiation occur parapatrically, I do not recognize proposed taxonomic revisions unless zones of contact have been carefully examined and the degree of gene flow between groups has been quantified. In cases in which zones of contact have not been examined in sufficient detail to apply criteria (2) and (3), I retain the older nomenclature and treat these as unresolved species complexes. Examples include the *Eurycea bislineata* complex and most members of the *Plethodon glutinosus* complex.

The common names of certain North American salamanders were unstable prior to 1960. However, the publication of field guides to North American amphibians and reptiles by Conant (1958, 1975) and Stebbins (1966, 1985) largely stabilized the use of English vernacular names applied to North American salamanders. Attempts have been made to formally standardize common name usage for North American salamanders (Collins 1990; Collins et al. 1978, 1982). However, many herpetologists have not fully adopted these names, in part because many of the proposed standard names differ from common names that traditionally have been applied to particular species. Other references for common names, such as the *Catalogue of American Amphibians and Reptiles* and the *Checklist of Vertebrates of the United States, the U.S. Territories, and Canada* (Banks et al. 1987, 1998), are inconsistent with the most recent checklist of standard names (Collins 1990). Instead, these generally follow the traditional names used in the published literature and by Conant (1958, 1975) and Stebbins (1966, 1985). Here, I retain common names that have historically been stable (e.g., red-backed salamander, black-bellied salamander) and that are consistent with the second edition of the *Checklist of Vertebrates of the United States, the U.S. Territories, and Canada* (Banks et al. 1998).

ACKNOWLEDGMENTS

This book is a tribute to the hundreds of scientists whose dedication to field studies produced the scientific literature upon which this synthesis is based. The quality of the work has been greatly enhanced through the generous help of professionals who reviewed species accounts, sent manuscripts in press, and freely shared their scientific expertise. These include Ronald Altig, Carl D. Anthony, Joe Bernardo, Alvin L. Braswell, Edmund D. Brodie, Jr., Richard C. Bruce, Ronald S. Caldwell, Carlos D. Camp, Carrie A. Carreno, Paul T. Chippindale, Paul V. Cupp, Jr., Ellen Dawley, C. Kenneth Dodd, Jr., Carl Ernst, Caitlin R. Gabor, Nelson G. Hairston, Sr., Reid N. Harris, Richard Highton, Joseph R. Holomuzki, Robert G. Jaeger, Thomas R. Jones, Lee B. Kats, Sandra Kilpatrick, Fred Kraus, Roy W. McDiarmid, Brian T. Miller, Paul E. Moler, Thomas K. Pauley, Charles R. Peterson, William H. Redmond, Travis J. Ryan, A. Floyd Scott, Raymond D. Semlitsch, Charles K. Smith, Stephen G. Tilley, Stanley E. Trauth, Robert Tucker, Paul A. Verrell, David B. Wake, Susan C. Walls, and George R. Zug.

I am deeply indebted to two outstanding nature photographers, William P. Leonard and Robert W. Van Devender, who provided most of the photographs used in the book. Special appreciation is also extended to Steven J. Arnold, Joe Bernardo, Edmund D. Brodie, Jr., Paul T. Chippindale, James P. Collins, Joseph T. Collins, Paul V. Cupp, Jr., Mario Garcia-

Paris, Klaus Haker, Robert W. Hansen, Grant Hokit, Joseph R. Holomuzki, Lee B. Kats, Barry Mansell, Wyman Meinzer, Ken Nemuras, Charles R. Peterson, and George R. Zug (R. W. Barbour collection and other photographs) for additional photographic contributions. Special appreciation is also extended to Robert R. Moody, who prepared range maps; to Daphne A. Thomas, who proofed the manuscript and drafted all pen-and-ink illustrations; and to Michael D. Stuart and Rebecca L. Elkin, who provided technical support and unending encouragement.

This book would not have been possible without the professional and courteous help of the staff of the Smithsonian Institution Press. I am particularly grateful to Peter Cannell, Duke Johns, Ruth Spiegel, and their associates for making a very complex process seem simple. My deepest appreciation also goes to Peter Strupp of Princeton Editorial Associates—a master editor who greatly improved every aspect of the book.

Finally, I wish to acknowledge three individuals who encouraged me to pursue natural history studies. While I was an undergraduate at Auburn University, Dr. George W. Folkerts spent an inordinate amount of time with undergraduates in the field; his comprehensive understanding of Alabama's diverse flora and fauna and endless enthusiasm for learning inspired me and many others who accompanied him in the field to pursue graduate studies in biology. His close friend, Dr. Robert H. Mount, had an encyclopedic knowledge of Alabama's herpetofauna. His love of the subject that he taught exceeded that of any laboratory biologist that I have known; he made me and other students understand why field biologists go about their work so passionately. While pursuing a doctoral degree at the University of Kentucky, I was greatly inspired by Dr. Roger W. Barbour, who still exhibited a youthful fascination with natural history after five decades of study. He strengthened my commitment to study natural history in an era when such endeavors were all too often discouraged by other members of the scientific community.

Introduction

Salamanders, frogs, and caecilians make up the class Amphibia. Members of this class are ectotherms ("cold-blooded" vertebrates) that have four limbs, two occipital condyles, no more than one sacral vertebra, and glandular skin that lacks scales, feathers, hairs, or other epidermal structures (Duellman and Trueb 1986). Although amphibians are considered to be tetrapods, many have limbs that have been reduced or lost evolutionarily.

Members of the three extant orders of amphibians can be readily identified by external morphology as well as by a variety of technical anatomical features. Caecilians (order Gymnophiona) are tropical amphibians that are specialized for burrowing. The adults have highly elongated bodies, no legs, ringed grooves around the body, and small eyes. Individuals either lack tails or have greatly reduced tails. Frogs (order Anura) lack tails and have bodies and limbs that are highly modified for jumping. Salamanders (order Caudata) have prominent tails and typically have two pairs of limbs that are about the same size.

The amphibians are an ancient group that arose some 360 million years ago. Many ancient forms reached their peak over 200 million years ago and have long since gone extinct. Today there are about 4600 described species of living amphibians, of which 8.5% are salamanders, 3.5% are caecilians, and 88% are frogs.

BIODIVERSITY

Salamanders are one of the least familiar groups of North American vertebrates because of their secretive nature and nocturnal habits. Of the approximately 400 species of salamanders that occur worldwide, a rich diversity of species occurs in the United States and Canada, including representatives of 9 families, 24 genera, and 127 species. The southeastern United States in many respects has the greatest diversity of salamanders in the world. No other area of comparable size is as taxonomically diverse in terms of the number of families (7), genera (19), and species (75+) of salamanders. Another global region of extraordinary salamander diversity is Mexico and Central America, where members of the tribe Bolitoglossini of the family Plethodontidae are represented by 11 genera and over 150 species. The ambystomatids are also well represented in Mexico.

ECOLOGICAL ROLES AND VALUE TO HUMANS

All salamanders are predators that feed primarily on invertebrates, but their ecological roles in natural communities are poorly understood. Salamanders are abundant in both aquatic and terrestrial habitats. Larval densities in optimal habitats often exceed 2–3 individuals/m^2, and adult densities may exceed 1–2 animals/m^2 of forest floor or streambed. Salamanders provide food for a variety of predators, including fishes, snakes, small mammals, woodland birds, and invertebrates. More important, they appear to play important roles in organizing many terrestrial and aquatic communities. Larvae of mole salamanders (*Ambystoma*) are top predators in vernal pond communities and influence the abundance and diversity of aquatic invertebrates as well as other amphibians. Giant salamanders (*Dicamptodon*) and dusky salamanders (*Desmognathus*) reach very high densities and biomasses in small streams that lack fish, and in many ways perform ecological roles similar to those of fish that occupy larger, permanent stream sections.

Salamanders are the most abundant vertebrates in many forest floor habitats and consume vast quantities of insects and other invertebrates. Studies of the role of woodland salamanders in controlling forest floor insects and influencing leaf litter decomposition and nutrient recycling in forest systems are in their infancy, but early evidence from studies in the eastern United States suggests that salamanders play important ecological roles in deciduous forest systems.

Widely used in scientific research, salamanders have proven to be valuable tools for examining a variety of conceptual and theoretical problems in evolution, ecology, animal behavior, physiology, genetics, and cell biology—to name only a few disciplines. For example, studies of salamanders have provided evolutionary biologists with a clearer understanding of how new species arise and of how genetic structure varies on different spatial scales within species. Ecologists have used salamanders to delineate rules of community assembly, to examine mechanisms of population regulation, and to understand and refine basic concepts of animal behavior. Salamanders have been used in medical research for understanding the basis for limb and tissue regeneration in vertebrates, for studying the inheritance of genetic disorders, for investigating the biochemistry and physiology of vision, and for numerous other purposes that will improve the human condition. Research is under way to explore whether the toxic skin secretions of salamanders have therapeutic value in treating human diseases such as cancer; environmental scientists are increasingly using salamanders as indicators of environmental health.

LIFE HISTORY DIVERSITY

Salamanders have evolved a variety of ecological life-styles and life history patterns that appear to be adaptive. Many species have biphasic life cycles consisting of an aquatic larval stage and a semiaquatic or terrestrial adult stage. Larvae that hatch from eggs may live in streams or ponds for a few weeks to several years before metamorphosing into juveniles. Metamorphosis involves many changes in morphology, the most conspicuous being the resorption of the gills and fins. After reaching sexual maturity, the adults return to aquatic breeding sites and lay their eggs in or near habitats used by the larvae.

Many species deviate from this biphasic life cycle. The terrestrial stage is often absent in species that live in permanent aquatic habitats such as rivers, lakes, and springs, particularly where the terrestrial habitat is unfavorable for survival. Species such as the Georgia blind salamander (*Haideotriton*) and the mudpuppy and waterdogs (*Necturus*) are permanently aquatic forms that mature into gilled adults. At the other extreme, many terrestrial species such as those in the genera *Plethodon, Aneides,* and *Ensatina* lack a larval stage and lay their eggs in decaying logs, subsurface cavities, or other terrestrial microhabitats. The embryos of these forms pass through a gilled stage within the egg capsule and lose the gills at or near the time of hatching. Most North American salamanders conform to one of these three life history patterns. Less common life history patterns, such as the presence of both gilled and transformed adults in populations of some salamanders, are discussed in detail in the individual accounts of species.

GENERAL NATURAL HISTORY

All species of salamanders in the United States and Canada lay eggs, and almost all species have well-defined breeding seasons. Egg laying is preceded by courtship and mating, but the gap between the seasonal time of mating and egg laying varies markedly among taxa. In some groups such as the mole salamanders (*Ambystoma*), females begin laying eggs within a few days after mating. In contrast, females of many lungless salamanders (Plethodontidae) are capable of long-term sperm storage and lay their eggs several months after mating.

Courtship behavior varies both among and within the nine families of North American salamanders, and in many cases certain aspects are rather stereotypic for a given taxon. One element of the courtship of all plethodontid salamanders, for example, is the tail-straddle walk. During this phase of courtship, the female places her chin on the dorsal base of the male's tail and is led forward as she straddles the male's tail (Fig. 1a). The males of many plethodontids develop elongated teeth during the mating season that are pulled or scraped along the female's dorsum (Figs. 1b and 3f). This action abrades the skin, provides tactile stimulation, and may facilitate the transfer of mental gland secretions into the female's circulatory system. Male newts of the genus *Notophthalmus* use their enlarged hindlimbs to amplex females in the region of the front limbs (Fig. 1c). In contrast, males of western newts (*Taricha*) amplex the females with the hindlimbs placed just in front of the female's hindlimbs. Torrent salamanders (*Rhyacotriton*) have a unique tail-waggle display, whereas many dusky salamanders (*Desmognathus*) swing the arms forward in a manner that resembles the butterfly stroke of swimmers.

Courtship behavior in almost all species culminates in the male's depositing a spermatophore on the substrate. Spermatophores vary in size, shape, and composition, but most consist of a gelatinous base that tapers toward the top and supports an apical sperm mass (Fig. 2). The shape of the spermatophore varies among species and may facilitate the transfer of sperm into the female's cloaca. In most instances a courting male deposits one or more spermatophores in front of a female. The female then moves forward (often while orienting to chemical secretions from glands of the male), aligns her vent above the spermatophore, and removes all or a portion of the sperm mass from the top. The sperm is stored in a special chamber (spermatheca) that leads off the cloaca, and the ova are fertilized internally shortly before they

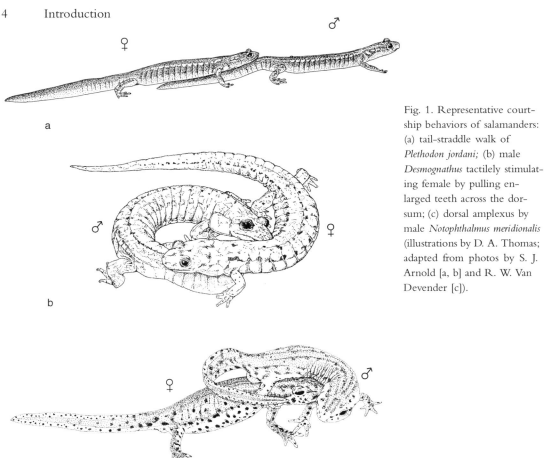

Fig. 1. Representative courtship behaviors of salamanders: (a) tail-straddle walk of *Plethodon jordani*; (b) male *Desmognathus* tactilely stimulating female by pulling enlarged teeth across the dorsum; (c) dorsal amplexus by male *Notophthalmus meridionalis* (illustrations by D. A. Thomas; adapted from photos by S. J. Arnold [a, b] and R. W. Van Devender [c]).

pass out of the cloaca. A few species deviate from this mode of reproduction. The hellbender (*Cryptobranchus*) has external fertilization and the male deposits milty secretions directly on the eggs as they are laid. Sirenids (*Siren, Pseudobranchus*) are also assumed to have external fertilization because females lack spermathecae and males do not produce spermatophores.

Male salamanders can often be distinguished from females by the presence of secondary sexual characteristics, such as hedonic glands, enlarged teeth, enlarged cloacal glands, nasal cirri, and pronounced tail fins. Most of these features develop seasonally in association with the onset of the breeding season. Males of most species have papillae on the walls and/or posterior margin of the cloaca, whereas females have pleated or folded cloacal walls (Fig. 3a,b). Male torrent salamanders have conspicuous vent lobes throughout the year (Fig. 3c), whereas sexually active male newts have cornified structures on the lower hindlimbs and/or digits that function as friction pads to facilitate amplexing females (Fig. 3d).

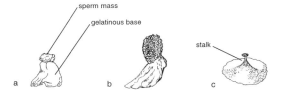

Fig. 2. Examples of spermatophores of salamanders: (a) *Ambystoma maculatum*; (b) *Eurycea bislineata*; (c) *Notophthalmus viridescens* (illustrations by D. A. Thomas, adapted from Bishop 1941a).

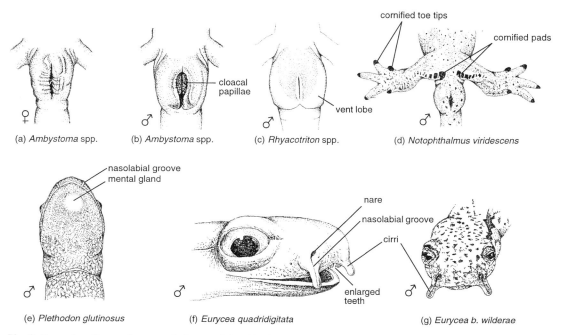

Fig. 3. Examples of secondary sexual characteristics: (a) pleated vent of female mole salamander (*Ambystoma*); (b) papillose vent of male mole salamander (*Ambystoma*); (c) vent lobe of male torrent salamander (*Rhyacotriton*) (adapted from photo by W. P. Leonard); (d) swollen vent and cornified limbs of male *Notophthalmus viridescens*; (e) mental gland of *Plethodon glutinosus* (adapted from photo by R. W. Van Devender); (f) enlarged teeth and nasal cirri of *Eurycea quadridigitata* (adapted from Dunn 1926); (g) nasal cirri of *Eurycea bislineata wilderae* (all illustrations by D. A. Thomas).

Many elements of salamander courtship are mediated by chemical signals that originate from special glands located on the skin or in the cloacal region. Many male plethodontids, for example, have rounded or oblong mental glands on or just posterior to the chin that are rubbed on the skin of the female during courtship (Fig. 3e). As noted earlier, the males of some species also develop elongated teeth that are used in courtship (Fig. 3f). Some male plethodontids develop elongated nasal cirri (Fig. 3f,g). These may be tapped on the substrate or placed on a female's dorsum, and they facilitate the transfer of water-borne chemicals to chemoreceptors in the nasal passageways.

Male newts have chemically laden genial glands along the sides of the head that are rubbed on the female's snout. In addition, glands in the cloacal region of male newts secrete chemicals that mediate courtship behavior. Males of certain newts waft these toward the female's nostrils by periodically fluttering their tails. Male mole salamanders (*Ambystoma*) have conspicuously swollen cloacae with glands that produce the spermatophores. Glandular secretions from the cloacae of both males and females may play a role in mediating species and sex recognition in these and many other salamanders.

Males abandon the females shortly after courting and, with rare exceptions, do not care for the young. Females have evolved a variety of reproductive strategies that correlate with the environments they inhabit. Depending on the species, females may lay their eggs directly in water or in moist microhabitats on land. The eggs are surrounded by a vitelline membrane

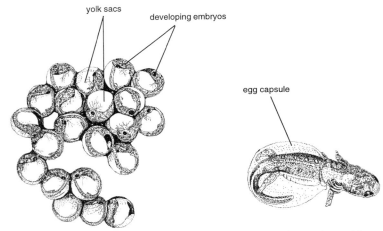

Fig. 4. Eggs and egg capsules of *Desmognathus*: (a) late-term embryos of *D. monticola* with conspicuous yolk reserves; (b) hatchling *D. quadramaculatus* emerging from the egg capsule (illustrations by D. A. Thomas).

and one or more jelly layers and/or associated membranes (Fig. 4). The jelly layers and membranes protect the eggs from desiccation, provide a defensive barrier against predators, and may function to protect the eggs and developing embryos from pathogens.

Pond-breeding species lay eggs that are small relative to those of stream-breeding species of similar size and females abandon their eggs immediately after ovipositing. Depending on the species, the eggs may be laid singly or in masses that contain aggregates of a few to more than 100 eggs. Salamanders that breed in running water often attach their eggs in monolayers or flattened masses to the undersides of rocks or logs, and the females often brood their eggs through hatching. Many species of dusky salamanders (*Desmognathus*) use semiaquatic microhabitats for nesting, and the females lay grapelike clusters of eggs in moss clumps, leaf packs, or other moist microhabitats. The larvae move into nearby streams or seepages after hatching.

Terrestrial breeders such as *Ensatina, Batrachoseps, Plethodon,* and *Aneides* deposit eggs in cryptic sites such as underground cavities, decaying logs, and moist rock crevices. The eggs are exceptionally large relative to female body size, and the females of most species brood their eggs through hatching. Brooding serves many purposes, the most important of which appears to be protection from predators.

Larval salamanders have morphological adaptations that correlate with the environments that the larvae inhabit (Noble 1927b). One dichotomy is that between species that typically breed in running versus standing water habitats. Larvae of species that characteristically breed in standing water ("pond-type" larvae) have long, bushy gills, dorsal fins that extend well onto the back, and nonfunctional limbs at hatching (Fig. 5a). Many pond-type hatchlings also have balancers, which are fleshy props that grow from the sides of the head. These are lost within a few days to several weeks after hatching as the limbs and digits develop. Larvae of species that characteristically breed in running water ("stream-type" larvae) typically have short, reduced gills, dorsal fins that terminate near the rear limbs, and fully formed, functional limbs at hatching (Fig. 5b,c). Certain stream forms such as *Necturus* deviate in having conspicuous bushy gills, whereas other forms such as *Dicamptodon* (Fig. 5d) have highly branched gills with

Fig. 5. Examples of larval morphology: (a) pond-form larva of *Ambystoma jeffersonianum*; (b) stream-form larva of *Eurycea multiplicata* (adapted from photo by R. W. Van Devender); (c) antlerlike gills of *Desmognathus quadramaculatus*; (d) bushy gills of *Dicamptodon tenebrosus* (adapted from photo by R. W. Van Devender) (illustrations by D. A. Thomas).

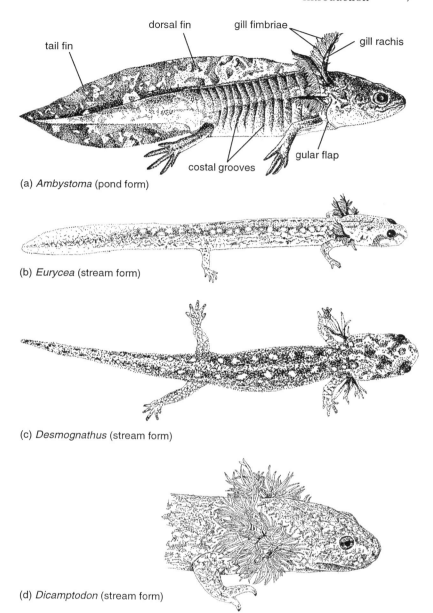

(a) *Ambystoma* (pond form)

(b) *Eurycea* (stream form)

(c) *Desmognathus* (stream form)

(d) *Dicamptodon* (stream form)

short gill rachises. Therefore the presence or absence of a dorsal fin along the back is the single most reliable trait for recognizing pond- versus stream-type larvae. Larvae that inhabit fast-flowing water often have cornified friction pads on the toe tips.

Valentine and Dennis (1964) recognized a third group of aquatic larvae ("mountain brook" larvae) that is perhaps better viewed as an extreme form of the stream-type morphology. Relative to typical stream breeders, this form shows even greater reduction in the gill structures and has a tail fin that terminates posterior to the cloaca. Here, I distinguish only between stream-type and pond-type larvae in the species accounts.

Larval salamanders are predators that feed primarily on aquatic invertebrates. Some of the larger species may cannibalize or feed on heterospecific larvae. Growth rates and length of the larval period vary markedly among species. Species that utilize seasonally ephemeral habitats have rapid growth rates and larval periods that often last only 2–6 months. Vernal ponds and other temporary habitats are productive sites that provide opportunities for rapid growth. However, they also are risky habitats, and the larvae of pond-breeders generally have higher mortality rates than the larvae of stream-breeding species. Species that use seasonally ephemeral habitats produce large numbers of small eggs. Selection pressures in these habitats favor rapid growth and development, and one of the associated costs is high mortality rates.

Stream-breeders tend to have inherently slower growth rates and correspondingly lower mortality rates than pond-breeders. Stream-breeding larvae often remain in hiding during the day and forage outside cover at night when visually oriented predators such as fish are less active. Stream-breeders generally have larger eggs, smaller clutch sizes, slower larval growth rates, and lower mortality rates than pond-breeders. From an energetic standpoint, stream larvae have a slower-paced life-style that invests less in rapid growth and more in predator avoidance than do larvae of pond-breeders. Length of the larval period varies markedly among stream-breeders with biphasic life cycles, from 1–2 months to several years.

The larvae of some salamanders are aggressive toward both con- and heterospecific larvae. Agonistic behaviors range from aggressive and submissive posturing to biting. The larvae of certain taxa such as *Ambystoma* and *Dicamptodon* are very aggressive and may cannibalize or prey upon heterospecifics. Cannibalistic morphs that have specialized morphology such as enlarged teeth and wider heads occur in two *Ambystoma* species. The larvae of some salamanders exhibit kin recognition and are generally less aggressive toward close relatives than distantly related individuals.

Juveniles of most pond-breeders disperse away from breeding sites following metamorphosis and live on land or in semiaquatic habitats. The juvenile stage is an important dispersal stage of the life cycle, but surprisingly little data are available on juvenile site fidelity or dispersal rates. Individuals frequently return to their natal ponds. However, newly constructed ponds are often quickly colonized by salamanders, indicating that many individuals are not philopatric to their natal pond.

The juveniles of most pond-breeders live in underground burrows and are only occasionally encountered on the ground surface. Juvenile newts are an exception and are often active on the ground surface during the day. After reaching sexual maturity, pond-breeders begin making seasonal migrations to and from breeding sites. Most species breed in late winter through early summer, and the embryos hatch as ponds begin their annual warm-up in the spring or early summer. Annual breeding patterns vary both within and among species. Males typically breed annually, but females may skip a year or two before returning to the ponds to breed. Females at low elevations also tend to breed more frequently than those at high elevations.

Most terrestrial salamanders are active at night on the ground surface, where they forage for invertebrates, defend territories, and seek mates. The majority of North American salamanders belong to the family Plethodontidae, commonly known as the lungless salamanders. Adult plethodontids lack both gills and lungs except for a few permanently gilled, aquatic species. Because their skin must be kept moist to facilitate gas exchange, plethodontids that live on the forest floor generally restrict their activity to moist microhabitats and are active on the ground surface at night when relative humidities are high. Individuals quickly dehydrate if microhabitats become dry, and they normally respond to dry conditions by moving underground or beneath moist, decaying logs on the forest floor. Mole salamanders (family Ambystomatidae) largely solve the dehydration problem by burrowing in the soil and feeding from burrow entrances. The nocturnal activity of most salamanders also functions to minimize predation risk from diurnally active species such as robins, thrushes, jays, and garter snakes.

Many plethodontids exhibit agonistic behaviors and appear to be territorial. Adults establish small seasonal home ranges on the forest floor and defend these from conspecifics. Detailed studies of the red-backed salamander (*Plethodon cinereus*) show that adults mark territories with fecal pellets and body secretions, and actively defend territories from adults of the same sex that trespass. Contests for territories may involve defensive posturing or escalate into chasing and biting that may occasionally result in significant injuries such as loss of part of the tail.

Larval and adult salamanders have evolved an array of defenses against predators. Larval salamanders tend to be most active at night and often remain in leaf litter or under cover objects during the day when diurnally active predators such as birds and fishes are most active. The larvae of some species respond to chemicals emitted by predators such as fishes by moving beneath cover, whereas those of other species such as newts produce toxic skin secretions. Fleeing is the most widespread antipredator behavior in larval salamanders.

Transformed animals employ a variety of defenses against predators. The juveniles and adults of most terrestrial species produce toxic or noxious skin secretions that are released from the skin during an attack, and certain species have bright coloration that functions to warn predators. Individuals that are exposed to an attacker may remain immobile, flee, or posture defensively. When posturing defensively, individuals typically hide the head or orient the head away from the predator, position the tail or other structures bearing noxious skin secretions toward the attacker, and maximally expose warning coloration. Some species will bite an attacker or smear it with gluey tail secretions. If a predator grasps the tail and escape is unlikely, many salamanders will autotomize their tails and flee.

Details of the general life history patterns and behaviors discussed previously are supplied in the accounts of individual species. Major reviews that provide more detailed summaries are included in the following publications: general life history (Duellman and Trueb 1986; Stebbins and Cohen 1995), courtship behavior and reproductive isolation (Arnold 1977; Arnold et al. 1993; Houck 1986; Houck and Verrell 1993; Salthe 1967; Salthe and Mecham 1974), antipredator defenses (Brodie 1977), modes of reproduction and life history evolution (Tilley and Bernardo 1993; Bruce 1996; Salthe 1969; Tilley 1977), social behavior and territoriality (Jaeger 1986; Jaeger and Forester 1993; Mathis et al. 1995), and population and community ecology (Hairston 1987).

Salamander Identification and Plan of the Book

Readers who are just embarking on learning to identify salamanders should take time to become familiar with the terminology and methodologies described in this section. Many salamander species are easy to identify because of their unique color patterning, unusual shape, large size, or unique morphological features such as the absence of limbs or toes. However, certain taxa, such as dusky salamanders (*Desmognathus*) or the unisexual mole salamanders (*Ambystoma*), are challenging to even professional herpetologists. In some cases these forms require genetic analyses for proper identification.

IDENTIFICATION, MEASUREMENT, AND MORPHOLOGY

If you are just getting started, focus on learning the families of salamanders and the major genera. With a little practice, most adult specimens can be readily assigned to the genus *Ambystoma*, *Aneides*, *Batrachoseps*, *Desmognathus*, *Dicamptodon*, *Eurycea*, *Necturus*, *Plethodon*, or *Rhyacotriton*. Once a specimen is narrowed to a particular genus, it can be identified to species using information on body size, habitat, collection locality, behavior, and an array of technical features. If in doubt, consult a qualified herpetologist or naturalist to verify identifications.

Technical features that are important in identifying larvae include the number of costal grooves, the size and shape of the head, fin morphology and coloration, gill morphology, the presence of bold lines or other patterning on the body, and the number and shape of digits on the limbs. Adult characteristics of importance include adult size, number of costal grooves, general body shape, number of limbs and digits, body coloration and patterning, tooth patterning, and the presence of unique structures such as nasolabial grooves or vent lobes.

The larvae and adults of most salamanders have costal grooves along the sides of the body that mark the position of the ribs (Fig. 6). Herpetologists use the number of costal grooves between the axilla and groin as a standard measure of costal groove number. When counting costal grooves, count the number of grooves that occur between the back margin of the front limb and the front margin of the rear limb, then add two to this number. Two is added to the count to include a groove for each limb. In some specimens, costal grooves near the limbs may be forked. If so, count each branch of the fork as a separate groove. The number of costal grooves between adpressed limbs is also useful in identifying many plethodontid salamanders. To determine this, press the limbs laterally along the body and count the number of costal grooves between the tips of the longest digit on each limb (Fig. 6b).

The two standard measurements of length used for larvae and adults are total length and

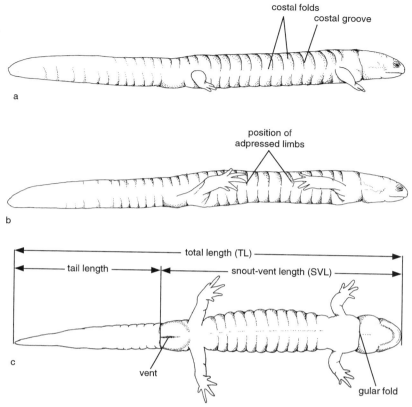

Fig. 6. Profiles of *Ambystoma* showing position of adpressed limbs and reference points for measuring length (illustrations by D. A. Thomas).

snout–vent length (Fig. 6c). Total length (TL) is the distance from the tip of the tail to the tip of the snout of a straightened animal. Snout–vent length (SVL) is the distance from the tip of the snout to the posterior margin of the vent.

The number of limbs and the number, relative length, and shape of the digits are useful in identifying both larval and adult salamanders. Adult salamanders and mature salamander larvae typically have four digits on each front limb and five digits on each rear limb. Deviations from this formula occur in certain taxa and are useful in identifying species or genera. The presence or absence of webbing between the toes and the occurrence of squared versus rounded digits are also helpful in identifying plethodontid salamanders (Fig. 7).

For preserved specimens, tooth patterning is useful for identifying specimens to family, and, in many cases, to genus or species. The teeth of adults are peglike, typically have two cusps, and are found on the bones of the upper and lower jaws as well as the roof of the mouth (Fig. 8). Monocuspid teeth are found in salamander larvae, in the males of certain sexually active plethodontids, and in permanently gilled forms. Almost all North American salamanders can be readily identified using external features. Therefore tooth characteristics are rarely used in identification except when attempting to identify poorly preserved specimens.

The identification of salamander larvae is more challenging and often requires an investigative approach that incorporates information on natural history and distribution, in addition to observations of technical features. Salamander larvae often undergo marked ontogenetic

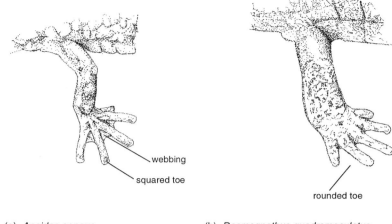

Fig. 7. Limb profiles of salamanders: (a) toe webbing and squared toe tips of *Aneides aeneus*; (b) rounded toe tip of *Desmognathus quadramaculatus* (illustrations by D. A. Thomas).

(a) *Aneides aeneus*   (b) *Desmognathus quadramaculatus*

changes in body coloration and toe and limb development, and identification is most reliable using older specimens. Conspicuous features that are useful in identifying larvae include banding and pigmentation patterns on the dorsum; the number of costal grooves; the size of the larvae; the presence or absence of dark pigmentation on the undersides; the general shape of the body; the size, shape, and coloration of the gills; and the presence of stream-type or pond-type morphology.

Each gill contains a gill arch with many gill rakers, and a fleshy gill rachis that supports numerous finely divided fimbriae where gas exchange occurs (Fig. 9). Pond-breeders have large, bushy gills with elongated rachises and numerous fimbriae (Fig. 5a). The rachises of stream-breeders may be greatly reduced (e.g., *Dicamptodon*) or absent (e.g., *Desmognathus*), and the number of fimbriae is often much lower relative to pond-breeders (Fig. 5c). *Necturus* is an exception and has conspicuous, bushy gills with long rachises and numerous fimbriae.

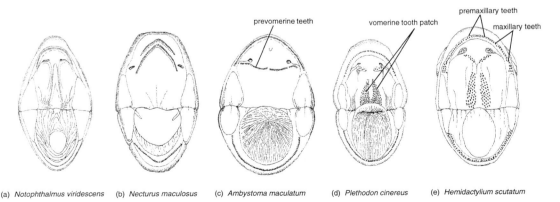

Fig. 8. Representative tooth patterns of salamanders: (a) *Notophthalmus viridescens*; (b) *Necturus maculosus*; (c) *Ambystoma maculatum*; (d) *Plethodon cinereus*; (e) *Hemidactylium scutatum* (illustrations by D. A. Thomas, adapted from Bishop 1941a).

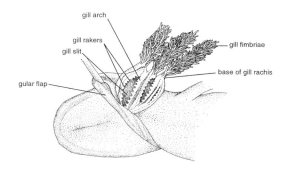

Fig. 9. Gills and associated structures of *Ambystoma tigrinum* (illustration by D. A. Thomas).

IMPORTANT INFORMATION FOR READERS

Information for the text comes primarily from over 2100 papers that have appeared in refereed scientific journals. This work includes a comprehensive survey of literature published through 1 January 1997. In addition, critical taxonomic works that appeared between 1 January and 1 September 1997 are included. Species accounts are presented alphabetically by family, tribe (Plethodontidae), genus, and species to facilitate use. A phylogenetic arrangement is not used because the phylogenetic relationships of many groups are poorly understood and controversial.

Range maps are based on information from the most current field guides (Conant and Collins 1991; Stebbins 1985), from recent state works, and from newly published range extensions. Filled patterns denote areas where populations occur more or less continuously across the landscape where suitable habitats occur, whereas an X denotes geographic isolates consisting of a single published record.

Data are expressed in metric units, and numeric values are rounded to the nearest whole number or decimal fraction that is biologically or statistically relevant. Measurements of variation are reported as ranges rather than standard deviations or standard errors. In a few instances in which values are not reported directly in the text of original papers, values are estimated from figures, as noted in the text.

The following abbreviations are used throughout the text: SVL (snout-vent length), TL (total length), mm (millimeters), cm (centimeters), m (meters), km (kilometers), g (grams), and kg (kilograms). Almost all researchers today measure SVL from the tip of the snout to the posterior margin of the vent, but in past decades some researchers measured SVL to the anterior margin of the vent (particularly for *Plethodon* spp.). All too often the older papers do not specify which margin of the vent was used to measure SVL. Here, all measurements are reported simply as SVL, with the realization that most are based on measurements to the posterior margin of the vent.

A confusing terminology has evolved over the years concerning sexually mature salamanders that retain larval traits such as gills or fins. Terms such as *neoteny, paedogenesis, paedomorphosis,* and *progenesis* have been used by various authors, but usage has been inconsistent and in many cases nonspecific and vague (Pierce and Smith 1979). Paedomorphosis is widely used in the literature today, but this is a phylogenetic term that has been misused. A derived larval form that is evolving away from an ancestral larval state and toward the adult condition would probably be described by ecologists as being paedomorphic, but many evolutionary biologists would deem this usage inappropriate. To avoid further confusion, all larviform adults are referred to as "gilled adults" or "perennibranchs."

Many species of North American salamanders have very restricted ranges or inhabit inaccessible habitats such as caves. Very little published literature on these forms is available compared with that on more widely distributed, surface-dwelling species. Because of the disparity in the amount of information about common and rare forms, two formats are used in the species accounts. For most species a full account is presented that includes the following sections: identification, systematics and geographic variation, distribution and adult habitat, breeding and courtship, reproductive strategy, aquatic ecology (if appropriate), terrestrial ecology (if appropriate), predators and defense, community ecology, conservation biology, and comments (optional). For the remaining species an abbreviated account is presented that includes sections on identification, systematics and geographic variation, and distribution. Depending on the data available, one or more of the additional sections listed previously may be included, or there may simply be a "comments" section that summarizes all that is known about the species.

Species accounts include a range map and one or more illustrations. The species are also illustrated in color in the plates section.

Conservation Biology of Amphibians

Salamanders and frogs are elements of a larger story involving a global decline in biodiversity associated with uncontrolled human population growth. The health of every major biome on Earth has deteriorated during the last two centuries in parallel with the rapid expansion of the human population. In particular, many North American amphibians have declined owing to environmental alteration from timber harvesting, agriculture, wetland drainage, urbanization, stream pollution and siltation, and the introduction of exotic predators (e.g., Ash 1988; Blaustein and Wake 1990, 1995; Bury 1983; Fisher and Shaffer 1996; Gamradt and Kats 1996; Gore 1983; Hayes and Jennings 1986; Orser and Shure 1972; Pechmann et al. 1991; Petranka et al. 1993, 1994; Pierce and Harvey 1987; Vial and Saylor 1993; Welsh and Lind 1988).

Although losses from direct environmental degradation are well documented, there is growing evidence that many amphibian populations are also declining in "pristine" areas that have not been conspicuously altered by human activity (Blaustein and Wake 1995; Drost and Fellers 1996; Stebbins and Cohen 1995; Vial and Saylor 1993). For example, the golden toad and other frogs have declined precipitously in virgin Costa Rican rain forests (Crump et al. 1992; Pounds and Crump 1994) as have several species of frogs in Australian rain forests. In western North America, the boreal toad (*Bufo boreas*), cascade frog (*Rana cascadae*), mountain yellow-legged frog (*R. muscosa*), Yosemite toad (*B. canorus*), and other frogs have disappeared or declined throughout many areas of their ranges (Blaustein et al. 1994; Bradford 1991; Drost and Fellers 1996; Fisher and Shaffer 1996; Sherman and Morton 1993). The federally endangered Wyoming toad (*B. hemiophrys baxteri*) has been reduced to a single population, and *R. tarahumarae* and *R. fisheri* have apparently been extirpated from the United States (deMaynadier and Hunter 1995). The extent to which salamander declines have paralleled those of frogs is largely unknown, although there is increasing evidence of similar patterns in salamanders (Blaustein and Wake 1995; Blem and Blem 1989, 1991; Fisher and Shaffer 1996).

The decline of amphibians in relatively undisturbed habitats around the world may be due to global anthropogenic factors such as acid precipitation or increased UV-B radiation (Blaustein 1994; Dunson et al. 1992). An alternative hypothesis is that the declines are components of natural population fluctuations. If so, then species that are currently in declining phases may eventually recover as part of natural population cycles related to drought, the outbreak of disease, or other natural causes (Blaustein et al. 1993; Pechmann and Wilbur 1994). Several agents have been hypothesized as responsible for the decline of regional populations of frogs and salamanders in the United States, including disease, predation, the introduction of exotic pests, changes in seasonal weather patterns, immunological stress, acid precipitation, UV-B radiation, and the widespread use of estrogen-mimicking pesticides (e.g., Blaustein et al. 1994; Blem and Blem 1989, 1991; Bradford 1991; Carey 1993; Crump et al. 1992; deMaynadier and Hunter 1995; Hayes and Jennings 1986; Long et al. 1995; Pechmann et al. 1991; Sherman and Morton 1993; Stone 1994). Studies of many of these factors are in their infancy, and future research will no doubt show that the relative importance of most varies among species and geographic regions (e.g., see Blaustein et al. 1994, 1996, and Grant and Licht 1995 concerning the importance of UV-B radiation).

Although long-term historical data are lacking, scientists agree that populations of most North American salamanders have declined markedly since European colonization. Environmental alteration from deforestation, habitat fragmentation, timber harvesting, agriculture, wetland drainage, urban growth, and stream pollution and siltation is the primary reason for salamander declines. Several studies indicate that many terrestrial salamanders are best adapted to conditions associated with late stages of forest succession (deMaynadier and Hunter 1995). Old-growth and mature forests provide cool, moist microenvironments that maximize the time salamanders can spend foraging on the forest floor. These conditions include a relatively thick leaf litter layer that acts as a sponge to retain moisture following rains, high levels of fallen debris that provide important surface cover during dry periods, and a multilayered, heavily shaded canopy that minimizes evaporative drying of the leaf litter during dry periods.

Many salamanders are sensitive to intensive forestry practices, such as clear-cutting, that greatly modify these conditions. Clear-cutting degrades forest-floor microhabitats for salamanders by eliminating shading and leaf litter, increasing soil surface temperature, and reducing leaf litter moisture. This process creates conditions that physiologically stress animals that rely on their moist skin as a respiratory organ. Several studies demonstrate that salamanders either decline following clear-cutting or are less abundant in young stands compared with mature forest stands (e.g., Ash 1988; Blymer and McGinnes 1977; Bury and Corn 1988a,b; deMaynadier and Hunter 1995; Dodd 1991; Means et al. 1996; Murphy et al. 1981; Petranka et al. 1993, 1994; Pough et al. 1987; Raymond and Hardy 1991; Welsh 1990; Welsh and Lind 1988, 1991). Some species are so sensitive to intensive forest management that they may be at risk of local or regional extinction.

Ambystomatid salamanders and other pond-breeding species have been adversely affected by the loss of wetlands and bottomland forests. Molecular and biogeographic evidence indicates that *Ambystoma* species rapidly colonized many regions of North America following postglacial retreats and other major climatic changes. This finding suggests that breeding sites dotted the landscape at sufficiently high densities to allow high levels of dispersal between local populations. Today, countless thousands of vernal ponds have been filled and millions of hectares of bottomland forests have been replaced with agricultural fields. Populations of pond-breeding salamanders have become increasingly isolated as a result, and in many areas are at the point at which the recolonization of ponds following local extinction is becoming increasingly unlikely.

Although amphibians in general appear to be more acid tolerant than many fishes (e.g., Andren et al. 1989; Gosner and Black 1957; Pierce 1985; Whiteman et al. 1995), the abundance and/or survival of some salamander species in eastern North America is inversely correlated with the acidity of breeding ponds (Clark 1986; Karns 1992; Pough 1976; Pough and Wilson 1977; Sadinski and Dunson 1992; however, see Cook 1983). Highly acidic solutions can kill embryos and hamper the ability of larvae to osmoregulate (Freda and Dunson 1985). Most larval salamanders do not show high embryonic mortality until pH drops below 4.5, although pHs of 4.5–5.5 can slow larval growth and development (Karns 1992; Ling et al. 1986; Pierce 1985; Sadinski and Dunson 1992).

Karns (1992) found that the blue-spotted salamander (*A. laterale*) is unable to reproduce successfully in bog habitats in Minnesota where the pH is <4.5. Sadinski and Dunson (1992) found that over half of the ponds studied in an area of Pennsylvania that receives relatively large amounts of acid deposition have a pH <4.5, and that complete egg mortality occurs in the Jefferson salamander (*A. jeffersonianum*) in ponds with low pH. Harte and Hoffman (1989) documented a long-term decline in populations of the tiger salamander (*A. t. nebulosum*) in the Colorado Rockies and surmised that acid rain may be responsible; however, more recent studies by Wissinger and Whiteman (1992) and Corn and Vertucci (1992) suggest that populations of *A. tigrinum* in this region are not being affected significantly by acid precipitation.

Because potentially toxic metals such as aluminum, copper, and manganese are more soluble at low pH, mortality associated with acidification of natural breeding sites may be due to a combination of high hydrogen ion concentration, high metal concentrations, and perhaps high concentrations of other compounds such as organic acids (Blem and Blem 1989, 1991; Karns 1992; Mathews and Morgan 1982; Whipple and Dunson 1993). Similar problems can arise when acid-bearing rocks are exposed during mining or road construction projects. For example, salamander larvae were eliminated from a stream section in the Great Smoky Mountains National Park after iron pyrite rock was used as road fill (Huckabee et al. 1975; Kucken et al. 1994; Mathews and Morgan 1982). Leachate from the rock reduces stream pH and increases metal concentrations to toxic levels.

We know little about the sensitivity of stream-breeding salamanders to environmental disturbance. Gore (1983) found that larvae of the dusky salamander (*D. fuscus*) are absent from many streams draining coal strip mines in eastern Kentucky and Tennessee. Stream siltation and high metal concentrations appear to be the primary factors reducing or eliminating *Desmognathus* from streams draining strip mines. Juveniles and adults of the black-bellied salamander (*D. quadramaculatus*) and seal salamander (*D. monticola*) can tolerate short-term exposure to water of pH 3.5–7.2 (Roudebush 1988). However, the feeding rates of animals after 3 weeks of exposure tend to decrease with ambient pH.

Recent research also suggest that acid precipitation can potentially affect terrestrial salamanders. Frisbie and Wyman (1991, 1992), Wyman (1988), Wyman and Hawksley-Lescault (1987), and Wyman and Jancola (1992) found that salamanders in New York are rare or absent in soils of very low pH. Their data suggest that extremely acidic soils may disrupt sodium balance and be potentially lethal to these and other terrestrial salamanders.

Although North American frogs and salamanders have been adversely affected by environmental degradation by humans, much can be done to improve habitats and minimize future losses. Populations of many pond-breeding salamanders such as the mole salamanders (*Ambystoma* spp.) have been severely depleted because of the loss of small wetlands. Maintaining viable populations will require protecting vernal ponds, swamps, and other small wetlands used for breeding, as well as protecting surrounding forests that provide habitats for adult salamanders. Small vernal ponds can be easily constructed in forested landscapes, and a growing number of forest managers, wetlands ecologists, and conservation biologists are designing and implementing plans to create vernal pond habitats for amphibians.

Methods of forest management strongly influence salamander diversity and biomass. In general, salamanders are most adversely affected by intensive management and harvesting techniques such as clear-cutting and plantation forestry (deMaynadier and Hunter 1995). Healthy populations of salamanders are best maintained by using partial harvest techniques and by maintaining forests on long-term rotation cycles. In the eastern United States, salamanders have been adversely affected by the conversion of mature hardwood forests into even-age pine forests that are maintained on short rotation cycles via clear-cutting. Management techniques that favor the retention of mature hardwood forests provide optimal habitats for these and many other forest vertebrates. Other management techniques, such as leaving 30- to 100-m forest buffer zones along streams and seepages, will benefit species that rely on aquatic habitats for reproduction. One of the greatest challenges for foresters in the next century will be to create innovative forest management plans that are ecologically sound, that provide a strong economic base, and that provide an adequate supply of forest resources to our growing population. Because most forested lands in North America are privately owned, a second challenge will be to educate and assist private landowners in adopting new approaches to land stewardship.

Key to Adult Salamanders of the United States and Canada

The following key is useful in identifying sexually mature transformed specimens, permanently gilled species, or species that rarely transform. The user should have a freshly caught or recently preserved specimen with coloration that is unfaded, and should be aware of the general collection locality. Unisexual species are best identified by genomic constitution and laboratory analyses of ploidy levels. Two unisexual biotypes (*A. platineum*; *A. tremblayi*) that have traditionally been formally recognized are included in the key. Other *Ambystoma* of hybrid origin are not included and are best identified using genetic or electrophoretic analyses.

1a. Rear limbs absent; external gills present . . 2
1b. Rear limbs present; external gills present or absent 6

2a. Four digits on each front foot; body unstriped 3
2b. Three digits on each front foot; body usually striped 5

3a. From states west of Alabama
. . . . *Siren intermedia* (in part)
3b. From states east of Mississippi 4

4a. Costal grooves 31–34; body uniformly colored or with small black spots; greenish yellow dashes and blotches absent on sides . . . . *Siren intermedia*
4b. Costal grooves 36–39; sides lighter than back; greenish yellow dashes and blotches on sides
. . . . *Siren lacertina*

5a. Lateral stripes indistinct (north Florida) or buff colored (south Florida); 32 chromosomes; restricted to all but northwestern portion of peninsular Florida, absent from panhandle
. . . . *Pseudobranchus axanthus*
5b. Lateral stripes distinct, yellow or tan; 24 chromosomes; from South Carolina through Florida panhandle and northwestern peninsular Florida . . . . *Pseudobranchus striatus*

6a. Aquatic, body eel-shaped; legs reduced and tiny; one to three digits per foot 7
6b. Aquatic or terrestrial, body not eel-shaped; legs not noticeably reduced; four or five digits per foot 9

7a. One digit per foot (check all feet)
. . . . *Amphiuma pholeter*
7b. Two or three digits per foot 8

8a. Two digits per foot (check all feet); light belly does not sharply contrast with darker dorsum; throat light; from South Carolina to extreme southeastern Louisiana . . . *Amphiuma means*
8b. Three digits per foot; light belly contrasts sharply with darker dorsum; dark throat patch present; from Texas to southeastern Alabama
. . . . *Amphiuma tridactylum*

9a. Head and body depressed, sides of body with wrinkled folds of skin; large, riverine species without bushy gills
. . . . *Cryptobranchus alleganiensis*
9b. Characters not as above; if large, riverine species, then with bushy gills 10

10a. Large, permanently gilled, lake-, stream-, or river-dweller; bushy gills; four digits on each foot; from eastern United States (*Necturus*)
. . . . 11
10b. Characters and range not as above 15

19

11a. Belly gray; dorsum typically uniformly brown, gray, or black (if from Cape Fear and Lumber River systems of North Carolina, fine spotting may be present on dorsum)
. . . . *Necturus punctatus*

11b. Belly not as above; body usually conspicuously spotted, and, if not, belly white . . 12

12a. Center of belly white and unspotted; dorsal spotting often absent; from western Georgia to eastern Mississippi . . . *Necturus alabamensis*

12b. Center of belly not white and usually conspicuously spotted; if white and unspotted then found west of Mississippi; dorsum usually spotted 13

13a. Dorsum rusty brown, both dorsum and venter spotted; restricted to Neuse and Tar River systems of North Carolina . . *Necturus lewisi*

13b. Dorsal coloration and degree of spotting variable; found outside Neuse and Tar River systems of North Carolina 14

14a. Dorsum dark brown with light tan intermeshed and heavily spotted; venter heavily spotted; restricted to southeastern Texas, southeastern Louisiana, and southern and central Mississippi . . . . *Necturus beyeri*

14b. Dorsum gray to rusty brown with scattered, indistinct spots; venter unspotted or with a few large spots; from northern Louisiana and northern Mississippi northward
. . . . *Necturus maculosus*

15a. Skin rough and granular; costal grooves absent or indistinct (newts) 16

15b. Skin smooth; costal grooves readily evident
. . . . 22

16a. Dorsum with red spots or lines encircled in black 17

16b. Dorsum not as above 18

17a. Back with red spots or broken lines
. . . . . . *Notophthalmus viridescens* (in part)

17b. Back with a continuous red line
. . . . *Notophthalmus perstriatus*

18a. Belly light yellow with fine black spotting; from eastern United States
. . . . *Notophthalmus viridescens*

18b. Belly orange or yellow-orange with black spotting; or from western United States . . 19

19a. Dorsum olive green; dorsum and venter with large scattered black spots; from Texas southward . . . . *Notophthalmus meridionalis*

19b. Dorsum brown to black; black spots on dorsum and venter normally absent; from western United States 20

20a. Eyes dark brown; belly tomato-red
. . . . *Taricha rivularis*

20b. Eyes yellow; belly yellow to orange 21

21a. Lower eyelid dark; eyes small (when viewed from above not meeting margin of the head); teeth in roof of mouth V-shaped
. . . . *Taricha granulosa*

21b. Lower eyelid light; eyes large (when viewed from above extending to or beyond margin of the head); teeth in roof of mouth Y-shaped
. . . . *Taricha torosa*

22a. Males with squarish cloacal lobes visible from above; nasolabial grooves absent; dorsum greenish brown to brownish gray; venter yellowish to orange; eyes conspicuously large; from western United States (*Rhyacotriton*)
. . . . 23

22b. Characters and range not as above 26

23a. Dorsum unspotted; venter with black spots; line of demarcation between dorsum and venter wavy and well defined; restricted to Olympic Mountains in northwestern Washington
. . . . *Rhyacotriton olympicus*

23b. Characters and range not as above 24

24a. Dorsum heavily blotched; venter unspotted; line of demarcation between dorsum and venter straight and well defined; restricted to western slopes of Cascades from Mt. St. Helens to Lane Co., Oregon . . *Rhyacotriton cascadae*

24b. Characters and range not as above 25

25a. Dorsum and venter both spotted; line of demarcation between dorsum and venter well defined and straight; from California to Little Nestucca River and Grande Ronde Valley in Polk, Tillamook, and Yamhill counties, Oregon
. . . . *Rhyacotriton variegatus*

25b. Dorsum and venter both unspotted (or weakly so); line of demarcation between dorsum and venter indistinct and straight; from coast ranges of Washington to Little Nestucca

River and Grande Ronde Valley in Polk, Tillamook, and Yamhill counties, Oregon
. . . . *Rhyacotriton kezeri*

26a. Specimen large and stout-bodied; costal grooves indistinct; no nasolabial grooves; adpressed limbs overlapping or nearly so; dorsum tan to reddish brown and often marbled on head or body; from western United States (*Dicamptodon*) 27

26b. Characters or range not as above 30

27a. Adults almost always gilled; dorsum not mottled and often with yellowish patches; light granular glands under yellow patches and at base of tail fin; maximum head width ≤ one-fifth SVL; costal folds zero to two between adpressed limbs . . . . *Dicamptodon copei*

27b. Adults often transformed; dorsum often mottled and lacks yellowish patches; dorsum lacks glands described above; maximum head width > one-fifth SVL; digit tips of adpressed limbs often overlap 28

28a. Dorsum and venter dark; dorsal mottling fine; restricted to eastern Oregon and Washington, northern Idaho, and western Montana
. . . . *Dicamptodon aterrimus*

28b. Dorsum and venter lighter colored; dorsum heavily mottled; restricted to coastal forests from southwestern British Columbia to central California 29

29a. From the San Francisco Bay area (southern Mendocino, Sonoma, and Napa counties southward to Santa Cruz Co., California)
. . . . *Dicamptodon ensatus*

29b. From north of Sonoma and Napa counties, California, to British Columbia
. . . . *Dicamptodon tenebrosus*

30a. Nasolabial grooves absent; lungs present; teeth in roof of mouth in a transverse row; relatively stout-bodied animals with rounded snouts (*Ambystoma*) 31

30b. Nasolabial grooves present (use hand lens); lungs absent; dentition includes a curved, nearly transverse row and an elongated patch that extends posteriorly between the orbits; body form variable but many species have relatively slender bodies and pointed snouts (Plethodontidae) 46

31a. Dorsum with dorsal stripes or bold bars, bands, or spots 32

31b. Dorsum overlain with lichenlike frosting, speckling, or flecking and lacking bold bars, bands, or spots 37

32a. Dorsum black with white or silvery crossbands that expand into hourglass shapes along the sides; venter black . . . *Ambystoma opacum*

32b. Dorsum not as above, often with cream, yellow, or greenish blotches, spots, or stripes; venter often light colored 33

33a. Dorsum grayish black with large, round, yellowish spots forming one or two irregular rows that extend onto tail; head often with either yellow or red spots; venter gray; from eastern United States
. . . . *Ambystoma maculatum*

33b. Characters or range not as above 34

34a. Dorsum black with cream to yellow oval spots; belly grayish with a few dull yellow spots or (rarely) yellow bands; digit tips pinkish; from Central Valley and lower elevations to the west in central California
. . . . *Ambystoma californiense*

34b. Characters and range not as above 35

35a. Dorsum gray or black with cream to yellow crossbands that often extend uninterrupted across back; venter light colored and unmarked; tail long and narrow; restricted to Ozark and Ouachita mountains of Ozark region; costal grooves 15
. . . . *Ambystoma annulatum*

35b. Characters and range not as above; costal grooves 12–13 36

36a. Dorsum highly variable, either (1) gray or black with large greenish to yellowish bars, blotches, or spots, (2) yellowish with black bars or blotches, or (3) brownish to gray with small dark spots or no markings; venter often spotted or blotched; usually > 17 cm TL
. . . . *Ambystoma tigrinum*

36b. Dorsum dark brown or black with a greenish, tan, or yellowish stripe, which is often interrupted into irregular rows of blotches; venter dark brown to black and lacking spots or blotches; usually < 15 cm TL
. . . . *Ambystoma macrodactylum*

37a. Snout short; head small relative to body size; plicae of tongue projecting from a median furrow; grayish black dorsum overlain with grayish lichenlike patches; venter black with tiny flecks or lichenlike mottling; costal grooves 14–15 38
37b. Characters not all as above 39

38a. Internarial distance usually ≥2.5 mm; teeth squared and not daggerlike; primarily from limestone regions of central Tennessee, central Kentucky, southwestern Ohio, and southeastern Indiana; eggs laid singly in streams (rarely in ponds) . . . . *Ambystoma barbouri*
38b. Internarial distance usually <2.5 mm; teeth daggerlike in specimens from Indiana, Ohio, Kentucky, and Tennessee; from areas outside range of 38a; eggs laid in small clusters in ponds (rarely in streams)
. . . . *Ambystoma texanum*

39a. Large, swollen parotoid gland behind each eye; rounded ridge along upper edge of tail; dorsum gray-brown to chocolate brown above; restricted to northwestern United States and western Canada
. . . . *Ambystoma gracile*
39b. Characters not as above; from eastern North America 40

40a. Costal grooves 13–16, usually 15; dorsum patterned with gray netlike pattern or frosted pattern; belly black with gray flecks; restricted to Coastal Plain from South Carolina to southeastern Mississippi . . . *Ambystoma cingulatum*
40b. Costal grooves 13 or fewer (usually 11–12); dorsum variable but often with flecks or lichenlike marks; belly grayish if specimen is from Coastal Plain; may or may not occur on Coastal Plain 41

41a. Head large relative to body size and with rounded snout; costal grooves 10–11; dorsum brown or gray with bluish white flecks; venter gray with light blotches; usually <10 cm TL
. . . . *Ambystoma talpoideum*
41b. Characters not all as above 42

42a. Adults ≤10 cm TL; head small relative to body size; costal grooves 13; dorsum brown to black with numerous light flecks; from southeastern Coastal Plain of North Carolina and eastern South Carolina . . . *Ambystoma mabeei*
42b. Adults >10 cm TL; head normal-sized relative to body size; costal grooves 12 or fewer; dorsal coloration variable; from areas other than Coastal Plain of North Carolina and South Carolina 43

43a. SVL usually ≤69 mm in males, 75 mm in females; dorsum blackish with scattered blue spots; venter only slightly paler than dorsum; vent region black 44
43b. SVL usually >69 mm in males, 75 mm in females; dorsum grayish or brownish with bluish flecks concentrated on lower sides; vent region grayish 45

44a. Bisexual species with 2n = 28 chromosomes; venter black with numerous gray spots; maximum TL of females ≤13 cm; perimeter of red blood cell enclosed area <850 μm^2
. . . . *Ambystoma laterale*
44b. Unisexual species with 3n = 42 chromosomes; venter often grayish with fewer blue marks; females often exceed 13 cm TL; perimeter of red blood cell enclosed area >850 μm^2
. . . JLL unisexuals (=*Ambystoma tremblayi*)

45a. Bisexual species with 2n = 28 chromosomes; bluish marking flecklike and often concentrated along lower sides and on limbs; perimeter of red blood cell enclosed area <850 μm^2
. . . . *Ambystoma jeffersonianum*
45b. Unisexual species (males very rare) with 3n = 42 chromosomes; bluish marking often larger and more widely distributed; perimeter of red blood cell enclosed area >850 μm^2
. . . JJL unisexuals (=*Ambystoma platineum*)

46a. Tail with a conspicuous constriction near the base 47
46b. Tail not constricted 48

47a. Belly white with black spots; from eastern United States . . . . . *Hemidactylium scutatum*
47b. Belly not white with black spots; from western United States . . . . *Ensatina eschscholtzii*

48a. Four digits on each foot, tail rounded in cross section; body often highly elongated, wormlike and with small limbs; from western United States (*Batrachoseps*) 49

48b. Five digits on each hind foot (or if four digits, then tail laterally compressed); body not wormlike, limbs normal in size; from eastern or western United States 57

49a. Belly black with conspicuous white blotches; from Cascades of Oregon
.... *Batrachoseps wrighti*

49b. Belly not as above; from California to southwestern Oregon 50

50a. Costal folds between digit tips of adpressed limbs >10; tail length often >1.5–2 × SVL
.... 51

50b. Costal folds between digit tips of adpressed limbs <9; tail length variable, often <1.5 × SVL
.... 52

51a. Underside of tail lighter than belly and often tinged with yellow; from central California northward ...... *Batrachoseps attenuatus*

51b. Underside of tail dark and not contrasting with belly; from central California southward
.... *Batrachoseps nigriventris*

52a. Dorsum heavily suffused with silvery or brassy flecks; underside of tail flesh-colored and contrasting with dark belly; endemic to Hidden Palm Canyon, Riverside Co., California
.... *Batrachoseps aridus*

52b. Characters and range not as above 53

53a. Dorsum with a bright coppery to orange-colored, diffuse stripe that breaks into patches on the tail; ratio of SVL to hind limb length 5–6; tail noticeably tapered; endemic to San Gabriel Mountains, Los Angeles Co., California
.... *Batrachoseps gabrieli*

53b. Characters and range not as above 54

54a. Costal folds between digits of adpressed limbs two to five; tail short, < three-quarters SVL; restricted to Inyo Mountains, Inyo Co., California .... *Batrachoseps campi*

54b. Characters and range not as above 55

55a. Costal folds between digits of adpressed limbs six to seven; costal grooves 18–19; feet strongly webbed with only one segment of digit free of webbing; restricted to Caliente Creek drainage in California
.... *Batrachoseps stebbinsi*

55b. Characters and range not as above 56

56a. Costal grooves 20–21; venter black with conspicuous white flecking; restricted to Kern River drainage in California
.... *Batrachoseps simatus*

56b. Costal grooves 18–20; coloration and morphology extremely variable; if from Kern River drainage, then venter whitish to light gray, and, if light flecking present, then missing from midventral area
.... *Batrachoseps pacificus*

57a. Rear limbs conspicuously larger than front limbs; pale diagonal line from eye to angle of jaw (may be absent on melanistic animals); costal grooves 14 (rarely 13); specimens often move by leaping (*Desmognathus*) 58

57b. Characters not as above 70

58a. Midpoint of tail rounded or oval in cross section; keel absent from posterior half of tail; adults generally ≤10 cm TL 59

58b. Midpoint of tail triangular in cross section; keel present on posterior half of tail; adults often >10 cm TL 64

59a. Adults usually <30 mm SVL; dorsum often with either herringbone pattern or dark midline branching into Y at eyes; dorsum of rear limbs usually with coppery coloration or spots
.... 60

59b. Adults usually ≥30 mm SVL; dorsum variable but typically lacking herringbone pattern or midline branching into Y at eyes; dorsum of rear limbs lacking coppery coloration or spots
.... 61

60a. Dorsum usually with conspicuous dark herringbone pattern; tail <45% TL; eyelids and dorsum of limbs coppery colored; top of head rugose; mental gland of males U-shaped with arms extending along dentaries; venter unpigmented ...... *Desmognathus wrighti*

60b. Dorsum without conspicuous dark herringbone pattern (faint pattern sometimes evident); tail about 50% TL; dark midline that branches into Y at eyes usually present and tan or light brown "portholes" often present on dorsum of each rear limb; top of head smooth; mental gland of males kidney-shaped and not extending onto dentaries; venter lightly mottled ...... *Desmognathus aeneus*

61a. Individuals with five to seven pairs of rounded, coalesced blotches on dorsum, or melanistic and brownish; restricted to Coastal Plain in Choctawhatchee, Chattahoochee, and Apalachicola river basins in Alabama, Georgia, and Florida panhandle
. . . . *Desmognathus apalachicolae*

61b. Dorsal pattern highly variable: may be striped or blotched and, if melanistic, often black; from Atlantic Piedmont northward 62

62a. From outside Great Smoky Mountains National Park, North Carolina and Tennessee
. . . . *Desmognathus ochrophaeus* complex (*D. ocoee, D. carolinensis, D. orestes,* and *D. ochrophaeus;* use range maps in text to distinguish between species)

62b. Specimen from Great Smoky Mountains National Park 63

63a. With red to yellow cheek patches
. . . . *Desmognathus imitator* (in part)

63b. Without red to yellow cheek patches (reliable identification not possible without electrophoretic analyses)
. . . . . . *Desmognathus ocoee* and *D. imitator*

64a. Internal nasal openings slitlike; belly dark gray to black with whitish central region (sometimes missing); dorsum variable but typically with two irregular rows of blotches
. . . . *Desmognathus marmoratus*

64b. Internal nasal openings circular; belly white, mottled, or black and if black, then lacking whitish central region; dorsum variable but often not blotched 65

65a. Dorsum brownish with weak spotting or patterning; restricted to Ouachita Mountains of Arkansas and Oklahoma
. . . . *Desmognathus brimleyorum*

65b. Dorsal patterning variable; found outside Ouachita Mountains of Arkansas and Oklahoma 66

66a. Digit tips with darkened friction pads and appearing grayish black to black 67

66b. Digit tips lacking friction pads and appearing white or light gray 69

67a. Belly black . . . *Desmognathus quadramaculatus*

67b. Belly white, light gray, or mottled with brown
. . . . 68

68a. Belly whitish to light gray and uniformly colored; dorsum often with bold, dark wormy marks or spots on a lighter ground color; tail rounded near base . . *Desmognathus monticola*

68b. Belly brownish and heavily mottled; dorsum often drab and lacking bold patterning; tail laterally compressed near base
. . . . *Desmognathus welteri*

69a. Venter dark brown to black and flecked with white; dorsum suffused with black and drab; sides of body with two rows of white (sometimes red) spots that continue onto tail
. . . . *Desmognathus auriculatus*

69b. Venter cream colored with varying amounts of brown mottling; dorsum often conspicuously blotched or striped and if unblotched, then suffused with brown; sides of body and tail lacking two rows of white spots
. . . . *Desmognathus fuscus*

70a. Dorsum and venter uniformly dark brown; costal grooves 20–22; costal folds between adpressed limbs 13–15; restricted to Red Hills of south-central Alabama
. . . . *Phaeognathus hubrichti*

70b. Characters and range not as above 71

71a. Dorsum light brown to dull yellow; series of dark parallel lines on sides (sometimes reduced to spots); belly yellow with scattered dark specks; restricted to Atlantic Coastal Plain from Virginia to Georgia
. . . . *Stereochilus marginatus*

71b. Characters and range not as above 72

72a. Cave form with pinkish white skin; eyes greatly reduced; gills red and bushy; snout rounded; restricted to southwestern Georgia and northwestern Florida . . . . . . *Haideotriton wallacei*

72b. Characters and range not as above 73

73a. Permanently gilled species restricted to Edwards Plateau region of Texas 74

73b. Character and range not as above 79

74a. Body white or yellowish white; costal grooves <13 75

74b. Body brownish to light yellow; costal grooves >12 (usually 14–17) 77

75a. Rarely >7 cm TL; snout moderately depressed at level of eyes . . . . . . *Eurycea tridentifera*
75b. Usually >7 cm TL; snout strongly depressed and truncated 76
76a. Legs extremely thin, and remnants of eyes present as dark spots; from underground streams in the Purgatory Creek system and along the San Marcos Springs Fault near San Marcos, Texas . . . . *Eurycea rathbuni*
76b. Similar to 75a but more robust; extremely rare, with only one specimen known from beneath bed of Blanco River, Hays Co., Texas . . . . *Eurycea robusta*
77a. Dorsum heavily blotched and mottled with dark pigment to produce salt-and-pepper effect; dorsum often with silvery-white pigmentation; eyes reduced; snout shovel-nosed; from within the city of Austin, Texas . . . . *Eurycea sosorum*
77b. Dorsum not heavily blotched with dark pigment; silvery-white pigmentation absent from dorsum; eyes appearing normal in size; snout not conspicuously shovel-nosed; from areas outside Austin, Texas 78
78a. ≤5.1 cm TL; dorsum plain brown; belly yellowish white; from Spring Lake, San Marcos, Texas . . . . *Eurycea nana*
78b. >5.1 cm TL; dorsum light brown to yellowish; belly whitish; from areas of the Edwards Plateau other than Spring Lake . . . . *Eurycea neotenes*
79a. From New Mexico 80
79b. From elsewhere than New Mexico 81
80a. Body noticeably elongated; costal grooves 19; restricted to Jemez Mountains in northern New Mexico . . . . . *Plethodon neomexicanus*
80b. Body not elongated; costal grooves 14–15; restricted to Sacramento, Capitan, and White mountains in southern New Mexico . . . . *Aneides hardii*
81a. From west of eastern edge of Rocky Mountains 82
81b. From east of Rocky Mountains 94
82a. Costal grooves 13; digits conspicuously webbed; tongue mushroom-shaped and unattached at front of mouth (*Hydromantes*) . . 83

82b. Costal grooves >13; digits not conspicuously webbed; tongue not mushroom-shaped and attached at front of mouth 85
83a. Dorsum uniformly brown; underside of tail yellowish; restricted to Lower Merced Canyon, Mariposa Co., California . . . . *Hydromantes brunus*
83b. Coloration and range not as above 84
84a. Dorsum mottled gray-green to reddish; tail yellowish to yellowish orange above; restricted to Mt. Shasta, Shasta Co., California . . . . *Hydromantes shastae*
84b. Dorsum mottled grayish; tail not yellowish above; from Sierra Nevada south of Mt. Shasta . . . . *Hydromantes platycephalus*
85a. Head triangular (particularly in males); front teeth in upper jaw conspicuously projecting beyond upper lip (feel by stroking gently with finger); body rather stout (*Aneides*) 86
85b. Head not conspicuously triangular; front teeth in upper jaw not conspicuously projecting beyond upper lip; body relatively slender (*Plethodon* [in part]) 88
86a. Dorsal ground color black; belly grayish black . . . . *Aneides flavipunctatus*
86b. Dorsal ground color brownish to grayish; belly whitish to brownish 87
87a. Dorsum brownish and often clouded with greenish gray, ash, or coppery coloration; costal grooves 16–17; from Oregon to northern California . . . . *Aneides ferreus*
87b. Dorsum grayish to brown, usually with cream or yellowish spots; costal grooves 15–16; from northern California to Baja California . . . . *Aneides lugubris*
88a. Costal grooves 17–18 89
88b. Costal grooves 14–16 (usually 14–15) . . 90
89a. Costal grooves usually 18; 6.5–7.5 costal grooves between adpressed limbs; dorsum brown or black; brown dorsal stripe sometimes present . . . . *Plethodon elongatus*
89b. Costal grooves usually 17; 4.0–5.5 costal grooves between adpressed limbs; dorsal chocolate or purplish brown, heavily speckled with whitish or yellowish marks; dorsal stripe absent . . . . *Plethodon stormi*

90a. Venter red to salmon pink; undersides of feet reddish . . . . *Plethodon larselli*
90b. Venter variably colored, but not reddish; undersides of feet variably colored, but not reddish 91
91a. Pale yellow throat patch present that contrasts with dark belly; parotoid gland well developed 92
91b. Pale yellow throat patch not evident; parotoid gland absent or inconspicuous 93
92a. Dorsal stripe yellow to tan and straight-edged; upper surface of limbs same color as dorsal stripe; restricted to western Washington . . . . *Plethodon vandykei*
92b. Dorsal stripe yellow, green, orange, or red and often wavy and narrow; upper surface of limbs dark; restricted to Idaho and Montana . . . . *Plethodon idahoensis*
93a. Costal grooves usually 16; dorsal stripe extending to tail tip; belly finely mottled to produce a salt-and-pepper pattern . . . . *Plethodon vehiculum*
93b. Costal grooves usually 15; dorsal stripe not extending to tail tip; belly slate gray and not salt-and-pepper . . . . *Plethodon dunni*
94a. Dorsum black to grayish with conspicuous blotches of green or yellow-green; digits square-tipped and expanded; body appearing flattened . . . . *Aneides aeneus*
94b. Characters not as above 95
95a. Cave form with pinkish to white skin; eyelids fused; eyes nonfunctional; adults transformed and without gills; restricted to Ozark Uplift . . . . *Typhlotriton spelaeus*
95b. Characters and range not as above 96
96a. Tail rounded or oval in cross section along entire length; dorsal ground color dark gray to black (often overlain with a dorsal stripe or light spotting or flecking); belly uniformly dark, with salt-and-pepper dark mottling, or dark with white blotching near chest region (*Plethodon* [in part]) 97
96b. Tail triangular in cross section (often only on posterior half); dorsal ground color often reddish, salmon, or brownish; belly usually light colored 118

97a. Costal grooves 16; dorsum frosted or blotched with gold or brassy coloration . . . . *Plethodon welleri*
97b. Costal grooves >16; or, if 16, then dorsum lacking gold or brassy blotching 98
98a. Costal grooves 16–17 (rarely 18); typically ≥10 cm TL 99
98b. Costal grooves 18–22; typically <10 cm TL 110
99a. From Ozark region of western Arkansas and eastern Oklahoma 100
99b. From elsewhere than Ozark region of western Arkansas and eastern Oklahoma 102
100a Throat dark and similar in color to belly . . . . *Plethodon glutinosus* (in part)
100b. Throat light and contrasting with darker belly 101
101a Dorsum with chestnut coloration or with two irregular rows of large whitish blotches down back; chest dark; restricted to areas immediately north of Caddo Mountains in southeastern Oklahoma and western Arkansas . . . . *Plethodon ouachitae*
101b. Dorsum lacking chestnut coloration and irregularly speckled with whitish markings; chest light; restricted to Caddo Mountains . . . . *Plethodon caddoensis*
102a Dorsum heavily marked with chestnut, reddish brown, or olive-brown blotching or banding 103
102b. Dorsum lacking chestnut, reddish brown, or olive-brown coloration 104
103a Dorsum reddish brown or olive-brown; chest and throat mottled with yellowish coloration; restricted to Pigeon Mountain in extreme northwestern Georgia . . . *Plethodon petraeus*
103b. Dorsum marked with chestnut coloration; chest and throat not mottled with yellowish coloration; restricted to Blue Ridge Mountains of North Carolina, eastern Tennessee, and southwestern Virginia . . . . *Plethodon yonahlossee*
104a Costal grooves typically 17–18; throat white or blotched with white and lighter than belly 105
104b. Costal grooves typically 16; throat pale gray

to gray and not conspicuously lighter than belly 106

105a. Dorsum with conspicuous brassy spots and flecks and scattered, fine white spots; red spotting sometimes present
. . . . *Plethodon wehrlei*
105b. Brassy spots and flecks on dorsum greatly reduced or absent, and white or yellow spots conspicuous on dorsum; red spotting absent
. . . . *Plethodon punctatus*

106a. Dorsum with abundant brassy spotting and flecking; chin lighter than belly; restricted to west slopes of Unicoi Mountains and vicinity in northeastern Polk and eastern Monroe counties, Tennessee, and northwestern Graham and northwestern Cherokee counties, North Carolina . . . . . . *Plethodon aureolus*
106b. Characteristics and range not as above
. . . . 107

107a. Dorsum with small white spots and larger lateral spots; restricted to areas west of French Broad River in southwestern North Carolina, and to adjacent areas of Tennessee, northwestern South Carolina, and Rabun Co., Georgia . . . . . . *Plethodon oconaluftee*
107b. Characteristics and range not as above
. . . . 108

108a. Dorsum dark gray, rarely with faint white spotting; legs or cheeks red in some populations; brassy frosting present on dorsum in other populations . . . . . . *Plethodon jordani*
108b. Dorsum blackish, typically marked with white spotting; no red coloration on legs or cheeks; brassy flecking present on dorsum in some populations 109

109a. Chin lighter than belly; dorsum with fine, scattered white spots and little brassy flecking; chin rather pointed; restricted to eastern Kentucky, southwestern Virginia, and southern West Virginia . . . . . *Plethodon kentucki*
109b. Chin not lighter than belly (within range of 109a); dorsum with conspicuous white spots or brassy flecking; chin rounded; broadly distributed in eastern United States
. . . . *Plethodon glutinosus*

110a. Costal grooves 18–19; dorsum marked with numerous brassy flecks; restricted to Cheat Mountains of southeastern West Virginia and Peaks of Otter region of Bedford and Rockbridge counties, Virginia 111
110b. Characters and range not as above . . . 112

111a. Costal grooves averaging 18; scattered brassy flecks on dorsum; restricted to Cheat Mountains of West Virginia . . . . *Plethodon nettingi*
111b. Costal grooves averaging 19; abundant brassy flecks often organized into blotches on dorsum or a weak dorsal stripe; restricted to Peaks of Otter region of Bedford and Rockbridge counties, Virginia
. . . . *Plethodon hubrichti*

112a. Costal grooves typically 20–22; dorsum usually with conspicuous brassy flecks . . . 113
112b. Costal grooves typically 18–19; dorsum often unmarked or with a red, yellowish, or brownish stripe 114

113a. Belly uniformly dark; throat mottled
. . . . *Plethodon richmondi*
113b. Belly dark with white mottling; throat mostly white . . . . *Plethodon hoffmani*

114a. Belly mottled with orange, white, and black; dorsal stripe (if present) noticeably wavy (straight-edged in specimens west of the Mississippi River) 115
114b. Belly not mottled or mottled and lacking orangish pigmentation; dorsal stripe (if present) straight or with serrated edges . . . 116

115a. From South Carolina, west-central Georgia, central and southern Mississippi, and central and southern Alabama; if from Jefferson Co., Alabama, then typically striped
. . . . *Plethodon websteri*
115b. From northern Alabama and extreme northern Mississippi and Georgia northward and westward; if from Jefferson Co., Alabama, then typically unstriped . . . *Plethodon dorsalis*

116a. Belly blackish with scattered flecks of white or yellowish pigment; costal grooves typically 18; restricted to Shenandoah National Park, northern Virginia . . . . *Plethodon shenandoah*
116b. Belly heavily mottled to produce a salt-and-pepper pattern; costal grooves 19 or more; or

from elsewhere than Shenandoah National Park, northern Virginia 117

117a. From west and south of French Broad River in western North Carolina; dorsal stripe of striped phase often serrated
. . . . *Plethodon serratus*
117b. From east and north of French Broad River in western North Carolina; dorsal stripe of striped phase straight . . . . *Plethodon cinereus*

118a. Dorsal coloration red, salmon, or orangish with dark spots or faint dark reticulations (may be rusty red to reddish brown in old animals) 119
118b. Dorsal coloration brown, yellowish, or pinkish white, with or without black spots or dark reticulations 122

119a. Body slender; costal grooves 14–15
. . . . *Eurycea lucifuga*
119b. Body stout; costal grooves 16–19 120

120a. Light line bordered by gray or black extending from eye to nostril (may be faint in some races); snout squarish; costal grooves 17–19
. . . . *Gyrinophilus porphyriticus*
120b. Light line bordered by gray or black not evident from eye to nostril; snout rounded; costal grooves 16–17 121

121a. Eyes yellow; coloration on sides and belly intergrades; dorsum covered with numerous black spots . . . . *Pseudotriton ruber*
121b. Eyes brown; coloration of sides and belly sharply contrasts; dorsum with widely scattered black spots . . . . *Pseudotriton montanus*

122a. Costal grooves 19–21; body slender; restricted to Ozark region 123
122b. Characters and range not as above . . . 124

123a. Adults gilled; belly pale colored; restricted to northeastern Oklahoma and adjacent corners of Arkansas and Missouri
. . . . *Eurycea tynerensis*

123b. Adults usually transformed or, if gilled, then south of range of 123a; belly yellowish or grayish; more broadly distributed in eastern Oklahoma, north-central Arkansas, and southwestern Missouri . . . . . *Eurycea multiplicata*

124a. Rear foot with four digits
. . . . *Eurycea quadridigitata*
124b. Rear foot with five digits 125

125a. Body slender; costal grooves 13–16 . . . 126
125b. Body stout; costal grooves 17–19 129

126a. Dorsum with yellowish brown to yellowish orange stripe bordered by two solid, broad bands that extend from eyes well onto tail
. . . . *Eurycea bislineata*
126b. Characters not as above 127

127a. Dorsum with yellowish brown stripe and three parallel dark bands
. . . . *Eurycea guttolineata*
127b. Characters not as above 128

128a. Tail >55% TL; either dorsum bordered by mottled bands on sides or tail with herringbone pattern on sides; broadly distributed in eastern United States . . . . *Eurycea longicauda*
128b. Tail <50% TL; dorsum either lacking broad mottled bands or with bands broken into irregular wavy lines; restricted to extreme southwestern North Carolina and adjoining portions of Tennessee . . . *Eurycea junaluska*

129a. Adults transformed; light line from eye to nostril indistinct; dorsum with faint reticulate pattern; restricted to General Davis Cave, Greenbrier Co., West Virginia
. . . . *Gyrinophilus subterraneus*
129b. Adults gilled (rarely transformed); light line from eye to nostril not evident; dorsum often spotted; restricted to cave systems in Tennessee and Alabama . . . . *Gyrinophilus palleucus*

Key to Larval Salamanders of the United States and Canada

This key is useful in identifying sexually immature larvae as well as larviform adults of species having both gilled and transformed adults. The key is based primarily on information provided in Altig and Ireland (1984), Pfingsten and Downs (1989), and Stebbins (1985).

Larval salamanders often undergo marked ontogenetic change from hatching to metamorphosis. This key is useful for identifying larvae that have fully differentiated rear limbs and digits (sirenids that lack rear limbs are also included). The key requires that the user know the general collection locality and have a freshly caught or recently preserved specimen showing natural coloration. Identification of specimens is best achieved with older larvae and by examining a series of animals from a collection site. Ontogenetic and geographic variation in coloration, color patterns, and morphology is poorly documented for larvae of most salamander species. The user should expect occasional deviations from the descriptions given here and use other evidence—such as the type of aquatic habitat, range, or presence of juveniles and adults collected at a field site—to aid in identifying larval forms. If necessary, large larvae should be raised through metamorphosis and the juveniles used to verify larval identifications.

1a. No rear limbs 2
1b. Rear limbs present 5

2a. Three digits on each front foot 3
2b. Four digits on each front foot 4

3a. 32 chromosomes; from extreme northeastern Florida southward to southern Florida and absent from panhandle region
. . . . *Pseudobranchus axanthus*
3b. 24 chromosomes; from southern South Carolina through Florida panhandle and northwestern portion of peninsular Florida
. . . . *Pseudobranchus striatus*

4a. Costal grooves 31–34; head with bold red or yellowish band (rarely absent) . . . *Siren intermedia*
4b. Costal grooves 36–39; head not with bold red or yellowish band . . . . *Siren lacertina*

5a. Body elongate and cylindrical; one to three digits per foot 6

5b. Body not elongate and cylindrical; four or five digits per foot 8

6a. One digit on each foot . . *Amphiuma pholeter*
6b. Two or three digits on each foot 7

7a. Two digits present on each foot
. . . . *Amphiuma means*
7b. Three digits present on each foot
. . . . *Amphiuma tridactylum*

8a. Snout broad and depressed; skin appearing loose along sides; digits flattened and fleshy; larvae ≤130 mm TL; from rivers and large streams . . . . . . *Cryptobranchus alleganiensis*
8b. Characters not as above 9

9a. From east of Continental Divide 10
9b. From west of Continental Divide 14

10a. Five digits on each rear foot 11
10b. Four digits on each rear foot 28

11a. Gills with rachises and with fimbriae through-

out length; three or four gill slits; tips of digits and soles of feet not keratinized 12
11b. Gills without rachises, short and branched from base; four gill slits; tips of digits and soles of feet often keratinized (desmognathine plethodontids) 64
12a. Dorsal fin extends onto body; rarely collected in running water 13
12b. Dorsal fin terminates on tail or near junction of tail and body; usually collected in running water (hemidactyliine plethodontids) . . . 34
13a. Four gill slits; keratinized jaw sheath absent; head somewhat pointed in dorsal view; body slender; skin of large specimens may be granular (Salamandridae) 53
13b. Three gill slits; keratinized jaw sheath usually present; head rounded in dorsal view; body stout; skin of large specimens smooth (Ambystomatidae) 55
14a. Three gill slits; skin smooth in large specimens; keratinized jaw sheath usually present; head rounded in dorsal view 15
14b. Four gill slits; skin granular in large specimens; keratinized jaw sheath absent; head angular in dorsal view 26
15a. Dorsal fin terminating on body; gills typically long and bushy with long rachises 16
15b. Dorsal fin terminating on tail or slightly anterior to pelvis; gills either small and not bushy, or bushy and with short rachises 19
16a. More than 15 gill rakers on third gill arch; parotoid gland and glandular ridge on tail absent; digits flattened and pointed 17
16b. Fewer than 15 gill rakers on third gill arch; parotoid gland and glandular ridge on tail sometimes present on large specimens; digits rounded 18
17a. Specimen from west-central California; gilled adults not known . . . *Ambystoma californiense*
17b. Specimen from elsewhere; gilled adults often present in populations
. . . . *Ambystoma tigrinum* (in part)
18a. 9-13 gill rakers on third gill arch; parotoid gland and glandular ridge on tail absent; costal grooves usually 12; ≤8 cm SVL
. . . . *Ambystoma macrodactylum*
18b. 7-10 gill rakers on third gill arch; parotoid gland and glandular ridge on tail usually evident on large larvae and gilled adults; costal grooves usually 11; >8 cm SVL
. . . . *Ambystoma gracile*
19a. Head flattened in lateral view; keratinized jaw sheath present; gills short but bushy; ≤30 cm TL (*Dicamptodon*) 20
19b. Head rounded in lateral view; keratinized jaw sheath absent; gills greatly reduced and not bushy; ≤6-7 cm TL (*Rhyacotriton*) 23
20a. Costal grooves between adpressed limbs zero to two; conspicuous granular glands on dorsum; 28-42 vomerine teeth; venter dark; transformed specimens rare . . . *Dicamptodon copei*
20b. Costal grooves between adpressed limbs 0.5, or tips of digits overlapping; granular glands absent on dorsum; 46-59 vomerine teeth; venter light; gilled adults rare in small streams, common in large streams 21
21a. From eastern Oregon and Washington, northern Idaho, and western Montana
. . . . *Dicamptodon aterrimus*
21b. From southwestern British Columbia to central California 22
22a. From San Francisco Bay area (southern Mendocino, Sonoma, and Napa counties southward to Santa Cruz Co., California)
. . . . *Dicamptodon ensatus*
22b. From north of Sonoma and Napa counties, California, to British Columbia
. . . . *Dicamptodon tenebrosus*
23a. Dorsum unspotted; black spots on venter; line of demarcation between dorsum and venter wavy and well defined; restricted to Olympic Mountains in northwestern Washington (traits only apply to large larvae; use range for small larvae of this and other *Rhyacotriton* spp.)
. . . . *Rhyacotriton olympicus*
23b. Characters and range not as above 24
24a. Dorsum blotched; venter unspotted; line of demarcation between dorsum and venter straight and well defined; restricted to western slopes of Cascades from Mt. St. Helens to Lane Co., Oregon . . . *Rhyacotriton cascadae*
24b. Characters and range not as above 25

25a.	Dorsum and venter both spotted; line of demarcation between dorsum and venter distinct and straight; from California to Little Nestucca River and Grande Ronde Valley in Polk, Tillamook, and Yamhill counties, Oregon . . . . *Rhyacotriton variegatus*
25b.	Dorsum and venter both unspotted (or very weakly spotted); line of demarcation between dorsum and venter indistinct and straight; from coast ranges of Washington to Little Nestucca River and Grande Ronde Valley in Polk, Tillamook, and Yamhill counties, Oregon . . . . *Rhyacotriton kezeri*
26a.	Dorsal fin terminating anteriorly before reaching insertion point of front limbs; from coastal streams of Sonoma, Mendocino, and south Humboldt counties, California . . . . *Taricha rivularis*
26b.	Dorsal fin extending forward to insertion point of front limbs; broadly distributed in western United States 27
27a.	Black dorsal stripe on either side of dorsal fin; gilled adults not known . . . . . *Taricha torosa*
27b.	Black stripe absent on either side of dorsal fin; large gilled adults occur in some high-elevation populations . . . . . . *Taricha granulosa*
28a.	Dorsal fin extending onto body; snout rounded; usually from lentic habitats 29
28b.	Dorsal fin terminating on tail; snout angular; usually from lotic habitats 30
29a.	Costal grooves 13–14; eye line present . . . . *Hemidactylium scutatum*
29b.	Costal grooves 14–17; eye line absent . . . . *Eurycea quadridigitata*
30a.	From Atlantic Coastal Plain drainage . . . 31
30b.	From Gulf Coast drainage 32
31a.	Dorsum uniformly brown to gray, and may have small dark spots if from Cape Fear and Lumber River systems of the Carolinas . . . . *Necturus punctatus*
31b.	Dorsum with light tan dorsal stripes flanked by darker pigment laterally; restricted to Neuse and Tar River systems of North Carolina . . . . *Necturus lewisi*
32a.	Center of dorsum dark and bordered laterally on each side by conspicuous yellow stripe . . . . *Necturus maculosus*
32b.	Yellow stripes not present on dorsum . . 33
33a.	Center of belly of large specimens white and unspotted, and dorsal spotting often absent; from western Georgia to eastern Mississippi . . . . *Necturus alabamensis*
33b.	Dorsum and venter of large specimens heavily spotted; restricted to southeastern Texas, southeastern Louisiana, and southern and central Mississippi . . . . *Necturus beyeri*
34a.	From south-central Texas along Balcones Escarpment and Edwards Plateau 35
34b.	From elsewhere in the eastern United States 40
35a.	Costal grooves ≤12; eyes conspicuously reduced 36
35b.	Costal grooves >12 (usually 14–17); eyes normal or slightly reduced in size 38
36a.	Dorsum very pale brown to brownish yellow; 4–15 pairs of dorsolateral light spots on body and base of tail . . . . . . *Eurycea tridentifera*
36b.	Dorsum white; dorsolateral light spots absent 37
37a.	Legs extremely thin, and remnants of eyes present as dark spots; from underground streams in the Purgatory Creek system and along the San Marcos Springs Fault near San Marcos, Texas . . . . *Eurycea rathbuni*
37b.	Similar to 37a but more robust; extremely rare, with only one specimen known from beneath bed of Blanco River, Hays Co., Texas . . . . *Eurycea robusta*
38a.	Dorsum heavily blotched and mottled with dark pigment to produce salt-and-pepper effect; dorsum often with silvery-white pigmentation; eyes reduced; snout shovel-nosed; from within the city of Austin, Texas . . . . *Eurycea sosorum*
38b.	Dorsum not heavily blotched with dark pigment; silvery-white pigmentation absent from dorsum; eyes appearing normal in size; snout not conspicuously shovel-nosed; from areas outside Austin, Texas 39

39a. Venter yellowish white to whitish with yellowish cast beneath tail; dark ring around lens of eye; dark bar extending from eye to nostril not evident; from Spring Lake, San Marcos, Texas . . . . *Eurycea nana*

39b. Throat, belly, and underside of tail whitish and translucent; dark ring not evident around lens of eye; dark bar extending from eye to nostril usually present; from areas of the Edwards Plateau other than Spring Lake . . . . *Eurycea neotenes*

40a. From Ozark and Ouachita mountains and vicinity, and costal grooves ≥16 41

40b. From east of Mississippi River or, if from Ozark and Ouachita mountains, then costal grooves ≤15 43

41a. Four to six costal grooves between adpressed limbs; eyes appearing small relative to body size . . . . *Typhlotriton spelaeus*

41b. Seven or more costal grooves between adpressed limbs; eyes appearing normal in size 42

42a. Venter gray or yellow; tail without brownish dorsal stripe; head slightly wider than neck . . . . *Eurycea multiplicata*

42b. Venter whitish; tail with brownish dorsal stripe; head same width as body . . . . *Eurycea tynerensis*

43a. Body whitish, pinkish, or flesh colored; eyes not evident or greatly reduced; from cave habitats 44

43b. Body appearing to be normally pigmented, often with conspicuous spots, blotches, or reticulations; eyes normal in size; usually from surface water habitats, but may be from caves 46

44a. Body silvery white; eyes missing or very small; costal grooves 12–13; from northern Florida and adjacent Georgia . . . *Haideotriton wallacei*

44b. Characters and range not as above 45

45a. Dorsum with distinct dark reticulations; two or three irregular rows of pale spots on sides of large larvae; from General Davis Cave, Greenbrier Co., West Virginia . . . . *Gyrinophilus subterraneus*

45b. Dorsum lacking distinct dark reticulations; irregular rows of pale spots not evident on sides of large larvae; from southeastern Tennessee, northern Alabama, and adjacent Georgia . . . . *Gyrinophilus palleucus*

46a. Costal grooves 13–16 47
46b. Costal grooves 17–20 50

47a. Dorsum with six to nine pairs of light spots 48
47b. Dorsum lacking pairs of light spots 49

48a. Dorsal ground color of large larvae light tan; body slender; margin of dark pigmentation on venter wavy; iridophores abundant on venter; widely distributed in eastern United States . . . . *Eurycea bislineata*

48b. Dorsal ground color of large larvae dark brown; body somewhat stout; margin of dark pigmentation on venter straight; iridophores absent on venter; restricted to Graham Co., North Carolina, and Sevier and Monroe counties, Tennessee . . . . . *Eurycea junaluska*

49a. Gular pigmentation extending medially immediately in front of first gill (missing in small larvae); ventral surface of rear feet pigmented; lateral pigment extending onto belly beyond area of limb insertions; costal grooves usually 15 . . . . . . *Eurycea lucifuga*

49b. Gular pigmentation not extending medially immediately in front of first gill (older larvae only); ventral surface of rear feet not pigmented (or weakly pigmented in large larvae); lateral pigment not extending onto belly beyond area of limb insertions; costal grooves usually 14 . . . . *Eurycea longicauda* complex (*E. longicauda* and *E. guttolineata;* use range maps in text to distinguish between species)

50a. Body slender; dorsum dark brown to dull yellow; series of fine, parallel dark lines or streaks on sides of body; venter yellowish . . . . *Stereochilus marginatus*

50b. Body stout; dorsum light brown to flesh colored (older larvae often marked with dark spots or faint reticulations); parallel lines or streaks not present on sides; venter whitish 51

51a. Dorsum of larger larvae lacking black spots;

supraotic lateral line pores arranged in an ellipse; costal grooves 17–19
. . . . *Gyrinophilus porphyriticus*

51b. Dorsum of larger larvae often with black spots; supraotic lateral line pores arranged in a circle; costal grooves 16–17 52

52a. Dorsum and sides of older larvae weakly mottled or with black streaks, and usually without distinct spots (exception: specimens in upper Piedmont of North or South Carolina often spotted and unstreaked); typically from small, clear streams . . . . *Pseudotriton ruber*

52b. Dorsum and sides of older larvae uniformly brown with widely scattered black spots (exception: specimens in upper Piedmont of North or South Carolina often streaked and unspotted); typically from muddy ponds, swamps, or sluggish bottomland streams (Kentucky populations an exception)
. . . . *Pseudotriton montanus*

53a. From southern Texas south of San Antonio River; diffuse dark midventral stripe present; small light spots along sides
. . . . *Notophthalmus meridionalis*

53b. Characters and range not as above 54

54a. Greatest diameter of eye equal to distance from eye to nostril; head wider at region of eyes than just behind eyes; scattered dusky spots on tail; restricted to southeastern Georgia and northern Florida
. . . . *Notophthalmus perstriatus*

54b. Greatest diameter of eye less than distance from eye to nostril; head same width at region of eyes as just behind eyes; conspicuous dark spots on tail; widely distributed in eastern United States, and gilled adults found in scattered populations, primarily in Coastal Plain . . . . *Notophthalmus viridescens*

55a. Chin and/or throat lightly to heavily pigmented 56
55b. Chin and throat immaculate 59

56a. Costal grooves 10–13 57
56b. Costal grooves 14–15 58

57a. Costal grooves 10–11; one or two longitudinal dark stripes on ventral surface of older larvae; body of older larvae with two conspicuous cream or dull yellow stripes on each side; gilled adults common in some populations
. . . . *Ambystoma talpoideum*

57b. Costal grooves 11–13; ventral surface of older larvae lacking longitudinal dark stripes; body of older larvae lacking dull yellow stripes on each side; gilled adults not known
. . . . *Ambystoma opacum*

58a. Typically from stream habitats (rarely ponds) in limestone regions of central Tennessee, central Kentucky, southeastern Indiana, and southwestern Ohio (geographic isolates in Livingston Co., Kentucky, and Wayne Co., West Virginia) . . . . *Ambystoma barbouri*

58b. Typically from ponds, ditches, and other standing water habitats in regions outside range of 58a . . . . *Ambystoma texanum*

59a. Dorsum with pale yellow to gold middorsal stripe and one or two continuous lateral stripes bordered by dark pigmentation; head with bold black stripe through each eye
. . . . *Ambystoma cingulatum*

59b. Dorsum not as above, or if light lateral stripes present, then broken or blotched; black stripes through eyes either absent or diffuse and poorly defined 60

60a. Dorsum brown to blackish above with two cream-colored stripes along either side of body; costal grooves 13; restricted to Coastal Plain of southeastern Virginia, North Carolina, and South Carolina . . *Ambystoma mabeei*

60b. Characters and range not as above 61

61a. Costal grooves 15; irregular light band present on sides of body and separating darker pigment above and below; restricted to Ozark region of Oklahoma, Missouri, and Arkansas
. . . . *Ambystoma annulatum*

61b. Costal grooves 12–13; light band usually absent from sides of body; broadly distributed in United States 62

62a. Digits flattened, broad at base; 14–24 gill rakers on third gill arch; larvae often ≥80 mm TL
. . . . *Ambystoma tigrinum* (in part)

62b. Digits rounded in cross section, not broad at base; <12 gill rakers on third gill arch; larvae usually <80 mm TL 63

63a. Head conspicuously wider than trunk; tail fin mottled with black; large dark dorsal blotches present in all but large larvae
...... *Ambystoma jeffersonianum* complex
63b. Head not conspicuously wider than trunk; tail fin moderately pigmented but not conspicuously mottled; dorsum drab and lacking large dark dorsal blotches .. *Ambystoma maculatum*

64a. From Ouachita Mountains of west-central Arkansas and adjacent Oklahoma
.... *Desmognathus brimleyorum*
64b. From elsewhere than Ouachita Mountains of west-central Arkansas and adjacent Oklahoma
.... 65

65a. Toe tips of older larvae black 66
65b. Toe tips of older larvae not black 69

66a. Dorsum of older larvae with series of paired light spots; body slender 67
66b. Dorsum of older larvae lacking paired light spots; body often stout 68

67a. Dorsum with four or five pairs of light spots between legs; rarely >16–17 mm SVL; total number of gill fimbriae per side 11–18; broadly distributed in Appalachian Mountains
.... *Desmognathus monticola* (in part)
67b. Dorsum with five to seven pairs of light spots between legs; often >18 mm SVL; total number of gill fimbriae per side 17–27; restricted to Cumberland Mountains and Plateau of southwestern Virginia, eastern Kentucky, and adjoining areas in northeastern Tennessee
.... *Desmognathus welteri*

68a. Dorsum light brown; sides of body lacking conspicuous light flecks; often >36 mm SVL
.... *Desmognathus quadramaculatus*
68b. Dorsum dark brown; sides of body with conspicuous light flecks; rarely >35 mm SVL
.... *Desmognathus marmoratus*

69a. Underside of tail diffusely blotched; angle of mouth falls within vertical lines from anterior to posterior margins of eyes; ventrolateral lateral-line pores partially or entirely outside pigment margin; dorsal spots distinct and without dark margins; rarely from Coastal Plain
.... *Desmognathus monticola*
69b. Characters and range not as above 70

70a. Gills bushy and darkly pigmented; primarily from Coastal Plain
.... *Desmognathus auriculatus*
70b. Gills not bushy (rachis shorter than fimbriae) and whitish; often from mountainous regions outside of Coastal Plain 71

71a. From Choctawhatchee, Chattahoochee, and Apalachicola River basins in Alabama, Georgia, and panhandle of Florida; dorsum with five to seven pairs of tan or reddish spots or blotches on back and tail
.... *Desmognathus apalachicolae*
71b. From elsewhere than in 71a; dorsal spots or blotches, if present, light tan 72

72a. From southeastern Tennessee and southwestern Virginia southward
.... *D. fuscus, D. orestes, D. carolinensis, D. ocoee,* and *D. imitator* (diagnostic traits for identifying larvae of southern Appalachian populations of these species have not been found; use range or elevation to eliminate species if possible; electrophoretic evidence may be required in some cases for positive identification)
72b. From southeastern Kentucky and southern Virginia northward............ 73

73a. Dorsum lacking spots and with pale middorsal stripe, or with two to seven pairs of spots, generally restricted to posterior half of body; tail fin poorly developed and occurring as keel on posterior half of tail; posterior region of head with dark spot (V- or Y-shaped pattern on older larvae) bordered by pair of round light spots on either side
.... *Desmognathus ochrophaeus*
73b. Dorsum with 5–10 pairs of spots; tail fin well developed and extending forward to base of tail; no central dark marking bordered by pale round spots on either side of head
.... *Desmognathus fuscus*

Family Ambystomatidae
Mole Salamanders

Members of this family are stout-bodied salamanders with short, rounded heads and conspicuous costal grooves. Nasolabial grooves are absent and transformed adults have lungs. Most species are less than 20 cm TL, but *Ambystoma tigrinum* may exceed 34 cm TL. Larvae have broad heads, three pairs of bushy gills with well-developed rachises, and broad caudal fins that extend well onto the back. The prevomerine teeth occur in a transverse row that crosses the palate near the posterior margins of the internal nares.

Fossil *Ambystoma* occur in lower Oligocene through Pleistocene deposits of North America and date back to around 30 million years ago (Gehlbach 1967a; Holman 1968; Rogers 1985; Tihen 1969). Pleistocene and Pliocene records of ambystomatids include specimens of several extant forms, such as *A. tigrinum, A. opacum,* and *A. maculatum.* Of these, *A. tigrinum* is particularly well represented in Pleistocene deposits (Holman 1975, 1977, 1996; Rogers 1985; Tihen 1942, 1955). A major radiation of Mexican forms is estimated to have occurred 10–12 million years ago, based on allozyme and geologic evidence (Shaffer 1984a). However, a later interpretation based on mitochondrial DNA evidence suggests that the Mexican radiation occurred less than 5 million years ago (Shaffer and McKnight 1996).

Recent treatments recognize two genera (*Ambystoma, Rhyacosiredon*) with about 30 living species. *Rhyacosiredon* consists of four species found in the mountains of central Mexico. Recent genetic and morphological evidence suggests that this genus should be contained within *Ambystoma* (Reilly and Brandon 1994; Shaffer 1984a,b, 1993). The remaining species consist of about 26 *Ambystoma* found through much of North America southward to south-central Mexico. Gilled adults are common in certain *Ambystoma,* and unisexual complexes of hybrid origin are found in southeastern Canada and the north-central and northeastern United States. Members of the latter group have semiclonal modes of reproduction and are treated in a special account on unisexual *Ambystoma.*

Major systematic studies include those of Brandon (1972, 1977), Jones et al. (1993a), Kraus (1988), Kraus and Petranka (1989), Reilly and Brandon (1994), Shaffer (1984a,b, 1993), Shaffer and McKnight (1996), Shaffer et al. (1991), and Tihen (1958). Significant disagreement exists concerning the systematic relationships of species in the United States and Canada because of the sometimes conflicting results of analyses using genetic variation, morphological characters, or a combination of the two (Jones et al. 1993a; Kraus 1988; Shaffer and McKnight 1996; Shaffer et al. 1991). Jones et al. (1993a), using a combination of electrophoretic and morphological characters, distinguished four major clusters: a *gracile-maculatum-talpoideum* cluster, a *jeffersonianum-opacum-laterale-macrodactylum* cluster, a *tigrinum-californiense* cluster, and an *annulatum-cingulatum-barbouri-texanum-mabeei* cluster (the *"Linguaelapsus"* group). Using mitochondrial DNA analyses, Shaffer and McKnight (1996) recognized eight major clades in *A. tigrinum* and other members of the *A. tigrinum* complex in Mexico. These groups are poorly differentiated from each other, a finding that suggests a very recent derivation.

Ambystoma annulatum Cope
Ringed Salamander
PLATE I

Fig. 10. *Ambystoma annulatum*; adult; Stone Co., Missouri (R. W. Van Devender).

IDENTIFICATION. The ringed salamander is a slender *Ambystoma* that is boldly marked with white, buff, or yellow crossbands on a grayish black ground color. The narrow crossbands extend from the neck to near the tail tip and are often broken into spots or bold streaks that partially encircle the dorsum. The belly varies from slate gray to buff yellow, depending on the extent to which lighter markings invade the ground color. There are usually 15 costal grooves between the limbs, and the tail is long and slender. Adults are 14–23.5 cm TL, and females are slightly larger than males on average (Anderson 1965). Males have a conspicuously swollen cloaca and papillose cloacal lining during the breeding season.

Larvae have a broad dorsal fin that extends forward to the head. Mature larvae have uniformly brownish to greenish brown pigmentation on the dorsal surfaces and sides, and have a pigment-free band that extends along the sides of the trunk from the gills to the basal third of the trunk (Bishop 1943). Hatchlings in two populations studied by Hutcherson et al. (1989) measured 6.9 and 7.6 mm average SVL. Trauth and Cartwright (1989) collected an albino larva in Arkansas.

Newly metamorphosed animals have olive to black backs, grayish white bellies, and a distinct row of dorsolateral light spots that extend from the forelimbs onto the tail. Juveniles acquire blotches of yellow coloration soon after metamorphosis and assume the adult pattern within 2 months after transforming (Hutcherson et al. 1989).

SYSTEMATICS AND GEOGRAPHIC VARIATION. *Ambystoma annulatum* has had a relatively stable nomenclature since it was described in 1886. No subspecies are recognized.

DISTRIBUTION AND ADULT HABITAT. Ringed salamanders are found in or near hardwood or mixed hardwood-pine forests in the Ozark Plateau and Ouachita Mountains of Arkansas, Missouri, and Oklahoma. Local populations are scattered throughout the range of the species where suitable breeding habitats exist. The adults are highly fossorial and live in subterranean retreats in forests throughout much of the year (Dowling 1956).

BREEDING AND COURTSHIP. Breeding occurs in fish-free habitats such as woodland pools, mud-bottomed farm ponds, and seasonally ephemeral ponds adjoining forests (Brussock and Brown 1982; McMillian and Wilkinson 1972; Peterson et al. 1991; Spotila and Beumer 1970; Trapp 1956). Adults use temporary pools both in low-lying habitats and along ridgetops and are capable of homing both uphill and downhill to ponds. Adults seem to prefer murky ponds used by livestock as breeding sites (Brussock and Brown 1982).

The ringed salamander tends to be an explosive breeder, and the adults normally engage in two to four major breeding bouts during the breeding season. Individuals begin migrating to breeding sites at night following the first heavy, sustained rains of late summer and early autumn. Migrations occur during or immediately after periods of rain, and the adults may require several weeks to reach the ponds. A few individuals may migrate on humid nights without rain (Brussock and Brown 1982).

Breeding occurs from about mid-September through November in most populations (Brussock and Brown 1982; Hutcherson et al. 1989; McDaniel and Saugey 1977; Noble and Marshall 1929; Peterson et al. 1991; Spotila and Beumer 1970; Trapp 1956). Records for Missouri ponds include observations of eggs in late September (Noble and Marshall 1929) and breeding adults from 18 September to 16 October (Hutcherson et al. 1989; Peterson et al. 1991). In Arkansas breeding

Range of *Ambystoma annulatum*

has been observed in early October (Trapp 1956) and on 14 February (Trauth et al. 1989a). Winter breeding in the latter study is atypical; observations at this site suggest that most adults breed during October.

Descriptions of courtship behavior are based primarily on the mass courtship of 43 adults that was observed in a laboratory tank (Spotila and Beumer 1970). Initially a male approaches a female, nudges her cloaca and sides briefly, then swims off a short distance and deposits one or more spermatophores. The male then returns to the female and repeats the sequence. While depositing spermatophores, a male grasps the substrate with his rear feet, presses his cloaca to the substrate, arches and curves his tail, then deposits the gelatinous base and sperm cap while slightly undulating the cloacal region. A male may deposit nine spermatophores in 2 minutes.

As additional males begin courting, they lose their specificity for particular females and begin nudging both males and females indiscriminately before depositing spermatophores. Females are generally unresponsive to males and do not pick up seminal fluid during this phase. Males may deposit spermatophores on rocks and gravel, on other spermatophores, and on the bodies and tails of other individuals. In natural habitats spermatophores are attached to decaying plant material or placed directly on the pond bottom.

Courtship observed by Spotila (1976) in a natural pond on 13 October is very similar to the laboratory observations of Spotila and Beumer (1970). Animals at this site gather in groups of 6–12 throughout the pond and engage in mass courtship. Males occasionally move away from groups to deposit spermatophores. As in the laboratory, females remain passive and do not pick up sperm caps while being actively courted by males. Spermatophores in the field are concentrated in areas of 100 cm^2 or less, and multiple spermatophores consisting of as many as 11 spermatophores are often present.

The spermatophore averages 3.4 mm in height and consists of a gelatinous base about 4 mm wide with a short stalk topped by a sperm cap. A slight sagittal fold runs across the top of each spermatophore and divides it in half (Spotila and Beumer 1970).

REPRODUCTIVE STRATEGY. Females begin ovipositing within 1–2 days after breeding, and the density of eggs in ponds can average as high as 638 eggs/m^2 (Peterson et al. 1992). Freshly laid ova are about 2 mm in diameter and are surrounded by a vitelline membrane and two jelly capsules.

Females in laboratory tanks will deposit eggs singly, in small strings, or in loose masses that hold 50 or more eggs (Spotila and Beumer 1970). In a Missouri population the females attach a small percentage of their eggs singly or in loose clumps to vegetation, but most are placed directly on the pond bottom (Noble and Marshall 1929). The number of eggs per mass at this site is 1–45 and averages 10, whereas in two Arkansas populations it is 2–31 and averages 14 and 8 (Trapp 1956). The only estimate of clutch size is that of Hutcherson et al. (1989), who found an average of 390 ± 16 (SE) mature ova in 16 females. As with many *Ambystoma*, clutch size is positively correlated with female SVL.

The embryonic period of *A. annulatum* is shorter than that of most *Ambystoma* because the eggs are laid in the autumn, when seasonal water temperatures are relatively high. Estimates by Trapp (1956) and Hutcherson et al. (1989) are 9–10 days and 16 days, respectively.

One risk of autumnal breeding is that the breeding ponds may not fill before the adults arrive to breed. In addition, ponds that do fill partially in early autumn may dry within weeks if rains do not maintain them. Hutcherson et al. (1989) observed one instance in which adults laid eggs in a dried pond that had no standing water. This unusual behavior is atypical of *A. annulatum*, but similar tendencies in ancestral forms of other species such as *A. opacum* and *A. cingulatum* may have provided a basis for the evolution of terrestrial breeding in *Ambystoma*.

The embryos in an Arkansas pond hatched on 12 October, but the pond dried within 2 weeks posthatching (Trapp 1956). This and other observations indicate that larvae occasionally incur heavy mortality from both premature pond drying and pond freezing (Hutcherson et al. 1989).

AQUATIC ECOLOGY. Ringed salamander larvae hatch in the fall, overwinter, then enter a major growth phase in late winter and spring as water temperatures increase. The larvae feed on a wide array of aquatic organisms. Small larvae at one site primarily eat cladocerans, copepods, and dipteran larvae, whereas larger animals feed mostly on dipteran larvae (Hutcherson et al. 1989). Other prey include ostracods, hemipterans, snails, and dragonfly and damselfly nymphs. None of the specimens contained conspecifics, even though cannibalism was observed in the field and laboratory. Larvae in a Missouri pond eat microcrustaceans, chironomids, beetles, snails, earthworms, frog eggs, and conspecific larvae (Nyman et al. 1993).

Cannibalism is more likely to occur when a local population contains a wide range of larval size classes. Cannibalistic larvae are typically larger than non-cannibalistic larvae and often transform at a larger size and an earlier age (Nyman et al. 1993). Cannibalistic larvae have slightly different head shapes, but it is uncertain if these are true genetic morphs, as seen in *A. tigrinum*.

Larvae may begin metamorphosing as early as April (Peterson et al. 1991), although at many sites metamorphosis begins 1–2 months later. At a Missouri site the larval period lasts 6–8.5 months, depending on the pond and year (Hutcherson et al. 1989). Larvae in most ponds transform in May when average SVL and mass at metamorphosis are 34–40 mm and 0.9–1.7 g, respectively. Larvae in one pond may not metamorphose until June or early July.

At another Missouri site metamorphosis occurs from late April to early June, and the larval period lasts 6.5–8 months (Nyman et al. 1993). Arkansas larvae studied by Trapp (1956) average 45 mm TL by 14 May and transform before 1 June. The larval period lasts about 7–7.5 months at this site.

As in most *Ambystoma*, the larval stage for this species is risky, and premetamorphic survival is often low. Peterson et al. (1991) estimate premetamorphic survival to be only 0.01% in one population. Age-specific survival is highest in young larvae and in larvae nearing metamorphosis. Mortality is minimal during the winter months, when larvae and their predators are less active.

TERRESTRIAL ECOLOGY. Almost no data are available on the terrestrial ecology of the juveniles and adults. Metamorphs primarily move away from ponds during rains, but some move during dry periods (Hutcherson et al. 1989). Juveniles presumably migrate to surrounding forests and live in underground burrows until sexually mature. Individuals do not appear capable of actively digging burrows but can enlarge existing holes and crevices (Semlitsch 1983c). Adults are rarely active on the ground surface outside the breeding season. All but 3 of 23 digestive tracts examined in one study were empty, and earthworms were the only identifiable prey (Hutcherson et al. 1989).

Length of the juvenile stage is not known, but based on data for other *Ambystoma* most juveniles probably become sexually mature within 2–3 years after transforming. Spotila and Beumer (1970) collected 155 adults that measured 74–112 mm SVL.

PREDATORS AND DEFENSE. Natural predators are not known but undoubtedly include owls, wood-

Fig. 11. *Ambystoma annulatum;* metamorph; Stone Co., Missouri (R. W. Van Devender).

Fig. 12. *Ambystoma annulatum;* defensive posturing; LeFlore Co., Oklahoma (E. D. Brodie, Jr.).

land snakes, shrews, skunks, raccoons, opossums, and other mammals. Defensive posturing in this species involves coiling the body while tucking the head underneath the base of the tail. Some individuals may weakly lash their tails when molested (Brodie 1977).

COMMUNITY ECOLOGY. Ringed salamander larvae often share ponds with other *Ambystoma* and with wood frog tadpoles (*Rana sylvatica*), but ecological interactions between community members have not been examined. Larvae can reach high densities and in all likelihood play important roles in structuring zooplankton communities.

CONSERVATION BIOLOGY. The ringed salamander has a relatively restricted range, and local populations are patchily distributed across the landscape where breeding habitats occur. Local breeding populations should be protected whenever possible by keeping ponds fish-free and by leaving a buffer of deciduous forests around breeding sites. The distance that adults move from ponds is not known, but a forest buffer of 200–400 m radius would probably provide suitable habitats for most adults. This spectacularly beautiful and docile animal would be an outstanding candidate for use in environmental education programs.

Ambystoma barbouri Kraus and Petranka
Streamside Salamander
PLATE 2

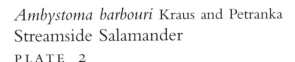

IDENTIFICATION. The streamside salamander is a sibling species of the small-mouthed salamander (*A. texanum*) and closely resembles the latter species. Adults have a grayish brown to grayish black dorsum that is heavily overlain by tan or light gray lichenlike markings. The venter is dark gray to grayish black with light grayish blotches, and the snout is wide and rounded. Adults are 11–17 cm TL and have 14–15 costal grooves. Males have conspicuously swollen cloacae during the breeding season and are slightly smaller on average than females. Adult females from southwestern Ohio populations are 8% longer in SVL on average than males (Downs 1989).

Hatchlings are similar in general color and patterning to *A. texanum* hatchlings, but they are more darkly pigmented and two to three times larger at any developmental stage. Larvae are the pond type and have broad dorsal fins that extend to the neck region. Small larvae are darkly pigmented with greenish brown above and usually have a series of three to six paired blotches or saddlelike markings down the middorsal region. The lighter markings become less conspicuous with age and are often missing in older larvae. Metamorphs are uniformly brown to brownish gray. Juveniles begin acquiring the adult color patterns within 3–6 weeks after transforming.

Fig. 13. *Ambystoma barbouri;* adult; Fayette Co., Kentucky (R. W. Barbour).

Fig. 14. *Ambystoma barbouri;* albino adult; Fayette Co., Kentucky (J. W. Petranka).

Adult *A. barbouri* are difficult to distinguish from *A. texanum* using external morphology, and the two species are best separated in areas of close geographic contact by tooth morphology and geographic range. In regions of parapatry, the maxillary and premaxillary teeth of *A. barbouri* have short, spatulate cusps compared with long, pointed cusps in *A. texanum* (Kraus and Petranka 1989). *Ambystoma barbouri* usually has a more rounded tail, a wider and more rounded snout, a greater distance between the nares, and more conspicuous costal grooves and folds than *A. texanum.*

SYSTEMATICS AND GEOGRAPHIC VARIATION. *Ambystoma barbouri* is a sibling species of *A. texanum* that was previously considered to be conspecific with *A. texanum* (Petranka 1982a). These species are mostly allopatric, but three narrow zones of contact are known. In one deforestation and the filling of wetlands have eliminated most populations. In the remaining two zones the adults occasionally hybridize, but there is no evidence of a smooth zone of intergradation (Kraus and Petranka 1989). *Ambystoma barbouri* was recently discovered in central Tennessee, and intensive collecting in this region will no doubt reveal a fourth area of parapatry with *A. texanum.* Electrophoretic data may be of limited value in distinguishing *A. barbouri* from *A. texanum* because both species show extreme heterozygosity of gene loci.

Ambystoma barbouri and *A. texanum* differ in preferred breeding habitat, mode of egg deposition, egg size, clutch size, larval morphology, larval behavior, and the shape of the teeth and premaxillas. These differences are maintained in populations that occur within 100 m of each other and provide strong evidence that gene exchange between species is minimal (Kraus and Petranka 1989).

I documented geographic variation in egg size, hatching stage, and larval pigmentation and morphology (Petranka 1982a). Variation in these traits is minor throughout the range of *A. barbouri,* and no subspecies are recognized.

DISTRIBUTION AND ADULT HABITAT. *Ambystoma barbouri* occurs in central Tennessee, western and central Kentucky, southeastern Indiana, southwestern Ohio, and in extreme western West Virginia. Geographic isolates occur in Livingston and Russell counties, Kentucky (Kraus and Petranka 1989); in Davidson, Jackson, and Rutherford counties, Tennessee (B. T. Miller and A. F. Scott, pers. comm.); and in Wayne Co., West Virginia. Specimens from Washington Co. in southeastern Ohio are either *A. texanum* or *A. barbouri-texanum* hybrids (F. Kraus, pers. comm.).

Adults occupy upland deciduous forests in regions with rolling topography. Populations are primarily restricted to limestone regions, and local populations are rarely found in stream sections where the surrounding forest has been eliminated.

BREEDING AND COURTSHIP. Adults usually breed in first- and second-order streams, although they occasionally breed in farm ponds, sinkhole ponds, and quarry ponds (Craddock and Minckley 1964; Petranka 1982a). Breeding streams are typically 2–5 m wide, have moderate gradients, and consist of alternating pools and riffles with large, flat limestone rocks. Breeding streams usually have small cascades or waterfalls along their lengths that act as barriers to the

Range of *Ambystoma barbouri*

upstream movement of predatory fish, and sections used by *A. barbouri* for breeding are usually seasonally ephemeral.

Ashton (1966) found freshly laid eggs that are presumably those of *A. barbouri* in streams in Davidson Co., Tennessee, from late December to early April.

Fig. 15. *Ambystoma barbouri;* breeding stream; Fayette Co., Kentucky (J. W. Petranka).

Adults that I monitored in central Kentucky migrate to breeding streams from late October through March (Petranka 1984a). Movements occur on rainy nights when air temperatures are 2–16°C and stop within 30 minutes after rainfall ceases. Males tend to arrive at breeding streams before females and outnumber females by about two to one. Breeding in this region occurs from late December to mid-April and lasts for about 3 months in any population.

Small numbers of sexually active adults can be found in streams throughout the breeding season. At one site in central Kentucky, breeding begins in early to mid-January, peaks in mid-March, and ends by mid-April. Water temperatures during the breeding season are 0–12°C, and adults are often sexually active when water temperatures are near freezing (Petranka 1984a).

Courtship has not been observed in nature, but observations of spermatophores and adults beneath rocks suggest that the adults often court in small groups beneath large rocks. Of 68 adults that I collected beneath rocks, 30 were alone and the remainder were in groups of two to five individuals (Petranka 1984b). The following account of courtship is based on my observations of small groups of adults maintained in laboratory aquaria (Petranka 1982b).

Males initially move slowly about and nudge the bodies of conspecifics. After encountering a female, a

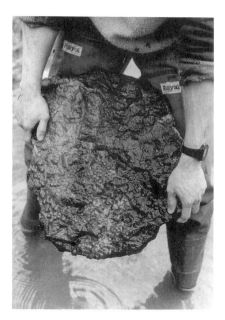

Fig. 16. *Ambystoma barbouri;* eggs on underside of limestone rock; Fayette Co., Kentucky (J. W. Petranka).

male begins to swim vigorously about with exaggerated undulations of the body and tail. Within minutes other males follow suit. Males then circle about in tight groups and vigorously nudge and prod others' bodies and cloacae. During this time, females normally remain motionless or move slowly about. After 1–18 minutes of mutual prodding, males begin depositing spermatophores on the substrate. When depositing a spermatophore, a male tightly grasps the substrate with his hindlimbs, presses his cloaca to the substrate, and slowly undulates his tail. Spermatophore deposition requires about 11 seconds. Males sometimes try to dislodge or disrupt spermatophore deposition by others, and they often deposit spermatophores on those of rival males.

Males may court females for one to several minutes after wandering from courting groups of males. Females eventually become responsive, move into spermatophore fields, and begin picking up spermatophores. Females may collect sperm from as many as 18–27 spermatophores.

When mounting a spermatophore a female grasps the substrate with her hindlimbs, then presses her cloaca to the spermatophore while undulating the tail slowly. After mounting spermatophores, females become docile and motionless; however, males may continue to court long after female sexual activity subsides. Females lay fertile eggs within 48 hours after courting.

REPRODUCTIVE STRATEGY. Females in streams usually attach eggs singly in a monolayer to the undersides of rocks (Kraus and Petranka 1989; Petranka 1982b). When ovipositing, a female flips over on her back and deposits a line of six to eight eggs spaced 1–4 cm apart (Petranka 1984b). The tail and dorsum are used to brace the female as she flips upside down. An egg is deposited every 15–45 seconds, and a female rests for several minutes before depositing a second line of eggs. Crisscrossing of lines of eggs ultimately produces a group of rather tightly packed eggs. Two or more females occasionally lay eggs under the same rock, so the number of eggs per monolayer can exceed the maximum clutch size of females. The number of eggs in 199 monolayers that I examined averaged 122 and varied from 8 to 1142 (Petranka 1984b). In general, larger rocks contain more eggs (Holomuzki 1991).

When breeding in ponds, females plaster eggs to the undersides of logs or rocks, or attach eggs singly to grasses or sedges. Eggs attached singly to grasses are concentrated within a 20- to 30-cm-diameter area rather than being scattered widely in the pond. In streams, egg densities are highest in pools, intermediate in runs, and lowest in riffles (Holomuzki 1991; Kats and Sih 1992). Females tend to avoid laying eggs in stream pools with green sunfish (*Lepomis cyanellus*), which, although they do not prey on the adults, are voracious predators on *A. barbouri* hatchlings. Maximum egg density in pools is about 26 eggs per square meter of habitat.

Ambystoma barbouri shows a strong tendency to lay eggs in cryptic sites, but its close relative, *A. texanum*, does not. To determine the adaptive value of this behavior, Sih and Maurer (1992) experimentally exposed *A. barbouri* eggs by flipping over rocks in a study stream. The results show that exposed eggs suffer extremely high mortality relative to control eggs that are

Fig. 17. *Ambystoma barbouri;* eggs on underside of stone; Fayette Co., Kentucky (J. W. Petranka).

maintained in their natural positions. Many of the exposed eggs wash loose from rocks, whereas others are exposed to air and desiccate as the stream subsides seasonally. This experiment demonstrates that cryptic egg laying is adaptive and explains why natural selection has favored cryptic egg deposition in *A. barbouri*. Ashton (1966) noted that eggs laid on the tops of rocks in central Tennessee often desiccated as the water levels in streams dropped.

The ova of *A. barbouri* are about twice as large as those of *A. texanum*. Recently deposited ova are 2.4–3.8 mm in diameter and are dark gray to brownish above and cream colored below. They are surrounded by two envelopes in addition to the vitelline membrane. The number of mature ova in 14 females that I examined varied from 184 to 397, averaged 262, and was positively correlated with female SVL (Petranka 1984a).

Most embryos in a central Kentucky population hatch during a 2- to 3-week period in April (Holomuzki 1991; Petranka 1984a). The incubation period for individual clutches is 29–82 days and is inversely related to the seasonal time of oviposition (Petranka 1984b). Survivorship to hatching is greatest in eggs deposited earliest in the season and averages 80–84% between years.

AQUATIC ECOLOGY. Larvae feed on benthic invertebrates as well as zooplankton. Larvae from several ponds and streams in central Kentucky that C. K. Smith and I examined primarily ate zooplankton, chironomid larvae, amphipods, and isopods. Cannibalism of smaller larvae occurs on very rare occasions (Smith and Petranka 1987). *Ambystoma barbouri* is a size-selective feeder on the isopod *Lirceus fontinalis* and cannot easily handle large individuals. Sparkes (1996) provides evidence that size-selective feeding on *Lirceus* by *A. barbouri* has favored the evolution of delayed maturity in local stream populations of *Lirceus*.

Larvae will often catch two or more prey before swallowing their catch. A. Sih and I examined optimal foraging by *A. barbouri* larvae and found that individuals do not conform to conventional theory (Sih and Petranka 1988), which assumes that predators cannot search for prey while handling a prey item. When a multiple-prey model is applied to larvae, individuals conform to optimal foraging theory. At low prey densities larvae are nonselective. At high prey densities larvae are nonselective when their mouths are empty but prefer larger prey when their mouths are partially full.

Fig. 18. *Ambystoma barbouri*; larva; Fayette Co., Kentucky (J. W. Petranka).

Larvae grow rapidly and transform 6–10 weeks after hatching, when they weigh 0.6–1.2 g on average (Petranka 1984b; Petranka and Sih 1986). Transformation usually occurs from late May to mid-July (Holomuzki 1991; Keen 1975; Petranka 1984b). In a comparative study of seven populations in central Kentucky, I found that growth rate, length of the larval period, and size at metamorphosis are highly variable. In general, larvae in ponds grow faster and are larger at metamorphosis than larvae in streams (Petranka 1984c). When larvae are raised in a common laboratory environment, differences among populations are slight. This finding suggests that environmental factors such as food availability and water temperature are primarily responsible for differences in growth rates between natural populations.

Fish are major predators of *A. barbouri* larvae, and local populations are primarily restricted to the upper sections of stream drainages that lack fish. I surveyed 90 streams and found that larvae and predatory fish rarely occur together in the same microhabitats within stream drainages (Petranka 1983). Green sunfish that wash into upstream pools from farm ponds rapidly eliminate local pool populations of larvae, and larvae occur at high densities in these same pools during years when fish are absent.

Fish occur sporadically in pools in transitional areas between highly ephemeral and permanent sections of streams. In such cases, larvae are often abundant in fishless pools but rare in fish pools. Sih et al. (1992) conducted a field experiment to examine how fish predation, larval drift, and modification of refuge use produce patterns of habitat segregation between fish and larvae in this transitional zone. When green sunfish are added to pools, larvae spend more time hiding beneath rocks, spend less time in the deeper sections of the pools, and increase their downstream drift at night. Overall, densities decrease more rapidly in fish pools than in control pools until almost no larvae remain in fish pools. Comparisons of immigration versus emigration rates indicate that fish predation per se is nearly completely responsible for the decline of larvae in fish pools. Overall, only 6–8% of the larvae that drift into fish pools survive to drift out.

More detailed studies show that larvae can use chemical cues to detect predatory sunfishes (Kats 1988; Kats et al. 1988; Petranka et al. 1987b; Sih and Kats 1994). Larvae normally respond to fish chemicals by quickly moving beneath rocks or other cover. However, in the absence of refuges they may use immobility (lack of active swimming) as a defense. Larvae are capable of responding to fish chemicals immediately after hatching, when they are most vulnerable to downstream drift into fish pools (Sih and Kats 1994). Although older larvae typically respond to fish chemicals by moving into refuges, they soon emerge and are quickly preyed upon by fish (Huang and Sih 1990; Sih et al. 1988). Predation rates are so high that long-term coexistence with fish in the same pool is impossible. Immobility in the presence of fish may allow a few larvae to drift downstream into pools that lack fish. In addition, larvae approaching metamorphosis may be able to survive in refuges beneath rocks until they transform and leave the stream.

Sih and Kats (1991) found that the behavioral response of larvae to fish chemicals is dependent upon the presence or absence of refuges. In trials that simulate larvae drifting into pools with fish, larvae tend to remain immobile after encountering fish chemicals if refuges are not present. When refuges are present, larvae quickly move beneath them. Thus, larvae can assess refuge availability and predation risk, and respond with behaviors that maximize the probability of surviving fish predation.

Other studies show that larvae are less likely to emerge from refuges in experimental pools with predatory fish when isopods are present (Huang and Sih 1990). The presence of isopods stimulates fish activity. This increase in activity, in turn, triggers strong avoidance responses from larvae in refuges. Huang and Sih (1991) document the direct and indirect interactions that occur between *A. barbouri*, an isopod (*Lirceus*), and green sunfish. Green sunfish and *A. barbouri* larvae both feed on isopods. Surprisingly, isopods actually benefit from the presence of fish because fish reduce overall predation rates of salamander larvae on the isopods.

During floods, larvae in upstream sections often wash downstream, where they are vulnerable to fish predation (Sih et al. 1992). Drift rates vary depending on the size of streams and current. Holomuzki (1991) found that 12% of larvae drift at least 10 m/day in a small headwater stream, and that about 1 month is required for complete turnover of populations in local pools.

A. Sih and I conducted a 6-year study of a stream population in central Kentucky to determine whether growth and survival of larvae are density dependent (Petranka 1983, 1984b,c; Petranka and Sih 1986). Larval survivorship from hatching to metamorphosis is low at this site and varies from 0.9 to 6.6% among years. Flooding, stream drying, and fish predation are the major sources of larval mortality. In some years severe flooding early in the larval period reduces larval densities to the point that competition for food is undetectable. Larvae in flood years are about the same size at metamorphosis as those fed ad libitum in the laboratory at similar temperatures. In years without catastrophic floods, hatchling densities remain relatively high throughout the larval period, the larval period is prolonged, and larvae metamorphose at a much smaller size than in flood years. Volume of food in the gut during years of high density is about half that during years when densities are relatively low. In years of high density, the mean size of larvae in local pools within the stream also varies inversely with density. These results suggest that density-dependent regulation in the larval stage may limit adult population size. Holomuzki (1991) notes that survival in a central Kentucky stream averages 15.9% in pools compared with only 3.5% in runs, where larvae often became trapped and die as these microhabitats dry seasonally.

TERRESTRIAL ECOLOGY. The metamorphs presumably disperse away from streams soon after transforming. I have collected adults on roads as far as 300–400 m from the nearest streams; many juveniles probably move similar distances away from streams. Juveniles are highly secretive and apparently remain beneath the ground until sexually mature. The adults are also highly fossorial and are rarely found on the ground surface except during the breeding season. No data are available on juvenile growth rates or the time required to reach sexual maturity.

PREDATORS AND DEFENSE. Although fish are major predators on *A. barbouri* larvae, crayfish, water snakes (*Nerodia*), and planarians (*Phagocotus gracilis*) also prey on small hatchlings and larvae (Holomuzki 1989a; Kats 1986; Petranka et al. 1987a). Larvae often con-

Fig. 19. *Ambystoma barbouri;* defensive posturing; Fayette Co., Kentucky (E. D. Brodie, Jr.).

centrate in mats of *Cladophora* algae, where food levels are high and risk from predation from crayfish and water snakes is relatively low (Holomuzki 1989a).

Sih and Moore (1993) demonstrated that *A. barbouri* embryos may delay hatching in response to a predator. When exposed to predatory planarians or planarian chemicals, the embryos hatch at a later developmental stage and larger size than embryos exposed to nonpredatory isopods. Because large hatchlings are less vulnerable to planarian predation, delayed hatching appears to be an antipredator adaptation. Experiments using green sunfish (*Lepomis cyanellus*) further show that embryos will delay hatching when exposed to chemicals from this predator (Moore et al. 1996). Hatching at a larger size and more advanced developmental stage could reduce the probability of hatchlings drifting into fish pools or enhance the probability of escape from fish attacks.

The natural predators of adults are poorly documented. One radioactively tagged adult was found in the stomach of an eastern garter snake (R. W. Barbour, pers. comm.). Adults defensively posture by raising the rear limbs, then raising and undulating the tail, which contains noxious secretions on its dorsal surface. If attacked, individuals may either tuck the head beneath the tail or coil the body and remain immobile with the head and tail raised and the hindlimbs extended posteriorly (Brodie 1977; Brodie et al. 1974b).

COMMUNITY ECOLOGY. *Ambystoma barbouri* larvae are important elements of certain headwater stream communities, where they replace fish as top predators. Headwater streams in central Kentucky have proven to be excellent systems for understanding how predation, competition, habitat instability, and organismal behavior interact to structure biological communities and influence patterns of species diversity along environmental gradients.

CONSERVATION BIOLOGY. Surveys that I have conducted show that adult populations are almost invariably associated with hardwood forests. *Ambystoma barbouri* has declined in many areas throughout its range where native forests have been destroyed and replaced with pastureland or residential areas. Several populations occur within small nature preserves in the Bluegrass region of Kentucky and are protected. However, additional protection of forested ravines is needed to protect local populations of this and other vertebrates in a region that is undergoing rapid urbanization.

COMMENTS. Several lines of evidence indicate that *A. barbouri* evolved from pond-breeding *A. texanum* stock that colonized streams. The most conspicuous is the fact that *A. barbouri* larvae have pond-type morphology. Delayed hatching, increased egg size, a reduction in clutch size, and the tendency to oviposit in monolayers beneath rocks parallel trends that occur in stream-breeding salamanders as a whole. Selection for increased egg size and delayed hatching appear to be closely tied to size-related mortality in hatchlings. Hatchlings that drift downstream are highly vulnerable to fish predation. My colleagues and I find that predation by planarians and susceptibility to stream drift decrease markedly with hatchling size (Petranka et al. 1987a). These factors may have favored delayed hatching and the production of larger eggs following the initial colonization of streams by *A. texanum* stock.

Ambystoma barbouri larvae are restricted to the ephemeral, upper sections of streams because of fish predation and often suffer heavy mortality from stream drying. In a comparative study with *A. texanum,* A. Sih and I demonstrated that *A. barbouri* has a shorter larval period and is smaller at metamorphosis than *A. texanum* when raised under the same conditions of food availability and temperature in the laboratory (Petranka and Sih 1987). Genetic crosses also demonstrate that these traits are influenced by parental genotypes and are responsive to natural selection. Mortality from stream drying appears to be the primary agent that has favored shorter larval periods in *A. barbouri*. The evolution of smaller size at metamorphosis may reflect the fact that these two variables are genetically linked and that selection for shorter larval periods has indirectly modified size at metamorphosis.

Ambystoma californiense (Gray)
California Tiger Salamander
PLATE 3

Fig. 20. *Ambystoma californiense;* adult; Monterey Co., California (R. W. Van Devender).

IDENTIFICATION. The California tiger salamander is a stocky salamander with a broad, rounded snout and small eyes. The dorsal ground color is lustrous black, and the salamander is marked with bold, rounded or irregular lemon yellow spots that are concentrated along the sides of the body. The belly is grayish and may contain a few small, dirty yellow spots. Albinism is rare but has been reported in Madera Co., California (Hensley 1959). Adults usually have 12 costal grooves and are 15-22 cm TL. The average SVL of males and females is similar; however, males have proportionately longer tails and a greater average TL. Males also develop swollen cloacae during the breeding season.

The larvae are yellowish gray and have broad, flat heads; truncated snouts; broad dorsal fins that extend well onto the back; and large, feathery gills. Hatchlings measured by Storer (1925) averaged 10.5 mm TL. California tiger salamanders only occasionally share breeding ponds with other salamanders, such as California newts, and the larvae are easy to identify based on their large size and head shape. Larvae transform prior to reaching sexual maturity even though several western subspecies of closely related *A. tigrinum* have gilled adults (Collins et al. 1980; Hensley 1964; Larson 1968).

SYSTEMATICS AND GEOGRAPHIC VARIATION. *Ambystoma californiense* is considered by some researchers (e.g., Gehlbach 1967a) to be a subspecies of *A. tigrinum*. However, most researchers now treat this form as a separate species because it is geographically isolated from *A. tigrinum* and differs in coloration and natural history from western subspecies of *A. tigrinum*. Recent genetic comparisons with subspecies of *A. tigrinum* indicate that *A. californiense* is well differentiated genetically from all *A. tigrinum* subspecies (Shaffer and McKnight 1996; T. R. Jones, pers. comm.) and provide further evidence for recognizing *A. californiense* as a separate species. Geographic variation in color patterning within *A. californiense* is rather uniform, and no subspecies are recognized.

DISTRIBUTION AND ADULT HABITAT. The California tiger salamander is restricted to the Central Valley of California and to lower elevations to the west where much of the original vegetation was grasslands and open oak woodlands. Some populations that once inhabited this region have been extirpated because of urbanization and the conversion of native grasslands and wetlands into agricultural lands. The remaining populations of *A. californiense* occur in scattered colonies in areas where there are suitable breeding sites (Fisher and Shaffer 1996).

BREEDING AND COURTSHIP. California tiger salamanders breed in fish-free, seasonally ephemeral ponds. Juveniles and adults are fossorial and are rarely seen on the surface except during the winter breeding season when the ground surface is moist. Migrations to and from the ponds occur from November through April, although most breeding occurs from December through March. Breeding pools often do not form until January or February when soils become saturated following the arrival of autumnal rains (Storer 1925). In drought years, ponds may not form at all and the adults cannot breed (Barry and Shaffer 1994).

Adults engage in mass migrations during a few rainy nights and are explosive breeders (Barry and Shaffer 1994). Males often outnumber females at the breeding sites (Twitty 1941), and the adults vacate ponds shortly after breeding. Records of breeding activity include observations of migrating or breeding adults from late November to March and observations of eggs from

Range of *Ambystoma californiense*

December to February (Holland et al. 1990; Loredo and Van Vuren 1996; Storer 1925; Twitty 1941).

The most detailed study of the reproductive ecology of a local population is by Loredo and Van Vuren (1996), who conducted a 2-year study of a population in Contra Costa Co. At this site, males may begin arriving at the breeding site in late November and early December following autumn rainstorms and subsequent pond filling. Most males arrive before mid-December, while the majority of females may not arrive until a week or two later. Data from pitfall trap captures suggest that the adults engage in one to three major breeding bouts during December through mid-January. Both the duration of stay (means for two consecutive years = 20 and 54 days for males, 14 and 6 days for females) and the sex ratio of males to females (0.8:1, 8.1:1) may vary markedly among years. The higher ratio in certain years may reflect the fact that females are less likely than males to undergo migrations to breeding sites in years when autumnal rainfall is low.

Detailed descriptions of courtship have not been published. Twitty (1941) observed males actively nudging the cloacal region of females, a behavior that suggests that courtship is similar to that of *A. tigrinum*. The spermatophore of *A. californiense* resembles that of *A. t. tigrinum* but is slightly larger. It consists of a broad gelatinous base that narrows to a neck. The apex flares into four raised horns that hold seminal fluid.

REPRODUCTIVE STRATEGY. Females attach their eggs singly, or in rare instances in groups of two to four, to twigs, grass stems, vegetation, or detritus (Storer 1925; Twitty 1941). Each ovum is surrounded by a vitelline membrane and three jelly coats. Freshly laid ova are pale yellowish brown and about 2 mm in diameter. The outer jelly coat is sticky and frequently covered with sediment, and the eggs hatch within 2–4 weeks after they are deposited (Storer 1925). Hatching may occur as early as 15 February in the western foothills of the Sierra Nevada (Storer 1925).

AQUATIC ECOLOGY. Hatchlings begin feeding within days after emerging from the egg capsules. Anderson (1968) found that hatchlings and small larvae feed mostly on zooplankton, whereas older larvae feed heavily on tadpoles. Prey items include cladocerans, copepods, ostracods, amphipods, midges, phantom midges, water boatmen, snails, and tadpoles of *Pseudacris regilla* and *Rana aurora*.

The larval period of *A. californiense* usually lasts 3–6 months since most breeding ponds dry up with the onset of the summer dry season. Specimens collected near Stockton during April varied from 47 to 58 mm TL (Storer 1925). A year later the ponds filled earlier and larvae collected on 11–14 March measured 35–48 and 28–47 mm TL at two study sites. Metamorphosing larvae were collected on 14 May.

Loredo and Van Vuren (1996) collected metamorphs between 1 June and 22 August during 3 years of trapping at their Contra Costa Co. site. A peak in emergence of metamorphs occurs from mid-June to mid-July, and the mean size of metamorphs may change significantly as the season progresses. Breeding in this population occurs primarily from early December to mid-January, suggesting an average larval period of about 4–5 months. The mean SVL of metamorphs differs among years (means = 58, 64, and 78 mm SVL for 3 years; $n = 3$ for the latter estimate), and metamorphs vary from 46 to 114 mm SVL. Per capita recruitment into the terrestrial population is highly variable among years and averages from 0.2 to 5.9 metamorphs/female. Estimates of total output of metamorphs per year vary from 3 to 1,248 individuals.

Very few data are available on larval behavior. Larvae often rest on the pond bottom in shallow water and stratify in the water column in deeper water. Stratifying larvae are wary and will dart into bottom vegetation when approached (Storer 1925).

TERRESTRIAL ECOLOGY. Metamorphosis occurs during the dry summer months, and metamorphs typically migrate from ponds at night during dry weather. Rare summer rains will stimulate relatively large numbers of metamorphs to emigrate from ponds (Loredo and Van Vuren 1996). The first night after leaving the breeding pond, metamorphs may move 6–57 m (mean = 26 m) from the pond (Loredo et al. 1996). Individuals settle in soil crevices or ground squirrel burrows at the end of nightly movements, and they presumably move over several nights before establishing permanent home ranges. The total distance that metamorphs move from the breeding ponds is not known. Adults at this site may move 8–129 m (mean = 36 m) the first night after leaving the pond, and about 83% settle in ground squirrel burrows. This and other studies suggest that both juveniles and adults rely rather heavily on the burrows of California ground squirrels (*Spermophilus beecheyi*) and valley pocket gophers (*Thomomys bottae*) as summer aestivation sites (Barry and Shaffer 1994; Loredo et al. 1996; Storer 1925). The adults are capable of burrowing through moist soil in plugged rodent burrows (Jennings 1996). Although ground squirrels are known to prey on amphibians, juvenile and adult *A. californiense* do not show preferences for occupied versus unoccupied ground squirrel burrows (Loredo et al. 1996).

Holland et al. (1990) observed an unusual movement of recently metamorphosed juveniles toward a breeding site in August following light rains. Hundreds of juveniles measuring 56–69 mm SVL were collected near a lake margin in September and October, and many were aggregated in clumps beneath debris; 29 individuals were found in a ball in a mouse nest beneath a plank. Most animals in the October sample apparently died from desiccation as the habitat dried.

PREDATORS AND DEFENSE. Almost no information is available on natural predators of *A. californiense*. Baldwin and Stanford (1987) observed a red-legged frog (*Rana aurora*) eat a larva.

COMMUNITY ECOLOGY. Tiger salamander larvae (*A. tigrinum*) play important roles in structuring many fish-free pond communities. *Ambystoma californiense* probably plays similar roles in structuring vernal pond communities in California, although these are undocumented.

CONSERVATION BIOLOGY. Habitat loss from agricultural and urban development along with other forms of human disturbance has affected *A. californiense* and many other amphibians in central California. A recent survey suggests that *A. californiense* is in the early stages of range contraction and fragmentation (Fisher and Shaffer 1996). This species is vulnerable to extinction if these trends continue (Barry and Shaffer 1994; Loredo et al. 1996).

One critical aspect of the species' natural history about which we know little is the distance that adults and juveniles move from ponds. Protection of both breeding sites and the surrounding habitats used by adults and juveniles is essential for maintaining viable local populations of this species. Based on studies of movements of other *Ambystoma* species, a 200- to 500-m radius of natural vegetation would probably provide suitable habitats for most adult populations.

The California tiger salamander appears to have a commensal relationship with ground squirrels and pocket gophers. California ground squirrels are currently controlled on over 4 million hectares in California, a practice that could potentially reduce critical microhabitats for *A. californiense* (Loredo et al. 1996). Detailed studies of ground squirrel–California tiger salamander interactions are needed to assess the importance of these rodents in managing populations of *A. californiense*.

Ambystoma cingulatum (Cope)
Flatwoods Salamander
PLATES 4, 5

Fig. 21. *Ambystoma cingulatum;* adult; Beaufort Co., Georgia (R. W. Van Devender).

IDENTIFICATION. The flatwoods salamander is a small, slender *Ambystoma* with light grayish lines or specks that form a reticulate or frosted pattern on a dark gray or black dorsum. Reticulation or frosting is absent in about 5% of specimens, and about 2% of specimens have light annuli on the dorsum instead of reticulations (Carr 1940; Martof 1968). The venter is dark gray with light gray flecks or spots. The head is small and only about as wide as the neck and shoulder region. Costal grooves vary from 13 to 16 and average 15. Adults measure 9–13.5 cm TL (Martof 1968; Palis 1996a). Sexually active males have a slightly enlarged cloacal region, but sexual dimorphism is not as pronounced as in most *Ambystoma*. Breeding males in a Florida population are 9% shorter in SVL than females on average (Palis 1997).

Hatchlings are uniformly dark brown above and pale brown below and measure 7.5–11.5 mm SVL and 10–19 mm TL (Anderson and Williamson 1976; Palis 1995). Soon after hatching, the larvae develop a yellow to gold vertebral stripe along the length of the body that is flanked by dark dorsolateral stripes (Anderson and Williamson 1976). Older larvae have a pale tan vertebral stripe and one or two pale cream lateral stripes that are bordered by dark pigmentation. Vivid, dark stripes are present on the head that extend from the snout through the eyes to the base of the gills. Juveniles may retain the midlateral stripe or remnants of the stripe for as long as 1 year after metamorphosing (Palis 1997). Mature larvae are illustrated in Ashton (1992), Hardy and Olmon (1974), Mecham and Hellman (1952), and Orton (1942).

SYSTEMATICS AND GEOGRAPHIC VARIATION. The flatwoods salamander is a well-defined species whose nomenclature has been stable for decades. Gulf coast populations tend to have brownish gray, reticulate markings on the dorsum compared with light gray, frosted markings on Atlantic coast specimens (Martof and Gerhardt 1965). However, numerous exceptions occur within these regions, and subspecies are not recognized.

DISTRIBUTION AND ADULT HABITAT. The flatwoods salamander lives in seasonally wet, pine flatwoods and pine savannas from southern South Carolina, southern Georgia, and northern Florida westward to southern Alabama (Goin 1950; Means et al. 1996). Populations are discontinuous throughout the range and may have been extirpated from Alabama (Palis 1996a). A record from Mississippi (Telford 1954) is probably misidentified *A. talpoideum* larvae.

The flatwoods salamander was found in sandy, seasonally wet, longleaf pine (*Pinus palustris*)–wiregrass (*Aristida stricta*) communities prior to European settlement. Most of these communities have been destroyed and replaced by slash pine (*P. elliottii*) or mixed slash pine–longleaf pine communities (Palis 1996a).

BREEDING AND COURTSHIP. The flatwoods salamander breeds in ephemeral habitats, including roadside ditches, borrow pits, marshy pasture ponds, swamps, and pond cypress (*Taxodium ascendens*) and

Range of *Ambystoma cingulatum*

black gum (*Nyssa sylvatica*) ponds (Anderson and Williamson 1976; Mecham and Hellman 1952). Breeding sites are typically in or near pine flatwoods or savannas that support longleaf pine, slash pine, and wiregrass. Although seasonally ephemeral, the breeding sites sometimes contain small fishes that enter during seasonal floods (Palis 1996a). At least 20 species of amphibians share breeding sites with *A. cingulatum* (Anderson and Williamson 1976).

Local breeding populations usually consist of 200–400 or fewer adults (Means et al. 1996; Palis 1997). Most adults migrate to breeding sites during periods of rainy weather from mid-October through early February (Anderson and Williamson 1976; Means 1972; Means et al. 1996; Palis 1996a, 1997). The adults usually move during heavy rains and most stop moving within 1–5 hours after the rains end. In rare instances, small numbers may move on rainless nights if soils are near saturation (Palis 1997). Although courtship has not been witnessed, individuals presumably court on land shortly after arriving at the breeding sites.

The only comprehensive study of adult movements is that of Palis (1997), who studied a Florida population for two seasons. Males and females in this population migrate in synchrony, and a peak in arrivals at the breeding site occurs in October and November. Small numbers of adults continue to arrive through February, but most individuals emigrate in December and January after spending an average of 38 days at the breeding site. Individuals tend to leave the breeding site where they enter, a behavior that suggests that they return to their home ranges after breeding. Gravid females lose about 37% of their weight after ovipositing.

The ratio of males to females for both yearling and older animals collected during two consecutive years was 0.6:1 and 0.9:1. When nonyearling females that did not breed are eliminated, the sex ratio of males to females is 2.4:1. Annual breeding patterns are poorly documented, but some females are known to breed annually. Courtship has not been described, but it would merit study since the flatwoods salamander and the marbled salamander are the only *Ambystoma* species that court on land.

REPRODUCTIVE STRATEGY. Most information on the nesting biology of *A. cingulatum* comes from Anderson and Williamson (1976), who studied populations in southern South Carolina and adjoining areas of Georgia. Females in these populations deposit eggs singly beneath logs, leaf litter, dead grasses, and sphagnum mats. Eggs are occasionally laid on bare soil and are often placed near the bases of bushes and small trees. In many instances eggs are deposited in or near the entrances to crayfish burrows. Females lay groups of 1–34 eggs in a linear or clumped fashion in the lowest elevations of depressions, then abandon the eggs immediately after ovipositing. Egg densities within pond beds are 0.44–2.28 eggs/m^2. Means et al. (1996) note that females in Florida normally oviposit in the grassy upper zones of flatwoods depressions.

Estimates of clutch size are based on counts of ovarian eggs. The number of mature ova in South Carolina and Georgia females ranges from 97 to 222 (mean = 163 ova) and is positively correlated with female SVL (Anderson and Williamson 1976). In Florida populations, the maximum clutch size is about 225 ova (Ashton 1992).

Eggs in early developmental stages have been found from early November through early December. Ova in early developmental stages are dark brown or black at the animal pole and pale brown to buff at the vegetal pole and measure 2.0–2.6 mm in diameter. The outermost capsule is soft and sticky. Embryos develop to the earliest hatching stages (Harrison stage 38–39) in about 2 weeks, then enter an arrested state much like the embryos of *A. opacum*. Embryos normally hatch within a few hours after they are inundated by rising water following heavy rains (Anderson and Williamson 1976).

AQUATIC ECOLOGY. Depending on seasonal rainfall patterns, hatching may be highly synchronous or staggered. Staggered hatching occurs whenever two or more major rains are required to fill the pond to capacity, and it can result in two or more cohorts hatching over a 3- to 6-week period (Anderson and Williamson 1976; Palis 1995). Hatching has been documented from November to late January.

At one Florida site larvae may hatch as two major cohorts, one in late November and a second in early January (Palis 1995). Under these conditions, larvae grow 1.7–1.9 mm SVL/week and the larval period lasts 15–18 weeks. At a second site larvae sometimes hatch synchronously in late January, grow an average of 2.5 mm SVL per week, and metamorphose 11–13 weeks later. The mean size of metamorphs in the laboratory is 39 mm SVL, compared with 42 mm SVL for animals in the field.

Hatchlings begin feeding on small zooplankton and other invertebrates soon after emerging from the eggs and grow rapidly. The larvae tend to concentrate around submerged vegetation and often float in the water column at night to forage on invertebrates (Palis 1995). The larval period lasts about 3–5 months in most populations, with transformation occurring from March to early May. In Florida, Means (1972) collected transforming individuals in late April, and Mecham and Hellman (1952) collected mature larvae as large as 78 mm TL (36 mm SVL) in March. The largest larva collected by Palis (1995) measured 47 mm SVL and 96 mm TL.

TERRESTRIAL ECOLOGY. Juveniles normally disperse from ponds shortly after metamorphosing, but they may stay close to ponds during seasonal droughts (Palis 1997). The adults move as far as 1000–1700 m from ponds (Ashton 1992; Ashton and Ashton 1988), and many regularly move 300–500 m (Means et al. 1996). The juveniles presumably disperse similar distances from ponds, then spend most of their time below ground until sexually mature.

Males in a Florida population become sexually mature when about 1 year old, but most probably do not mate initially until the following year (Palis 1997). Females mature and reproduce for the first time when 2 years old. Females become gravid when >52 mm SVL whereas males mature sexually when >43 mm SVL.

The adults are highly fossorial and spend much of their time in crayfish burrows or root channels, where they feed on earthworms (Ashton 1992; Goin 1950). A few adults may make short movements to and from the breeding site after the breeding season has terminated, but the majority disperse away from ponds and live sedentary, subterranean lives. Adults have activity ranges of about 1500 m^2 (Ashton 1992).

PREDATORS AND DEFENSE. Larvae often show evidence of attacks from aquatic invertebrates or conspecifics, and as many as 77–84% of older larvae may have damaged tails (Palis 1995). Defensive posturing in adults typically involves coiling the body and tucking the head underneath the base of the tail. In rare instances, individuals may lash weakly with their tails (Brodie 1977).

COMMUNITY ECOLOGY. *Ambystoma cingulatum* and *A. opacum* are the only *Ambystoma* species that breed on land, and *A. cingulatum* is one of only a few terrestrially breeding salamanders in North America that abandons its eggs after ovipositing. Terrestrial reproduction allows larvae to hatch immediately after pond formation. This strategy maximizes the time available for larval growth and may reduce the susceptibility of larvae to predators that colonize pools. In addition, larvae that hatch early may gain a competitive advantage over conspecifics that share breeding pools (Anderson and Williamson 1976).

CONSERVATION BIOLOGY. The flatwoods salamander is a relatively rare *Ambystoma* that consists of small, widely scattered local populations. Numerous populations have been destroyed or drastically reduced in size by land clearing, ditching, filling of wetlands, and the conversion of native longleaf pine forests to intensively managed tree farms (Ashton and Ashton 1988; Palis 1996a; Vickers et al. 1985). In the most detailed long-term study to date, Means et al. (1996) documented the decline of a local population following the conversion of a native longleaf pine forest into a bedded slash pine plantation maintained on a short rotation cycle.

Ambystoma cingulatum is currently a potential candidate for federal listing (Palis 1996a, 1997), and populations appear to be declining throughout the range of the species. The flatwoods salamander is currently considered to be endangered in South Carolina and rare in Alabama, Georgia, and Florida. Any additional populations that are discovered should receive high priority for protection by land managers and forest ecologists. Proper management will require protecting both the aquatic and the terrestrial habitats.

Ambystoma gracile (Baird)
Northwestern Salamander
PLATES 6, 7

Fig. 22. *Ambystoma g. gracile;* adult; Thurston Co., Washington (W. P. Leonard).

Fig. 23. *Ambystoma g. gracile;* head with parotoid glands; Skamania Co., Washington (R. W. Van Devender).

IDENTIFICATION. The northwestern salamander occurs as both gilled and transformed adults. Most transformed adults are uniformly dark brown, gray, or black; however, individuals in northern populations often have irregularly shaped white or yellowish flecks on a dark background. Conspicuous oval parotoid glands occur immediately behind the eyes, and the tail ridge appears roughened because of concentrations of granular glands (Brodie and Gibson 1969). Transformed males become darker than females during the breeding season and have conspicuously enlarged cloacae. Sexual dimorphism is similar in gilled adults, except that males have hypertrophied hindlimbs and feet and are less spotted, and the glandular ridge on the tail is enlarged (Snyder 1956). Terrestrial adults measure 14–22 cm TL, whereas gilled adults may reach 13 cm SVL and 26 cm TL (Boundy and Balgooyen 1988). Specimens usually have 11 costal grooves. Males and females examined by Taylor (1984) do not differ significantly in SVL, but in many populations females are slightly larger than males on average.

Hatchlings measure about 8 mm SVL on average (Licht 1975) and have concentrated dark pigment along the base of the dorsal fin. Older larvae vary from deep brown to olive green or light yellow. They often have sooty blotches or spots on the dorsum, and they may have yellow flecks or spots along the sides (Stebbins 1985). Parotoid glands and dorsal tail glands are readily evident on larvae of >50 mm SVL and are less conspicuous on gilled adults than on transformed animals (Licht and Sever 1993). Gilled adults are brownish to olive green and are mottled with yellow and black. Distinct yellow spots sometimes occur along the sides and tail. The venter varies from cream to pale or dark gray, and it darkens after metamorphosis. Metamorphs are uniformly light olive brown and tend to darken as they age. An albino specimen was found near Portland, Oregon (Dyrkacz 1981).

SYSTEMATICS AND GEOGRAPHIC VARIATION. Two subspecies have traditionally been recognized. The brown salamander (*Ambystoma g. gracile*) is a southern race that differs from the British Columbia salamander (*A. g. decorticatum*) in having three rather than four phalanges in the fourth toe, more prominent parotoid glands, more uniformly dark brown coloration, and teeth on the prevomers in groups of four rather than two (Dunn 1944). Titus (1990) found that these distinctions do not always hold, and that genetic variation does not correlate well with recognized subspecies. Thus, subspecific recognition may not be warranted. Six populations from Oregon and British Columbia differ in genetic distances (*D*) from 0.005 to 0.22 and mean heterozygosity ranges from 0.005 to 0.17. Heterozygosity is much lower in *A. g. decorticatum* than in *Ambystoma g. gracile*.

Populations with gilled adults are widespread and increase in frequency with elevation. Adults at very high elevations are predominantly gilled, whereas those at low to intermediate elevations are almost entirely terrestrial (Eagleson 1976; Snyder 1956, 1963; Sprules 1974a,b; Titus and Gaines 1991). A study of allozyme variation in a cluster of coastal populations consisting mostly of terrestrial adults, and a similar cluster of montane populations consisting almost entirely of gilled adults, indicates that isolated local populations located only a few kilometers apart show little genetic differentiation in both cases (F_{st} = 0.006 and 0.01, respectively). Given that gene flow among local demes of gilled adults in montane areas is relatively low, the small degree of genetic differentiation between these populations is surprising.

DISTRIBUTION AND ADULT HABITAT. The northwestern salamander is found in mesic habitats along the Pacific coast from northern California through British Columbia to extreme southeastern Alaska. Terrestrial adults live in habitats ranging from grasslands to mesophytic forests and occur from sea level to near 3,110 m in elevation. The terrestrial adults are mostly fossorial except when migrating to and from breeding sites, but they are occasionally found beneath logs and other surface objects outside the breeding season.

BREEDING AND COURTSHIP. Adults breed in permanent or semipermanent habitats ranging from small, shallow ponds to large, deep lakes. Individuals are occasionally observed in sluggish pools of slowly-flowing streams. The breeding season varies substantially depending on latitude and elevation. Egg deposition in lowland populations in the United States and British Columbia generally occurs from January through April (Brown 1976; Henry and Twitty 1940; Licht 1969, 1975; Slater 1936; Snyder 1956; Watney 1941), and at any given site breeding lasts 1-7 weeks (Eagleson 1976; Snyder 1956). Adults in high-elevation populations on Mt. Rainier begin breeding shortly after the lake surface becomes ice-free (Snyder 1956), but breeding may also take place while lakes are still partially frozen (Eagleson 1976).

Researchers have observed eggs from late June to late August in Oregon lakes at elevations of 1372-1676 m (Snyder 1956); from March through May in Marion Lake, British Columbia (Efford and Mathias 1969); and from March through August in southwestern British Columbia lakes at elevations of 100-1200

Fig. 24. *Ambystoma g. gracile*; head of adult; Benton Co., Oregon (G. Hokit).

Range of *Ambystoma gracile*

m (Eagleson 1976). Yearly breeding patterns are poorly documented for the terrestrial adults and presumably vary with elevation. Gilled adult females breed annually in lowland lakes, but they breed biennially in montane lakes (Nussbaum et al. 1983).

Field and laboratory observations suggest that courtship behavior varies among regional populations. Licht (1969) observed five amplexed pairs of courting *A. gracile* in a shallow pond in Vancouver, British Columbia, and a single pair of adults in a laboratory tank. In this population a male first straddles a female after approaching her from either side. The male then clasps the female with his forelimbs positioned directly behind the female's forelimbs and his hindlimbs wrapped around the female's hindlimbs or body. The male repeatedly pumps the hindlimbs forward while vigorously rubbing his chin from side to side over the female's snout. At the same time, the male lashes his tail slowly from side to side and rubs his cloaca over the dorsum of the female's tail. Licht (1969) did not witness spermatophore pickup by the female.

Knudsen (1960) observed a slightly different courtship pattern in *A. gracile* from Pierce Co., Washington. In this population a male first approaches and straddles a female, then clasps the female about the forelimbs with his hindlimbs. The pair remains in this position for an extended period. The male periodically lashes his tail back and forth across the female's back, then dismounts and attempts (unsuccessfully) to deposit a spermatophore. This sequence is repeated in quick succession. The male next moves from head to tail across the back of the female while dragging his cloaca along the back and dorsal edge of the female's tail. The male then grasps the female's tail with his hindlimbs while forcing the tail deep within his cloaca. He dismounts and deposits a spermatophore. The female moves to the point of spermatophore deposition and picks up the sperm cap.

Fig. 25. *Ambystoma g. gracile;* egg mass; Thurston Co., Washington (W. P. Leonard).

REPRODUCTIVE STRATEGY. Transformed females attach egg masses to tree limbs, cattails, and other sturdy support structures in ponds within a few days after mating, frequently attaching the egg masses 0.5–1 m below the water surface (Licht 1975). The jelly coat quickly swells to produce large, firm egg masses that resemble those of *A. maculatum*. Females in a Washington population produce egg masses that are 5–15 cm long and 5–8 cm wide, and that contain 30–270 ova (Slater 1936). In populations in British Columbia, the egg masses contain from as many as 106 (range = 74–126; Licht 1975) to as few as 47 (Lindsey 1966) ova on average. Recently fertilized ova are surrounded by two jelly envelopes, are 1.5–3 mm in diameter, and have brown to black animal poles and cream gray to white vegetal poles. As egg masses age, they are often invaded by symbiotic algae, which provide oxygen to the developing embryos (Carl and Cowan 1945; Patch 1922).

Snyder (1956) notes that gilled adults in small cirque ponds near Mt. Rainier produce smaller egg masses of a much looser consistency than those laid by transformed adults. Ovum number in a series of egg masses from one pond ranges from 15 to 143. Many of the masses are laid directly on the pond bottom, unattached to support structures. Henry and Twitty (1940) also report flimsy egg masses of *A. gracile* from the same locales near Mt. Rainier, but Knudsen (1960) observed gilled adults laying firm masses elsewhere. Knudsen (1960) suggests that Snyder (1956) mistook eggs of *A. macrodactylum* for those of *A. gracile*.

Length of the embryonic period, which is positively correlated with water temperature, is 2–9 weeks. In Washington, the incubation period lasts 2–4 weeks at several sites studied by Slater (1936) and an estimated 62 days at a site studied by Brown (1976). Embryos in a pond in Vancouver, British Columbia, hatch in mid-May, about 6 weeks after the first eggs are deposited (Watney 1941). The embryonic period lasts about 4 weeks in a lowland population in British Columbia (Licht 1975), with hatching occurring in late April. Data on embryonic survival are not available.

AQUATIC ECOLOGY. Hatchlings begin feeding on zooplankton soon after emerging from the eggs. As larvae grow, they incorporate larger prey, such as amphipods and insect larvae, into their diet. Larvae in a lowland population in British Columbia feed on annelids, mollusks, cladocerans, ostracods, amphipods, anostracans, isopods, copepods, mites, dipterans, and a variety of other insect larvae (Licht 1975). Larvae in different size classes have high dietary overlap, even though large larvae incorporate large prey such as insect larvae into their diets. Larvae in this population feed during the winter but grow little during this time of year.

Larvae in Marion Lake, British Columbia, feed heavily on amphipods. A variety of other prey—including aquatic and terrestrial insects, copepods, ostracods, snails, and flatworms—are eaten in much smaller quantities (Efford and Tsumura 1973). Individual larvae and gilled adults often specialize on benthic prey or zooplankton (Henderson 1973) and prefer frog tadpoles over toad tadpoles that produce skin toxins (Peterson and Blaustein 1991).

Larvae in lowland populations grow rapidly and reach 50–90 mm TL after about 1 year (Eagleson 1976; Licht 1975; Snyder 1956; Watney 1941). Most transform the following spring when 12–14 months old. Larvae begin metamorphosing when around 50 mm SVL (75–90 mm TL), but a few remain in ponds and become sexually mature during their second year of growth after reaching 70–75 mm SVL. Efford and Mathias (1969) estimate that larvae in Marion Lake, British Columbia, reach sexual maturity late in their second year of growth. Larvae in montane habitats grow more slowly than those in lowland habitats and do not transform until their third year of life (Eagleson 1976).

Walls et al. (1996) studied agonistic behavior in larvae from three populations in Oregon. Their data suggest that larvae from each of the ponds cannot distinguish conspecifics based on the genetic related-

Fig. 26. *Ambystoma g. gracile;* gilled adult; Lewis Co., Washington (W. P. Leonard).

Fig. 27. *Ambystoma g. gracile;* defensive posturing; Tillamook Co., Oregon (E. D. Brodie, Jr.).

ness or familiarity of individuals from the same pond. Larvae can discriminate between nonrelatives from the same versus a different pond, but only when they have had previous experience with their opponents. Levels of aggressive and submissive behavior also differ significantly between pond populations.

TERRESTRIAL ECOLOGY. The terrestrial adults are most active on the ground surface during the breeding season and during periods of fall rain. Length of the terrestrial juvenile stage is not known. However, both lowland and montane animals become sexually mature when they reach about 70–75 mm SVL. This finding suggests that juveniles mature after 1–2 years of growth. Gilled adults in Marion Lake live about 5 years (Efford and Mathias 1969).

The percent of gilled adults in a population generally increases with altitude (Eagleson 1976; Snyder 1956; Sprules 1974a,b), a phenomenon that is probably due to geographic variation in both genotype and environmental factors such as water temperature. Licht (1992) demonstrated that high food levels increase larval growth rates and reduce the time to metamorphosis. However, food level does not affect size at metamorphosis or the percent of animals that become gilled adults. Eagleson (1976) conducted common garden experiments that suggest that genetic differences between lowland and montane populations influence growth rate, time to and size at metamorphosis, and the percent of salamanders that become gilled adults.

PREDATORS AND DEFENSE. Natural predators are poorly documented, but they include trout and beetle larvae (Nussbaum et al. 1983). Northwestern salamanders have large concentrations of granular glands on the parotoid glands and tail ridge, which produce toxic secretions (Brodie and Gibson 1969). When molested, animals may emit a ticking sound (Licht 1973) and posture themselves with the glands oriented toward the predator. Head butting and tail lashing help bring the secretions into contact with the predator. Larvae and gilled adults typically respond to simulated attacks by fleeing (Licht and Sever 1993).

COMMUNITY ECOLOGY. Although most *Ambystoma* species utilize fish-free habitats for breeding, *A. gracile* often coexists with predatory fish, which in turn strongly affect the behavior, microhabitat use, and competitive interactions of *A. gracile* larvae in montane lakes (Efford and Mathias 1969; Sprules 1974a; Taylor 1983 a,b, 1984). Larvae in lakes that lack fish are active during the day, whereas those in lakes with fish are almost exclusively nocturnal. In the presence of fish, sexually immature larvae tend to concentrate in vegetated shallows, which serve as refuges (Neish 1971; Taylor 1983a). Larger larvae and gilled adults, which regularly encounter fish, are more likely to flee toward shallower water when threatened than larvae that rarely encounter fish (Taylor 1983b). These data suggest that larvae regularly monitor fish and respond with appropriate behaviors that reduce predation risk.

Despite these behavioral adaptations, fish do occasionally prey on *A. gracile* (Efford and Tsumura 1973) and may maintain larval populations below competitive levels. In lakes lacking predatory fish, *A. gracile* populations may reach sufficiently high densities to compete with conspecifics and rough-skinned newts (Taylor 1984). There is limited evidence that *A. gracile* larvae also influence the composition of zooplankton communities in breeding ponds (Giguere 1979).

CONSERVATION BIOLOGY. Studies of habitat requirements of *A. gracile* have produced conflicting data on the affinity of this species for old-growth forests. Aubry and Hall (1991) found *A. gracile* to be far less abundant in young forests compared with older forests in the Washington Cascades. In contrast, Corn and Bury (1991) found little correlation of salamander abundance with stand age, although no animals were found on recent clear-cuts. Leaving a forest buffer of 200 to 250 m radius around small ponds and lakes used as breeding sites by terrestrial adults would provide optimal conditions for maintaining viable local populations of this species.

Blaustein et al. (1995) found that embryos of *A. gracile* suffer higher mortality when exposed to ambient UV-B light (study site 183 m in elevation) than do controls that are shielded from UV-B. Eggs in this experiment were immersed in 5–10 cm of pond water and exposed to direct sunlight. Since females often deposit eggs at greater depths in ponds, and sometimes use sites with murky water, the extent to which embryonic mortality from UV-B light is negatively affecting natural populations is not fully known.

Ambystoma jeffersonianum (Green) Jefferson Salamander
PLATE 8

Fig. 28. *Ambystoma jeffersonianum*; adult; Pike Co., Ohio (R. W. Van Devender).

IDENTIFICATION. The Jefferson salamander is a long, slender, dark gray to brownish gray *Ambystoma* with 12–13 costal grooves and elongated limbs and toes. Light bluish gray or silvery flecks occur on the limbs and along the lower sides of the body and tail. The tail is nearly as long as the body and laterally compressed. Males in breeding condition have conspicuously swollen vents and more compressed tails than females. Males also often have a dull yellowish brown ridge along the top of the tail. Females examined by Uzzell (1964a) measured 13% longer than males on average. Adults range from 65 to 95 mm SVL.

Hatchlings measure 10–14 mm TL (Bishop 1941a) and are olive green to brown above with tinges of yellow on the sides of the neck, head, and dorsal fin. Mature larvae have grayish bodies with heavily mottled, broad dorsal fins. The head is very broad, and the toes are elongated and tapered. The belly lacks obvious pigmentation and is silvery or white. Metamorphs are rather nondescript and are uniformly gray or brownish above with inconspicuous brownish yellow specks on the sides. Both albino adults and larvae have been collected on rare occasions (Collins 1965; Dyrkacz 1981).

SYSTEMATICS AND GEOGRAPHIC VARIATION. *Ambystoma jeffersonianum* is one of four *Ambystoma* species that have hybridized to form unisexual species complexes involving diploid, triploid, tetraploid, and pentaploid unisexual forms (see special account of unisexual *Ambystoma*). Weller et al. (1978) report localities of this and other members of the complex in Ontario.

DISTRIBUTION AND ADULT HABITAT. The Jefferson salamander inhabits deciduous forests from New England south and southwestward to Indiana, Kentucky, West Virginia, and Virginia. A geographic isolate occurs in east-central Illinois. Populations are patchily distributed across the landscape and are restricted to sites where suitable breeding ponds occur. This species shows a stronger affinity for upland forests than do many other *Ambystoma* species. Adults are frequently collected in and around breeding ponds in

Range of *Ambystoma jeffersonianum*

late winter and early spring, but they largely reside beneath the ground at other times of the year. Juveniles are active on the ground surface in autumn through early spring in northern sections of the range, but they are rarely collected above ground in southern areas. Individuals have been collected from several West Virginia caves (Green and Brant 1966).

BREEDING AND COURTSHIP. Adults typically breed in seasonally ephemeral woodland pools and farm ponds, but occasionally use permanent habitats (Bishop 1941a; Douglas and Monroe 1981). Populations often breed in upland ponds along ridges and at the tops of stream drainages, but they also utilize ponds in woodlots and floodplains (Thompson and Gates 1982; Thompson et al. 1980; Weller et al. 1978). Abandoned, fishless farm ponds with rank growths of cattails and other vegetation are good sites to find breeding populations.

The Jefferson salamander is one of the earliest seasonal breeders. In instances in which one or more *Ambystoma* species share a pond with *A. jeffersonianum*, this species is typically the first to breed (Brodman 1995). Migrations in New York often occur when ponds are still ice-covered and before the ground is completely thawed (Bishop 1941a). Individuals normally migrate to and from ponds on rainy nights when temperatures are moderate, and males begin arriving at the breeding ponds in advance of the females (Douglas 1979; Douglas and Monroe 1981; Downs 1989). A small percentage of individuals may migrate during the day during overcast weather (Bishop 1941a), and in some instances males and females may migrate in synchrony on nights without rain (Brodman 1995).

As in most eastern *Ambystoma* species, northern populations of *A. jeffersonianum* breed later in the year than southern populations. Bishop (1941a) observed adult migrations and breeding in March and early April in New York following periods of rain or heavy snowmelt. Eight dead females that apparently froze while trying to reach open water were found around the margin of one pond. Other breeding observations include eggs in early April in New York (Smith 1911a), adult males in an Ontario pond on 30 March and 30 April (Weller and Sprules 1976), breeding adults in central Pennsylvania on 25-28 March (Mohr 1931), and breeding adults from 5 February to 3 April (Uzzell 1964a) and from early to mid-March (Brodman 1995) in Ohio populations.

Breeding in southern populations may begin as early as January, and in some populations adults begin migrating to ponds during the autumn. Adults first breed from 27 December to 28 January in southern Indiana (Downs 1989) and as early as mid-February in southwestern Ohio (Collins 1965). At a central Kentucky site adults begin migrating to ponds in late October and November (Douglas and Monroe 1981). At another central Kentucky site eggs have been found as early as 3 January (Smith 1983). Breeding at the latter site lasts 3-11 weeks in any pond. Adults breed as early as 2 February in West Virginia, although breeding at some sites does not begin until early March (Wilson and Friddle 1950).

Individuals in northern populations tend to breed more synchronously than those in populations farther south. At any one site in the north the breeding season may last for only a few days (Bishop 1941a; Brodman 1995), whereas at more southern locales breeding entails two or three major bouts that are interrupted by periods of cold weather. Males at a Kentucky site begin arriving at ponds before females and outnumber females by more than two to one on average (Douglas 1979). Operational sex ratios of males to females within ponds are usually greater than 3:1. Average duration of the stay varies from 16 to 30 days for males and from 19 to 21 days for females. Other studies indicate that males frequently outnumber females in the breeding ponds by 1.5:1 or more (Collins 1965; Downs 1989).

After mating and ovipositing, the adults migrate from breeding ponds into the surrounding forest, where they reside in underground retreats. In a Kentucky population individuals tend to enter and exit breeding sites from the same point and require about 45 days to move to permanent sites that are situated an average of 250 m from the shoreline (Douglas and Monroe 1981). These observations suggest that adults generally return to the same areas of the forest after mating. The maximum distance that adults move from ponds seems to be greater in *A. jeffersonianum* than in most *Ambystoma* species. Individual specimens have been found as far as 625 and 1600 m from breeding sites (Bishop 1941a; Downs 1989). The available data suggest that males breed annually, but females often skip one or more years before returning to breed (Downs 1989).

Aspects of courtship have been observed by Bishop (1941a) and Mohr (1930, 1931) in the field and by Uzzell (1964a) in the laboratory. Breeding activity begins when groups of two to four adults gather in scattered localities in a pond (Bishop 1941a). A male typically approaches a female, then amplexes her dorsally with his forelimbs positioned just behind those of the female. Pairs may remain amplexed for an extended period before additional courtship occurs. Eventually the male begins rubbing his snout on the top and sides of the female's head while undulating his tail. The male sometimes rocks his body and cloaca back and forth along the back of the female. The male may lash about vigorously during this time. After rubbing activity reaches a peak, the male moves forward along the female and dismounts. The male strongly undulates the tail and posterior body while moving past the female, then deposits one or more spermatophores while clasping the substrate with his rear legs. The female follows the male, often nudging his cloaca, then moves forward and picks up seminal fluid from the top of the spermatophore with her cloacal lips. Bishop (1941a) observed unpaired females butting and nudging amplexed pairs in a pond where females greatly outnumbered males. This behavior may reflect sexual competition among female *A. jeffersonianum* and unisexual *Ambystoma* for males.

The spermatophores are attached to detritus, vegetation, or stones (Bishop 1941a; Mohr 1931; Uzzell 1969). Each spermatophore has a clear gelatinous base, is strongly tapered at the top, and has four quadrangularly spaced projections at the top, which form a receptacle for the seminal fluid (Mohr 1931; Uzzell 1969). The entire structure is about 5-6 mm tall, 7 mm wide at the base, and 1-2 mm wide at the top of the stalk (Mohr 1931). *Ambystoma jeffersonianum* spermatophores are roughly twice the size of those produced by *A. laterale*. A male may deposit over 20 spermatophores, but averages of 4-12 per breeding bout are typical. Males produce about twice as many spermatophores when courting conspecifics as when courting JJL unisexuals (Uzzell 1964a, 1969). This behavioral difference may reflect the fact that male *A. jeffersonianum* can chemically discriminate between female *A. jeffersonianum* and JJL unisexuals (Dawley and Dawley 1986).

Unisexual *Ambystoma* (JJL) that share breeding ponds with *A. jeffersonianum* or *A. laterale* are dependent on diploid males of these or other *Ambystoma* species for successful reproduction. Female unisexuals often greatly outnumber diploid males in local populations (Nyman et al. 1988; Uzzell 1964a; Wilbur 1971). Diploid males produce fewer spermatophores when courting triploid females compared with diploid females, and diploid females have larger clutches of eggs than their triploid counterparts. These factors lessen the probability of triploids extirpating local populations of diploids (Uzzell 1964a, 1969).

REPRODUCTIVE STRATEGY. Females begin laying eggs within 1–2 days after mating and deposit small masses of eggs on grasses, twigs, fallen tree branches, and other support structures in ponds. As the eggs are laid, the female strongly arches her back while raising and undulating her tail. Several minutes are required to lay a mass of eggs (Bishop 1941a). Freshly laid ova are 2–2.5 mm in diameter and are surrounded by a vitelline membrane and three jelly envelopes.

When using fallen tree branches and other firm support structures, females often attach their masses in close proximity to one another to produce elongated, sausage-shaped groups. When laid on blades of grass or vegetation, individual egg masses are scattered about the ponds. Older egg masses frequently appear green from the large numbers of algae (*Oophila*) that live in the jelly and egg membranes. Colonization of the egg masses by algae occurs after the eggs are laid and does not involve the transmission of algal cells from the female to the egg masses (Gatz 1973).

Several workers (e.g., Bishop 1941a; Bleakney 1957; Piersol 1910a; Smith 1911a; Stille 1954) have described egg masses of *A. jeffersonianum*, but the descriptions in some cases may represent mixtures of *A. jeffersonianum*, *A. laterale*, or their hybrids (Uzzell 1964a). In central Kentucky, where other members of the complex are absent, the egg masses measure 39 mm wide and 43 mm long on average, and the number of eggs per mass is 2–67 (average = 23 eggs/mass; Smith 1983). In New York, females produce cylindrical egg masses that reach an average of 5 cm in length and 2.5 cm in diameter and contain 7–60 eggs (average = 16 eggs/mass; Bishop 1941a). Other literature reports that appear to describe *A. jeffersonianum* include a mean of 14 eggs/mass for a population near Syracuse, New York (Smith 1911a), a single egg mass with 35 eggs from West Virginia (Wilson and Friddle 1950), 22 eggs per mass (range = 8–55 eggs/mass) in southern Ohio populations (Seibert and Brandon 1960), and 30–33 eggs/mass in an Ohio population (Brodman 1995). Average egg densities are 57 eggs/m^2 in Ohio (Brodman 1995) to 123 eggs/m^2 in Indiana (Cortwright 1988).

Estimates of clutch size are based mostly on counts of mature ovarian eggs. Ovarian complements of females examined by Uzzell (1964a) vary from 140 to 280 eggs (estimated from figure in text), whereas averages for two Ohio populations are 183 and 212 ova. Estimates for New York populations include 166, 173, 215, and 286 eggs laid by four females maintained in captivity (Bishop 1941a) and 107, 123, 180, and 216 mature ova in four preserved specimens.

The embryonic period may last as long as 14 weeks depending on the seasonal time of egg deposition. Embryos in New York hatch in late April and early May after an incubation period of 30–45 days (Bishop 1941a), whereas embryos in central Kentucky hatch after 3–14 weeks depending on the seasonal time of oviposition (Smith 1983). Hatching in Kentucky occurs from late March to mid-April. Eggs in an Ohio population require an average of 28 days to hatch, and the incubation period is inversely proportional to the incubation temperature (Brodman 1995).

Data from several populations indicate that embryonic survival is moderately high. Mean embryonic survival to hatching is 87% in a northern Ohio population (Brodman 1995) and is positively correlated with egg mass size. Mean embryonic survival in five Massachusetts ponds is 60–88% and is independent of pond pH (Cook 1983). Median embryonic survival in ponds in Pennsylvania is 71–96% (Rowe and Dunson 1993).

AQUATIC ECOLOGY. Larvae begin feeding on small zooplankton soon after hatching and have voracious appetites. The dietary items of larvae from ponds in central Kentucky that C. K. Smith and I studied included nematodes, water mites, cladocerans, copepods, ostracods, collembolans, mosquito larvae, chironomid larvae, snails, and assorted insects (Smith

Fig. 29. *Ambystoma jeffersonianum;* larva; Jessamine Co., Kentucky (J. W. Petranka).

Fig. 30. *Ambystoma jeffersonianum;* juvenile; Hocking Co., Ohio (R. W. Van Devender).

and Petranka 1987). Individuals are sometimes cannibalistic and also feed on small *A. maculatum* larvae (Bishop 1941a; Brandon 1961; Smith and Petranka 1987).

Larvae grow rapidly and may complete the larval period in as few as 2 months. In New York populations, the larval period lasts 2–4 months and larvae transform from early July to mid-September (Bishop 1941a). By early November small juveniles may reach 52–78 mm TL and the average size is 62 mm. In an Ohio population the larval stage lasts an average of 66–80 days depending on conditions in the pond from year to year (Downs 1989). Metamorphs have been collected in Indiana as early as 22 June (Minton 1954).

Most aspects of larval behavior are poorly documented. In New Jersey ponds that contain mixed populations of *A. jeffersonianum* and JJL unisexuals, *A. jeffersonianum* larvae emerge from vegetation at dark and either move about on the substrate, swim through the water column, or float at or near the water surface while feeding (Anderson and Graham 1967). Hatchlings and larvae nearing metamorphosis spend proportionately more time on the pond bottom at night and are less likely to stratify than larvae of intermediate sizes.

Larvae are very aggressive and will cannibalize smaller individuals or bite off and consume the limbs, gills, and fins of conspecifics. The extent to which injuries result in death from secondary infections is not known. Although this behavior is poorly documented, invertebrate predators undoubtedly consume many hatchlings and larvae. Data from two studies suggest that premetamorphic survival is very low. Estimates of premetamorphic survival in an Ohio population are 0.08% and 0.7% for 2 years (Downs 1989), compared with 0% in a Maryland pond (Thompson et al. 1980).

TERRESTRIAL ECOLOGY. Very little is known about the ecology of juveniles and adults outside the breeding season. Juveniles in one population may move 3–247 m (mean = 92 m) away from the breeding site during a 3- to 10-day period (Downs 1989). Juveniles presumably spend most of their time in subsurface retreats, where they feed on soil invertebrates. The smallest mature males and females examined by Uzzell (1967a) measured 62 and 76 mm SVL, respectively. Downs (1989) reports values of 68 and 78 mm SVL, respectively, for Ohio populations. Juveniles in this region require 3 years or more to become sexually mature.

Adults enter and leave ponds near the same area and migrate to adjoining forests, where they live in rodent burrows and other subsurface retreats (Douglas and Monroe 1981). The adults are rarely seen outside the breeding season and presumably feed on earthworms and other invertebrates when underground. In an Ohio population only about 10% of adults survive for 3 years (Downs 1989).

PREDATORS AND DEFENSE. Predation on juveniles and adults is poorly documented. Adults are vulnerable to predators such as owls, striped skunks (*Mephitis mephitis*), and raccoons (*Procyon lotor*) during migrations to and from breeding ponds. Shrews and woodland snakes probably prey on the juveniles and adults. Defensive posturing in *A. jeffersonianum* is indistinguishable from that of *A. laterale* and consists of

Fig. 31. *Ambystoma jeffersonianum;* defensive posturing (note tail secretions); locality unknown (E. D. Brodie, Jr.).

raising and undulating the tail, lashing the tail, and coiling the body (Brodie 1977). Small animals typically raise the tail vertically and rapidly undulate it; larger animals hold the tail either straight or arched to reveal the tail secretions.

When exposed to a smooth earth snake (*Virginia valeriae*), individuals may elevate, undulate, and arch the tail or engage in body flipping, fleeing, or biting (Dodd 1977). Exposure to a single tongue flick by an eastern garter snake (*Thamnophis sirtalis*) may trigger individuals to undulate the tail from side to side for as long as 4 minutes (Ducey and Brodie 1983). The intensity of the tail undulating response differs among individual salamanders when they are confronted with tongue flicks from either predatory or nonpredatory snakes (Brodie 1989). However, the overall response to tongue flicking is not influenced by whether the snakes prey upon salamanders.

COMMUNITY ECOLOGY. *Ambystoma jeffersonianum* larvae frequently coexist with other *Ambystoma* larvae in breeding ponds and function as top predators in these systems. Cortwright (1988) examined interactions between *A. opacum, A. jeffersonianum,* and *A. maculatum* larvae in experimental cages. Marbled salamander larvae prey upon *A. jeffersonianum* larvae and significantly decrease their survival, but *A. jeffersonianum* do not lower the survival of *A. maculatum. Ambystoma jeffersonianum* larvae that survive predation by *A. opacum* have shorter larval periods than controls but do not differ in mean mass.

Rowe and Dunson (1995) examined how variable hydroperiods affect species interactions between *A. jeffersonianum, A. maculatum,* and wood frog tadpoles (*Rana sylvatica*) in experimental tanks set up in a forest in Pennsylvania. None of the species metamorphosed in the shortest hydroperiod (56-day treatment), and only *Rana* transformed in the 84-day treatment. Average instantaneous growth rates of *A. jeffersonianum* decreased with the duration of the hydroperiod, but growth rate, size at metamorphosis, and time to metamorphosis were independent of the initial densities of all species.

CONSERVATION BIOLOGY. Although amphibians in general are rather tolerant of acid conditions, *A. jeffersonianum* appears to be more vulnerable to habitat acidification than most salamanders in eastern North America. Rowe and Dunson (1993) found that the number of eggs that adults deposit (an indicator of relative adult population size) is positively correlated with pond pH and alkalinity, and negatively correlated with aluminum levels. Over half of the ponds examined in an area of Pennsylvania that receives relatively large amounts of acid deposition have a pH <4.5 (Sadinski and Dunson 1992). Complete egg mortality occurs in ponds with low pH, and water of pH <4.5 is often lethal to larvae.

Ambystoma laterale Hallowell
Blue-spotted Salamander
PLATE 9

IDENTIFICATION. The blue-spotted salamander is a small, slender *Ambystoma* with relatively short legs. Adults have large bluish white blotches and flecks on a dark gray to grayish black dorsum. The venter is dark gray or brownish gray with scattered light spots on the belly and throat. Mature adults measure 7.5–13 cm TL and have 13 costal grooves.

Males are on average smaller than females and have tails that are longer relative to body length. Females collected by Lowcock et al. (1991) at five Ontario sites

Fig. 32. *Ambystoma laterale;* adult; Schoolcraft Co., Michigan (R. W. Van Devender).

measured 0.3–15.9% longer than males, whereas females examined by Uzzell (1964a) from museum collections were 9% longer than males on average. Sexually active males have laterally compressed tails and swollen vents that are absent in females. This species is easily confused with LLJ and other unisexual *Ambystoma.* Identification based on cell size, molecular evidence, or chromosome number is required where populations of unisexuals occur.

Hatchlings have dark brown dorsums with yellowish transverse bars and measure 8–10 mm TL on average. Older larvae are brownish with a yellowish lateral stripe that becomes less distinct with age. The venter is generally unmarked except for small amounts of dark pigment that are often present along the margins of the lower jaw. The dorsal fins of older larvae are heavily blotched and mottled with black. Recently metamorphosed juveniles have yellowish spotting on the dorsum and venter.

SYSTEMATICS AND GEOGRAPHIC VARIATION. *Ambystoma laterale* is one of four parental species from which unisexual *Ambystoma* that are of hybrid origin have been derived. Details of unisexual species complexes are presented under the special account of unisexual *Ambystoma.* No subspecies of *A. laterale* are recognized.

DISTRIBUTION AND ADULT HABITAT. *Ambystoma laterale* occurs in formerly or recently glaciated regions across a broad swath of southern Canada and the extreme northern United States from the Gulf of St. Lawrence to Minnesota and Iowa. This species is found primarily in deciduous or mixed deciduous-conifer forests and is often locally abundant in areas with sandy or loamy soils (Downs 1989).

BREEDING AND COURTSHIP. Adults breed in a variety of fish-free habitats, including pools along lakeshores, springs in pastures, quarry ponds, marshes, roadside ditches, and both seasonally ephemeral and permanent woodland pools (Anderson and Giacosie 1967; Bleakney 1957; Piersol 1910a; Stille 1954; Van Buskirk and Smith 1991; Weller et al. 1978). Migrations to breeding sites occur during the late winter or early spring on rainy nights, and males begin arriving at breeding ponds shortly before females (Downs 1989; Lowcock et al. 1991; Uzzell 1969).

As with many *Ambystoma,* there is a seasonal progression of breeding time from south to north. Migrating adults, breeding adults, or freshly laid eggs have been found from late March to mid-April in southern Michigan (Clanton 1934), in mid-April in Wisconsin (Vogt 1981), and in March and early April in northern Illinois (Stille 1954; Uzzell 1969). On Isle Royale in northwestern Lake Superior breeding may not occur until after snowmelt in early May (Van Buskirk and Smith 1991).

In northern areas breeding has been documented from 30 April to 5 June in southern Canada (Bleakney 1957), from early to late April in central Ontario (Lowcock et al. 1991), and from early April to mid-May in Nova Scotia (Gilhen 1974). In the latter study the ratio of males to females for 240 adults collected over a 5-year period was 1.7:1. A small percentage of females in these samples is likely to be unisexual *Ambystoma* (LLJ).

The blue-spotted salamander tends to be an explosive breeder. Breeding in local populations typically consists of one to three major breeding bouts depending on weather patterns and local conditions. Adults in a Massachusetts population breed for only 2–3 days annually (Talentino and Landre 1991); however, a breeding season of 2–3 weeks is more typical for this species (Clanton 1934; Lowcock et al. 1991; Uzzell 1969).

Courtship behavior is described by Kumpf and Yeaton (1932) and Storez (1969) and closely resembles that of *A. jeffersonianum.* The following account comes mostly from the slightly more detailed account of Storez (1969).

Males swim about in more or less random fashion until they encounter conspecifics. As a male ap-

Range of *Ambystoma laterale*

proaches a second animal, it lightly pushes its snout into the side of the animal. This behavior is initially directed toward both sexes, but later it is directed only at females. After nudging a female, the male moves anteriorly and amplexes her, usually behind the forelegs (Kumpf and Yeaton 1932; Storez 1969; Uzzell 1964a). Amplexus may last as long as 12 minutes, and nearby males will often attempt to break apart amplexed pairs.

After initial amplexus, the male shifts anteriorly and rubs his chin from side to side over the female's snout while undulating his body over the back of the female. During this phase, his legs are often moved in a rapid stroking motion along the sides of the female. A male sometimes moves forward, turns about, and walks down the back of the female until his body is even with her tail. He then turns and returns to the clasping position. The male ultimately releases the female, strides forward while vibrating the body, then deposits a spermatophore while the tail is elevated, arched to one side, and undulated.

The male clasps the substrate with the hindfeet and presses his cloaca to the substrate while depositing a spermatophore. The female then follows the male forward while nudging his cloaca. As she moves over the spermatophore, she picks up seminal fluid from the spermatophore with her cloacal lips. The male then moves forward again and deposits additional spermatophores. The female responds and continues to remove seminal fluid from these. Female unisexual *Ambystoma* that share breeding ponds may use sperm from male *A. laterale* to facilitate reproduction (Downs 1978).

Males deposit an average of 8–35 spermatophores per breeding bout; the spermatophores are similar to those of *A. jeffersonianum* but about half the size (Arnold 1977; Kumpf and Yeaton 1932; Uzzell 1964a, 1969). Each is about 3 mm tall and consists of a clear gelatinous base that strongly tapers to the top. Four gelatinous horns at the top form a receptacle for the seminal fluid.

REPRODUCTIVE STRATEGY. Females attach eggs singly or in poorly defined masses of 2-10 (rarely more) eggs to leaf petioles, twigs, detritus, or rocks within a few days after courting (Bleakney 1957; Gilhen 1974; Stille 1954; Uzzell 1964a; Wilbur 1977). Most eggs are laid singly or in strings of two to four eggs. They are usually attached to decaying vegetation or the bases of rocks, but in some instances they may be scattered directly on the pond bottom. Talentino and Landre (1991) record an average of 19 eggs per mass in Massachusetts—an exceptionally high value for this species.

Freshly laid ova are 1.5-1.6 mm in diameter and have darkly pigmented animal poles (Clanton 1934; Minton 1972). The embryos hatch 3-4 weeks after the eggs are deposited (Minton 1972; Smith 1961; Talentino and Landre 1991). Hatchlings emerge from the eggs in May and early June on Isle Royale (Van Buskirk and Smith 1991), in late April and early May in northern Illinois (Smith 1961), and in early May in Massachusetts (Talentino and Landre 1991).

Reported clutch sizes based on the number of mature ova in females are 120-300 ova (average = 216 ova) for a large series of females from southern Michigan (Clanton 1934), an average of 196 ova for 23 females from the same region (Wilbur 1977), and 82-489 ova for Indiana specimens (Minton 1972). In Nova Scotia, females contain 123-538 ova (estimated mean = 250 ova based on figure in text), and clutch size is positively correlated with female SVL (Gilhen 1974).

AQUATIC ECOLOGY. Little information is available on the larval ecology of *A. laterale*. Larvae feed on a variety of aquatic organisms, but detailed studies of larval diet have not been published. Zooplankton and dipteran larvae probably form the bulk of the diet, as in many other *Ambystoma* species.

The larval period is brief and typically lasts 2-3 months. Larvae in northern Illinois transform in late June through mid-August after 8-10 weeks of growth (Smith 1961), whereas those in Massachusetts transform in July after a 9- to 11-week larval period (Talentino and Landre 1991). Other estimates of length of the larval period are 84-100 days for animals in experimental enclosures (Wilbur 1971), and a mean

Fig. 33. *Ambystoma laterale;* larva; Washtenaw Co., Michigan (R. W. Van Devender).

of 87 days for laboratory animals (Lowcock 1994). Metamorphs have been found in Wisconsin in mid-September (Edgren 1949) and in Nova Scotia from late July to early September (Gilhen 1974). Estimates of mean size at metamorphosis in natural populations are 34 mm SVL (range = 30-38 mm) in Canada (Lowcock 1994), 30 mm SVL in Massachusetts (estimated from figure in Talentino and Landre 1991), and 24-38 mm SVL in Wisconsin (Edgren 1949).

The aquatic larval stage probably functions as an ecological bottleneck that sets an upper limit on adult population size. Van Buskirk and Smith (1991) studied factors that could regulate adult population size in 36 larval populations inhabiting pools along the rocky shores of Lake Superior on Isle Royale. Median densities of hatchlings in their two study areas are 15 and 12 hatchlings/m^2, although as many as 158 hatchlings/m^2 can occur in pools. Growth rate, survival, and percent of individuals with intact tails are inversely related to larval density in both experimental and natural populations. However, larvae do not appear to be food-limited and have little impact on the available food supply. Intraspecific larval aggression is the primary mechanism resulting in density-dependent growth and survival. Although survival was not tracked to metamorphosis, approximately 17% of the populations suffered catastrophic mortality because of premature pool drying and wave-washing.

TERRESTRIAL ECOLOGY. Unlike most *Ambystoma*, the blue-spotted salamander is often active on the ground surface during the warmer months of the year (Downs 1989; Minton 1972; Vogt 1981). Dietary items from 29 postmetamorphic *A. laterale* from Ontario, Canada, include snails, earthworms, isopods, centipedes, spiders, and numerous insects (Judd 1957). Other recorded prey include beetles, centipedes, and roaches in Indiana specimens (Minton 1972) and snails, earthworms, beetles, beetle larvae, springtails, and spiders in 25 Wisconsin specimens (Vogt 1981). About 80% of 243 adults from Nova Scotia lack food in the stomachs (Gilhen 1974). The remainder contain both terrestrial and aquatic prey, including spiders, centipedes, springtails, mayfly nymphs, caddisfly and midge larvae, bugs and beetles, earthworms, snails, slugs, moth larvae, and miscellaneous plant debris.

The juveniles grow rather rapidly and in Michigan become mature when about 2 years old (Wilbur 1977). Maturation in 2 years is probably typical for the species as a whole. The smallest mature females in a Pelee Island, Lake Erie, population measure around 52 mm SVL (Licht 1989), whereas the smallest mature males and females from general localities measure 42 mm and 51 mm SVL, respectively (Uzzell 1967b). Males and females in Nova Scotia reach 45–70 mm SVL and 51–77 mm SVL, respectively (Gilhen 1984).

PREDATORS AND DEFENSE. Rand (1954) observed defensive posturing in *A.* "*jeffersonianum*" (presumably *A. laterale;* see Minton 1972) from northern Indiana. When touched roughly, juveniles and small adults raise their tails vertically, spread and brace their hindlimbs, raise their vents off the ground, lower their heads, and undulate their tails rapidly while secreting a sticky, white substance from the tail. Depending on which part of the body is attacked, individuals may also engage in tail lashing or body coiling (Brodie 1977). Defensive posturing by *A. laterale* appears to be nearly identical to that of *A. jeffersonianum* and *A. macrodactylum*.

COMMUNITY ECOLOGY. *Ambystoma laterale* often shares breeding ponds with diploid, triploid, and tetraploid unisexuals. Larvae of these species are similar in morphology and diet but often differ slightly in length of the larval period, size at metamorphosis, and the seasonal time at which individuals use a pond. Mechanisms that allow unisexual complexes to coexist in local ponds are not fully understood, nor is the role of larval competition in maintaining mixtures of *A. laterale* and unisexuals in local populations.

Lowcock (1994) conducted a detailed life history study of a population consisting of mixtures of mostly diploid, triploid, and tetraploid forms of the *A. laterale-*

Fig. 34. *Ambystoma laterale;* defensive posturing; Calhoun Co., Michigan (E. D. Brodie, Jr.).

jeffersonianum complex in southern Ontario. These forms differ in seasonal migratory patterns of adults, growth and development rates of larvae, and age at maturity of juveniles. Similar differences are evident in unisexual complexes in central Ontario (Lowcock et al. 1991). The extent to which these contribute to the stable coexistence of the biotypes that share breeding ponds is unknown.

CONSERVATION BIOLOGY. The blue-spotted salamander is a relatively common species in many areas of its range. Nonetheless, this species has declined as native forests have been replaced by urban and agricultural regions and thousands of natural breeding sites have been destroyed. *Ambystoma laterale* is unable to reproduce successfully in bog habitats in northern Minnesota where pH is <4.5 (Karns 1992). Continued acidification of wetlands through acid deposition could potentially reduce the reproductive success of this species in many areas of its range.

COMMENTS. Unisexuals that originated through hybridization of *A. laterale* and *A. texanum* occur on Kellys Island, Pelee Island, and the Bass Islands of Lake Erie, and in northwestern Ohio and southeastern Michigan. Males are uncommon in these populations, and diploid, triploid, and tetraploid unisexuals can occur in the same population. Unisexuals produce fewer and larger eggs than either parental species, and unisexual females are generally larger than *A. laterale* females (Licht 1989; Licht and Bogart 1989b). Tetraploids metamorphose later, exhibit delayed maturation, and are generally larger than diploids throughout the larval and juvenile stages. Details of these and other unisexual complexes are described under the account of unisexual *Ambystoma*.

Ambystoma mabeei Bishop
Mabee's Salamander
PLATES 10, 11

Fig. 35. *Ambystoma mabeei*; adult; Scotland Co., North Carolina (J. W. Petranka).

IDENTIFICATION. Mabee's salamander is a relatively small *Ambystoma* with a dark brownish gray to black dorsum overlain with light gray or whitish flecks. The light flecks are concentrated on the sides, and the venter is light gray or grayish brown with a few scattered, light flecks (Hardy and Anderson 1970). The tail makes up only about 40% of the total length and there are usually 13 costal grooves. Adults measure 8–12 cm TL. Mabee's salamander superficially resembles the small-mouthed salamander, *A. texanum*, and has been misidentified on several occasions as this species even though *A. texanum* does not occur in the Atlantic Coastal Plain.

Larvae are the pond type with bushy gills and broad dorsal fins that extend onto the back. Hatchlings have a single yellow stripe on either side of the body and lack conspicuous mottling on the dorsal and ventral fins. A hatchling illustrated by Hardy (1969b) measures 9 mm TL. Older larvae are brown to blackish above with two cream-colored stripes along either side of the body that are often broken or blotched. The venter is inconspicuously marked and tends to be flesh-colored. The dorsal fin is moderately developed and becomes heavily mottled with black along the tail as larvae mature. Older larvae generally resemble *A. cingulatum* larvae, but on *A. mabeei* the light lateral stripes are usually broken or blotched and the black stripes through the eyes are either absent or diffuse and poorly defined (they are bold and conspicuous in *A. cingulatum*). Recently metamorphosed animals are black or dark gray above with little or no light flecking.

SYSTEMATICS AND GEOGRAPHIC VARIATION. This is a well-defined species that shows no evidence of racial variation.

DISTRIBUTION AND ADULT HABITAT. Mabee's salamander inhabits Coastal Plain habitats from extreme southeastern Virginia to southern South Carolina. Adults are active on the ground surface during the breeding season and have been collected beneath surface objects in sandy pinewoods or near cypress-tupelo stands in the vicinity of breeding ponds (Mitchell and Hedges 1980; Mosimann and Rabb 1948).

BREEDING AND COURTSHIP. Adults breed in fish-free habitats, including semipermanent farm ponds, foxholes filled with water, vernal ponds in river bottomlands, Carolina bays, and cypress-tupelo ponds in pinewoods (Hardy and Anderson 1970). In North Carolina adults often use ephemeral, acidic ponds in or near extensive stands of pine, but they may occasionally use ponds in open, grassy fields (Hardy 1969a).

Many aspects of the breeding biology of this species are not documented. Adults migrate to breeding ponds and breed during periods of heavy, warm rains in winter or early spring. In North Carolina adults typically breed from early February to late March, but breeding may begin as early as mid-January in some years (Hardy 1969a,b). Courtship behavior has not been described.

REPRODUCTIVE STRATEGY. Females presumably begin ovipositing within a day or two after mating. The eggs are attached singly or in loose strings of two to six eggs to vegetation and detritus in ponds and are scattered widely throughout the breeding site. Each ovum is surrounded by a vitelline membrane and two jelly capsules. The embryonic period lasts 9–14 days in North Carolina populations (Hardy 1969b).

AQUATIC ECOLOGY. Larvae begin feeding on zooplankton and other invertebrates in ponds shortly

Range of *Ambystoma mabeei*

after hatching and are capable of rapid growth. The larval period lasts only a few months, and there is no evidence that larvae overwinter or develop into gilled adults. Larvae in a North Carolina population transform in early to mid-May after 2 months of growth (Hardy 1969a). The smallest transforming larvae in this population measure 55–60 mm TL. Other reports of size at metamorphosis include a 63-mm TL transforming larva collected in May from North Carolina (Bishop 1928), five metamorphs collected in late April and mid-May in Virginia that measured 37–46 mm SVL and 60–79 mm TL (Mitchell and Hedges 1980), and larvae collected from a South Carolina pond on 24 March that transformed 8–12 days later at 50 mm TL (Jobson 1940).

TERRESTRIAL ECOLOGY. The movement of adults after breeding is poorly documented, but limited evidence suggests that some adults remain in close proximity to breeding ponds, whereas others disperse widely into surrounding forests and fields. Adults at a North Carolina site can be found throughout most of the year beneath surface cover in or near the dried bottoms of breeding ponds (Hardy 1969a). Other adults in the same population inhabit open fields and adjoining forests during the summer months. Adults are active on the ground surface on rainy nights during the breeding season, but it is uncertain if this behavior reflects migratory activity.

Fig. 36. *Ambystoma mabeei*; larva; Scotland Co., North Carolina (J. W. Petranka).

Fig. 37. *Ambystoma mabeei*; juvenile; Scotland Co., North Carolina (R. W. Van Devender).

70   Ambystomatidae

Fig. 38. *Ambystoma mabeei;* defensive posturing; Orangeburg Co., South Carolina (E. D. Brodie, Jr.).

Very little is known about the ecology of the juveniles. Large numbers of juveniles ($n = 91$) have been found in late May over 800 m from the nearest known body of water (Hardy 1969a).

PREDATORS AND DEFENSE. *Ambystoma tigrinum* larvae regularly feed on the larvae (Hardy 1969a), and pond invertebrates such as odonates and dytiscid beetles undoubtedly eat many larvae. Defensive posturing of juvenile and adult *A. mabeei* is similar to that of *A. opacum* and *A. annulatum,* and consists primarily of coiling the body and tucking the head beneath the base of the tail (Brodie 1970). Specimens may also elevate and undulate their tails when contacted by snakes (Dodd 1977).

COMMUNITY ECOLOGY. *Ambystoma mabeei* larvae inhabit breeding sites that are shared by a diverse array of salamanders and anuran larvae. Ecological interactions of *A. mabeei* with other members of these communities have not been studied.

CONSERVATION BIOLOGY. Mabee's salamander breeds in fish-free habitats, and many breeding sites have been destroyed by the ditching and draining of wetlands and the conversion of forests into croplands. Comprehensive studies of adult and juvenile habitat requirements and the distribution of extant populations are needed to assess the conservation status of this species.

Ambystoma macrodactylum Baird
Long-toed Salamander
PLATES 12, 13

Fig. 39. *Ambystoma m. columbianum;* adult; Custer Co., Idaho (R. W. Van Devender).

IDENTIFICATION. The long-toed salamander is a small, dark gray to black *Ambystoma* with an uneven yellow, tan, or olive green dorsal stripe that is sometimes broken into blotches (Nussbaum et al. 1983). The sides have white speckling, and the venter is gray or black. Males have proportionately longer limbs and tails than females, and they develop swollen, papillose vents during the breeding season (Beneski et al. 1986; Nussbaum et al. 1983). Adults have 12–13 costal grooves and measure 10–17 cm TL. Females examined by Beneski et al. (1986) are 6% longer in SVL than males on average.

Hatchlings measure 6–12 mm TL (Anderson 1967a; Howard and Wallace 1985) and the larvae have pond-type morphology with bushy gills and a dorsal fin that extends to near the forelimbs. The dorsum is olive gray to brownish gray above and is mottled with brown and black. Both cannibalistic and typical morphs occur at a high-elevation site in Oregon (Walls et al. 1993a,b), but the cannibalistic morph has not been found elsewhere. The cannibalistic morph has larger vomerine teeth and a broader, longer head than the typical morph. In many respects it is similar to the cannibal-

Fig. 40. *Ambystoma m. macrodactylum;* adult; Dechutes Co., Oregon (G. Hokit).

istic morphs of *A. tigrinum*. Very young juveniles of all subspecies have bright yellow to buff dorsal stripes (Ferguson 1961).

SYSTEMATICS AND GEOGRAPHIC VARIATION. Five subspecies are currently recognized that differ primarily in dorsal banding patterns and vomerine tooth counts (Ferguson 1961). The eastern long-toed salamander (*A. m. columbianum*) has a bright yellow to tan dorsal stripe, and the head, snout, and eyelids have spots. The southern long-toed salamander (*A. m. sigillatum*) has fine spots on the head and a narrow, bright yellow dorsal stripe that is usually interrupted by blotches. The Santa Cruz long-toed salamander (*A. m. croceum*) has a black ground color overlain by yellowish orange to orange blotches and is disjunct from the remaining subspecies. The northern long-toed salamander (*A. m. krausei*) has a continuous, narrow, yellow dorsal stripe that continues onto the head and is widest just behind the eyes. The western long-toed salamander (*A. m. macrodactylum*) has a grayish ground color with a dull green to yellow dorsal stripe that is reduced to diffuse flecks on the head, snout, and sides of the back.

A study of microgeographic variation in allozymes in *A. m. columbianum* from northeastern Oregon and western Idaho indicates that the degree of genetic differentiation among regional populations is higher in this species than in most other vertebrate groups (Howard and Wallace 1981). Genetic distance (D) varies from 0.01 to 0.10 and heterozygosity is 0.10 on average.

DISTRIBUTION AND ADULT HABITAT. The long-toed salamander occurs from near sea level to about 2700 m in elevation and lives in a wide variety of habitats, including bottomland forests, coniferous forests, alpine meadows, sagebrush communities, and dry woodlands (Anderson 1967a; Ferguson 1961; Howard and Wallace 1985; Nussbaum et al. 1983). Populations range from southeastern Alaska southward to northern California, north-central Idaho, and western Montana. Geographic isolates occur in Santa Cruz and Monterey counties, California, and in south-central and southeastern Oregon.

BREEDING AND COURTSHIP. The life history of *A. macrodactylum* is highly variable depending on elevation and climate. Adults breed in both seasonally ephemeral and permanent lakes, ponds, and flooded meadows, although a small percentage of populations breed in slowly-moving streams (Beneski et al. 1986). At low elevations, the adults may migrate and begin breeding as early as October or November. At high elevations, breeding may not occur until late spring or early summer following snowmelt. As with most *Ambystoma*, the peak arrival of males typically precedes that of females by one to several days. The most comprehensive comparisons of the life histories of lowland and montane populations are by Anderson (1967a), Howard and Wallace (1985), and Kezer and Farner (1955).

Anderson (1967a) compared the life history of a lowland population of *A. m. croceum* with that of two montane populations of *A. m. sigillatum* in California. Adult *A. m. croceum* migrate to ponds during periods of heavy rain from October through February and breed explosively. Breeding occurs from mid-January through late February. In any year the breeding season

Fig. 41. *Ambystoma m. macrodactylum;* adult male (above) and female (below); Pierce Co., Washington (W. P. Leonard).

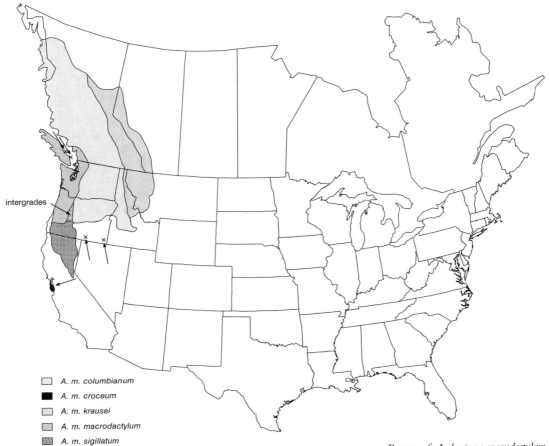

Range of *Ambystoma macrodactylum*

lasts about 1 week and rainfall is the major factor stimulating breeding. Montane *A. m. sigillatum* emerge and migrate to ponds as soon as the first thaw reduces snow cover and melts the ice cover on ponds and lakes. Adults begin courting and mating immediately upon reaching the breeding sites. Breeding occurs during May at 1980 m and from late May to late June at 2450 m. Adults in a high-elevation population in Oregon breed explosively during a 1- to 2-day period, and runoff from snowmelt triggers breeding (Walls et al. 1993b).

Howard and Wallace (1985) document variation in life history characteristics of *A. m. columbianum* at different elevations in Oregon and western Idaho. Adults breed at a lowland site (420 m) from early February to mid-April, at a middle-elevation site (1140 m) in April, and at a high-elevation site (2470 m) during June and July. Breeding occurs at night and frequently begins while ice and snow still cover the ground.

Observations at sites in Oregon, Washington, and British Columbia also indicate a general seasonal progression in breeding from low- to high-elevation populations. Populations in valleys below 610 m in Oregon breed between mid-February and mid-March (Fitch 1936), whereas populations near Crater Lake, Oregon (elevation 1550-1900 m) breed in July (Kezer and Farner 1955). Additional information on breeding activity includes observations of eggs on 9 March on Vancouver Island, British Columbia (Carl 1942), eggs in May near Victoria, British Columbia (Carl and Cowan 1945), eggs and gravid females on 6 April in the Blue Mountains in northeastern Oregon (Ferguson 1954), and eggs close to hatching near Pullman, Washington, on 12 April (Svihla and Svihla 1933).

The migratory period in a large Idaho population lasts at least 58 days (Beneski et al. 1986). This is a minimum estimate since the authors did not begin monitoring drift fences until after migration began.

Adults at this site enter the pond daily during the 16-day migratory period, and males and females use the same migratory routes. Individuals enter and exit the pond along the same routes. Emigration is both continuous and episodic, and it may last 50 days before ending in late April. The largest migrations occur on days with precipitation and above-zero temperatures, but many animals move on days with no precipitation. Males tend to enter the ponds before females and exit the ponds later than females. Males spend an average of at least 28 days in the ponds compared with 18 days for females. Males lose an average of 4% of their wet mass during the breeding season compared to 18% for females. The sex ratio of males to females is 1.7:1.

Accounts of courtship include those of Knudsen (1960) and Slater (1936), but the most complete account is that of Anderson (1961). The following summary is based on Anderson's account and indicates that courtship behavior of *A. macrodactylum* is very similar to that of *A. jeffersonianum*.

A male rapidly approaches a female, then mounts and dorsally amplexes her with his forelimbs clasped immediately behind the female's forelimbs and his hindlimbs stretched outward. The female often tries to swim away and dislodge the male when initially amplexed but soon becomes more passive. The male then forcefully begins rubbing movements in which his chin is moved back and forth over the head of the female. The tempo of these movements gradually increases as courtship progresses, and the female becomes quiet and motionless.

The male next moves forward over the head of the female, raises his tail vertically, and slowly waves the tail tip. If sufficiently aroused, the female follows and nudges the male's tail and cloaca with her snout. The male continues to walk forward with the tail undulating, and the female follows. During this time, the male strongly undulates his pelvic region while dragging his cloaca along the substrate. The male then stops and deposits a spermatophore. The female moves forward, arches her tail, and picks up seminal fluid from the spermatophore. The male may deposit a second spermatophore, but the female often departs after picking up the first.

A male can deposit up to 15 spermatophores over a 5-hour period. The spermatophore has a gelatinous base, is pyramidal in shape, and is topped by a small white to buff sperm cap. The apex is flared slightly to hold the sperm cap. The entire structure is 3–9 mm high and 6–10 mm in diameter at the base. Zalisko and Larsen (1989) found that males leaving breeding ponds often contain large quantities of sperm that are evacuated from the cloaca.

REPRODUCTIVE STRATEGY. Females begin ovipositing within a few days after mating and attach eggs either singly or in loose clumps to vegetation, twigs, or detritus. A female may oviposit on rocks or directly on the pond bottom when support structures for eggs are lacking (Howard and Wallace 1985; Slater 1936). The egg masses are very sticky and are not firm as in *A. maculatum* and *A. gracile*. Freshly laid ova are 2.1–2.7 mm in diameter and have a black animal pole and a grayish or white vegetal pole (Howard and Wallace 1985; Slater 1936). Each egg is surrounded by a vitelline membrane and two distinct jelly envelopes.

The mode of egg deposition varies geographically. Female *A. m. croceum* attach eggs singly to vegetation in shallow water within 5–8 cm of the water surface. In contrast, *A. m. sigillatum* in the Sierra Nevada of California attach eggs singly or in loose, linear clusters to the undersides of logs or large branches in deep water (Anderson 1967a). Field observations suggest that eggs of the northern subspecies are typically laid in small masses, although some Oregon populations of *A. m. columbianum* attach the eggs singly in large groups to the undersides of rocks (Walls et al. 1993b). The number of eggs per mass in *A. m. columbianum* populations sampled in Oregon and western Idaho varies from 2 to 57 and averages 15–17 (Howard and Wallace 1985). Egg masses of *A. m. macrodactylum* near Corvallis, Oregon, contain 5–25 eggs and average 10–12 (Ferguson 1961).

Fig. 42. *Ambystoma m. macrodactylum*; egg mass; Thurston Co., Washington (W. P. Leonard).

Fig. 43. *Ambystoma macrodactylum;* larva; locality unknown (R. W. Van Devender).

Clutch size is geographically variable, but there is little evidence of strong elevational or latitudinal trends. Number of mature ova in females from three Oregon populations varies from 90 to 167 and is positively correlated with female SVL (Howard and Wallace 1985). Other estimates are 85–345 ova (mean = 178 ova) for 11 Oregon females (Ferguson 1961) and 215–411 ova (mean = 307 ova) for five large *A. m. croceum* (Anderson 1967a). The incubation period lasts 2–5 weeks depending on average pond temperatures, and it generally increases with elevation.

AQUATIC ECOLOGY. Hatchlings begin feeding on small invertebrates soon after emerging from the egg masses and are capable of rapid growth when water temperatures and food levels are high. Larvae feed on a wide variety of aquatic organisms, particularly copepods, cladocerans, ostracods, chironomid larvae, oligochaetes, and tadpoles of *Pseudacris regilla* (Anderson 1968). Small larvae are mostly sit-and-wait foragers but will make small lunges at prey. Older larvae tend to stalk prey on the pond bottom and make longer lunges.

Growth rate, length of the larval stage, and size at metamorphosis vary markedly among populations. Larvae at high-elevation sites generally transform at a larger size and have slower growth and development rates than larvae at low-elevation sites. Proximate factors such as water temperature, length of the growing season, food level, larval densities, and seasonal drying of ponds influence these growth parameters.

Howard and Wallace (1985) found that the larval period at a site at 1140 m lasts only 3 months and that larvae transform on average at about 36 mm SVL. In contrast, larvae at 2470 m transform after 26 months when their average SVL is 47 mm. Length of the larval period is highly variable in ponds and lakes around Crater Lake, Oregon (Kezer and Farner 1955). In ponds below 1676 m, all larvae transform during the summer after hatching at 60–80 mm TL. In permanent bodies of water above 1829 m, the eggs are deposited in summer and the larvae transform during the following summer or later at 70–90 mm TL. In temporary ponds at high elevations, larvae that measure 40–50 mm TL transform during their first year of growth. Smaller larvae are often killed as the ponds dry seasonally.

Length of the larval period of *A. m. croceum* is a function of the amount of rainfall each season. In years of high rainfall, adults breed early and the ponds retain water longer than in years of low rainfall (Anderson 1967a). The larval period usually lasts 3–4.5 months and larvae transform when they reach 1.0–1.5 g and 60–90 mm TL (average SVL about 38 mm). In montane populations of *A. m. sigillatum* the larval period lasts 14 months, but larvae are about the same size at metamorphosis as those of *A. m. croceum*. Larvae on Vancouver Island have been observed metamorphosing during late July at 23–26 mm SVL (Carl 1942). The larval period at this site lasts about 4 months.

Older larvae are aggressive and will cannibalize smaller individuals or bite the limbs, tails, and gills off conspecifics (Anderson 1967a; Walls et al. 1993b). Although cannibalistic morphs appear to be extremely rare, they occur in at least one population in Oregon and have broader heads and larger vomerine teeth than typical larvae of similar body size (Walls et al. 1993a,b). When fed different combinations of conspecifics, brine shrimp, and tadpoles, larvae develop different head shapes. However, none assumes the shape of cannibalistic *A. tigrinum* morphs and none develops enlarged teeth.

The few observations of diurnal activity suggest that larvae of *A. m. croceum* are secretive and remain in shallow water beneath litter or in aquatic vegetation during the day. In contrast, *A. m. sigillatum* of intermediate size are often active in open areas of ponds during the day (Anderson 1967a). In general, larvae in most populations are more secretive during the day and are more likely to be active outside cover during the night.

TERRESTRIAL ECOLOGY. Metamorphs disperse away from the breeding sites soon after transforming unless conditions are dry. Juveniles sometimes respond to water stress by forming tight aggregates in the soil, a behavior that reduces water loss (Alvarado 1967). Individuals become sexually mature when they reach 48–56 mm SVL, and the juvenile stage lasts from <1 to 2 years depending on elevation and size at metamorphosis (Anderson 1967a; Howard and Wallace 1985).

The juveniles and adults are fossorial in lowland areas or in dry habitats, but metamorphs may be active on or near the ground surface during periods of rainfall prior to dispersing away from ponds. At higher elevations, juveniles and adults are often active on the ground surface during the summer breeding season (Anderson 1967a; Ferguson 1961; Fitch 1936; Kezer and Farner 1955). In laboratory trials Ducey (1989) found that both resident and intruder animals will bite each other and engage in aggressive posturing. In addition, individuals tend to avoid one another when housed together in laboratory chambers. These observations suggest that adults actively defend burrows in terrestrial habitats.

The stomach contents of 27 *A. macrodactylum* around Crater Lake, Oregon, consisted of fragments of ants, beetles, and flies as well as aquatic larvae of beetles, flies, and caddisflies (Farner 1947). Because numerous fragments of dead arthropods occur beneath rocks where the salamanders reside, the salamanders probably scavenge for arthropod remains. Isopods and other arthropods are the dietary staple of adult *A. macrodactylum* in certain California populations (Anderson 1968). Females apparently do not feed while breeding, but males collected from ponds sometimes contain a few aquatic insects.

PREDATORS AND DEFENSE. Natural predators are poorly documented although garter snakes (*Thamnophis* spp.) and bullfrogs feed on the adults (Beneski et al. 1986; Ferguson 1961; Nussbaum et al. 1983). Defensive posturing involves undulating and lashing the tail, and coiling the body to protect the head (Brodie 1977). In a laboratory trial, Anderson (1963) found that a California mole (*Scapanus latimanus*) would not attack *A. macrodactylum* even though it would readily feed on other organisms. Chivers et al. (1996) found that adults will avoid areas that are treated with chemicals from injured conspecifics. Individuals appear to use chemical signals to detect areas where predators may be foraging.

COMMUNITY ECOLOGY. *Ambystoma macrodactylum* larvae function as upper-level predators in freshwater habitats in the western United States, but their role in structuring these communities has not been investigated experimentally.

CONSERVATION BIOLOGY. The long-toed salamander is a relatively common salamander throughout its range, but wetland breeding habitats have been destroyed in many urban and agricultural regions along the Pacific coast. *Ambystoma m. croceum* has a very limited range and consists of a few scattered populations around Monterey Bay that are threatened with extinction (Bury et al. 1980). This subspecies is federally endangered and is currently receiving high priority for protection by local citizens, state officials, and the U.S. Fish and Wildlife Service.

Bradford et al. (1994) discuss the potential effects of acid precipitation on populations in the Sierra Nevada and test the sensitivity of *A. macrodactylum* to acidity and aluminum toxicity. The continued introduction of fish to fish-free, higher-elevation lakes is adversely affecting many local populations of *A. macrodactylum*. The impact of fish introductions on this and other western *Ambystoma* species must be critically evaluated.

Ambystoma maculatum (Shaw)
Spotted Salamander
PLATES 14, 15

Fig. 44. *Ambystoma maculatum;* adult; Macon Co., Alabama (J. W. Petranka).

IDENTIFICATION. The spotted salamander is a relatively large *Ambystoma* that has two irregular rows of large yellow or yellowish orange spots on a steel gray or black dorsum. The spots extend from the head to the tip of the tail, and individuals in some populations have bright orange spots on the head that contrast with yellow spots on the body and tail. Adults that lack spotting (Easterla 1968; Husting 1965; Mount 1975) as well as albinos and partial albinos occur in low frequencies in certain populations (Brandt 1952; Dyrkacz 1981; Hensley 1959; Smith and Michener 1962). There are 11-13 costal grooves and adults measure 15-25 mm TL.

Males in breeding condition have conspicuously swollen vents. Outside the breeding season males are best identified by the presence of a series of parallel ridges on the inside of the cloaca that run perpendicular to the cloacal slit. Breeding females are typically slightly larger on average than breeding males, a phenomenon that may be due to more rapid growth and delayed maturity in females (Flageole and Leclair 1992; Hillis 1977; Peckham and Dineen 1954).

Hatchlings are dull olive green, lack conspicuous markings, and measure 12-17 mm TL (Bishop 1941a; DuShane and Hutchinson 1944). Older larvae are also dull greenish and are not strongly marked. Larvae at all stages have white or light bellies that are generally free of dark pigmentation. The tail fin is finely stippled or lightly mottled with dark pigmentation near the tip.

Juveniles are similar to adults but have lighter venters and less prominent spotting on the dorsum.

SYSTEMATICS AND GEOGRAPHIC VARIATION. The presence or absence of orange spotting on the head often varies among local and regional populations of spotted salamanders, but trends are rather chaotic across the range and no subspecies are recognized. Phillips (1994) analyzed patterns of variation in mitochondrial DNA and identified two divergent, geographically separate lineages that differ by a minimum of 19 restriction sites. Both groups contain paired disjuncts in either the eastern and western, or northern and southern, portions of the range. Many populations in the Ozarks are genetically uniform, a finding that suggests that *A. maculatum* only recently invaded this area.

DISTRIBUTION AND ADULT HABITAT. The spotted salamander occurs in portions of southern Canada and throughout much of the eastern United States. Adults are most common in bottomland forests in or adjoining floodplains, but they occur sporadically in upland forests and in mountainous regions where suitable breeding sites are available (Thompson and Gates 1982). Mature deciduous forests with vernal ponds offer optimal habitats for this species, but local populations also inhabit coniferous and mixed coniferous-deciduous forests.

BREEDING AND COURTSHIP. Adults normally breed in seasonally ephemeral, fish-free habitats such as vernal ponds, swamps, roadside ditches, and flooded tire tracks, but they may occasionally use permanent ponds (Figiel and Semlitsch 1990; Harris 1984; Husting 1965). A New York population breeds in a large, fish-free lake (Bahret 1996); however, egg masses have never been reported from lakes with fish. In the Atlantic Coastal Plain adults frequently use sloughs and backwater areas of streams that contain predatory fish (Semlitsch 1988) even though the larvae are palatable to fish and lack any obvious chemical defenses against these predators (Kats et al. 1988).

Adults migrate to breeding ponds during the winter

Range of *Ambystoma maculatum*

and early spring on rainy or foggy nights when temperatures are moderate. In some instances adults will migrate in the absence of rain if the humidity is very high and temperatures are warm (Hillis 1977; Whitford and Vinegar 1966). In northern regions adults may migrate following periods of heavy snowmelt. Several researchers (Baldauf 1952; Bishop 1941a; Blanchard 1930; Wright and Allen 1909) report that adults do not engage in mass migrations unless air temperatures are 10–12°C. However, Missouri adults will migrate whenever mean air temperatures over a 3-day period are >5.5°C (Sexton et al. 1990), whereas Indiana populations will migrate when air temperatures are 5.5–11.0°C (Peckham and Dineen 1954).

Individuals begin breeding as early as December in the southern portion of the range but may not begin until March or April farther north. Breeding has been documented in late December and early January in Mississippi (Walls and Altig 1986), from late December through early February in Alabama (Mount 1975), from late January through late March in central North Carolina (Brimley 1921; Gray 1941; Harris 1980), and from mid-January through February in southeastern Tennessee (King 1939).

The initial date of breeding in a Missouri pond varies substantially among years, but most adults breed in February and March (Sexton et al. 1986). Adults in central Kentucky and West Virginia begin breeding in mid-February through early March (Green 1956; Keen 1975; Welter and Carr 1939). Breeding occurs in March and April in Connecticut (Woodward 1982), Indiana (Minton 1972; Peckham and Dineen 1954), Maryland (Hardy 1952; Worthington 1968, 1969), Massachusetts (Moulton 1954; Stangel 1988; Talentino and Landre 1991), Michigan (Blanchard 1930; Dempster 1930; Husting 1965), New Jersey (Nyman 1987, 1991), New York (Bishop 1941a; Wright and Allen 1909), Ohio (Brodman 1995; Seibert and Brandon

1960), Pennsylvania (Baldauf 1952; Pawling 1939), and Rhode Island (Whitford and Vinegar 1966). Females in some areas of Nova Scotia may not breed until early May (Pinder and Friet 1994).

The breeding season may last from 3 days to over 2 months and is usually restricted to two or three major breeding bouts that occur during or immediately following periods of rain (Harris 1980; Mount 1975; Peckham and Dineen 1954; Sexton et al. 1986). The breeding season is estimated to last 37 days for females and 46 days for males in a Missouri population (Sexton et al. 1986), compared with 56–73 days for a North Carolina population (Stenhouse 1985a). Individuals in the latter study remained for 19–48 days at the breeding pond. At a Michigan site the annual breeding season over a 5-year period lasts 9–29 days (Husting 1965), whereas in Indiana adults remain an average of 25 days at one breeding site (Peckham and Dineen 1954).

In some northern populations, breeding is highly synchronized and restricted to one major breeding bout. The respective breeding seasons at Massachusetts (Talentino and Landre 1991) and Ohio (Brodman 1995) sites last only 2–3 and 3–5 days.

Males are typically 1.5–3.5 times more numerous than females in breeding ponds (Downs 1989; Flageole and Leclair 1992; Hillis 1977; Husting 1965; Peckham and Dineen 1954; Sexton et al. 1986; Stenhouse 1985a; Whitford and Vinegar 1966) and arrive slightly earlier than females (Blanchard 1930; Downs 1989; Hillis 1977; Peckham and Dineen 1954; Sexton et al. 1990; Shoop 1974). Annual sex ratios documented by Sexton et al. (1986) were 1.42–2.55 males/female over a 6-year span. Sex ratios differ significantly between years at this site but are unpredictable from year to year. Sex ratios in a population studied by Peckham and Dineen (1954) over 2 years were 2.9 and 2.7 males/female. The sex ratio of surface-active terrestrial adults in a southern Illinois population is around 1:1 (Parmelee 1993).

Studies of yearly patterns of breeding have revealed marked variation among populations. In some populations most adults breed annually, whereas in others males and females breed erratically and may skip a year or two before breeding again. About 90% of adults at a Rhode Island site breed annually (Whitford and Vinegar 1966). Similarly, Douglas and Monroe (1981) found that six of eight tagged individuals bred annually. The remaining two adults migrated toward ponds but stopped 70 m short and did not breed. Phillips and Sexton (1989) found that 38% of marked females and 30% of marked males return to breed 1 year after leaving a Missouri breeding pond.

Husting (1965) tracked a marked population for 5 years and documented erratic breeding patterns. Each year large numbers of both males and females can be collected that are unmarked. A small number of individuals breed annually nearly every year, but most skip one or more years before returning to breed. Females studied by Sexton et al. (1986) lost an average of 38% of their weight during breeding compared to only 13% for males.

Adults often leave the pond near the same point where they initially entered (Douglas and Monroe 1981; Phillips and Sexton 1989; Sexton et al. 1986; Shoop 1965a, 1968; Stenhouse 1985a) and tend to arrive in the same order from year to year (Stenhouse 1985a). These observations strongly suggest that individuals migrate to and from ponds along specific pathways and return to their respective home ranges in the adjoining forest after breeding. In fact, some adults return to the same burrow from which they initially migrated (Downs 1989).

When captured individuals are released next to foreign ponds, they bypass these and return to their home ponds (Shoop 1968). Animals released as far as 130–500 m from a pond may return in as few as 11 days after being released (Shoop 1968; Whitford and Vinegar 1966). The mechanisms used for orientation during migrations are poorly understood. McGregor and Teska (1989) found that intercepted adults prefer substrates with home pond odors compared with foreign pond odors; this observation suggests that olfaction is important in orienting to ponds.

Adults may move 200 m or more from ponds after breeding. At a Missouri site adults move as far as 172 m from a breeding pond, but most stay within 100 m of the pond margin (Sexton et al. 1986). Other estimates are that adults move 6–220 m (mean = 150 m; $n = 6$) from a Kentucky pond (Douglas and Monroe 1981), compared with 157–249 m (mean = 192 m; $n = 6$) from a Michigan pond (Kleeberger and Werner 1983). At both sites, adults became sedentary and migrations ceased by the first of June. Unpublished data from other studies indicate that adults often move <100 m from the breeding ponds (Stenhouse 1985a).

Courtship of *A. maculatum* is described by Wright and Allen (1909) and Arnold (1976). The following description is from Arnold's more detailed account.

Fig. 45. *Ambystoma maculatum;* spermatophores; Orange Co., North Carolina (J. W. Petranka).

When courting in individual pairs, the male contacts the female repeatedly with his snout, swings his head back and forth over her dorsum, and lifts his head under her chin. The male circles around the female repeatedly during this period, and the female turns and nudges the male with her snout whenever she is contacted. The male moves a short distance away from the female in a shuffling motion that maintains the vent in frequent contact with the substrate, then deposits a spermatophore either on the substrate or on another spermatophore. While depositing a spermatophore, the male arches the midbody upward slightly and slowly undulates the tail while holding it level with the substrate. The male moves forward and shuffles his vent against the substrate. If another spermatophore is encountered he immediately deposits a spermatophore on top of this. Otherwise, the male returns to the female and repeats the nudging and circling behavior before depositing additional spermatophores.

The female searches for spermatophores by stepping side to side with her hindlimbs while moving slowly forward. When a female finds a spermatophore, she orients tactually with her body, then squats upon the spermatophore and removes seminal fluid with her cloacal lips. Her posture while picking up a spermatophore is similar to that of the male while depositing, except that the tail is not undulated. As the female moves off the spermatophore she arches her tail base. The female usually mounts 15–20 spermatophores before ending courtship.

Males that have not previously courted produce an average of about 40 spermatophores (range = 10–81 spermatophores) before ceasing courtship. A male may court again on succeeding nights following the initial courtship, but the number of spermatophores produced drops precipitously. Over the entire breeding season a male probably produces fewer than 100 spermatophores. The spermatophore is about 6–8 mm tall with a gelatinous base that is 6–9 mm in diameter at its widest point. The base weakly tapers toward the top and has a four-pronged concave summit that holds the seminal fluid (Arnold 1976; Mohr 1931; Smith 1910).

Adults normally court in groups that are scattered about shallow regions of the pond. A group may contain from 3 to 50 or more adults (Breder 1927), and males typically outnumber females in local breeding aggregates. Scattered spermatophore fields that often contain several hundred spermatophores in an area <1 m^2 can be seen in ponds following breeding bouts. Adults sometimes make audible sounds when courting (Wyman and Thrall 1972).

Males actively nudge members of both sexes when courting en masse. The presence of other males lowers the reproductive success of an individual, either because a given individual's spermatophores make up a smaller proportion of the total number in a spermatophore field, because its spermatophores are covered by those of others, or because it is prevented by other males from successfully courting the female (Arnold 1976, 1977). Multiple spermatophores constitute about 6% of spermatophores at a natural breeding site in Missouri (Sexton et al. 1986). Because of intense sexual competition, rapid courtship before other males gather about females is probably a key courtship strategy used by males.

REPRODUCTIVE STRATEGY. Within 2–3 days after mating, females begin depositing eggs on twigs, fallen tree branches, and aquatic plants. The eggs are occasionally laid directly on the pond bottom (Sexton et al. 1986) and are deposited in firm, ovoid to cylindrical masses that are 5–15 cm wide and 5–25 cm long (Bishop 1941a). The egg masses are either scattered individually about the pond or deposited in dense, communal aggregates that may contain 50 or more egg masses. The outer jelly membrane surrounding the eggs is colonized by a unicellular green alga (*Oophila amblystomatis*) that increases oxygen supply to the developing embryos, particularly those near the center of a mass that are not adequately supplied with oxygen by diffusion from the surrounding water (Bachmann et al. 1986; Gatz 1973; Gilbert 1942, 1944; Hammen and Hutchison 1962; Hutchison and Hammen 1958; Pinder and Friet 1994).

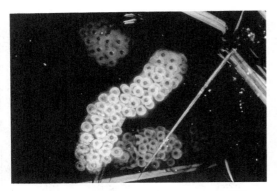

Fig. 46. *Ambystoma maculatum;* clear egg masses; Wake Co., North Carolina (J. W. Petranka).

Fig. 47. *Ambystoma maculatum;* clear egg mass; Breckenridge Co., Kentucky (J. W. Petranka).

The egg masses of spotted salamanders consist of clear, white, and intermediate color morphs depending on the presence of proteins in the outer jelly layers (Hardy and Lucas 1991; Ruth et al. 1993). Local pond populations often lay only the clear egg masses; however, masses that are milky white in appearance can make up a high percentage of egg masses in some populations. Intermediate forms are relatively uncommon.

Ruth et al. (1993) conducted a comprehensive study of egg morph variation among populations and found no obvious adaptive advantage of white versus clear masses. The frequency of clear masses in populations correlates with the levels of certain nutrients in ponds. However, this relationship varies among years. In an Ohio population white egg masses contain an average of 103 eggs compared with 66 eggs in clear masses (Brodman 1995).

A female normally deposits two to four egg masses that hold from 1 to 250 eggs each (Bishop 1941a; Shoop 1974). Mean numbers of eggs per mass reported from different localities are 88 (Stangel 1988) and 155 (Talentino and Landre 1991) in Massachusetts, 134 (Bishop 1941a) in New York, 99 in Nova Scotia (Pinder and Friet 1994), and 60–90 (Seibert and Brandon 1960) and 58 (Downs 1989) in Ohio. Mean number of eggs per mass may vary from 72 to 79 over a 5-year period in an Ohio population (Brodman 1995). In a central North Carolina pond, mean number of eggs per mass varies seasonally from 65 to 104 and decreases as the season progresses (Harris 1980). Embryos from eggs deposited in late January have a lower probability of surviving to hatching than those in egg masses deposited later in the season. The only estimate of egg density is 21–250 eggs/m^2 (mean = 123) in 13 Indiana ponds (Cortwright 1988).

When depositing eggs, a female grasps a support structure with her rear legs and arches her body upward with the forelimbs often free and the tail pointed downward (Bishop 1941a; Wright 1908). Egg masses are 2–6 cm long when first laid but quickly swell to a much larger size. Females require 30–60 minutes to deposit a single mass (Wright and Allen 1909). In two instances eggs have been found on land that were laid singly beneath boards or logs near water (Brimley 1921; Smith 1921). These may have been deposited by ovulating females that were forced to leave ponds because of severe freezing.

Mean clutch size is fairly similar throughout the range of *A. maculatum*. Females at two breeding sites in Connecticut contain 144–370 eggs and have median complements of 182 and 223 eggs (Woodward 1982). Clutch size is positively correlated with female SVL at one site but not at the second. Ovarian com-

Fig. 48. *Ambystoma maculatum;* white egg masses; Graham Co., North Carolina (J. W. Petranka).

plements of 22 Massachusetts females are 92–328 and average 224 ova (Shoop 1974). Other reported values are an average of 207 ovarian eggs in 40 Virginia specimens (Ireland 1989) and a mean of 172 ova in 10 females from Michigan (Wilbur 1977).

Embryos in blastula or preblastula stages average from 1.8 to 2.8 mm in diameter (DuShane and Hutchinson 1944; Woodward 1982). Each embryo is surrounded by two jelly capsules in addition to the vitelline membrane. Hatchling size, developmental stage, and length of the embryonic period are negatively correlated with the temperature at which embryos develop (Voss 1993a). Embryos from the Chicago region are smaller than those from Pennsylvania and New Jersey sites (DuShane and Hutchinson 1944), and embryos from larger eggs develop faster and are larger at hatching than those from smaller eggs. Larval growth rate is independent of ovum and hatchling size in a Mississippi population (Walls and Altig 1986).

The spotted salamander has a relatively long incubation period that normally lasts for 4–7 weeks. Talentino and Landre (1991) compared the developmental period of *A. maculatum* and *A. laterale* embryos in Massachusetts and found that the latter species has much faster development rates. Embryos of *A. laterale* hatch 9–10 days before those of *A. maculatum*. This may give *A. laterale* a competitive advantage since hatchlings gain an early growth advantage over *A. maculatum* hatchlings. At an Ohio site *A. maculatum* has an incubation period (mean = 50 days) that is 3 weeks longer than that of syntopic *A. jeffersonianum* eggs laid at the same time (Brodman 1995).

Bishop (1941a) reported an incubation period of 4–6 weeks in New York populations, with hatching occurring from mid-May to early June. Average incubation periods reported for other localities are 34 days for central Kentucky populations (Keen 1975), 39 days for a Massachusetts population (Stangel 1988), and 31 days for a Missouri population (Sexton et al. 1986). Egg masses deposited at the beginning of the breeding season in a New Jersey pond hatch after 42–43 days (Nyman 1991).

Within a single population, the embryonic period may be highly variable both within and between years. During one year the incubation period in a Maryland pond studied by Worthington (1968, 1969) varied from 38 to 52 days depending on the seasonal time of oviposition and whether eggs were laid in shady or sunny portions of the pond. In the following year when adults bred about 2 weeks later than usual, the incubation period was only 20–26 days in the sunny portion of the pond. Hatching occurred around mid-April in both years. Hatching in New Jersey typically occurs in late April through May (Freda 1983; Nyman 1991). The hatching period is more compressed than the breeding season and may last only 8–14 days (Shoop 1974).

Embryonic mortality is often highly variable both among and within ponds. Estimates of embryonic mortality in seven North Carolina ponds range from 0 to 100% and 0 to 40% for 2 years (Stenhouse 1987), compared with 38–65% over a 5-year period in Virginia populations (Ireland 1989). In a Massachusetts pond, mean embryonic mortality is 3% (Stangel 1988), compared with 2–30% in 13 other Massachusetts populations (Cook 1983). In an Ohio population annual embryonic mortality is 28–40% for 4 years and is independent of egg mass size (Brodman 1995). In central Ontario, mortality in 16 ponds varies from 1 to 100% and averages 45% (Clark 1986), whereas in New York mortality varies from 1 to 73% in ponds surveyed by Pough (1976).

Both the seasonal time of deposition and the site of egg deposition within a pond can affect embryonic mortality. Embryonic mortality is highest in eggs laid early in the season in a North Carolina pond (Harris 1980). Experiments conducted by Brodman (1995) show that eggs placed in groups at shallow depths have higher survival than ungrouped eggs placed at relatively deep depths. Eggs deposited at shallow depths in ponds may occasionally suffer significant mortality from freezing (Ireland 1989).

AQUATIC ECOLOGY. Larvae are gape-limited predators and show a general tendency to incorporate a broader array of prey into their diets as they grow. Zooplankton is the dietary staple of larvae of all size classes. In experimental ponds, larvae are more likely to reduce populations of isopods and amphipods than zooplankton (Harris 1995).

Larvae in a small temporary pond in New Jersey feed predominantly on cladocerans and copepods, but also occasionally take isopods, ostracods, odonates, beetles, and trichopterans (Freda 1983). Larvae of the red-spotted newt are a major dietary item of mature larvae, but there is no evidence of cannibalism. Small larvae in New Jersey ponds feed almost entirely on zooplankton, whereas older larvae feed on zooplankton as well as chironomids, chaoborids, and isopods (Nyman 1991). Zooplankton are also the major dietary items of larvae in Mississippi ponds (Branch and Altig 1981).

Larvae grow rapidly and most transform within 2-4 months after hatching. Larvae in a Maryland population begin transforming in mid-June, about 2 months after hatching, when they reach 29-32 mm SVL (Worthington 1968). Larvae in New Jersey populations begin to transform in late June through early July after a larval period of 6-8 weeks (Freda 1983; Nyman 1991). Metamorphosing larvae have been collected in mid-June in Oklahoma (Dundee 1947).

Bishop (1941a) reports a larval period of 9-16 weeks in New York ponds with larvae metamorphosing between mid-July and mid-September at an average of 43-53 mm TL. Larvae in a Massachusetts population metamorphose in August after 3 months of growth when they reach 27-28 mm SVL (Talentino and Landre 1991), whereas those in a southern Michigan pond transform in mid-August after 13 weeks of growth (Dempster 1930). Average size at metamorphosis ranges from 49 to 60 mm TL. Ball (1937), Ling et al. (1986), and Stangel (1988) document several instances in which ponds dried before any larvae metamorphosed.

Although most larvae transform from June to August, slowly growing larvae in some populations overwinter and transform the following spring or summer. Overwintering larvae have been found in a cold, spring-fed pond in Rhode Island (Whitford and Vinegar 1966) and in ponds in Arkansas (Ireland 1973), Massachusetts (Stangel 1988), Missouri (Phillips 1992), and Nova Scotia (Bleakney 1952). At the Arkansas site most larvae metamorphose in June, but a few overwinter and transform before early March.

At the Missouri site overwintering patterns vary annually. In some years most larvae transform during the summer, but in three out of five years on average the majority overwinter (Phillips 1992). Some overwintering larvae transform in April and May and may reach an average TL of 69 mm by early May. In years when larvae do not overwinter, size at metamorphosis is positively correlated with the date of capture in drift fences and varies from 45 to 75 mm TL.

Experimental studies by Wilbur (1972) and Wilbur and Collins (1973) demonstrate that growth rate and size at metamorphosis are phenotypically plastic traits of this and other *Ambystoma* species. When established at high densities larvae grow more slowly, have higher mortality, and are smaller at metamorphosis; these density-dependent responses could function to limit adult population size. The extent to which initial density affects larval growth and development in natural ponds depends on additional factors such as food levels and predator densities.

Ambystoma larvae are very aggressive and in laboratory cultures will cannibalize or dismember each other unless food is supplied in excess. Spotted salamander larvae caught in the field often have missing tail tips, a finding that may reflect aggressive behavior among conspecifics. Agonistic behavior in *A. maculatum* involves looking toward, lunging at, and biting conspecifics. Smaller individuals often respond to these behaviors by moving away from attackers and tend to be less aggressive once attacked (Walls and Semlitsch 1991).

Stratification, a behavior in which a larva floats in a stationary position in the water column, has been documented in some spotted salamander populations and may function both to enhance foraging efficiency and to lessen the vulnerability of smaller larvae to attacks from larger conspecifics. Anderson and Graham (1967) did not observe stratification of larvae in New Jersey ponds, but Branch and Altig (1981) did in Mississippi ponds. At another New Jersey site, late-stage larvae stratify but younger larvae remain on the pond bottom or in vegetation in shallow areas of the pond (Nyman 1991).

Premetamorphic survival of *A. maculatum* is usually <10% and often much lower. Estimates of annual survivorship from oviposition to metamorphosis are 1-12.6% for a 4-year study of a Massachusetts population (Shoop 1974). Survivorship at this site is highest in years when relatively few eggs are deposited in the ponds. However, these are also years when ponds dry late in the season, so the cause for the inverse relationship between density and survivorship is uncertain. Premetamorphic survival in populations in North Carolina is typically <1%; the high mortality is due in part to intense predation from *A. opacum* larvae (Stenhouse 1985b, 1987). Premetamorphic survival is 0-3% in a Massachusetts pond (Stangel 1988), and complete mortality may occur in years when the pond dries prematurely. Rowe and Dunson (1993) record catastrophic mortality of embryos because of premature

Fig. 49. *Ambystoma maculatum;* metamorph; Watauga Co., North Carolina (R. W. Van Devender).

pond drying. Premetamorphic survival in a fish-free Virginia pond varies from 4.2% to 12.5% among years, but the total output of juveniles does not differ significantly among years (Ireland 1989). Larvae have type II survivorship, and age-specific survival is nearly constant throughout the larval period.

TERRESTRIAL ECOLOGY. Metamorphs leave the ponds within a few weeks after transforming and disperse into the surrounding forest during rainy weather. Emigrating metamorphs have been observed on 31 July in Maryland that measure 50 mm TL on average (Hardy 1952), and from late July through early October in Massachusetts that measure from 36 to 46 mm SVL on average (Shoop 1974).

Metamorphs often move beneath rocks surrounding the margins of a pond in New York (Pough and Wilson 1970). On sunny days they are exposed to high temperatures that approximate their critical thermal maxima. Individuals often aggregate beneath rocks during these physiologically stressful periods, and severely dehydrated animals cannot right themselves. Metamorphs that burrow into mud near the pond margin may also die from desiccation if there is a prolonged period without rain (Shoop 1974).

Juveniles that successfully disperse away from ponds presumably spend most of their time burrowed in the soil and do not return to the breeding ponds until sexually mature. Few data are available on the length of the juvenile stage. Although detailed data are lacking, individuals in more southern localities may require only 2-3 years to reach sexual maturity (Bishop 1941a; Minton 1972). In Michigan, males and females reach maturity after 2-3 and 3-5 years, respectively (Wilbur 1977). Most females in a Quebec population mature by age 7 when they reach >78 mm SVL, whereas males mature when 2-6 years old and >63 mm SVL (Flageole and Leclair 1992). Most of the animals in this population are 2-18 years old, but some live as long as 32 years. The peak in age frequency is 6-8 years. Spotted salamanders can live 22-25 years in captivity (P. H. Pope 1928, 1937).

Nonmigrating adults at a Kentucky site live in subterranean rodent burrows and restrict their activity to 12-14 m^2 of the forest floor (Douglas and Monroe 1981). At other locales the adults may move an average of 14 m over a 6- to 7-week period and have average home ranges of about 10 m^2 (Kleeberger and Werner 1983). Animals at this site are found 72% of the time beneath the ground surface, 21% of the time in or under decaying hardwood logs, and 7% of the time beneath wet leaf litter. On rainy nights individuals are occasionally found on the forest floor and may feed while protruding their heads from burrow entrances (Gordon 1968). A small percentage of individuals in a southern Illinois population can be found beneath large logs on the forest floor in spring, but the adults move beneath ground with the onset of summer weather (Parmelee 1993). Individuals do not appear capable of actively digging burrows, but they can enlarge existing holes and crevices and may live as deep as 1.3 m below the ground surface (Gordon 1968; Semlitsch 1983c).

Adults feed on a wide variety of forest floor invertebrates, except during the breeding season when they apparently do not feed (Smallwood 1928). Prey of adults includes earthworms, mollusks, spiders, millipedes, centipedes, and a wide variety of larval and adult insects (Bishop 1941a; Pope 1944). The adults are aggressive and may defend home burrows or feeding areas from conspecifics (Ducey and Heuer 1991; Ducey and Ritsema 1988). Because adults typically occur at low densities in forest habitats, it is uncertain if defense of local burrows would act as an effective density-dependent regulating mechanism.

Ducey et al. (1994) found that *A. maculatum* is aggressive toward *Plethodon cinereus;* in staged laboratory encounters *A. maculatum* may kill and eat 9% of the *P. cinereus* intruders. Aggression in this instance may reflect predation more than defense of territories. Recently metamorphosed juveniles rarely will bite *A.*

maculatum or *A. talpoideum* intruders of similar age (Walls 1990). However, residents do spend more time moving toward both conspecific and heterospecific intruders. This behavior is directed more toward conspecifics than heterospecifics.

Yearly survivorship of adults is very high. At least 90% of breeding adults at a Rhode Island site return to breed the following year (Whitford and Vinegar 1966), and 100% of eight tagged animals at a Kentucky site may survive from one breeding season to the next (Douglas and Monroe 1981). Husting (1965) tracked survivorship and breeding returns in a marked population for four consecutive years and found that annual survivorship is >70–80% in most years. Females have higher mortality rates than males, and this finding may explain in part why males typically outnumber females at breeding ponds.

Severe winter weather can be a significant source of mortality in some populations. Harris (1980) notes that 38 adults were killed when a pond in North Carolina froze; all but one of the salamanders was female. About 50,000 eggs were laid during the breeding season at this site. If we assume that each female lays 200 eggs (Woodward 1982), about 15% of the female breeding cohort was killed during the freeze.

PREDATORS AND DEFENSE. Spotted salamander egg masses have thick, firm outer jelly coats that function both as a defense against predators and as a means to prevent desiccation of eggs that become stranded when ponds recede during dry periods (Nyman 1987). The jelly coat reduces predation on embryos from leeches, centrarchid fish, and other predators (Cory and Manion 1953; Semlitsch 1988; Ward and Sexton 1981). Egg mortality in fish versus fish-free ponds in Virginia does not differ significantly (Ireland 1989). This finding, along with a study by Semlitsch (1988), suggests that *Lepomis* and other centrarchids have difficulty eating *A. maculatum* eggs.

Despite the protective value of the jelly coat, spotted salamander eggs are preyed upon by many aquatic species. Adult red-spotted newts are effective predators on the eggs of this and other amphibians (Hamilton 1932; Wood and Goodwin 1954), as are certain caddisfly and midge larvae (LeClair and Bourassa 1981; Murphy 1961; Stout et al. 1992). Midge larvae (*Parachironomus*) prey heavily on the developing embryos in populations in Quebec, and embryonic mortality may reach 70–100% in certain ponds where infestations are very high (LeClair and Bourassa 1981).

The caddisfly *Ptilostomis postica* can be a major predator on spotted salamander embryos and hatchlings (Rowe et al. 1994). In some years >90% of *A. maculatum* embryos are killed prior to hatching, and clear and white egg masses are equally vulnerable. *Ptilostomis* is extremely acid tolerant and reaches relatively high densities in acidic ponds, perhaps because of release from competition with acid-intolerant competitors. Thus, populations of spotted salamanders breeding in highly acidic (pH <4.5) ponds often suffer heavy embryonic mortality from a combination of both *Ptilostomis* predation and physiological stress associated with low pH. In comparable Pennsylvania ponds with higher pH and without high densities of *Ptilostomis*, median egg survival is 94–100% (Rowe and Dunson 1993). Another larval caddisfly, *Banksiola dossuaria*, also is a major predator on spotted salamander eggs (Stout et al. 1992).

In the southern Appalachians *A. maculatum* embryos often suffer heavy mortality from predatory wood frog (*Rana sylvatica*) tadpoles. Experiments that my colleagues and I conducted in both artificial mesocosms and natural habitats indicate that white egg masses are less vulnerable to predation than clear egg masses, and that egg predation rates are higher when food levels are low (Petranka et al. 1988). Wood frog tadpoles will also attack and kill late-term embryos that escape from the egg capsules as the tadpoles feed. Differential tadpole predation on clear versus white egg masses appears to favor the white egg morph in southern Appalachian populations of *A. maculatum*.

Larvae are eaten by a wide array of aquatic predators, including aquatic insects, *Ambystoma* larvae, and fish. Stangel (1983) observed least sandpipers (*Calidris pusilla*) feeding on larvae in a Massachusetts pond, and Ling and Werner (1988) found that hatchlings in the laboratory often died from protozoan (*Tetrahymena*) infections. Huheey and Stupka (1967) found three dead adults along with the remains of numerous wood frogs that were attacked and eaten by raccoons around a breeding site in eastern Tennessee. Around the perimeter of a breeding pond in western North Carolina, Beachy (1991) discovered 21 adults that had been killed and partially eaten by an unidentified predator. This observation underscores the risk associated with migrating to and from breeding ponds.

Fig. 50. *Ambystoma maculatum;* defensive posturing; Macon Co., North Carolina (E. D. Brodie, Jr.).

Like many other salamanders, adult *A. maculatum* produce noxious skin secretions that repulse mammalian, avian, and reptilian predators (Barach 1951; Brodie et al. 1979; Howard 1971). The bright spotting on this species presumably functions as warning coloration. Animals that are attacked by garter snakes or other predators will often secrete noxious chemicals from the tail that ward off attacks. A spotted salamander may respond to an approaching smooth earth snake (*Virginia valeriae*) by arching the body and biting (Dodd 1977). A butting defensive posture that protects the head is typical of this and several other *Ambystoma* species (Brodie 1977). When attacked by shrews (*Blarina*), individuals respond by butting with the head if attacked from the front or lashing with the tail if attacked from the rear. Individuals of all sizes may also vocalize (Brodie et al. 1979).

COMMUNITY ECOLOGY. Spotted salamander larvae often share breeding ponds with congeners such as *A. jeffersonianum, A. opacum,* and *A. texanum*. Several researchers have examined size-specific interactions among these species, particularly between *A. maculatum* and *A. opacum*. In many ponds *A. opacum* larvae are nearing metamorphosis when *A. maculatum* larvae hatch. At this stage *A. opacum* is sufficiently large to eat *A. maculatum* (Stenhouse et al. 1983; Stewart 1956). In North Carolina ponds premetamorphic survivorship of *A. maculatum* is <1% in ponds with *A. opacum* and in some cases appears to be 0% because of intense predation (Stenhouse 1985b). In addition, densities of *A. maculatum* are negatively correlated with densities of *A. opacum* across study ponds. Laboratory trials show that *A. maculatum* is less prone to predation when alternate prey is available for *A. opacum* and refuges are available for *A. maculatum* larvae.

At other sites predation of *A. opacum* larvae on *A. maculatum* larvae does not appear to be as severe. Worthington (1968) collected substantial numbers of *A. maculatum* larvae nearing transformation from a Maryland pond that supports a large population of *A. opacum*. In cage experiments, Cortwright (1988) found that the presence of *A. opacum* or *A. jeffersonianum* does not significantly affect *A. maculatum* survival, growth, or developmental rate.

Despite the low premetamorphic survivorship of *A. maculatum,* large populations of *A. maculatum* often coexist locally with *A. opacum*. Dispersal from surrounding populations, high adult survivorship, and selection for long life spans may be important features that allow *A. maculatum* to maintain healthy populations in areas where *A. opacum* is a major predator (Husting 1965; Stenhouse 1985b).

Nyman (1991) examined dietary overlap between *A. maculatum* and members of the *A. jeffersonianum* complex in New Jersey ponds. At these sites *A. maculatum* embryos hatch later than members of the *A. jeffersonianum* complex and are frequently preyed upon by *A. jeffersonianum* and JJL unisexuals. Although larvae of all *Ambystoma* species at this site utilize zooplankton rather heavily, differences are evident in the utilization of insects and isopods by *A. maculatum* and members of the *A. jeffersonianum* complex. Dietary differences parallel differences in size and stratification behavior. Noble (1931) also noted that *A. jeffersonianum* often prey upon *A. maculatum* when the two are crowded together.

Brodman (1996) conducted a series of experiments in laboratory aquaria and field enclosures in a northeastern Ohio pond to examine interactions between *A. maculatum* and *A. jeffersonianum*. The results indicate that high larval densities reduce survival to metamorphosis in both species. This reduced survival occurs when larvae are grown either alone or with members of the other species. However, the adverse effects of interspecific interactions at high densities are about the same as those of intraspecific interactions at the same densities. In this regard, the two species appear to be equal in competitive ability.

Other data show that growth and survival to metamorphosis are inversely correlated with the seasonal duration of ponds, and that larvae suffer heavy mortality in the absence of cover. Both species exhibit similar microhabitat use when alone but tend to partition habitats when together. Habitat partitioning is due to *A. maculatum* shifting microhabitat use in the presence of larger, predatory *A. jeffersonianum* larvae. The behavioral flexibility of *A. maculatum* appears to be a key trait that allows larvae to successfully exploit habitats utilized by larger *A. jeffersonianum* larvae. The final preferred temperature of *A. maculatum* is warmer than that of *A. jeffersonianum* (35°C versus 25°C), suggesting that these species may spatially segregate late in the season along thermal gradients in ponds (Stauffer et al. 1983).

Rowe and Dunson (1995) examined how variable hydroperiods affect species interactions between *A. jeffersonianum, A. maculatum,* and *Rana sylvatica* in experimental tanks set up in a forest in Pennsylvania. Under these conditions, growth rate, time to metamorphosis, and mass at metamorphosis of *A. maculatum* larvae are independent of both the hydroperiod and the initial densities of the three species. The survival of wood frog tadpoles is negatively correlated with the growth rate of *A. maculatum,* a finding that presumably reflects predation of *Ambystoma* on *Rana.*

Wilbur (1972) examined density-dependent interactions of *A. maculatum* larvae with several other *Ambystoma* species and found that crowding generally decreases the growth rate, survival, and size at metamorphosis of most species. Walls and Jaeger (1987) found that *A. maculatum* larvae inhibit the growth of both conspecifics and *A. talpoideum* larvae, perhaps via exploitative competition for food. However, *A. talpoideum* larvae are more aggressive to both conspecifics and *A. maculatum.* This trade-off between superiority in exploitative versus interference competition may explain why neither species seems capable of driving the other extinct locally.

In a related study Walls (1995) examined how the presence of predatory *A. opacum* larvae could mediate ecological interactions between *A. maculatum* and *A. talpoideum* larvae. *Ambystoma maculatum* is more vulnerable to predation by *A. opacum* than is *A. talpoideum.* *Ambystoma talpoideum* increases its use of leaf litter and deep-water refugia in the presence of *A. opacum,* but *A. maculatum* does not. These different responses to *A. opacum* may reduce habitat overlap and reduce competitive interactions between the two prey species.

Semlitsch (1988) examined interactions between *A. maculatum* and *A. talpoideum* on the Atlantic Coastal Plain, where the two species almost never share the same breeding pond. A field transplant experiment coupled with competition experiments suggests that neither species is capable of excluding the other from sites via competition or predation. The allotopic distribution of these species may be related more to the habitat requirements of adults than to larval interactions.

In contrast to findings at Atlantic coast sites, *A. maculatum* and *A. talpoideum* frequently coexist in local breeding ponds on the Gulf coast. Semlitsch and Walls (1993) compared geographic variation in the competitive abilities of specimens collected from Mississippi and South Carolina and found that the geographic origins of conspecific and heterospecific competitors have little effect on competitive outcomes. Overall, *A. maculatum* is the inferior competitor, presumably because *A. talpoideum* is more aggressive. However, *A. maculatum* does not suffer increased mortality in the presence of *A. talpoideum.*

Several researchers have examined predator-prey interactions between fish and spotted salamanders. Centrarchid fish such as the sunfishes (*Lepomis* spp.) cannot feed on the eggs because of the thick jelly coat that surrounds them, but they will readily eat hatchlings and older larvae (Semlitsch 1988). Spotted salamanders rarely breed in fish ponds in most areas outside the Atlantic Coastal Plain. This observation suggests either that fish predation is sufficiently intense to eliminate *A. maculatum* from fish ponds or that adults actively avoid ovipositing in ponds with fish.

Sexton et al. (1994) found that naive adult spotted salamanders do not respond to fish chemicals by reducing the number of egg masses oviposited. Thus, the use of fish-free habitats does not appear to be strongly tied to the active avoidance of sites with predatory fish. At a site in Virginia where *A. maculatum* uses both a fish-free and a fish pond for breeding, *A. maculatum* larvae do not hide from fish and none survives to metamorphosis (Ireland 1989). In an adjoining fish-free pond, larvae do survive to metamorphosis. These observations suggest that *A. maculatum* is largely restricted to fish-free habitats because of direct predation on larvae.

Other studies have been conducted in Atlantic Coastal Plain habitats where *A. maculatum* often coexists with predatory fish. Based on the results of a seminatural field experiment, Semlitsch (1988) concluded that *A. maculatum* larvae are less vulnerable to fish predation that *A. talpoideum*. In a related experiment, Figiel and Semlitsch (1990) found that *A. maculatum* larvae shift their microhabitat use from open water to leaf litter when placed with predatory bluegills (*Lepomis macrochirus*) in experimental tanks. These results suggest that populations that regularly coexist with fish have evolved behavioral adaptations that allow coexistence. Surprisingly, larvae did not suffer higher mortality in the presence of fish, even though they did in a similar experiment conducted by Semlitsch (1988).

Gray treefrogs (*Hyla chrysoscelis*) actively avoid ovipositing in experimental ponds with *A. maculatum* larvae, which are potential predators on their offspring (Resetarits and Wilbur 1989). Thus, spotted salamander larvae may indirectly affect community structure by altering habitat use by other community members.

CONSERVATION BIOLOGY. Woodland vernal ponds are the primary breeding sites of spotted salamanders. Protecting and preserving a 200- to 250-m radius of deciduous forest around vernal ponds are essential for maintaining healthy local populations of spotted salamanders. Local populations of this and other *Ambystoma* species are becoming increasingly isolated from each other as habitat fragmentation, deforestation, and loss of vernal ponds reduce gene flow among demes. Inbreeding depression could eventually become a significant problem in some of the smaller, isolated populations. In addition, colonists may not be able to reach ponds where *Ambystoma* populations have suffered local extinctions. Maintaining forest corridors between local populations and constructing breeding sites to allow island hopping would enhance the long-term survival of regional populations of this and other *Ambystoma* species.

In addition to direct habitat loss, conservationists are increasingly concerned that acid deposition in the eastern United States is resulting in the long-term acidification of seasonally ephemeral breeding sites used by this and many other amphibians. Spotted salamanders have been used extensively in acid tolerance studies, and data indicate that the tolerance of eggs and larvae to acid conditions varies markedly among local and regional populations. Sadinski and Dunson (1992) found that survival is markedly depressed when eggs and larvae from Pennsylvania are maintained at a pH ≤ 4.3 but is much higher at a pH >4.5. In some populations a pH in the 4.5–5.5 range can reduce hatching success and larval growth and development (Brodman 1993; Clark 1986; Gosner and Black 1957; Ling et al. 1986; Pough 1976; Pough and Wilson 1977; Saber and Dunson 1978). Values within this range are not uncommon in many areas of the eastern United States and southern Canada (Clark 1986; Cook 1983; Ling et al. 1986; Sadinski and Dunson, 1992). Currently, there is little evidence that local populations of *A. maculatum* have evolved tolerance to acidity (Clark 1986; Tome and Pough 1982).

Spotted salamander populations have declined in eastern Virginia during the last decade, and decreases in pH and increases in metal concentrations associated with acid deposition may have contributed to the declines (Blem and Blem 1989, 1991). In some ponds, hatching success is inversely related to total and inorganic aluminum concentrations (Clark and Hall 1985). The number of eggs present in Pennsylvania ponds is positively correlated with pH and pond size, and negatively correlated with total cations and silica (Rowe and Dunson 1993). In central Ontario, hatching success is positively correlated with pond pH, whereas the number of eggs per pond is positively correlated with pond alkalinity (Clark 1986). In addition to directly killing the eggs and larvae, low pH can affect larval survival and growth rates and alter predator-prey interactions between *A. maculatum* and *A. jeffersonianum* larvae (Brodman 1993). In contrast to the previously mentioned findings, Cook (1983) surveyed 13 ponds in Massachusetts and found no relationship between pond pH and embryonic mortality. Because the spotted salamander is so widespread and well studied, it would make an excellent focal species for use in long-term amphibian monitoring programs.

Ambystoma opacum (Gravenhorst)
Marbled Salamander
PLATE 16

Fig. 51. *Ambystoma opacum*; male (above) and female (below); Wake Co., North Carolina (J. W. Petranka).

IDENTIFICATION. The marbled salamander is a stout, medium-sized *Ambystoma* that has a black ground color overlain by conspicuous white or light gray crossbands across the head, back, and tail. The crossbands often run together or are interrupted and are broader on the sides of the body. In rare instances the crossbands may be absent to produce a morph having longitudinal stripes (Trauth and Richards 1988). Males have silvery white crossbands whereas those of females are silvery gray. The venter is black and lacks crossbands. Sexually active males can be readily distinguished from females by their white crossbands and swollen cloacal glands. Adults measure 77–127 mm TL.

Hatchlings are blackish and rather drab and have pond-type morphology with bushy gills and dorsal fins that extend almost to the front limbs. Hatchlings usually measure 10–14 mm TL (Kaplan 1980), but more advanced embryos may reach 19 mm TL at hatching (Bishop 1941a; Noble and Brady 1933). The larvae are drab brown or blackish and have a ventrolateral series of light spots that form a line just below the level of the limb insertions (Anderson 1967b; Trauth et al. 1989b). Older larvae often develop mottling on the body and have varying degrees of light yellowish green coloration (Trauth et al. 1989b). The throat is uniformly stippled with dark pigment, and scattered melanophores occur on the lateral portions of the venter. Because of fall breeding, larvae collected in the early spring are normally much larger than other *Ambystoma* larvae that share breeding ponds.

Recently transformed juveniles are brown or black with light flecks. The flecks become more pronounced and form lichenlike patterns within 1–3 weeks after metamorphosis. Juveniles begin developing the adult pattern within 1–2 months after transformation (Bishop 1941a). Albino larvae were reported by Dyrkacz (1981).

SYSTEMATICS AND GEOGRAPHIC VARIATION. This is a well-defined species whose nomenclature has been stable for decades. Geographic variation has not been extensively analyzed, but variation in color patterns appears to be chaotic and no subspecies are recognized.

DISTRIBUTION AND ADULT HABITAT. The marbled salamander occurs throughout much of the eastern deciduous forest from southern New England to northern Florida and westward to the tallgrass prairie. Floodplain forests with numerous oxbows and cut-off stream channels often abound with this species. Adults also occur in upland forests that have suitable breeding sites. Bishop (1941a) noted that marbled salamanders in New York are often collected outside the breeding season in drier habitats than other *Ambystoma* species.

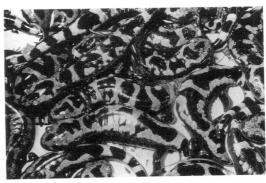

Fig. 52. *Ambystoma opacum*; color pattern variation in females; Orange Co., North Carolina (J. W. Petranka).

Range of *Ambystoma opacum*

BREEDING AND COURTSHIP. This is one of only two *Ambystoma* species that mates and oviposits on land. Females nest in the dried beds of temporary ponds or along the margins of reduced ponds. Adults begin migrating to the breeding sites in late summer or autumn and move at night during rainy weather. Until recently, mating was thought to occur in the immediate vicinity of nesting sites. Krenz and Scott (1994), however, found that 31–49% of females that are intercepted while migrating toward a breeding pond lay fertilized eggs when held in captivity. These data provide strong evidence that males often court females before arriving at the nesting sites.

Adults in northern populations tend to breed earlier than those in southern populations—a trend that is the reverse of that in other *Ambystoma* species in the eastern United States (Anderson and Williamson 1973). Migrating adults or nests have been found in August and November in Rhode Island (Shoop and Doty 1972), and in September in New Jersey (Anderson and Williamson 1973), New York (Deckert 1916; Noble and Brady 1933), Pennsylvania (Pawling 1939), Virginia (Dunn 1917), and West Virginia (Green 1956). Breeding in October or November has been observed in Alabama (Mount 1975), Arkansas (Trauth et al. 1989b), Georgia (Anderson and Williamson 1973; Kaplan and Crump 1978), southern Illinois (O'Donnell 1937), Louisiana (Doody 1996; Walls and Roudebush 1991), Mississippi (Walls and Altig 1986), North Carolina (Bishop 1924; Brimley 1920a; Petranka 1990), South Carolina (Anderson and Williamson 1973; Krenz and Scott 1994), and Tennessee (Gentry 1955; Huheey and Stupka 1967; King 1939).

Males arrive at the ponds from a few days to more than 2 weeks before females (Noble and Brady 1933). In South Carolina 64% of males enter a study site before the first females arrive (Krenz and Scott 1994). Females arrive an average of 9 days later than males, and the operational sex ratios within a drift fence boundary vary from 6:1 to 85:1. Overall, more than four times as many males as females enter the site during the breeding season. Stenhouse (1987) reports a 3.3:1 ratio of males to females, whereas the sex ratio of a sample of 18 adults collected from terrestrial habitats in southern Illinois is about 1:1 (Parmelee 1993).

Adults in North Carolina and Rhode Island populations tend to exit breeding ponds very close to their point of entry (Shoop and Doty 1972; Stenhouse 1985a). This finding suggests that adults follow the same pathways when moving to and from ponds and return to familiar territory after breeding. The proximate mechanisms that adults use to detect sites that will later flood are not known.

Data on annual breeding patterns are largely lacking, but the available evidence suggests that many females breed biennially. Pechmann et al. (1991) document a marked increase in the number of breeding females over an 11-year period in a South Carolina pond. Marbled salamanders did not reduce breeding during dry years even though syntopic *A. talpoideum* and *A. tigrinum* did. A more recent analysis of trends over a 16-year period indicates that the number of adults that breed annually is positively correlated with the amount of rainfall during the breeding season (Semlitsch et al. 1996). The number of breeding adults is also correlated positively with the number of metamorphs produced in previous years.

Descriptions of courtship behavior are given in Bishop (1941a) and Noble and Brady (1933). The following is a composite summary based on observations made in both the field and the laboratory.

A male initiates courtship when he begins rapidly moving about and nudging and lifting other males or females, particularly near the tail and cloacal region. Other males soon become active and begin butting and lifting other individuals while moving about. Males frequently butt or nudge both males and females, and females often respond to male nudging and butting by nudging the cloacal regions of males. When paired, a male and female often move in a circular fashion as they mutually nudge their cloacal regions. The male eventually moves forward along the body of the female and, while undulating his tail and raising his body, deposits a spermatophore on the substrate. If responsive, the female positions her chin and body over the spermatophore while being led forward by the male. The female then positions her cloaca over the sperm cap and picks up seminal fluid from the top of the spermatophore. In some cases the female does not follow the male forward and simply noses about until a spermatophore is encountered. She then positions her cloaca above the spermatophore and picks up seminal fluid.

The spermatophores may be either single or multiple. A multiple spermatophore is produced when a male deposits a spermatophore on top of an existing spermatophore. This behavior presumably functions to eliminate the sperm of rival males (Arnold 1976). Single spermatophores are about 4.0–5.5 mm tall, 6 mm wide at the base, and 2 mm wide at the top of the

Fig. 53. *Ambystoma opacum;* female with eggs (leaf litter removed); Durham Co., North Carolina (J. W. Petranka).

stalk. The top has four quadrangularly spaced, raised knobs, which delineate a slightly concave receptacle that holds the seminal fluid (Noble and Brady 1933).

REPRODUCTIVE STRATEGY. After mating, each female selects a site in the dried or partially dried bed of a temporary pond or ditch and constructs a shallow nest. Females construct nests by burrowing ovoid to oblong cavities in the soil surface immediately below the leaf litter or surface cover (Noble and Brady 1933). They then lay their eggs singly within the depressions, coil around their clutches, and occasionally move about and turn the eggs while brooding (Bishop 1941a; Noble and Brady 1933; Petranka and Petranka 1981a). Nests are usually constructed singly, but I have observed communal nests holding from two to seven clutches (Petranka 1990; Petranka and Petranka 1981b). At an Indiana site as many as 59% of adults may nest communally in nests that contain from 163 to 1813 eggs (Palis 1996b). This site is an artificial pond that contains almost no leaf litter or cover when most eggs are deposited. The clumping of adults may be an artifact of these unusual conditions.

Clutch size typically increases with female SVL (Petranka 1990; Walls and Altig 1986) and has been estimated largely from counts of eggs in nests. Estimated means and ranges of clutch sizes include 95 (48–200) for 44 nests in Alabama (Petranka and Petranka 1981a), 107 (87–131) for 10 nests in Arkansas (Trauth et al. 1989b), 110 (51–169) for 12 nests in Indiana (Palis 1996b), 150 (75–232) for 15 nests in New York (Noble and Brady 1933), 130 (99–172) for six nests in eastern Tennessee (King 1939), and 87

(37–130) for 11 nests in West Virginia (Green 1956). Other reported values include 50–150 eggs per nest in Indiana populations (McAtee 1907), a mean of 79 eggs per nest in two North Carolina populations (Stenhouse 1987), a mean of 97 eggs in a South Carolina pond (Scott 1990), and two individual nests near Raleigh, North Carolina, with 73 and 102 eggs (Bishop 1924). I surveyed 11 central North Carolina ponds and found a grand mean of 92 eggs in nests attended by females versus 76.5 eggs in abandoned nests (Petranka 1990).

Nests are most often constructed in bare soil or rodent burrows immediately below the leaf litter, but they have also been found at the bases of grass clumps, in moss mats near the bases of trees, and beneath bark, logs, and stones (Bishop 1941a; Brimley 1920a; Doody 1996; Jackson et al. 1989; Noble and Brady 1933; Petranka and Petranka 1980, 1981a). Freshly laid ova measure 1.9–2.6 mm in diameter and increase in size as development proceeds (Kaplan 1979). The eggs have a tough, sticky outer membrane and often appear black from the soil and debris that stick to them. The egg capsules may shrink or swell markedly depending on their state of hydration; those measured by Trauth et al. (1989b) averaged 5.4 mm in diameter.

J. G. Petranka and I studied nest site selection in a pond in Alabama and found that most nests are placed at intermediate depths in the pond bed (Petranka and Petranka 1981b). In ponds that fill slowly over an extended period, the placement of eggs with respect to depth determines the time of hatching relative to conspecifics as well as congeners that share breeding ponds. Larvae from eggs laid near the bottoms of partially filled ponds may have a competitive advantage because they are the first to emerge; however, the hatchlings can be killed if subsequent rains do not occur and the pond dries (King 1935; Petranka and Petranka 1981b). Eggs laid at the margins of ponds hatch last, are more subject to freeze damage, and may not be inundated. In addition, the larvae may be at a disadvantage because they are smaller than conspecifics that hatch earlier and have less time to exploit food resources before the pond dries in summer.

The extent to which these and other factors influence nest site selection varies geographically. In a South Carolina pond, nest density does not differ significantly in three depth zones even though as many as 44% of the nests may not flood during dry years (Jackson et al. 1989). In contrast, females in 11 North Carolina ponds strongly avoid laying eggs in pond bottoms (Petranka 1990). When breeding in relatively deep ponds, females at these sites most frequently nest at intermediate elevations as observed in Alabama (Petranka and Petranka 1981b). When using shallow ponds (<40 cm deep), females most frequently select the shallowest areas near pond margins.

Figiel and Semlitsch (1995) exposed gravid females to three substrate types (leaves, wood, and grass clumps) and two moisture regimes in outdoor experimental chambers. Under these experimental regimes, almost all females lay eggs beneath grass clumps and only 4% oviposit when conditions are dry; this finding suggests that a minimum threshold moisture level is required to stimulate egg laying. Marangio and Anderson (1977) found that breeding animals do not select specific soil moistures when placed on a soil moisture gradient.

Many females brood their eggs until their nests are flooded, but nests without females are frequently found (Bishop 1941a; Jackson et al. 1989; Kaplan and Crump 1978; Noble and Brady 1933; Petranka 1990; Petranka and Petranka 1981a). Females often brood nests in which all eggs are dead and often desert nests in which all eggs are viable (Petranka 1990). In the Piedmont of North Carolina, 25% of the nests in unflooded ponds may be abandoned within 2–4 weeks after the eggs are deposited. Mean egg viability and clutch size are significantly lower in unattended nests compared with attended nests. Abandoned nests at an Alabama site also contain significantly fewer eggs than those with females (Petranka and Petranka 1981a). In South Carolina, embryonic survival is positively cor-

Fig. 54. *Ambystoma opacum;* hatchlings trapped in drying pond—one of the costs of depositing eggs in the deepest portions of ponds; Macon Co., Alabama (J. W. Petranka).

related with the length of time that females brood their eggs (Jackson et al. 1989). Collectively, these observations strongly suggest that brooding enhances offspring survival. Brooding probably functions to protect the eggs against predators, to minimize fungal attacks, and to reduce desiccation.

Females that remain with their eggs through hatching may brood for 2 weeks to several months depending on the seasonal pattern of pond filling. In northern areas with severe winters, adults are restricted to using ponds that fill in the autumn or early winter, and the brooding period is relatively short (Bishop 1941a). In coastal regions and the Deep South, where the eggs are less likely to freeze, the brooding period may be longer because adults can utilize habitats that do not flood until late winter. Kaplan and Crump (1978) found that there are no detectable energetic costs of brooding over a 9-day period. However, it is likely that females that brood for many weeks or even months incur significant costs.

It is estimated that females remain with their eggs for about 1 month in West Virginia (Green 1956), whereas the maximum time spent at a nest in the study by Jackson et al. (1989) was 44 days. Worthington (1968) records a brooding period of at least 2 months in Maryland, and Hassinger et al. (1970), a brooding period of about 3 months in New Jersey. Some pools that J. G. Petranka and I monitored in Alabama may not fill until February, about 3-4 months after females oviposit (Petranka and Petranka 1981b). Viable eggs have been found as late as March (Noble and Brady 1933).

Embryos develop to the hatching stage within 9-15 days after oviposition, but do not hatch until the eggs are flooded (Kaplan and Crump 1978; King 1935). My colleagues and I found that hypoxia triggers hatching (Petranka et al. 1982). When covered with water the embryos become oxygen stressed because their metabolic demands exceed the supply of oxygen entering from the water that surrounds the egg. This situation triggers the release (from hatching glands on the snout) of digestive enzymes that dissolve the egg capsule and allow the embryo to escape. Because hatching is environmentally induced, the size at hatching varies markedly depending on the developmental stage that larvae reach prior to pond filling and egg submergence.

Kaplan (1980) examined the relationship between ovum size and development and growth rate in *A. opacum*. In general, there is no relationship between initial ovum size and development rate to hatching, but a strong positive correlation exists between ovum size and hatchling size. Larger larvae at a common hatching stage tend to reach the feeding stage sooner and are larger early in the larval period. Large larvae that develop from large eggs may have a competitive advantage over smaller larvae in certain circumstances, but at a cost to the female in terms of reduced clutch size.

Limited data indicate that catastrophic mortality often occurs in individual clutches. Embryos in only 36% of the clutches studied by Stenhouse (1987) survive to hatching; however, it is uncertain how disturbance from repeated checking of nests contributes to embryonic mortality. In South Carolina, as few as 20% of clutches may hatch during drought years when many nests do not flood (Jackson et al. 1989).

AQUATIC ECOLOGY. Embryos hatch within a few hours to 1-2 days after a nest is submerged and begin feeding almost immediately on zooplankton. The larvae feed predominantly on macrozooplankton, although large larvae eat amphibian larvae and eggs. Cladocerans, copepods, and ostracods are the most numerically important prey of larvae (Branch and Altig 1981; Petranka and Petranka 1980; Stewart 1956). Larvae also feed on a variety of aquatic insects, as well as isopods, mites, snails, and oligochaetes. Small larvae feed mostly on cladocerans, copepods, and ostracods. As larvae grow, they incorporate larger prey into their diets but continue to take large numbers of small zooplankters. Large larvae sometimes feed rather heavily on caterpillars that fall from overhanging trees (Petranka and Petranka 1980). They also feed on *A. maculatum* hatchlings and chew the legs and tails off conspecifics (Stenhouse 1985b; Stewart 1956).

Large larvae in natural populations eat chorus frog (*Pseudacris triseriata*) tadpoles, wood frog (*Rana sylvatica*) embryos, and hatchling tiger salamander larvae (Stine et al. 1954; Walters 1975). They also eat the eggs and embryos of freeze-damaged *A. jeffersonianum* egg masses, but they are unable to prey on intact egg masses. In the laboratory, larvae will eat eggs or hatchlings of *A. maculatum*, *A. texanum*, *A. jeffersonianum*, and gray treefrogs (*Hyla versicolor*). Large larvae will also feed on wood frog and gray treefrog (*H. chrysoscelis*) tadpoles in experimental cages (Cortwright 1988). J. G. Petranka and I did not find evidence of predation on other *Ambystoma* species that shared breeding ponds (Petranka and Petranka 1980), but others have (Branch and Altig 1981).

Hatchlings and small larvae tend to congregate in leaf litter in warm, shallow water during the day. At night they disperse more evenly throughout the ponds and feed in the water column. Larvae in New Jersey emerge from vegetation and leaf litter at dark and either move about on the substrate, swim through the water column, or float at or near the water surface while feeding (Anderson and Graham 1967). This latter phenomenon, termed stratification, is most evident during the first 3–4 hours after sunset (Branch and Altig 1981; Petranka and Petranka 1980). I have observed larvae in murky habitats stratifying during the day and larvae in clear water stratifying on overcast days.

In New Jersey ponds stratification is more prevalent in larvae of intermediate sizes; hatchlings and older larvae nearing metamorphosis spend proportionately more time on the pond bottom at night (Anderson and Graham 1967; Hassinger et al. 1970). This behavioral shift is not as pronounced in populations in Alabama (Petranka and Petranka 1980) and Mississippi (Branch and Altig 1981), where both hatchlings and older larvae frequently stratify. Anderson and Graham (1967) and Hassinger et al. (1970) considered stratification to be a restricted feeding phase; however, Branch and Altig (1981) and Petranka and Petranka (1980) provided evidence that suggests that larvae feed continuously throughout the day and night.

Larvae grow rapidly as water temperatures increase seasonally in late winter and spring, and they often achieve a mass of 2 g or more before metamorphosing. Metamorphosis occurs primarily in May and June in the central and northern areas of the range, and in March and April in the South. Length of the larval period varies depending on climatic patterns and the time of hatching. Northern populations generally have longer larval periods than southern populations because of prolonged winter weather that greatly reduces growth. The larval period in many New York populations lasts about 8–9 months, with most larvae transforming from May through early July (Bishop 1941a; Deckert 1916). Thirty-one transforming larvae collected by Bishop (1941a) on 23 June measured 72 mm TL, and six metamorphs from a nearby site measured 66 mm TL on average.

Larvae in a Maryland pond transform in mid- to late May when they reach 33–34 mm SVL (Worthington 1968, 1969). In Arkansas, individuals that hatch in early February transform 3 months later, in early May, when they measure about 30 mm SVL and 49 mm TL on average (Trauth et al. 1989b). Larvae in Indiana populations transform in June when they reach 48–51 mm TL (Minton 1972), whereas those in central Kentucky transform in mid-May at an average SVL of 34 mm (Keen 1975). Hassinger et al. (1970) estimate the larval period to last 4.5 months in New Jersey ponds; larvae begin transforming in early June when they measure 38 mm SVL and 72 mm TL on average. Larvae from central North Carolina ponds metamorphose from late March to mid-June and average 45–58 mm TL. The larval period lasts 5–7 months depending on ambient water temperatures and food levels (Stewart 1956).

Observations from farther south suggest that the larval period is relatively brief. Doody (1996) tracked growth in a Louisiana population where larvae hatched synchronously on 2 December after the pond formed and reached full capacity following a single rainfall event. Under these conditions, larvae grow 0.23 mm SVL/day, growth rates are relatively constant and linear, and variation in larval size is much less than in populations with staggered hatching. Transformation begins in mid-April after 130 days of growth, when larvae measure about 32–33 mm SVL on average. Larvae elsewhere in Louisiana may metamorphose as early as mid-March at 53–56 mm TL (Liner 1954). Larvae in a pond in central Alabama transform in March and April, about 3 months after hatching, when they reach an average of 33 mm SVL and 49 mm TL (Petranka and Petranka 1980).

Density-dependent regulation of local populations of marbled salamanders appears to occur primarily during the larval stage. Data from several studies indicate that larval growth, size at metamorphosis, and survival are often inversely related to larval densities in ponds. Stenhouse (1985b) found an inverse correlation between the density of larvae and mean growth rate in ponds in North Carolina. Growth rates are highly variable among ponds and years depending on larval density and pond temperature.

I tested for density-dependent growth and survival by dividing natural ponds in North Carolina in half and manipulating larval density (Petranka 1989). Larvae at natural densities grow more slowly, have lower survival, and are smaller at the initiation of metamorphosis than larvae established at reduced densities. Larval density does not affect the biomass of zooplankton in ponds, but larval growth is positively correlated with zooplankton biomass across ponds. Most larvae

Fig. 55. *Ambystoma opacum;* larvae showing tail damage from conspecifics; Durham Co., North Carolina (J. W. Petranka).

suffer tail damage from attacks by conspecifics, and the percentage of injured larvae is positively correlated with larval density. My data suggest that intraspecific aggression is more important than exploitative competition for food in explaining density-dependent growth and survival in *A. opacum* larvae.

Using large field enclosures in natural ponds in South Carolina, Scott (1990) found that larvae at high densities grow more slowly, are smaller at metamorphosis, and have longer larval periods and lower survival compared with larvae at low densities. When ponds dry prematurely, larvae metamorphose sooner, are smaller, and have lower survival than in years when ponds dry later in the season. Thus, both density and hydroperiod are important in influencing growth and survival of larvae.

Marbled salamander larvae are aggressive and will eat both conspecific and heterospecific salamander larvae in ponds. In addition, individuals may bite off and eat the limbs, tails, or gills of live conspecifics. Smith (1990) conducted experiments that indicate that larger larvae are not competitively superior to smaller larvae when exploitative competition occurs for macrozooplankton. When allowed to physically interact, however, larger larvae often gain a competitive advantage by biting off the limbs and tails of smaller animals. Many of the victims die following repeated attacks by larger, more aggressive larvae.

Walls and her colleagues conducted a series of experiments to examine how social and environmental factors mediate the expression of kin discrimination in *A. opacum* larvae (Hokit et al. 1996; Walls and Blaustein 1994, 1995; Walls and Roudebush 1991).

When individuals of the same size are paired, larvae are less aggressive toward siblings than nonsiblings; however, they make no distinction between familiar and unfamiliar individuals (Walls and Roudebush 1991). Kin discrimination by larvae could enhance an individual's inclusive fitness by minimizing aggressive encounters with close kin that could lead to serious injury to one or both individuals.

When large and small larvae are paired, large larvae may cannibalize siblings more than nonsiblings (Walls and Blaustein 1995). This seemingly maladaptive behavior may actually be adaptive if the indirect loss in inclusive fitness to the cannibal is more than compensated for by the nutritional benefits. In a study that more directly measures the ecological significance of kin recognition, Walls and Blaustein (1994) found that kinship does not affect the growth and survival of *A. opacum* larvae established at different experimental densities in artificial ponds.

These experiments examined single factors but failed to examine interactions that can potentially occur in nature. Hokit et al. (1996) implemented a multifactorial experiment that examined the main and interactive effects of food level, initial larval size, kinship, and degree of physical interactions among larvae. This study demonstrates that the frequency of aggression between larvae depends on interactions between kinship, food level, and relative larval size. The main effects and interactions of these factors can in turn alter growth rates, size at metamorphosis, and other parameters that affect individual fitness. Collectively, these studies show that larval aggression is context dependent and is influenced by food level, the size of interacting individuals, the mechanism of competition, and kinship.

In natural populations premetamorphic survival is usually low, and most individuals in a cohort die before metamorphosing. Survival to metamorphosis in the South Carolina pond varies markedly between years depending on the length of time that the pond holds water (Pechmann et al. 1991). Larvae may suffer complete mortality in years when this or other ponds dry prematurely (e.g., Stenhouse 1985b).

Stenhouse (1987) estimated premetamorphic survival in a North Carolina population to be 8.8% and near 0% during two consecutive years of study. Larval survival was 24.2% for one year and <1% for a second. Survival from hatching to the initiation of metamorphosis in 11 ponds that I studied was 1–44% and averaged 15% for larvae at natural densities (Petranka 1989).

Fig. 56. *Ambystoma opacum;* metamorph; locality unknown (Ken Nemuras; courtesy of R. W. Van Devender).

TERRESTRIAL ECOLOGY. Very little is known about the ecology of juveniles. Recently transformed individuals can be found beneath litter and debris around the margins of breeding ponds, but the metamorphs disperse from ponds during rainy weather soon after transforming (Stenhouse 1987). Laboratory experiments indicate that metamorphosed juveniles are less aggressive toward both siblings and familiar individuals, and that sibling recognition is enhanced by familiarity (Walls 1991). Kin and neighbor recognition by juveniles may act to space out kin and minimize competition for resources among relatives.

The juveniles and adults are fossorial throughout much of the year but are sometimes active on the ground surface following rains in the summer and autumn. Specimens have been found nearly 1 m deep in the ground (Pike 1886). Adults tend to become more active on the ground surface during or shortly before the breeding season.

A breeding population at Rainbow Bay at the Savannah River Site increased markedly during the first 8 years of monitoring, then stabilized (Semlitsch et al. 1996). The initial increase in breeding population size may reflect a colonization event. The number of metamorphs produced per female at this site is independent of initial larval densities, and there is little evidence of strong density-dependent interactions with other larval salamanders that share the pond.

Scott (1994) examined the effects of density-dependent larval growth on adult traits. He grew larvae at low and high densities, marked and released the metamorphs, then examined the life-history characteristics of adults that returned to breed. This experiment indicates that age at first reproduction varies depending on the year in which animals are released and the density treatment. Individuals return to breed for the first time when 1–5 years old. In general animals reared at low densities are larger at metamorphosis and breed at an earlier age and larger size than those raised at high densities. Low-density metamorphs also have a higher probability of surviving to first reproduction. Scott's data strongly suggest that density-dependent growth during the larval stage plays an important role in regulating adult populations.

A second factor that could potentially play a role in controlling adult populations is territoriality. Adults live in rodent burrows or in burrows that they construct themselves. Individuals do not appear capable of actively digging burrows, but they can enlarge existing holes and crevices (Semlitsch 1983c). In some instances marbled salamanders are aggressive toward conspecifics; this observation suggests that they may defend burrows or feeding territories from intruders. Ducey (1989) found that resident animals often bite and engage in threat posturing during initial encounters with conspecific intruders. If maintained for 8 days together, however, paired animals stop avoiding one another and will often share tunnels. Individuals also ignore substrates from cages of conspecifics. Adults are more likely to be aggressive when food levels are low, suggesting that aggressive behavior may function in defense of feeding areas (Ducey and Heuer 1991).

Food intake is an important variable that can potentially affect traits related to individual fitness. Scott and Fore (1995) found that females that are fed high levels of food grow faster, have higher lipid levels, produce larger eggs, and have larger clutches than females fed relatively low levels of food. Females receiving large amounts of food are also more likely to produce a clutch of eggs. In the only experimental study to date of density dependence in the terrestrial stages of *Ambystoma,* Pechmann (1995) found that doubling the density of metamorphs of either conspecifics or *A. talpoideum* has no detectable effect on survival to first reproduction of *A. opacum* in experimental pens.

PREDATORS AND DEFENSE. Marbled salamanders are undoubtedly preyed upon by owls, raccoons, skunks, snakes, and other woodland predators. Liner (1954) notes that a ribbon snake (*Thamnophis sauritus*) regurgitated two metamorphs. When molested, adults

Fig. 57. *Ambystoma opacum;* defensive posturing; Oktibbeha Co., Mississippi (E. D. Brodie, Jr.).

secrete copious amounts of milky secretions from the tail that repel predators. An individual that is attacked will often engage in defensive posturing that involves lowering the head, raising the rear limbs, and lashing the raised tail at the attacker (Brodie 1977; DiGiovanni and Brodie 1981). Nesting females will often posture when initially uncovered.

COMMUNITY ECOLOGY. Marbled salamanders almost never share breeding habitats with fish, perhaps because the larvae are palatable to fish and lack any obvious defenses against these predators (Kats et al. 1988). The larvae are important predators on both invertebrates and vertebrates in vernal pond communities. Because this species is usually the first amphibian to hatch after vernal ponds fill with water, the larvae are often major predators on other amphibian larvae and may compete for food with other *Ambystoma*. Details of ecological interactions of *A. opacum* with *A. maculatum* and *A. jeffersonianum* are presented under the accounts of these species.

Morin (1995) found that *A. opacum* larvae and adult *Notophthalmus viridescens* are functionally equivalent in artificial ponds. Both species nearly eliminated *Bufo americanus* and *Hyla andersonii* tadpoles from tanks, but had far less impact on those of *Pseudacris crucifer*. The effects on *Pseudacris* were nonadditive (higher survival than expected if effects were additive), perhaps because the growth of *A. maculatum* was reduced when it was placed with *Notophthalmus*.

CONSERVATION BIOLOGY. The greatest environmental threat to this species is loss of bottomland hardwoods and associated vernal pond habitats. Thousands of local populations of marbled salamanders have already been eliminated by habitat loss, and more will be lost in the future. Landowners who wish to provide breeding sites for these and other amphibians can do so by constructing artificial vernal ponds. Habitats 0.5–1.0 m deep and the size of a large living room will support a surprising array of native frogs and salamanders and can be constructed at minimal cost.

Ambystoma talpoideum (Holbrook)
Mole Salamander
PLATES 17, 18

Fig. 58. *Ambystoma talpoideum;* juvenile; Buncombe Co., North Carolina (R. W. Van Devender).

IDENTIFICATION. Mole salamanders occur as both gilled and terrestrial adults. Terrestrial adults are short, stocky salamanders with short tails, large limbs, and prominent, rounded heads. The dorsum may be light brown, light bluish gray, dark gray, or blackish. The dorsum and sides have light gray specks, which are inconspicuous in many specimens. Juveniles and adults have parotoid glands that produce noxious secretions. Sexually active males have swollen vents and rough, granular regions on the dorsolateral regions of the tail. Louisiana males and females studied by Raymond and Hardy (1990) did not differ significantly in SVL. Adults reach 8–12 cm TL.

Hatchlings have bold, alternating black and yellow

blotches along the midline of the back. Conspicuous patches of yellow and black also occur on the tail fin (Volpe and Shoop 1963). Hatchling size is geographically variable and appears to be smaller on average in Atlantic coast populations. Louisiana hatchlings measure 10 mm SVL on average (Raymond and Hardy 1990), whereas South Carolina hatchlings measure 4.5–8 mm SVL (Semlitsch 1985b).

Larvae in middle to late developmental stages have two conspicuous cream or dull yellow stripes on each side that break into blotches near the tail. The upper stripe occurs midlaterally and may be indistinct or broken in some specimens. The lower stripe occurs at the junction of the side and venter and is normally distinct. A characteristic dark band extends along the midline of the belly in older larvae, gilled adults, and recently transformed juveniles (Orton 1942; Volpe and Shoop 1963), but it may be absent, poorly formed, or split into two smaller bands in some specimens. Gilled adults often lack the midlateral stripe and are generally darker than sexually immature larvae. Recently transformed juveniles are brownish green and may retain the midlateral stripe for a few weeks after transforming. Juveniles begin developing adult patterning such as gray flecking within 2–4 weeks after metamorphosis. Sexually active males of both adult types can be distinguished from females by their swollen cloacal glands.

SYSTEMATICS AND GEOGRAPHIC VARIATION. Atlantic coast and Gulf coast populations differ in mode of egg deposition and other life history parameters (see "Reproductive Strategy"), suggesting that these groups may be genetically differentiated. Detailed studies of geographic variation in morphology and genetic variation are not available, and no subspecies are recognized.

DISTRIBUTION AND ADULT HABITAT. Mole salamanders occur throughout much of the Coastal Plain of the southeastern United States from South Carolina to eastern Texas northward to southern Illinois. Geographic disjuncts occur outside the Coastal Plain in Alabama, Arkansas, Georgia, Kentucky, North Carolina, South Carolina, Tennessee, and Virginia. Terrestrial adults are most commonly found in extensive floodplain forests, often in areas near gum and cypress ponds (Semlitsch 1981; Shoop 1960). Populations outside the Coastal Plain usually inhabit upland hardwood forests or mixed pine-hardwood forests surrounding vernal ponds or other breeding sites. A local population may consist entirely of terrestrial adults or of a mixture of gilled and terrestrial adults. Terrestrial adults predominate in areas with temporary ponds that dry annually, whereas gilled adults predominate in permanent or semipermanent ponds (Heintzel and Rossell 1995; Scott 1993; Semlitsch and Gibbons 1985; Semlitsch et al. 1990). Detailed studies by R. Semlitsch and his colleagues (see "Aquatic Ecology") indicate that both genotype and environmental factors influence the percentage of gilled adults found in local populations.

BREEDING AND COURTSHIP. Adults breed in a wide variety of temporary and permanent habitats, including roadside ditches, gravel pits, woodland pools, and Carolina bays (Mount 1975; Patterson 1978; Semlitsch 1981; Shoop 1960; Smith 1961). This species shows a strong tendency to used fish-free habitats (Semlitsch 1988) and is rarely found in ponds with large predatory fish.

Shoop (1960) found that breeding is most intense during periods with heavy, sustained rains and cold temperatures. Results of an 11-year study of a South Carolina population by Pechmann et al. (1991) indicate that the number of breeding females is positively correlated with cumulative rainfall during the breeding season. Many females do not migrate to breed in years when rainfall is low and larval mortality from premature pond drying is catastrophic.

The most comprehensive information on the breeding biology of *A. talpoideum* comes from Patterson (1978), Semlitsch (e.g., 1981, 1985a,b, 1987a), and other researchers at the Savannah River Site in South Carolina. Adult salamanders at the Savannah River Site migrate on rainy nights to breeding ponds from September through late March (Patterson 1978; Semlitsch 1981, 1985a). They most frequently move along corridors of hardwood vegetation and avoid open, grassy areas. A peak in arrival occurs from November through January depending on annual rainfall patterns. Adults are stimulated to migrate by rainfall and cool temperatures of late autumn and early winter, and males tend to arrive before females and to leave the ponds later. Emigration from the ponds occurs in March.

Where a local population consists of both gilled and terrestrial adults, gilled adults at the Savannah River Site normally breed 6 weeks or more before the terrestrial adults, and the larvae gain a size advantage (Krenz and Sever 1995; Patterson 1978; Scott 1993).

Range of *Ambystoma talpoideum*

Most gilled adults mate in early November, whereas the peak in arrival of migrating adults is from mid-November to early January depending on annual rainfall patterns. This pattern results in partial reproductive isolation between the forms and may contribute to maintaining mixed populations of gilled and transformed adults in populations. Gilled adults have sperm in their spermathecae throughout the breeding season (Trauth et al. 1994). Although individuals can store sperm for as long as 5 months, they cannot retain viable sperm into the next breeding season.

Analyses of annual breeding patterns over periods of 6 (Semlitsch 1987a) and 16 (Semlitsch et al. 1996) years indicate that breeding population size is positively correlated with the amount of rainfall during the breeding season, but not with the number of metamorphs produced during previous years. The arrival time of specific individuals relative to others is unpredictable from year to year, suggesting that environmental variation is the primary factor affecting when individuals return to ponds to breed.

Observations from localities other than the Savannah River Site indicate that breeding occurs primarily from December through February. In southeastern Louisiana, Shoop (1960) found that breeding of terrestrial adults lasts only 7–15 days in any year and occurs from early December to mid-February. Hardy and Raymond (1980), however, report a more prolonged breeding season (99–108 days) in a population in northwestern Louisiana, with breeding activity lasting from December through March. Individuals remain in or about this pond for 8–108 days. Males arrive at the pond earlier than females and remain in the pond longer; however, both emigrate at approximately the same time. Breeding occurs in either late December, January, or February in Alabama (Mount 1975), Arkansas (Trauth et al. 1993a, 1995), Florida (Carr 1940), Mississippi (Allen 1932; Walls and Altig 1986), and Tennessee (Gentry 1955). Smith (1961) observed an unusual breeding aggregate on 8 November in Illinois.

Data for Louisiana and South Carolina indicate that most adults breed annually (Raymond and Hardy 1990; Semlitsch et al. 1993). Virtually all adults marked by Raymond and Hardy (1990) return to the same pond to breed from year to year, although a few individuals switch ponds. The sex ratio at this site does not differ significantly from 1:1 when data for 5 years are combined. However, both males and females may significantly outnumber the opposite sex in certain years.

Shoop (1960) studied courtship behavior of terrestrial adults in a large outdoor tank, and the following is a summary of his observations. After bumping a female for about 10 seconds, the male runs his head along her sides to the cloacal region. The female begins nosing the cloacal region of the male and both push with their heads. This results in a circular movement or waltz of one or two revolutions. The male next breaks contact with the female's cloacal region and straightens his body. The female slides her head down to the tip of the male's tail. The male then begins wagging the pelvic region and proximal portion of his tail while keeping the posterior part of the tail undulating and often touching the female's head. At this time the cloacal aperture opens and the spermatophore stalk may begin to form. The male moves forward slowly while continuing the wagging motion. The female follows, keeping her head in close proximity to the male's tail.

After the female has followed the male's tail for 1–11 minutes, she pushes her snout along the tail until it reaches the cloacal region or hindlimbs. She then bumps the male in the cloacal region one or two times. The male ceases his pelvic movements and deposits a spermatophore. If he does not stop the pelvic movements, she returns to the distal end of his tail. Immediately after depositing a spermatophore, the male resumes the pelvic motions. The female moves over the spermatophore and picks up the entire spermatophore or the sperm cap with her cloacal lips. She then wanders away from the male. If the female is unsuccessful, she continues to follow the male until she picks up a spermatophore. Females are often courted by several males, and one male may break up courting pairs to lead the female away from a rival. The entire courtship lasts from 7 to 37 minutes.

Males can produce as many as 15 spermatophores. The spermatophore has a wide circular base and a thin stalk. The top of the stalk is oval to ellipsoidal and is divided in half by a fold that runs down the middle. Spermatophores are 6–8.5 mm tall with a base that is 5–8 mm wide. The stalk narrows near the top and is 1–2.6 mm in diameter at the apex.

REPRODUCTIVE STRATEGY. Mode of egg deposition and clutch size differ among Gulf Coastal Plain and Atlantic Coastal Plain populations. Most terrestrial females in Gulf coast populations lay their eggs in small clusters that are often placed on the same twig, whereas females along the Atlantic Coastal Plain scatter their eggs singly in ponds (Semlitsch and Walls 1990). Each egg is darkly pigmented above and dull white below and is surrounded by a vitelline membrane and two jelly envelopes.

Estimates of number of eggs per mass in Louisiana populations are 4–20 eggs (Bishop 1943), 3–50 eggs (mean = 17 eggs) for 38 masses (Raymond and Hardy 1990), and a mean of 18 eggs for 107 masses (Raymond and Hardy 1990). Egg masses in the former study were 30–57 mm in length and 24–34 mm in width. Egg masses in Arkansas ponds contain an average of 41 eggs (range = 12–99 eggs), although gilled adults lay eggs singly (Trauth et al. 1995).

Eggs in Louisiana are most frequently laid on small twigs or other support structures oriented at an angle to the water's surface (Shoop 1960). When ovipositing on a twig, a female grasps it with her hind legs and often turns the twig under before depositing eggs along its length. Females usually oviposit at night and require as long as 3 nights to lay their entire complement. Gilled adults from South Carolina lay about 35 eggs per night and require several days to deposit their entire complement (Krenz and Sever 1995).

Mosimann and Uzzell (1952) observed egg masses in South Carolina that were 15–58 mm in length and 12–25 mm in width and contained an average of 21 eggs. Based on their description of the egg masses, these are in all likelihood eggs of *A. tigrinum* rather than *A. talpoideum*. All other accounts of South Carolina and North Carolina populations indicate that the eggs are laid singly (Semlitsch and Walls 1990).

Clutch size varies geographically and averages higher in Atlantic coast populations. Seven terrestrial adults from Louisiana contain 208–504 ova (mean = 331 ova; Raymond and Hardy 1990) compared with 226–401 ova (mean approximately 300 ova) in 14 other Louisiana females (Shoop 1960). In South Carolina, estimates of clutch size are 481–696 ova (mean = 590 ova) in eight terrestrial females (Raymond and Hardy 1990) and respective mean clutches of 248, 275, 445, 516, and 668 ova for transformed females that are 1, 2, 3, 4, and ≥5 years old (Semlitsch 1985b).

In the population studied by Semlitsch (1985b) clutch size and egg size increase with female SVL. Terrestrial females are substantially larger than gilled females, and they lay more eggs, and larger eggs, than the latter (mean = 173 ova for gilled adults). When adjusted for body size differences, the two morphs produce similar numbers of eggs, but terrestrial females produce larger ova. Krenz and Sever (1995) also report a positive correlation between clutch size and female SVL in gilled adults.

In a related study, Semlitsch and Gibbons (1990) found that 4-year-old females produce larger eggs (mean diameter = 1.38 mm) than 1-year-old females (mean diameter = 1.19 mm), and that egg size is positively correlated with hatchling size. When raised in outdoor tanks, larvae from large eggs have higher growth rates early in the larval period than larvae from small eggs. When water levels are held constant, larvae from large eggs have higher survival to the initiation of metamorphosis than larvae from small eggs. However, exposure to an artificial pond-drying regime does not affect larval survival. These data indicate that egg size may sometimes affect the number of clutchmates that are recruited into the adult population. Walls and Altig (1986) report that larval growth rate is positively correlated with ovum and hatchling size.

AQUATIC ECOLOGY. Larvae feed mostly on pond invertebrates, with zooplankton and dipteran larvae forming the bulk of the diet. Taylor et al. (1988) compared the diet of *A. talpoideum* larvae with that of *N. viridescens* and *E. quadridigitata* that share a Carolina bay in South Carolina. *Ambystoma talpoideum* and *N. viridescens* at this site have strong dietary overlap. Both species feed rather heavily on small zooplankton when young and incorporate larger prey such as chironomid larvae into their diets as they grow. Copepods are underrepresented in the diet, probably because their rapid, darting movements make them difficult to capture. Ostracods, cladocerans, and chironomid larvae are the major dietary items, and larvae selectively feed on the larger individuals of cladocerans. Dietary overlap between *A. talpoideum* and *E. quadridigitata* is much lower than that between *A. talpoideum* and *Notophthalmus*. These differences are probably due to size variation between larvae and the fact that larvae feed in the water column at night.

In Illinois the larval period lasts only 3–4 months and all larvae apparently transform when sexually immature (Smith 1961). Metamorphs at a southern Illinois site first appeared in late July and reached a peak in September (Parmelee 1993). At the Savannah River Site, larvae have very plastic life histories (Semlitsch 1985b). Larvae grow rapidly after hatching in late winter or early spring. They begin to mature sexually as gilled adults by September when they reach >30 mm SVL and first breed in December or January

Fig. 59. *Ambystoma talpoideum;* mature larva; Buncombe Co., North Carolina (J. W. Petranka).

when about 1 year old. Females at this time typically measure >34–36 mm SVL. Gilled adults in a southern Appalachian population measure >44 mm SVL (Heintzel and Rossell 1995).

Larvae at the Savannah River Site that do not become gilled adults normally metamorphose during the first year from May through September. Metamorphosed animals become sexually mature within a few months after transforming and breed during their first year of life when females measure >43–48 mm SVL (Semlitsch 1985b). A small percentage of larvae may overwinter as immature forms and either transform the next summer or mature into gilled adults. Large immature and sexually mature larvae can metamorphose at any time if ponds dry during the summer, and experimental studies indicate that pond drying triggers early metamorphosis (Semlitsch and Gibbons 1985; Semlitsch and Wilbur 1988; Semlitsch et al. 1990).

According to Patterson's (1978) study larvae that hatch in mid-March grow very rapidly during the summer and mature sexually by September or October. Gilled adults breed during this time and the majority transform the following April and May when 12–15 months old. A few larvae remain in the ponds and never transform. Larvae in ponds that dry in late summer transform prior to maturing sexually.

Larvae can apparently transform at any time of the year if their SVL is >32–38 mm. Recently metamorphosed animals collected by Raymond and Hardy (1990) averaged 35–38 mm SVL. Larvae in a pond studied by Semlitsch et al. (1988) for 8 years vary in mean size at metamorphosis from 41 to 50 mm SVL depending on when the pond dries annually. The minimum and maximum sizes of larvae are 33 and 68 mm SVL, respectively.

Several researchers have examined environmental factors that influence growth parameters. Larvae grow fastest when at low densities, with access to high food levels, and in nondrying ponds (Semlitsch 1987b). Faster-growing larvae mature sexually at a larger size and produce more ova than slower-growing larvae. Thus, factors that affect larval growth ultimately influence the fitness of gilled adults.

Simulated pond drying, low larval density, and low food levels increase the percentage of larvae that metamorphose, whereas constant water conditions and low densities increase the percentage of larvae that become gilled adults (Semlitsch 1987c). Surprisingly, the tendency to become gilled adults does not correlate closely with the seasonal duration of natural ponds that local populations use as breeding sites.

Although environmental factors can trigger metamorphosis, the propensity to metamorphose also varies among local populations subjected to the same experimental regimes (Semlitsch and Gibbons 1985; Semlitsch et al. 1990). These and other studies indicate that the tendency to metamorphose has a genetic basis in some populations (Harris et al. 1990). The percentage of animals that become either gilled or terrestrial adults can be altered significantly through artificial selection in as few as four generations (Semlitsch and Wilbur 1989). This finding demonstrates a heritable component to the gilled versus terrestrial adult phenotypes found in South Carolina populations.

Data collected by Semlitsch (1987a) on three pond populations over a 6-year period provide important clues to factors affecting the population dynamics of this species. Survival to metamorphosis and length of the larval period tend to increase with the number of days that ponds contain water. Larvae remain longer in ponds of long duration but do not metamorphose at a larger size. Although the number of eggs laid per pond is highly variable from year to year, size at metamorphosis and survivorship to metamorphosis do not correlate with the total number of eggs oviposited. Survivorship in ponds that last <145 days is at or near 0% because of early pond drying. In most ponds, and in most years, survival is <3%, but it may range as high as 4.1%. Pechmann et al. (1991) also report that larval survival varies markedly between years and strongly correlates with annual rainfall and pond hydroperiod.

Results of a detailed 8-year study of a natural population involving one of the ponds mentioned previously (Rainbow Bay) indicate that the modal time of metamorphosis and mean size at metamorphosis are positively correlated with the date that the pond dries (Semlitsch et al. 1988). In years when the pond dries late in the year, two cohorts of roughly equal size emerge from the pond. Individuals that emerge early are larger at first and second reproduction than those that emerge late in the year. Regardless of the time of emergence, size at metamorphosis is positively correlated with size at first reproduction. Early metamorphosis allows many females to breed a year earlier than those that emerge late in the year; however, similar trends are not evident among male cohorts. Neither time of metamorphosis nor size at metamorphosis influences survival to first reproduction. These data further demonstrate that larval growth and development can influence adult fitness.

Researchers at the Savannah River Site have now tracked populations at Rainbow Bay for 16 years (Semlitsch et al. 1996). The most recent analysis of long-term trends indicates that breeding populations fluctuate markedly from year to year and that recruitment into the adult population is episodic. The amount of seasonal rainfall and the season duration of Rainbow Bay appear to be the most critical factors affecting the number and size of offspring recruited into the adult population. Nonetheless, larval competition also appears to be important, and the number of metamorphs produced per female is inversely related to initial larval density.

Studies of larval behavior have focused on feeding behavior and aggression. Larvae exhibit diurnal shifts in microhabitat use that involve hiding in leaf litter during the day and stratifying in the water column at night. In natural ponds larvae of all sizes float in the water column at night and feed on zooplankton (Anderson and Williamson 1974; Branch and Altig 1981). In studies in outdoor tanks, vertical migrations peak within 1–2 hours after dark when temperatures are 25–27°C (Stangel and Semlitsch 1987). Stratification does not correlate with the movement of zooplankton into the upper water column and may be a means of thermoregulating. Larvae do not migrate above the leaf litter into the upper water column at night when fish are present in experimental ponds.

Larvae are aggressive toward conspecifics and other *Ambystoma*. Larval agonistic behavior in this species primarily involves looking toward, lunging at, and biting conspecifics (Walls and Semlitsch 1991).

TERRESTRIAL ECOLOGY. Semlitsch (1985a,b) found that metamorphs in South Carolina ponds begin emigrating as early as May. Dispersal away from ponds continues until the ponds dry. In more permanent ponds, juvenile emigration may continue through October. Estimates are that juveniles move a median distance of only 47 m from the ponds compared with 156 m for females and 213 m for males (Semlitsch 1981). Adults spend the summer in one to six small activity centers of 0.02–0.21 m^2 that are separated from each other by a distance of 0.3–5.3 m. Home ranges are established in both pine and hardwood forests. Adults live during the summer in underground burrows and tunnels that on average are 5 cm beneath the soil surface. Individuals are capable of both actively digging burrows and enlarging existing cracks and crevices (Semlitsch 1983c).

In Louisiana populations studied by Raymond and Hardy (1990), most postmetamorphic growth occurs during the first year after transformation. Individuals first breed when about 2 years old and 52–55 mm in SVL. Growth after sexual maturity is very slow and typically averages 1–2 mm per year. The mass of very old individuals may actually decline from year to year. Recaptures indicate that some individuals live for at least 8 years. Juveniles studied by Shoop (1960) become sexually mature when they reach 44–48 mm SVL.

Transformed animals exhibit relatively weak agonistic behavior that involves residents occasionally biting and aggressively posturing at intruders (Ducey 1989; Ducey and Heuer 1991). Recently metamorphosed juveniles exhibit little agonistic behavior toward either conspecific or heterospecific intruders (Walls 1990).

Very few data are available on adult survival. Based on recaptures at drift fences, Patterson (1978) estimates that 45% of the adults die before leaving the breeding ponds. Annual adult mortality in a population studied by Raymond and Hardy (1990) varies from 16% to 37%, averages 26%, and does not differ between sexes. Semlitsch et al. (1993) noted that some individuals live for 6–8 years.

PREDATORS AND DEFENSE. Juveniles and adults have well-developed parotoid glands on the head and granular glands along the tail that produce noxious secretions. Individuals often exhibit a head-down defensive posture when attacked and may lash their tails at potential predators (Brodie 1977). Test animals confined with a smooth earth snake (*Virginia valeriae*)

Fig. 60. *Ambystoma talpoideum;* defensive posturing; Johnson Co., Illinois (E. D. Brodie, Jr.).

respond to approaches by head butting, biting, and body flipping, followed by immobility or movement away from the snake.

COMMUNITY ECOLOGY. Mole salamanders are important members of vernal pond communities, where they prey on zooplankton and may potentially compete with other salamanders for food. Ecological interactions between *A. maculatum* and *A. talpoideum* (Semlitsch 1988; Semlitsch and Walls 1993; Walls and Jaeger 1987) are discussed in detail under the species account of *A. maculatum*. Walls and Jaeger (1989) found no evidence that chemical interference competition occurs between these species.

Predatory fish such as the bluegill (*Lepomis macrochirus*) are voracious predators on the eggs and larvae of *A. talpoideum* (Semlitsch 1988), an observation that may explain why *A. talpoideum* almost always breeds in fish-free habitats. Jackson and Semlitsch (1993) found that exposure of larvae to caged fish (i.e., fish chemicals and visual stimuli) decreases the proportion of larvae that become gilled adults. Larvae respond to fish chemicals in the laboratory by moving into refuges; however, larvae become habituated when chronically exposed to fish chemicals in experimental ponds.

CONSERVATION BIOLOGY. Many local populations of *Ambystoma* have been lost in the eastern United States as native forests with seasonally ephemeral wetlands have been converted into agricultural and urban areas. Intensive timbering practices also may be affecting some *Ambystoma* species. Raymond and Hardy (1991) found that survival of adult *A. talpoideum* decreased following the clear-cutting of a mixed pine-hardwood forest adjoining a breeding pond in Louisiana.

Ambystoma texanum (Matthes)
Small-mouthed Salamander
PLATE 19

Fig. 61. *Ambystoma texanum*; adult; Warren Co., Iowa (R. W. Van Devender).

IDENTIFICATION. The small-mouthed salamander is a medium-sized *Ambystoma* with a conspicuously small head and mouth. The dorsum of adults is brownish gray to grayish black and is overlain with light gray speckles or lichenlike markings, particularly along the lower sides of the body. The ventral ground color is similar, but is not strongly speckled with gray. Adults reach 11–19 cm TL and have an average of 15 costal grooves. Adult females average 3–11% longer in SVL than males (Downs 1989; Plummer 1977), and males have conspicuously swollen, papillose vents during the breeding season.

Hatchlings and small larvae are brownish to olive green with three to six light yellowish to olive green saddles or paired blotches along the midline of the back. Hatchlings vary from 7 to 14 mm TL (Liner 1954; Minton 1972; Smith 1934). Older larvae are light brown and often have inconspicuous or diffuse saddles and a weak broken stripe along each side of the body. Metamorphs are nondescript and brownish to brownish gray. Juveniles begin acquiring the light gray flecks and lichenlike patterns of the adults within a few weeks after transforming. Albinos are reported by Allyn and Shockley (1939), Dyrkacz (1981), and Jones (1991).

SYSTEMATICS AND GEOGRAPHIC VARIATION. Populations in central Kentucky and adjoining areas that were previously treated as *A. texanum* are now recognized as a separate species, *A. barbouri* (Kraus and Petranka 1989). Details of the systematics of this group are given under the account of *A. barbouri*. Unisexual *Ambystoma* that are of hybrid origin between *A. texanum* and other diploid *Ambystoma* species occur in the northeastern and north-central portion of the range of *A. texanum*. A detailed description of this unisexual complex is presented under the special account of unisexual *Ambystoma*.

The degree of gray speckling varies substantially among local *A. texanum* populations. Populations in Texas and Oklahoma tend to be more heavily speckled than those farther north, but considerable local variation exists. I described geographic variation in egg size, egg mass size, and habitat use in populations in western Kentucky and adjoining regions (Petranka 1982a). No subspecies of *A. texanum* are recognized.

DISTRIBUTION AND ADULT HABITAT. The small-mouthed salamander occurs from Ohio westward to eastern Kansas, and southward to eastern Texas, Louisiana, Mississippi, and Alabama. Scattered geographic isolates occur in southeastern Indiana, southern Ohio, and western West Virginia. One specimen from Washington Co. in southeastern Ohio (indicated by question mark on range map) is an *A. barbouri-texanum* hybrid (F. Kraus, pers. comm.).

Populations are most abundant in bottomland forests and associated wetlands in or adjoining floodplains (Bragg 1949; Downs 1989; Minton 1972; Petranka 1982a; Strecker and Williams 1928). In the western portion of the range adults have been collected in upland ponds and sluggish streams in tallgrass prairies and in oak-hickory-pine forests in the Ouachita Uplift (Bragg 1949; Collins 1993).

BREEDING AND COURTSHIP. Adults breed in seasonally ephemeral lentic habitats, including woodland ponds, oxbow ponds, roadside ditches, borrow pits, flooded fields, prairie ponds, and swamps (Bailey 1943; Petranka 1982a; Ramsey and Forsyth 1950). On rare occasion adults may breed in sluggish streams or pools in headwater tributaries. Larvae or eggs have been collected from streams in southern Illinois (Cagle

Range of *Ambystoma texanum*

1942), Indiana (Kraus and Petranka 1989), Kansas (Collins 1993; Gloyd 1928), Kentucky (Petranka 1982a), and Texas (Burt 1938).

Small-mouthed salamanders often select shallower breeding sites than members of other *Ambystoma* species and may utilize pools that are only a few centimeters deep (Bragg 1949; Minton 1972; Petranka 1982a). Adults tend to be explosive breeders. Migrations to breeding sites occur at night during rainy weather, and breeding is usually confined to one to three brief bouts during or immediately following rains in winter or early spring (Kraus and Petranka 1989; Petranka 1984a). Adults in Iowa will migrate to ponds and breed in the absence of rainy weather during dry springs (Bailey 1943). Reported sex ratios of adults are about 1:1 for populations in southern Illinois (Parmelee 1993) and Kansas (Plummer 1977).

Adults breed in late winter through early spring and show a weak seasonal progression of breeding from south to north. Breeding has been documented in Alabama in February (Brandon 1966a), in Texas in mid-January through February (Ramsey and Forsyth 1950), in Oklahoma from February through mid-March (Bragg 1949), in Indiana from late January to late March (Hay 1892; Minton 1972), in Illinois and Ohio from late February through March (Cagle 1942; Downs 1989; Smith 1961), and in Iowa from mid-February through mid-April (Bailey 1943; Camper 1990). Hatchlings have been observed in a Louisiana pond on 8 February, suggesting that most females oviposit in early to mid-January (Doody 1996). In Kansas, eggs have been found in late January through April (Gloyd 1928; Smith 1934), but most breeding occurs in late February and March (Plummer 1977).

Two contradictory accounts of courtship have been described for *A. texanum*. Wyman (1971) observed males from north-central Illinois dorsally amplex females before leading them forward to mount spermatophores. Garton (1972), however, reported that males from southern Illinois do not dorsally amplex females or lead them forward to mount spermatophores. I observed courtship in closely related *A. barbouri* (Petranka 1982b) that essentially conformed to Garton's (1972) description. I also observed spatial arrangements of spermatophores in numerous *A. texanum* breeding ponds that indicate that adults engage in nonamplexing mass courtship similar to that of *A. maculatum*. Additional field observations of spermatophore deposition by Downs (1978) are consistent with my observations.

Licht and Bogart (1990) observed courtship between male *A. texanum* and diploid, triploid, and tetraploid unisexual *Ambystoma* on Pelee Island, Ontario. Courtship behavior of the males is similar to that described by Garton (1972), and males do not dorsally amplex or lead females. Unisexuals pick up sperm caps of *A. texanum* males even though they do not engage in amplexus or leading behavior. The description by Garton (1972) that follows is presumed to be typical of *A. texanum* throughout its range.

After encountering a female, a male begins to nudge the female, especially around the pelvis, cloaca, and tail. The nudging activity becomes more intense and the male begins moving with exaggerated undulations of the tail and pelvic regions. During this phase the male may repeatedly move over and under the female. The female ignores the male and may move slowly about during this period. The male then moves away from the female and deposits one or more spermatophores while pressing and undulating his cloaca against the substrate.

After 6–10 spermatophores are deposited, the female begins to nudge the male. Mutual nudging intermixed with spermatophore deposition occurs during this phase of courtship, but the female does not actively follow the male when he moves away to deposit a spermatophore. When a female encounters a spermatophore she positions her cloaca above the spermatophore, then squats and picks up seminal fluid. A female may collect sperm from eight or more spermatophores, and a male may deposit 30–40 spermatophores before ending courtship. *Ambystoma texanum* confined with unisexual *Ambystoma* may produce 4–19 spermatophores (Downs 1978).

When courting in groups males often court other males in a manner similar to that seen when courting females. Patterns of spermatophore deposition indicate that adults gather in scattered groups in ponds and engage in mass courtship. Garton (1972) found clusters of 72–100 spermatophores in small, isolated areas in a ditch where adults bred. I made similar observations in numerous ponds in Kentucky and southern Illinois (Petranka 1982b).

Males deposit spermatophores on leaves, twigs, and other spermatophores. When depositing a spermatophore, the hindfeet are held against the cloacal walls, the back is arched slightly, and the tail is elevated and undulated. Deposition requires 10–20 seconds, depending on temperature (Labanick and Davis 1978). The spermatophores are about 5 mm high and have a slightly oval base about 5 mm in diameter. The stalk

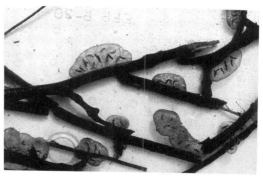

Fig. 62. *Ambystoma texanum;* egg masses; Livingston Co., Kentucky (J. W. Petranka).

tapers rapidly from the base and is slightly flared at the apex. The sperm mass sits on four lateral horns at the apex (Labanick and Davis 1978).

REPRODUCTIVE STRATEGY. Females deposit eggs either singly, in loose clusters, or in small masses on twigs, leaf petioles, leaves, grasses, and other vegetation and detritus in ponds (Minton 1972; Petranka 1982a). The mode of egg deposition varies substantially among 50 populations that I observed in Illinois, Indiana, Kentucky, Ohio, and Tennessee (Petranka 1982a). In five of the populations almost all eggs are deposited singly, but in the remaining populations the eggs are laid in small masses that average 2–15 eggs. Females on Pelee Island in Lake Erie also deposit eggs singly (Licht 1989). Two Iowa populations sampled by Bailey (1943) and Camper (1990) produce masses with an average of 11 and 15 (range = 4–19) eggs per mass, whereas egg masses from Arkansas average 13 eggs per mass (Trauth et al. 1990).

Large egg masses are sausage shaped and rather flimsy. Each egg is surrounded by a vitelline membrane and two outer envelopes (Minton 1972). The animal pole is dark brown to black, and the vegetal pole is cream to white. Freshly laid ova are 1.6–2.5 mm in diameter (Licht 1989; Minton 1972; Petranka et al. 1987a; Plummer 1977). The incubation period lasts 2–8 weeks depending on water temperature (Minton 1972). Hatchlings or eggs at the hatching stages have been observed on 20 January in Louisiana (Liner 1954), in late February through mid-March in Oklahoma (Bragg 1949; Dundee 1947), and in early April in Indiana (Hay 1892).

The average clutch size of the small-mouthed salamander is relatively large compared to those of other

Ambystoma of similar size. Reports include an average of 543 eggs in three specimens from Iowa (Camper 1990); 694 (Smith 1934) and 550 (Cagle 1942) ova in two individuals from Kansas and Illinois, respectively; and an average of 545 ova in eight Arkansas specimens (Trauth et al. 1990). The mean number of ova for 38 Kansas females is 658 and is positively correlated with female SVL (Plummer 1977).

AQUATIC ECOLOGY. Hatchlings begin feeding on small zooplankton soon after emerging from the egg masses and incorporate larger prey into their diet as they grow. Older larvae may feed both from the water column and while resting on the pond bottom (McWilliams and Bachmann 1989b). When on the bottom substrate, individuals alternate between crawling and resting motionless in an attack position. When floating, larvae often alternate between swimming and floating motionless. Air gulping serves to increase the buoyancy of animals that are floating. Attacks on prey are preceded by a rapid burst of swimming if a larva is suspended in water, or a lunge forward with the limbs if a larva is on the bottom. In general, larger larvae tend to spend more time foraging on the bottom than in the water column.

Larvae in an Indiana pond feed mostly on *Daphnia* and ostracods when very young and on larger prey such as isopods when older (Whitaker et al. 1982). Other prey include chironomid larvae, corixids, dytiscid larvae, and trichopterans. Larvae at an Iowa site feed mostly on isopods and ostracods, but also eat chironomids, amphipods, cladocerans, copepods, gastropods, odonates, and coleopterans (McWilliams and Bachmann 1989a). Larvae at this site feed continuously throughout the day and night, and appear to prefer ostracods and cladocerans over chironomids, amphipods, and (when the larvae are younger) isopods. Isopods become more important in the diet as the season progresses, and in general there is a shift from pelagic to benthic prey as larvae grow. Small larvae do not feed on benthic prey because the larvae lack well-developed limbs and can only handle very small prey such as zooplankton. Larger larvae do not stratify. Instead, they tend to forage on the surface of the leaf litter at night and move into the leaf litter during the day. On rare occasion larger larvae may cannibalize smaller conspecifics (Minton 1972).

Few data are available on larval growth rates or size at metamorphosis. Bragg (1949) collected larvae with an average TL of 48 mm from a pond in Oklahoma on 7 May and observed metamorphosing larvae in June and July in Oklahoma. Hay (1892) and Minton (1972) found larvae nearing metamorphosis in late May-early June in Indiana. In southern Illinois, Cagle (1942) collected metamorphs on 19 May, Parmelee (1993) first collected metamorphs in late May, and Rossman (1960) found several metamorphs during mid-June that measured about 61 mm TL. Smith (1961) found transforming larvae from late May through July at various sites in Illinois. These observations indicate a larval period of 2-4 months in most populations.

TERRESTRIAL ECOLOGY. Juveniles move away from breeding ponds within a few weeks after metamorphosing and reside in forest floor retreats until sexually mature. At a floodplain site in southern Illinois, juveniles can be found beneath cover objects for a month or two following metamorphosis, but then they move underground with the onset of hot weather (Parmelee 1993). Juveniles tend to utilize substrates with leaves more so than adults. Limited observations suggest that juveniles and adults of this species do not move as far away from breeding ponds as do members of many other *Ambystoma* species; many adults appear to remain within 50-60 m of their breeding ponds (Parmelee 1993). Individuals become sexually mature when they reach 60-70 mm SVL, but the duration of the juvenile stage is not known.

Adults are sometimes active on rainy nights outside the breeding season but spend most of their time in underground burrows (Minton 1972). Individuals are frequently plowed up by farmers and excavation crews, and they also live in crayfish burrows (Cagle 1942; Minton 1972; Parmelee 1993; Strecker and Williams

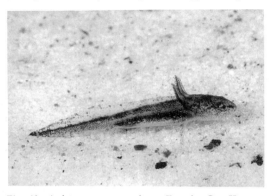

Fig. 63. *Ambystoma texanum;* larva; Douglas Co., Kansas (R. W. Van Devender).

1928). During the breeding season the adults reside beneath logs or litter around pond margins.

In a floodplain forest in southern Illinois, adults and juveniles can be found beneath moist logs and other cover objects from March to June. Surface activity peaks in March during the breeding season and progressively declines with the arrival of hot summer weather (Parmelee 1993). Individuals prefer large cover objects, and adults generally select larger cover objects than juveniles. Within reproductive classes, the size of cover occupied by an individual is independent of the resident's SVL.

Data from animals recaptured at this site suggest that individuals remain in the same activity area for extended periods of time (maximum distance moved = 20 m by an adult female). Individuals frequently share cover objects with conspecifics, and infrequently share cover objects with other *Ambystoma* species. Males are far less likely to be found with another male than with a female; however, females show no evidence of avoiding members of the same sex. The adults do not appear to be territorial (Martin et al. 1986; Parmelee 1993). Individuals do not respond to substrates marked by conspecifics and are no more aggressive to conspecifics than to control surrogates.

Whitaker et al. (1982) found that earthworms and shed skin are the primary dietary items of both juveniles and adults in Indiana. Other prey include centipedes, lepidopteran larvae, elaterid larvae, isopods, and *Chauliognathus* larvae. Adults feed very little while in the breeding ponds, but a small percentage may contain aquatic organisms such as isopods in their guts. Other documented prey include earthworms in the stomachs of four Illinois males (Cagle 1942) and beetles, earthworms, centipedes, a spider, a weevil, and a mayfly in the stomachs of Indiana and Kansas specimens (Minton 1972; Plummer 1977).

PREDATORS AND DEFENSE. Aquatic insects and tiger salamander larvae eat *A. texanum* larvae (Wilbur 1972), and garter (*Thamnophis*) and water (*Nerodia*) snakes eat the adults. When attacked by snakes or other predators, the adults posture defensively by lowering the head, curling the body, and raising and waving the tail, which is covered with noxious secretions (Dodd 1977; Hay 1892; Minton 1972).

COMMUNITY ECOLOGY. The small-mouthed salamander shows very strong affinities for fish-free breeding sites. This species is vulnerable to fish predation because the larvae are palatable to fish and do not use chemical cues to monitor fish presence (Kats et al. 1988). Larvae often reach high densities in shallow vernal ponds, but their role in organizing vernal pond communities has not been studied. Wilbur (1972) conducted field enclosure experiments to determine the impact of *A. tigrinum* larvae on *A. laterale*, JLL unisexuals, and *A. texanum*. All three species have lower survival in the presence of *A. tigrinum*.

Adults often share breeding ponds with other *Ambystoma* species and may co-occur locally in forest floor habitats outside the breeding season. A detailed study of surface cover selection at a southern Illinois site where *A. texanum* coexists with *A. maculatum*, *A. opacum*, and *A. talpoideum* suggests that adults of these species spatially segregate to some extent, primarily according to the amount of substrate moisture beneath cover objects (Parmelee 1993). In general, *A. texanum* and *A. talpoideum* use moister microhabitats than the other species. Juveniles of the same species show less evidence of spatial segregation.

CONSERVATION BIOLOGY. Populations of small-mouthed salamanders have been eliminated throughout their range as floodplain forests have been cleared and converted into agricultural fields. Maintaining viable populations of this and many other amphibians will require protecting bottomland forests and vernal ponds that provide critical habitats for the larvae and adults.

Ambystoma tigrinum (Green)
Tiger Salamander
PLATES 20, 21, 22, 23, 24

Fig. 64. *Ambystoma t. melanostictum*; adult; Wyoming (R. W. Van Devender).

Fig. 65. *Ambystoma t. tigrinum*; adult; Minnesota (R. W. Barbour).

IDENTIFICATION. *Ambystoma tigrinum* is a wide-ranging species that exhibits marked geographic variation in morphology, behavior, and life history patterns. As many as eight subspecies are currently recognized by some scientists (*diaboli, californiense, mavortium, melanostictum, nebulosum, stebbinsi, tigrinum,* and *valasci*; see detailed descriptions under "Systematics and Geographic Variation"). In addition, local populations may contain from two to six morphs that include cannibalistic forms and gilled or transformed adults (Collins et al. 1980). Sexually immature larvae occur as either "typical" or "cannibalistic" morphs. The cannibalistic morph is usually larger than the typical morph and has a broader, flatter head; a larger skull; and an extra row of prevomerine teeth (Collins et al. 1980; Gehlbach 1969; Powers 1907; Rose and Armentrout 1976). The teeth are more elongated and conspicuous than in the typical morph and in some populations are recurved. The cannibalistic morph has been documented in certain populations of *A. t. mavortium, A. t. melanostictum, A. t. nebulosum,* and *A. t. tigrinum* (Collins et al. 1993; Reilly et al. 1992). However, the morphological differences between typical and cannibalistic forms are not absolute and intermediate forms occur in some populations (Pedersen 1993).

Four adult morphs are recognized: typical metamorphosed adults, cannibalistic metamorphosed adults, typical gilled adults, and cannibalistic gilled adults. Gilled adults occur in all subspecies of *A. tigrinum*; however, they are rare in *A. t. tigrinum* and are currently known only from Michigan populations (Collins et al. 1980; Hensley 1964; Jones et al. 1993b; Larson 1968). In addition to these six general morphs, two types of typical adults occur in western Texas that differ in color pattern and clutch size (Rose and Armentrout 1976). "Small" morphs inhabit seasonally ephemeral ponds and transform at a relatively small size when immature, whereas "large" morphs inhabit highly eutrophic playas and typically transform when sexually mature.

Ambystoma tigrinum is one of the largest terrestrial salamanders in North America, rivaled only by *Dicamptodon*. Terrestrial adults can reach 33–35 cm TL and are stout-bodied animals with conspicuously broad heads and relatively small eyes (Langebartel 1946; Smith 1949; Smith and Reese 1968). Coloration and patterning are highly variable depending on subspecies but generally involve light yellow, brown, or yellowish green crossbands or blotches on a dark brown or black background. Albino larvae and adults occur at very low frequencies in a few populations (Dyrkacz 1981).

Terrestrial males have laterally compressed tails and swollen cloacae during the breeding season. Males also have proportionately longer tails than females and tend to have about the same or slightly smaller average body lengths (Downs 1989; Semlitsch 1983a; Sever and Dineen 1978).

Hatchlings and older larvae have broad heads with

rounded snouts, conspicuous bushy gills, and broad dorsal fins. Rudimentary balancers that are lost just before hatching are present on the late-term embryos (Nicholas 1925). Reported hatchling sizes are 9–10 mm TL for *A. t. mavortium* (Webb and Roueche 1971), 9–14 mm TL for *A. t. nebulosum* (Tanner et al. 1971), and 13–17 mm TL for *A. t. tigrinum* (Bishop 1941a).

Hatchlings of *A. t. mavortium* have alternating dark and light middorsal blotches and a pale lateral stripe along the body (Webb and Roueche 1971). Hatchlings of *A. t. tigrinum* have paired blotches along the dorsum and an unpigmented venter (Bishop 1941a). Older larvae of all subspecies are dull green to dusky colored above and white beneath and often have black dorsal spotting. In some populations larvae begin to develop the adult color patterns during or immediately before metamorphosis (e.g., Webb and Roueche 1971). Juvenile *A. t. tigrinum* acquire the adult coloration within 1 month after transforming (Engelhardt 1916a).

SYSTEMATICS AND GEOGRAPHIC VARIATION. The tiger salamander is a wide-ranging, polytypic species that shows marked geographic and ontogenetic variation in coloration, color patterning, and morphology (e.g., Fernandez and Collins 1988; Gehlbach 1967a,b; Jones et al. 1988). Scientists disagree about the number of subspecies in the complex (Jones et al. 1988; Shaffer and McKnight 1996). The most widely used classification is that of Gehlbach (1967a), which recognizes seven subspecies that differ in adult coloration and patterning, gill raker count, and tendency toward developing into gilled adults (see names and detailed descriptions later in this section). Most workers now recognize one of these subspecies (*A. t. californiense*) as a separate species, and some recognize a Mexican subspecies (*A. t. valasci*) as a separate species. Populations in Arizona are considered by some researchers to represent three subspecies (*A. t. nebulosum, A. t. stebbinsi, A. t. utahense*), but Gehlbach (1967a) synonymized these with *A. t. nebulosum*. Populations of *A. t. mavortium* also have been widely introduced into southeastern Arizona for use as fish bait.

Populations referable to *A. t. stebbinsi* in the San Rafael Valley in southern Arizona have low genetic heterozygosity (mean heterozygosity = 0.0015). This group is very similar to *A. t. mavortium* in terms of adult coloration and allozyme patterns, but it has mitochondrial DNA that is identical to that of *A. t. nebulosum* (Collins et al. 1988; Jones et al. 1988). The reticulate pattern that is diagnostic of *A. t. stebbinsi* is a juvenile trait that often disappears with age. Nonetheless, individuals referable to *A. t. stebbinsi* often have color patterning that is unlike that of both *A. t. mavortium* and *A. t. nebulosum*. Mitochondrial DNA analyses suggest that populations referable to *A. t. stebbinsi* are remnants of populations that occupied a former hybrid zone between *A. t. nebulosum* and *A. t. mavortium* (Jones et al. 1995). These populations are now distinct from either of the parental types from which they were likely derived; have a unique color pattern, geographic range, and genetic composition; and warrant formal recognition (Collins et al. 1988).

Here, I follow Gehlbach's classification with the exception of recognizing *A. californiense* as a separate species and *A. t. stebbinsi* as a subspecies. All Mexican populations of *A. tigrinum* are assigned to *A. t. valasci*. The following descriptions of subspecies are based on adult coloration and patterning. More detailed data on subspecific differences are given in Bishop (1941a), Collins (1981), Gehlbach (1967a), and Lowe (1954), and in the life history accounts that follow. Ranges of subspecies are based on data provided by Conant and Collins (1991), Shaffer and McKnight (1996), and Stebbins (1985).

The eastern tiger salamander (*A. t. tigrinum*) has a dark brown to grayish black dorsal ground color that is overlain with numerous, large, irregular, dirty brown to brownish yellow spots or vertical streaks. The belly is marked with irregular pale yellow blotches on a darker background. The gray tiger salamander (*A. t. diaboli*) is light gray or yellowish brown above and below. Small, rounded black spots overlay the lighter ground color. The barred tiger salamander (*A. t. mavortium*) has a black dorsal ground color that is overlain with broad, bold, vertical yellow bars that extend from the lower sides and belly to near the middorsal region. A few rounded or irregular yellow markings are often interspersed between the vertical bands. The Arizona tiger salamander (*A. t. nebulosum*) has an olive green to dark grayish dorsum with small, scattered black dots on the back, tail, limbs, and head. The venter is lighter and is often mottled with darker coloration. The Huachuca tiger salamander (*A. t. stebbinsi*) has a dark brownish gray to grayish black dorsum that is overlain with a branching network of light yellow lines. The venter is black and may have a few scattered yellow spots. Small yellow spots may also occur on the dorsum and along the sides of the body and tail. The reticulate pattern is strongly developed in juveniles that are >2-3 months postmetamorphosis. Adults may have a similar pattern, or a pattern that resembles that of *A. t. mavortium*. Thus, this subspecies is best identified by a combination of range and mitochondrial DNA data. The blotched tiger salamander (*A. t. melanostictum*) has a dorsum that is heavily mottled with irregular, dirty yellow and blackish blotches. The venter varies from being light to nearly as heavily mottled as the dorsum. The plateau tiger salamander (*A. t. valasci*) has yellow to olive spots or blotches scattered irregularly over the back and sides of the body. In many ways this subspecies resembles *A. t. tigrinum* and is best identified by its range. It is restricted to the highlands of northern and central Mexico and gilled adults are common in many populations. Individual, ontogenetic, and interpopulational variation occur in all subspecies, so one should expect deviations from the foregoing descriptions.

The taxonomic status and phylogeny of forms currently assigned to *A. tigrinum* continue to be investigated using molecular approaches, but much work remains before a comprehensive analysis is completed. Routman (1993b) examined mitochondrial DNA and isozyme variation in populations of *A. t. tigrinum* and *A. t. mavortium* across a broad region extending from Illinois to Colorado. Evidence from mitochondrial DNA and protein variation indicates that these two forms are well differentiated and probably evolved independently of each other before coming into recent contact and forming a hybrid zone. Local populations within each subspecies are well differentiated, and gene flow between demes appears to be very low. Genetic diversity of allozymes is the same for populations with gilled adults and those with terrestrial adults both within (*A. t. mavortium*) and between subspecies. However, genetic diversity for mitochondrial DNA is higher in populations with gilled adults. Additional analyses of these populations provide evidence of historical fragmentation of populations as well as range extensions of certain forms across previously glaciated regions (Templeton et al. 1995).

In other regional comparisons, Pierce and Mitton (1980) found substantial electrophoretic differences in a survey of polymorphic loci between *A. t. mavortium* and *A. t. nebulosum* in Colorado, New Mexico, and Texas. An abrupt change in allelic frequencies occurs across the Front Range in Colorado, where montane populations of *A. t. nebulosum* contact Great Plains populations of *A. t. mavortium*. Gene exchange between forms in this area is nonexistent, presumably because of the rapid elevational transition from grassland to montane forest communities.

Jones and Collins (1992) document an extensive hybrid zone between *A. t. nebulosum* and *A. t. mavortium* in west-central New Mexico, where changes in topographic relief are not as abrupt. Gene flow in this region is largely unidirectional from *A. t. mavortium* to *A. t. nebulosum* and occurs over a broad geographic area. These studies suggest that evolutionary relationships between subspecies of *A. tigrinum* are complex and can only be resolved by examining several contact zones from throughout the range of adjoining subspecies.

Studies of the evolutionary relationships between *A. tigrinum* and closely related members of the *A. tigrinum* complex in Mexico have been conducted using both electrophoresis of proteins (Shaffer 1983, 1984a) and mitochondrial DNA (Shaffer and McKnight 1996). The latter study sampled most of the subspecies of *A. tigrinum* using a broadscale approach, and only one specimen was usually analyzed from each population. The data show surprisingly little differentiation among recognized forms, even though many species were sampled over broad geographic regions using a fast-evolving segment of DNA. In this set of molecular data, subspecies of *A. tigrinum* often cluster more closely with other species than with conspecific subspecies, and the phylogeny of the major clades within the complex cannot be resolved.

Although Shaffer and McKnight (1996) question the validity of recognizing *A. tigrinum* as a polytypic species, it is perhaps best to continue treating *A. tigrinum* as such until more comprehensive genetic data become available. Nomenclatural changes at this point would do little to clarify evolutionary relationships among members of the complex and would undoubtedly be unstable. The subspecies of *A. tigrinum* are best viewed as members of a species complex that consists of geographic groups that have diverged evolutionarily to levels ranging from subspecies to semispecies or (perhaps) full species (Collins et al. 1980; Pierce and Mitton 1980). In addition to the genetic variation described previously, unisexual *Ambystoma* that contain genomes of *A. tigrinum* and other diploid *Ambystoma* species occur in the northeastern and north-central United States. Details of these and other unisexual complexes are presented under the special account on unisexual *Ambystoma*.

Other conspicuous geographic variation in *A. tigrinum* involves the occurrence of cannibalistic morphs, which, when present, usually make up a small proportion (<1–30%) of the total larval population (Pfennig et al. 1991). These morphs occur sporadically in *A. t. mavortium, A. t. melanostictum,* and *A. t. nebulosum* populations, and rarely in *A. t. tigrinum* (Collins 1981; Lannoo and Bachmann 1984; Rose and Armentrout 1976). They have not been reported for other subspecies.

DISTRIBUTION AND HABITATS. *Ambystoma tigrinum* is the most widely distributed salamander in North America. Populations occur in scattered locales throughout much of the United States and west-central Canada south to the southern Sierra Madre Occidentalis in Mexico. Populations are found from sea level to about 3350 m in elevation. The terrestrial adults are found in a wide array of habitats, including bottomland deciduous forests, conifer forests and woodlands, open fields and brushy areas, alpine and subalpine meadows, grasslands, semideserts, and deserts (Bishop 1941a; Collins 1981; Duellman 1954a; Sexton and Bizer 1978). On rare occasion adults have been found in streams in the western portion of the range (Collins 1981; Duellman 1955). Areas with sandy or otherwise friable soils and abundant breeding sites offer optimal habitats for this species in the eastern United States. Tiger salamander larvae are used as fish bait, and commercial bait collectors have introduced nonnative subspecies into many regions of the western United States (Collins 1981; Espinoza et al. 1970). These introductions have hampered efforts to determine the original distribution of the species, particularly in the southwestern United States.

BREEDING AND COURTSHIP. Adults breed in temporary and permanent habitats, including ditches, vernal ponds, farm ponds, quarry ponds, cattle tanks, subalpine lakes, and (on rare occasion) sluggish streams (Bishop 1941a; Collins 1981). In one instance larvae were found in a cave, but it is uncertain whether this was the site of oviposition (Black 1969). A breeding population of gilled adult *A. t. mavortium* inhabits a cave in New Mexico (Thompson and Jones 1992). Although *A. tigrinum* often breeds in permanent habitats in the western United States, adults rarely use sites with predatory fish (Blair 1951; Brandon and Bremer 1967; Carpenter 1953; Werner and McPeek 1994; Woodbury 1952).

Breeding has been documented in almost every month of the year, depending on subspecies, latitude, and elevation. *Ambystoma t. tigrinum* breeds from November through May and adults migrate to ponds at night during rainy weather (Semlitsch and Pechmann 1985; Sever and Dineen 1978). At relatively northern latitudes breeding in noncoastal regions typically occurs from late February through April, but in coastal regions breeding may begin as early as January. Breeding adults or freshly laid eggs have been found in March in Ohio (Downs 1989) and New York (Bishop 1941a), in February in Missouri (Seale 1980) and Arkansas (Trauth et al. 1990), in January and February in New Jersey (Anderson et al. 1971; Hassinger et al. 1970), from mid-January through late March in Maryland (Cooper 1955; Stine et al. 1954), and from late February through April in southern Illinois (Brandon and Bremer 1967), Indiana (Peckham and Dineen 1954; Sever and Dineen 1978), and northwestern Iowa (Lannoo and Bachmann 1984). Populations in the southeastern United States breed in both autumn and winter. Breeding may occur as early as November in Tennessee (Gentry 1955), and from November through February in the Sandhills of North Carolina (Morin 1983a).

At the Savannah River Site in South Carolina, the adults migrate during or immediately following rains, and males arrive at the breeding ponds 2–8 weeks before females (Semlitsch 1983a). Peak migrations to the ponds occur in January and February, whereas peak migrations from the ponds occur from January through March. One radioactively tagged male moved 162 m from the breeding pond on 6 March, then made short movements of a few meters over the next 5–6 weeks (Semlitsch 1983b). He returned to the same breeding pond the following fall. Detailed analyses after 11 and 16 years of monitoring a population at Rainbow Bay at the Savannah River Site reveal that the number of breeding females varies markedly from year to year (Pechmann et al. 1991; Semlitsch et al. 1996). Breeding activity is positively correlated with rainfall, and few or no females migrate to ponds during years with low rainfall.

At an Indiana site, male *A. t. tigrinum* arrive at a breeding pond at about the same time as females and adults remain from 10 to 29 days (Peckham and Dineen 1954). Males stay longer on average than females and most females breed annually (Sever and Dineen 1978). The minimum size of the breeding

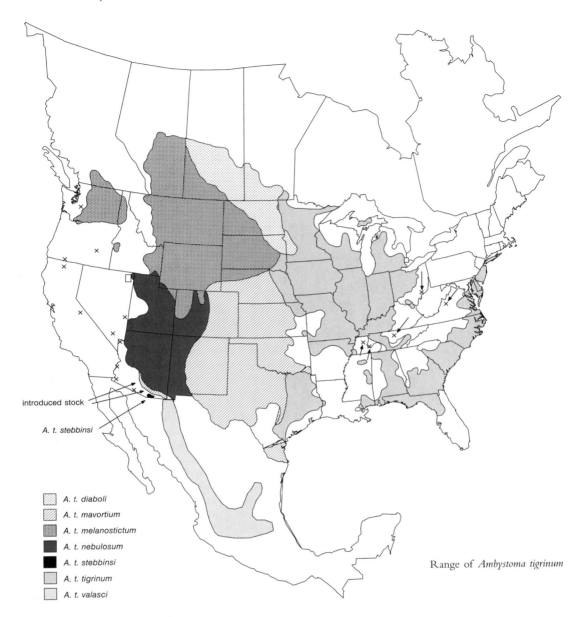

Range of *Ambystoma tigrinum*

population is 1100 adults (probably closer to 1500–2000), compared with 540 adults in a New Jersey population (Anderson et al. 1971). The breeding season in New Jersey lasts about 2 months (Hassinger et al. 1970).

The large morphs of *A. t. mavortium* in Texas breed from January to May, but small morphs breed sporadically throughout the year whenever rainfall is sufficient to fill playas (Rose and Armentrout 1976). Two metamorphosed female small morphs collected from below ground in October laid fertile eggs when injected with pituitary extract. Others maintained in tanks without males metamorphosed and produced fertile eggs. Both observations indicate that females in some populations are capable of long-term sperm storage (Rose and Armentrout 1976).

Dundee (1947) found sexually active gilled and terrestrial adults of *A. t. mavortium* in an Oklahoma pond in April and early June. A population of *A. t. mavortium* in a cattle tank in New Mexico breeds between

September and April, but annual breeding patterns vary markedly depending on patterns of seasonal rainfall and temperature (Webb and Roueche 1971). Adults breed during the fall if heavy rains occur and during the winter if the weather is mild. Otherwise, they may delay breeding until late winter or early spring. Populations in southern New Mexico often breed in mid- to late summer depending on weather patterns (Jones and Collins 1992).

Ambystoma t. nebulosum in high-elevation ponds in Colorado breed from June to early March, with terrestrial adults breeding in June (Whiteman et al. 1995; Wissinger and Whiteman 1992). In high-elevation sites in Utah, *A. t. nebulosum* may not breed until July or August (Worthylake and Hovingh 1989), but at lower elevations breeding may begin in May and continue sporadically for several months during periods of heavy rainfall (Tanner et al. 1971). In these and other high-elevation populations in the West, terrestrial adults normally begin breeding during or just following snowmelt. Eggs of *A. t. nebulosum* have been found in western New Mexico and Arizona ponds in early spring immediately after snowmelt (Jones and Collins 1992; Pfennig et al. 1991).

In a hybrid zone between *A. t. nebulosum* and *A. t. mavortium* in New Mexico, breeding varies from early spring in the most *nebulosum*-like populations to late spring or early summer in the most *mavortium*-like populations (Jones and Collins 1992). *Ambystoma t. stebbinsi* breeds from mid-February to early May, but most breeding occurs from mid-March to late April (Collins et al. 1988). Populations of *A. tigrinum* in the Pacific Northwest breed in March and April (Leonard and Darda 1995; Nussbaum et al. 1983).

Arnold (1976) and Kumpf (1934) described courtship of *A. t. tigrinum,* but detailed descriptions of courtship of other subspecies are not available. The following generalized account is from Arnold (1976).

Adults most frequently gather in widely dispersed groups of two to three individuals in breeding ponds and court. Upon encountering a female, the male moves along the length of the female while nudging, swinging his head, and lifting the female's venter. The male then pushes his snout against the lateral or ventral portions of the female's body and shoves her several centimeters to a meter. The male soon initiates a tail-nudging walk in which he moves a short distance from the female while maintaining contact by tapping the tail on her dorsum. The female moves forward and contacts the male's cloaca with her snout—this stimulates the male to move forward again. The male moves forward with his tail raised and deposits a spermatophore. During this process he stretches out and raises his limbs above the substrate and rapidly undulates his tail. The female follows and moves forward until her vent contacts the spermatophore. She then shuffles her vent laterally over the spermatophore and inserts it into her cloaca. In the process, she presses her hindlimbs against her tail base, arches her body slightly, and weakly undulates her raised tail. This process is similar to the action taken by males when depositing spermatophores. If the female remains responsive, the male may continue moving forward while monitoring the female's position with his tail and may deposit additional spermatophores.

A male will often engage in sexual interference or sexual defense by shoving a female away from a second male. A male may also intrude between a female and a rival male, and, while mimicking the female's behavior, cover up the rival's spermatophores with his own. The number of spermatophores deposited per male in a single night ranges from 8 to 37 and averages 21.

Tiger salamanders produce large spermatophores relative to other *Ambystoma* species, and the sperm cap is held in a quadrangularly shaped apex of the neck that has four raised horns. The neck slants forward, a feature that may facilitate the insertion of the sperm cap into the cloaca of the female. Male *A. t. tigrinum* in a New Jersey pond attach spermatophores to bottom debris, twigs, and willow leaves. The spermatophore is 8–10 mm tall. The base is 8–12 mm wide and broadly tapers to form a neck 2–3 mm in diameter (Anderson 1970).

Males typically outnumber females in breeding populations. Reported sex ratios of males to females are about 2:1 in South Carolina (Semlitsch 1983a) and 1:1, 3.2:1, and 5.3:1 in Indiana (Peckham and Dineen 1954; Sever and Dineen 1978).

REPRODUCTIVE STRATEGY. Mode of egg deposition varies among subspecies and may involve laying eggs singly or in masses. *Ambystoma t. tigrinum* deposits eggs in masses on twigs, weed stems, and other support structures in ponds. The ova are 2–3 mm in diameter, pigmented on the animal pole, and surrounded by three envelopes (Collins et al. 1980; Kaplan 1980). Freshly laid egg masses are firm and globular or oblong in shape and average about 55 x 70 mm. By the end of the embryonic period the masses swell to a much larger size and become very loose and flimsy. Eggs in New Jersey ponds are usually laid in water >20 cm deep (Hassinger et al. 1970), but I have often observed egg masses in ponds in Indiana, Kentucky, North Carolina, and Ohio in water >0.5–1 m deep.

Egg number per mass is fairly consistent geographically. Nine masses observed by Bishop (1941a) in New York averaged 38 eggs (range = 23–76), whereas 14 other masses from Long Island averaged 52 eggs per mass. Values for other sites include a mean of 59 eggs per mass in Maryland (Stine et al. 1954), a mean of 44 eggs per mass (range = 5–122) for several New Jersey ponds (Anderson et al. 1971; Hassinger et al. 1970), and a range of 13–52 eggs per mass for North Carolina ponds (Morin 1983a). Two masses in Arkansas contained 96 and 165 eggs (Trauth et al. 1990).

Length of the incubation period varies depending on local pond conditions and the seasonal time of oviposition. Reported incubation periods for different *A. t. tigrinum* populations are 40–45 days in Arkansas (Trauth et al. 1990), 19–31 days in Indiana (Couture and Sever 1979; Sever and Dineen 1978), 36 days in Maryland (Stine et al. 1954), 40–50 days in New Jersey (Hassinger et al. 1970), 30–40 days in New York (Engelhardt 1916b), and 29–36 days in North Carolina (Morin 1983a).

Female *A. t. mavortium*, *A. t. melanostictum*, and *A. t. nebulosum* deposit eggs singly or in very small clusters or linear strings on twigs, vegetation, or detritus. Eggs in early developmental stages are 1.7–2.5 mm in diameter, grayish or brownish above, and surrounded by three envelopes (Collins et al. 1980; Larson 1968; Tanner et al. 1971; Webb and Roueche 1971). When laying eggs, a female *A. t. nebulosum* grasps a support structure with the hindlimbs, elevates and waves the tail, and shuffles the cloaca from side to side against the substrate immediately prior to ovipositing (Hamilton 1948). The reported incubation period of *A. t. mavortium* is 8.5 days at 25°C (Webb and Roueche 1971). The incubation period of *A. t. nebulosum* from Utah is 6.5 days at 19°C and about 2–3 weeks at a natural breeding site (Tanner et al. 1971).

The tiger salamander shows the greatest variation in clutch size of any salamander. Wilbur (1977) reports a mean clutch size of 421 ova for 14 *A. t. tigrinum* from Michigan. In contrast, the average clutch sizes of gilled adult *A. t. mavortium* in western Texas are 2385 and 5670 for small and large morphs, respectively (Rose and Armentrout 1976). The maximum recorded clutch size of the large morph is 7631 eggs—the largest of any salamander. Estimates of average clutch sizes of transformed animals in this same population are 625 and 805 for small and large morphs, respectively. Thus, metamorphosis results in an 86% reduction in annual reproductive output for the large morph.

AQUATIC ECOLOGY. Larvae feed on a wide variety of aquatic organisms, including nematodes, insects, snails, clams, cladocerans, copepods, ostracods, fairy shrimp, crayfish, amphibian eggs and larvae, and the eggs and larvae of conspecifics (Black 1969; Brophy 1980; Brunkow and Collins 1996; Collins and Holomuzki 1984; Dodson and Dodson 1971; Lannoo and Bachmann 1984; Rose and Armentrout 1976). Large larvae will eat bullfrog and green frog tadpoles in captivity even though these tadpoles are noxious to fish (Werner and McPeek 1994). *Ambystoma t. nebulosum* larvae in Arizona shift to higher trophic levels as they mature or differentiate into the cannibalistic morph (Holomuzki and Collins 1987). Larvae measuring <30 mm SVL primarily eat cladocerans and dipteran larvae, whereas those over 30 mm in SVL incorporate a greater variety of prey into the diet, including zooplankton, snails, tiger salamander eggs, and a wide variety of aquatic insects. Cannibals occupy the highest trophic level, with 84% of their diet consisting of conspecific larvae.

Although large larvae take large prey, small zooplankton may constitute the bulk of the diet in certain populations. Large larvae (88–103 mm SVL) in an irrigation canal in Oklahoma may feed mostly on tiny ostracods and cladocerans (Tyler and Buscher 1980), as may larvae in ponds in Illinois (Brophy

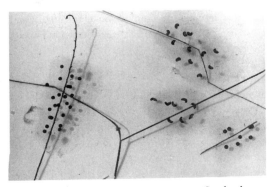

Fig. 66. *Ambystoma t. tigrinum*; egg masses; Scotland Co., North Carolina (J. W. Petranka).

Fig. 67. *Ambystoma t. tigrinum;* larva; Scotland Co., North Carolina (R. W. Van Devender).

1980) and Maryland (Lee and Franz 1974). Large morphs studied by Rose and Armentrout (1976) specialize on cladocerans, but small morphs often eat larger prey, such as water striders, water boatmen, and conspecific larvae.

Large gilled adults of *A. t. mavortium* in a Colorado pond feed primarily on chironomid larvae and pupae, although the abundance of these prey varies seasonally (Norris 1989). Other prey include water striders, water boatmen, beetle larvae, and the eggs of conspecifics. Cannibalism of larvae has not been documented in this population even though marked size differences occur between gilled adults and young larvae.

In an Indiana population the diet consists of chironomid larvae, clam shrimp, cladocerans, copepods, ostracods, rotifers, insects, water mites, amphipods, oligochaetes, gastropods, leeches, and American toad (*Bufo americanus*) tadpoles (Sever and Dineen 1978). Wood frog (*Rana sylvatica*) tadpoles and other *Ambystoma* larvae are major prey items of Michigan larvae (Wilbur 1972), and spadefoot toad (*Scaphiopus*) tadpoles and assorted invertebrates are important prey in New Mexico (Little and Keller 1937). Dietary items in specimens examined by Gehlbach (1965) include aquatic insects, gastropods, leeches, and eggs of *Hyla arenicolor.*

Dodson (1970) and Dodson and Dodson (1971) conducted a comprehensive dietary analysis of tiger salamander larvae in Colorado. Larvae in these ponds rarely cannibalize and eat a wide array of prey that consists mostly of arthropods. Very small larvae (<2 cm SVL) feed almost entirely on zooplankton, whereas larger larvae eat zooplankton along with larger prey such as amphipods, insects, and mollusks. Most prey are consumed roughly in proportion to their abundance. However, large prey such as *Chaoborus* and large zooplankton are preferred, whereas other prey such as adult beetles, notonectids, caddisfly larvae, sponges, and copepods are rarely eaten. Larvae feed primarily during the warmer hours of the day and are inactive at night. In a related study in Colorado, Sprules (1972) found that gilled adults are restricted to deep subalpine ponds that do not completely freeze. Predation by gilled adults eliminates large prey such as *Daphnia pulex* and the fairy shrimp, *Branchinecta shantzi,* from ponds.

Larvae have a variety of foraging strategies that include floating in the water column at night (Anderson and Graham 1967), resting motionless on the bottom substrate then lunging at prey that approach, and actively crawling through vegetation or bottom debris in search of prey. Leff and Bachmann (1986) examined ontogenetic changes in foraging behavior of Iowa *A. tigrinum* larvae. Small larvae crawl less and lunge more at prey than larger larvae. Larvae feed mostly on the substrate rather than in the water column. Small larvae show a preference for *Daphnia* and are less successful than larger larvae at capturing large prey such as damselfly larvae and water boatmen. Small larvae in New Jersey ponds tend to float in the water column, whereas large larvae more often swim through the water or crawl over the substrate in search of prey (Hassinger et al. 1970). These ontogenetic shifts in foraging behavior correlate with the development of functional limbs. Larvae from different microhabitats have different diets that correspond to the different prey types available in the microhabitats (Zerba and Collins 1992).

Cannibals tend to eat larger prey than typical morphs, including other salamander larvae, and often consume larvae nearly as large as themselves (Collins and Holomuzki 1984; Holomuzki and Collins 1987; Lannoo and Bachmann 1984; Rose and Armentrout 1976). Comparisons of cannibalistic and typical morphs in *A. t. melanostictum* show that the two morphs are equally effective at catching small fish or small salamander larvae. However, the cannibalistic morph is more efficient at catching large salamander larvae (Reilly et al. 1992) and consumes higher quantities of chorus frog tadpoles (*Pseudacris triseriata*) than do noncannibals (Loeb et al. 1994).

As for other *Ambystoma* species, growth rate, length of the larval period, and size at metamorphosis are phenotypically plastic traits that are influenced by environmental factors such as food level, density, and the seasonal time of pond drying (Wilbur and Collins 1973). In pen experiments involving *A. t. nebulosum* in Arizona, Brunkow and Collins (1996) found that larvae at a high experimental density (15 larvae/m^3) grow and develop more slowly than those at a lower density (8 larvae/m^3). However, survivorship to the initiation of metamorphosis is independent of density. When variability in initial larval size is reduced experimentally, larval growth rate and diet are not affected, but survivorship decreases. Development rate also increases when larvae are at low density. Overall, initial larval density appears to be more important than initial size variability in affecting growth and development rates.

Natural populations vary markedly in growth rate, length of the larval period, and size at metamorphosis. Observations from several locations in the Midwest and northeastern United States indicate that the larval period of *A. t. tigrinum* typically lasts 2.5-5 months, with most larvae metamorphosing between June and August (e.g., Bishop 1943; Engelhardt 1916a; Lannoo and Bachmann 1984; Seale 1980; Sever and Dineen 1978; Stine et al. 1954; Trauth et al. 1990). Eastern tiger salamanders hatch in March and April in New York and often exceed 10 cm TL after 2-3 months of growth (Bishop 1941a). In one study metamorphosing larvae averaged 104 mm TL (Bishop 1943).

Embryos in an Indiana population hatch in April and the larvae may metamorphose in early to mid-July, when they measure 49-63 mm SVL, if the pond begins to dry (Sever and Dineen 1978). The larval period in a Missouri population lasts 3.5-4 months and transformation occurs from early July through August when larvae reach 60-75 mm SVL (Seale 1980). In a New Jersey population, the larval period lasts an average of 2.5 months (Hassinger et al. 1970) compared with 3.7 months in Maryland larvae (Stine et al. 1954). At the former site pond drying may stimulate larvae to metamorphose in early June, when they average only 40 mm SVL and 70 mm TL.

Ambystoma t. tigrinum larvae in permanent habitats may overwinter and transform the following year at exceptionally large sizes. Larvae transform in June in the majority of populations in southern Illinois, but larvae in some ponds may overwinter and reach 17-21 cm TL by the following April (Brandon and Bremer 1967). Despite their large size, overwintering larvae at these sites are sexually immature. Gilled adult *A. t. tigrinum* are rarely encountered, but they have been found in some Michigan populations (Collins et al. 1980; Hensley 1964; Jones et al. 1993b; Larson 1968).

The larval period in southern populations of *A. t. tigrinum* may be relatively long if adults breed in the autumn and the larvae overwinter. Larvae in Louisiana have been found on 23 December that measure 56-90 mm TL and on 6 April that measure 60-125 mm TL (Dundee 1974). These presumably hatched from eggs laid in the autumn.

In montane regions of western North America, duration of the larval period and size at metamorphosis vary markedly depending on elevation, pond hydroperiod, and correlated variables such as pond temperature. Larvae of *A. t. nebulosum* in high-elevation lakes in Utah hatch in July or August and grow an average of 1.2 mm/day (Worthylake and Hovingh 1989). Some may transform within 2-3 months if they exceed 40 mm SVL. Most, however, overwinter and transform the following year. At lower elevations, most larvae transform within 2-3 months after hatching, and a small percentage overwinter and transform the following year (Tanner et al. 1971). Larvae of *A. t. nebulosum* in two Colorado ponds transform either the summer after hatching (elevation 2880 m) or the following summer (3097 m; Dodson and Dodson 1971). Members of the latter group transform when they measure 70-80 mm SVL and appear to reach sexual maturity by the following spring; members of the former group transform when they measure around 50 mm SVL and require an additional summer of growth before reaching sexual maturity. At even higher elevations in Colorado, permanent ponds often contain three or four age classes of larvae as well as gilled and metamorphosed adults (Wissinger and Whiteman 1992). *Ambystoma t. nebulosum* on the north and south rims of the Grand Canyon in Arizona average 53 and 65 mm SVL (Gehlbach 1967b), respectively.

Sexton and Bizer (1978) surveyed 60 ponds from 2600-3650 m elevation in Colorado and found that length of the larval period of *A. t. nebulosum* is inversely correlated with the average temperature of the ponds (range = <11.1-16.6°C). Bizer (1978) further showed that growth rates of first-year larvae are positively correlated with average pond temperatures. For populations having an average larval period of 1 year, average mass at metamorphosis decreases with average pond temperature. In the warmest ponds sexually immature larvae usually transform during their first year of growth, but in cooler, permanent ponds (11.2-16.6°C) transformation does not occur until the second or third year of life. Most animals metamorphose at 60-80 mm SVL, but a few reach nearly 90 mm SVL before metamorphosing. Gilled adults are found in high-elevation ponds (>3300 m) with average temperatures below 13°C, and they typically vary from 76 to 96 mm SVL (Sexton and Bizer 1978).

Fig. 68. *Ambystoma t. diaboli;* gilled adult; Douglas Co., Minnesota (R. W. Barbour).

In Colorado, gilled adults occur only at high elevations in the coldest ponds (Sexton and Bizer 1978). In contrast, the proportion of gilled adults in larval populations does not vary strongly with elevation in Arizona habitats (Collins 1981). Larvae in high-elevation populations typically do not become sexually mature until their third or fourth year of growth. Most gilled adults do not transform, and those that do often suffer high mortality.

Ambystoma t. stebbinsi in southern Arizona breeds mostly in March and April. The embryos hatch within a few weeks and larvae grow to about 60 mm SVL by mid-July. An estimated 17–40% of larvae metamorphose from late July to early September, but most remain in ponds and mature into gilled adults. Individuals usually exceed 100 mm SVL within 8–10 months after hatching and breed for the first time when 1 year old (Collins et al. 1988).

Large gilled adults of *A. t. mavortium* may reach 25–30 cm TL (Knopf 1962). Small morphs in Rose and Armentrout's (1976) western Texas ponds transform at an average size of 80–100 mm SVL, whereas large morphs transform at 140–150 mm SVL. Larvae metamorphose during all months of the year, but a peak occurs in the autumn and late winter months. Most gilled adults are incapable of metamorphosing. Gehlbach (1965) notes that larvae of *A. t. nebulosum* in New Mexico transform if ponds begin to dry, but otherwise develop into gilled adults. Carpenter (1953) collected transforming *A. t. melanostictum* in August in Wyoming that measured 135 mm TL on average.

Tiger salamander larvae have inherently faster growth rates than larvae of other *Ambystoma* species (Keen et al. 1984) and are capable of extremely fast growth in warm habitats. Recently transformed individuals in a Chihuahuan Desert cattle tank often reach

Fig. 69. *Ambystoma t. nebulosum;* gilled adults; Arizona (J. R. Holomuzki).

sexual maturity 2.5 months after hatching when about 15 cm TL, whereas gilled adults reach sexual maturity 5 months after hatching when about 19 cm TL (Webb and Roueche 1971). A 25-cm TL gilled adult *A. t. mavortium* was collected from a farm pond in Oklahoma 12–14 months after the pond was constructed and filled with water (Glass 1951).

The growth rates above are dwarfed by those of larvae in warm playa lakes of western Texas (Rose and Armentrout 1976). Larvae from eggs laid in late spring can reach an average of 82 mm SVL within 19 days posthatching, a size that is normally achieved in 1–2 years in many montane populations. Individuals may reach sexual maturity and transform when only 5–6 weeks old. If eggs are laid in the autumn, larvae may require 6–7 months to transform and reach sexual maturity. The genetic basis for variation in individual growth is poorly understood. Pierce and Minton (1982) found a weak positive correlation between allozyme heterozygosity and growth rate during the early larval period in some, but not all, *A. tigrinum* populations in Colorado and New Mexico.

The cannibalistic morph occurs sporadically in populations, and its frequency of occurrence is influenced by both genotype and local environmental factors. The tendency for larvae to develop into cannibalistic morphs

Fig. 70. *Ambystoma t. nebulosum;* scanning electron micrograph showing enlarged teeth of cannibalistic morph; Arizona (J. P. Collins; courtesy of J. R. Holomuzki).

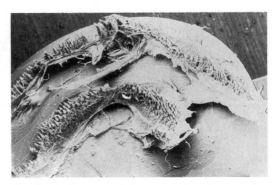

Fig. 71. *Ambystoma t. nebulosum;* scanning electron micrograph showing teeth of typical morph; Arizona (J. P. Collins; courtesy of J. R. Holomuzki).

in laboratory tanks is influenced by larval density, but not by food level (Collins and Cheek 1983; Pfennig et al. 1991). The size structure of the population and the relative sizes of larvae can also influence the development of cannibals. Large larvae are more likely to become cannibals than smaller individuals, and cannibals are more likely to develop when variation in body size is high within a population (Maret and Collins 1994).

Lannoo et al. (1989) found that typical morphs of *A. t. tigrinum* that cannibalize have faster growth rates and shorter larval periods than typical morphs that feed on other prey. Nonetheless, size at metamorphosis is not affected by experimental diet. Loeb et al. (1994) fed *A. t. nebulosum* larvae a variety of prey including tadpoles of *Pseudacris triseriata* and found that cannibals only develop when they eat conspecifics.

Regional differences occur in the percentage of cannibalistic morphs in natural populations of *A. t. nebulosum;* however, the proportion of cannibals in a population is not correlated with larval density in ponds (Pfennig et al. 1991). Cannibalistic morphs have significantly more parasitic nematodes and are more likely to acquire a lethal disease by cannibalizing diseased conspecifics. Higher percentages of cannibalistic morphs occur in ponds that have relatively low incidences of disease. Thus, susceptibility to disease may be one factor selecting against cannibalistic morphs in certain populations.

Both cannibalistic and noncannibalistic *A. t. nebulosum* larvae are able to distinguish genetic relatedness via chemical cues and are less likely to prey upon close relatives than distantly related larvae (Pfennig et al. 1994). Cannibals are also less likely to eat cannibalistic versus typical morphs that are siblings. Other studies indicate that larvae reared with mixed-brood groups are more likely to develop into cannibals—and to develop into cannibals sooner—than are larvae reared with siblings (Pfennig and Collins 1993). This may be an adaptive developmental response that increases inclusive fitness.

Tiger salamander larvae switch microhabitat use in response to temperature, predation risk, and zooplankton migrations. Diurnal movements of *A. tigrinum* larvae to and from the edges of ponds are thought to play a role in thermoregulation (Heath 1975; Holomuzki and Collins 1983; Prosser 1911; Whitford and Massey 1970). Larvae in a New Mexico pond move into warm water near the pond margin in the daytime, then retreat to deeper, relatively warmer, waters at night (Whitford and Massey 1970). Diurnal movements to and from the shore do not occur during the warmer months of early summer when temperatures exceed 15°C, nor do they occur on days when stiff winds homogenize water temperature throughout the pond. Similar patterns occur in ponds in Colorado (Heath 1975). Recaptures of marked animals indicate that larvae in ponds along the rim of the Grand Canyon return to the same general vicinity of the shoreline each day following daily migrations to and from deeper water (Gehlbach 1967b).

Larvae in Arizona exhibit both vertical and horizontal diurnal movements (Holomuzki 1989b; Holomuzki and Collins 1983). Larvae move both inshore and to the upper water column during the day but are most abundant on the pond bottom in deeper water at night. Thermoregulation is at a premium in many high-elevation ponds because of the limited growing season and cool pond temperatures. Lannoo and Bachmann (1984) found that both pelagic prey availability and darkness stimulate *A. t. tigrinum* larvae to move into and float in the water column.

Ambystoma t. nebulosum larvae in Arizona shift microhabitat use in response to the predatory diving beetle, *Dytiscus dauricus* (Holomuzki 1986a). Tiger salamander larvae can make up as much as 15% of the diet of these beetles (Holomuzki 1985a,b), which hunt in vegetated shallows at night. In the absence of *Dytiscus*, tiger salamander larvae move during the day into vegetated shallows, where temperatures are warmer, food levels are higher, and opportunities for growth are greater than in deep water (Holomuzki 1986c). In ponds with *Dytiscus*, however, larvae leave the vegetated shallows at night and move into open water. This study demonstrates that larvae are able to balance predation risk and growth opportunities in an adaptive way.

In a related study, Holomuzki (1986b) found that small first-year larvae do not avoid second-year cannibalistic or typical morphs. This seemingly maladaptive behavior may be explained by the sit-and-wait foraging behavior of older larvae which makes them difficult to detect. Skelly (1992) found that tadpoles of the gray treefrog (*Hyla versicolor*) increase their use of refuges when placed with *A. tigrinum* larvae. This behavior in turn slows tadpole growth.

Data from a variety of sources suggest that larval tiger salamander populations often suffer mass mortality from catastrophic events such as pond drying or disease (e.g., Sever and Dineen 1978). Populations of larvae and gilled adult *A. t. stebbinsi* sometimes suffer complete mortality in ponds in Arizona (Collins et al. 1988), and egg predation by eastern newts may in some cases be sufficiently high to virtually exclude *A. t. tigrinum* from ponds (Morin 1983a). Survival during the first year of growth of *A. t. nebulosum* larvae in Colorado is much higher in permanent ponds than in semipermanent ponds that are susceptible to drying (Wissinger and Whiteman 1992). Larvae in drying ponds often die from desiccation and predation by gray jays. In certain years local populations in Arizona and Utah suffer major die-offs that appear to be due to epizootic bacteria, including *Acinetobacter* (Holomuzki 1986c; Pfennig et al. 1991; Worthylake and Hovingh 1989). Larvae sometimes have heavy leech infestations, but it is uncertain if these cause mortality (Carpenter 1953; Holomuzki 1986c).

Long-term studies at the Savannah River Site indicate that the production of *A. t. tigrinum* juveniles is episodic (Pechmann et al. 1991; Semlitsch et al. 1996). Larvae often suffer complete mortality from pond drying in years with low rainfall. The number of metamorphs produced per female is not significantly affected by the initial larval densities in the pond, and there is only weak evidence of negative interactions of *A. tigrinum* larvae with other salamander larvae that share the pond.

Production of *A. t. tigrinum* metamorphs ranges from 0 to 24 per female in populations studied by Semlitsch (1983a). Only 6 of 1041 juveniles that left one site returned to breed 2 years later, although 52 were observed during the third year. Premetamorphic survivorship in a study by Anderson et al. (1971) varied from 0 to 8.7%, and roughly 60% of the premetamorphic mortality occurred prior to hatching.

TERRESTRIAL ECOLOGY. Duration of the terrestrial juvenile stage is highly variable depending on subspecies and local environmental conditions. Under favorable conditions (low larval densities, high food levels, permanent ponds) larvae of many subspecies do not metamorphose until they approach sexual maturity. In some populations, metamorphs disperse hundreds of meters from ponds, whereas in others they remain near ponds (e.g., Gehlbach et al. 1969). Metamorphs studied by Gehlbach (1967b) move as far as 229 m from their home ponds. However, Webb and Roueche (1971) collected juveniles with tadpoles and other aquatic prey in their guts, a finding that suggests that individuals in some populations remain in or near ponds after transforming. Nonbreeding terrestrial adults and metamorphs also may forage in seasonally ephemeral and semipermanent Colorado ponds (Whiteman et al. 1994).

Eastern populations often have a longer juvenile stage than western populations, and the juveniles disperse away from ponds and return to breed when sexually mature. Emigration from an Indiana pond occurs in July during rainy weather when temperatures are relatively low (Sever and Dineen 1978). In South Carolina the smallest adults at the Savannah River Site measure 81–85 mm SVL, and individuals return to breed for the first time when 2 years old (Semlitsch 1983a).

Adults of the western subspecies often live in or about mammal burrows and are active on the ground surface in the summer and autumn during periods of rain (Calef 1954; Carpenter 1955; Collins 1993; Gehlbach 1967b; Hamilton 1946). Over 200 adults and juveniles were killed by automobiles as individuals moved about on rainy nights in October and November in southeastern Michigan (Duellman 1954a). There was no evidence of unidirectional migration at the time.

Adult tiger salamanders will actively construct burrows by digging with the front limbs (Gruberg and Stirling 1972; Semlitsch 1983c). One radioactively tagged specimen of *A. t. tigrinum* studied by Semlitsch (1983b) was found an average of 12 cm below ground in tunnels averaging 28 mm in diameter. Gehlbach (1965) reported that an adult was excavated from nearly 2 m below ground. An adult *A. t. melanostictum* was found overwintering in a rattlesnake den in Alberta, Canada (Stark 1986).

Relatively few data are available on the diets of adults. Dietary items include insects, worms, and young field mice in New York specimens (Bishop 1941a); insects and a snail in specimens from Alberta, Canada (Moore and Strickland 1955); and a hatchling racerunner (*Cnemidophorus sexlineatus*) in an Iowa specimen (Camper 1986). Captive specimens will eat small frogs and snakes (Camper 1986; Duellman 1948), but it is uncertain if they prey on these in nature. Webb and Roueche (1971) found *Scaphiopus* tadpoles and nonaquatic insects in aquatic, transformed subadults of *A. t. mavortium*.

Terrestrial adult *A. t. nebulosum* in high-elevation Colorado ponds breed in both semipermanent and permanent ponds. After breeding, some adults move back onto land, but others remain during the summer in ponds, where they feed on both aquatic and terrestrial organisms (Whiteman et al. 1994). Adults in permanent ponds often move to semipermanent ponds, where opportunities for feeding are greater and the combined densities of terrestrial adults, gilled adults, and large larvae are lower. Nonbreeding terrestrial individuals may also move from land to feed in semipermanent ponds during the summer. All adult movements on land occur during summer rainstorms. Terrestrial adults in ponds feed heavily on fairy shrimp, but they also eat other zooplankton, benthic insects, and terrestrial insects that land on the water surface.

Very few data are available on adult survivorship. Harte and Hoffman (1989) document a 65% decline in an adult population in Colorado over 7 years that parallels a decline in larval recruitment. Semlitsch et al. (1996) document a similar long-term decline in a South Carolina population. Adult *A. t. nebulosum* in Utah are sometimes subject to catastrophic die-offs that are due to epizootic bacteria in breeding ponds (Worthylake and Hovingh 1989). Inputs of nitrogen from the breakdown of the feces of grazing sheep and from atmospheric sources may trigger the bacterial blooms that result in mass mortality of both larvae and adults. Gilled adult and terrestrial adult *A. tigrinum* may live as long as 25 and 16 years, respectively, in captivity (Nigrelli 1954).

PREDATORS AND DEFENSE. Caddisflies (Dalrymple 1970), newts (Morin 1983a), and tiger salamander larvae (Rose and Armentrout 1976) often feed on the eggs of *A. t. tigrinum*. Predators of larvae include dragonfly naiads, caddisfly larvae, diving beetles, marbled salamander larvae, and garter snakes (Anderson et al. 1971; Holomuzki 1985a,b; Holomuzki and Collins 1987; Sever and Dineen 1978; Stine et al. 1954). Larvae are capable of matching their color to existing backgrounds, an ability that may reduce predation from birds and other sight-oriented predators (Fernandez and Collins 1988).

Birds such as killdeers, bitterns, grackles, and gray jays, and mammals such as bobcats (*Lynx rufus*) and coyotes (*Canis latrans*), may eat larvae trapped in nearly dried ponds (Sever and Dineen 1978; Webb and Roueche 1971; Wissinger and Whiteman 1992). Owls, eastern hognose snakes (*Heterodon platirhinos*), and badgers (*Taxidea taxus*) eat the adults (Collins 1993).

When defensively posturing, terrestrial adult *A. t. melanostictum* spread and arch their hindlimbs, raise their tails nearly vertically, and wave them back and forth (Carpenter 1955). Similar defensive behavior has

Fig. 72. *Ambystoma t. melanostictum*; defensive posturing by adult; Douglas Co., Washington (W. P. Leonard).

Fig. 73. *Ambystoma t. mavortium;* defensive posturing (note milky tail secretions); western Texas (E. D. Brodie, Jr.).

been documented for *A. t. nebulosum* (Gehlbach 1967b) and *A. t. tigrinum* (Smith 1985). Animals prodded by Brodie (1977) lashed their tails and in one case threw tail secretions over 1 m onto the researcher's face, shoulder, and upper arm.

COMMUNITY ECOLOGY. Tiger salamander larvae are important predators in fish-free ponds and can influence the structure of both vertebrate and invertebrate communities. Holomuzki et al. (1994) examined the role of tiger salamander larvae as community organizers in montane Arizona ponds. In experimental enclosures, predation by larvae cascades downward to decrease herbivorous zooplankton biomass. This effect, in turn, increases phytoplankton biomass and lowers orthophosphate levels. Weaker cascading effects occur in a natural breeding pond, where the abundances of a few prey such as chironomid larvae and ostracods are correlated with salamander densities.

Larval tiger salamanders may also affect the density, size structure, and spatiotemporal distribution of *Daphnia pulex* in experimental enclosures (Holomuzki 1989b). In the presence of predatory salamanders, *Daphnia* occur at lower densities, are smaller on average, and are more evenly distributed throughout the water column. Densities of *Chaoborus* and *Diaptomus* are also reduced in treatments with salamanders. Data presented by Dodson (1970), Dodson and Dodson (1971), and Sprules (1972) suggest that larvae may exclude large prey such as *Daphnia* and fairy shrimp from ponds in Colorado.

Wilbur (1972) conducted experiments to determine the effect of *A. tigrinum* larvae on *A. laterale, A. texanum,* and *A. tremblayi* and found that all three species have lower survival in the presence of *A. tigrinum*. Adding wood frog tadpoles decreases the survival of *A. laterale, A. texanum,* and *A. tremblayi,* presumably because *A. tigrinum* can grow sufficiently large after feeding on tadpoles to prey on congeneric *Ambystoma* larvae. Brophy (1980) found very similar diets between *A. t. tigrinum* and *Notophthalmus viridescens* larvae that coexist in a pond in southern Illinois. *Ambystoma tigrinum* larvae can completely eliminate assemblages of tadpoles in experimental cattle tank communities (Morin 1983b).

CONSERVATION BIOLOGY. Populations in the southeastern United States have been affected by deforestation and loss of wetland habitats and appear to be declining in many areas. Harte and Hoffman (1989) document a long-term decline of *A. t. nebulosum* populations in the Colorado Rockies and surmise that acid rain may be responsible. Studies by Corn and Vertucci (1992) and Wissinger and Whiteman (1992), however, indicate that acid precipitation is not the cause of these declines. Whiteman et al. (1995) found that males in some populations avoid acidic water and prefer water with a neutral pH. About 50% of the embryos suffer mortality at pH 4.2 whereas >90% of embryos survive at pH >5.0. Low pH reduces the size of hatchlings and the length of the embryonic period.

Kiesecker (1996) found that *A. t. nebulosum* larvae maintained in the laboratory at pH between 4.5 and 7.0 have reduced growth and longer larval periods at the lower pH levels. Because growth and development rates of *Pseudacris triseriata* tadpoles are not affected, *A. tigrinum* larvae are less effective gape-limited predators at the lower pH levels. These results suggest that environmentally induced changes in pH could alter community interactions in natural systems.

The introduction of predatory fish into formerly fishless lakes and ponds is affecting *A. tigrinum* throughout the West (e.g., Carpenter 1953; Collins et al. 1988). The consequences of these introductions on population dynamics and community structure require study. The ecological effects of fish introductions on native amphibians should be carefully considered by fish and wildlife managers when deciding whether or not to stock natural, fish-free habitats.

Unisexual *Ambystoma* Biotypes
PLATES 25, 26

Unisexual *Ambystoma* of hybrid origin occur throughout the Great Lakes region to New England and the Maritime Provinces of Canada. These forms are often polyploid, consist almost entirely of females, and contain genetic complements of *A. jeffersonianum, A. laterale, A. texanum,* and *A. tigrinum*. A given individual may contain genetic complements of as many as three of these four species. Triploid unisexuals are common in most populations, and tetraploid and pentaploid unisexuals occur at low frequencies in some populations (e.g., Bogart 1989; Bogart and Licht 1986; Kraus 1985a; Lowcock and Murphy 1991; Lowcock et al. 1992). In addition, diploid unisexuals containing genomes of *A. laterale* and *A. texanum* are common in many populations in northwestern Ohio and southeastern Michigan.

Some of the unisexual biotypes have been formally named (i.e., *A. nothagenes, A. platineum, A. tremblayi*), but most current treatments simply refer to all unisexual forms by their genomic complements, such as "TTTL" for a tetraploid that has three sets of genes from *A. texanum* and one set of genes from *A. laterale*. Standard abbreviations used in the treatment that follows are J (*A. jeffersonianum*), L (*A. laterale*), T (*A. texanum*), and Ti (*A. tigrinum*). Genomic combinations that have been positively identified include JL, LT, JJL, JLL, JLT, JLTi, LLT, LTT, LTTi, TTT, JJLT, JLLL, LLLL, LLLT, LLTT, LTTT, LTTTi, and JLLLL (Bogart and Licht 1986; Bogart et al. 1985, 1987; Downs 1978; Kraus et al. 1991; Licht and Bogart 1987, 1989a; Lowcock 1994; Lowcock and Murphy 1991; Lowcock et al. 1991, 1992; Morris and Brandon 1984; Uzzell 1964a).

The first unisexual complex that was discovered involved *A. jeffersonianum, A. laterale,* and two triploid unisexuals with genomes from these parental species. The two unisexuals are JJL and JLL, which are referred to in older literature as *A. platineum* and *A. tremblayi*, respectively. These four forms have traditionally been referred to as the "*Ambystoma jeffersonianum* complex" (Clanton 1934; Uzzell 1963, 1964a) and consist of two diploid, bisexual species (*A. laterale, A. jeffersonianum*; 2n = 28), two triploid unisexuals (JJL, JLL; 3n = 42), and occasional unisexuals of other ploidy levels and/or genomic combinations.

Ambystoma jeffersonianum and *A. laterale* are genetically well-differentiated species that presumably arose via geographic speciation (Lowcock 1989). Earlier analyses of morphological, serological, and distributional data suggest that JJL and JLL arose from hybridization of *A. jeffersonianum* and *A. laterale* ancestral stock (Uzzell 1964a; Uzzell and Goldblatt 1967). However, more recent analyses using mitochondrial DNA indicate that the derivation of JJL and JLL is more complicated than previously thought.

Two diploid-triploid pairs of the complex (pair 1: *A. jeffersonianum* and JJL; pair 2: *A. laterale* and JLL) often occur in close geographic proximity to each other, but members of opposing pairs almost never share the same breeding pond (Nyman et al. 1988; Uzzell 1964a; Weller and Menzel 1979; Weller et al. 1978). Adults of the two diploid species can be distinguished from each other using external morphological features; however, the triploids have intermediate morphologies and often cannot be distinguished from either of their diploid counterparts using external morphology (Uzzell 1964a). Like other unisexuals, JJL and JLL are nearly always female and have larger cells and nuclei, larger eggs, and smaller clutch sizes than their diploid counterparts (Uzzell 1964a). Nonetheless, identification based on a combination of chromosome number, morphology, and allozymes is most reliable (Austin and Bogart 1982).

Additional polyploid unisexuals that are of hybrid origin occur in the lower Great Lakes region and vicinity. Morris and Brandon (1984) document unisexual JJLT in Illinois, and Morris (1985) reports a single 4n individual of unconfirmed identity from Indiana. Kraus et al. (1991) describe JLT and JLTi unisexuals from Michigan and Ohio, and Kraus (1985b) describes LTTi from Kellys Island in Lake Erie. Allozyme studies indicate that Kellys Island contains diploid, triploid, and tetraploid biotypes with *laterale, texanum,* and *tigrinum* genomes (Bogart et al. 1987). Unisexual diploids, triploids, and tetraploids that contain genomic complements of *A. laterale* and *A. texanum* occur on Pelee Island and the Bass Islands of Lake Erie (Bogart and Licht 1986; Bogart et al. 1985; Downs 1978; Kraus 1985a; Licht and Bogart 1987, 1989a). Unisexuals with *A. texanum* and *A.*

laterale genomes are also common in northwestern Ohio and southeastern Michigan (e.g., Kraus 1985a). These and other unisexuals are almost entirely female, have low embryonic survival, and appear to reproduce via both gynogenesis and syngamy (Bogart and Licht 1986; Bogart et al. 1989; Kraus 1991; Lowcock 1994). Reconstitution of viable diploids from triploids is unlikely since larvae having molecular characteristics of reconstituted diploids do not survive to maturity (Taylor 1992).

One hypothesis concerning the formation of the *A. jeffersonianum* complex and other unisexual complexes is that one or a small number of hybridization events occurred between the diploid species and was followed by backcrossing with the diploid parental species to form unisexual polyploids. According to this interpretation, the unisexuals (JJL, JLL, and others) are sexually dependent upon diploid, bisexual species (Macgregor and Uzzell 1964; Uzzell 1964a,b; Wilbur 1971) and produce unreduced ova that are stimulated to divide when a sperm of a diploid male penetrates the egg membrane. Fertilization does not ensue, so the male's genes are not incorporated into the genome of the resulting embryo. This process is known as gynogenesis.

Another hypothesis is that unisexuals have arisen on multiple occasions rather than during one or a small number of hybridization events, and that continual gene exchange and hybridization are occurring between diploids and unisexuals, albeit at low rates. Several lines of evidence provide partial support for this hypothesis and are not consistent with a strictly gynogenetic mode of reproduction. Lowcock and Bogart (1989) found a wide range of genotypes occurring in triploids, and Bogart (1982) found that both diploid JL and triploid JJL and JLL can arise from the same egg mass. In addition, JJL females can produce diploid *A. jeffersonianum* offspring (Sessions 1982). Tetraploids occur in some hybrid populations, and even pentaploids have been discovered (e.g., Lowcock 1994; Lowcock and Murphy 1991; Lowcock et al. 1991, 1992; Spolsky et al. 1992b).

These patterns indicate that meiotic reduction and syngamy occur under certain circumstances. Tetraploids and pentaploids that are present in some populations are produced through ploidy elevation. This is a process by which sperm is "accidentally" incorporated into an unreduced egg to produce offspring of a higher ploidy level. Through ploidy elevation, triploid females may produce tetraploid female offspring. Similarly, tetraploid females may on very rare occasion produce pentaploid female offspring. Triploid offspring that differ genetically from their triploid mothers are probably produced by meiotic reduction followed by syngamy.

Two important lines of research that involve temperature-dependent reproductive modes and mitochondrial DNA analysis have helped clarify some of the results of past research. In one, Bogart et al. (1989) and Elinson et al. (1992) found that triploid offspring are more likely to be produced via gynogenesis when eggs are incubated at relatively low temperatures that approximate those in natural ponds. In contrast, tetraploid offspring that are derived sexually are likely to develop at relatively high temperatures. Thus, reproductive mechanisms in unisexual complexes can potentially include the production of reduced and unreduced eggs and involve mixtures of both gynogenesis and syngamy depending on temperature (Bogart et al. 1989; Licht 1989).

A second line of evidence involving mitochondrial DNA supports clonal reproduction as the major mechanism of reproduction in unisexuals. Kraus (1989) conducted a mitochondrial DNA analysis of the *A. texanum-laterale* complex and found surprising genetic uniformity among unisexuals, but not among *A. texanum*, which is the presumed maternal parental species. His data suggest that one or a very small number of hybridization events produced this complex, and that gene exchange via syngametic mechanisms has not occurred at the high rate postulated by researchers supporting the multiple origin hypothesis. Studies by Spolsky et al. (1992a,b) reconfirm Kraus's finding of uniformity in mitochondrial DNA in unisexuals and point to ancient unisexual lineages that may be millions of years old. These data, together with electrophoretic markers, provide strong evidence that gynogenesis is the predominant form of reproduction in JJL.

Kraus and Miyamoto (1990) analyzed JLL using both allozymes and mitochondrial DNA. Their electrophoretic data show, as expected, that the nuclear genome consists of two complements of *A. laterale* alleles and one complement of *A. jeffersonianum* alleles. Surprisingly, however, the mitochondrial DNAs are derived solely from a third species, *A. texanum*. These data suggest strongly that unisexuals in the *A. jeffersonianum* complex are secondarily derived from an original *A. texanum-laterale* lineage and provide incontrovertible evidence for genomic recombination in the unisexuals.

Hedges et al. (1992) used a similar approach to compare mitochondrial DNA and nuclear DNA patterns in four sexual species and their corresponding unisexual forms collected over a broad geographic area. The mitochondrial data are consistent with those of Kraus (1989), Kraus and Miyamoto (1990), and Spolsky et al. (1992a,b) and confirm that the unisexuals have very similar mitochondrial genomes. Surprisingly, however, none is similar to any of the four bisexual species other than *A. texanum*. As with other mitochondrial DNA studies, there is strong evidence of ancient unisexual lineages.

Collectively, research to date indicates that unisexuals are derived from one or a few ancient ancestral hybridization events between *A. texanum* and *A. laterale*, and that the wide array of unisexual combinations has evolved in the context of a single hemiclonal lineage (i.e., genetic background). The data do not support the hypothesis that unisexuals are continually being produced through hybridization between diploid bisexual species (i.e., *A. jeffersonianum, A. laterale, A. texanum, A. tigrinum*). Gynogenetic reproduction in which unisexual females are dependent upon males of bisexual species for successful reproduction appears to be the primary mechanism of reproduction in most local populations. Nonetheless, unisexuals do regularly acquire nuclear material from bisexual species through syngametic mechanisms, leading to ploidy elevation and greater genetic diversity. In addition, alleles from unisexuals may occasionally introgress back into the diploid parental species (Kraus 1985a, 1989). There is no evidence that unisexuals reproduce by parthenogenesis (Bogart and Licht 1986).

SUMMARY OF LIFE HISTORIES OF UNISEXUALS. Much remains to be learned about the ecology and natural history of unisexual *Ambystoma*. All of the unisexuals breed in late winter or spring and their general life histories are similar. Because unisexuals are dependent on diploid males for successful reproduction, diploids are almost always found in ponds with unisexual populations. In populations with few or no *A. jeffersonianum* or *A. laterale*, females depend upon other diploid *Ambystoma* species, particularly *A. texanum*, for successful reproduction (Downs 1978; Morris and Brandon 1984; Peckham and Dineen 1954; Spolsky et al. 1992b).

Fig. 74. JJL unisexual (*Ambystoma platineum*); adult; locality unknown (R. W. Barbour).

Female unisexuals often greatly outnumber diploid males in local populations (Nyman et al. 1988; Uzzell 1964a; Wilbur 1971). Diploid males produce fewer spermatophores when courting triploid females compared with diploid females, and diploid females have larger clutches of eggs than their triploid counterparts. These factors may lessen the probability of triploids being extirpated from local populations of diploids (Uzzell 1964a, 1969).

The following is a summary of life history information for the three most widespread and well-studied unisexual groups. Information on the life histories of other forms is very incomplete.

JJL (formerly *Ambystoma platineum*; silvery salamander). This biotype is so similar morphologically to *A. jeffersonianum* that the two cannot be reliably distinguished using external morphology. Adults are brownish or brownish gray above with scattered pale blue spots and blotches that are concentrated on the sides, tail, and venter. The venter is more lightly colored than the dorsum and there are usually 13 costal grooves. JJL unisexuals are best identified by a combination of factors: sex, chromosome number, electrophoretic protein patterns, and the size of cells and nuclei.

Hatchlings and older larvae closely resemble those of *A. jeffersonianum*. Mature larvae have grayish bodies with heavily mottled, broad dorsal fins. The head is very broad, and the toes are elongated and tapered. The belly is whitish.

The distribution of JJL is poorly documented because laboratory analyses are required to make positive identifications. Populations are known from scattered

Range of JJL unisexuals (*Ambystoma platineum*)

locations in east-central Illinois, northern Kentucky, northern and central Indiana, northwestern Ohio, southern Michigan, northern New Jersey, western Massachusetts, and southern Ontario (Morris and Brandon 1984; Spolsky et al. 1992b; Uzzell 1964a; Weller and Menzel 1979; Weller et al. 1978). In many areas JJL inhabits deciduous upland forests along with *A. jeffersonianum*. In areas outside the range of *A. jeffersonianum*, populations are often found in bottomland forests (Downs 1989).

Because of the difficulty of identifying specimens in the field, little unequivocal information is available on the life history of this form. Females primarily depend upon *A. jeffersonianum* males for reproduction but also utilize *A. texanum* and *A. tigrinum* (Morris and Brandon 1984). In Illinois and Indiana, *A. texanum* is the primary host of most JJL populations (Spolsky et al. 1992b). Because JJL relies on diploid species to facilitate gynogenetic reproduction, the breeding season generally parallels that of *A. jeffersonianum* or other diploid *Ambystoma* species in local ponds. Adults breed in late winter or early spring in woodland ponds or other fish-free bodies of water.

Male *A. jeffersonianum* can chemically discriminate between female conspecifics and female JJL (Dawley and Dawley 1986) and are more likely to court and produce relatively large numbers of spermatophores when experimentally paired with conspecifics than with JJL (Uzzell and Goldblatt 1967). This finding suggests that female JJL sometimes use excess spermatophores that are deposited as males of other *Ambystoma* species mate with conspecifics. Female JJL lay about 140–200 eggs, and the mode of egg deposition and egg mass size resemble those of *A. jeffersonianum* (Uzzell 1964a). Eggs measured by Downs (1989) from a single JJL female that laid 137 eggs measure an average of 2.7

Fig. 75. JLL unisexual (*Ambystoma tremblayi*); adult; locality unknown (R. W. Van Devender).

mm in diameter (range = 2.7–2.8 mm). Each egg is surrounded by three envelopes and the number of eggs per mass is on average 19. Morris and Brandon (1984) report an average of 10.5 eggs per mass (range = 2–50) from Illinois.

Breeding occurs from mid-February to the end of March in Ohio (Downs 1989) and has been observed in mid-March in New Jersey (Nyman 1991). In the latter study, eggs laid early in the breeding season began hatching on 12 April after an incubation period of 28 days. Metamorphosing larvae were first observed on 21 June, and metamorphosis continued through July. Larvae often float in the water column, where they feed on copepods, cladocerans, ostracods, insect larvae (chironomids, chaoborids, coleopterans, anisopterans, and lepidopterans), and *A. maculatum* larvae.

JLL (formerly *Ambystoma tremblayi*; Tremblay's salamander). JLL is a triploid, unisexual form that is of hybrid origin between *A. jeffersonianum* and *A. laterale*. This biotype closely resembles *A. laterale* and is a small, slender *Ambystoma* with small bluish white blotches and flecks on a dark gray to brownish gray dorsum. Tremblay's salamander cannot be reliably distinguished from *A. laterale* using external morphology. In general, adult JLL tend to be larger than adult *A. laterale*, with a lighter body color. JLL lay fewer but larger eggs and have larger larvae at maturity than *A. laterale*. Unfortunately, none of these traits is diagnostic. Sex, allozyme data, chromosome number, and the size of cells and nuclei are the most useful characters for separating the two species. Adults reach 9.5–16 cm TL (Downs 1989; Uzzell 1967c), and there are usually 13 costal grooves.

Hatchlings and older larvae are the pond type and closely resemble those of *A. laterale*. Older larvae are brownish above with a yellowish lateral stripe that becomes less distinct with age. The dorsal fins of older larvae are heavily blotched and mottled with black.

JLL occurs sporadically throughout much of the range of *A. laterale* (Morris and Brandon 1984; Uzzell 1964a; Weller et al. 1978). Positive identifications have been made for specimens from northern Indiana, Maine, Massachusetts, southern Michigan, Nova Scotia, northwestern Ohio, southern Quebec and Ontario, and Wisconsin (Downs 1989; Uzzell 1964a). JLL is primarily dependent on *A. laterale* for successful reproduction, and its habitat requirements and annual breeding cycle are similar to those of the host species. JLL occurs primarily in forested habitats with sandy soils.

Adults migrate to breeding ponds following the arrival of warm weather in late winter or early spring and breed in a variety of habitats, including ponds in open, grassy areas next to forests, as well as woodland ponds and roadside ditches (Downs 1989). Breeding occurs from mid-March to early April in Michigan (Clanton 1934; Uzzell 1964a) and in April and early May in Ontario (Lowcock 1994). In an Ontario population, *A. laterale*, JLL, and JLLL differ in time of arrival at the breeding sites (Lowcock 1994; Lowcock et al. 1991). Male *A. laterale* tend to arrive first, followed by JLL, JLLL, and female *A. laterale*.

Courtship behavior has not been described for heterospecific matings. Females are primarily dependent upon *A. laterale* males to provide sperm for gynogenetic reproduction. However, the absence of *A. laterale* from sites with JLL-like triploids suggests that females use sperm from other *Ambystoma* species that share breeding ponds (Downs 1989).

Male *A. laterale* may be a limited resource in many populations because female JLL and *A. laterale* both rely on males for reproduction. Unisexuals almost always produce only female offspring, but female *A. laterale* produce offspring of both sexes. All else being equal, a consequence of this characteristic is that populations could become heavily dominated by unisexual females if females of both forms were equally successful at courtship and mating (Uzzell 1964a, 1969; Wilbur 1971). Breeding experiments conducted by Uzzell (1969) indicate that this is not the case. In paired matings, male *A. laterale* are more successful at courting and inseminating female *A. laterale* than female JLL, as measured by the number of spermatophores and fertilized egg masses produced.

Range of JLL unisexuals (*Ambystoma tremblayi*)

Despite the fact that male *A. laterale* discriminate against JLL, many mixed populations of *A. laterale* and JLL consist mostly of females (Clanton 1934; Lowcock 1994; Uzzell 1964a, 1967c; Wilbur 1971). Egg mortality is often high when females greatly outnumber males because females are not successfully inseminated and lay eggs that subsequently fail to develop. Females that do not find spermatophores may leave the ponds and resorb their ova (Clanton 1934).

JLL females normally lay their eggs in small, sausage-shaped masses attached to twigs and leaves, although single eggs may be deposited occasionally. The average number of eggs in 200 masses observed by Wilbur (1971) was about five. Published average clutch sizes vary from 136 to 142 eggs (Clanton 1934; Wilbur 1977), but individual females may produce from 50 to 337 eggs (Clanton 1934; Gilhen 1974; Uzzell 1967c). Freshly laid eggs are 1.8–2.2 mm in diameter and are pigmented both above and below (Clanton 1934). Data provided by Uzzell (1964a) suggest that the incubation period lasts about 1 month.

Little is known about the larval stage, but the larval ecology of JLL is presumably similar to that of other *Ambystoma* in the eastern United States. Larvae transform after an average of 81–99 days in field pens (Wilbur 1971, 1972) and after 88 days in the laboratory (Lowcock 1994). The smallest juvenile observed by Minton (1972) measured 31 mm SVL, whereas metamorphs from a natural pond studied by Lowcock (1994) reached an average of 37 mm SVL (range = 32–41 mm).

JLL often shares breeding ponds with other unisexual biotypes. Mechanisms that allow JLL to coexist in local ponds with these forms are not fully understood, nor is the role of larval competition in maintaining mixtures of unisexuals in local populations.

Range of *Ambystoma laterale-texanum* unisexuals (LT, LLT, LTT, LLLT, LLTT, LTTT)

Wilbur (1971, 1972) compared the competitive abilities of *A. laterale* and JLL larvae grown in experimental cages under natural conditions by growing larvae both in isolation and in mixed groups. In one set of experiments JLL had lower survivorship and a slightly longer larval period than *A. laterale* when grown in isolation at high densities (Wilbur 1971). However, the reverse finding was documented in a second set of experiments (Wilbur 1972). When mixed populations are grown at different densities, contradictory results have been obtained in some cases. In general, crowding of larvae results in longer larval periods, slower growth, and smaller size at metamorphosis for both species. Three species systems with *A. laterale, A. maculatum,* and JLL also provide evidence of density-dependent interactions among these species.

Lowcock (1994) conducted a detailed life history study of a population in southern Ontario consisting of mixtures of mostly *A. laterale,* JLL, and JLLL. These forms differ in the seasonal migratory patterns of adults, the growth and development rates of larvae, and the age at maturity of juveniles. Similar patterns are evident in mixed populations of unisexuals in central Ontario (Lowcock et al. 1991). The extent to which these differences contribute to the stable coexistence of diploid sexual species and unisexuals through time is not understood.

The smallest sexually mature JLL examined by Uzzell (1967c) measured 56 mm SVL and 93 mm TL. Nova Scotia specimens examined by Gilhen (1984) measured 67–85 mm SVL and 120–155 mm TL. Females probably mature sexually within 2–3 years after transforming (Downs 1989).

Ducey (1989) and Ducey and Heuer (1991) found that resident animals are aggressive toward conspecific intruders and frequently bite and aggressively posture during initial encounters. When food levels are low, residents are far more likely to pursue and vigorously bite intruders. This observation suggests that individuals may protect feeding areas from conspecifics.

Ambystoma laterale-texanum. Unisexuals of hybrid origin between *Ambystoma texanum* and *A. laterale* are common in northwestern Ohio and southeastern Michigan, and on Kellys Island, Pelee Island, and the Bass Islands of Lake Erie (Bogart et al. 1987; Kraus 1985a). Most individuals in local populations are diploids (LT) or triploids (LLT, LTT), although tetraploids are found at low frequencies in some populations. Licht (1989) provides measures of clutch sizes for *A. laterale-texanum* unisexuals collected from ponds on Pelee Island, but most specimens had already begun to oviposit prior to being collected and clutch size estimates are therefore unreliable. Twenty-five LLT females collected before breeding contained 2–284 ova (mean = 140) with an average diameter of 2.35 mm. The female with only two eggs was presumably spent.

Mean egg size differs among forms, and tetraploids have larger eggs on average than triploids. Mating experiments indicate that the different genotypes are equally successful at being inseminated and have similar fertilization (40–78%) and hatching (most <25%) rates. Male *A. laterale* will court LLT females, and females will follow males that lead them forward. However, spermatophore deposition has never been documented. LT females in this population reach sexual maturity at about 68 mm SVL. Most LLT females that have been collected measure >70 mm SVL.

Populations of *A. laterale-texanum* unisexuals are largely restricted to clay-based soils and breed from late February to mid-April, with a peak between mid-March and early April. The eggs are attached singly in close proximity to one another to limbs or other substrates, and egg mortality is often extremely high (F. Kraus, pers. comm.).

CONSERVATION BIOLOGY. Some unisexuals, such as JLT, JJLT, and LTTi, have very restricted ranges and are vulnerable to extinction. Kraus (1995) discusses some of the problems regarding the conservation of these and other unisexual vertebrate populations. Current federal and state laws are inadequate for protecting unisexual *Ambystoma* because of the specific legal wording of legislation. Because of their complex genetic systems, involving elements of both clonal and syngametic reproduction, these forms do not conform precisely to either the biological or the evolutionary species concept, upon which current legislation is based. Despite the fact that some unisexual *Ambystoma* are both rare and unique elements of biodiversity, they are currently receiving no protection from public agencies or conservation organizations. Kraus (1995) argues for a more process-oriented view of biodiversity that would accommodate groups that fall outside the realm of biological species.

Family Amphiumidae
Amphiumas

The Amphiumidae contains three extant species of *Amphiuma* that are largely restricted to the Coastal Plain of the southeastern United States. The fossil record indicates that amphiumids were widely distributed in the United States from the upper Cretaceous until the upper Miocene and became restricted to the Coastal Plain in the Pleistocene (Brattstrom 1953; Duellman and Trueb 1986; Salthe 1973c). Adults are elongated, cylindrical salamanders that may reach 1 m or more in length. The pelvic and pectoral girdles are reduced and the limbs are vestigial. Adults lack eyelids, have a single pair of gill slits, possess lungs, and lack external gills. Adults inhabit marshes, swamps, and other aquatic habitats. Females lay long strings of eggs and guard their clutches. Amphiumas have numerous colorful colloquial names in the South, including ditch eel, fish eel, conger eel, congo eel, congo snake, lamprey, lamp-eater, lamp eel, and lamprey eel (Baker 1937; Brimley 1920b; Meade 1934; Schmidt 1920).

Amphiuma means Garden
Two-toed Amphiuma
PLATE 27

Fig. 76. *Amphiuma means;* adult (note tiny front limbs); Orange Co., Florida (R. W. Van Devender).

IDENTIFICATION. The two-toed amphiuma is an elongated, slimy, eel-like salamander with four tiny limbs and two toes per foot. Adults have lungs, and there is a single pair of gill slits but no external gills (Salthe 1973a). The tail is laterally compressed and makes up 20-25% of the body length. The number of costal grooves is 57-60 and the average is about 58. The dorsum is nondescript and varies from black to dark brown or gray. The venter is similar, but lighter colored, and does not sharply contrast with the darker dorsum. Albinistic animals have been found in South Carolina (Hensley 1959). The two-toed amphiuma is one of the largest salamanders in the world, with adults reaching 46-116 cm TL.

Hatchlings reach an average of around 55 mm TL, are black above, and have tan bellies, grayish throats, and white gills (Weber 1944). All four limbs are functional at hatching, and juveniles are similar in color to the adults. Albino specimens are known from South Carolina (Schwartz 1957).

SYSTEMATICS AND GEOGRAPHIC VARIATION. Most herpetologists currently recognize three species of amphiumas. *Amphiuma means* and *A. tridactylum* are a closely related species pair whose ranges overlap in portions of Alabama, Mississippi, and Louisiana. Electrophoretic analyses of proteins by Karlin and Means (1994) indicate a high level of genetic similarity between the two species (Nei's $D = 0.12$). In contrast, *A. pholeter* is an ancient evolutionary offshoot that is very distinct genetically from the remaining species. No conspicuous geographic variation in the morphology of *A. means* is evident and no subspecies are recognized.

DISTRIBUTION AND ADULT HABITAT. Two-toed amphiumas occur in Coastal Plain habitats from southeastern Virginia to eastern Louisiana. Juveniles and adults occur in or near swamps, cypress bays, ditches, temporary ponds, sloughs, and sluggish streams. Individuals are often found in crayfish burrows and in root masses of water hyacinths and other aquatic plants (Carr 1940; Duellman and Schwartz 1958). Specimens are most easily collected by dip-netting, seining, or trapping with turtle and minnow traps.

BREEDING AND COURTSHIP. Many aspects of the breeding biology of this species have not been adequately documented. Breeding is apparently confined to the winter months and occurs in Louisiana from December through February, based on seasonal changes in reproductive structures (Rose 1967). Brooding females with eggs at early developmental stages have been found in February in Florida (Weber 1944). Courtship behavior has not been described, but fertilization is internal (Rose 1967).

REPRODUCTIVE STRATEGY. Females deposit eggs in cavities that they construct, then remain coiled about their eggs through hatching (Salthe 1973a; Weber 1944). The eggs are laid in rosarylike fashion and have 1-mm-diameter constrictions of the outer sheaths that separate adjoining ova by a distance of 5-10 mm. The eggs and outer sheaths are about 10 mm in diameter (Weber 1944). Very few data are available on clutch size. The estimated mean clutch size of a southern Louisiana population is 201 (range = 106-354; $n = 22$ females), and clutch size is positively correlated with female SVL (Rose 1966c).

Most nesting females have been found on land, but in most cases females probably lay in water, and their nests are exposed as water levels recede seasonally. Seyle (1985) found a 29-cm SVL female coiled around

Range of *Amphiuma means*

33 late-term eggs on 12 September. The eggs were in a hollow log in the dried bed of a cypress-tupelo swamp in Georgia. Weber (1944) found a female coiled around a cluster of 49 eggs in southern Florida in early February. The nest cavity was 8 cm wide, 13 cm long, and 2.5 cm deep. At other Florida sites four groups of eggs were found in July beneath logs in the partially dried beds of ponds (Brimley 1910). In Florida the nests are occasionally found in the nest mounds of American alligators. The alligator nests are usually constructed in June, and *Amphiuma* eggs can be found in July (P. E. Moler, pers. comm.).

The incubation period is exceptionally long and may last 5–6 months. Eggs estimated by Weber (1944) to have been deposited in the middle of January hatched in late June. Others have collected eggs in July that were probably laid in winter (Bishop 1943).

AQUATIC ECOLOGY. Larvae resorb their gills almost immediately after hatching, and juveniles as small as 60 mm TL have been collected (Neill 1947). Hatchlings and small juveniles presumably feed on small invertebrates, but little is known about their diet. Adult two-toed amphiumas are large, powerful predators and can handle both vertebrate and invertebrate prey. Crayfish, salamanders, and small frogs are important

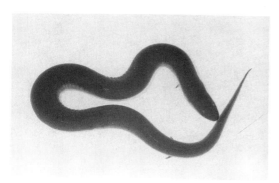

Fig. 77. *Amphiuma means;* adult; Orange Co., Florida (R. W. Van Devender).

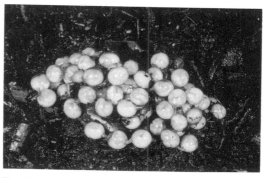

Fig. 78. *Amphiuma means;* eggs; Florida (Barry Mansell; courtesy of P. E. Moler).

food items in Florida populations (Carr 1940). Other documented prey include maggots, odonates, aquatic beetles, crayfish, and a fish in southern Florida specimens (Duellman and Schwartz 1958), and insects, frogs, tadpoles, water snakes, turtles, lizards, mosquitofish, sunfish, bass, crayfish, snails, and spiders in adults from southern Florida and South Carolina (Hamilton 1950).

Juveniles and adults construct burrows in or near aquatic habitats and have been excavated by farmers and construction crews from >1 m below the ground surface (Brimley 1920b; Knepton 1954). Individuals are primarily active at night and may actively prowl about in shallow water in search of prey or remain in burrows and snap at prey that come within striking distance (Carr 1940; Funderburg 1955).

Two-toed amphiumas are occasionally taken in terrestrial pitfall traps and have been seen on rare occasion crossing roads on rainy nights. These findings suggest that individuals are capable of dispersing short distances overland (Gibbons and Semlitsch 1991). During seasonal droughts individuals aestivate in burrows, and the adults may live 1–3 years without food (Gunter 1968; Rose 1966b). Almost no data are available on survival and longevity in nature. Captive specimens may live as long as 27 years (Nigrelli 1954).

PREDATORS AND DEFENSE. Amphiumas are preyed upon by water snakes (*Nerodia*), mud and rainbow snakes (*Farancia*), cottonmouths (*Agkistrodon piscivorus*), and large wading birds (Duellman and Schwartz 1958; Dye 1982; Funderburg 1955; Harper 1935; Telford 1952). Defensive behavior primarily involves biting. Large specimens can deliver a painful bite and should be handled with care. Carr (1940) reports that an 18-kg (40-lb) test line was broken twice while trying to dislodge hooked animals from burrows. He notes in a mild understatement that "An enraged forty inch *Amphiuma* is unpleasant to handle."

COMMUNITY ECOLOGY. Two-toed amphiumas often function as top predators in freshwater systems, but their importance in influencing community composition and species diversity is not known.

CONSERVATION BIOLOGY. The widespread loss of wetlands in the southeastern United States has undoubtedly eliminated many local populations of this species. Two-toed amphiumas are still common in many permanent and semipermanent habitats in the South, but long-term population trends have not been documented.

Amphiuma pholeter Neill
One-toed Amphiuma
PLATE 28

IDENTIFICATION. The one-toed amphiuma is an elongated, eel-shaped salamander that has one toe on each of its greatly reduced front limbs. The limbs and head are proportionately shorter than those in other amphiumas, the eyes are very small, and there is a single gill opening on each side of the head (Means 1992a; Neill 1964). The dorsum and venter are dark grayish or grayish brown, and the dorsum does not contrast noticeably with the venter. This is the smallest species of *Amphiuma*; the adults reach only 22–33 cm TL. Males and females do not exhibit conspicuous sexual dimorphism in external traits. The larvae have never been found and it is uncertain whether a larval stage occurs (Means 1992a).

Fig. 79. *Amphiuma pholeter*; adult; Florida (R. W. Van Devender).

Range of *Amphiuma pholeter*

SYSTEMATICS AND GEOGRAPHIC VARIATION. Electrophoretic analyses of proteins by Karlin and Means (1994) indicate that *A. pholeter* is an ancient evolutionary offshoot that is very distinct genetically from *A. means* (Nei's $D = 0.90$) and *A. tridactylum* ($D = 0.73$). Detailed studies of geographic variation in morphology have not been conducted. However, no conspicuous trends are evident, and no subspecies are recognized.

DISTRIBUTION AND ADULT HABITAT. One-toed amphiumas occur in extreme southwestern Georgia and adjoining portions of the panhandle of Florida. A possibly disjunct population occurs in extreme southwestern Alabama near Mobile (Carey 1984). Populations are patchily distributed and very localized. Specimens have been taken from a dried streambed (Stevenson 1967), from creeks (Karlin and Means 1994), and from areas with small, muck-bottomed ponds and intermittent streams (Neill 1964). The preferred habitat is deep, organic, liquid muck in floodplain swampy terrace streams and in alluvial swamps of low-gradient second- or third-order streams (Means 1992a).

COMMENTS. Little is known about the biology of this rare and highly secretive species. Adults court in winter or spring and the eggs are laid in late spring or summer (Means 1992a). Young raised in the laboratory may reach adult size in about 2 years. Groups of as many as four to five individuals have been excavated from burrows at the interface of the muck layer and the underlying substrate. During cold winter weather individuals move deeper into the muck and are torpid when uncovered. Dietary items of Florida specimens include sphaeriid clams, earthworms, larvae of dipterans and odonates, and terrestrial beetles (Means 1992a).

The one-toed amphiuma should receive high priority for protection because of its small range and general rarity. A comprehensive analysis of its distribution, ecology, and natural history is needed.

Amphiuma tridactylum Cuvier
Three-toed Amphiuma
PLATE 29

Fig. 80. *Amphiuma tridactylum*; adult; Jefferson Co., Texas (R. W. Van Devender).

IDENTIFICATION. The three-toed amphiuma is a large, elongated, salamander with four tiny limbs that each have three toes. Adults measure 46–106 cm TL and are black, slate gray, or brownish above and light gray beneath. A dark patch is present on the throat and the dorsal and ventral colors contrast sharply. The tail is laterally compressed and makes up about 20–25% of the body length. There are no external gills, but a single gill slit occurs on each side of the body. The average number of costal grooves is 57–60. Albinos are extremely rare but have been found in Mississippi and Louisiana (Cagle 1948; Hensley 1959).

Individuals can be sexed by their cloacal anatomy (Baker 1937). Females have dark cloacal walls that are smooth, whereas males have whitish walls with papillose oval patches on the lateral walls. Males also develop swollen cloacae during the breeding season (Cagle 1948). Hatchlings measure 43–64 mm TL (Bishop 1943; Cagle 1948; Hay 1888). They are dark brown above and light below and have short, whitish gills and well-developed legs and toes.

The three-toed amphiuma closely resembles the two-toed amphiuma, but it has three toes per foot and a distinctly bicolored body (Salthe 1973b). Other less diagnostic differences are summarized by Baker (1947). One or more toes may be missing from a foot owing to injury, so all four limbs should be checked when identifying specimens.

SYSTEMATICS AND GEOGRAPHIC VARIATION. *Amphiuma means* and *A. tridactylum* constitute a closely related species pair whose ranges overlap in portions of Alabama, Mississippi, and Louisiana. Electrophoretic analyses of proteins by Karlin and Means (1994) show a high level of genetic similarity between the two species (Nei's $D = 0.12$).

DISTRIBUTION AND ADULT HABITAT. Three-toed amphiumas occur in Coastal Plain habitats from eastern Texas to western Alabama and northward to southeastern Missouri and extreme western Kentucky. Redmond and Scott (1996) consider the sight record for Benton Co., Tennessee (?), to be questionable. This species prefers semipermanent or permanent habitats with abundant vegetation and inhabits drainage ditches, swamps, sloughs, sluggish streams, and semipermanent ponds (Baker 1945; Cagle 1948; Chaney 1951).

BREEDING AND COURTSHIP. *Amphiuma tridactylum* has an extended breeding season. In southern Louisiana breeding occurs primarily during the winter and spring months. Males have swollen cloacae from mid-January to mid-May, and peak breeding activity occurs in March following periods of heavy rain (Cagle 1948). Wilson (1940, 1941a) reports that male cloacae reach their maximum seasonal development from January to March. Sperm is present in male cloacae in

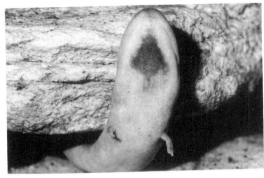

Fig. 81. *Amphiuma tridactylum*; adult (note dark throat patch); Jefferson Co., Texas (R. W. Van Devender).

Range of *Amphiuma tridactylum*

decreasing amounts from March through August and in female cloacae as early as March. Caged specimens bred in July in Tennessee (Baker et al. 1947).

Data from several studies indicate that females oviposit on biennial cycles, whereas males reproduce annually. About 42% of mature females examined by Cagle (1948) contained enlarged ova in a given year compared with 43–48% of mature females examined by Wilson (1941b, 1942). Sex ratios are near 1:1 (Cagle 1948).

Baker et al. (1947) observed courtship in late July involving one male and two females maintained in an outdoor pen. When courting, the animals swim rapidly about in a spiral fashion and occasionally make sudden twists of the body or dart rapidly through the water. Courtship activity is greatest early in the morning and in late afternoon. At the termination of these sessions the female rubs her snout along the body of the male. As the male rests near the shore with his head protruding from the water, the female glides underneath, rolls over, and positions herself so that the cloacae are in contact. The pair may remain in this manner for 20 minutes before separating.

Cagle (1948) observed over 100 individuals that were courting in a shallow pool on 25 March. The animals swam rapidly through the water and often flipped their tails out of the water. Several were lying in pairs, but cloacal contact was not observed. All of the males had lacerated bodies from bites from other males.

Males produce flattened, flakelike spermatophores that are directly transferred to the females (Cagle 1948). Females are capable of storing viable sperm for at least 6–8 months after mating (Kreeger 1942).

REPRODUCTIVE STRATEGY. Females deposit eggs in rosarylike strings similar to those of *A. means* and remain coiled about them until hatching (Cagle 1948; Hay 1888). Mature ova are 6–7 mm in diameter, and freshly laid ova are surrounded by jelly envelopes that are separated by thin cords 5–12 mm long. Reported clutch sizes are a mean of 98 (range = 42–131) enlarged ova in 26 Louisiana females (Cagle 1948), 80 and 127 enlarged ova in two Arkansas specimens (Trauth et al. 1990), and a clutch of about 150 eggs in Arkansas (Hay 1888).

Brooding females have been found coiled about their eggs beneath logs and other cover objects along the margins of ponds or swamps. Field observations suggest that the eggs are laid in water, then exposed as the water line recedes. Nesting records include a clutch of late-term eggs in Louisiana in mid-November (Bishop 1943), a female with late-term eggs about

9 mm in diameter in Arkansas in late August (Hay 1888), a clutch with an attending female in Alabama on 9 November (Ultsch and Arceneaux 1988), and nests with brooding females in western Tennessee in August, September, and midwinter beneath logs within 3.7 m of the water's edge (Parker 1937).

Cagle (1948) surmised that most females at his study site lay eggs in burrows, since he found eggs in open water only once. He collected one female with eggs protruding from her cloaca in late June, and another female on 28 September with egg cases in her stomach. Females in this population develop enlarged ova in May and June when oviposition presumably occurs. Most embryos hatch in October and November after a 4- to 5-month incubation period, although small numbers of hatchlings can be found from March to May. Observations from this and other studies suggest that oviposition can occur throughout most of the warmer months of the year, with a peak in late spring or early summer.

AQUATIC ECOLOGY. The three-toed amphiuma has a very abbreviated larval stage, and individuals lose their gills within 3 weeks after hatching (Ultsch and Arceneaux 1988). Individuals reach 10–14 cm TL during their first year of growth, and females first develop large ova when they measure about 33 cm SVL (Cagle 1948). Individuals probably require 3–4 years to reach sexual maturity.

In a Louisiana population individuals remain in their burrows during the day and emerge at night to forage (Cagle 1948). Peak activity occurs 3–4 hours after sunset, and individuals are most active after heavy rains when temperatures exceed 5°C. During the twilight hours they frequently attempt to capture prey while remaining in their burrows with the head or upper body extended from the entrance.

Crayfish and earthworms make up most of the diet of southern Louisiana specimens. Other prey include fish, aquatic and terrestrial insects, spiders, snails, ground skinks, and small amounts of vegetable matter (Chaney 1951). Prey of other specimens from Louisiana include crayfish and the common snapping turtle, *Chelydra serpentina* (Fontenot and Fontenot 1989).

When feeding on crayfish, amphiumas partially emerge from their burrows, capture the prey, then pull them into the burrows to eat. Amphiumas often twist violently after capturing prey, a behavior that may reduce the probability of being injured by large prey.

The adults are capable of burrowing through soft muck and probably enlarge crayfish burrows for their own use. During droughts, individuals remain in underground burrows and can live for many months without feeding. Individuals generally remain in a fairly restricted area; however, individuals are capable of dispersing at least 396 m from their original site of capture (Cagle 1948). Animals often move overland on nights with heavy rains and have been observed on land 12 m from the water's edge.

PREDATORS AND DEFENSE. Liner (1954) observed a cottonmouth (*Agkistrodon piscivorus*) consuming a large adult, but a more important predator is the mud snake, *Farancia abacura* (Curd 1950; Ernst and Barbour 1989; Meade 1934). Like *Amphiuma means*, three-toed amphiumas are slimy-skinned animals that are difficult to handle and can deliver a painful bite. Amphiumas sometimes produce a whistling sound when provoked (Baker 1937). Biting and fleeing appear to be the major defensive mechanisms of this species.

COMMUNITY ECOLOGY. Three-toed amphiumas function as upper-level predators in many freshwater systems. Although undocumented, their role in structuring communities is probably similar to that of large predatory fishes.

CONSERVATION BIOLOGY. Three-toed amphiumas are locally common in many areas of their range and do not appear to be in immediate need of protection. Because this species can coexist with fish, the creation of canals and permanent ponds has compensated for the loss of swamps and other natural wetlands that are used as foraging and breeding sites.

Family Cryptobranchidae
Hellbender and Giant Salamanders

This family consists of only two extant genera (*Cryptobranchus, Andrias*) and three extant species. *Cryptobranchus* is restricted to eastern North America, and fossils have been found from the upper Paleocene of Saskatchewan, Canada, and the Pleistocene of Maryland (Holman 1977; Nickerson and Mays 1973a). *Andrias* occurs in central China and Japan, and fossils have been found from the upper Oligocene and Pliocene of Europe, the Miocene of North America, and the Pleistocene of Japan (Duellman and Trueb 1986; Estes 1981).

The cryptobranchids are gigantic, stream-dwelling salamanders with depressed bodies, flattened heads, and conspicuous dermal folds of skin. Living *Andrias* may exceed 1.5 m TL, and certain extinct forms exceed 2.3 m TL (Estes 1981). *Cryptobranchus* reaches a maximum length of about 74 cm TL. Female cryptobranchids lay their eggs in paired strings and fertilization is external. All species are perennibranchs that undergo varying degrees of metamorphosis. Adults lack eyelids, and in *Cryptobranchus* one pair of gill slits is usually present. Sexually mature *Cryptobranchus* retain the larval dentition in which the prevomerine teeth form a curved series that is parallel to the premaxillary and maxillary (upper jaw) bones.

Cryptobranchus alleganiensis (Daudin)
Hellbender
PLATE 30

Fig. 82. *Cryptobranchus alleganiensis;* adult; Tazewell Co., Virginia (R. W. Van Devender).

IDENTIFICATION. This unmistakable species is a large, slimy, aquatic salamander that reaches 30–74 cm TL. The body and head are strongly flattened dorsoventrally and the eyes are small. A wrinkled, fleshy fold of skin occurs along each side of the body. The dorsum is greenish, yellowish brown or slate gray, with varying amounts of black spotting or blotching. Albinos and morphs with orangish or reddish ground colors have occasionally been found (Dyrkacz 1981; Fauth et al. 1996; Nickerson and Mays 1973a). Adult males develop swollen cloacal glands during the breeding season and on average are about the same TL as females. Larvae undergo partial metamorphosis so that adults have a single pair of circular gill openings; these are sometimes missing on one or both sides of the head.

Hatchlings reach 25–33 mm TL, are uniformly colored above, have conspicuous yolk sacs, and lack functional limbs at hatching (Bishop 1941a; Peterson 1988; Smith 1912b). One-month-old larvae are uniformly dark above and have white venters, whereas 6-month-old larvae have large, dark spots and slightly pigmented venters (Smith 1912b). Dorsal spots and blotches are most prominent in larvae 1–2 years old and in young adults (Bishop 1941a; Smith 1907).

SYSTEMATICS AND GEOGRAPHIC VARIATION. The hellbender shows remarkably little variation in allozymes throughout its range. Merkle et al. (1977) found that only 2 of 24 loci were variable in samples from a broad geographic region (mean heterozygosity = 0.007), whereas Shaffer and Breden (1989) found similarly low genetic variation in two other populations ($H = 0.02$). Nonetheless, Routman (1993a) found high variation in mitochondrial DNA, which suggests that *Cryptobranchus* may have experienced a population bottleneck in the recent past. In more detailed studies involving mitochondrial DNA, Routman et al. (1994) found that populations north of the Ohio River are very similar genetically to populations in the northern Ozarks. This finding suggests that invasion from one region to the other occurred relatively recently. Their data also suggest that both subspecies are paraphyletic.

Two subspecies are recognized that are best identified by geographic range and color patterns (Dundee and Dundee 1965). The hellbender (*C. a. alleganiensis*) occurs in central Missouri and from southern New York southwestward to northern Alabama and extreme northeastern Mississippi. Adults of this subspecies typically have small black spots on the dorsum and a uniformly colored chin; however, adults in southern populations are sometimes blotched. The Ozark hellbender (*C. a. bishopi*) occurs in southern Missouri and adjoining portions of Arkansas. The adults have large blotches on the dorsum and a chin that is mottled with dark pigmentation.

Fig. 83. *Cryptobranchus alleganiensis;* adult; Tazewell Co., Virginia (R. W. Van Devender).

Range of *Cryptobranchus alleganiensis*

DISTRIBUTION AND ADULT HABITAT. Hellbenders inhabit large, rocky, fast-flowing streams at elevations below 762 m and are frequently taken beneath large rocks in shallow, rocky rapids (Hillis and Bellis 1971; Nickerson and Mays 1973a; Taber et al. 1975). This species occurs from southern New York to northern Alabama and extreme northeastern Mississippi, and westward to central and southern Missouri and northern Arkansas.

BREEDING AND COURTSHIP. Length of the breeding period is variable among eastern and western populations. Many populations outside the Ozarks have short breeding seasons that extend from mid- or late August through mid-September. Breeding occurs at this time in New York and Pennsylvania (Bishop 1941a; Smith 1907, 1912a; Swanson 1948), Ohio (Pfingsten 1990), Tennessee (Fitch 1947; Huheey and Stupka 1967), and West Virginia (Green 1933). Breeding in an Indiana population occurs from 7 September to 11 October (Kern 1986), and adults in breeding condition can be collected in Alabama in late September and early October (Mount 1975).

Populations in the Ozark Uplift have more variable breeding seasons. The spawning season in Missouri of *C. a. alleganiensis* lasts from early September to mid-November (Dundee and Dundee 1965). Breeding occurs in late September in a Missouri population, with spermatogenesis occurring in summer (Ingersol et al. 1991). Vitellogenic activity is greatest from May to mid-September. Nests have been found in Missouri between 13 September and 8 October (Nickerson and Mays 1973a; Nickerson and Tohulka 1986).

The breeding season of *C. a. alleganiensis* in the Niangua River, Missouri, lasts from the second week of September through the first week of October (Peterson 1988). In contrast, *C. a. bishopi* in the Spring

Fig. 84. *Cryptobranchus alleganiensis*; typical stream habitat; South Fork of Holston River, Virginia (J. W. Petranka).

River of Arkansas breeds primarily in January, when populations elsewhere have ceased breeding (Peterson et al. 1989a). In an earlier study of the Spring River population, Baker (1963) did not find males in breeding condition after 15 December.

Sex ratios in most Ozark populations are about 1:1 (Nickerson and Mays 1973a; Peterson 1987); however, males frequently outnumber females in eastern populations. Pfingsten (1990) reports a 3:1 ratio of males to females in Ohio samples; in Pennsylvania sex ratios vary from 1.6:1 to 3:1 (Hillis and Bellis 1971; Smith 1912a).

Unlike most salamanders, *Cryptobranchus* employs external fertilization, and males produce milty secretions rather than spermatophores. Males and females in Pennsylvania become restless shortly before mating (Smith 1907). Adults move about the streambed during the day and prod with their noses beneath crevices under rocks. In some instances as many as 6-12 adults may congregate on the stream bottom. Aggregates of as many as 13 animals, 11 of which are males, have been observed in a Missouri stream in mid-September (Peterson 1988).

Shortly before mating, each male moves to a brooding site. This is normally an excavated, saucer-shaped depression beneath a large rock, log, or plank, with the entrance positioned out of direct current on the downstream side of the depression (Alexander 1927; Bishop 1941a; Kern 1986; Nickerson and Mays 1973a). In Missouri nests are occasionally constructed in rock crevices with the entrances pointing in a direction other than downstream (Nickerson and Tohulka 1986). Males may also nest at the base of mud-gravel banks (Peterson 1988).

Males remain posed with their heads emerging from entrances in wait for females. As a gravid female approaches an entrance, the male often guides or drives her into his burrow. The male may aggressively prevent the female from leaving until she oviposits (Nickerson and Mays 1973a).

The male becomes attracted to the female soon after eggs begin to emerge from her cloaca (Smith 1907). As eggs are laid by the female, the male positions himself alongside or slightly above the female. He then sprays the eggs with seminal fluid that is often in stringy cords. During this process the male sways the posterior of his body laterally and lowers and raises his hindlimbs. These motions disperse sperm over the egg mass. Peterson (1988) notes that two males were attracted to a submerged sack that contained a female that had released eggs. This finding suggests that chemical communication is important in mediating courtship.

REPRODUCTIVE STRATEGY. Females lay their eggs during a 2- to 3-day period in rosarylike strings in depressions within the males' burrows (Nickerson and Mays 1973a; Smith 1907). The egg strings ultimately become twisted about each other to form a heap in the bottom of the depression. Freshly laid ova are pale to light yellow, 5-7 mm in diameter, and surrounded by two gelatinous envelopes (Bishop 1941a; Nickerson and Mays 1973a). The largest ovarian eggs are about 6 mm in diameter (Dundee and Dundee 1965). Freshly laid eggs with the surrounding envelopes are 18-20 mm in diameter and are separated from adjoining eggs by short cords 5-10 mm in length (Bishop 1941a; Smith 1912a). By the time of hatching, the outer capsules are often 25-30 mm in diameter (Bishop 1941a).

During or shortly after oviposition, males and females may prey upon their own and other individuals' clutches. Most hellbenders examined during the breeding season contain between 15 and 25 eggs in their stomachs (Smith 1907). Males frequently regurgitate eggs (King 1939; Pfingsten 1990), and females sometimes eat their own eggs while ovipositing them (Nickerson and Mays 1973a).

Females are apparently driven away from nests soon after they oviposit, but males guard the eggs for an undetermined amount of time. Brooding males will often rock their bodies back and forth and undulate their lateral folds, increasing oxygen supply to both the eggs and the adult (Bishop 1941a). However, the primary value of brooding is probably nest defense from conspecifics. Smith (1907) witnessed a rather violent fight in which a large male successfully defended its clutch against a marauding female and smaller male. All three animals had eggs in their stomachs. Bishop (1941a) also witnessed a male driving off an intruder male.

The number of eggs in nests is often much lower than the number of enlarged ovarian eggs, a discrepancy that may reflect heavy egg cannibalism. Egg counts for individual nests are 153 eggs in an Indiana nest (Kern 1986), 138 eggs for a Missouri nest (Dundee and Dundee 1965), 319 and 334 eggs for two New York nests (Bishop 1941a), and 250 eggs in an Ohio nest (Pfingsten 1990). A female collected on 1 September in Tennessee oviposited 200 eggs in captivity the following day (Fitch 1947). Joint nests in which several females oviposit in the same depression have been observed on several occasions. Joint nests may contain as many as 1946 eggs (Bishop 1941a).

The number of enlarged ovarian eggs in 21 females from Missouri varies from about 150 to 750 and averages between 400 and 500 (Topping and Ingersol 1981). Clutch size is positively correlated with female TL in both subspecies. However, only about 76% of the mature ova of Missouri females are laid; the remainder are reabsorbed after the breeding season. Mean number (and range) of mature ova in females of the two subspecies examined by Peterson et al. (1988) from four rivers in Arkansas and Missouri are 365 (215–452), 429 (95–481), 450 (159–687), and 480 (296–908). Number of mature ova is positively correlated with female TL and age in all of these populations, and *C. a. bishopi* produces more ova than *C. a. alleganiensis* of similar size.

The incubation period is 68–75 days in Pennsylvania and New York populations (Bishop 1941a) and around 45 days in a Missouri population (Peterson 1988). Late-term embryos or hatchlings have been found on 7–23 November (Bishop 1941a) and 2 November (Peterson 1988).

AQUATIC ECOLOGY. Hatchlings have conspicuous yolk sacs and presumably rely on yolk as a major source of energy for the first few months after hatching (Smith 1912b). Hatchlings in New York populations grow 35–40 mm during their first year of life and reach 68–70 mm TL when 1 year old (Bishop 1941a). Four young of the year collected by Smith (1912b) in August varied from 6.4 to 7.7 cm TL. Larvae partially transform and lose their external gills after reaching 100–130 mm TL and 1.5–2 years of age (Bishop 1941a; Nickerson and Mays 1973a; Smith 1907, 1912b).

Hatchlings and larvae are less abundant than adults in collections. This finding suggests that larvae either have very high mortality rates relative to adults or use microhabitats that are rarely searched. Larvae and small transformed individuals often live beneath small stones in gravel beds or shallow-water habitats (Nickerson and Mays 1973a; Smith 1912b). Very small larvae are rarely collected and probably use cryptic habitats such as interstitial areas in gravel.

Individuals mature sexually when 5–8 years old (Bishop 1941a; Dundee and Dundee 1965), and males normally mature at a smaller size and younger age than

Fig. 85. *Cryptobranchus alleganiensis;* larva; Watauga Co., North Carolina (R. W. Van Devender).

females. Females in the Eleven Point River in Missouri become sexually mature after reaching 30 cm TL (Peterson et al. 1988). However, in most Missouri populations most females mature sexually when they reach 37–39 cm TL and are 6–8 years old (Dundee and Dundee 1965; Nickerson and Mays 1973a; Peterson et al. 1983, 1988; Taber et al. 1975; Topping and Ingersol 1981). In contrast, most males become sexually mature when they reach about 30 cm TL and are 5–6 years old (Taber et al. 1975). Males and females in western Pennsylvania become sexually mature when they measure between 33 and 35 cm TL (Smith 1907).

Growth rates decline steadily with age after metamorphosis. Animals at 18 months grow about 60–70 mm TL per year, whereas those approaching 25–30 years of age grow only 1 mm per year (Peterson et al. 1983, 1988; Taber et al. 1975). Males and females grow at about the same rate. Data on annual growth rates indicate that some animals may live for more than 25–30 years in the wild. Adults may survive as long as 29 years in captivity (Nigrelli 1954).

Males appear to have relatively constant age-specific mortality throughout life (Peterson et al. 1983; Taber et al. 1975), but females are more variable. In one population age-specific mortality of both males and females is relatively low during the first 5–10 years of life but remains relatively constant thereafter (Peterson et al. 1983). In a second population, age-specific mortality tends to decrease slightly with age throughout the lifetime of an average individual (Taber et al. 1975). In both instances females have higher survivorship than males. These estimates assume stable age distributions in the populations.

Hellbenders are primarily active on the stream bottom at night (Noeske and Nickerson 1979; Smith

1907; Swanson 1948). Adults remain beneath rocks or other cover during the day, sometimes with their heads protruding (Hillis and Bellis 1971; Smith 1907). At dusk, they emerge and forage over the stream bottom. During the breeding season and on overcast days, adults may emerge from cover and move about the stream during the day (Nickerson and Mays 1973a).

Transformed *Cryptobranchus* in a Pennsylvania stream tend to favor habitats with large, flat rocks in swift, shallow waters (Hillis and Bellis 1971). Hellbenders are less abundant in deeper waters or stream sections with rounded rocks or boulders embedded in the substrate. Marked animals at this site rarely move more than 10-20 m from their point of release. Estimates of the average and median home ranges are 346 m^2 and 113 m^2, respectively. In the Niangua River in Missouri the respective average and median home ranges are 28 m^2 and 13 m^2 for females and 81 m^2 and 77 m^2 for males (Peterson and Wilkinson 1996). The size of the home range in this population is independent of adult SVL, and the home ranges of both males and females frequently overlap. Although adults sometimes move long distances (up to 3500 m), most appear to remain within a restricted home range (Peterson 1987). Displacement experiments suggest that animals can home both upstream and downstream to their home territories.

Nonbrooding adults defend rocks they occupy from conspecifics and rarely share home rocks with other individuals (Hillis and Bellis 1971; Nickerson and Mays 1973a; Peterson and Wilkinson 1996). The North Fork of the White River in Missouri supports an estimated 428 mature hellbenders per kilometer of stream and a biomass of 156 kg/km of stream (Nickerson and Mays 1973b). As many as 10 animals per 80-100 m^2 of stream bottom occur in optimal riffle habitats. Peterson et al. (1988) estimate densities of 1-6 hellbenders per 100 m^2 of stream in four populations in Arkansas and Missouri.

Crayfish and small fish (*Campostoma, Catostomus, Cottus, Ichthyomyzon, Notemigonus, Notropis, Percina, Salmo*) are the dietary mainstay of hellbenders. Bishop (1941a) lists crawfish, small fish, mollusks, worms, and insects as dietary items. Crayfish and fish are the primary dietary items of hellbenders in Pennsylvania (Netting 1929; Smith 1907; Swanson 1948) and West Virginia (Green 1933, 1935). Missouri specimens sometimes contain large numbers of lampreys (Nickerson et al. 1983); however, crayfish often form the bulk of the diet of *C. A. bishopi* (Nickerson and Mays 1973a).

Peterson et al. (1989b) found that crayfish and fish make up more than 90% of the diet and that the diet changes little seasonally. Other items reported in hellbender guts include snails, tadpoles, worms, insect larvae, a toad, aquatic reptiles, and a small mammal (Green 1935; Nickerson and Mays 1973a). Adults also eat their shed outer skin (Smith 1907).

PREDATORS AND DEFENSE. Hellbenders produce very slimy skin secretions that are noxious to some predators (Brodie 1971a; Nickerson and Mays 1973a). Nonetheless, this species is occasionally preyed upon by large fish, turtles, and water snakes (Barbour 1971; Huheey and Stupka 1967; Minton 1972; Nickerson and Mays 1973a; Surface 1913). The eggs and larvae are frequently eaten by conspecifics. Native Americans have traditionally used this species as a food source (McCoy 1982), and hellbenders are frequently collected by fishers on baited hooks.

COMMUNITY ECOLOGY. Although the natural history of *Cryptobranchus* is well documented, researchers still do not understand the extent to which this species affects species diversity and patterns of resource use in freshwater streams. Its appetite for crayfish suggests that *Cryptobranchus* may play an important role in regulating crayfish populations in some streams.

CONSERVATION BIOLOGY. Because of its large size and unusual appearance, the hellbender is one of our most impressive native salamanders. This species was undoubtedly abundant in most large streams within its range prior to European colonization. It has declined throughout much of its range during this century because of stream impoundment, pollution, and siltation (Bury et al. 1980; Dundee 1971; Gates et al. 1985; McCoy 1982; Smith and Minton 1957; Trauth et al. 1992; Williams et al. 1981).

Family Dicamptodontidae
Pacific Giant Salamanders

This family contains four extant species of *Dicamptodon*. Members of this genus are restricted to southwestern British Columbia, California, Idaho, Montana, Oregon, and Washington, and fossils are known from the lower Pliocene of California (Nussbaum 1976). The terrestrial adults are large, stout salamanders that may reach 34 cm TL. Large adults are capable of delivering a painful bite and should be handled with care. Gilled adults are common in many local populations and often outnumber transformed individuals. Adults have compressed, bladelike teeth and lay large, unpigmented eggs. Larvae are the stream type with tail fins that extend forward only to the insertion of the hindlimbs. The gills are short, bushy, and dull red. Heavy black mottling is often present on the tail fins.

The systematic status of the genus *Dicamptodon* has been debated by numerous researchers. Until recently, this genus was included in the family Ambystomatidae (Tihen 1958) or in a separate family (Estes 1981). After reviewing morphological and biochemical evidence used in classifying *Dicamptodon, Rhyacotriton,* and other genera that have been included in the Ambystomatidae, Good and Wake (1992) recommended that *Dicamptodon* and *Rhyacotriton* be placed in separate families. Here I follow this recommendation and recognize the Dicamptodontidae, Rhyacotritonidae, and Ambystomatidae.

Dicamptodon aterrimus (Cope)
Idaho Giant Salamander
PLATE 31

Fig. 86. *Dicamptodon aterrimus;* gilled adult; Idaho (C. R. Peterson).

IDENTIFICATION. The Idaho giant salamander is a stout-bodied salamander with a dark brown to nearly black dorsum that is overlain with fine brown spotting or marbling. The head is depressed in front of the eyes and the posterior half of the tail is laterally compressed. This species closely resembles *D. tenebrosus,* but it is darker above and below and has finer dorsal mottling (Nussbaum 1976). Specimens in some populations have a dark brown middorsal region that lacks mottling. Transformed adults have 12–13 indistinct costal grooves and measure 17–25 cm TL.

Larvae are the stream type with tail fins that extend forward only to the insertion of the hindlimbs. The gills are short, bushy, and dull red. Young larvae are darker than those of *D. tenebrosus* of similar size and have little dorsal mottling. They have a faint stripe behind the eye, and the tail tip is only slightly darker than the rest of the tail. Larvae >60 mm SVL lack a dark tail tip and have a dark bluish gray venter and purplish brown dorsum with few or no markings (Nussbaum 1976). Maughan et al. (1976) collected a gilled adult with large ovarian eggs. Both gilled and terrestrial adults occur in some populations.

SYSTEMATICS AND GEOGRAPHIC VARIATION. The genus *Dicamptodon* was previously considered to consist of a single species, but detailed morphological

Range of *Dicamptodon aterrimus*

and electrophoretic studies indicate that *Dicamptodon* is polytypic. Daugherty et al. (1983) examined genetic variation in specimens referable to *Dicamptodon ensatus* and found strong genetic divergence between Rocky Mountain and coastal populations of *D. ensatus* (Nei's $D = 0.52$). They recommended that Rocky Mountain populations be recognized as a separate species, *D. aterrimus*. Good (1989) verified that Rocky Mountain populations are strongly differentiated genetically from coastal populations ($D = 0.50$) and also recognized *D. aterrimus*.

DISTRIBUTION AND ADULT HABITAT. *Dicamptodon aterrimus* is found in north-central Idaho and in a small adjoining portion of extreme western Montana. A southern isolate occurs in the Salmon River drainage near Warm Lake, Idaho. Larvae are common in Idaho in small streams above 975 m in elevation (Maughan et al. 1976). The adults inhabit stream and streamside habitats in moist forests.

COMMENTS. Most aspects of the life history of *D. aterrimus* have not been studied. This species is closely related to *D. tenebrosus*, and many aspects of the natural history are similar. A sample of larvae collected by Nussbaum and Clothier (1973) from the Palouse River, Idaho, contains individuals measuring 50–180 mm TL. Larvae in this population metamorphose during their third year of growth when they reach about 140–180 mm TL. In some populations larvae mature into gilled adults when as small as 107 mm SVL (Nussbaum 1976). Gilled adults have enlarged ova 4.5–5.0 mm in diameter (Maughan et al. 1976).

The only breeding record is for a terrestrial female collected from the base of a small waterfall in Idaho on 1 September that laid 185 eggs in an aquarium on 1 October (Nussbaum 1969b). The eggs lack pigmentation and are attached singly to the substrate by short pedicels. The ova are 6.0–6.3 mm in diameter.

The diet of Idaho *Dicamptodon* includes snails, nematomorphs, spiders, stoneflies, caddisflies, beetles, mayflies, true flies, lepidopteran larvae, orthopterans, and tailed-frog tadpoles (Metter 1963). Individuals often consume bark, twigs, small rocks, and other debris while feeding. In general, the taxonomic diversity of prey consumed increases with body size.

Dicamptodon copei Nussbaum
Cope's Giant Salamander
PLATES 32, 33

Fig. 87. *Dicamptodon copei*; gilled adult; Mason Co., Washington (R. W. Van Devender).

IDENTIFICATION. *Dicamptodon copei* is a sibling species of *D. tenebrosus* that very rarely metamorphoses even when treated with thyroxin. The few terrestrial adults that have been collected usually have dorsal mottling. Larvae are the stream type with tail fins that extend forward to the insertion of the hindlimbs. The gills are short, bushy, and dull red. The dorsum is brown with little mottling and often has yellowish tan patches. Light granular glands occur on the dorsum and are most conspicuous under the yellow patches and at the base of the tail fin. The tail has little mottling, lacks the black tip as in *D. tenebrosus*, and is similar in color to the body. The venter of young specimens is white and that of older specimens bluish gray. Descriptions of sexual dimorphism in external traits are not available.

This species has a slimmer head and body than other *Dicamptodon* species. The maximum head width is less than or equal to one-fifth of the SVL, and there are 0–2 costal folds between the adpressed limbs. The number of maxillary and premaxillary teeth varies from 33 to 44 and the number of vomerine teeth from 28 to 42. Palatopterygoid teeth are lacking. Size at

Dicamptodontidae

Range of *Dicamptodon copei*

sexual maturity is smaller in *D. copei* than in *D. tenebrosus*; the smallest gilled adult *D. tenebrosus* is normally larger than the largest gilled adult *D. copei*.

SYSTEMATICS AND GEOGRAPHIC VARIATION. *Dicamptodon copei* was not recognized for many years because of its close resemblance to *D. tenebrosus*. Nussbaum (1970, 1976, 1983) lists characters that distinguish *D. copei* from *D. tenebrosus*, and Daugherty et al. (1983) provide electrophoretic evidence that *D. copei* is reproductively isolated from *D. tenebrosus*. Body size, tooth counts, gill raker counts, and other traits

Fig. 88. *Dicamptodon copei*; gilled adult; Mason Co., Washington (R. W. Van Devender).

vary slightly among local and regional populations of *D. copei* (Nussbaum et al. 1983). In addition, specimens from the Columbia River gorge tend to be darker than those from elsewhere. No subspecies are recognized.

DISTRIBUTION AND ADULT HABITAT. *Dicamptodon copei* inhabits cool mountain streams with temperatures of 8-14°C in humid coniferous forests in the Coast Range of western Washington and extreme northwestern Oregon (Nussbaum 1983). Individuals live under stones or flood debris in streams. Specimens have been found as high as 1372 m in elevation and have occasionally been taken from clear, cold mountain lakes and ponds (Nussbaum et al. 1983).

BREEDING AND COURTSHIP. The mating season is poorly documented and descriptions of courtship behavior are not available.

REPRODUCTIVE STRATEGY. Eggs are laid throughout the year except during the winter (Nussbaum et al. 1983). Females deposit their eggs singly on the undersides of rocks or other structures in underground retreats and guard their clutches through hatching. Freshly laid ova are white, measure about 5.5 mm in diameter on average and are surrounded by several jelly envelopes that superficially appear as two capsules. Each egg is suspended by a short pedicel to a rock, log, or other support structure, and the eggs are clustered in close proximity to one another.

Brooding females are highly aggressive toward marauding conspecifics that feed on the eggs. Brooding females and other salamanders near nests often have bite scars on their bodies and tails. Eggs are almost always missing from the nests, as evidenced by the remains of the pedicels, and much of the mortality may be due to nest piracy by conspecifics. Animals collected near nests often contain eggs in their stomachs (Nussbaum et al. 1983).

Clutch size varies from 25 to 115, with an average of about 50 eggs per clutch (Nussbaum et al. 1983). Females probably oviposit every other year because of the exceptionally long incubation period. Hatchlings have conspicuous yolk sacs and do not begin feeding until they reach 34 mm TL. Embryos incubated at 8°C require 240 days to reach the feeding stages.

AQUATIC ECOLOGY. Larvae live beneath rocks, bark, and other cover in streams and feed primarily on aquatic organisms. Antonelli et al. (1972) compare the stomach contents of a mixture of 132 *D. tenebrosus* and *D. copei* larvae (individuals could not be identified to species) with those of benthic invertebrates collected from a Washington stream. Both species appear to be opportunistic benthic feeders and eat mayflies, stoneflies, caddisflies, flies, beetles, homopterans, hymenopterans, lepidopterans, orthopterans, clams, snails, spiders, mites, nematodes, and small trout and sculpins. In addition to eating fish eggs and small fish, individuals occasionally eat the eggs and tadpoles of tailed frogs, along with smaller larvae of conspecifics and *D. tenebrosus* (Nussbaum et al. 1983).

Cope's giant salamander is largely aquatic, but larvae sometimes emerge at night following heavy rains and crawl on wet rocks and vegetation in or near the stream (Nussbaum et al. 1983). Average size at maturity is 65–77 mm SVL and varies among populations. In extreme cases larvae may not mature until they reach 114 mm SVL.

Fig. 89. *Dicamptodon copei*; larva; Grays Harbor Co., Washington (W. P. Leonard).

TERRESTRIAL ECOLOGY. The terrestrial adults are extremely rare. Only six specimens have been reported to date (Loafman and Jones 1996; Nussbaum 1976), and no data are available on the life history of the terrestrial phase.

PREDATORS AND DEFENSE. Cope's giant salamander is preyed upon by garter snakes (*Thamnophis* spp.), northern water shrews (*Sorex palustris*), and *D. tenebrosus* (Loafman and Jones 1996; Nussbaum 1970; Nussbaum et al. 1983).

COMMUNITY ECOLOGY. *Dicamptodon copei* often coexists with *D. tenebrosus*, but competitive and predatory interactions between these species have not been examined. Substantial dietary overlap occurs between *Dicamptodon*, rainbow trout, and slender sculpins, suggesting that *Dicamptodon* may compete for food with fishes (Antonelli et al. 1972).

CONSERVATION BIOLOGY. Logging that results in stream siltation can deplete populations of this and other *Dicamptodon* species. Details of studies that have examined the impact of timbering on *Dicamptodon* are presented under the account of *D. tenebrosus*.

Dicamptodon ensatus (Eschscholtz)
California Giant Salamander
PLATE 34

Fig. 90. *Dicamptodon ensatus;* adult; Santa Clara Co., California (M. Garcia-Paris).

IDENTIFICATION. The California giant salamander is one of the largest salamanders in the Pacific Northwest. In most populations larvae greatly outnumber transformed individuals. In addition, gilled adults are common in many populations. The ground color of the dorsum of transformed animals is often light reddish brown and is overlain with copper-colored, coarse marbling. Marbling often extends onto the chin, throat, and undersides of the forelimbs and pectoral girdle (Nussbaum 1976). Young metamorphs have bright, golden marbling. The head is depressed in front of the eyes, and the posterior half of the tail is laterally compressed. Transformed adults have 12-13 indistinct costal grooves and measure 17-30.5 cm TL (Nussbaum 1976).

Larvae are the stream type with tail fins that extend forward only to the insertion of the hind limbs. The gills are short, bushy, and dull red. Larvae are light brown above with white to yellowish white venters. The tail tip is not as conspicuously mottled with black, white-edged blotches as in *D. tenebrosus*. A short, yellow stripe is usually present behind each eye, and the tips of the digits are black and cornified. The snout appears depressed when viewed from the side.

SYSTEMATICS AND GEOGRAPHIC VARIATION. The genus *Dicamptodon* was previously considered to consist of a single species, but detailed morphological and electrophoretic studies indicate that this group is polytypic. Good (1989) provided evidence that coastal members of *Dicamptodon* in the San Francisco Bay region are sufficiently distinct to be recognized as a species (*D. ensatus*) separate from populations to the north (*D. tenebrosus*). Details of these and other studies on the systematics of *Dicamptodon* are given under the account of *D. tenebrosus*.

No subspecies of *D. ensatus* are currently recognized. Transformed animals south of San Francisco Bay have more mottling on the chin and throat and a greater number of maxillary, premaxillary, and vomerine teeth than individuals to the north (Nussbaum 1976).

DISTRIBUTION AND ADULT HABITAT. California giant salamanders are found in and about semipermanent and permanent streams in mesic coastal forests from Sonoma and Napa counties north of San Francisco Bay to Santa Cruz Co. to the south (Anderson 1969; Good 1989; Nussbaum 1976). A geographic isolate also occurs in Monterey Co., farther south. The terrestrial adults are far less abundant than the aquatic larvae. Larvae are most abundant in permanent, small to medium-sized mountain streams. The adults are occasionally found under rocks, logs, and other surface cover, or under stones in streams during the breeding season.

BREEDING AND COURTSHIP. Information on the mating season and courtship behavior is not available.

REPRODUCTIVE STRATEGY. The few nests discovered to date have been in subterranean habitats in running water. Nesting records include accounts of about 70 eggs discovered on 19 June on the underside of a large timber submerged in a creek (Henry and Twitty 1940) and about 100 eggs and the remains of two adults that were washed out of a drilling excavation about 6 m into a spring-fed embankment on 20 March (Dethlefsen 1948).

Eggs in early developmental stages are pure white, measure about 5.5 mm in diameter (including the outer membrane), and are attached singly to rocks or logs by pedicels about 6 mm long. The outer capsules of older eggs may measure around 8.3 mm in average diameter (Henry and Twitty 1940). The embryos are

Range of *Dicamptodon ensatus*

surrounded by six envelopes, but these appear as two when viewed with the naked eye (Dethlefsen 1948). This species apparently has one of the longest incubation periods of all salamanders. Embryos in the early tail bud stage require nearly 5 months to develop to the point at which yolk supplies are nearly exhausted (Henry and Twitty 1940).

AQUATIC ECOLOGY. Data on the diet of larvae are not available. The larvae presumably have diets similar to that of *D. tenebrosus* and shift to larger prey such as large aquatic insects as they age. Kessel and Kessel (1943a,b, 1944) studied growth and length of the larval period of *Dicamptodon* in a small stream. Early in the year when water flow is rapid, small larvae occur in slowly moving water near the banks. As flow drops seasonally, first-year larvae move toward deeper pools and into the main stream channel where older larvae live. Larvae grow about 8–12 mm TL per month during the warmer months of their first year of life. They reach 100 mm TL within 1 year after hatching, and most transform the following June–August when they reach 130–140 mm TL.

TERRESTRIAL ECOLOGY. Very little information is available on the terrestrial ecology of *D. ensatus*.

Juveniles and adults forage on the forest floor on rainy nights and have been found as far as 12 m from nearby streams. J. D. Anderson (1960) collected a 211-mm TL adult that regurgitated a smaller *D. ensatus* that measured 130 mm TL.

PREDATORS AND DEFENSE. Conspecifics are the only known predators, although shrews, birds, and other vertebrates undoubtedly prey on the juveniles and adults. Antipredator posturing has not been studied, but it is presumably similar to that of *D. tenebrosus*. Individuals often emit a noise resembling a bark when molested, and *D. ensatus* is one of the few salamanders that appear to truly vocalize (Stebbins 1951).

COMMUNITY ECOLOGY. No information is available on the community ecology of *D. ensatus*. The larvae often reach high densities in streams and probably play important roles in organizing invertebrate communities.

CONSERVATION BIOLOGY. The California giant salamander has a small range and is at greater risk from stream siltation and urban development than other *Dicamptodon* species. Detailed studies of its life history and habitat requirements are needed.

Dicamptodon tenebrosus (Baird and Girard)
Pacific Giant Salamander
PLATES 35, 36, 37

Fig. 91. *Dicamptodon tenebrosus;* adult; Multnomah Co., Oregon (W. P. Leonard).

Fig. 92. *Dicamptodon tenebrosus;* adult; Douglas Co., Oregon (R. W. Van Devender).

IDENTIFICATION. The Pacific giant salamander is the largest salamander in the Pacific Northwest and one of the largest terrestrial salamanders in the world. In most populations larvae greatly outnumber transformed individuals. In addition, gilled adults are common in many populations. The ground color of the dorsum of transformed animals is dark brown to nearly black and is overlain with light brown spotting or marbling. Young metamorphs have bright golden marbling that becomes diffuse and less conspicuous with age. Very old animals may become patternless except on the head. The degree of marbling may vary from very fine to coarse. The venter is normally white to light gray but is dark in some specimens. The head is depressed in front of the eyes, and the posterior half of the tail is laterally compressed. Transformed adults have 12-13 indistinct costal grooves and vary in length from 17 to 34 cm TL; the largest gilled adult on record measured 35.1 cm TL (Nussbaum 1976).

Hatchlings measure about 18-19 mm SVL and 33-36 mm TL on average (Nussbaum and Clothier 1973). Larvae are the stream type with tail fins that extend forward to the insertion of the hindlimbs. The gills are short, bushy, and dull red. Larvae measuring <55 mm SVL are light to dark brown above with immaculately white venters. The tail tip is black and the upper portion of the caudal fin is heavily mottled with black, white-edged blotches. A short, yellow stripe is usually present behind each eye, and the tips of the digits are black and cornified (Nussbaum 1976). The snout appears depressed when viewed from the side. Older larvae and gilled adults in many populations are mottled and have dark ventral surfaces and coppery gold color on the dorsum, especially the head. The dark blotches on the tail and the black tail tip become less conspicuous with age. Albino and partial albino larvae are reported by Jones and Bury (1986) and Nussbaum (1976). Detailed descriptions of sexual dimorphism in the transformed adults are not available.

SYSTEMATICS AND GEOGRAPHIC VARIATION. The genus *Dicamptodon* was previously considered to consist of a single species, but detailed morphological and electrophoretic studies indicate that this group is polytypic. Nussbaum (1976) recognized four major groups based primarily on morphometric and life history data and recognized two species, *D. copei* and *D. ensatus*. The former species coexists locally with *D.* "*ensatus*" (=*tenebrosus*) without interbreeding.

Daugherty et al. (1983) examined genetic variation in specimens referable to *D. ensatus* and found strong genetic divergence (seven fixed allele differences; Nei's $D = 0.52$) between Idaho and coastal populations. Good (1989) verified that Idaho populations are strongly differentiated from coastal populations and found that coastal populations that were formerly recognized as *D. ensatus* consist of two well-differentiated genetic groups (Nei's $D = 0.50$). These groups inter-

Fig. 93. *Dicamptodon tenebrosus*; profile of head of adult; Lane Co., Oregon (G. Hokit).

Fig. 94. *Dicamptodon tenebrosus*; profile of head of larva; Lane Co., Oregon (G. Hokit).

breed over a 4.7-km contact zone near Anchor Bay, Mendocino Co., in north coastal California. Gene flow between local populations in this region is very low because most populations inhabit streams that flow directly into the Pacific Ocean and do not interconnect.

Good (1989) recognized these forms as separate species (*D. ensatus* to the south; *D. tenebrosus* to the north) because the zone of hybridization is relatively narrow, the two groups are strongly differentiated genetically, there is a deficiency of heterozygotes and hybrid genotypes in the hybrid populations, and there is little evidence of introgression outside the hybrid zone. Reproductive closure between these forms appears to be nearly complete, and the two appear to be at a level between semispecies and full species. Here, I treat *D. tenebrosus* as a full species.

No subspecies of *D. tenebrosus* are currently recognized. Nussbaum (1976) presents a comprehensive summary of geographic variation in *D. tenebrosus* and notes that both adult and larval populations are highly variable in coloration. Older larvae in Trinity and Siskiyou counties, California, are unusual in being boldly marked with white or yellow spots, blotches, streaks, and bars. In certain populations, such as those near Mt. Rainier, the venter is more lightly colored in older larvae than in populations elsewhere. In general, geographic variation in morphological traits does not correlate strongly with patterns of genetic variation (Good 1989).

DISTRIBUTION AND ADULT HABITAT. Pacific giant salamanders are found in and about semipermanent and permanent streams in mesic coastal forests from southwestern British Columbia to lower Sonoma Co. in north coastal California (Good 1989; Maughan et al. 1976; Nussbaum 1976). They are most frequently encountered at elevations <960 m but have been observed as high as 2160 m (Anderson 1969; Nussbaum 1976). This is one of the most abundant aquatic salamanders in the Pacific Northwest, although the terrestrial adults are far less abundant than the aquatic larvae. Larvae sometimes inhabit large creeks, rivers, clear mountain lakes, and ponds, but they are most abundant in permanent, small to medium-sized mountain streams (Nussbaum and Clothier 1973). Transformed individuals are found in moist coniferous forests under rocks, logs, and bark. Adults can be collected under stones in streams during the breeding season (Nussbaum et al. 1983; Stebbins 1951). Adults are typically found within 50 m or so of streams but have been found as far as 400 m from streams in Oregon (McComb et al. 1993b).

BREEDING AND COURTSHIP. Courtship occurs during the spring and fall (Nussbaum et al. 1983). Field observations suggest that the adults court in nest chambers in crevices or beneath logs and stones; however, courtship behavior has yet to be observed. Males produce up to 16 spermatophores that average 7.6 mm in height and 5.7 mm in maximum basal diameter.

REPRODUCTIVE STRATEGY. Most females appear to oviposit during early to mid-May then guard their eggs through hatching (Nussbaum et al. 1983). The few nests discovered to date have been in subterranean habitats in running water. Nussbaum (1969b) found two nests in Oregon on 17 and 31 May. The first contained 146 eggs at the blastula stage that were submerged in running water and guarded by a female. The second nest was in running water at the base of

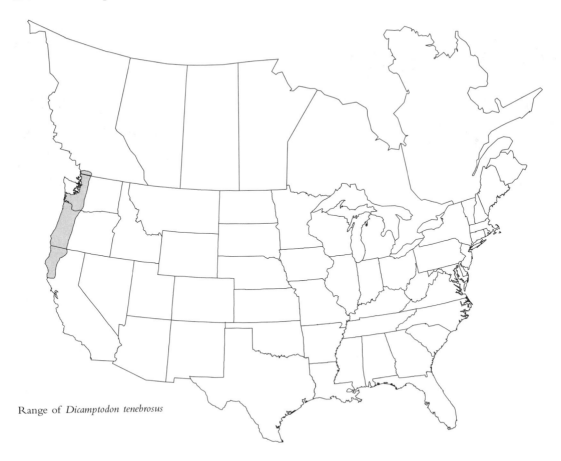

Range of *Dicamptodon tenebrosus*

a natural dam of rocks and logs and contained 83 eggs at the neural plate stage. The eggs were attached to a stick and two small rocks about 1 m within the base of the dam and were guarded by a female.

Eggs in early developmental stages measure about 5.5–7 mm in diameter, are pure white, and lack pigmentation. They are surrounded by six envelopes, but these appear as two when viewed with the naked eye (Nussbaum 1969b). The eggs are attached singly by short pedicels to substrates such as rocks or logs. This species has one of the longest incubation periods of all salamanders. Hatchlings do not appear in streams until December or January in coastal regions, about 6–7 months after oviposition (Nussbaum 1969b). Hatchlings measure 18–19 mm SVL and 33–36 mm TL on average and can survive on their yolk reserves until they reach 45–51 mm TL and 24–28 mm SVL (Nussbaum and Clothier 1973).

AQUATIC ECOLOGY. Larvae feed primarily on stream invertebrates, although large larvae are capable of handling vertebrate prey. Schonberger (1944) found amphipods, caddisflies, and a sowbug in the stomachs of five larvae from California. Larvae in a California lake consume caddisflies, beetles, odonates, mayflies, amphipods, isopods, chironomids, hemipterans, spiders, and *Ambystoma gracile* larvae (Johnson and Schreck 1969). Larvae readily prey upon conspecifics and *A. gracile* in laboratory aquaria, but they will not eat *Taricha granulosa* larvae.

Antonelli et al. (1972) compared the stomach contents of 132 larval *Dicamptodon* (*tenebrosus* and *copei*) with benthic invertebrates collected from a Washington stream and concluded that larvae are opportunistic benthic feeders. Prey items in the diet include mayflies, stoneflies, caddisflies, flies, and beetles, as well as several miscellaneous items including sculpins and rainbow trout. Substantial dietary overlap occurs between *Dicamptodon*, rainbow trout, and slender sculpins, which suggests that *Dicamptodon* may compete for food with these fish.

Parker (1994) presents a detailed account of the

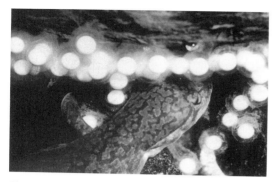

Fig. 95. *Dicamptodon tenebrosus;* female with eggs; Benton Co., Oregon (E. D. Brodie, Jr.).

Fig. 96. *Dicamptodon tenebrosus;* larvae; Humboldt Co., California (R. W. Van Devender).

feeding ecology of a larval population in a northern California stream. Larvae emerge from cover during the night, but there is no consistent diel pattern of feeding activity based on gut contents. Aquatic insects such as mayflies, caddisflies, stoneflies, and dipterans are the most important prey. Other prey include a variety of terrestrial invertebrates, ostracods, mites, nematomorphs, turbellarians, and rainbow trout. Four instances of cannibalism were documented.

As larvae grow, they include larger prey in the diet but continue to consume large numbers of small prey. Larvae in different size classes do not strongly partition food resources. Mayflies and other large, mobile prey tend to be overrepresented in the diet, whereas cryptic prey and caddisflies are underrepresented. In addition, larvae prefer large mayflies over smaller ones. The tendency of larvae to select larger prey is due in part to the fact that large larvae are more likely to attack large prey and that encounter rates and reactive distances increase with prey size (Parker 1993).

Dicamptodon larvae are often the most abundant vertebrate predators in headwater streams in the Pacific Northwest. Larvae reach their highest densities where stones cover streambeds that have little silt. They may make up >90% of total predator biomass in streams in western Oregon and northern California (Murphy and Hall 1981). Total *Dicamptodon* biomass ranges from 0.2 to 51.5 g/m^2 and averages 14.3 g/m^2 in old-growth sites and 19.5 g/m^2 in clear-cut streams. Hawkins et al. (1983) estimate a mean density and biomass of 0.73 individuals/m^2 and 6.9 g/m^2 in seven streams; larvae are absent from 13 other streams that contain fish. Corn and Bury (1989) report an average density of 2.3 larvae per square meter and an average biomass of 8.2 g/m^2 in Oregon streams. Larval density is positively correlated with stone density in a California stream and varies from 0.6 to 5.1 larvae per square meter of habitat (Parker 1991).

Length of the larval stage and size at maturity vary widely depending on environmental conditions, particularly on whether streams are permanent. Nussbaum and Clothier (1973) found that most larvae in small permanent or semipermanent streams transform 18–24 months after hatching in June and July. A few second-year larvae overwinter and transform the following summer. In dry years, larvae inhabiting semipermanent streams often suffer high mortality from either cannibalism or desiccation as they concentrate and are trapped in drying pools in streambeds.

In permanent streams of intermediate size, stream flow is usually adequate to provide cover and food year round, and larvae often do not transform until 2–3 years old. In cases in which the growing season is short and water temperatures are cool, larvae may not transform until >3 years old. Second-year larvae may transform as late as November, and gilled adults are occasionally found (Nussbaum and Clothier 1973).

In large streams, many animals remain as gilled adults, and distinct size classes may not be evident. Ponds, lakes, and large permanent streams often have the highest percentages of gilled adults. Although they are presumed to be aquatic, one gilled adult was collected in a pitfall trap located 3 m from a small stream in northern California (Welsh 1986).

Larvae in a California stream consist of 1- and 2-year-olds (Parker 1991). First-year larvae in an Oregon population grow 1–4 mm TL per month (Nussbaum and Clothier 1973). At two other sites, average growth rates for the year are about 3–4 mm TL per month.

In some populations all larvae metamorphose when sexually immature. Metamorphosis can occur when animals reach 53-93 mm SVL and 92-166 mm TL, but most transform when they reach 110-150 mm TL (Nussbaum and Clothier 1973). Transforming animals have been collected in June and July in Oregon streams (Graf et al. 1939; Nussbaum and Clothier 1973). Other populations contain mixtures of transformed and gilled adults. Larvae usually do not mature into gilled adults until they measure 115 mm SVL, but some individuals in scattered populations throughout the range mature when they measure 85-107 mm SVL (Nussbaum 1976).

Gilled adults may grow as large as 205 mm SVL and 351 mm TL (Nussbaum and Clothier 1973), but most are considerably smaller. Transformed individuals can grow almost as large (Nussbaum 1976) and mature sexually when they reach >115 mm SVL. At a site in northwestern Oregon an estimated 43% of larvae survive their first year of life (Nussbaum and Clothier 1973). Cannibalism and predation appear to be the primary causes of death.

TERRESTRIAL ECOLOGY. Terrestrial adults emerge from logs and other surface cover on warm, rainy nights and forage on the forest floor. Individuals may forage as high as 1.5-2 m above the ground on tree trunks (Nussbaum et al. 1983; Stebbins 1951). Large individuals can handle large prey such as small mammals. Dietary records include a small microtine rodent that was probably *Microtus longicaudus* (Wilson 1970), a shrew in the stomach of the holotype (Nussbaum and Clothier 1973), and nine mammals, a fence lizard, two slender salamanders, and several groups of terrestrial invertebrates in the guts of 12 terrestrial adults (Bury 1972).

Graf (1949) observed a Pacific giant salamander that was 18 cm in TL biting the head of a garter snake that was 62.5 mm long. Diller (1907) made a similar observation of a 23-cm-long adult seizing the head of a 600-cm-long garter snake. In both cases, it is uncertain whether the salamander was attempting to forage or defend itself from attack. The latter explanation seems more likely given the sizes of the snakes and the fact that garter snakes prey upon *Dicamptodon* (Fitch 1936).

PREDATORS AND DEFENSE. Natural predators include fish, common garter snakes (*Thamnophis sirtalis*), northern water shrews (*Sorex palustris*), river otters (*Lutra canadensis*), weasels (*Mustela*), and conspecifics (Nussbaum and Maser 1969; Nussbaum et al. 1983). Large larvae and terrestrial adults often have eggs of conspecifics in their stomachs, and the larvae are sometimes taken on hook and line (Graf 1949).

Terrestrial forms have a host of antipredator defenses. When approached by a predator, an individual usually postures with its body strongly arched and raised off the substrate. The tail, which produces noxious secretions, may be lashed at an attacking predator. Adults will also head-butt and can bite viciously (Brodie 1977). When molested, terrestrial forms from the redwood forest of California often produce a rattling or growling sound while snapping their jaws and lashing their tails (Bogert 1960; Maslin 1950; Nussbaum 1983).

COMMUNITY ECOLOGY. Feminella and Hawkins (1994) found that tailed-frog tadpoles (*Ascaphus truei*) will move beneath cover when exposed to chemical cues from predatory *Dicamptodon* larvae. Thus, the presence of *Dicamptodon* can indirectly alter community structure. The high frequency of occurrence of small mammals in the diet of large adults suggests that these salamanders may play a role in structuring shrew and mice communities.

CONSERVATION BIOLOGY. Several studies indicate that local populations of *D. tenebrosus* often decline after old-growth forests are logged. This species is far more abundant in unsilted compared with heavily silted streams (Hawkins et al. 1983). The short-term response of *D. tenebrosus* to clear-cutting varies depending on stream gradient (Murphy and Hall 1981; Murphy et al. 1981). In high-gradient streams, populations may increase for the first 5-17 years after logging because removal of the forest canopy increases primary productivity until the canopy reforms. In low-gradient streams, populations normally decline because silt deposition eliminates microhabitats required by this species. After canopy closure, populations appear to decline below levels typically seen in old-growth forests. Bury and Corn (1988b) and Corn and Bury (1989) found similar patterns in cut and uncut forests in Oregon and conclude that stream siltation following timbering markedly reduces densities of local populations of this species. Pacific giant salamanders in this region are nine times more abundant in old-growth forests compared with previously logged sites with 14- to 40-year-old regenerating forest.

Family Plethodontidae
Lungless Salamanders

The Plethodontidae is the largest family of salamanders and contains 27 genera and about 240 species. Two subfamilies are recognized based on skeletal features and head musculature: the Desmognathinae, which contains *Desmognathus* and *Phaeognathus,* and the Plethodontinae, which contains all remaining genera (Larson 1984; Wake 1966). The desmognathines are also characterized by their relatively small genome sizes (Larson 1984). Three tribes are recognized within the Plethodontinae: (1) the Hemidactyliini with aquatic larvae (*Eurycea, Gyrinophilus, Haideotriton, Hemidactylium, Pseudotriton, Stereochilus,* and *Typhlotriton*); (2) the Plethodontini, which lack aquatic larvae and have large ossified second basibranchials (*Aneides, Ensatina,* and *Plethodon*); and (3) the Bolitoglossini, which lack aquatic larvae and second basibranchials (all other plethodontine genera). Most North American plethodontids have a diploid number of 28 chromosomes compared with 26 in neotropical plethodontids. The Bolitoglossini is a very large tribe that contains roughly 40% of all known salamander species (Wake 1970; Wake and Lynch 1976). This tribe has undergone adaptive radiation in the tropics and includes both fossorial and arboreal species.

Systematic relationships among many genera and species are poorly understood, although much progress has been made in understanding the evolutionary relationships of certain groups such as *Plethodon*. Major systematic treatments of plethodontids along with summaries of interspecific variation in anatomy, development, morphology, genome size, chromosome characteristics, and molecular data that relate to classification are provided in Highton (1962a, 1972, 1991, 1995), Highton and Larson (1979), Highton et al. (1989), Larson (1983a,b, 1984, 1991), Larson and Highton (1978), Larson et al. (1981), Lombard and Wake (1977), Macgregor et al. (1973), Maxson and Maxson (1979), Maxson et al. (1979, 1984), Mizuno and Macgregor (1974), Mizuno et al. (1976), Sessions and Kezer (1987), Sessions and Larson (1987), Titus and Larson (1996), Wake (1966, 1993), Wake and Brame (1969), Wake and Larson (1987), Wake and Lynch (1976, 1982), and Wake and Marks (1993).

Studies of members of the tribe Plethodontini indicate that the eastern and western *Plethodon* belong to widely divergent lineages that separated about 40 million years ago (Highton and Larson 1979). Evolutionary stasis is common in this genus, and molecular techniques have proven to be most reliable in understanding systematic relationships. The eastern and western *Plethodon* differ markedly in immunological distance, electrophoretic protein variants, genome size (about twice as large in western *Plethodon*), and several morphological traits (Highton and Larson 1979; Maxson and Maxson 1979; Maxson et al. 1979; Mizuno and Macgregor 1974; Sessions and Larson 1987; Wake 1966). Evidence suggests that *Aneides* may have been derived from western *Plethodon* subsequent to the separation of eastern and western *Plethodon* (Highton 1991; Highton and Larson 1979). If this hypothesis is proven, a reclassification of these three groups may be required.

Both eastern and western *Plethodon* contain several well-defined clusters (Highton 1991; Highton and Larson 1979; Maxson et al. 1979; Mizuno and Macgregor 1974). The eastern *Plethodon* consists of the *P. glutinosus* group (*aureolus, caddoensis, glutinosus, jordani, kentucki, oconaluftee, ouachitae, petraeus, yonahlossee*), the *P. wehrlei* group (*punctatus, wehrlei*), the *P. welleri* group (*dorsalis, websteri, welleri*), and the *P. cinereus* group (*cinereus, hoffmani, hubrichti, nettingi, richmondi, serratus, shenandoah*). The western *Plethodon* consists of four additional groups: the *P. elongatus* group (*elongatus, stormi*), the *P. vehiculum* group (*dunni, vehiculum*), the *P. vandykei* group (*idahoensis, vandykei*), and the *P. neomexicanus* group (*larselli, neomexicanus*).

Despite the diversity of life forms within this family, all plethodontids lack lungs and possess nasolabial grooves, which are slitlike channels that extend from the margin of the upper lip to the lateral corner of each nostril. The nasolabial grooves function to transport waterborne chemicals from the substrate to the vomeronasal organ and are important in facilitating chemically mediated behaviors (Dawley 1992a,b; Dawley and Bass 1989). Sexually active males of most species have short protuberances (cirri) or nasal swellings associated with the nasolabial grooves, mental glands that are used in courtship, and papillose cloacal lips. The number of vomerine teeth in most species increases with age. Fully terrestrial species typically have rounded tails, whereas stream-breeding species tend to have triangular, keeled tails. Highly fossorial forms often show tendencies to have reduced limbs, whereas highly arboreal forms tend to have prehensile tails and either square-toed or webbed feet.

Plethodontids are widely distributed in eastern and western North America, Mexico, and Central America. Two genera occur in South America and two species of *Hydromantes* also occur in southern Europe and on Sardinia. Fossils of six genera are known from the lower Miocene to the Pleistocene of North America (Duellman and Trueb 1986).

SUBFAMILY DESMOGNATHINAE

Desmognathus aeneus Brown and Bishop
Seepage Salamander
PLATE 38

Fig. 97. *Desmognathus aeneus;* adult; southwestern North Carolina (J. Bernardo).

IDENTIFICATION. This slender, tiny salamander has a yellow to reddish brown dorsal stripe that is bordered laterally by a dark band. The dorsal stripe can be either straight or wavy. Specimens often have a light patch or circular mark on the dorsal surface of each thigh and a Y-shaped mark immediately behind the eyes that may continue as a faint middorsal stripe or line of dots down the back. A faint herringbone pattern is often evident on the back. The venter is mottled and light colored, and the top of the head is smooth rather than rugose as in *D. wrighti*. The tail is round in cross section and makes up less than half of the total length. The toes lack cornifications. Adults measure 38–57 mm TL and have 13–14 costal grooves.

Hatchlings measure about 6–7 mm SVL and have proportionately shorter tails and larger heads than the adults. Juveniles resemble the adults but are more brightly colored. Adult males have a small, bean-shaped mental gland on the chin.

SYSTEMATICS AND GEOGRAPHIC VARIATION. Genetic variation in this species has not been analyzed in detail. Studies of mitochondrial DNA suggest that the lineages leading to *D. aeneus* and *D. wrighti* originated relatively early in the evolution of desmognathine salamanders (Titus and Larson 1996). Geographic variation in color patterning is chaotic, and no subspecies are recognized (Harrison 1992).

DISTRIBUTION AND ADULT HABITAT. Seepage salamanders are spotily distributed in deciduous forests from extreme southeastern Tennessee and southwestern North Carolina to central Alabama. A major disjunct occurs in western populations in the Fall Line Hills of Alabama that is separated from eastern populations in the Blue Ridge and Piedmont. Two smaller disjuncts are present in the Piedmont of northeastern Georgia (Harrison 1992) and in Oconee Co., South Carolina. Populations have not been found to date north of the Little Tennessee River and are restricted to elevations of 30–1340 m.

Seepage salamanders are most frequently encountered in and around seepages or in terrestrial habitats adjoining small streams (Bishop and Valentine 1950; Hairston 1986; Harrison 1967; Jones 1982; Mount 1975). This species frequents moist or wet leaf litter but is occasionally found beneath logs, moss mats, and other surface objects. Individuals apparently spend much of their time in thick leaf litter where they forage for tiny invertebrates.

BREEDING AND COURTSHIP. Published information on the mating season of *D. aeneus* is not available. Like many *Desmognathus,* this species probably mates in both autumn and spring. Females oviposit annually, in contrast to the biennial pattern of some *Desmognathus* (Harrison 1967).

The seepage salamander has a biting phase during courtship that substitutes for the pulling, snapping, and butterfly motions that are used by most *Desmognathus* species. The following account of courtship is from Promislow (1987), who observed numerous courtships in laboratory animals.

Initially a male approaches a female and touches his snout to her body or rubs his chin on her back. He then bites the female and maintains a hold that can last for several hours. The male most frequently bites the

Range of *Desmognathus aeneus*

female near the midbody or proximal part of the tail. During this phase, the male pulls from side to side or away from the female at a rate of about 1-2 tugs per second, and the female attempts to escape the grasp of the male. Biting creates a wound that allows mental gland secretions to enter the circulatory system of the female. Struggling by the female diminishes during the first 30 minutes, at which point she becomes quiescent. Thereafter, the female may bite the male or move her head over the dorsal part of his tail in a rhythmic motion at a rate of one pass every 5-6 seconds.

During this phase of courtship the animals form a full circle. The male eventually releases the female and undulates his tail. The female rests her head on his tail and follows him in a tail-straddle walk that may last 0.5-2.0 hours. In many instances males resort to biting and repeat the sequence just described before resuming one or more tail-straddle walks. Near the termination of the tail-straddle walk, the male stops moving and begins undulating the tail from side to side while the female keeps her head on the tail. The male then deposits a spermatophore and the female head-swings vigorously. The male then moves forward about one body length with the distal part of the tail flexed laterally at approximately 45°. The female continues forward with her chin positioned on the distal part of the tail until her cloaca lines up with the spermatophore. The female then picks up the sperm mass with her cloaca while undulating the proximal part of her tail.

During sperm transfer, the male undulates his tail beneath the chin of the female and then becomes quiescent. Sperm transfer is usually successful, and the male deposits only a single spermatophore per night. If the female is unsuccessful in picking up the sperm mass, a second spermatophore may be deposited.

REPRODUCTIVE STRATEGY. Females usually nest in clumps of *Thuidium, Mnium,* or other mosses near seepages or streams, but they may use other cover, such as rotting logs, for nesting (Harrison 1967; Jones 1982; Mount 1975). Nesting females sometimes aggregate in seepages, and as many as 12 nests may occur within a 0.2-0.3 m^2 area (Harrison 1967).

Females lay eggs in grapelike clumps and remain beneath or curled about the eggs until the embryos hatch (Bishop and Valentine 1950; Brown and Bishop 1948). The eggs are creamy white and are surrounded by two envelopes. Each egg has a short pedicel, and the pedicels of adjoining eggs are wrapped about each other to form a common support that binds the cluster together. Average diameters of freshly laid eggs vary among local populations from 1.9 to 2.7 mm, and the entire egg capsule varies from 2.4 to 3.0 mm in diameter (Beachy 1993b; Brown and Bishop 1948; Harrison 1967).

Geographic variation in average clutch size is slight; clutches range from 11 to 14 eggs. The number of mature ova in 34 females from the southern Appalachians was 6–17, averaged 12, and was positively correlated with SVL (Harrison 1967). The mean size of 27 clutches found in the field varied from 6 to 18 and averaged 11. The means (and ranges if reported) for other populations are 11 (5–12) for Alabama (Bishop and Valentine 1950), 12 (Beachy 1993b) and 14 (3–19) for North Carolina (Collazo and Marks 1994), and 12 (8–15) for southeastern Tennessee (Jones 1982).

Females oviposit in late April and early May in Georgia, North Carolina, and Tennessee (Harrison 1967; Jones 1982), and as early as February in western Alabama (Valentine 1963c). Brooding females have been found near Tuscaloosa, Alabama, in late March (Bishop and Valentine 1950) and during the spring and from July to October in eastern Alabama (Mount 1975).

Hatching occurs from late May to early August in Georgia and North Carolina populations (Harrison 1967) and as early as late June in Tennessee (Jones 1982). Based on observations of brooding females, hatching in Alabama populations probably occurs during both the spring and autumn months. The incubation period lasts 6–9 weeks (Harrison 1967; Jones 1982; Valentine 1963c).

AQUATIC ECOLOGY. Seepage salamanders sometimes hatch with gills, but they are lost a few days or weeks after hatching. Thus, there is no true larval feeding stage as in most other *Desmognathus* species (Harrison 1967). Larvae from Alabama measure 10.2–12 mm TL and transform when they reach 11.5–14 mm TL (Valentine 1963c).

TERRESTRIAL ECOLOGY. Very little information is available on the ecology of the juveniles and adults. They appear to be more secretive than those of other *Desmognathus* species and are less likely to emerge at night to forage on the ground surface for invertebrates or to climb on vegetation (Hairston 1987). Because of its small size, this species may be more vulnerable to predators than larger *Desmognathus* species. Hatchlings become sexually mature in 2 years after reaching 18–19 mm SVL (Harrison 1967). Many females, however, may not breed until nearly 3 years old.

Donovan and Folkerts (1972) found a wide variety of dietary items in 133 specimens collected in Alabama and Georgia. Food items include isopods, amphipods, mites, spiders, myriopods, earthworms, terrestrial snails, nematodes, shed salamander skins, numerous insects (particularly beetles, flies, and springtails), and a conspecific. Mites and collembolans are the dietary staple of seepage salamanders in a Tennessee population. Other prey include nematodes, snails, isopods, various insects, millipedes, centipedes, spiders, and pseudoscorpions (Jones 1982). Individuals appear to forage mostly beneath litter and are rarely observed above ground at night.

PREDATORS AND DEFENSE. No data are available on natural predators. Individuals often remain immobile when uncovered, a behavior that may make them less conspicuous to predators (Dodd 1990b).

COMMUNITY ECOLOGY. In the southern Appalachians, seepage salamanders live in diverse salamander communities in which all other members are larger on average. Because smaller animals are often at a competitive disadvantage and are more vulnerable to predators, the ability of this species to thrive under such conditions is surprising.

Desmognathus aeneus is similar in size to *D. wrighti*, suggesting that the two species may compete for limited resources. These species are mostly allopatric, but they occur microsyntopically in Macon Co., North Carolina (Bruce 1991; Rubin 1971; Tilley and Harrison 1969).

CONSERVATION BIOLOGY. Southern populations are strongly associated with seepages in forested habitats and in all likelihood are vulnerable to intensive management practices, such as clear-cutting, that eliminate leaf litter and shading. Whenever possible, forest buffers should be left around seepages and headwater streams in areas scheduled for timbering, since these are the major breeding sites of this and many other salamander species.

Desmognathus apalachicolae Means and Karlin
Apalachicola Dusky Salamander
PLATE 39

Fig. 98. *Desmognathus apalachicolae;* adult; Liberty Co., Florida (R. W. Van Devender).

IDENTIFICATION. The Apalachicola dusky salamander is closely allied to other members of the *D. ochrophaeus* complex and resembles these species in general morphology. The base of the tail is rounded in cross section, and the tail narrows to a fine, laterally compressed point (Means 1993). *Desmognathus apalachicolae* is larger on average than *D. ocoee* from northern Georgia and is more robust than regional populations of *D. fuscus conanti* (Means and Karlin 1989). Individuals normally have 10–14 rounded, coalesced blotches on the dorsum, and melanistic adults tend to be brownish rather than black. The belly superficially appears to be nearly immaculate, but actually has finely stippled melanophores. The toe tips lack cornifications. Adults measure 8.5–10 cm TL, and there are 14 costal grooves. Adult males have a sinuate jaw commissure as in other members of the *D. ochrophaeus* complex.

Larvae have 10–14 large round or oval blotches on the dorsum between the front and rear limbs. The blotches are usually tan or reddish and either alternate or occur as 5–7 opposite pairs along the back and tail. The gills are white; the head is rounded with large, dark eyes; and the rear limbs are conspicuously larger than the front limbs. Juveniles and adult females are boldly patterned with 10–14 dorsal blotches between the limbs that are fringed laterally by black or brown pigment. These often form a scalloped or zigzag pattern. The pattern in adult males is similar but is often obscured by dark dorsal pigmentation (Means 1993).

SYSTEMATICS AND GEOGRAPHIC VARIATION. Karlin and Guttman (1986) compared protein variation in *D. fuscus*-like specimens from Florida and elsewhere and found that Florida specimens consist of two distinct species, *D. apalachicolae* and *D. fuscus,* that have fixed or nearly fixed differences in six loci. *Desmognathus apalachicolae* is actually more similar genetically to some southern Appalachian populations of the *D. ochrophaeus* complex ($D = 0.20$) than to *D. fuscus* ($D = 0.35$). Verrell (1990c) attempted to mate *D. apalachicolae* with congeners and found that this species is sexually incompatible with *D. fuscus, D. ocoee,* and *D. orestes* from the southern Appalachians. However, *D. apalachicolae* did readily court *D. carolinensis* from Mt. Mitchell, North Carolina.

The only conspicuous geographic variation in *D. apalachicolae* is that members of local populations from the Tallahassee Red Hills and from limestone ravines often are larger than those from steephead populations along the Apalachicola or Ochlockonee rivers.

DISTRIBUTION AND HABITATS. *Desmognathus apalachicolae* has a very restricted distribution and occurs as two geographic isolates that are found in the Choctawhatchee, Chattahoochee, and Apalachicola river basins in Alabama, Georgia, and the panhandle of Florida. The northern group occurs from near the fall line at Columbus, Georgia, southward to the northern edge of the Dougherty Plain physiographic province; the southern group occurs from near the confluence of the Flint and Chattahoochee rivers southward along the Apalachicola River's eastern valley wall. This species occurs in mesic ravines with steep sidewalls or near permanent seepages at the heads of first-order streams that dissect uplands. Large rocks, leaf litter, and leaf packs in or near streams are important microhabitats for the juveniles and adults.

Our knowledge of the natural history of *D. apalachicolae* comes mostly from works by Means (1974) and Means and Karlin (1989). Except where noted, the life history account that follows is primarily from Means and Karlin (1989).

BREEDING AND COURTSHIP. No detailed data are available on the breeding season of *D. apalachicolae.*

Range of *Desmognathus apalachicolae*

Captive animals breed in late summer and autumn, but it is uncertain if courtship continues into the winter and spring months. Courtship is similar to that of other medium-sized *Desmognathus* species, such as *D. fuscus* and *D. orestes* (Verrell 1990c, 1994d). A male initially pursues a female and engages in bouts of head rubbing. These may last for 5 minutes or more if the female does not flee. The male may also butterfly one or both forelimbs when close to the female, or he may stroke her with a forelimb. The male may occasionally interrupt rubbing bouts, move posteriorly, and pull the head across the female's dorsum. The male may also vigorously snap the head while lying more or less parallel to the female, or when positioned in front of the female with the body curled backward in a C shape. Pulling and snapping behaviors lacerate the skin and inoculate the female with male mental gland secretions. The female will often move away from the male during this phase of courtship, with the male in pursuit.

The male eventually slides forward beneath the female's chin while lifting his body and undulating his tail. As the female straddles the male's tail, the pair may move forward with the female's chin near the base of his tail. At some point during this tail-straddle walk, the male stops and deposits a spermatophore. The female remains on his tail with her chin at the base. He then leads her forward and flexes his tail to one side. The female moves over the spermatophore and inserts the sperm-filled cap into her cloaca. She often rapidly undulates her pelvic region when being led forward over the spermatophore. This behavior may facilitate detecting the spermatophore through tactile stimulation. Males typically produce a single spermatophore per night. In Verrell's (1994d) laboratory trials the period from the initiation of head rubbing to the start of the tail-straddle walk lasted 37–117 minutes, and the tail-straddle walk lasted 4.0–7.5 minutes.

REPRODUCTIVE STRATEGY. Females lay their eggs in clusters much like those of *D. fuscus* and *D. ochrophaeus*. The eggs are normally deposited beneath stones, small logs, and other cover or along the flowing rills of seepage sites. Females remain with their eggs until hatching and presumably guard them from predators. Most females oviposit in May and June, and eggs at the hatching stages have been found in October and November (Means and Karlin 1989).

AQUATIC ECOLOGY. Embryos hatch from midsummer through the autumn, and larvae transform before the following April when about 14 mm in SVL. The maximum length of the larval period is 9–10 months (Means 1993). The larvae presumably feed on small aquatic invertebrates, but the foraging ecology of this species has yet to be studied.

TERRESTRIAL ECOLOGY. Data on the juvenile and adult stages are scant. Juveniles reach sexual maturity when they reach 33–40 mm SVL. Sexually mature males measure 40–52 mm SVL, whereas females measure 33–47 mm SVL (Means and Karlin 1989).

This species is closely tied to streams and seepages and rarely moves far from these habitats. Individuals hide beneath leaf litter, rocks, or friable streambanks during the day and emerge at night to feed. Specimens from sites with dark substrates tend to be much darker than those from sites with lighter substrates, a phenomenon that may in part reflect the ability of individuals to change color to match their backgrounds.

PREDATORS AND DEFENSE. Natural predators are not known and defensive mechanisms have not been investigated.

COMMUNITY ECOLOGY. *Desmognathus apalachicolae* is allopatric with *D. fuscus* and *D. monticola* but occurs parapatrically with *D. auriculatus*. Means (1975) provided distributional evidence suggesting that *D. auriculatus* is excluded from local habitats by *D. apalachicolae*. In northern Florida, *D. auriculatus* occupies a broad range of habitats, including steepheads, small tributaries, and lowland habitats such as swamps and mucky seeps. This species is restricted to downstream, lowland habitats whenever *D. apalachicolae* occupies steepheads.

CONSERVATION BIOLOGY. Maintaining mature hardwoods around the mesic ravines that *D. apalachicolae* inhabits is the single most important action that will assure the long-term survival of this species. *Desmognathus apalachicolae* is one of many endemic species of organisms that live in the Apalachicola River drainage, and it provides another reason why the region should receive high priority for protection.

Desmognathus auriculatus (Holbrook)
Southern Dusky Salamander
PLATES 40, 41

Fig. 99. *Desmognathus auriculatus;* adult; Carteret Co., North Carolina (R. W. Van Devender).

IDENTIFICATION. *Desmognathus auriculatus* is a medium-sized salamander that has a drab, dark brown or black dorsum with one or two rows of light spots along each side of the body. The venter varies from grayish brown to black and usually has fine white specking. The tail is laterally compressed and keeled, and the toe tips lack cornifications. Adults reach 7.5–16 cm TL and have an average of 14 costal grooves.

Hatchlings from clutches from Florida average 18 mm TL (Goin 1951). Larvae are dusky brown to black and have small, pale dorsal spots. The darkly pigmented gills are bushier than those of other, sympatric *Desmognathus* species and have 30 or more gill fimbriae (Valentine 1963d). Juveniles have six to seven pairs of dorsal spots between the limbs that may fuse to form a jagged middorsal stripe, which disappears with age.

SYSTEMATICS AND GEOGRAPHIC VARIATION. This species was previously considered to be a subspecies of *D. fuscus,* but Means (1974) and Valentine (1963d) presented evidence that *D. auriculatus* is a valid species. Genetic data also indicate that *D. auriculatus* is well differentiated (Nei's $D > 0.50$) from *D. apalachicolae, D. brimleyorum,* and *D. fuscus* (Karlin and Guttman 1986). Evidence suggests that *D. auriculatus* and *D. fuscus* hybridize in portions of their range (Mount 1975). However, the degree of gene exchange be-

Range of *Desmognathus auriculatus*

tween these species is very limited (Means 1974; Valentine 1963d).

Specimens in Louisiana and Mississippi often are lighter colored and more likely to have dorsal spots than those from the remainder of the range (Cook and Brown 1974; Means 1974; Valentine 1963d). No subspecies are recognized.

DISTRIBUTION AND ADULT HABITAT. *Desmognathus auriculatus* is largely restricted to the Coastal Plain of the eastern United States from southeastern Virginia to eastern Texas. Disjunct populations occur in northeastern Georgia and north-central Mississippi. This species is characteristically found in or around the margins of slowly moving or stagnant bodies of water with mucky, acidic soils. Specimens have been found in springs, cypress swamps, sloughs, mud-bottomed pools in floodplains, and slowly moving muddy streams (Means 1974; Neill and Rose 1949; Robertson and Tyson 1950; Rossman 1959). Populations are mostly confined to deciduous forest habitats in Alabama (Mount 1975).

BREEDING AND COURTSHIP. Published information on the mating season is not available. The following summary of courtship behavior is for South Carolina populations studied by Verrell (1997).

A male initiates courtship by orienting to the female. He may repeatedly follow the female as she moves about and may butterfly with the forelimbs, especially when close to the female. The female eventually stops moving away from the male, and the male provides tactile stimulation by nudging the female, rubbing her body and head with his chin or head, or pulling. The latter behavior entails moving the chin across the female's dorsum, using one or more short strokes as the head is bent down. The male may also butterfly during this period.

The male eventually slides his body beneath the female's chin, and the female rests her chin on the dorsal surface of the male's tail. From this position, the male undulates his tail from side to side and curves his body backward in a C shape. The couple may "waltz" about in a circular pattern as the male undulates his tail. During a waltz the male may pull against the female's dorsum or snap his chin against the female with sufficient force to throw himself several centimeters from his partner. If the female moves away from the male during this tactile period, the male pursues her.

The male next slides beneath the female's chin, and she places her chin on the undulating dorsal base of the male's tail as she straddles the tail. From this position, the pair walks forward in a straight line in a tail-straddle walk as the male continues to undulate his tail. The male eventually stops moving forward, undulates his tail more intensely, and squats and deposits a spermatophore on the substrate. The male then moves forward one body length, swings his undulating tail out to one side, and stops forward movement. He vigorously pumps his pelvic region and the proximal portion of his tail up and down by extending and flexing his hindlimbs, and the female sways her pelvic region back and forth laterally in search of the spermatophore while keeping her chin on the base of the male's tail. The female picks up the spermatophore in her cloaca, and the pair separates soon thereafter.

Fig. 100. *Desmognathus auriculatus;* larva; New Hanover Co., North Carolina (R. W. Van Devender).

REPRODUCTIVE STRATEGY. Females have been found guarding late-term embryos or hatchlings in mid-June in southern Georgia (Neill and Rose 1949), from early September to late October in North Carolina (Eaton 1953; Robertson and Tyson 1950), and on 28 October in Florida (Goin 1951). Nests with attending females have also been found from 4 September to 12 October in Alabama (Mount 1975) and on 27 August in southeastern Virginia (Wood and Clarke 1955). Females lay the eggs in grapelike clusters in cavities or small depressions, although a few eggs may occasionally be placed individually on rootlets or other structures. Nests have been found in sphagnum moss mats, in cypress logs and stumps, and beneath logs and bark within 1-2 m of a water source. Clutch size estimates 9-19 eggs in 10 clutches from Florida (Goin 1951), 14-20 eggs for six North Carolina females (Robertson and Tyson 1950), and 26 eggs in a single nest in Virginia (Wood and Clarke 1955).

Females can home to their nests when displaced as far as 6 m away (Rose 1966a). Individuals released upstream or downstream from their nests are generally more successful at homing than those placed at right angles to the stream.

AQUATIC ECOLOGY. Little information is available on the larval period. Transforming larvae collected by Valentine (1963d) measured 22-32 mm SVL, whereas the average size of metamorphs collected by Means (1974) in Florida was 21 mm SVL.

TERRESTRIAL ECOLOGY. Surprisingly little information is available on the natural history of the juveniles and adults. Dietary items in Florida specimens include beetle larvae, earthworms, tabanid flies, lycopsid spiders, and crane flies (Carr 1940). Most males in central Florida become mature when they reach >32 mm SVL, but the length of the juvenile stage is not known precisely (Rossman 1959).

PREDATORS AND DEFENSE. No data are available on predators or defense.

COMMUNITY ECOLOGY. Means (1975) provided distributional evidence suggesting that *D. auriculatus* is excluded from local habitats by *D. apalachicolae*. In northern Florida, *D. auriculatus* occupies a broad range of habitats, including steepheads, small tributaries, and lowland habitats such as swamps and mucky seeps. The southern dusky salamander is restricted to downstream, lowland habitats whenever *D. apalachicolae* occupies steepheads. Means (1975) did not find these species to be syntopic, but Valentine (1963d) reports numerous locations in Mississippi where *D. auriculatus* and *D. fuscus* co-occur.

CONSERVATION BIOLOGY. Bottomland hardwood forests offer optimal habitats for this species. The retention of mature hardwood forest buffers around small streams, swamps, and marshes will help maintain high-quality habitat for this and many other vertebrates in southeastern Coastal Plain communities.

Desmognathus brimleyorum (Stejneger)
Ouachita Dusky Salamander
PLATE 42

Fig. 101. *Desmognathus brimleyorum;* adult; Oklahoma (R. W. Van Devender).

IDENTIFICATION. This is one of the largest and least-studied *Desmognathus* species. Mature adults tend to be uniformly dark brown above with faint spotting or patterning on the dorsum. The venter varies from whitish in young animals to very light grayish brown in older specimens. The tail is laterally compressed and keeled, and a light line extends from the eye to the angle of the jaw. The Ouachita dusky salamander is most easily identified by its range, since no other *Desmognathus* coexists with this species. Adults measure 7–18 cm TL and have 14 costal grooves.

Juveniles have five to six pairs of light spots on the dorsal surface of the body that become diffuse and tend to disappear with age. The dorsal ground color is light brown and the tail is keeled. Older juveniles and young adults may have varying degrees of wormy patterning on the back, particularly near the base of the tail.

SYSTEMATICS AND GEOGRAPHIC VARIATION. This species is well differentiated genetically from *D. apalachicolae, D. auriculatus,* and *D. fuscus* (Karlin and Guttman 1986). Karlin et al. (1993) examined protein variation in populations on different mountain ranges in Arkansas and Oklahoma and found moderate genetic differentiation between eastern and western groups (Nei's $D = 0.22$). Their data suggest that valleys between mountains are acting as weak barriers to gene flow and that gene flow between populations inhabiting the same mountain range is moderate to high.

DISTRIBUTION AND ADULT HABITAT. *Desmognathus brimleyorum* is endemic to the Ouachita Mountains of Arkansas and Oklahoma at elevations of 120–790 m. Like other large *Desmognathus* species, this species is found in the immediate vicinity of running water. Adults often reside under large rocks lying in or at the edge of rocky, gravelly streams (Means 1974). Juveniles are abundant in gravel and rubble along streamsides, especially where seepages or freshets drain.

BREEDING AND COURTSHIP. No data are available on the breeding season and courtship behavior. Both sexes breed annually (Taylor et al. 1990).

Fig. 102. *Desmognathus brimleyorum;* adult; Ouachita Mountains, Arkansas (J. W. Petranka).

REPRODUCTIVE STRATEGY. Females oviposit between March and September, but most clutches are laid during midsummer with a peak in July. Clutch size estimates include an average of 24 mature ova in 21 females (Taylor et al. 1990), a mean of 22 eggs for 14 clutches observed in the field in July and August (Taylor et al. 1990), and an average clutch size of 27 mature ova (Trauth et al. 1990). Clutch size is positively correlated with female SVL, as in most *Desmognathus* species.

Trauth (1988) reports finding clutches of 20, 28, and 29 eggs with brooding females on 25 July and 15 August. The eggs can be found about 0.5 m deep in seepages near the permanent water line and are suspended both singly and in grapelike clusters from the ceilings of brooding chambers constructed in

Fig. 103. *Desmognathus brimleyorum;* larva; LeFlore Co., Oklahoma (R. W. Van Devender).

mud. Egg clutches have also been found attached to the undersides of rocks in a mine shaft (Heath et al. 1986). The largest mature ova are 4.5–5 mm in diameter (Trauth et al. 1990).

AQUATIC ECOLOGY. Small larvae (<21 mm SVL) can be found throughout the winter and spring months, suggesting that some females oviposit during this period (Trauth et al. 1990). The smallest metamorphs from two studies measured 24 mm SVL (Means 1974) and 27–31 mm SVL (Trauth et al. 1990), and the larval period lasts about 1 year.

TERRESTRIAL ECOLOGY. Transformed animals are aquatic or semiaquatic and spend most of their time in the stream proper or in streambank habitats. No information is available on diet, length of the juvenile stage, growth rates, and many other aspects of the natural history of the juveniles and adults. The minimum size of adults appears to be greater in this species than in most *Desmognathus* species. The smallest gravid females found by Means (1974) measured 70 mm SVL, whereas the smallest males with two lobes per testis measured nearly 80 mm SVL. In a second study the minimum size of females with yolked ovarian eggs was 63 mm SVL (Trauth et al. 1990).

PREDATORS AND DEFENSE. Natural predators probably include garter snakes and other semiaquatic snakes, small birds, and shrews. The only documented predator is the speckled king snake (*Lampropeltis getula holbrooki;* Trauth and McAllister 1995).

COMMUNITY ECOLOGY. *Desmognathus brimleyorum* does not share streams with other dusky salamanders and is often the most abundant salamander in rocky, perennial streams within its range. The extent to which this species plays a role in organizing stream communities has not been investigated.

Range of *Desmognathus brimleyorum*

CONSERVATION. Siltation of streams following timbering operations probably poses the greatest potential threat to *D. brimleyorum*. Hardwood forest buffers maintained around small woodland streams will help assure high-quality habitats for this and many other species of woodland salamanders.

Desmognathus carolinensis Dunn
Carolina Mountain Dusky Salamander
PLATE 43

Fig. 104. *Desmognathus carolinensis;* adult; Buncombe Co., North Carolina (J. W. Petranka).

IDENTIFICATION. The Carolina mountain dusky salamander is part of the *D. ochrophaeus* complex, which consists of *D. apalachicolae* and five sibling species previously referred to as *D. ochrophaeus* (Tilley and Mahoney 1996). Members of the complex are so similar in size, general morphology, coloration, and color patterns that they are best identified by geographic range and molecular data. The five species in the complex that were previously treated as *D. ochrophaeus* (*D. carolinensis, D. imitator, D. ocoee, D. orestes,* and *D. ochrophaeus,* sensu Tilley and Mahoney 1996) are medium-sized salamanders that have pigmented testes and rounded, unkeeled tails. The tail is slightly longer than the body, and the toe tips lack cornifications. Dorsal color patterning ranges from straight-edged stripes to wavy, blotched patterning (Martof and Rose 1963; Tilley 1969). Dorsal stripes and blotches occur in a wide variety of colors, including greenish gray, light brown, rusty red, yellowish brown, and bright red. The venter is light colored in young animals but becomes light gray to grayish black with age. Older animals are often melanistic, and males are more likely to be melanistic than females. Dyrkacz (1981) reports a partial albino adult.

Relative to females, males of all five species of the complex have well-developed premaxillary teeth and large jaw musculature. Males also possess small mental glands, are on average slightly larger than females, have a mouth line that appears sinuate when viewed from the side, and tend to lose the vomerine teeth with age. Males of the *D. ochrophaeus* complex average 6–20% longer in SVL than females (Bruce 1993; Martof and Rose 1963; Orr 1989). Adults reach 7–11 cm TL, and there are 14 costal grooves. Hatchlings measure 13–18 mm TL and have prominent yolk reserves (Bishop 1941a; Bishop and Chrisp 1933; Huheey and Brandon 1973; Orr 1989).

The dorsal patterning of *D. carolinensis* is highly variable; the species includes both straight-edged and wavy-striped individuals (Tilley 1969). Hatchlings and larvae have rounded snouts and a series of five to six pairs of alternating spots along the dorsum, but these may be absent or very faint in individuals that develop straight dorsolateral lines as adults. Juveniles and young adults typically have reddish or yellowish dorsal stripes or blotches. Older adults in high-elevation populations are usually melanistic and are often dark gray to brownish gray with reduced dorsal patterning. Males in local populations are on average 10–20% larger than females and are more likely to be melanistic (Martof and Rose 1963).

SYSTEMATICS AND GEOGRAPHIC VARIATION. Populations that have traditionally been referred to as *D. ochrophaeus* in the eastern United States exhibit marked geographic variation in body size, coloration, color patterning, size at sexual maturity, and other traits. The complexity of geographic variation in this group has led to an unstable nomenclature and much debate about the taxonomic status of many forms. Three forms were described as separate species or subspecies based on color patterning and morphology (Dunn 1916; Neill 1950; Nicholls 1949), but these taxa were later relegated to a single monotypic species (Conant 1975; Huheey and Brandon 1973; Martof and Rose 1963). The use of molecular techniques to study genetic variation led to the discovery of *D. imitator*. This sibling species coexists with a second member of the *D. ochrophaeus* complex without interbreeding (Tilley et al. 1978; Verrell 1990b; Verrell and Tilley 1992). Additional studies of protein variation and ethological isolation indicated that the remaining populations of *D. ochrophaeus* consist of several geographic subgroups that are well differentiated genetically and show partial to nearly complete reproductive isolation (Arnold et al. 1996; Herring and Verrell 1996; Tilley and Mahoney 1996; Tilley et al. 1978, 1990).

In a detailed study that sampled populations from throughout the range of *D. ochrophaeus*, Tilley and Mahoney (1996) documented five genetic subgroups (A, B, C, D, and E). Using interpretations based on an evolutionary species concept, they recommended that four species (*D. carolinensis*, *D. ochrophaeus*, *D. ocoee*, and *D. orestes*) be recognized. These forms are parapatrically distributed, or nearly so, and most are strongly differentiated genetically, with fixed or nearly fixed differences for several alleles. The northernmost form (*D. ochrophaeus*; group A) is geographically isolated from *D. orestes* by a low valley in southwestern Virginia. Populations on either side of the valley are well differentiated genetically and show no evidence of recent gene exchange.

Desmognathus carolinensis (group D) is found immediately southwest of *D. orestes* (groups B and C). These forms are parapatric and are strongly differentiated, with eight fixed or nearly fixed allele differences (Nei's $D = 0.53$). Along a transect established in the western portion of the contact zone, where groups C, B, and D replace each other from north to south, the frequencies of alleles for three of six marker loci for groups B and C increase gradually over a distance of 20–30 km, whereas alleles for four marker loci for groups B and D change abruptly and concordantly over a 1.2-km zone. Tilley and Mahoney (1996) interpret this pattern as evidence of intergradation between groups B and C (both considered to be conspecific and to constitute *D. orestes*) and as evidence of greatly restricted gene flow between *D. carolinensis* and *D. orestes*.

A second transect through the eastern portion of the contact zone reveals different patterns in loci showing fixed or nearly fixed differences between forms. Alleles at three of eight loci gradually increase in frequency over a distance of approximately 50–80 km, whereas those at the remaining five loci change abruptly in frequency over a 12-km zone in Burke and Mitchell counties, North Carolina. Populations within this 12-km zone have not been sampled, and the steepness of the transition in allele frequencies is not known.

Tilley and Mahoney (1996) interpret groups B, C, and D as representing two evolutionary species that are evolving independently of one another. Their interpretation is based on a synthesis using information on genetic distance, evidence for gene flow in contact zones, and ethological isolation. They interpret abrupt changes in alleles as evidence of independent evolutionary trajectories and restricted gene flow among forms. However, this interpretation is not completely satisfactory because it does not explain the gradual increase in frequency of other alleles over larger geographic distances.

Group E is treated as a fourth species, *D. ocoee*. This species is a highly heterogeneous group that consists of populations that occupy major mountain ranges in the southern Appalachians southwest of the Pigeon River. Genetic distances between local populations of group E are relatively high (mean $D = 0.24$; range = 0.01–0.44; $F_{st} = 0.65$), and certain populations have developed ethological isolation that appears to reflect isolation by distance. There is no single allozyme locus that is unique to *D. ocoee*. However, this form typically has one fixed or nearly fixed allozyme that will distinguish it in paired comparisons with other members of the complex. *Desmognathus ocoee* intergrades extensively with *D. carolinensis* as far east as the Blue Ridge Divide east of Asheville, North Carolina. Patterns of genetic variation in this zone of intergradation are currently being investigated in detail (S. G. Tilley, pers. comm.).

Tilley and Mahoney's (1996) recognition of four species could be readily challenged depending on one's taxonomic philosophy. Those who endorse phylogenetic species concepts would recognize many more species, whereas those who follow a strict interpretation of the biological species concept might recognize only one or two species. One of the major problems in applying any species concept to this complex is that local populations in many cases appear to have evolved ethological isolation by distance. That is, the probability of individuals successfully courting tends to decrease with the geographic distance separating local populations. This pattern tends to hold both within and between many genetic subgroups in the southern portion of the range of members of the complex (Herring and Verrell 1996; Houck and Schwenk 1984; Houck et al. 1988; Tilley et al. 1990; Verrell and Arnold 1989). As a result, distant populations within the same genetic group may exhibit partial to almost complete ethological isolation that reflects isolation by distance.

We currently lack a species concept that can adequately represent the complex patterns of genetic variation and reproductive isolation in this group. Here I follow Tilley and Mahoney's (1996) classification as a tentative working model. As more detailed information on gene exchange in contact zones becomes available, additional revisions may be required.

Desmognathus carolinensis exhibits marked geographic variation in body size, coloration, size at sexual reproduction, and other traits. Body size generally increases with elevation, but local populations at similar elevations may differ markedly in average size (Hairston 1949; Martof and Rose 1963; Tilley 1974).

DISTRIBUTION AND ADULT HABITAT. The Carolina mountain dusky salamander is confined to mountainous, forested habitats in the southwestern Blue Ridge physiographic province, including the Blue Ridge, Black, Bald, and Unaka mountains. Its range extends from a region between Linville Falls and McKinney Gap on the Blue Ridge Divide (Burke and Mitchell counties, North Carolina) and the valley of the Doe River on the North Carolina–Tennessee border, southwestward to the valley of the Pigeon River in Haywood and Buncombe counties, North Carolina. This species is abundant on wet rockfaces, in seepage areas, and in forest-floor habitats in and about the vicinities of streams at elevations of 300–2000 m (Tilley 1980). Populations at relatively low elevations are concentrated in or near seepages or streams, whereas those at higher elevations (generally >1370 m) are often found on the forest floor far from running water.

BREEDING AND COURTSHIP. Mating of this and other members of the *D. ochrophaeus* complex in the southern Appalachian Mountains occurs during the spring, late summer, and autumn (Huheey and Brandon 1973; Organ 1961a), but seasonal patterns for *D. carolinensis* are poorly documented. Females mate and oviposit annually (Martof and Rose 1963; Tilley 1973b; Tilley and Tinkle 1968), and courtship involves a stereotypical tail-straddle walk typical of plethodontids (Houck et al. 1985a; Uzendoski and Verrell 1993). Details of a comparative study of courtship behavior of *D. carolinensis, D. ocoee,* and *D. orestes* are presented by Herring and Verrell (1996). The courtship sequence of all three of these members of the *D.*

Range of *Desmognathus carolinensis*

ochrophaeus complex appears to be nearly identical and involves an orientation phase followed by persuasion and sperm transfer phases (for a detailed description, see the *D. orestes* species account). Evidence of multiple inseminations (Houck et al. 1985b; Tilley and Hausman 1976) and long-term sperm storage (Houck and Schwenk 1984) suggests that some females court several times per year and that sperm competition occurs among males.

Males normally deposit only one spermatophore per night, even when presented with multiple partners (Verrell 1988a). In addition, females are not responsive to males for 2-4 days after successful insemination. This interval corresponds to a time when the female's cloaca is physically blocked by the sperm cap (Verrell 1991b). Despite the use of the tail in courtship, tailless males are as successful as males with intact tails in courting and inseminating females (Houck 1982). The energetic cost of courtship is minimal and probably does not constrain the number of times that animals can court (Bennett and Houck 1983). Although larger females are more fecund, the mating frequency of females is independent of female SVL (Shillington and Verrell 1996). The size of courtship arenas and the complexity of the physical environment also appear to have little effect on mating frequency (Verrell 1988c).

REPRODUCTIVE STRATEGY. Females deposit their eggs in small, grapelike clusters in hollowed depressions beneath rocks, decaying logs, and leaf litter and in moss mats in or near springs, seeps, and small streams (Martof and Rose 1963; Tilley 1973b). Each egg has a short, gelatinous stalk and is attached to a common point to form a cluster (Martof and Rose 1963). The maximum diameters of mature ova measured by Tilley (1973b) were 2.9-3.2 mm.

Females often gather in large congregations in seepages or other wet microhabitats and construct their nests within a few centimeters or decimeters of each other (Tilley 1973b). The mean clutch sizes of eight populations studied by Tilley (1973a) were 13-26 ova and increased with elevation, whereas the mean clutch sizes of two populations studied by Martof and Rose (1963) were 12.5 and 28 ova. Clutch size is positively correlated with female body size (Tilley 1973a).

Females brood their clutches through hatching. Brooding females protect the embryos from predators, remove dead eggs from nests, reduce egg desiccation, and reduce fungal attacks (Tilley 1972). Females may consume prey that wander near nests, but brooding necessitates that females greatly reduce their foraging activity (Martof and Rose 1963). Brooding females may lose 12% of their weight over a 1-month period, but much of that loss may be due to desiccation (Tilley 1972).

The seasonal time of oviposition varies among regional populations. Populations studied by Tilley (1973b) at high elevations oviposit from 20 May to 10 July compared with 1 June-30 July for low-elevation populations. Females have been found brooding hatchlings in early March in western North Carolina, suggesting that some females oviposit in winter (Tilley 1973b; Tilley and Tinkle 1968).

The time of hatching varies markedly depending on the egg-laying schedule of females in local populations. Tilley (1973b) found hatchlings with yolk in both late summer and early March in North Carolina, although in any population hatching occurs in either the spring or late summer and early autumn. Embryos that hatch in the spring may be from females that oviposit in late autumn and overwinter with their clutches. Most embryos appear to hatch in August and September (Tilley 1972), and the incubation period is 58-69 days (Tilley 1972, 1973b).

AQUATIC ECOLOGY. The larvae live in seepages; in shallow, slow sections of headwater streams; and on wet rockfaces. The larval period may last 2-8 months depending on the time of oviposition. Hatchlings from eggs laid in the late winter or early spring typically transform in 2-3 months, whereas those from eggs laid in the summer may overwinter and transform 4-8 months later (Tilley 1973b). Published dietary studies of larvae are not available.

TERRESTRIAL ECOLOGY. The juveniles and adults appear to be generalist feeders that eat snails, earthworms, beetles, flies, and other terrestrial arthropods (Hairston 1949). Female *D. carolinensis* sometimes feed on their own eggs and hatchlings. Females will preferentially eat dead eggs introduced into their clutches and will sometimes eat healthy eggs from their own broods immediately after consuming a dead egg (Tilley 1972). Hatchlings are sometimes cannibalized, and in some instances females of this and other members of the *D. ochrophaeus* complex appear to eat their own young (Forester 1981; Martof and Rose 1963; Wood and Wood 1955).

Growth rates vary depending on age, elevation, and the microhabitats frequented by the adults. Juvenile growth rates of 8–10 mm SVL per year appear typical for most populations (Tilley 1973a). Age and size at first reproduction, maximum adult size, mean clutch size, and adult survivorship also increase with elevation. The median size at first reproduction increases with elevation from 32–34 mm SVL in low-elevation populations to 40–42 mm SVL in high-elevation populations (Tilley 1973a). Differences in size at maturity, growth rate, and mean clutch size may also occur between rockface and woodland populations at the same locality (Tilley 1974). These may reflect differences in food availability between rockface and woodland habitats, but genetic causes cannot be ruled out. Maximum longevity of animals in the field is around 10–11 years.

Carolina mountain dusky salamanders are nocturnally active and remain beneath cover objects or in recesses in rockfaces during the day. They emerge at night and actively move about in search of prey or mates. On overcast days, a small percentage of animals may be active outside cover (Hairston 1949). This and other dusky salamanders will retreat into cover if individuals dehydrate only slightly (Feder and Londos 1984). This behavior may explain why nocturnal surface activity is minimal during periods of dry weather and why surface activity often declines within a few hours after dark. On rainy nights, individuals frequently climb plants, where they sometimes perch 1 m or more above the ground. The adults are aggressive and will actively defend space from conspecifics. Bennett and Houck (1983) found that the energetic cost of aggression is minimal.

Females apparently migrate short distances to breeding habitats each year, and young juveniles disperse from these sites into the surrounding forest. Individuals are active on the ground surface except during the coldest winter months. Adults and juveniles congregate in seepages or underground retreats during the winter months and disperse from winter retreats to surrounding forests in the spring (Tilley 1973b).

PREDATORS AND DEFENSE. *Desmognathus carolinensis* is undoubtedly preyed upon by many vertebrates inhabiting mesic forests. Woodland birds, small snakes, and large salamanders such as *D. monticola, D. quadramaculatus,* and *Gyrinophilus porphyriticus* are major predators on other members of the *D. ochrophaeus* complex (Bruce 1979; Coker 1931; Formanowicz and Brodie 1993; Hairston 1986; Whiteman and Wissinger 1991) and undoubtedly feed on *D. carolinensis.*

COMMUNITY ECOLOGY. *Desmognathus carolinensis* often shares streamside habitats with *D. monticola, D. quadramaculatus,* and *D. wrighti* (Hairston 1949). Ecological interactions between these forms have not been examined in detail, but studies of other members of the *D. ochrophaeus* complex suggest that *D. carolinensis* may be an important food source that is competed for by *D. monticola* and *D. quadramaculatus* (see the account under *D. ocoee*).

CONSERVATION BIOLOGY. *Desmognathus carolinensis* is one of the most common species in the Appalachian Mountains and is in minimal need of protection. This species reaches its highest densities in mature hardwood forests, and local populations may decline markedly following clear-cutting (Petranka et al. 1993). The Carolina mountain dusky salamander would be a good candidate for use as an indicator species of environmental quality because it is widespread, occurs over a wide altitudinal distribution, and has both aquatic and terrestrial stages that may be susceptible to environmental disturbance.

Desmognathus fuscus (Rafinesque)
Dusky Salamander
PLATES 44, 45

IDENTIFICATION. The dusky salamander is a medium-sized *Desmognathus* that is one of the most common species in North America. The basal third of the tail is laterally compressed and the tail is keeled and somewhat triangular in cross section. A light line extends from the eye to the angle of the jaw and there are 14 costal grooves. The toe tips lack cornifications.

Adults in the northern portion of the range often

Fig. 105. *Desmognathus f. fuscus*; adult; Knott Co., Kentucky (R. W. Barbour).

have a relatively uniform, light dorsal stripe that extends to the anterior portion of the tail. However, retention of the larval spotting pattern occurs to varying degrees and is often conspicuous in more southern populations. Old adults throughout the range of the species are often melanistic and are dark above with little spotting. In addition, albinos or partial albinos have occasionally been reported (e.g., Channell and Valentine 1972; Dyrkacz 1981).

Populations in the southern Appalachians are the most variable, particularly where *D. f. fuscus* and *D. f. santeetlah* intergrade. Depending on the locale, individuals can range from being rather uniformly yellowish brown to being boldly marked with dorsal patterns that resemble those of members of the *D. ochrophaeus* complex (Tilley 1988).

The venter appears cream colored (except in melanistic individuals) and the underside of the tail is often weakly washed in yellow. Melanophores on the venter are clumped to produce a weakly reticulate or peppered pattern that becomes more conspicuous as individuals age. Melanophores are usually present on the dorsal lining of the mouth, and the mouth line in most populations is not strongly sinuate when viewed from the side.

Adults range from 6 to 14 cm TL. The maximum size of males of this and other *Desmognathus* species exceeds that of females, a feature that may reflect greater average longevity in males (Organ 1961a). The average SVL of males in Ohio specimens is 7% larger than that of females (Karlin and Pfingsten 1989). Adult males have a small mental gland on the apex of the lower jaw, enlarged premaxillary teeth, and papillose cloacal lips. The cloacal lips of females have smooth folds. Characters useful in separating *D. fuscus* from *D.*

monticola and *D. welteri* are provided by Caldwell and Trauth (1979) and Juterbock (1978, 1984).

Hatchling size is geographically variable and ranges from about 8 to 12 mm SVL and from about 12 to 20 mm TL. Hatchlings have prominent yolk reserves and whitish gills (Bishop 1941a; Krzysik 1980a; Organ 1961a; Wood and Fitzmaurice 1948). Young larvae and many juveniles have five to eight pairs of even or alternating dorsal blotches or spots between the front and rear limbs that continue onto the tail. Blotches and spots are separate in hatchlings but become partially fused or develop into a dorsal stripe in older larvae and juveniles. The sides of the body typically have a dark dorsolateral stripe that can be either wavy or straight, and the dorsum may be either uniformly colored or conspicuously blotched with brown to reddish pigments.

SYSTEMATICS AND GEOGRAPHIC VARIATION. Tilley (1981) proposed the name *D. santeetlah* for high-elevation populations of *fuscus*-like animals taken from the Unicoi, Great Smoky, and Great Balsam mountains in the southern Appalachians. *Desmognathus santeetlah* differs from nearby populations of *D. fuscus* in reaching sexual maturity at a slightly smaller size, in being generally darker above with inconspicuous markings, in having slightly different body proportions, and in having fewer dorsal spots on the larvae. In addition, electrophoretic analyses show that the two taxa are fixed for alternate electromorphs at three protein loci. In the Unicoi Mountains, populations of these two forms are parapatric and appear to be separated by a slight elevational gap.

Detailed genetic studies indicate that the two forms are moderately differentiated genetically with Nei's distances (D) of 0.20-0.35 (Karlin and Guttman 1986;

Fig. 106. *Desmognathus f. fuscus*; head profile of adult; Wilkes Co., North Carolina (R. W. Van Devender).

Tilley and Schwerdtfeger 1981). In addition, the two forms appear to interbreed extensively where they come into geographic contact in a broad geographic area along the northwest escarpment of the Great Smoky Mountains (Tilley 1981, 1988). This complex probably reflects derivatives of an ancestral form that became geographically isolated and diverged evolutionarily. The two forms have since come into contact to form a secondary hybrid zone. Despite genetic and life history differences that occur between the forms, there is little evidence that the forms are reproductively isolated from each other in areas where there are elevational overlap and opportunities to interbreed (Maksymovitch and Verrell 1993; Tilley 1988; Verrell 1990d). The lack of evidence of hybrids in the area of parapatry in the Unicoi Mountains could reflect the development of partial or complete reproductive isolating mechanisms, the lack of opportunities to interbreed because of altitudinal separation, or a combination of both.

This species pair is yet another example of parapatric forms of salamanders that are at a level of evolutionary divergence that falls between the subspecies and full species level. My view is that *D. santeetlah* is closer to the former, and therefore I treat this form as a subspecies of *D. fuscus*.

Karlin and Guttman (1986) examined genetic variation in *D. fuscus* and found that northern *D. f. fuscus* and southern *D. f. conanti* populations are well differentiated (Nei's $D = 0.47$). In addition, southern populations consist of several moderately differentiated subgroups. The authors suggested that the southern and northern populations should be considered as separate species but did not name them formally because zones of contact between northern and southern groups have yet to be examined in detail. One group that was thought to be a member of the *D. fuscus* group has since been shown to be more closely related to members of the *D. ochrophaeus* complex. This form is now recognized as *D. apalachicolae* (Means and Karlin 1989). Titus and Larson (1996) raised *D. f. conanti* to the level of a full species based on very limited evidence from mitochondrial DNA analyses. Because they examined only two populations from outside the contact zone with *D. f. fuscus*, I have not adopted their interpretation. I concur with Karlin and Guttman (1986) that detailed studies of contact zones should be made before revisions to the nomenclature are proposed.

Hybrids between *D. fuscus* and *D. ochrophaeus* occur in several populations in Ohio and Pennsylvania (Karlin and Guttman 1981) and in the Chateauguay River drainage of Quebec, Canada (Sharbel and Bonin 1992; Sharbel et al. 1995). However, gene exchange between these species is very limited and reflects introgressive hybridization (Houck et al. 1988).

Local and regional populations of *D. fuscus* often vary markedly in color patterns, body size, and life history traits and consist of several genetically differentiated subgroups (Danstedt 1979; Davic 1983; Means 1974; Tilley 1988). Whether any or all of these should be recognized as separate species is open to debate. Here, I treat this group as a single species consisting of three subspecies.

The northern dusky salamander (*D. f. fuscus*) occurs from northern Maine to south-central Kentucky, eastern Tennessee, and western North Carolina. It has a strongly keeled tail and the adults have a relatively uniform, light dorsal stripe that extends onto the anterior portion of the tail. Retention of the larval spotting pattern occurs to varying degrees and may be conspicuous in populations in the southern Appalachians.

The Santeetlah dusky salamander (*D. f. santeetlah*) occurs in the Unicoi, Great Smoky, and Great Balsam mountains in the southern Appalachians. Adults of this subspecies have a moderately keeled tail, a drab brownish or greenish dorsum with poorly defined dorsolateral stripes, and a yellowish wash to the underside of the tail. Juveniles often have four or five pairs of chestnut-colored spots or blotches on the dorsum that may fuse to form a wavy dorsal stripe. Adults are smaller and less fecund on average than populations of *D. f. fuscus* (Jones 1986; Tilley 1988). *Desmognathus fuscus santeetlah* and *D. f. fuscus* intergrade extensively along the northwest escarpment of the Great Smoky Mountains.

The spotted dusky salamander (*D. f. conanti*) occurs from southern Illinois southeastward to northeastern Georgia and southward to Florida, Alabama, Mississippi, and Louisiana. This subspecies has a moderately keeled tail and has bolder marking on the dorsum. Dorsal spots and blotches on the dorsum are red to golden and are more prominent in older animals. Ashton and Ashton (1988) indicate two disjunct populations in central Florida whose identities have not been confirmed by genetic studies. Conant and Collins (1991) indicate a broad zone of intergradation between *D. f. fuscus* and *D. f. conanti* in western Kentucky, central and southeastern Tennessee, northeastern

Range of *Desmognathus fuscus*

Georgia, and South Carolina. However, the extent to which populations in this zone reflect intergradation between northern and southern forms has not been clearly delineated. This region is treated here as a "contact zone" where subspecific and genetic variation is poorly documented.

DISTRIBUTION AND ADULT HABITAT. The dusky salamander has an exceptionally wide latitudinal distribution, ranging from northern Maine and adjoining portions of Canada to extreme southern Alabama and Mississippi. It is generally found east of the Mississippi River except for isolates in northern Louisiana and Arkansas. This species is rare in the Ridge and Valley Physiographic Province of Tennessee and Virginia, and in much of the southwestern Blue Ridge Physiographic Province.

The dusky salamander occurs in a variety of aquatic or semiaquatic forested habitats throughout much of the eastern United States. At lower elevations individuals are often abundant in or about seeps, and in or along the margins of small streams with rocks, logs, mosses, and other cover. Hom (1988) found that females on Chihowee Mountain in eastern Tennessee prefer rocks as cover, except during the nesting season when they move from streams onto streambanks and nest in moss clumps. In mountainous regions *D. f. fuscus* and *D. f. santeetlah* are often found several meters from the water's edge, while in the Coastal Plain *D. f. conanti* is sometimes found around the margins of swamps and sluggish, muddy streams (Mount 1975). *Desmognathus fuscus* is generally rare above 1200 m in elevation except in the Great Smoky Mountains and vicinity.

BREEDING AND COURTSHIP. Because of the extensive latitudinal range of this species, it is no surprise that the breeding season, seasonal activity patterns, annual breeding schedules, and other aspects of the natural history vary geographically. Adult *D. f. fuscus* mate during both autumn and spring in New York and Virginia (Bishop 1941a; Organ 1961a), and females breed every year in many populations (Danstedt 1975; Spight 1967). Females in mountainous regions of southwestern Virginia appear to have biennial reproductive cycles and the sex ratio is about 1:1 (Organ 1961a).

Courtship behavior has been described by several authors and appears to be virtually identical for *D. f. santeetlah* and *D. f. fuscus* (Maksymovitch and Verrell 1992; Verrell 1995 a,b). The following is a general description of courtship based primarily on observations by Organ (1961a) and supplemented with those of Maksymovitch and Verrell (1992), Uzendoski and Verrell (1993), and Verrell (1995b).

When a male encounters a receptive female, he engages in a "butterfly walk" (Uzendoski and Verrell 1993) in which he approaches while synchronously rotating the forelimbs in a manner that is similar to the butterfly stroke of swimmers. The male may repeatedly jerk his body to and fro and undulate his tail during this period. After reaching the female, the male rubs his head on the head region of the female and often undulates his tail (Maksymovitch and Verrell 1992; Uzendoski and Verrell 1993).

The male eventually places his snout on the female's back and raises and lowers his forelimbs simultaneously (Organ 1961a). Holding his snout in contact with the female's back, the male arches his body upward and raises his belly and forelimbs clear of the ground. The hindlimbs and tail are kept pressed to the substrate, and the snout is pressed downward with considerable force against the female's back. This action not only brings the hedonic glands into contact with the female, but also causes tactile stimulation from the enlarged premaxillary teeth of the male. This arched position is held for a short period of time, then the male violently snaps his body straight. The force with which the male snaps often propels him 5–10 cm or more from the female. He then returns and repeats the entire sequence several times while slowly moving toward the female's head. Perhaps more commonly, the male positions himself such that his undulating tail is beneath the chin of the female. From this position, he reaches back and snaps the female's dorsum or neck. The snapping action abrades the skin and presumably vaccinates the female with secretions from the mental gland (Arnold 1977; Uzendoski and Verrell 1993).

In addition to snapping, a male may slowly pull his teeth across the female's dorsum to vaccinate her. The female often moves away during this persuasion phase, and the male may return repeatedly to rub and snap her until she becomes fully receptive. This phase of courtship often lasts 1–2 hours.

Once the female is fully receptive, the male forces his head under her chin, then moves forward while holding his back against her chin. He undulates his tail as it passes under her chin. The female places her forelimbs astride the male's tail and the courting pair moves forward in a tail-straddle walk. With his undulating tail in contact with the belly of the female, the male secretes a colorless fluid from the dorsum of his tail.

Immediately prior to spermatophore deposition, the male begins a series of lateral pelvic rocking movements and the female moves her head in synchronization with, but counter to, his movements. The male then lowers his vent to the substrate and deposits a spermatophore. Immediately thereafter, he arches the base of the tail upward and bends the tail sharply to one side. He then violently undulates the tail in a manner similar to the wiggling movements of an autotomized salamander tail. The female holds her chin against the base of the bent tail and the courting pair moves forward quickly. As the female passes over the spermatophore, she picks up the sperm cap with her cloacal lips. The animals decouple shortly thereafter. The spermatophore has a greatly expanded, flangelike base that narrows to a slender tip (Bishop 1941a).

Males often attempt to court other males and are often attacked by rival males in the process. Rival males that become stimulated and undulate their tails may in turn be attacked by the courting male. Males may also attack rival males that are courting females. When males are presented with a choice, they are more likely to court large females than small females (Verrell 1994a, 1995a). When presented with females sequentially, however, males are equally likely to court females that differ in body size. Males that choose larger females may enhance their own fitness because fecundity is positively correlated with female SVL. The extent to which males have choices and are selective in natural populations has not been determined.

Uzendoski and Verrell (1993) compared the courtship behavior of *D. orestes* and *D. fuscus* and found that certain aspects of courtship differ significantly between these species. Nonetheless, differences in courtship do not appear to be the primary reason for reproductive isolation between these species. Instead, males typically ignore females of the opposite species from the point of initial contact. Differences in female chemical cues may be the primary reason for the lack of responsiveness exhibited by males.

Maksymovitch and Verrell (1993) compared the sexual compatibility of individuals from three populations of *D. f. santeetlah* in the southern Appalachians and found statistically significant differences in levels of sexual compatibility among populations. These data are consistent with other studies of salamanders that indicate that aspects of mate recognition systems can diverge among conspecific populations in the absence of geographic isolation.

Verrell (1990d) conducted courtship trials between *D. f. fuscus* from a population in the Chilhowee Mountains in Tennessee and *D. f. santeetlah* from a population about 40 km away at Indian Gap in the Great Smoky Mountains. The results indicate that *D. f. fuscus* males are just as likely to court and inseminate female *D. f. fuscus* as female *D. f. santeetlah* when paired. However, when caged with females of both subspecies, males are more likely to court members of the same population and subspecies compared with members of the opposing population and subspecies, regardless of whether they are given simultaneous or sequential choices of mates. Courtship trials using *D. f. santeetlah*, *D. imitator*, and *D. ocoee* from Indian Gap demonstrate that all three species are sexually isolated from one another (Verrell 1990b).

REPRODUCTIVE STRATEGY. Females lay their eggs in cryptic microhabitats in or near aquatic sites. Female *D. f. fuscus* may attach their eggs to the undersides of rocks in streams, in cavities in and under rotting logs, and in leaf mats or clumps of moss near streams and seeps (Bishop 1941a; Dennis 1962; Hom 1987; Krzysik 1980a; Noble and Evans 1932; Organ 1961a; Wood and Fitzmaurice 1948). Eggs of *D. f. santeetlah* are often placed beneath moss mats (Jones 1986). At any given site, females often tend to prefer one microhabitat type strongly over another.

The eggs of *D. fuscus* are deposited in a globular or grapelike cluster that is often suspended from the roof of a cavity or attached to mosses or fibers. The outer envelope of each egg is drawn out into a short stalk that is attached either to support structures or to other

Fig. 107. *Desmognathus f. fuscus*; female with eggs; Watauga Co., North Carolina (R. W. Van Devender).

eggs or stalks. Nests are typically located within 1–2 m of the stream edge.

When nesting in soft substrates, a female may actively construct a small depression in which to lay eggs (Bishop 1941a; Dennis 1962; Hom 1987; Juterbock 1987; Krzysik 1980a; Wilder 1913). Females normally deposit their clutches within a 24-hour period (Dennis 1962; Wilder 1913), and freshly laid eggs are 3–4.5 mm in diameter. The ova are cream colored to whitish and 2.5–3.5 mm in diameter, and are surrounded by three envelopes (Bishop 1941a).

Females coil about their eggs for the duration of the incubation period but sometimes leave their nests during the night (Brode 1961). Females returning to their nests apparently use chemical cues to recognize their clutches (Forester 1986). Brooding females in Ohio will aggressively defend their nests from conspecifics that are potential predators of the embryos (Dennis 1962). Gravid females will often abandon their nesting cavities if disturbed immediately prior to ovipositing, but brooding females rarely abandon their nests when disturbed (Juterbock 1987). One cost of brooding is that foraging rates are reduced and females feed only opportunistically (Krzysik 1980b; Montague and Poinski 1978).

Dennis (1962) found that female *D. f. fuscus* displaced as far as 32 m from their nests will return and resume brooding. In one experiment in which females were released from locations 2.4–10.6 m from their nests, 8 out of 13 individuals successfully returned. Females rarely abandon their eggs, and those that are abandoned usually suffer catastrophic mortality (Dennis 1962; Hom 1987).

Nesting dusky salamanders are relatively easy to locate, and numerous researchers have provided information on clutch size and dates of oviposition and hatching. With rare exceptions, females lay their eggs in late spring or early summer. In Tennessee *D. f. santeetlah* oviposits from as early as mid-May through early July, whereas *D. f. fuscus* at lower elevations oviposits from late June through early August (Hom 1987; Jones 1986).

In Ohio most *D. f. fuscus* oviposit during the last 2 weeks of July and the embryos hatch from early September to mid-October (Dennis 1962; Juterbock 1986; Orr and Maple 1978; Wood and Fitzmaurice 1948). Populations in eastern Kentucky and southeastern Ohio almost always oviposit during July, and the embryonic period lasts from 46 to 61 days (Juterbock 1986). The embryos hatch from late August to mid-October. The median incubation period is 47 days in Tennessee populations (Hom 1987).

Records from other areas of the range indicate similar times of oviposition. Eggs have been found from mid-July to mid-October in Alabama (Mount 1975), from 28 June to 5 October in Indiana (Minton 1972), from early June through early August in Maryland (Danstedt 1975, 1979), from 11 June to 24 September in Massachusetts (Wilder 1913), from July through September in southwestern Pennsylvania (Krzysik 1980a), and from late June to mid-August in Virginia (Organ 1961a). Hatchlings at the Maryland and Pennsylvania sites appear in July and August. Hatching mostly occurs in August and September in New York (Bishop 1941a).

Mean clutch size differs among local populations, but geographic variation appears to be chaotic. Reported clutch sizes are as follows: 13-24 eggs per nest for populations in Alabama (Mount 1975), 5-33 eggs per nest in nine nests in Massachusetts (Wilder 1913), an average of 27 mature ova in 18 females from western North Carolina (Spight 1967), an average of 26 eggs in seven nests in Ohio (Wood and Fitzmaurice 1948), an average of 21 ova (range = 11-36) in 17 Pennsylvania females (Hall 1977), and 22-34 eggs per nest in 44 Pennsylvania nests (Pawling 1939). The mean for six Maryland populations studied by Danstedt (1975) ranges from 24 to 33 ova. Clutch size is positively correlated with body size in all six of these populations, but mean egg size does not differ significantly. Hom (1987) also found a positive correlation between female SVL and the number of eggs in nests.

The clutch size of *D. f. santeetlah* populations in the Great Smoky Mountains averages 17-20 eggs and is significantly smaller than that of nearby *D. f. fuscus* (Tilley 1988). Clutch sizes of *D. f. santeetlah* × *fuscus* are slightly smaller than the clutch sizes of *D. f. santeetlah*. In general, clutch size positively correlates with adult body size in these populations. The clutch size of *D. f. fuscus* in the Unicoi Mountains averages 23 ova compared with 21 in *D. f. santeetlah* (Jones 1986). At a given SVL, *D. f. fuscus* produces about three eggs fewer on average than *D. f. santeetlah*. Clutch size is positively correlated with SVL in both subspecies. Another clutch size estimate for *D. f. santeetlah* is 20 eggs in 29 nests (Beachy 1993b).

Reproductive success is variable but low in eastern Tennessee populations studied by Hom (1987). The percentage of nests in which at least one embryo survives to hatching varies from 24% to 49% among years, and large females hatch a greater percentage of their eggs than small females.

AQUATIC ECOLOGY. Yolk absorption by hatchlings may take over 2 months (Orr and Maple 1978), but larvae begin feeding long before absorption is complete (Montague 1979; Wilder 1913). Larvae feed on small prey such as copepods, chironomid larvae, plecopteran nymphs, collembolans, mites, and fingernail clams (Burton 1976, Wilder 1913).

Larvae in four populations in northeastern Kentucky and southern Ohio grow very little during the fall and winter following hatching (Juterbock 1990). Growth rates increase beginning in mid- to late April and continue to increase until larvae metamorphose in June or July after about a 9-month larval period. In general, larvae in southern populations studied by Juterbock (1990) grow faster, transform slightly sooner, and are larger at metamorphosis than larvae farther north. This trend may be due in part to temperature differences associated with climate. Larvae in springs and seeps typically have less tail damage, but lower fat stores, than those in streams.

Size at metamorphosis varies among local and regional populations and ranges from 9 to 20 mm SVL. Most larvae in Juterbock's (1990) populations transform when they reach 15-20 mm SVL. The larval period of both *D. f. santeetlah* and *D. f. fuscus* is slightly less than 1 year in eastern Tennessee, and the smallest metamorphs measure 9-12 mm SVL (Jones 1986). Transforming larvae that are an estimated 8-10 months old have been collected in Massachusetts in June (Wilder 1913). Maryland larvae transform in late spring and early summer when 9–12 months old (Danstedt 1975), whereas larvae in southwestern Virginia transform in July–October after 11–14 months of growth (Organ 1961a).

TERRESTRIAL ECOLOGY. The juveniles and adults feed predominantly on terrestrial or semiterrestrial invertebrates. As individuals grow, they incorporate larger prey into the diet but continue to consume large numbers of small prey (Krzysik 1979; Sites 1978). Terrestrial and aquatic insects, centipedes, amphipods, spiders, and snails are eaten by *D. fuscus* in eastern Kentucky (Barbour and Lancaster 1946). Individuals in Tennessee feed on terrestrial insects during late summer and early autumn but shift toward isopods and other aquatic insects during the spring (Sites 1978). Prey measuring <1 mm TL are rarely eaten.

Lepidopteran larvae and adult and larval flies are the dietary mainstay in a Pennsylvania population (Krzysik 1979). New Hampshire populations eat a variety of aquatic and terrestrial invertebrates, particularly flies and beetles (Burton 1976). Large animals occasionally cannibalize both larvae and juveniles. Records of cannibalism include *D. fuscus* larvae in adults uncovered in underground winter retreats (Hamilton 1943), *Desmognathus* larvae in the stomachs of two adults (Wilder 1913), and a juvenile *D. fuscus* in the stomach of a larger female (Carr 1940).

Brooding females have greatly reduced foraging rates relative to nonbrooding animals (Krzysik 1980b; Montague and Poinski 1978). Brooding females will eat their own eggs, as well as the eggs of other females (Baldauf 1947; Bishop 1941a; Jones 1986). The extent to which females eat live embryos is not fully understood, but there is at least one recorded instance of a wild-caught, gravid female with three viable embryos in her gut (Bauldauf 1947). A female maintained by Bishop (1941a) in the laboratory consumed her entire clutch. Eating dead eggs would be adaptive since these are often attacked by microorganisms that can spread and destroy viable embryos.

Growth rates are fairly constant during the first 3–4 years of life, then begin to slow. Juveniles grow about 9–10 mm SVL per year (Danstedt 1975), but growth rates undoubtedly vary depending on local conditions. Most males and females in Kentucky, Maryland, North Carolina, Ohio, Pennsylvania, and Virginia populations become sexually mature when they reach 38–44 mm SVL (Danstedt 1979; Davic 1983; Hall 1977;

Juterbock 1978; Organ 1961a; Spight 1967). Males in certain populations in Maryland and Virginia reach sexual maturity about 2 years after hatching, but females require an additional year to mature (Danstedt 1975, 1979). At high-elevation sites in southwestern Virginia, males become sexually mature 3.5 years after hatching (Organ 1961a). Females require an additional year and first oviposit when 5 years old.

Jones (1986) found that respective sizes at sexual maturity for *D. f. santeetlah* and *D. f. fuscus* in Tennessee are 30 and 35 mm SVL for males, and 35 and 40 mm SVL for females. Males of both subspecies reach sexual maturity when 2 years old, whereas females require an additional year of growth.

Patterns of survival may vary in local populations. Males in southwestern Virginia exhibit type I survivorship, in which the probability of dying increases slightly with age (Organ 1961a). Sexually mature females have lower survivorship than males, which may reflect the costs of brooding. In other populations, survivorship is slightly lower during the first 3 years of life than in later years (Danstedt 1975). Nonetheless, age-specific survivorship tends to be relatively constant throughout life, particularly for males.

Dusky salamanders can reach relatively high densities, which often exceed 1–2 animals/m^2 in optimal habitats. Using the Lincoln-Petersen index, Danstedt (1979) estimates the densities of two populations in the Piedmont as 2.8 and 0.4 salamander/m^2. Other estimates include a peak density of 2.48 salamanders/m^2 in a Piedmont population (Orser and Shure 1975), 0.4–1.4 salamanders/m^2 in a North Carolina streambed (Spight 1967), and 0.75–0.78 salamander/m^2 at a Pennsylvania site (Hall 1977).

Nonbreeding adults are rather sedentary and rarely move more than a few meters during a 24-hour period (Barbour et al. 1969). Home ranges of five animals in eastern Kentucky are 25–114 m^2, and peak diurnal activity occurs in early morning and shortly before midnight. Individuals tracked by Hall (1977) move an average of 2.7 m between captures. Ashton (1975) reports the average home range of 16 adults in Ohio to be 1.4 m^2 over 6.5 months, whereas Barthalmus and Bellis (1972) estimate that the home range encompasses <3 m of stream length in Pennsylvania.

Comparisons made over a 2-year period by Hom (1987) show that most adults remain within a 15-m stretch of stream between years. Animals will return to their home ranges after being displaced as far as 30 m (Barthalmus and Bellis 1969; Barthalmus and Savidge 1974). The time before animals are recaptured at home is positively correlated with the distance animals are displaced, and blinded animals home as efficiently as animals that are normal.

Stewart and Bellis (1970) found that animals are randomly distributed beneath cover boards placed along a streambank and that individuals normally move to other cover within 1–2 days after being uncovered. *Desmognathus ochrophaeus* and *D. fuscus* are also distributed randomly with respect to each other.

Several researchers have examined factors that affect diurnal activity patterns and spatial patterning. *Desmognathus fuscus*, *D. ochrophaeus*, and *Eurycea bislineata* that inhabit streamside habitats in Ohio are most active on the ground surface during the first hour after sunset (Holomuzki 1980). This interval corresponds to a time when prey are most active on the ground surface. Keen (1984) studied the relationship between substrate moisture, nocturnal activity pattern, and size of substrate used for cover. Under low-moisture conditions, nocturnal activity is curtailed during the latter half of the night and animals prefer large rather than small rocks for cover. Thus, moisture stress could cause local clumping of animals beneath moist substrates. In laboratory trials, individuals show a preference for sites they have previously occupied (Keen et al. 1987). There is no evidence, however, that individuals use chemical cues or pheromones to mark or recognize home sites.

Adults are active on the ground surface at night throughout the warmer months of the year but move into underground retreats during the winter months (Ashton 1975; Hamilton 1943; Organ 1961a; Orser and Shure 1975). Adults studied by Ashton (1975) move into winter retreats 12–25 cm deep when the stream temperature drops below 7°C and occasionally gather in dense aggregates. As many as 10–15% of the specimens that Hamilton (1943) uncovered in winter retreats in New York were dead. Dietary analyses suggest that individuals feed while in winter retreats (Ashton 1975; Hamilton 1943).

PREDATORS AND DEFENSE. Uhler et al. (1939) found that dusky salamanders are occasionally taken by water snakes (*Nerodia sipedon*) and garter snakes (*Thamnophis sirtalis*). Other predators—such as raccoons, skunks, and birds—also undoubtedly eat them.

The spring salamander (*Gyrinophilus porphyriticus*) is an important predator on this and other salamanders in the Appalachian Mountains.

Dusky salamanders lack obvious chemical defenses against predators and use other mechanisms such as fleeing or biting to reduce predation. Individuals often bite and autotomize their tails when attacked by garter snakes (Whiteman and Wissinger 1991); 33% of 239 museum specimens examined by Wake and Dresner (1967) had missing portions of tails. Individuals may also remain immobile when uncovered, making themselves less conspicuous to predators (Dodd 1990b).

COMMUNITY ECOLOGY. Dusky salamanders are important invertebrate predators in headwater stream communities and often share microhabitats with other salamanders. Southerland (1986d) found that animals from a Piedmont site that lacks congeners have a broader preference of substrate types than individuals from a Blue Ridge population that coexist with members of three other *Desmognathus* species. This finding suggests that niche specialization with respect to microhabitat use may occur when salamander communities are packed with *Desmognathus* species. Organ (1961a) presents distributional data suggesting that *D. fuscus* may be excluded from some stream habitats where both *D. monticola* and *D. quadramaculatus* co-occur. Additional studies of ecological interactions of *D. fuscus* with congeners by Keen (1982), Keen and Sharp (1984), Krzysik (1979), and Means (1975) are discussed under the accounts of *D. monticola*, *D. ochrophaeus*, and *D. auriculatus*.

CONSERVATION BIOLOGY. Like many aquatic organisms, dusky salamanders are sensitive to stream pollution and siltation. *Desmognathus fuscus* larvae are absent from many streams draining coal strip mines in eastern Kentucky and Tennessee (Gore 1983). Stream siltation and high metal concentrations appear to be the two primary factors in reducing or eliminating *Desmognathus* from these streams. Densities of *D. fuscus* in streams near Atlanta, Georgia, are inversely proportional to the degree of stream disturbance associated with urbanization (Orser and Shure 1972). Stream scouring, siltation, and loss of ground cover are the most likely reasons for low densities of *D. fuscus* in urban areas.

Desmognathus imitator (Dunn)
Imitator Salamander
PLATE 46

Fig. 108. *Desmognathus imitator;* adult; Swain Co., North Carolina (J. W. Petranka).

Fig. 109. *Desmognathus imitator;* adult with red cheek patch; Swain Co., North Carolina (J. W. Petranka).

IDENTIFICATION. The imitator salamander is a member of the *D. ochrophaeus* complex and is restricted to the Great Smoky, Balsam, and Plott Balsam mountains of eastern Tennessee and western North Carolina (Tilley 1985; Tilley et al. 1978). As in other members of the complex, the venter is gray and the tail is rounded and lacks a prominent keel. The dorsum often has a wavy dorsolateral stripe, but melanistic individuals that are mostly grayish black above are common. The toe tips lack cornifications. Adults measure 7–11 cm TL and there are normally 14 costal grooves.

Desmognathus imitator occupies stream and streamside habitats and often coexists locally with *D. ocoee.* Most local populations of *D. imitator* consist of color morphs both with and without yellowish, orangish, or reddish cheek patches. Because morphs of *D. ocoee* with colored cheek patches apparently do not occur within the range of *D. imitator,* any specimen collected with colored cheeks within *D. imitator*'s range is *D. imitator.* In many populations, specimens of *D. imitator* that lack colored cheeks are morphologically indistinguishable from *D. ocoee* and can only be identified reliably using molecular evidence. At sites along the crest of the Great Smoky Mountains, the two species can be distinguished with some degree of reliability by color pattern and the presence of dorsal bands. *Desmognathus imitator* typically has strongly undulating dorsolateral stripes and rarely has the straighter dorsal bands typical of sympatric *D. ocoee* (Tilley 1985; Tilley et al. 1978). Melanistic specimens are common in both species and are difficult to identify to species.

SYSTEMATICS AND GEOGRAPHIC VARIATION. Although *D. imitator* closely resembles *D. ocoee,* evidence from both courtship behavior (Verrell 1990b; Verrell and Tilley 1992) and electrophoretic analyses of proteins (Tilley et al. 1978) shows that these species are reproductively isolated from one another. Geographic variation in color patterns and protein patterns is described in detail in Tilley et al. (1978). Rockface populations at Waterrock Knob in the Balsam Mountains are genetically differentiated from populations that inhabit streamside and forest floor habitats elsewhere and will likely be described as a separate species (S. G. Tilley, pers. comm.).

DISTRIBUTION AND ADULT HABITAT. The imitator salamander is found in mountainous, forested habitats in the Great Smoky, Balsam, and Plott Balsam mountains at elevations of 900–2024 m. The range extends from the Smokies to just east of Soco Gap on Balsam Mountain. This species uses microhabitats similar to those used by *D. ocoee* but is generally found closer to streams and seepages (Tilley et al. 1978). The imitator salamander is also more common at lower elevations than is *D. ocoee.*

Range of *Desmognathus imitator*

BREEDING AND COURTSHIP. The mating season has not been established for *D. imitator*, but it probably involves both autumn and spring breeding, as is the case for many small *Desmognathus*. Courtship is indistinguishable from that of other members of the *D. ochrophaeus* complex and involves a period of sexual persuasion followed by a tail-straddle walk (Verrell 1994e; Verrell and Tilley 1992). The most detailed observations are those of Verrell (1994e) for specimens from Waterrock Knob in the Balsam Mountains of North Carolina. The systematic affinity of this population is currently under review, and the form may eventually be described as a sibling species.

Courtship behavior is initiated when a male begins pursuing a female to provide tactile and chemical stimulation. The male may butterfly the forelimbs during this phase when close to the female. Eventually the female stops fleeing and the male engages in head-rubbing bouts that may last as long as 6 minutes, accompanied by occasional butterflying. The male interrupts head-rubbing periodically, moves to the female's back region, and pulls or forcefully snaps with his snout. The male may also slide beneath the female's chin while undulating his tail, then curl the body backwards and forcefully snap the snout on the female's back. Snapping and pulling lacerate the skin and presumably function to deliver mental gland secretions to the female's circulatory system. Verrell (1994e) observed two instances in which a male seized the female in his jaws for 1–18 minutes. This behavior has also been recorded in *D. wrighti* and *D. aeneus*, and presumably functions to lacerate the skin and facilitate the transfer of mental gland secretions to the female.

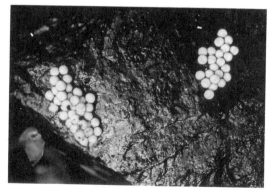

Fig. 110. *Desmognathus imitator*; eggs on underside of rock removed from seepage area; Swain Co., North Carolina (J. W. Petranka).

The female eventually straddles the male's tail and the pair engages in a tail-straddle walk, which is followed by spermatophore deposition and pickup. The period from the initiation of head-rubbing to the initiation of the tail-straddle walk lasts 180-215 minutes.

Courtship trials indicate that males of *D. ocoee* and *D. imitator* rarely initiate courtship with females of the opposite species (Verrell 1989a). Olfactory discrimination appears to play an important role in determining whether sexually active males are initially attracted to females of the same or different species.

COMMENTS. Most aspects of the life history of this species have not been studied. The general life cycle is presumably similar to that of *D. ocoee,* with mating occurring during the fall and spring months and oviposition occurring in late spring and early summer. S. G. Tilley (pers. comm.) found a clutch of eggs that were in all likelihood those of *D. imitator* on 21 June beneath the surface of a spring in North Carolina. C. K. Smith and I found two clutches of eggs with attending females that appeared to be *D. imitator.* Most eggs were attached singly in a tight monolayer on the underside of a rock that was embedded in a seepage area.

The imitator salamander is a Batesian mimic of red-cheeked *P. jordani* in the Great Smoky Mountains. Morphs of *D. imitator* with colored cheeks are common in most populations where *P. jordani* occurs and may comprise 15-20% of local populations (Tilley et al. 1978). Details of the mimetic relationships between *P. jordani* and members of the *D. ochrophaeus* complex are provided in the account of *P. jordani*. Antipredator defenses include remaining immobile when uncovered, followed by fleeing (Dodd 1990b).

Desmognathus imitator resides almost entirely within national park boundaries and most populations are protected. The greatest long-term threat to this species may be acidification of waters and soils from acid precipitation falling on high-elevation sites in the southern Appalachian Mountains.

Desmognathus marmoratus Moore
Shovel-nosed Salamander
PLATES 47, 48

Fig. 111. *Desmognathus marmoratus;* adult; Watauga Co., North Carolina (R. W. Van Devender).

IDENTIFICATION. The shovel-nosed salamander is a large desmognathine salamander that is closely related to *Desmognathus quadramaculatus* (Titus and Larson 1996). The dorsal ground color of adults is dark brown to black and is frequently overlain with two rows of irregular, yellowish brown or grayish blotches (Martof 1962). Two rows of light spots usually occur along each side of the body, but in some populations the adults are dark dorsally with little evidence of spotting. The venter is light colored in subadults and dark gray with a lighter central area in adults. In some populations the belly may be uniformly dark and lack a lighter central area, or be uniformly patterned with dark mottling. The tail is laterally compressed and strongly keeled above, and the toe tips are dark. The internal nares form slits rather than round pores as in other *Desmognathus* species. Adult males and females often lack vomerine teeth or have fewer than five vomerine teeth.

Males grow slightly larger than females; the largest males in five populations examined by Martof (1962) are 6-13% greater in average SVL than the largest females. Males also have relatively wide heads and fewer vomerine teeth (when they are present at all in local populations), enlarged maxillary teeth, papillate cloacae (smooth in females), and an inconspicuous mental gland on the chin. Albinistic individuals compose as much as 9% of some populations. Adults measure 8-15 cm and there are 14 costal grooves. The shovel-nosed salamander is easily confused with *D. quadramaculatus.* Martof (1962) listed a series of char-

Fig. 112. *Desmognathus marmoratus;* ventral view of adult; Sevier Co., North Carolina (J. W. Petranka).

acters that are useful in separating the two species, the most reliable being the shape of the internal nares.

Hatchlings measure 11 mm SVL on average. They have two rows of light spots on the dorsum and closely resemble hatchling *D. quadramaculatus*. Older larvae resemble *D. quadramaculatus* larvae but are darker, have conspicuous light flecks along the sides of the body, are more slender, have longer legs, and have more spatulate tails (Martof 1962).

SYSTEMATICS AND GEOGRAPHIC VARIATION.
Like most desmognathines, shovel-nosed salamanders show marked geographic variation in color patterns. As many as five subspecies have been proposed for *D. marmoratus,* but none is currently recognized. Regional populations vary from melanistic with little dorsal marking to brightly blotched (Martof 1956). Based on electrophoretic comparisons of protein variants, Voss et al. (1995) found that populations in the Tennessee drainage are very different from populations in the Chattahoochee and Savannah drainages to the south (Nei's $D = 0.76$). Degree of genetic divergence among groups is correlated with the time that populations in major drainages have been isolated from each other. This species has traditionally been placed in a separate genus (*Leurognathus*); however, mitochondrial DNA evidence does not support this interpretation (Titus and Larson 1996).

DISTRIBUTION AND ADULT HABITAT.
Shovel-nosed salamanders inhabit cool, well-oxygenated streams at elevations of 300–1680 m from southwestern Virginia to northern Georgia. Populations are patchily distributed and appear to be absent from some of the drainages in this region. This species is generally more common in second- and third-order streams in the southern Appalachians at elevations below 1220 m. Populations may also inhabit headwater streams with gentle gradients (Martof 1962; Pope and Hairston 1947). Ideal microhabitats for adults are shallow areas in streams having angular rocks, loose gravel, and moderate to fast-flowing water. Bait dealers have introduced this species into some areas outside its natural range (Martof 1953, 1962).

Range of *Desmognathus marmoratus*

BREEDING AND COURTSHIP. Information on the breeding season and courtship behavior is not available. Martof (1962) believed that females breed every other year, but this hypothesis has not been substantiated.

REPRODUCTIVE STRATEGY. Females oviposit in late spring or early summer and remain with their clutches through hatching. The eggs are attached to the undersides of rocks in fast-flowing water. At a site in northern Georgia, the eggs are attached to the undersides of relatively large rocks in water that is 8–37 cm deep. The eggs are attached either singly or in groups of two to four in tight clusters that average 4–5 cm in diameter. At a western North Carolina site, eggs have been found attached singly in two layers to the edge of a submerged stone in fast-flowing water (C. H. Pope 1924).

Freshly laid eggs are unpigmented, about 4 mm in diameter, and surrounded by a transparent capsule attached by a short pedicel (Martof 1962; C. H. Pope 1924). Egg masses can be found in a northern Georgia stream between 22 May and 10 August (Martof 1962). Data on developmental stages indicate that most females at this site oviposit from May to early July.

Mean clutch size often varies markedly between populations, a phenomenon that appears to be due primarily to differences in average female body size. Number of mature ova in specimens from several populations in northern Georgia and North Carolina varies from 20 to 65 and is positively correlated with female SVL (Martof 1962). Mean clutch size ranges from 24 ova in a North Carolina population to 40–42 ova in several Georgia populations. C. H. Pope (1924) reports a clutch of 28 eggs with fully developed embryos found on 6 August in western North Carolina.

In northern Georgia the embryonic period lasts 10–12 weeks and the embryos hatch from mid-August to mid-September (Martof 1962). Hatching in western North Carolina occurs in August and September (Bruce 1985a; Pope 1924).

AQUATIC ECOLOGY. Hatchlings are relatively weak swimmers and are washed downstream into pools or other microhabitats with low current (Martof 1962). The larvae and adults are secretive during the day and often remain beneath rocks with the head slightly protruding. The larvae feed mostly on insects and are more active outside cover at night.

Fig. 113. *Desmognathus marmoratus;* larva; Watauga Co., North Carolina (R. W. Van Devender).

Bruce (1985a) estimates the larval period to be about 3 years in a population in western North Carolina. Larvae grow 7–10 mm SVL per year and reach a maximum size of 37–38 mm SVL. At one Georgia site metamorphs reach an average SVL of 30–36 mm and are present from May through October (Martof 1962). Average size at metamorphosis declines seasonally in this population, and the larval period lasts 10–20 months. In other populations, average size at metamorphosis is 26–33 mm SVL.

Juveniles remain in the stream proper, where they occupy habitats similar to those of the adults. Males at a Georgia site become sexually mature at 43–50 mm SVL and females at 55–59 mm SVL (Martof 1962). In rare instances larvae with enlarged gonads have been collected that appear to be nearing sexual maturity.

Adult *D. marmoratus* are more aquatic than adult *D. quadramaculatus* and are frequently found beneath submerged stones in fast current. In laboratory aquaria, adults remain beneath cover during the day but emerge after dark and move about. Adults reach their highest densities in small to medium-sized trout streams with broken rocks and loose gravel. Specimens are most abundant in shallow rapids and riffles, where population density sometimes exceeds 6 salamanders/m^2 (Martof 1962). The adults move slowly away when disturbed, in contrast to the rapid fleeing characteristic of *D. quadramaculatus* (Bishop 1924; Martof 1962; C. H. Pope 1924). This behavioral difference is often useful in identifying specimens in the field. Southerland (1986e) collected eight adults in pens immediately next to a study stream, indicating that adults may move 1–2 m away from streams on rainy nights.

Martof and Scott (1957) document a variety of

prey in 150 specimens from Georgia and North Carolina. Mayflies, caddisflies, and true flies are the most numerically important prey. In addition, specimens eat stoneflies, beetles, hymenopterans, crayfishes, mites, snails, and salamanders, including *Eurycea bislineata* and conspecifics. Adults continue feeding through late autumn after water temperatures drop. Comparison of stomach contents with stream samples of invertebrates suggests that *D. marmoratus* is a generalist feeder that will attempt to capture most prey it encounters. Stoneflies with hard stone cases seem to be less preferred than soft-bodied species such as mayflies.

PREDATORS AND DEFENSE. Both trout and conspecifics feed on shovel-nosed salamanders. Martof (1962) made an unusual field observation regarding the food chain in one stream. He collected a large water snake (*Nerodia sipedon*) that was swallowing a trout. In the trout's throat was a large *D. marmoratus*, which, in turn, had eaten a conspecific. In the stomach of the conspecific were several aquatic insects. Mathews (1982) found that large nymphs of the stonefly *Acroneuria* attack and kill small *D. marmoratus* in laboratory aquaria. The extent to which predation from stream invertebrates occurs in nature is not known.

COMMUNITY ECOLOGY. *Desmognathus marmoratus* often coexists in streams with *D. quadramaculatus*. These species have very similar life-styles, but competitive and predatory interactions between the two species have not been examined.

CONSERVATION BIOLOGY. Dams have eliminated many populations of *D. marmoratus*. Martof (1962) notes that *D. marmoratus* is often absent from streams that are heavily silted and that contain bedrocks embedded in the substrate. Forestry and agricultural practices that minimize siltation of streams would benefit this and most other aquatic vertebrates.

Desmognathus monticola Dunn
Seal Salamander
PLATES 49, 50

Fig. 114. *Desmognathus monticola*; adult; Knott Co., Kentucky (R. W. Barbour).

IDENTIFICATION. The seal salamander is a relatively large *Desmognathus* with a tail that is rounded on the anterior third and laterally compressed and keeled on the posterior two-thirds. The dorsum often has reticulate, mottled, or wormy, dark brown or blackish markings on a light brown or grayish background. In many populations the wormy markings and dark mottling are absent or greatly reduced. The venter is whitish in young animals and whitish to light gray in older individuals. Melanophores on the venter are uniformly spaced so that the belly lacks any conspicuous mottling. A single line of white spots often occurs along the sides between the legs. The toe tips have cornified friction pads and are usually darker than other portions of the toes.

Adult males have enlarged premaxillary teeth, an inconspicuous mental gland on the chin, and papillose cloacal lips. The maximum adult size of males generally exceeds that of females, although in one sample males on average are only 2% larger in SVL than females (Bruce 1993). Adults measure 7.5–15 cm TL and there are 14 costal grooves.

Hatchlings measure on average about 11–12 mm SVL and the hatchlings and older larvae have four to five pairs of light dorsal spots between the limbs. The toe tips of older larvae usually have black cornifications. Young juveniles have four or five pairs of chestnut-colored spots that become fused and invaded by darker pigment with age.

SYSTEMATICS AND GEOGRAPHIC VARIATION. Genetic variation has not been examined, but future research will likely reveal much regional differentiation, as is typical of most *Desmognathus* species. Hoffman (1951) noted that specimens in the Blue Ridge Mountains of Virginia have less conspicuous dark dorsal patterning than specimens elsewhere (*D. m. monticola*) and described these as a different subspecies (*D. m. jeffersoni*). However, populations that I have observed in southwestern Virginia and western North Carolina are highly variable and often conform in dorsal patterning more to *D. m. jeffersoni* than to *D. m. monticola*. Given the variability in color patterning exhibited among local and regional populations of *D. monticola*, the recognition of subspecies based on the degree of dorsal patterning is not warranted.

DISTRIBUTION AND ADULT HABITAT. The seal salamander ranges from southwestern Pennsylvania to southern Alabama and extreme western Florida. This species is largely confined to the Appalachian Mountains, but populations occur in several Coastal Plain localities in southern Alabama and extreme western Florida (Means and Longden 1970; Mount 1975; Rose and Dobie 1963). Geographic isolates are also known from east-central and north-central Georgia. Seal salamanders are most common in hardwood forests in or adjoining small to moderate-sized, rocky streams with well-aerated, cool water (Hairston 1949; Krzysik 1979; Mount 1975). Although populations range in elevation as high as 1555 m, they are most abundant at elevations below 1219 to 1372 m (Hairston 1949; Organ 1961a). At relatively high elevations in mountainous regions, *D. monticola* often burrows in streambanks and avoids the stream proper, where *D. quadramaculatus* lives (Hairston 1949, 1986; Organ 1961a). At low elevations, *D. monticola* often lives beneath rocks and other cover directly in the streambed (Krzysik 1979), and in some populations it is not spatially segregated from *D. quadramaculatus*.

BREEDING AND COURTSHIP. Information on the breeding season is very limited. Courtship has been observed in September, October, and April in Virginia (Organ 1961a), suggesting that mating occurs in the spring, late summer, and autumn. Organ (1961a) concluded that females reproduce biennially in Virginia, but Tilley (1968) questioned this interpretation and suggested an annual breeding cycle.

The seal salamander engages in a stereotypical tail-straddle walk that is similar for all *Desmognathus* species. The following account is based on observations by Brock and Verrell (1994) and Organ (1961a). The latter authors note that geographic variation occurs in some aspects of the behaviors described below.

Sexually active males seek out and remain in close proximity to females. When approaching a female, a male will often "butterfly," a behavior that involves making circular strokes of the front limbs that resemble the butterfly stroke of swimmers. In some instances only a single limb is butterflied, whereas in others both limbs are rotated simultaneously. The male may also stroke the neck and head of the female with his forearms. Males in certain populations may also jerk the body to and fro when approaching females.

After what is often a prolonged period of pursuit and orientation, the male initiates a new phase of courtship that involves rubbing his head along the female's head. Butterflying of the limbs typically accompanies this phase of courtship, and individual episodes of head-rubbing may last more than 10 minutes. The male eventually slides his body under the female's chin and undulates his tail. Once the female's head is positioned on the dorsal surface of the base of the tail, the male may curve his body backward toward the female's neck or dorsum and pull or snap with the teeth to scratch the female's skin. This action transfers mental gland secretions into the female's circulatory system and can also occur when a male is positioned alongside a female. During this phase, the female often moves away from the male, and the male pursues and continues to stimulate her.

The pair eventually engages in a tail-straddle walk in which the male laterally undulates his tail and slides under the female's head until her chin rests on the dorsal base of the tail. The male moves forward in a straight line while undulating his tail laterally, and the female follows as she straddles his tail. Eventually the male increases the intensity of tail undulating and deposits a spermatophore. He then leads the female forward, and she picks up the sperm cap in her cloaca while laterally undulating her pelvic region. Courting males are often aggressive toward each other and will attack and bite potential competitors that could interfere with courtship.

REPRODUCTIVE STRATEGY. Females attach their eggs in cryptic sites that are in or near running

Range of *Desmognathus monticola*

water; they are frequently found with the eggs (Bruce 1996; Organ 1961a; C. H. Pope 1924). The eggs are attached singly to a support structure to form a monolayer, or in some instances a loose group that may be two to three egg layers thick (C. H. Pope 1924). Each egg is attached by an elastic stalk so that movements of the attending female or water currents rock the eggs back and forth. Organ (1961a) found most nests about 30 cm below ground in stream banks where water freely percolated through the surrounding substrate. Bruce (1990, 1996) found them buried in rock, gravel, and mud in the bed of a small stream, or attached to small rocks buried in the streambed or streambank. In some cases a female may attach her eggs to several adjacent small stones or nearby rootlets. I have found eggs in western North Carolina in thick leaf packs in streams. Two clutches discovered by Bruce (1990) contained ova that averaged 4.1 mm in diameter.

Eggs in early developmental stages have been found from mid-June to early August. Nesting records for North Carolina include two nests with 21 and 30 eggs in early embryonic stages in late July and early August (C. H. Pope 1924) and five clutches (mean = 25; range = 18–27 ova) found in early to mid-July (Bruce 1990, 1996). Number of eggs per nest varies from 16 to 39 and averages 27 in southwestern Virginia populations (Organ 1961a). Eggs in these populations are in early developmental stages in June and near hatching in early September. Hatchlings first appear in September, suggesting an incubation period of about 2 months.

Estimates of clutch sizes based on counts of enlarged ovarian eggs are similar to or slightly higher than egg counts from nests. The number of mature ova in specimens examined by Tilley (1968) from the southern Appalachians varies from 16 to 40, averages 27, and is independent of female SVL. Estimates for the average number of ova in two western North Carolina populations vary from 24 to 30 depending

Fig. 115. *Desmognathus monticola*; eggs on underside of rock; Macon Co., North Carolina (J. W. Petranka).

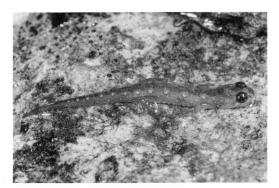

Fig. 116. *Desmognathus monticola;* larva; Henderson Co., North Carolina (R. W. Van Devender).

on the sample size and site (Bruce 1995, 1996; Bruce and Hairston 1990). SVL is positively correlated with number of ovarian eggs at one site, but not at the second.

AQUATIC ECOLOGY. Very few data are available on the ecology or life history of the larvae. Embryos hatch in late summer or early fall. The larvae overwinter, then undergo a relatively brief period of growth before transforming the following year. Organ (1961a) collected hatchlings in September and recently transformed larvae in June and July, suggesting a larval period of 10-11 months. Juterbock (1984) estimates a larval period of 8-9 months in Kentucky populations, with transformation occurring in May and June. The mean sizes of individuals in a western North Carolina population that are 1, 2, 3, and >4 years old are 16.7, 28.9, 40.4, and 55.3 mm SVL, respectively (Bruce 1990).

TERRESTRIAL ECOLOGY. Males in southwestern Virginia become sexually mature when 3.5 years old (Organ 1961a). Females become mature when 4.5 years old and first reproduce the following year when 5 years old. Males and females in western North Carolina populations first reproduce when 4-5 and 5-7 years old, respectively, and when they have reached 46-48 and 52-53 mm SVL (Bruce 1989, 1990, 1995; Castanet et al. 1996). Data for other populations indicate that males and females first become sexually mature when they have reached >41 and >47 mm SVL (Juterbock 1978; Means and Longden 1970; Organ 1961a). Females undergo little growth after maturing sexually and do not grow as large as males. Skeletochronological aging of a western North Caro-

lina population shows that growth rate declines to near zero as individuals reach 60-65 mm SVL and 7-8 years of age (Castanet et al. 1996). Some individuals may live as long as 11 years.

Metamorphosed *D. monticola* remain beneath rocks or logs or in burrows in streambanks during the day. They emerge at night shortly after dark and forage for invertebrates and other salamanders (Brandon and Huheey 1971; Hairston 1986; Shealy 1975). At night adults either sit with their bodies partially extended from their burrow entrances waiting for prey or actively move about in search of prey. Adults in the southern Appalachians sometimes climb 1-2 m above ground on tree trunks on rainy nights. Nocturnal surface activity peaks around midnight, with a secondary burst of activity near dawn (Hairston 1949, 1986; Shealy 1975).

The juveniles and adults feed on both aquatic and terrestrial invertebrates. Individuals in Pennsylvania incorporate larger prey into their diets as they grow (Krzysik 1979). Animals of all size classes eat a variety of invertebrates, including stoneflies, caddisflies, bugs, lepidopterans, beetles, flies, wasps, millipedes, and earthworms. Individuals in North Carolina populations feed mostly on terrestrial insects (Hairston 1949).

Seal salamanders will readily defend cover sites from conspecific intruders regardless of the sex of the intruder (Keen and Sharp 1984). The experimental addition of food to a cover site typically increases the site specificity of animals and reduces the amount of time spent outside refuges at night (Keen and Reed 1985). Individuals often show microhabitat segregation with respect to size, which may reflect ontogenetic shifts in the size of cover objects preferred as well as the active avoidance of larger individuals by smaller conspecifics. Adults prefer larger substrates than juveniles (Krzysik and Miller 1979), but juveniles also avoid cover objects occupied by larger conspecifics (Southerland 1986c).

Colley et al. (1989) found that juveniles and adults typically use different microhabitats that vary in substrate size, moisture level, and coarseness. Adults prefer large cover objects with complex microhabitats and high moisture. These differences appear to reflect the innate preferences of individuals, as well as juveniles actively avoiding adults. Microhabitat segregation may benefit juveniles by lessening predation risks from large adults and by reducing interference competition.

Individuals will often move temporarily out of streams and onto streambanks or the forest floor after heavy rains (Kleeberger 1984, 1985). However, as a general rule resident animals have small home ranges and rarely move long distances. A study of small-scale movements of *D. monticola* on a rocky bank shows that most adults remain in the same microhabitat for extended periods of time (Brandon and Huheey 1971). In another study one animal that was intensively monitored over a 57-day period remained sedentary for long periods of time, but moved 9 m during one period of heavy rain. The estimated home range is 8.4 m^2 in a Kentucky population (Hardin et al. 1969), compared with 0.07–0.45 m^2 for populations studied by Kleeberger (1984, 1985) in western North Carolina.

Kleeberger (1984) estimates densities to be 0.72–1.4 individuals/m^2 in optimal habitats in a study stream in western North Carolina. Average lipid content in experimental animals decreases slightly relative to controls when density is increased experimentally in natural habitats. Experimental plots in this study contain both resident and foreign animals collected from nearby watersheds, whereas control plots contain only resident animals. Thus it is uncertain whether reduction in lipid content is related to stress from being transported to unfamiliar habitats or to some density effects such as competition for food or space. When cover is experimentally added to plots, surface densities of salamanders increase (Kleeberger 1984, 1985); this finding indicates that cover objects are limiting and are a potential source of competition. Data provided by Bruce (1995) suggest densities on the order of 2.5–3.5 animals/m^2 in optimal stream habitats.

Surface activity in a South Carolina population peaks in April and generally declines throughout the remainder of the year. A second peak may occur in winter following warming trends (Shealy 1975). Surface activity is most strongly correlated with soil temperature.

Age-specific survivorship is fairly constant in transformed males throughout life but tends to increase slightly as females become sexually mature (Organ 1961a). The mean annual survival rate is about 50% during the early years of life for males and immature females. The annual rate of survival of brooding females is only 30%. These estimates assume that females breed biennially. In contrast to Organ's findings, Tilley (1968) found that age-specific survivorship decreases with age in western North Carolina populations. Annual survival of 50–60% is required to maintain stable numbers of females in a study population in North Carolina (Bruce 1995).

PREDATORS AND DEFENSE. Very few data are available on natural predators. Spring salamanders and black-bellied salamanders undoubtedly prey upon juvenile and larval *D. monticola* in the southern Appalachians. Wake and Dresner (1967) report that 11% of a sample of museum specimens have broken tails.

COMMUNITY ECOLOGY. Ecological interactions of *D. monticola* with *D. quadramaculatus*, *D. ochrophaeus*, and *D. ocoee* have been studied by several researchers and are presented in detail under the accounts of these species. Kleeberger (1984) manipulated salamander density and cover objects in a stream in western North Carolina and concluded that intraspecific competition for food and cover objects may be important for juvenile and adult *D. monticola*.

Keen (1982) compared activity patterns and a variety of environmental factors relating to microhabitat use in *D. monticola* and *D. fuscus*. The results indicate that the two species differ in their preferred substrate texture, substrate moisture, size of cover objects, and diel activity patterns. However, the presence or absence of one species does not significantly influence the microhabitat choice or activity pattern of the second.

In field enclosures *D. fuscus* tends to use forest floor and streambank habitats more frequently and to be less active when in the presence of *D. monticola*. However, *D. monticola* does not shift habitat use or level of activity when co-occurring with *D. fuscus*. The density of salamanders used in Keen's field experiments (4.3/m^2) was higher than that reported for natural populations of *D. fuscus* and *D. monticola*, so the results should be interpreted cautiously.

Keen and Sharp (1984) studied the agonistic responses of resident *D. monticola* to *D. monticola* and *D. fuscus* intruders. Residents are more likely to be aggressive toward smaller intruders in general and toward *D. fuscus* than *D. monticola*. In some instances residents will bite off portions of the tails of intruders. The extent to which aggression occurs in natural populations is unknown. Southerland (1986c) found no evidence of aggression in *Desmognathus* species maintained in outdoor cages even though animals are aggressive in the laboratory. Collectively, however, the work of Keen (1982) and Keen and Sharp (1984) suggests that *D. monticola* may influence the spatial distribution of *D. fuscus* when the two coexist at high densities.

192 Plethodontidae: Desmognathinae

CONSERVATION BIOLOGY. We know little about the sensitivity of adult *Desmognathus* to stream acidification. Juvenile and adult *D. quadramaculatus* and *D. monticola* that are exposed for 3 weeks to water of pH 3.5–7.2 do not suffer any mortality (Roudebush 1988). However, feeding rates tend to decrease with increasing acidity.

Desmognathus ochrophaeus Cope
Allegheny Mountain Dusky Salamander
PLATE 51

Fig. 117. *Desmognathus ochrophaeus;* adult; eastern Kentucky (R. W. Barbour).

IDENTIFICATION. The Allegheny Mountain dusky salamander is a member of the *D. ochrophaeus* complex, which consists of *D. apalachicolae, D. carolinensis, D. imitator, D. ocoee, D. orestes,* and *D. ochrophaeus* (Tilley and Mahoney 1996). Members of the complex are sibling species that are best identified by geographic range and molecular data. The five species in the complex that occur outside the Coastal Plain are medium-sized *Desmognathus* that have pigmented testes and rounded, unkeeled tails. The tail is slightly longer than the body, and the toe tips lack cornifications. Dorsal color patterning ranges from straight-edged stripes to wavy, blotched patterning that may be brownish, yellowish, or reddish (Martof and Rose 1963; Tilley 1969). The venter becomes light gray to grayish black with age and lacks conspicuous mottling. Older animals are often melanistic, particularly males.

Relative to females, males of all five species have well-developed premaxillary teeth and large jaw musculature. Males also possess small mental glands, are on average slightly larger than females, have a mouth line that appears sinuate when viewed from the side, and tend to lose the vomerine teeth with age. Males of the *D. ochrophaeus* complex are on average 6–20% longer in SVL than females (Bruce 1993; Martof and Rose 1963; Orr 1989). Adults measure 7–11 cm TL and there are 14 costal grooves. Hatchlings measure 13–18 mm TL and have prominent yolk reserves (Bishop 1941a; Bishop and Chrisp 1933; Huheey and Brandon 1973; Orr 1989).

Desmognathus ochrophaeus is the northernmost member of the complex and ranges from eastern Tennessee and southwestern Virginia to New York and extreme southern Quebec. This species is somewhat distinguished from other members of the complex in that most individuals have relatively straight dorsolateral stripes (Tilley and Mahoney 1996). Specimens also often have a row of middorsal melanophore patches that are frequently chevron shaped. Hatchlings and larvae often have straight dorsolateral stripes and either lack alternating spots along the dorsum or have them restricted to the posterior portion of the trunk. Juveniles and young adults have reddish or yellowish dorsal stripes, whereas older males are often melanistic and lack a dorsal stripe. Males from northeastern Ohio are on average 6% longer in SVL than females (Orr 1989).

SYSTEMATICS AND GEOGRAPHIC VARIATION. Populations that have traditionally been referred to as *D. ochrophaeus* in the eastern United States exhibit marked geographic variation in body size, coloration, color patterning, size at sexual maturity, and other traits. Different interpretations of geographic variation in this group have led to an unstable nomenclature and much debate about the taxonomic status of many forms. Tilley and Mahoney (1996) conducted a detailed analysis of geographic variation in protein variants and split populations referred to *D. ochrophaeus* into four species (see detailed summary under the account of *D. carolinensis*). The name *D. ochrophaeus* is now restricted to northern populations that extend from eastern Ten-

Fig. 118. *Desmognathus ochrophaeus;* adult; Herkimer Co., New York (J. W. Petranka).

nessee and southwestern Virginia northward to New York and extreme southern Quebec.

The Allegheny mountain dusky salamander is the least variable member of the *D. ochrophaeus* complex in terms of color patterning and genetic variation. This species appears to have been derived from a southern form that rapidly expanded its range northward following the last glacial period. Populations distributed over a broad geographic region are surprisingly uniform, with Nei's genetic distance (*D*) between populations averaging only 0.06 (Tilley and Mahoney 1996). Individuals typically have a straight-edged dorsal stripe that is most frequently reddish to light brown, and geographic variation in color patterning is not as pronounced as in southern members of the complex.

DISTRIBUTION AND ADULT HABITAT. The Allegheny mountain dusky salamander is confined to forested habitats from the northern edge of the Adirondack Mountains in Quebec, Canada (Sharbel and Bonin 1992), southward to eastern Tennessee and extreme southwestern Virginia. Populations extend through the Allegheny Mountains and Plateau southward and occupy the higher ridges of the Ridge and Valley Province, including the Brumley, Clinch, Walker, and Potts mountains of southwestern Virginia (Tilley and Mahoney 1996). Scattered populations occur in the Cumberland Plateau of eastern Kentucky and eastern Tennessee, but the southern limit in eastern Tennessee is poorly defined. An isolate occurs in central Vermont. Populations are concentrated in or near seepages or streams but at relatively high elevations may be found on the forest floor away from running water.

BREEDING AND COURTSHIP. Mating occurs during the spring, late summer, and autumn. Adults have been found in New York with sperm caps in their cloacae in October and May (Bishop 1941a; Bishop

Range of *Desmognathus ochrophaeus*

and Chrisp 1933). Most females reproduce annually in populations in Ohio (Fitzpatrick 1973; Keen and Orr 1980) and presumably elsewhere. Courtship involves a stereotypical tail-straddle walk characteristic of all *Desmognathus* species (Houck et al. 1985a). Courtship behavior has not been described in detail, but it is presumably nearly identical to that described under the account of *D. orestes*.

REPRODUCTIVE STRATEGY. *Desmognathus ochrophaeus* shows a stronger tendency to nest underground than do southern members of the *D. ochrophaeus* complex (Keen and Orr 1980). In Ohio females often nest in seepage banks (Orr 1989), whereas in southwestern Virginia nests have been found in a dried, muddy streambed far from running water (Wood and Wood 1955). Females at the latter site oviposit in mud crevices and depressions beneath logs and stumps that are embedded in the mud. Eggs or recently hatched larvae have been found in Pennsylvania and New York beneath logs and stones embedded in hillside seeps and springs (Bishop and Chrisp 1933).

The eggs are deposited in small, grapelike clusters in hollowed depressions. Each egg has a short, gelatinous stalk and is attached to a common support to form a cluster. Eggs are sometimes laid singly or in small numbers next to the main cluster (Bishop 1941a). Mature ova measure an average of 2.5–3.0 mm in diameter and developing embryos may be 4 mm in diameter (Bishop 1941a; Keen and Orr 1980). Females remain with their clutches through hatching. Most females have a negative energy budget and lose weight over the brooding season. An estimated 16% of the annual energy budget is used during brooding (Fitzpatrick 1973), and females may prey upon eggs or hatchlings.

Most females in Ohio oviposit during late spring and early summer, but a small percentage lay eggs in late winter and early spring in underground retreats (Keen and Orr 1980). With rare exceptions, oviposition begins in March, reaches a peak in mid-May, and continues through September (Orr 1989). The seasonal time of oviposition is poorly documented in other areas of the range, but it appears to occur primarily from May to July. The number of ovarian eggs in a Pennsylvania population varies from 8 to 24, averages 16, and is positively correlated with female SVL (Hall 1977). Additional estimates of clutch size are 11–14 eggs for seven nests in New York (Bishop 1941a) and 10–37 eggs (mean = 20) for 20 nests in Virginia (Wood and Wood 1955).

The time of hatching varies markedly depending on the egg-laying schedule of females in local populations. Most embryos in Ohio populations hatch between 14 September and 6 October, with a peak in hatching in early October (Keen and Orr 1980; Orr 1989). Clutches at or near hatching have been found in March, September, and October in New York (Bishop 1941a), whereas a small percentage of hatchlings appear in Ohio in mid-April (Orr 1989). Late-term clutches found in New York in March are presumably from eggs laid in late fall or early winter.

AQUATIC ECOLOGY. Larvae live in seepages or sluggish portions of streams and feed on small invertebrates. Length of the larval period varies among local populations and may last from 2–3 weeks to as long as 8 months (Bishop and Chrisp 1933; Keen and Orr 1980). Larvae that transform within a few weeks after hatching presumably rely on yolk reserves for energy and may not feed before metamorphosing.

Factors such as the time of oviposition and seasonal drying patterns of streams and seepages may affect the duration of the larval period. Larvae from eggs laid in the late winter or spring typically transform the following summer or early autumn. In contrast, those from eggs laid in the summer may overwinter and transform the following year. Transforming larvae have been collected from March through October. Estimates of mean size at metamorphosis are 10 mm SVL in Ohio (Bishop 1924; Keen and Orr 1980) and 18 mm TL in Pennsylvania (Bishop and Chrisp 1933).

TERRESTRIAL ECOLOGY. Allegheny mountain dusky salamanders are nocturnally active and remain beneath cover objects or in recesses during the day. They emerge soon after sunset and actively move about in search of prey or mates. Peak surface activity in an Ohio population occurs shortly after dark (Holomuzki 1980). The juveniles and adults appear to be generalist feeders. Pennsylvania specimens eat a diverse array of terrestrial invertebrates, and the average prey size increases with SVL (Krzysik 1980b). The most important prey are larval lepidopterans and adult and larval dipterans. Specimens from northeastern Ohio consume snails, isopods, mites, earthworms, and both larval and adult insects (Fitzpatrick 1973; Keen 1979). The number of prey eaten is positively correlated with the level of precipitation and declines when minimum daily temperatures are <5°C (Keen 1979).

Female *D. ochrophaeus* sometimes feed on their own eggs. Wood and Wood (1955) note that one female consumed four hatchlings and a second female, four eggs. In both instances the eggs or hatchlings were at the same developmental stages as eggs or hatchlings found in nests. Females of other members of the *D. ochrophaeus* complex will eat dead eggs in their nests, and female *D. ochrophaeus* presumably do the same. Thus the extent to which oophagy reflects the attacking of healthy embryos versus the adaptive removal of dead or diseased eggs from nests is not fully understood.

Young juveniles in an Ohio population grow about 7–9 mm SVL per year (Keen and Orr 1980). Females in this population first oviposit when 3–4 years old and 30–34 mm in SVL. The smallest females with mature ovarian eggs in a series collected by Hall (1977) in Pennsylvania measure >30 mm SVL.

Adults are aggressive and will actively defend space from both conspecifics and heterospecifics. *Desmognathus ochrophaeus* is aggressive toward *Plethodon cinereus*, and individuals can frequently displace even larger redbacked salamanders from cover objects (Smith and Pough 1994). In contrast, intruder red-backed salamanders are rarely able to displace resident *D. ochrophaeus* regardless of relative size. Spatial patterning of males, females, and juveniles appears to be mediated to some extent by chemical markers. When presented with choices of substrates marked with either their own or a conspecific's body chemicals, adult females exhibit no preferences, adult males prefer substrates marked with the chemicals of other males, and juveniles prefer substrates marked with their own chemicals (Evans and Forester 1996). The attraction of males to substrates marked by other males may be a mechanism by which males locate and drive off rivals, and the seeming avoidance of adult males by juveniles may be an adaptive behavior that reduces competition for food and minimizes injury or the risk of cannibalism.

Despite evidence of aggression and chemically mediated behavior in the laboratory, the extent to which adults establish and defend permanent seasonal habitats is poorly understood. In studies of spatial patterning, Stewart and Bellis (1970) found that the distribution of *D. ochrophaeus* beneath cover boards placed along a streambank is random and that individuals normally move to other cover within 1–2 days after being uncovered. *Desmognathus ochrophaeus* and *D. fuscus* were also distributed randomly with respect to each other. Holomuzki (1982) found that a minimum of 25% of animals displaced 30 m from their home range return home.

Estimates of *D. ochrophaeus* densities for stream populations are 0.96–1.20 animals/m^2 in optimal habitats in Ohio (Orr 1989) and 0.62–1.07 salamanders/m^2 along a Pennsylvania stream (Hall 1977). The mean home range of animals in Ohio is <1 m^2 (Holomuzki 1982). Individuals are active on the ground surface except during the coldest winter months. Adults and juveniles congregate in seepages or underground retreats during the winter months. In Ohio individuals emerge from winter retreats in late March through April and are active on the ground surface through October (Keen 1979).

PREDATORS AND DEFENSE. *Desmognathus ochrophaeus* is undoubtedly preyed upon by many vertebrates inhabiting mesic forests. Birds and small snakes are known predators (Coker 1931; Whiteman and Wissinger 1991). *Desmognathus ochrophaeus* will avoid skin extracts of *D. brimleyorum*, *Notophthalmus viridescens*, and conspecifics, but not those of *Plethodon richmondi* (Lutterschmidt et al. 1994). The authors surmised that this behavior represents an alarm response that reduces predation risk.

Individuals will readily autotomize their tails if the tail is grasped. Like almost all other plethodontids, these salamanders cannot voluntarily autotomize the tail, but they can break the tail loose from the body if the tail is grasped by a predator. Individuals may also remain immobile when uncovered, making themselves less conspicuous to predators (Dodd 1990b).

When attacked by garter snakes, *D. ochrophaeus* uses body flipping, biting, and tail autotomy to aid in escape (Whiteman and Wissinger 1991). *Desmognathus ochrophaeus* and *D. fuscus* from Pennsylvania rely more heavily on biting than does *Eurycea b. bislineata* (Whiteman and Wissinger 1991), and individuals will often bite the mouth or face of an attacking snake when seized by the body or tail. *Desmognathus ochrophaeus* also avoids substrates that are marked with the odors of the ringneck snake (*Diadophis punctatus*; Cupp 1994).

Dodd et al. (1974) found that the adults are noxious to avian predators and hypothesized that *D. ochrophaeus* is a Batesian mimic of *P. hubrichti*.

COMMUNITY ECOLOGY. Krzysik (1979) compared microhabitats of *D. monticola, D. fuscus,* and *D. ochrophaeus* in Pennsylvania and in most instances found significant differences in the substrate used, the cover used, and the mean distance of each species from running water. Prey size utilization is similar in interspecific comparisons of similarly sized groups (e.g., mature *D. ochrophaeus* versus immature *D. monticola*), but generally different when groups that differ substantially in size are compared.

Interspecific aggression may alter patterns of resource use in streamside salamander communities. *Desmognathus ochrophaeus* can successfully defend cover objects from *P. cinereus* intruders, and *D. ochrophaeus* intruders will drive resident *P. cinereus* from cover objects (Smith and Pough 1994). In the presence of *D. fuscus, D. ochrophaeus* shifts away from surface water, whereas in the presence of *D. monticola, D. fuscus* selects smaller cover objects. Krzysik (1979) hypothesized that interspecific interference competition (aggression) has been the driving force in the evolution of body size and terrestrialism in *Desmognathus,* since his dietary analysis shows no evidence of interspecific predation.

CONSERVATION BIOLOGY. *Desmognathus ochrophaeus* is a common species in the Appalachian Mountains and is in minimal need of protection throughout much of its range. Isolated mountaintop populations in eastern Kentucky and Tennessee may require closer watch.

Desmognathus ocoee Nicholls
Ocoee Salamander
PLATES 52, 53

Fig. 119. *Desmognathus ocoee;* adult; Macon Co., North Carolina (J. W. Petranka).

IDENTIFICATION. The Ocoee salamander is a member of the *D. ochrophaeus* complex, which consists of *D. apalachicolae, D. imitator,* and four sibling species that until 1996 were referred to as *D. ochrophaeus* (Tilley and Mahoney 1996). Members of the complex are so similar in size, general morphology, coloration, and color patterns that they are best identified by geographic range and molecular data. Five species in the complex that are found in the Appalachian Mountains (*D. carolinensis, D. imitator, D. ochrophaeus, D. ocoee,* and *D. orestes;* Tilley and Mahoney 1996) are made up of medium-sized individuals that have pigmented testes and rounded, unkeeled tails. The tail is slightly longer than the body, and the toe tips lack cornifications. Dorsal color patterning ranges from straight-edged stripes to wavy, blotched patterning (Martof and Rose 1963; Tilley 1969). The dorsal patterning in juveniles and young adults varies from mottled shades of brown or gray to bright shades of yellow or red. The venter is light colored in young animals but becomes light gray to grayish black with age. Older animals are often melanistic, particularly males. A partial albino adult is reported by Dyrkacz (1981).

Relative to females, males of all five species have well-developed premaxillary teeth and large jaw musculature. Males also possess small mental glands, are on average slightly larger than females, have a mouth line that appears sinuate when viewed from the side, and tend to lose the vomerine teeth with age. Males of the *D. ochrophaeus* complex are on average 6–20% longer in SVL than females (Bruce 1993; Martof and Rose 1963; Orr 1989). Adults measure 7–11 cm TL and there are 14 costal grooves. Hatchlings measure 13–18 mm TL and have prominent yolk reserves (Bishop 1941a; Bishop and Chrisp 1933; Huheey and Brandon 1973; Orr 1989).

The dorsal coloration and patterning of *D. ocoee* are highly variable. Individuals may have straight-edged dorsal stripes, but more typically they have four to six pairs of light blotches between the limbs that are separate in young animals. These fuse to varying degrees in older animals to form a wavy dorsal stripe that is highly variable in color (Martof and Rose 1963; Tilley 1969; Tilley et al. 1978). Hatchlings and larvae have rounded snouts and four to six conspicuous pairs of alternating light spots along the dorsum. Juveniles and young adults typically have reddish to brownish, wavy dorsal stripes. Older adults in high-elevation populations are often melanistic and have little dorsal patterning. Males in local populations are on average 3–15% larger than females and are more likely to be melanistic (Martof and Rose 1963).

SYSTEMATICS AND GEOGRAPHIC VARIATION. Populations that have traditionally been referred to as *D. ochrophaeus* in the eastern United States exhibit marked geographic variation in body size, coloration, color patterning, size at sexual maturity, and other traits. The complexity of geographic variation in this group has led to an unstable nomenclature and much debate about the taxonomic status of many forms. Tilley and Mahoney (1996) summarize the taxonomic history of the group and present detailed analyses of geographic variation in protein variants. They split populations previously referred to as *D. ochrophaeus* into four species (see detailed summary under the account of *D. carolinensis*). Populations south of the Pigeon River in western North Carolina, which until recently were treated as *D. ochrophaeus,* are now recognized as *D. ocoee.*

Desmognathus ocoee is the most geographically variable member of the *D. ochrophaeus* complex in terms of color patterning and genetic variation. Body size generally increases with elevation, but local populations at similar elevations may differ markedly in average size (Martof and Rose 1963; Tilley 1974). Individuals with straight dorsolateral stripes are absent or occur at very low frequencies in most local populations, with the exception of the Great Smoky Mountains and adjacent parts of the Great Balsam Mountains, where straight-striped morphs often predominate (Tilley 1969; Tilley and Mahoney 1996; Tilley et al. 1978). Specimens in the Nantahala Mountains and adjoining areas of the Blue Ridge sometimes have conspicuous red to yellowish coloration on the cheeks and/or upper portions of the legs and appear to be Batesian mimics of *Plethodon jordani* (Labanick 1983). Details of studies addressing mimicry are included in the account of *P. jordani*.

Populations on major mountain ranges within the range of *D. ocoee* are well differentiated genetically (average Nei's distance = 0.24; range = 0.01–0.44; F_{st} = 0.65), and certain populations have developed ethological isolation that appears to reflect isolation by distance. There is no single allozyme locus that is unique to *D. ocoee,* but this species has one fixed or nearly fixed allozyme that will distinguish it in paired comparisons with other members of the *D. ochrophaeus* complex.

DISTRIBUTION AND ADULT HABITAT. The Ocoee salamander occurs as two or possibly three allopatric groups. The main group occurs in the southwestern Blue Ridge Physiological Province and includes populations in the Balsam, Blue Ridge, Cowee, Great Smoky, Nantahala, Snowbird, Tusquitee, and Unicoi mountains, along with low-elevation populations in the gorges of the Hiwasee, Ocoee, and Tugaloo rivers (Tilley and Mahoney 1996). A second group occurs in the Appalachian Plateau of northeastern Alabama.

In the southern Appalachians, *D. ocoee* is widespread and abundant on wet rockfaces, in seepage areas, and in forest floor habitats in and about the vicinities of streams. Populations at relatively low elevations are concentrated in or near seepages or streams, whereas those at higher elevations (generally >1370 m) are often abundant on the forest floor far from running water. This species ranges over a greater elevational span than any other *Desmognathus* species and extends from low-lying gorges to the highest mountaintops in the Great Smoky Mountains.

BREEDING AND COURTSHIP. Mating occurs during the spring, late summer, and autumn, and females reproduce annually in most populations (Forester 1977; Huheey and Brandon 1973; Martof and Rose 1963). Specimens have been collected with sperm in the vasa deferentia from September through June (Huheey and Brandon 1973). Details of a comparative study of courtship behavior of *D. ocoee, D. carolinensis,* and *D. orestes* are presented by Herring and Verrell (1996). The courtship sequences of all three of these members of the *D. ochrophaeus* complex appear to be nearly identical (for a detailed description, see the *D. orestes* species account). The basic sequence involves a period of persuasion followed by a tail-straddle walk and spermatophore deposition. During courtship, the male may repeatedly snap or pull his

Range of *Desmognathus ocoee*

chin across the female's dorsum. This behavior acts to scratch the female's skin and transfer mental gland secretions into the female's circulatory system (Houck 1986). Mental gland secretions increase the sexual responsiveness of female *D. ocoee* to males (Houck and Reagan 1990). Evidence of multiple inseminations (Houck et al. 1985a; Labanick 1983) suggests that some females court several times per year, and that sperm competition occurs among males.

Several researchers have examined whether phenotypic traits of potential mates influence mating success. Large and small males are equally successful in courting females when a single male is paired with a female (Houck 1988). However, when large and small males are paired in trios with females, larger males outcompete smaller males for mates by chasing smaller males away from females prior to mating. Pairs of adults are less likely to court successfully when in the presence of other males or predatory spring salamanders (Uzendoski et al. 1993). Age does not influence the mating success of males (Houck and Francillon-Vieillot 1988), and inexperienced males are as successful as experienced males in courting experienced females (Verrell 1991a). When males are presented a choice between small and large females, they prefer large females that are more fecund (Verrell 1989a).

Houck et al. (1985a) found that certain males are more successful than others at inseminating females. In addition, certain females are more successful than others at eliciting spermatophore deposition from males. This variation is important because it allows natural selection to occur in each sex. Verrell and Donovan (1991) observed male homosexual behavior that may be a form of sexual interference.

Males normally deposit only one spermatophore per night, even when presented with multiple partners (Verrell 1988a). In addition, females are not responsive to males for 2–4 days after successful insemination. This interval corresponds to a time when the female's cloaca is physically blocked by the sperm cap (Verrell 1988b, 1991b). Courtship success between males and females that are repeatedly exposed to each other in laboratory trials tends to decrease after five or six encounters relative to similar trials using unfamiliar pairs (Donovan and Verrell 1991). The relevance of this finding to natural populations is uncertain, since males probably encounter several females during the mating season. The size of courtship arenas and the complexity of the physical environment also appear to have little effect on mating frequency (Verrell 1988c). Tilley et al. (1990) found that the strength of sexual isolation among populations is statistically independent of Nei's genetic distance, although both of these variables correlate with the geographic distance between populations.

REPRODUCTIVE STRATEGY. Females move to nesting sites as long as 2-3 weeks before ovipositing (Forester 1981). The eggs are deposited in small, grapelike clusters in hollowed depressions. Nesting females have been found beneath rocks and decaying logs, and in leaf litter and moss mats in or near springs, seeps, and small streams (Bruce 1990; Forester 1977; Martof and Rose 1963; C. H. Pope 1924). Each egg has a short, gelatinous stalk and is attached to either an overhead support or moss fibers and rootlets at a common base (Martof and Rose 1963). The stalks become twisted about each other to form a single cluster that the female broods. Mature ova measure 2-3 mm in diameter on average (Bruce 1990; Huheey and Brandon 1973; Martof and Rose 1963).

Nests are usually constructed at or slightly above the water surface. Females often gather in large congregations in seepages or other nesting sites and construct their nests within a few centimeters or decimeters of each other (Forester 1977; Martof and Rose 1963; C. H. Pope 1924). In the Great Smoky Mountains nesting *D. ocoee* are rarely found in seepages and other sites that are occupied by *D. f. santeetlah*.

Females remain with their clutches through hatching, but it is uncertain whether they periodically move short distances away from their clutches to forage. Forester (1979b) found that 73% of females displaced 2 m from their nesting cavities return to their nests within 24 hours. Females are usually able to recognize their own eggs even when these are only a few centimeters from other nests, and they appear to use chemical cues for clutch recognition (Forester 1986; Forester et al. 1983). Brooding functions to protect the embryos from predators, to remove dead eggs from nests, to reduce egg desiccation, and to reduce fungal attacks (Forester 1979a, 1984; Tilley 1972). Brooding females may even help hatchlings escape from nesting cavities (Huheey and Brandon 1975).

When females are experimentally removed from their clutches, predators such as carabid beetles and other salamanders quickly eat the eggs (Forester 1979a; Tilley 1972). In one case a large male and a subadult *D. ocoee* were found at the nests with eggs in their stomachs. In laboratory trials, females can successfully defend their nests against predaceous beetles and conspecifics, but will make only feeble attempts to defend their nests from *Gyrinophilus porphyriticus, Desmognathus monticola,* and ringneck snakes (Forester 1978). Females will lunge at and bite some intruders. Aggression increases during the brooding season, and females that have invested most heavily in parental care (i.e., those with late-term embryos) are more likely to stand and defend their nests (Forester 1983).

Females will consume prey that wander near nests, but brooding necessitates that females greatly reduce their foraging activity (Forester 1981; Martof and Rose 1963; Tilley 1972). Most females have a negative energy budget and lose weight over the brooding season. Brooding females may lose 12% of body weight over a 1-month period, but much of this loss may be due to desiccation (Tilley 1972).

Egg laying occurs in July and early August in most populations, but in some instances freshly deposited eggs have been found in early September (Martof and Rose 1963). Eggs have been found in northern Alabama in August and September (Mount 1975). Estimates of clutch size based on ovarian complements include means of 17 (range = 12-28; n = 12) and 16 (range = 9-22; n = 17) eggs for two western North Carolina populations (Bruce 1996), modal clutch sizes of 12 and 16 eggs for two additional populations in western North Carolina (Bernardo 1994), and means of 8-20 eggs for 21 populations from throughout the southern Appalachians (Martof and Rose 1963). The only estimate from nests is a mean of 13 eggs (range = 5-23) for 29 nests at a western North Carolina site (Bruce 1990, 1996). Clutch size in most populations is positively correlated with female body size and increases with elevation (Bruce 1996; Martof and Rose 1963; Tilley 1968, 1980).

Clutches at or near hatching have been found from August to late September (Martof and Rose 1963; Noble 1927a; C. H. Pope 1924; Tilley 1972). Estimated incubation periods are 57-74 days (Forester 1977) and 52-69 days (Tilley 1972) for North Carolina populations. Embryonic survivorship from early developmental stages to hatching is near 80% in clutches examined by Tilley (1972).

AQUATIC ECOLOGY. The larval period lasts 9-10 months in western North Carolina, with larvae hatching in August-September and metamorphosing in May and June when they have reached 11-15 mm SVL (Bruce 1989). The larval period lasts for at least 9 months in rockface populations (Huheey and Brandon

1973). Mean size at metamorphosis in two populations studied by Bernardo (1994) is 13-14 mm SVL.

Both food level and temperature can affect larval growth and development rates (Beachy 1995). High temperature and high food levels accelerate larval growth. Size at metamorphosis increases with food level, but length of the larval period does not. When maintained at high temperatures, larvae metamorphose earlier and are smaller on average than larvae maintained at low temperatures.

TERRESTRIAL ECOLOGY. The Ocoee salamander appears to be a generalist feeder. Rockface populations take a wide variety of invertebrates, particularly flies and beetles (Huheey and Brandon 1973). On rare occasion individuals eat *Desmognathus* larvae. Prey of other populations include larval and adult insects, spiders and mites, nematodes, salamander eggs and juveniles, and shed skin (Forester 1981).

Female *D. ocoee* sometimes feed on their own eggs and hatchlings (Bruce 1990; Forester 1981; Martof and Rose 1963). The extent to which oophagy reflects the attack of healthy embryos versus the removal of dead or diseased eggs from nests is not fully understood. Females will eat dead eggs introduced into their clutches, and will sometimes eat healthy eggs from their own broods immediately after consuming a dead egg (Tilley 1972). Brooding females that are disturbed will consume healthy eggs from their own clutches (e.g., Bruce 1990). Hatchlings are sometimes cannibalized, and in some instances females appear to eat their own young (Forester 1981; Martof and Rose 1963). Collectively, these data suggest that females regularly remove dead eggs from their clutches but occasionally eat healthy eggs and hatchlings. Perhaps movement of hatchlings or late-term embryos stimulates hungry females to prey upon their offspring.

Growth rates vary depending on age, elevation, and the microhabitats frequented by the adults. Males of two rockface populations reach their maximum seasonal growth rates from June through August in North Carolina (Tilley 1977, 1980). Growth is about 5-7 mm SVL per year in males until they reach 4-5 years of age, then it slowly declines. Growth rates are similar in females but begin to slow after females reach 3-4 years of age.

Individuals in western North Carolina and extreme southeastern Tennessee often become sexually mature when they measure <30 mm SVL (Huheey and Brandon 1973; Martof and Rose 1963). In one population females and males mature when they reach 29-30 mm and ≥28 mm SVL, respectively (Bruce 1990). Both sexes reach sexual maturity when 3 years old, and females first oviposit when 4 years old. Skeletochronological aging of a western North Carolina population suggests that males mature when 3-4 years old and females when 4-5 years old, although the majority of individuals mature when either 3 (males) or 4 (females) years old (Castanet et al. 1996). Some individuals in this population may live as long as 10 years, but most do not live >7 years.

Age and size at first reproduction, maximum adult size, mean clutch size, and adult survivorship tend to increase with elevation (Tilley 1977, 1980). In a detailed study of two rockface populations, Tilley (1980) estimated annual adult survival rates to be 74% and 63%, respectively, for high- (1448 m) and low- (1024 m) elevation populations. However, the probability of surviving from hatching to first oviposition is lower at the high-elevation site (6% versus 9%) because females first oviposit during their fifth summer, 1 year later than females at the low-elevation site. Tilley (1977, 1980) proposed that variation in life history features reflects adaptive life history tactics. That is, higher survivorship in high-elevation populations favors delayed reproduction, and subsequently larger adult body size and larger average clutch size. However, he was unable to determine the extent to which additive genetic variance, nonadditive genetic variance, and the environment contribute to phenotypic variation among populations.

Bernardo (1994) conducted experiments that indicate that the difference in age at first reproduction between low- and high-elevation populations has a genetic component. Supplementally fed juveniles from both populations grow at similar rates, but the lowland population channels more energy into gonadal production and matures faster than the upland population.

Population densities on rockfaces usually exceed those in streamside habitats, perhaps because of reduced predation from larger salamanders. Estimated densities on rockfaces in western North Carolina are 6-7 adults/m^2 (Tilley 1980), 10-12 salamanders/m^2 (Bernardo 1994), and 11-22 and 7-19 salamanders/m^2. The latter values are the maximum adult and juvenile densities estimated by Huheey and Brandon (1973).

Ocoee salamanders are nocturnally active and remain beneath cover objects or in recesses in rockfaces during the day. They emerge at night and actively move about in search of prey or mates. On overcast days a small percentage of animals may be active outside cover. Peak surface activity in a South Carolina population occurs around midnight (Shealy 1975). In the southern Appalachians, individuals frequently climb plants on rainy nights.

Adults are aggressive and will actively defend space from conspecifics. Large males tend to dominate smaller males and can exclude them from food-rich patches (Jaeger 1988). Males are aggressive toward other males, even in the absence of females or patchy food resources (Verrell and Donovan 1991). A male initiates interactions by approaching another male. This approach is often followed by biting or gaping the mouth at any opponent that does not flee when approached. Bites are often directed to the tail, and in rare instances a portion of the tail may be lost. When bitten, an opponent will often bite back.

Females apparently migrate short distances to breeding habitats each year, and young juveniles disperse from these sites into the surrounding forest. Individuals are active on the ground surface except during the coldest winter months. Adults and juveniles congregate in seepages or underground retreats during the winter months (Shealy 1975), and the degree of winter activity declines with elevation. Rockface populations may be active on the surface as late as November and December (Huheey and Brandon 1973). In the southern Appalachians, individuals disperse from winter retreats to surrounding forests in the spring.

PREDATORS AND DEFENSE. Birds, small snakes, and large salamanders such as *Gyrinophilus porphyriticus*, *D. monticola*, and *D. quadramaculatus* are major predators (Bruce 1979; Formanowicz and Brodie 1993; Hairston 1986). Experimental data from feeding trials indicate that vulnerability of *D. ocoee* to *G. porphyriticus* and *D. quadramaculatus* decreases with prey size because of gape limitations (Formanowicz and Brodie 1993). Laboratory trials further suggest that *D. ocoee* is far more likely to flee from *G. porphyriticus* than *D. quadramaculatus* (Hileman and Brodie 1994). In addition, individuals are less likely to flee from *Gyrinophilus* in the summer, which corresponds to a time when the two species are more spatially segregated.

Individuals will readily autotomize their tails when attacked by birds (Labanick 1984). Like almost all plethodontids, individuals of this species cannot voluntarily autotomize the tail, but they can break the tail loose from the body if the tail is grasped by a predator. Chickens used in Labanick's experiments often ate wiggling autotomized tails before attacking the salamanders. The wiggling tail directs attention away from the body and sometimes allows salamanders time to escape. Individuals may also remain immobile when uncovered, which makes them less conspicuous to predators (Dodd 1990b).

When attacked by garter snakes, *D. ocoee* uses body flipping, biting, and tail autotomy to aid in escape (Brodie et al. 1989). Individuals may bite the mouth or face of an attacking snake when seized by the body or tail. Salamanders that are bitten on the tail and cannot flip free typically respond by biting the snake. If this does not induce the snake to release its grip, individuals may autotomize their tails and escape.

COMMUNITY ECOLOGY. In the Appalachian Mountains, *D. quadramaculatus*, *D. monticola*, members of the *D. ochrophaeus* complex, and *D. aeneus* or *D. wrighti* form a series of species with habitats ranging from aquatic to terrestrial (Hairston 1949; Organ 1961a). In some regions *D. fuscus* is also present and occupies a position in the series similar to that of members of the *D. ochrophaeus* complex. Correlated with this ecological progression from surface water to land is a series of morphological traits, including a negative correlation between average body size and distance from surface water (Hairston 1980c, 1986).

In Pennsylvania, *D. fuscus*, *D. monticola*, and *D. ochrophaeus* share streams and often differ in the mean distance of each species from running water. These species also differ in the substrate and cover used. Prey size utilization is similar in interspecific comparisons of similar-sized groups (e.g., mature *D. ochrophaeus* versus immature *D. monticola*) but generally differs between groups that differ significantly in size (Krzysik 1979). Patterns of microhabitat use suggest that species shift microhabitat use when coexisting with other species. In the presence of *D. fuscus*, *D. ochrophaeus* tends to be displaced farther from surface water, whereas in the presence of *D. monticola*, *D. fuscus* tends to select smaller cover objects. Krzysik (1979) hypothesized that interspecific interference competition (aggression) has been the driving force in the evolution of body size and terrestrialism in *Desmognathus*, since his dietary analysis shows no evidence of interspecific predation.

Hairston (1986) experimentally lowered densities of *D. monticola* and *D. ocoee* on study plots to determine if competition or predation (Tilley 1968) is the major organizing force in a *Desmognathus* community composed mostly of *D. monticola*, *D. ocoee*, and *D. quadramaculatus*. When *D. ocoee* is removed from plots, the abundance of *D. monticola* and *D. quadramaculatus* decreases relative to that in control plots. Removal of *D. monticola* results in an increase in *D. ocoee*. *Desmognathus ocoee* also shifts its distribution toward streams when larger congeners decrease in numbers. Hairston (1986) concluded that predation rather than competition is the major force that has driven the evolution of body size and habitat preference in the genus. The conclusions of Krzysik (1980b) and Hairston (1986) are conflicting, but both may be correct depending on geographic location.

CONSERVATION BIOLOGY. *Desmognathus ocoee* is one of the most common species in the southern Appalachian Mountains and is in minimal need of protection. Populations in northern Alabama may be more vulnerable to environmental disturbance, and their conservation status may merit study.

Desmognathus orestes Tilley and Mahoney
Blue Ridge Dusky Salamander
PLATE 54

Fig. 120. *Desmognathus orestes;* adult; Avery Co., North Carolina (J. W. Petranka).

IDENTIFICATION. The Blue Ridge dusky salamander is a member of the *D. ochrophaeus* complex, which consists of *D. apalachicolae*, *D. carolinensis*, *D. imitator*, *D. ochrophaeus*, *D. ocoee*, and *D. orestes* (Tilley and Mahoney 1996). Members of the complex are so similar in size and external appearance that they are best identified by geographic range and molecular data. Five species in the complex that were previously treated as *D. ochrophaeus* are medium-sized dusky salamanders that have pigmented testes and rounded, unkeeled tails. The tail is slightly longer than the body, and the toe tips lack cornifications. Dorsal color patterning grades from straight-edged stripes to wavy, blotched patterning and varies from light brown through shades of yellow to bright red (Martof and Rose 1963; Tilley 1969). The venter is light colored in young animals but is uniformly light gray to grayish black in adults. Older animals are often melanistic, particularly males. A partial albino adult is reported by Dyrkacz (1981).

Relative to females, sexually active males of all five species have well-developed premaxillary teeth and large jaw musculature. Males also possess small mental glands, are on average slightly larger than females, have a mouth line that appears sinuate when viewed from the side, and tend to lose the vomerine teeth with age. Males of the *D. ochrophaeus* complex are on average 6-20% longer in SVL than females (Bruce 1993; Martof and Rose 1963; Orr 1989). Adults measure 7-11 cm TL and there are 14 costal grooves. Hatchlings measure 13-18 mm TL and have prominent yolk reserves (Bishop 1941a; Bishop and Chrisp 1933; Huheey and Brandon 1973; Orr 1989).

The dorsal patterning of *D. orestes* is highly variable and includes both straight-edged and wavy-striped individuals (Tilley 1969). Hatchlings and larvae have not been described but are presumably polymorphic in color patterning and similar to those of *D. carolinensis*. Juveniles and young adults typically have brownish, reddish, or yellowish dorsal stripes or blotches. Older adults (primarily males) are often melanistic with little dorsal patterning. This species is easily confused with other members of the *D. ochrophaeus* complex, and specimens are best identified by the collection locality and/or molecular data.

Range of *Desmognathus orestes*

SYSTEMATICS AND GEOGRAPHIC VARIATION. Populations that have traditionally been referred to as *D. ochrophaeus* in the eastern United States exhibit marked geographic variation in body size, coloration, color patterning, size at sexual maturity, and other traits. Experts have disagreed on the taxonomic status of several forms that were described as separate species or subspecies (Dunn 1916; Nicholls 1949; Neill 1950); these were considered junior synonyms of *D. ochrophaeus* in later treatments. Tilley and Mahoney (1996) analyzed geographic variation in protein variants and split populations previously referred to as *D. ochrophaeus* into four species (see detailed summary under the account of *D. carolinensis*).

Like most other members of the *D. ochrophaeus* complex, local populations of *D. orestes* often differ markedly in dorsal coloration, color patterning, and body size, with average body size increasing with elevation (Tilley 1969). Geographic variation in protein variants is discordant, and there are no strong geographic trends within the species (Tilley and Mahoney 1996).

DISTRIBUTION AND ADULT HABITAT. *Desmognathus orestes* is confined to mountainous, forested habitats in the Blue Ridge Physiographic Province from Mt. Rogers and vicinity in Floyd Co., southwestern Virginia, southwestward to somewhere between Linville Falls and McKinney Gap in Burke and Mitchell counties, North Carolina. Farther west, the species extends to the headwaters of Toms Creek and Clark Creek about 1.5 km northeast of Iron Mountain Gap on the North Carolina–Tennessee border in Avery and Carter counties (Tilley and Mahoney 1996). This species inhabits wet rockfaces, seepage areas, and forest floor habitats in and about the vicinities of streams and seepages. Populations occur to the tops of the highest mountain peaks within the range of the species.

BREEDING AND COURTSHIP. The adults mate during the spring, late summer, and autumn (Organ 1961a). Females reproduce annually in most populations, and courtship involves a stereotypical tail-straddle walk (Organ 1961a; Uzendoski and Verrell 1993). The following account of courtship behavior is from Herring and Verrell's (1996) study of *D. carolinensis, D. ocoee,* and *D. orestes* and Uzendoski and Verrell's (1993) study of *D. orestes.*

The male initially attempts to position himself next to the female where he can stimulate her. When approaching or following a female, the male may jerk his body to and fro or rotate the front limbs in a circular

Fig. 121. *Desmognathus orestes;* courting adults (the male on the right is leading the female forward with the tail displaced to his side; note the spermatophore beneath the female's chest region); Grayson Co., Virginia (S. J. Arnold).

motion, either separately (pawing) or in synchrony (butterflying). The female may initially move away from such approaches but eventually stops moving. A male that encounters a female often initiates the persuasion phase by rubbing his head over all portions of the female's head. This behavior is interspersed with head-lifting in which the male repeatedly lifts and lowers the female's head. Other behaviors such as jerking of the body or pulling the snout across the female's dorsum may occur occasionally during this period.

If the female turns toward the male during rubbing, he may slide his body under her chin and undulate his tail. Once the female's head is positioned on the dorsal surface of the base of the tail, the male may curve his body backward toward the female's neck or dorsum. With his tail undulating, the male presses his chin against the dorsum of the female and snaps his head backward with a rapid movement of the body (this violent action may flip the male several centimeters from the female). In addition to rapid snapping, the male may slowly pull his depressed chin across the female's dorsum. The male may also snap and pull when positioned alongside a female. This behavior acts to scratch the female's skin and transfer mental gland secretions into the female's circulatory system (Houck 1986).

During this persuasion phase, the female often moves away from the male and the male pursues and continues to stimulate her. The pair next engages in a tail-straddle walk in which the male laterally undulates his tail and slides under the female's head until her chin rests on the dorsal base of the tail. The male moves forward in a straight line while undulating his tail laterally, and the female follows as she straddles his tail. Eventually the male ceases moving forward, increases the intensity of tail undulation, and deposits a spermatophore. The spermatophore is about 2 mm high and has a clear gelatinous base about 2.5 mm in diameter.

The male then leads the female forward about one body length and swings his tail from underneath the female and to one side. With the female's head resting on the base of his tail, the male vigorously and repeatedly raises and lowers his pelvic region. The female responds by picking up the sperm cap in her cloaca as she laterally moves her own pelvic region to and fro. The pair splits apart within a few minutes after courtship is completed. Evidence of long-term sperm storage (Houck and Schwenk 1984) suggests that sperm competition occurs among males.

Males normally deposit only one spermatophore per night, even when presented with multiple partners (Verrell 1988a). In addition, females are not responsive to males for 2-4 days after successful insemination. This period corresponds to a time when the female's cloaca is blocked by the sperm cap (Verrell 1991b). The size of courtship arenas and the complexity of the physical environment appear to have little affect on mating frequency (Verrell 1988c).

Desmognathus orestes often coexists with other *Desmognathus* species, but individuals will rarely attempt to court members of other species. In a comparison of interspecific courtships, Uzendoski and Verrell (1993) found that certain aspects of courtship differ significantly between *D. orestes* and *D. fuscus.* Nonetheless, these differences do not appear to be the primary reason for reproductive isolation. Instead, males typically ignore females of the opposite species from the point of initial contact. Differences in female chemical cues may be the primary reason for the lack of responsiveness exhibited by males.

Verrell (1994b) raised *D. orestes* in isolation to adulthood and found that sexually naive, laboratory-reared animals are as unresponsive to *D. f. fuscus* females as field-collected males. Thus experience does not appear to be important for the maintenance of sexual incompatibility between syntopic populations of these species. Mating success of *D. orestes* is significantly reduced when paired conspecifics are in the presence of a female *D. fuscus* compared to when they are alone (Verrell 1994c). This observation suggests that a slight reproductive cost may be incurred by *D. orestes* that are sympatric with *D. fuscus*.

REPRODUCTIVE STRATEGY. Females deposit their eggs in grapelike clusters in depressions beneath logs, rocks, leaf litter, or moss mats in or immediately next to springs, seeps, and small streams. Each egg has a short, gelatinous stalk and is attached to a common base to form a cluster (Martof and Rose 1963). Nests are usually constructed at or slightly above the water surface, and females remain with their clutches through hatching. Although reproductive behavior is poorly documented, most females presumably oviposit in May–July and most embryos hatch in late summer or early fall after an incubation period of about 2 months.

The mean clutch sizes of two North Carolina populations are 14 and 16 mature ova (Martof and Rose 1963). Clutch size is positively correlated with female body size in this and most other *Desmognathus* species (Martof and Rose 1963; Tilley 1968).

AQUATIC ECOLOGY. Larvae live in seepages, rivulets, and the sluggish portions of small streams, where they feed on small invertebrates. Organ (1961a) collected metamorphs in April and May in Virginia and estimated an 8- to 9-month larval period for offspring hatching from eggs laid the previous June. Tilley (1973b) and Tilley and Tinkle (1968) questioned this interpretation because some of the metamorphs may have been from winter-brooding females.

TERRESTRIAL ECOLOGY. No information is available on diet. Organ (1961a) found that males and females in southwestern Virginia become sexually mature in the autumn when 3.5 and 4.5 years old. Females first reproduce the following year when about 5 years old. Age-specific survivorship of males decreases with age. Brooding females have higher mortality rates than immature females, a fact that may reflect the costs of breeding.

Blue Ridge dusky salamanders are nocturnally active and remain beneath cover during the day. They emerge at night and actively move about in search of prey or mates. On overcast days a small percentage of animals may be active outside cover. Females migrate short distances to breeding habitats each year, and the young disperse from these sites into the surrounding forest. Individuals are active on the ground surface except during the winter months. Adults and juveniles congregate in seepages or underground retreats during the winter months and disperse from winter retreats to surrounding forests in the spring (Organ 1961a,b).

PREDATORS AND DEFENSE. Natural predators have not been identified, but birds, shrews, small snakes, and large salamanders undoubtedly prey upon *D. orestes*. Individuals cannot voluntarily autotomize the tail but can break the tail loose from the body if the tail is grasped by a predator.

COMMUNITY ECOLOGY. *Desmognathus orestes* often coexists with *D. quadramaculatus, D. monticola,* and *D. wrighti,* forming a series of species with habitats ranging from aquatic to terrestrial (Hairston 1949; Organ 1961a). *Desmognathus fuscus* is also present in some communities and occupies a position between that of *D. monticola* and *D. orestes*. Ecological interactions of *D. orestes* with these forms have not been investigated, but studies of interactions between *D. ocoee, D. monticola,* and *D. quadramaculatus* suggest that *D. orestes* is a common food resource and source of competition between the latter two species (Hairston 1980c, 1986).

Members of the *D. ochrophaeus* complex have been implicated as being mimics of several other species of salamanders, including *Eurycea bislineata, Plethodon cinereus, P. nettingi,* and *P. welleri* (Brodie 1981; Brodie and Howard 1973). Virtually all of the common color morphs of this species complex have at one time or another been interpreted as being mimics of other salamanders. Much work remains to be done to test the validity of these hypotheses.

CONSERVATION BIOLOGY. *Desmognathus orestes* is a common species at higher elevations throughout its range and is in minimal need of protection. Local populations are often severely depressed after clear-cutting, and low-elevation populations may take many decades to recover following intensive timbering (Petranka et al. 1994).

Desmognathus quadramaculatus (Holbrook)
Black-bellied Salamander
PLATE 55

Fig. 122. *Desmognathus quadramaculatus;* adult; Macon Co., North Carolina (J. W. Petranka).

Fig. 123. *Desmognathus quadramaculatus;* ventral view of adult; Buncombe Co., North Carolina (J. W. Petranka).

IDENTIFICATION. The black-bellied salamander is a large, stocky *Desmognathus* with a black belly and one or two rows of light dots along each lower side of the body. The tail is laterally compressed and keeled, and the toe tips are black and cornified. The dorsal ground color is black or dark brown and is overlain with dull greenish yellow to rusty blotches. Specimens in some areas are heavily suffused with light golden or brassy flecks. Old individuals are often black above and below with inconspicuous markings. This is the largest *Desmognathus* species. Adult males have slightly enlarged premaxillary teeth, papillose cloacal lips, and a small mental gland, and they are on average larger than females. Adult males in one western North Carolina population average 3% larger in SVL than adult females (Bruce 1993). Adults measure 9-21 cm TL and there are 14 costal grooves.

The hatchlings are light brown, have six to eight pairs of light spots between the limbs, and measure 11-16 mm SVL. Older larvae are much larger than larvae of other *Desmognathus* species and have whitish gills and dark toe tips. Young juveniles are stocky with varying amounts of blotching on the dorsum. The bellies of very young juveniles are whitish and acquire dark pigmentation within several months to a year after metamorphosis.

SYSTEMATICS AND GEOGRAPHIC VARIATION. Populations to the southwest and northeast of the French Broad River in western North Carolina show evidence of genetic and phenotypic differentiation. The heads, limbs, and dorsal tail surface of specimens in northern Georgia and southwestern North Carolina often appear slightly lighter than the body, and specimens from this region are often larger on average than those from more northern regions. These groups were once recognized as separate subspecies but are now considered to be members of a single, monotypic species. Hinderstein (1971) provides very limited evidence of genetic differentiation between southern and northern populations based on studies of lactate dehydrogenase. Southern and northern populations may also differ in their mode of egg deposition (Smith et al. 1996a).

DISTRIBUTION AND ADULT HABITAT. The black-bellied salamander occurs in mountainous regions from southern West Virginia and south of the Tennessee Valley Divide in the Allegheny Mountains in Virginia southward to northern Georgia. This species occupies a wide variety of habitats ranging from headwater tributaries to large trout streams with rapidly flowing water. The black-bellied salamander is often found in swifter current than most *Desmognathus* species and reaches its highest densities in small, perennial, unsilted streams with numerous rocks and cobbles (Davic and Orr 1987; C. H. Pope 1924). The black-bellied salamander is most abundant at elevations from 490 to 1676 m (Hairston 1949; Organ 1961a),

Range of *Desmognathus quadramaculatus*

but it may occur at lower elevations if local conditions are suitable. Specimens are typically found directly in or near flowing water, but large adults sometimes forage on the forest floor adjoining streams on rainy nights. This species is widely used as fish bait and has been introduced into several areas in Georgia where it is not native (Camp 1989; Martof 1953).

BREEDING AND COURTSHIP. Information on the breeding season is very limited. Organ (1961a) found females in September with spermatophores in their vents, suggesting that breeding occurs in the late summer and autumn. It is not certain whether adults mate in the spring as do some other *Desmognathus* species. Courtship has not been described in detail but presumably is similar to that of other large *Desmognathus* species.

REPRODUCTIVE STRATEGY. The mode of egg deposition appears to vary geographically. In northwestern North Carolina, Virginia, and West Virginia, eggs are attached singly to the undersides of rocks, stones, or other support structures in small clusters (Organ 1961a; C. H. Pope 1924; T. K. Pauley and R. W. Van Devender, pers. comm.). When in flowing water, the eggs are bounced about and constantly aerated by the water current. The eggs appear as a monolayer when attached to flat surfaces and often as bilayers when attached to concave or irregular surfaces. Each egg is suspended by a short, elastic pedicel that is about 3 mm long and 1.5 mm in diameter (C. H. Pope 1924).

In southwestern North Carolina, eggs have been found attached in large globular clusters rather than in monolayers. My colleagues and I collected two clutches from the Smoky and Nantahala mountains of North Carolina that were laid in this manner (Smith et al. 1996a). Each mass contains several subclusters of two to eight eggs that are attached to a common stalk and that resemble the grapelike egg clusters of *D. ochrophaeus*.

Fig. 124. *Desmognathus quadramaculatus;* egg clutch; Macon Co., North Carolina (J. W. Petranka).

Fig. 125. *Desmognathus quadramaculatus;* hatchlings; Macon Co., North Carolina (J. W. Petranka).

Fig. 126. *Desmognathus quadramaculatus;* larva; Caldwell Co., North Carolina (R. W. Van Devender).

Females oviposit in May and June and the embryos hatch from July through September. Breeding records include 10 nests in western North Carolina between 13 June and 6 August in moderate to advanced stages of development (C. H. Pope 1924), three other western North Carolina nests found in June (Bruce 1996), seven clutches in Virginia on 22 June in early to moderate stages of development (Organ 1961a), and several clutches at or near hatching in early August (Orr and Maple 1978). Austin and Camp (1992) found eggs in May and hatchlings measuring 16 mm SVL in July in Georgia.

The eggs are typically placed in fast-flowing water on the undersides of rocks and have often been found near cascades or small waterfalls (Organ 1961a; C. H. Pope 1924). Nests are occasionally found above the water line in or near the streambed. In most instances, females guard their egg masses. Estimates of clutch size are means of 32 (range = 26–43; n = 10) and 45 (range = 38–55; n = 3) for nests in North Carolina (Bruce 1996; C. H. Pope 1924), a mean of 31 (range = 21–43; n = 7) for nests in Virginia (Organ 1961a), and an average of 54 eggs (range = 52–56) for five females that oviposited in both the field and the laboratory (one after hormonal injection; Collazo and Marks 1994). Estimates from counts of mature ova in gravid females from two western North Carolina sites are 54 (range = 38–69; n = 5) and 52 (range = 42–62; n = 13). Clutch size is positively correlated with female SVL in some, but not all, populations of this species (Bruce 1996, Tilley 1968).

AQUATIC ECOLOGY. Hatchlings have conspicuous yolk masses and live off yolk reserves for 1–2 months after hatching. Older larvae often inhabit fast-flowing sections of streams beneath rocks or debris in the streambed. Larvae transform in Virginia in August to October after a 24- to 26-month larval period (Organ 1961a). Most larvae in this region transform when they reach 34–40 mm SVL, but a small percentage of individuals may transform when as large as 44 mm SVL.

Bruce (1985a) estimates the larval period of individuals in a population in western North Carolina to be approximately 3 years. In more detailed studies Bruce (1988a, 1989) found that three populations in western North Carolina differ significantly in length of the larval period and size at metamorphosis. Individuals metamorphose after 2–4 years of growth, but most appear to have a 3- or 4-year larval period. Most larvae in these populations appear to metamorphose between 35 and 42 mm SVL, and skeletochronological aging is consistent with estimates based on analyses of size distributions (Castanet et al. 1996). At a low-elevation site in Georgia, the larval period lasts nearly 3 years and larvae transform in May and June when they measure 40–43 mm SVL on average (Austin and Camp 1992). At a higher-elevation site the larval period lasts as long as 4 years and individuals metamorphose at an average of 54 mm SVL.

Size distributions of larvae from small tributaries and a larger stream suggest that reproduction is concentrated in the smaller tributaries (Bruce 1985a). Hatchlings and small larvae drift downstream as they age. Estimates of total densities of larval, juvenile, and adult *D. quadramaculatus* in streams in western North Carolina are 5.6–11.7 individuals/m^2 of streambed (Davic and Orr 1987).

TERRESTRIAL ECOLOGY. Organ (1961a) concluded that males in Virginia become sexually mature when 3.5 years old and females when 4.5 years old.

Females first reproduce the following year when about 5 years old. Bruce (1988a, 1989) questioned this interpretation and estimated the minimum age at first reproduction in western North Carolina populations to be 6 years for males and 7 years for females. Skeletochronological aging is consistent with the latter interpretation and shows that growth rate declines to near zero as individuals approach 80 mm SVL and 9-10 years of age (Castanet et al. 1996). Most individuals >70 mm SVL are sexually mature, and some individuals may live as long as 13 years. Organ (1961a) found that age-specific survivorship is relatively constant for metamorphosed males. In contrast, mortality rates increase after females reach sexual maturity. This species has higher mortality rates than other *Desmognathus* species.

Juveniles and adults are most abundant in streambeds, seepages, and other aquatic habitats. Juveniles often reach high densities in seepages with moderate water flow and numerous small stones that serve as cover. Adults are more likely to be found in streams under large rocks or logs. Individuals emerge from their daytime retreats at dusk and often sit with the head and upper body projecting from burrows or cover objects, where they forage on invertebrates and smaller salamanders. Although this species is strongly associated with water, large adults often forage several meters from streams, particularly on rainy nights. About 30% of adults along a stream in the Great Smoky Mountains forage at night >5 m from the stream, and several animals can be found >10 m from the stream's edge (Kucken et al. 1994).

Camp and Lee (1996) conducted a detailed analysis of movements and territoriality in a population in the upper Piedmont of Georgia. Animals <70 mm SVL are mostly found in the stream, but larger animals live in rock crevices or burrows in the streambank. The tendency of animals to wander at night decreases with size, and 90-97% of individuals >60 mm SVL feed from refuges where they ambush passing prey. Individuals often shift to new refuges but remain in home ranges with mean minimal areas of 1207 cm^2 (range = 138-4903 cm^2). Behavioral trials conducted in the field indicate that individuals aggressively defend burrows from conspecifics. The size of home ranges is independent of SVL. Home ranges are randomly distributed and those of neighbors often overlap. The median distance moved between captures is 0.74 m (range = 0.1-19.3 m), and the maximum distance moved by an individual is independent of SVL. Animals that move long distances often return to their original home ranges.

Camp (1996) analyzed patterns of bite scars in populations in northern Georgia and found that 9% of individuals have scars. Adult males and females do not differ in the proportion of individuals scarred and have a disproportionately high number of scars on the head compared with other areas of the body. In contrast, juveniles incur a higher proportion of bites on the hip. A higher proportion of adults have scars compared with juveniles, reflecting the greater levels of aggression exhibited by large adults. Over 40% of the animals lack intact tails.

Davic (1991) documents ontogenetic shifts in prey in a population from the southern Appalachians. Most (82%) of the diet of larvae at this site consists of aquatic prey, including mayflies, stoneflies, craneflies, chironomids, nematodes, collembolans, crayfish, and salamander larvae. In contrast, the highly aquatic juveniles feed primarily on aerial prey; only 36% of prey are aquatic organisms. The adults feed mostly on terrestrial species. Terrestrial organisms eaten by juveniles and adults include hymenopterans, lepidopterans, beetles, centipedes, and spiders. Hairston (1949) found a variety of aquatic and terrestrial insects in specimens from North Carolina, including stoneflies, flies, ants, collembolans, and bugs. Black-bellied salamanders in northern Georgia consume stoneflies, mayflies, caddisflies, true flies, beetles, hymenopterans, and other salamanders (Martof and Scott 1957).

Black-bellied salamanders are abundant in southern Appalachian streams and often prey on other salamanders (Formanowicz and Brodie 1993). Dietary studies suggest that *D. quadramaculatus* does not specialize on salamanders as much as *Gyrinophilus* (Bruce 1979). However, black-bellied salamanders will readily cannibalize when presented juveniles (Camp and Lee 1996), and regularly include *Desmognathus, Eurycea,* and *Plethodon* in their diets. Bishop (1924) found four salamanders, including a *Plethodon,* in five specimens. Predation of *D. quadramaculatus* on *D. ocoee* is size dependent, and vulnerability of prey to attack or injury decreases with prey size (Formanowicz and Brodie 1993).

PREDATORS AND DEFENSE. Black-bellied salamanders are large and can successfully defend themselves against relatively large predators such as shrews

(Brodie 1978). When approached by *Blarina brevicauda*, individuals open their jaws to expose the white lining, then lunge forward while snapping the jaws. If further provoked, they viciously bite the body or head of the shrew. Black-bellied salamanders use two defenses against garter snakes (*Thamnophis*). The first involves writhing and twisting the body in an attempt to flip free; the second is biting. Biting of the head or body of the snake often results in the snake releasing its grip and the salamander escaping (Brodie et al. 1989).

COMMUNITY ECOLOGY. Dusky salamanders (*Desmognathus* spp.) are important predators in headwater streams, where they feed on aquatic and terrestrial invertebrates and other salamanders. Microhabitat segregation of *Desmognathus* species in and along streams in the Appalachian Mountains is influenced by many factors, including size-specific preferences for cover objects, predation, aggressive defense of foraging sites, and the active avoidance of potential predators and competitors. Kleeberger (1984) reports little evidence of intra- or interspecific competition for food or cover objects in *D. quadramaculatus*. Addition of conspecifics to experimental plots does not significantly alter the density or lipid content of either conspecifics or *D. monticola*. In contrast, Hairston (1986) provides experimental evidence that suggests that *D. monticola* and *D. quadramaculatus* compete for prey, including *D. ocoee* (see the *D. ocoee* account for details).

The composition of *Desmognathus* assemblages in North Carolina streams correlates to some extent with both abiotic factors and the abundance of other *Desmognathus* species (Southerland 1986a–c). Field enclosure experiments indicate that *D. monticola* has lower growth and survival when confined with equal-sized *D. orestes* and *D. quadramaculatus*. These effects are presumably due to interference competition among species. Large *D. quadramaculatus* also appear to prey on smaller *D. monticola*.

Additional field studies (Southerland 1986a) show that *D. quadramaculatus* can displace *D. monticola* from streamside habitats and that small individuals of all species are usually underrepresented in microhabitats occupied by larger *Desmognathus* species. Burrows of large salamanders are rarely in close proximity to one another, an observation that suggests that aggression between individuals affects local dispersion patterns. The addition of surface cover to plots increases local densities of salamanders, particularly *D. monticola*.

Fig. 127. *Desmognathus quadramaculatus*; adult attempting to bite garter snake (*Thamnophis sirtalis*); locality unknown (E. D. Brodie, Jr.).

These results suggest that the amount of surface cover mediates predator-prey interactions; however, experimentally increasing *D. monticola* densities has no effect on *D. quadramaculatus*.

Adult *D. monticola* become more active on the surface and avoid the preferred microhabitats of *D. quadramaculatus* when confined with the latter species. Juveniles also avoid *D. quadramaculatus* microhabitats, but become less active in the presence of this potential predator (Southerland 1986c). *Desmognathus monticola* decreases its surface activity when present with *D. quadramaculatus* (Carr and Taylor 1985). In addition, *D. monticola* alters microhabitat use by *D. ocoee*. Seal salamanders will avoid substrates with *D. quadramaculatus*, and laboratory experiments indicate that *D. monticola* uses chemical cues to detect *D. quadramaculatus* (Roudebush and Taylor 1987a,b). *Desmognathus monticola* avoids mucous gland secretions of *D. quadramaculatus* but is attracted to feces and cloacal gland extracts (Jacobs and Taylor 1992). Additional details of the interactions of *D. quadramaculatus* with other species are discussed under the accounts of *D. monticola* and *D. ocoee*.

In a section of stream in the Great Smoky Mountains that is contaminated with acidic drainage, my colleagues and I compared streamside salamander communities with those in uncontaminated control sections (Kucken et al. 1994). Chronic stream degradation for >30 years has eliminated almost all *D. quadramaculatus* and *Eurycea bislineata wilderae* from the community and reduced the densities of members of the *D. ochrophaeus* complex. Small prey such as *D. wrighti* and juvenile *Plethodon jordani* are far more abundant on contaminated plots than control plots, a pattern that is most likely due to the elimination of predatory *D. quadramaculatus* from the community.

Using a field experiment, Beachy (1993a) found that large *D. quadramaculatus* larvae prey heavily on both smaller conspecifics and *E. b. wilderae* larvae. There is no evidence of either intra- or interspecific competition. In a related experiment, Beachy (1994) found that *Eurycea* has lower survival in the presence of predatory *Gyrinophilus porphyriticus* and *D. quadramaculatus* larvae, and that *Gyrinophilus* is the more effective predator. There is no evidence that *Gyrinophilus* and *D. quadramaculatus* larvae compete for limited food resources.

CONSERVATION BIOLOGY. My colleagues and I found that *D. quadramaculatus* and other salamanders with aquatic larval stages are absent from a stream in the Great Smoky Mountains that has been chronically exposed to acidic leachate from exposed Anakeesta rocks (Kucken et al. 1994). This rock formation is commonly associated with coal deposits, so drainage from coal mines could have similar effects on salamanders. Roudebush (1988) reported 100% survival in juvenile and adult *D. quadramaculatus* and *D. monticola* that are exposed for 3 weeks to water of pH 3.5–7.2. However, the feeding rate of animals after 3 weeks of exposure tends to decrease at low pH.

Black-bellied salamanders are often harvested from local streams to sell as bait to fishers, and overharvesting could potentially affect local populations.

Desmognathus welteri (Barbour)
Black Mountain Dusky Salamander
PLATE 56

Fig. 128. *Desmognathus welteri*; adult; Bell Co., Kentucky (J. W. Petranka).

Fig. 129. *Desmognathus welteri*; adult (note mottling on the belly); Bell Co., Kentucky (J. W. Petranka).

IDENTIFICATION. The Black Mountain dusky salamander is light brown above with varying amounts of irregular dark markings. The tail is laterally compressed near the base and has a distinct keel on the posterior half. Most specimens have dark toe tips and cornified pads. The belly is often heavily mottled with light grayish or brownish blotches. In many populations the dorsum of the tail is tan and lighter colored than the body. Adult males have papillate vent lips, enlarged premaxillary teeth, and a small mental gland on the chin. Females have pleated vent lips that lack papillae. The adults measure 7.5–17 cm TL and there are 14 costal grooves.

Hatchlings reach 11–13 mm SVL, and larvae have five to eight pairs of dorsal light spots between the limbs that disappear soon after metamorphosis. The venter is light colored and cornified pads occur on the toes. Young juveniles have a series of five to eight pairs of light blotches on the dorsum that fuse and become obscure with age. The venter is whitish or lightly mottled in young animals and heavily mottled in older animals.

This species often coexists locally with *D. fuscus* and *D. monticola* and can be difficult to identify. Large *D. fuscus* often resemble *D. welteri*, but they lack dark toe tips and often have a yellowish wash on the underside of the tail. The belly of the seal salamander lacks

Range of *Desmognathus welteri*

conspicuous mottling and the base of the tail is rounded or oval in cross section. Caldwell (1980), Caldwell and Trauth (1979), and Juterbock (1984) summarize characters that are helpful in separating these species in areas of sympatry.

SYSTEMATICS AND GEOGRAPHIC VARIATION. Genetic variation has not been analyzed. Local populations often vary with respect to body size and coloration, but detailed studies of geographic variation have not been conducted. No subspecies are currently recognized.

DISTRIBUTION AND HABITATS. *Desmognathus welteri* occurs in the Cumberland Mountains and Cumberland Plateau in eastern Kentucky and adjoining portions of Virginia and Tennessee (Juterbock 1984; Redmond 1980; Redmond and Scott 1996). This species is typically found along mostly permanent, small to moderate-sized streams flowing through mesophytic forests. Streams with steep to moderate gradients and coarse gravel or rocky substrates are optimal habitats (Redmond 1980). Large rocks offer ideal cover for the adults.

BREEDING AND COURTSHIP. The breeding season is not known and courtship has not been described.

REPRODUCTIVE STRATEGY. Few data are available on the nesting biology of this species. My colleagues and I found four clutches of eggs with three attending females between 8 July and 3 August in eastern Kentucky (Smith et al. 1996b). All nests were inside leaf packs in the main channel of a stream, 5–20 cm above the water line. Each egg has a transparent pedicel that is 4–12 mm long. Pedicels are attached to and wound about each other so that the eggs form a tight, grapelike cluster similar to that of *D. fuscus*. Ten eggs from one clutch averaged 4.5 mm in diameter (range = 4.4–4.6 mm). Number of eggs per clutch varied from 18 to 33 and averaged 26. Based on the

Fig. 130. *Desmognathus welteri;* adult with egg clutch; Bell Co., Kentucky (J. W. Petranka).

Fig. 131. *Desmognathus welteri;* egg cluster; Bell Co., Kentucky (J. W. Petranka).

developmental stages of eggs in July samples, we estimate that females at this site oviposit during the first or second week of June. Juterbock (1984) found hatchlings mostly in September, a finding that further suggests that oviposition occurs in early summer.

COMMENTS. Very little is known about the natural history of *D. welteri.* The larval period lasts about 20–24 months and most larvae appear to transform in May and June when they measure >21 mm SVL (Juterbock 1984). Juveniles become sexually mature when they reach 50–55 mm SVL (Juterbock 1978, 1984). Based on the length of larval period and growth rates of other *Desmognathus* species, sexual maturity is probably reached when individuals are 4–5 years old. Earlier life history accounts of *D. fuscus welteri* (e.g., Barbour and Hays 1957) refer to as many as three species of *Desmognathus* (Juterbock 1984) and are therefore not included in this account.

Desmognathus wrighti King
Pygmy Salamander
PLATE 57

Fig. 132. *Desmognathus wrighti;* adult; Graham Co., North Carolina (R. W. Van Devender).

Fig. 133. *Desmognathus wrighti;* adult; Grayson Co., Virginia (J. W. Petranka).

IDENTIFICATION. The pygmy salamander is a tiny species that has a reddish brown to coppery bronze dorsal stripe with a dark herringbone pattern down its center. The top of the head and snout is rugose, and the venter is flesh colored and lacks dark mottling. The eyelids are often coppery colored, which is a good field trait for distinguishing this species from other small, terrestrial *Desmognathus* species. The tail is rounded in cross section and shorter than the body. The toe tips lack cornifications. Adults measure 3.7–5.1 cm TL and there are 13–14 costal grooves. Males have a U-shaped mental gland, and at least one albino has been collected (Dyrkacz 1981).

Hatchlings have four to five pairs of light dorsal spots and very short tails. The mean SVL and TL of hatchlings vary from 7.0 to 8.1 and from 10.0 to 11.2 mm, respectively (Organ 1961a). The gills are lost at or shortly before hatching and there is no aquatic larval stage. Juveniles resemble miniature adults.

SYSTEMATICS AND GEOGRAPHIC VARIATION. No detailed published data on geographic variation in

Range of *Desmognathus wrighti*

genetic or phenotypic traits are available, and no subspecies are recognized. Studies of mitochondrial DNA suggest that the lineage leading to *D. wrighti* originated very early in the evolution of desmognathine salamanders (Titus and Larson 1996).

DISTRIBUTION AND ADULT HABITAT. *Desmognathus wrighti* is found at elevations of 838–1981 m in mountainous habitats from Whitetop Mountain and Mt. Rogers in southwestern Virginia, southwestward to the Nantahala, Cowee, and Great Smoky mountains in western North Carolina and eastern Tennessee (Bruce 1977; Rubin 1971; Tilley and Harrison 1969). Although this species is often characterized as a high-elevation species that is strongly associated with spruce-fir forests, populations often occur at relatively high densities in mature mesophytic cove forests at lower elevations (Hairston 1949; Tilley and Harrison 1969). Adults inhabit the forest floor and often occur far from running water.

BREEDING AND COURTSHIP. Mating occurs primarily during the fall and spring months. Virginia adults maintained in the laboratory will court from September to October and April to May (Organ 1961a,b). Organ (1961a) considers *D. wrighti* to be a biennial breeder, but Tilley (1968) argues for an annual breeding cycle.

Detailed descriptions of courtship are not available. A male that encounters a female may initially clamp his jaws to her tail, body, or head, and in some cases restrain the female for several hours (Houck 1980). This biting behavior is similar to that of *D. aeneus* (Promislow 1987) and may function to abrade or cut the skin and introduce mental gland secretions into the female's circulatory system. Adults eventually engage in a tail-straddle walk that may be preceded by the male placing his snout on the female's back and snapping. The tail-straddle walk is essentially the same as that of *D. orestes* and *D. fuscus* and culminates in the male depositing a spermatophore and the female picking up the sperm cap. The spermatophore is similar to that of *D. fuscus* but is only about 2.5 mm high (Organ 1961b). The base and stalk are clear and support a milky white sperm cap. The base is heart shaped and flared, and it tapers to a slender, conical stalk. Males will chase and bite rival males that attempt to court females.

REPRODUCTIVE STRATEGY. Females lay clusters of eggs similar to those of *D. fuscus* and remain coiled about their eggs with their heads resting upon the cluster or thrust into its center. The only nesting record

Fig. 134. *Desmognathus wrighti*; courting adults (note spermatophore being deposited by the male on the left); Grayson Co., Virginia (S. J. Arnold).

is for six clusters in late developmental stages found on 16 October in Virginia beneath the bank of a headwater stream (Organ 1961a,b). The eggs are 3.5 mm in diameter and are suspended by a single pedicel formed from the outer capsules of one or two eggs. Number of eggs per clutch varies from three to eight and averages six. Females at high-elevation sites that are far removed from running water or seepages presumably oviposit in underground retreats.

Bruce (1996) reports a mean of 9 ova (range = 8–10; $n = 4$) in females from southwestern North Carolina. The number of mature ova in females examined by Tilley (1968) from southwestern Virginia varies from 7 to 14, averages around 10, and is positively correlated with SVL (estimated from Fig. 1 in text). Eggs from four clutches maintained in the laboratory hatched between 19 and 25 October.

AQUATIC ECOLOGY. Pygmy salamander embryos resorb their gills shortly before hatching and there is no larval stage. This and *D. aeneus* are the only *Desmognathus* species that lack a larval feeding phase.

TERRESTRIAL ECOLOGY. In southwestern Virginia, males become sexually mature when they are 3.5 years old and measure >24 mm SVL. Females become sexually mature when 4.5 years old and first oviposit when 5 years old (Organ 1961a). Age-specific mortality tends to increase with age in *D. wrighti*, and annual survivorship is higher than that in other *Desmognathus* species. Mean annual survivorship during the first few years of life is estimated to be 91% for males and nonbrooding females.

In Virginia, large aggregates of *D. wrighti* can be found in streambanks from October through mid-May, but few are present in mid-June (Organ 1961a,b).

Most males disperse from hibernacula to forest floor habitats after the arrival of warm weather in late spring. Females either remain near winter hibernacula all year long or vacate forest floor microhabitats in early summer and move to streambanks and seepages to nest. Bruce (1977) collected 16 individuals from the wet banks of a seepage on 15 February.

At high-elevation sites the adults and juveniles can be found widely dispersed on the forest floor far from streams or seepages. It is not known if these animals engage in long-distance dispersal to and from distant streams to hibernate or breed, or simply remain in their home ranges and move beneath ground seasonally. Pygmy salamanders readily climb vegetation. On wet nights the majority of surface-active animals may perch on plants, where they forage on small invertebrates and avoid larger, predatory salamanders that forage from the leaf litter below.

PREDATORS AND DEFENSE. *Gyrinophilus porphyriticus* and carabid beetles are known to feed on pygmy salamanders (Huheey and Stupka 1967). Individuals often do not flee when uncovered. Instead, they remain immobile, a behavior that may make them less conspicuous to predators (Dodd 1990b). Wake and Dresner (1967) report that 7% of a sample of museum specimens have broken tails.

COMMUNITY ECOLOGY. In the southern Appalachians, *Desmognathus* species exhibit niche partitioning with respect to the distance that different species live from streams and seepages. The pygmy salamander is the most terrestrial of all *Desmognathus* species and avoids areas next to streambanks where larger, predatory species such as *D. monticola* and *D. quadramaculatus* abound. Detailed accounts of ecological interactions in

Desmognathus communities are presented under the accounts of *D. monticola, D. ocoee,* and *D. quadramaculatus.*

CONSERVATION BIOLOGY. Red spruce–Fraser fir forests are declining in many high-elevation sites of the southern Appalachians. The extent to which pygmy salamanders are being affected as these forests disappear locally is not known. At relatively low elevations, I have found *D. wrighti* to be common in old-growth deciduous forests but uncommon in young forest stands.

Phaeognathus hubrichti Highton
Red Hills Salamander
PLATE 58

Fig. 135. *Phaeognathus hubrichti;* adult; Alabama (R. W. Van Devender).

IDENTIFICATION. The Red Hills salamander is a large, dark, elongated desmognathine salamander with 20–22 costal grooves. The entire body is dark brown except for the snout, jaws, and soles of the feet, which are light brown. The limbs are short relative to body length, and individuals lack a light line from each eye to the angle of the jaws, as seen in most other desmognathine salamanders (Highton 1961; Valentine 1963a). Females measured by Schwaner and Mount (1970) average 3% larger in SVL than males. Adults measure 10–26 cm TL and there is no aquatic larval stage.

SYSTEMATICS AND GEOGRAPHIC VARIATION. Based on a very limited number of specimens, McKnight et al. (1991) found no variation in mitochondrial DNA among 14 individuals collected from 13 populations. In contrast, electrophoretic data suggest that two distinct groups exist within the range of the species.

DISTRIBUTION AND ADULT HABITAT. The Red Hills salamander is a federally threatened species that is confined to the Tallahatta and Hatchetigbee formations of the Red Hills Province in south-central Alabama. Scattered colonies occur between the Alabama and Conecuh rivers in moist, cool, mesic ravines of claystone in mixed deciduous forests (Dodd 1991; French and Mount 1978; Schwaner and Mount 1970). Older juveniles and adults live in oval-shaped burrows that have smoothly rounded rims. North-facing slopes with mature hardwood canopies provide optimal habitats (Dodd 1989b, 1991; Schwaner and Mount 1970). The burrows are typically constructed on ravine slopes that are too steep to hold leaf litter, but are occasionally found on nearly level land with leaf litter (French and Mount 1978; Schwaner and Mount 1970; Valentine 1963b). Many populations are fragmented because of the patchy distribution of ravines with claystone (Dodd 1991).

BREEDING AND COURTSHIP. Information is not available on the seasonal time of mating or courtship behavior.

REPRODUCTIVE STRATEGY. Most aspects of the nesting ecology of this species are poorly documented. Females presumably nest deep within underground burrows, although eggs have not been discovered in nature. Estimates of clutch size are 4–7 ova (mean = 5.5) for six specimens (K. Dodd, pers. comm.), 8 and 9 mature ova in two specimens (105 and 109 mm SVL; Brandon 1965b), and a clutch of 16 eggs laid by a 115-mm SVL captive female (Brandon and Maruska 1982). The pigmentless eggs are attached in a cluster from a common stalk to an overhanging support. The eggs are 7–7.2 mm in diameter and the outer envelopes are 8.5 mm in diameter.

Most females appear to oviposit in late winter or

Range of *Phaeognathus hubrichti*

spring. Spent females or females with large (>5.0 mm in diameter) ovarian eggs can be found in February through May (Brandon 1965b; Schwaner and Mount 1970). One spent female was collected in September.

TERRESTRIAL ECOLOGY. Small juveniles are underrepresented in samples, suggesting that they spend most of their time in underground retreats. Males mature sexually after reaching 80 mm SVL; however, females do not mature until they reach 93-98 mm SVL (Brandon 1965b; Schwaner and Mount 1970). Age at first reproduction is not known precisely, but juveniles probably require 5-6 years to reach sexual maturity given the large size of the smallest adults (Parham et al. 1996). A sample of 11 specimens that were aged using skeletochronology contained adults that were 5-11 years old.

At sites with mature forest canopies, individuals typically construct burrows on the upper two-thirds of the slope of a ravine (Dodd 1991). The entrances often emerge beneath and to the sides of claystone rocks, and the burrows are extensive and deeply coursing (Brandon 1965b). Adults cannot burrow readily into firmly packed soil, but can enlarge existing depressions or cavities (Brandon 1965b). In a survey of 10 sites with relatively high densities of *Phaeognathus*, Dodd (1990a) found an average of 2.6-9.4 burrows per 100 m^2 of bank habitat.

Individuals normally remain in subsurface retreats during the day, but emerge partially from their burrows at night to forage on insects and other small invertebrates (Brandon 1965b; Schwaner and Mount 1970). Individuals sometimes perch at burrow entrances during the day when cloudy, humid weather prevails (Dodd 1991). The tail is prehensile and facilitates moving within burrows (Blair 1967a).

Prey items in 13 specimens included snails, millipedes, mites, beetles, ants and other hymenopterans, flies, and insect larvae (Brandon 1965b). The diet of *Phaeognathus* is very similar to that of *P. glutinosus* inhabiting the same habitats.

PREDATORS AND DEFENSE. Dodd (1991) reported that feral pigs (*Sus*) and armadillos (*Dasypus novemcinctus*) frequently root about in *Phaeognathus* habitats and may be important predators on this species.

COMMUNITY ECOLOGY. Almost no information is available on the community ecology of the Red Hills salamander. *Eurycea b. cirrigera, E. guttolineata,* and *Notophthalmus viridescens* have been observed in *Phaeognathus* burrows (K. Dodd, pers. comm.), but it is uncertain if these species compete with *Phaeognathus* for space.

CONSERVATION BIOLOGY. The Red Hills salamander is a federally protected species and should receive high priority for protection because of its restricted range and specialized microhabitat requirements. French and Mount (1978) and Jordan and Mount (1975) report that many of the habitats suitable for *Phaeognathus* have been severely degraded by paper corporations employing forestry techniques involving clear-cutting and mechanical site preparation in large-scale operations. Dodd (1989b, 1991) found that Red Hills salamander burrows were common at about 65% of sites that had not been logged since 1976 compared with only 18% of sites that had been logged.

Clear-cutting of slopes and mechanical site preparation are highly detrimental to local populations and should be avoided. Slopes with *Phaeognathus* should be left uncut with a 100- to 200-m buffer of hardwoods maintained on either side, and timbered sites should be allowed to regenerate into mature hardwood forest. Paper companies are now taking an active role in protecting *Phaeognathus,* and cooperative agreements are in place to protect critical habitats of this Alabama endemic.

SUBFAMILY PLETHODONTINAE
Tribe Bolitoglossini

Batrachoseps aridus Brame
Desert Slender Salamander
PLATE 59

Fig. 136. *Batrachoseps aridus;* adult; Riverside Co., California (M. Garcia-Paris).

IDENTIFICATION. The desert slender salamander is a relict species that was described by Brame (1970) from Deep Canyon in Riverside Co., California. This species can be distinguished from all other *Batrachoseps* species by its distinctive ventral coloration. The belly and throat region is blackish maroon and contrasts markedly with the underside of the tail, which is flesh colored. The dorsum is also blackish maroon and is usually overlain by brassy flecking with larger patches of metallic, golden-orange colors. Males reach about the same SVL on average as females (Brame 1970). Relative to those of most other *Batrachoseps* species, the limbs are long and the head broad (Marlow et al. 1979). The tail is roughly about the same length as the body and there are 16–19 (usually 18) costal grooves. Fourteen adults described by Brame (1970) measure 30–48 mm SVL and average 37 mm SVL. The young are black to dark brown above and often have greatly reduced brassy flecking (Stebbins 1985).

Range of *Batrachoseps aridus*

DISTRIBUTION AND ADULT HABITAT. This species is known only from Hidden Palm Canyon in Riverside Co., California (Brame 1970). The type locality is a desert canyon at around 850 m that supports cacti, creosote bush, manzanita, and other xeric vegetation on the slopes, along with willow, Washington palms, mesquite, grasses, mosses, maidenhair ferns, and other vegetation on the clifflike walls of the canyon. The canyon walls contain seepages that provide wet microhabitats for the salamanders. Specimens have been collected beneath sheets of limestone and beneath rocks at the lower levels of the cliffs. A nearby population in Guadalupe Canyon probably belongs to this species (Stebbins 1985).

COMMENTS. The desert slender salamander is a federally endangered species that is protected by law. Its only known habitat is now an ecology reserve that has been purchased by the California Department of Fish and Game and can only be entered by permit (Bury et al. 1980; Stebbins 1985). Almost no published data are available on the ecology and natural history of this species. Data in Brame (1970) suggest that both males and females become sexually mature when they reach about 30-31 mm SVL. Immobility is the primary antipredator defense of this species (Brodie 1977).

Batrachoseps attenuatus (Eschscholtz)
California Slender Salamander
PLATE 60

Fig. 137. *Batrachoseps attenuatus;* adult; Del Norte Co., California (W. P. Leonard).

IDENTIFICATION. The California slender salamander is an elongated, slender species with four digits on each foot. The limbs are very short relative to body length and are about as long as the body is wide. The tail is very long and often makes up 55-65% of the total length of an individual. The ground color is dark brown to blackish and is usually overlain by a brown, pale yellow, or reddish tan dorsal stripe (Nussbaum et al. 1983; Stebbins 1951, 1985). A herringbone pattern is often present on the back. The venter is sooty to blackish with fine white stippling, and the underside of the tail is often lighter than the belly and appears tinged with yellow from underlying fat deposits. Adults measure 7.5-14 cm TL and there are 19-22 costal grooves (usually 20 or 21). Sexually active males have enlarged premaxillary teeth that project just beyond the closed mouth, and a somewhat broader, more truncated, snout than females. These traits are often difficult to distinguish, and individuals are best sexed by examining specimens for eggs or other reproductive structures.

Hatchlings measure 14-19 mm TL (Burke 1911; Storer 1925). Juveniles are similar to adults but have proportionately shorter tails and longer limbs.

SYSTEMATICS AND GEOGRAPHIC VARIATION. Slender salamanders (*Batrachoseps* spp.) are a taxonomically difficult group that exhibits marked geographic variation in color patterns, body proportions, morphological features, and genetic characteristics. Earlier treatments of this genus based on analyses of morphology and color patterns (e.g., Brame and Murray 1968; Hendrickson 1954) recognized from as few as two to as many as seven species in the United States. Yanev (1980) conducted extensive biochemical analyses of isozymes and documented numerous genetically distinct groups that contained three sibling species. She recommended recognizing seven species and treated *B.*

Range of *Batrachoseps attenuatus*

pacificus as a superspecies consisting of six semispecies that are near the species level of differentiation. One additional species that was not examined is *B. aridus,* a well-differentiated form with a very restricted distribution. Another species (*B. gabrieli*) was described from southern California (Wake 1986). D. B. Wake and his colleagues at the University of California at Berkeley are currently working on a revision of the genus *Batrachoseps* based on a more comprehensive set of molecular data. Several undescribed species will soon be described, and some forms that were previously considered to be subspecies may be raised to species rank. Here I follow Yanev's taxonomic classification for *Batrachoseps* and recognize nine species. As many as 18 species of *Batrachoseps* may eventually be recognized, and major revisions of the genus can be expected in the next few years (D. B. Wake, pers. comm.).

Hendrickson (1954) summarizes geographic variation in color patterns of this and other attenuate species of *Batrachoseps*. Populations in the redwoods region of northern California often have a higher frequency of individuals with reddish or reddish brown dorsal stripes than those to the south.

DISTRIBUTION AND ADULT HABITAT. *Batrachoseps attenuatus* ranges from southern Oregon to central California and into the foothills of the Sierra Nevada in east-central California. This species inhabits a variety of habitats, including grasslands, chaparral, coniferous forests, and interior live oak woodlands (P. K. Anderson 1960; Stebbins 1951; Yanev 1980). Specimens can be readily collected from beneath logs and rocks during the wetter months of the year and are common in urban areas. In the northern portion of its range, *B. attenuatus* shows strong affinities for mature and old-growth forests.

Fig. 138. *Batrachoseps attenuatus;* adult; northern California (J. W. Petranka).

BREEDING AND COURTSHIP. Males that surface in the autumn with the arrival of rains have viable sperm in their ducts, as do those collected throughout most of the rainy season (Stebbins 1951). Mating may conceivably occur in the summer when animals are in underground retreats. Most females oviposit annually, and are able to acquire sufficient energy to do so because they store fat in the large tail and do not brood (Maiorana 1976, 1977a). Courtship behavior has not been described.

REPRODUCTIVE STRATEGY. Most females appear to oviposit within a few weeks after they emerge in October and November from underground summer retreats following the arrival of autumn rains (P. K. Anderson 1960; Maslin 1939; Stebbins 1951). Time of emergence and initiation of breeding are strongly geared to annual rainfall patterns (Hendrickson 1954; Maiorana 1976; Storer 1925). The small number of eggs that have been found by researchers suggests that most females oviposit in underground cavities that are inaccessible to collectors. Clutches of 7 and 12 eggs have been found 30 cm below the ground surface, although most records are for surface nests (Stebbins 1951).

Females often lay eggs in joint nests, as evidenced by the large number of eggs found in groups relative to average clutch sizes. Nesting records include a group of 74 eggs and an ovipositing female found on 5 November beneath a strip of tin in a damp depression (Maslin 1939), 53 eggs in advanced stages of development found under a plank on 7 March (Storer 1925), and groups of 10 and 21 eggs in advanced stages found in small pockets in the ground under a log on 5 January (Burke 1911). Numbers of eggs in these nests generally exceed the average ovarian complements of females, which range from 8-9 (range = 5-15; P. K. Anderson 1960) to 12 (range = 6-21; Maslin 1939).

Freshly laid ova are unpigmented, about 3-4 mm in diameter, and surrounded by two envelopes (Maslin 1939; Snyder 1923). The outer capsule averages about 6 mm in diameter. The eggs are attached to each other by narrow cords of jelly about 12 mm in length (Maslin 1939; Snyder 1923). These form a rosarylike chain, but adjoining eggs often break apart during the course of development (Stebbins 1951). Females have been collected near eggs, but do not appear to actively brood the eggs.

The exact length of the incubation period under natural conditions is not known. Eggs incubated at 17.5 and 21°C hatch after an average of 64 days (Anderson 1958). Incubation in the field may be longer where soil temperatures are cooler. Advanced embryos have trilobed, filamentous gills that are lost shortly before hatching (Snyder 1923; Stebbins 1951; Storer 1925). Hatching typically occurs from mid-January through early April (Burke 1911; Snyder 1923).

TERRESTRIAL ECOLOGY. California slender salamanders are active on the ground surface during periods of wet weather, but retreat underground whenever the soil dries or the air temperature drops below freezing (Hendrickson 1954; Stebbins 1951). Individuals remain underground throughout the dry summer months and surface with the return of wet weather in the autumn. Hatchlings and small juveniles are often underrepresented in samples, suggesting that they spend more time beneath ground than do the adults. Maiorana (1976) believed that the underrepresentation of juveniles reflects higher mortality rates in juveniles relative to adults. Juveniles have higher metabolic rates and lower fat reserves in the tail than adults. These characteristics reduce their chances of surviving through the summer when feeding opportunities are limited.

Juveniles studied by Hendrickson (1954) grew 3-6 mm SVL/year, but growth declined to 1-3 mm SVL/year after individuals became sexually mature. Most animals became sexually mature when 2-4 years old, and some live as long as 10 years. Data provided by Maiorana (1976) and P. K. Anderson (1960) indicate that females become mature when they are 2.5-3.5 years old and they have reached about 34 mm SVL. Using skeletochronology, Wake and Castanet (1995) estimate that specimens measuring >40 mm SVL are 4-8 years old. Their data indicate that body size is a poor predictor of an individual's age.

P. K. Anderson (1960) compared life history traits of island and mainland populations of *B. attenuatus* in the San Francisco Bay area, California. Island populations occur at much higher densities than mainland populations; in favorable habitats densities may reach 1.1-1.7 salamanders/m^2. Population size structure also differs among island and mainland groups. Island populations have higher proportions of older individuals than mainland populations, and females delay maturity and often fail to reproduce at sites with high salamander densities. These factors help explain the underrepresentation of younger animals in many island populations.

California slender salamanders appear to have very low dispersal rates. In one population animals rarely move >1.5–2 m between captures and most remain beneath the same cover object from one year to the next (Hendrickson 1954). A few individuals are more mobile, but most have home ranges of <12–16 m^2. On rainy nights animals are more active and tend to move greater distances (P. K. Anderson 1960; Hendrickson 1954).

Adults have extremely long tails that can be autotomized at any segment and regrown. Maiorana (1977b) found several *Batrachoseps* in the field that appeared to be dying from heat stress or desiccation. Most of the stressed animals had autotomized their tails. During the dry season females rarely feed and use fat stored in the tail for egg production. Autotomy may function to distract predators, but it can also reduce clutch size, delay sexual maturation, or reduce reproductive output because of its drain on energy supplies (Maiorana 1977a; Hendrickson 1954). Tail regeneration may take 1–3 years depending on the length that is autotomized (Hendrickson 1954).

The diet consists entirely of invertebrates, and individuals are generalist, sit-and-wait predators. In central California populations, *B. attenuatus* eats snails, collembolans, mites, dipterans, aphids, and thrips (Maiorana 1978b). Individuals largely eat what is available, but in some instances tend to select larger prey or specific taxa. In other populations, collembolans, ants, thrips, beetles, lepidopterans, flies, isopods, arachnids, and snails are eaten, with snails, mites, beetles, and collembolans being the most numerically important (Adams 1968). Specimens in northern California feed predominantly on mites, springtails, small spiders, flies, and snails (Bury and Martin 1973).

PREDATORS AND DEFENSE. Natural predators of *Batrachoseps* include the arboreal salamander, small snakes, and white-footed mice (Stebbins 1954). Individuals will readily autotomize their tails if picked up by the tail (Storer 1925) or attacked and seized by garter (*Thamnophis*) and ringneck (*Diadophis*) snakes (Hubbard 1903). *Batrachoseps attenuatus* exhibits two antipredator behaviors. When initially uncovered, animals often coil into a tight spiral and remain immobile. If molested, they may rapidly coil and uncoil their bodies and, in springlike fashion, fling themselves as far as 60 cm away from potential predators (Brodie 1977; Brodie et al. 1974b; Stebbins 1951; Storer 1925). Anecdotal observations suggest that *Batrachoseps* may produce toxic skin secretions (Cunningham 1960; Hubbard 1903), but detailed studies have not been conducted to determine if toxic or noxious secretions are used to repel potential predators.

COMMUNITY ECOLOGY. Maiorana (1978a) studied patterns of resource use in *B. attenuatus* and syntopic *Aneides lugubris* and found that the spatial distribution of *Batrachoseps* is better explained by the distribution of cover rather than that of food resources. Although competition for food may sometimes occur between these species, the abundance of cover seems to be the primary factor regulating both populations. Lynch (1985) compared prey items of four sympatric salamanders, including *B. attenuatus*, and found species-specific differences in diet that are related to morphology and gape width.

CONSERVATION BIOLOGY. Although *B. attenuatus* is often found in nonforested regions in the southern portion of its range, populations in northern California and southwestern Oregon appear to reach their highest densities in mature forests. Welsh and Lind (1988, 1991) found that *B. attenuatus* is more abundant in old-growth forests in the Coast Range of California and Oregon than in young forests regenerating after timbering. Bury (1983) found *B. attenuatus* to be 10 times more abundant on old-growth redwood plots than on matched plots in 6- to 14-year-old, regenerating forests.

Batrachoseps campi Marlow, Brode, and Wake
Inyo Mountains Salamander
PLATE 61

Fig. 139. *Batrachoseps campi;* adult; Inyo Co., California (M. Garcia-Paris).

Fig. 140. *Batrachoseps campi;* adult; Inyo Co., California (R. W. Van Devender).

IDENTIFICATION. The Inyo Mountains Salamander is a large, robust species of *Batrachoseps* with a relatively short tail, broad head, and large feet (Marlow et al. 1979). Tail length represents <50% of the TL, and there is no dorsal stripe as in many *Batrachoseps*. Sixteen males from the type locality measure 41–53 mm SVL (mean = 45.3) and 23 females measure 32–61 mm SVL (mean = 45.5). The dorsal ground color

Range of *Batrachoseps campi*

is dark brownish black and is overlain with silvery to greenish gray patches of iridophores. In most populations the silvery patches cover the dorsum, but in some they are restricted to the eyelids, head, and front of the body (Yanev and Wake 1981). The eyes are relatively large and moderately protuberant, and there are 16-18 costal grooves.

SYSTEMATICS AND GEOGRAPHIC VARIATION. Yanev and Wake (1981) examined protein variation in *B. campi* and found relatively high levels of genetic differentiation among local and regional populations (Nei's $D = 0.01$–0.23; mean = 0.12). Geographic isolation and small population sizes may be responsible for the degree of differentiation documented in this group. Electrophoretic evidence further indicates that *B. campi* is not closely related to any of the known species of *Batrachoseps*, although it is most closely related to *B. wrighti* ($D = 0.71$). *Batrachoseps campi* is a primitive, generalized species that is a relict of what was formerly a more widely distributed ancestral form (Marlow et al. 1979).

DISTRIBUTION AND ADULT HABITAT. This species is restricted to spring-fed, desert canyons supporting mesic vegetation in the Inyo Mountains of Inyo Co. in southern California. Populations occur along a band about 32 km long and 10.5-13.5 km wide across the Inyo Mountains at elevations of 550-2620 m (Yanev and Wake 1981). Specimens have been collected beneath rocks in spring-fed, moist microhabitats (Marlow et al. 1979; Yanev and Wake 1981).

COMMENTS. The Inyo Mountains salamander was described in 1979 and most aspects of its life history remain unknown. Because of its limited range and restricted microhabitats, populations should be protected whenever possible when designing land use policies.

Batrachoseps gabrieli Wake
San Gabriel Mountain Slender Salamander

IDENTIFICATION. The San Gabriel Mountain slender salamander is a very recently described species that is known only from two localities in the San Gabriel Mountains in southern California (Wake 1996). This species is a medium-sized, slender *Batrachoseps* species with a relatively broad head, long limbs, large feet, and a markedly tapered tail. The dorsum has a broad, irregular band that is bright coppery bronze or orange colored. Coloration is most intense in the shoulder and pelvic regions, and the band breaks into diffuse patches along the tail. The venter is black with scattered iridophores that are more concentrated near the pelvic region. White spots occur in the throat region. The tail is moderately short relative to the tails of other attenuate *Batrachoseps* and averages 1.18 and 1.24 times the SVL for females and males, respectively. Coastal grooves average 18-19 and there are 5.5-7.5 costal folds between adpressed limbs. Sexually active males have slightly swollen nasolabial protuberances and swollen tissue around the anterior portions of the jaws and snout, but lack mental hedonic glands. Respective mean lengths for 16 adult females and 8 adult males were 46.1 (range = 41–50) and 42.4 (range = 40–46) mm SVL.

SYSTEMATICS AND GEOGRAPHIC VARIATION. *Batrachoseps gabrieli* is a relict species that is strongly differentiated genetically from all other *Batrachoseps*. The lowest value for Nei's genetic distance (D) is 0.65 for comparisons with certain populations of *B. p. major* and *B. stebbinsi*, and there are fixed differences for 12 of 26 loci sampled. Evidence from allozyme variation and mitochondrial DNA sequencing suggests that this lineage has been separated from other lineages in the genus for perhaps 8-13 million years.

DISTRIBUTION AND ADULT HABITAT. *Batrachoseps gabrieli* is currently known from only two localities in the San Gabriel Mountains in Los Angeles Co., California. The type locality is a northwest-facing talus slope at 1550 m that is shaded by a mixture of hardwood and conifer trees, including oaks (*Quercus* spp.), pines (*Pinus* spp.), white fir (*Abies concolor*), California laurel (*Umbellularia californica*), Oregon big-leaf maple (*Acer macrophyllum*), and incense cedar (*Calocedrus decurrens*). Specimens live primarily in the talus and reside under large rocks, rotting logs, and bark when active near the surface. The second known population occurs about 1 km to the south at 1158 m,

Range of *Batrachoseps gabrieli*

where two specimens have been collected within 15 m of a stream. *Batrachoseps nigriventris* occurs at this site but has not been found at the type locality.

COMMENTS. We currently have very few data on the natural history of *B. gabrieli*. Wake (1996) notes that surface activity is probably limited to a few months in the winter and early spring. Individuals are relatively common on the surface in February and March even when the soil surface temperature is only a few degrees centigrade about freezing. As many as three adults have been found sharing the same rock, an observation that suggests that adults are not aggressive or strongly territorial. During the dry months of summer and early fall, individuals presumably move deep within the talus, where conditions are cooler and moister. When first uncovered, individuals form a tight coil, but they then relax. Individuals are rather agile and are capable of springing or leaping away from pursuers.

The San Gabriel Mountain slender salamander is a relict species that occupies a habitat that is both restricted and fragile. The known habitat consists of two small areas about 1 hectare in extent. Efforts should be made immediately to protect the species, and herpetologists should refrain from removing any specimens from the wild.

Batrachoseps nigriventris Yanev
Black-bellied Slender Salamander
PLATE 62

IDENTIFICATION. The black-bellied slender salamander is a slender, short-limbed, long-tailed species of *Batrachoseps* that has 18–21 costal grooves and 9–15 costal folds between the toe tips of adpressed limbs. The tail of adults is often twice as long as the body in mainland populations, but it may be only about body length in adults on Santa Cruz Island. The dorsum typically has a stripe that is brown, reddish, or beige (Stebbins 1985). The venter is black or grayish black with fine white specks that cover the entire ventral surface.

Fig. 141. *Batrachoseps nigriventris;* adult; Los Angeles Co., California (R. W. Van Devender).

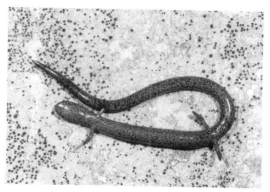

Fig. 142. *Batrachoseps nigriventris;* ventral view of adult; Los Angeles Co., California (R. W. Van Devender).

SYSTEMATICS AND GEOGRAPHIC VARIATION. Yanev (1980) conducted biochemical analyses of isozymes and documented several genetically distinct groups of *Batrachoseps*, including *B. nigriventris*. Subgroups currently assigned to *B. nigriventris* in the Coast Range and Sierra Nevada are genetically distinct from each other and are best viewed as members of a semispecies or superspecies complex. Some of these will probably be described as full species based on new molecular evidence (D. B. Wake, pers. comm.).

DISTRIBUTION AND ADULT HABITAT. This species occurs in three isolated areas: on Santa Cruz Island, in the southern coastal mountains, and along the western slope of the central and southern Sierra Nevada. The coastal mountain populations extend from around San Simeon in Monterey Co. southward to the Palos Verdes Hills and Santa Ana Mountains. The Sierra Nevada populations extend from the Merced River drainage south through the Techachapi Mountains toward the coastal populations. The adults occur in scat-

Range of *Batrachoseps nigriventris*

tered colonies and specimens can be found under rocks, logs, and bark and in termite channels in damp locations (Stebbins 1985). This species chiefly inhabits oak woodlands but can also be found in oak-pine forests, streamside habitats, and arid grasslands.

COMMENTS. Very few data are available on the life history of *B. nigriventris*. In southern California eggs are laid in the winter and hatch in winter and early spring. Eggs have been found in early December at 430 m in the Coast Range and in May at 1310 m in the Sierra Nevada (Stebbins 1985). Three females examined by Stebbins (1951) from the Santa Monica Mountains had four, six, and nine large ovarian eggs.

Batrachoseps pacificus (Cope)
Pacific Slender Salamander
PLATE 63

Fig. 143. *Batrachoseps pacificus major;* adult; California (M. Garcia-Paris).

IDENTIFICATION. The Pacific slender salamander is a geographically variable group that was treated by Yanev (1980) as a superspecies consisting of six semispecies. Isolated groups within this complex have undergone regional differentiation and semispecies now exist that differ in size, body proportions, and color patterns. There is no single set of traits that will distinguish this species from other *Batrachoseps* species (Stebbins 1985). However, many of the semispecies can be recognized in areas where they coexist locally with other *Batrachoseps* species; other subspecies can only be reliably diagnosed using biochemical evidence. Readers should refer to the individual subspecies discussions that follow for detailed descriptions of regional populations.

This species typically has 16-21 costal grooves, and there are 5-13 costal folds between adpressed limbs. Adults measure 7.5-16 cm TL and females are on average slightly larger than males. The average SVL of females in six populations examined by Brame and Murray (1968) is 1-9% greater than that of males. Albino adults have been found on rare occasions (Dyrkacz 1981).

Hatchlings of the different semispecies have not been adequately described. Hatchling *B. p. major* have darker bellies and a much more conspicuous dorsal stripe than older animals (Cunningham 1960). Reports of hatchling size are an average of 10 mm SVL and 16 mm TL (Grant 1958) and 21-23 mm TL (Cunningham 1960).

SYSTEMATICS AND GEOGRAPHIC VARIATION. D. B. Wake and his colleagues at the University of California at Berkeley are currently working on a revision of the genus *Batrachoseps* based on a comprehensive analysis of allozyme and mitochondrial DNA data. Several previously undescribed species will soon be described, and some forms that were previously considered to be subspecies may be raised to the species rank. Populations currently assigned to *B. pacificus* contain several well-differentiated subgroups that make up a semispecies or superspecies complex. In the most recent published treatment, Yanev (1980) recognizes six semispecies of *B. pacificus* based on electrophoretic analyses of protein variation. One occurs in Mexico and the remaining five in the United States. Three of the U.S. forms have been formally recognized as subspecies, whereas two others are simply referred to by their common names. The following descriptions of major geographic groups in the United States currently assigned to *B. pacificus* are from Stebbins (1985).

The Channel Islands slender salamander (*B. p.*

pacificus) reaches a relatively large size and has a broad head, long legs, and a relatively short tail. Specimens are brownish to pinkish above with a broad dorsal stripe with poorly defined borders. The venter and sides are pale and are weakly speckled or streaked with melanophores. There are 18–20 costal grooves. This subspecies is restricted to San Miguel, Santa Rosa, Santa Cruz, and the Anacapa Islands off the coast of Santa Barbara, California. Black-bellied slender salamanders that occur in sympatry on South Channel Island are smaller, darker, and slimmer and usually have white specks on the underside of the tail.

The garden slender salamander (*B. p. major*) is a more elongated form with relatively short limbs, usually a longer tail, and a narrower head than *B. p. pacificus*. Specimens are typically light brown, light tan, pink, or light gray above and frequently have rust on the tail, snout, and shoulders. Stebbins (1985) noted that some populations are dark in uplands of the Peninsular Ranges, in the San Pedro Martir Mountains, and on Todos Santos Island. In addition, those from near El Rosario, Baja California, are very pale. The dorsal stripe often has diffuse borders or is absent, and the belly is light gray with a weak speckling or streaking of melanophores. Dark populations have a sooty colored belly with heavier melanophore streaking. There are 17–21 costal grooves and 9–12 costal folds between the toes of adpressed limbs. The tail is elongated and may be twice as long as the body.

This subspecies is found in southern California from the base of the San Gabriel and San Bernardino mountains south to the San Pedro Martir Mountains in Baja California. Populations are also present on Catalina, Los Coronados, and Todos Santos islands. *Batrachoseps nigriventris* is sympatric with *B. p. major* in Los Angeles and Orange counties and can be distinguished by its darker coloration, slimmer and smaller body, and shorter limbs and the presence of a dorsal stripe in some individuals.

The relictual slender salamander (*B. p. relictus*) has a relatively broad head and long legs like *B. p. pacificus*, but is smaller (rarely exceeding 45 mm SVL) and usually dark gray to black above and below. The dorsal stripe is reddish, yellowish brown, or dark brown and is often obscure in large specimens. There are 16–20 costal grooves and 7–9 costal folds between the toe tips of adpressed limbs, but regional populations often differ in the mean number of costal grooves (Brame and Murray 1968). The tail is moderately elongated and may be as much as 1.75 times as long as the body. This subspecies is patchily distributed along the western slope of the southern Sierra Nevada from the American River drainage to the Kern River Canyon. It occurs in mixed coniferous forests and is typically found at higher elevations than *B. nigriventris* in the same region.

The Gabilan slender salamander is an undescribed race that resembles *B. attenuatus*. This race has a well-defined dorsal stripe, short legs, and a long tail. There are typically 20–21 costal grooves and 10–12 costal folds between the toes of adpressed limbs. This form is found mostly in oak woodland habitats along the east slope of the Santa Lucia Mountains and the Gabilan and southern Diablo ranges, from the vicinity of Soquel and the San Benito River south to the Polonia Pass area in San Luis Obispo Co., California. The Gabilan slender salamander is mostly isolated from *B. attenuatus* to the north and *B. nigriventris* to the south and is best identified by its range. In areas of overlap this form cannot be reliably distinguished from *B. nigriventris* without using biochemical data (Stebbins 1985).

The Santa Lucia slender salamander has relatively large hindlimbs and wide feet and is darker above and below than *B. p. relictus*. There are 19 costal grooves and 8–9 costal folds between the toes of adpressed limbs. The tail is about 1.75 times as long as the body. This race is found in deep leaf litter of moist canyons in redwood and mixed evergreen forests along the western slopes of the Santa Lucia Mountains from the Monterey area to northern San Luis Obispo Co., California. Like the Gabilan slender salamander, this race is mostly parapatrically distributed with *B. nigriventris* and is best identified using its range. In areas of overlap in the extreme southern range of this race, the two species can only be reliably distinguished by using biochemical data (Stebbins 1985).

DISTRIBUTION AND ADULT HABITAT. The Pacific slender salamander occurs as several geographically isolated populations that range from central California southward to northern Baja California. Members of this complex are found in a variety of habitats, including oak woodlands, coniferous and mixed deciduous-coniferous forests, chaparral, and grasslands. Specimens have also been collected along washes in salt marshes and under beach driftwood (Stebbins 1985). The relictual pacific salamander often inhabits

Range of *Batrachoseps pacificus*

seepages in addition to woodland habitats (Brame and Murray 1968).

BREEDING AND COURTSHIP. There are few data available on breeding dates or courtship of *B. pacificus*.

REPRODUCTIVE STRATEGY. Adults emerge from underground summer retreats with the onset of autumn rains and begin ovipositing shortly after emergence. Stebbins (1951) notes that two gravid females collected on 17 December contained 15 and 20 mature ova. Breeding records include a clutch of 15 eggs of *B. p. relictus* found on 8 January in a log lying in river-washed debris (Grinnell and Storer 1924) and a clutch of 18 eggs of *B. p. major* in advanced developmental stages found on 20 December beneath a wooden box (Davis 1952). The eggs are laid in beadlike fashion and are connected by narrow cords of jelly but often break apart into single or paired eggs with time. Mature ova are 3-4 mm in diameter and are surrounded by two jelly envelopes. The outer capsule is 5.8-6.0 mm in diameter and the embryos have trilobed gills. Hatchlings have been found from 20 January to 1 February (Cunningham 1960; Davis 1952; Grant 1958).

TERRESTRIAL ECOLOGY. Like members of most other California *Batrachoseps* species, individuals are mostly active on the ground surface during the rainy season from autumn through spring. However, *B. p. relictus* may be active year round in seepage habitats (Brame and Murray 1968). The diet consists of small invertebrates, but detailed dietary studies are not available.

Cunningham (1960) conducted a life history study of a population inhabiting a residential lot in Santa Monica, California, that is occasionally disturbed by mowing and gardening activities. Individuals at this site remain in communal burrows during the day and surface at night to forage on invertebrates. They may climb as high as 60 cm above ground on vegetation and occasionally use their tails in a prehensile manner. Individuals are most active on the surface during periods of wet weather and may sometimes die from severe desiccation on windy, dry nights. The distance that an animal moves between captures is positively correlated with body size. Individuals often move >1 m between captures. Some animals make extended excursions away from their home ranges and then return, whereas others make long-range movements (≤10 m) and establish new home ranges, particularly following disturbance of the habitat. Many juveniles in the population appear to be floaters that are not strongly philopatric.

Individuals often share old earthworm burrows and show no evidence of being aggressive (Cunningham 1960). A tight cluster of six aestivating salamanders was unearthed about 91 cm below ground in a cavity in moist soil. Congregations of 15–20 *B. p. major* have been found beneath a single board, further suggesting that the adults are not aggressive or territorial (Stebbins 1951).

PREDATORS AND DEFENSE. About a third of the salamanders that Cunningham (1960) studied for several months lost their tails, presumably while being attacked by predators. Jerusalem crickets (*Stenopelmatus*) are frequently found under shelters used by *Batrachoseps* and may be major predators.

CONSERVATION BIOLOGY. *Batrachoseps pacificus* is surprisingly tolerant of urbanization and is often common in residential areas. Except for island populations, this species needs minimal protection.

COMMENTS. Studies of *Batrachoseps* species by Cunningham (1960) and Davis (1952) are based on observations made in the zone of overlap between *B. pacificus* and *B. nigriventris*. These are tentatively assumed to be accounts of *B. pacificus* based on brief descriptions of the hatchlings or adults given by the authors.

Batrachoseps simatus Brame and Murray
Kern Canyon Slender Salamander
PLATE 64

Fig. 144. *Batrachoseps simatus*; adult; Kern Co., California (M. Garcia-Paris).

IDENTIFICATION. The Kern Canyon slender salamander is a relict species that is known only from the Kern River drainage in the southern Sierra Nevada. Relative to other *Batrachoseps* species, this species has long limbs, a narrow head, and a slightly flattened head and body (Brame and Murray 1968). The ground color of the dorsum, sides, and venter is black and is overlain with patches of bronze or reddish pigment that may produce a faint dorsal band that extends onto the tail. The venter is conspicuously speckled with light flecks and the tail is slightly longer than the body. There are 20–21 costal grooves and 7–9 costal folds between adpressed limbs. Adults measure 9–13 cm TL.

This species coexists with *B. p. relictus* and can be readily distinguished from the latter by the presence of light flecking on the black venter and more costal grooves (typically only 16–20 in *B. p. relictus*). Males are smaller than females on average and have enlarged premaxillary teeth that protrude from the lip (Brame and Murray 1968). In a sample of nine adults, males were on average 5% smaller in SVL than females.

Range of *Batrachoseps simatus*

DISTRIBUTION AND ADULT HABITAT. This species is known only from the Kern River drainage from the canyon mouth to near Fairview at elevations ranging from 430 to 1920 m. Specimens have been found beneath pine, oak, and chaparral scrub logs, and from beneath rocks and talus on north-facing slopes of the canyon (Brame and Murray 1968). Isolated colonies of this species are found along stream courses in shaded, narrow, tributary canyons with willows and cottonwoods. Colonies also occur on ridges and hillsides supporting live oak, canyon oak, and pines (Stebbins 1985).

COMMENTS. Most aspects of the life history of *B. simatus* are unknown, and a detailed life history study of the species is needed. Because of its highly restricted range, this species should receive high priority in management decisions that could adversely affect local populations. This species is currently protected by California state law.

Batrachoseps stebbinsi Brame and Murray
Tehachapi Slender Salamander
PLATE 65

IDENTIFICATION. The Tehachapi slender salamander is a large species of *Batrachoseps* that has long legs, toes that are expanded near the tips, and strongly webbed feet with only one segment on each toe free from the webbing. This species normally has 18-19 costal grooves and 6-7 costal folds between adpressed limbs. The dorsum may be dark red, brick red, or brown. Light beige patches and blotches may sometimes form an indistinct dorsal band (Brame and Murray 1968). Light flecking is reduced or absent from the midline of the belly, and the tail is slightly shorter than the body length. Adults measure 9-13 cm TL, and males have enlarged premaxillary teeth that protrude from the lip. In one sample, males averaged 4% smaller in SVL than females (Brame and Murray 1968). *Batrachoseps stebbinsi* is best distinguished from sympatric *B. nigriventris* by the former's relatively long legs, large feet, and broad head.

Range of *Batrachoseps stebbinsi*

Fig. 145. *Batrachoseps stebbinsi;* adult; Kern Co., California (R. W. Hansen; courtesy of J. T. Collins and the Center for North American Amphibians and Reptiles).

DISTRIBUTION AND ADULT HABITAT. *Batrachoseps stebbinsi* is known from the Caliente Creek drainage in the Piute Mountains at the southern end of the Sierra Nevada in Kern Co., California (Stebbins 1985). Populations occur from around 610 to 1400 m in moist canyons and ravines in oak or mixed pine-oak woodland. Individuals can be collected beneath rocks in or near talus slopes and from beneath rotting logs, especially where there is much leaf litter. Brame and Murray (1968) collected the type specimens from beneath rocks on a wet talus slope with thick leaf litter. The surrounding woody vegetation consists of oak, sycamore, horse chestnut, mule fat, and digger pine.

COMMENTS. Virtually no data are available on the life history of this species, and a detailed life history study is needed. Because of its highly restricted range, this species should receive high priority for protection when designing land management policies.

Batrachoseps wrighti (Bishop)
Oregon Slender Salamander
PLATE 66

Fig. 146. *Batrachoseps wrighti*; adult; Linn Co., Oregon (W. P. Leonard).

Fig. 147. *Batrachoseps wrighti*; adult; Linn Co., Oregon (W. P. Leonard).

IDENTIFICATION. Like many *Batrachoseps* species, this is a slender, elongated salamander with shorter limbs than most, and only four toes on the hindfeet. The toes are long compared to those of other *Batrachoseps* species, and there are 16–17 costal grooves. *Batrachoseps wrighti* can be distinguished from other species of *Batrachoseps* by its range and its distinctive dark or black venter, which is heavily marked with large whitish flecks and blotches. The dorsum is black with scattered metallic flecks and has a reddish brown stripe from the head to the tail that is often faded in older animals. Females on average are about 12% greater in SVL than males, and adults measure 8.5–12 cm TL.

Hatchlings and small juveniles have stouter bodies, shorter tails, and longer limbs than adults (Nussbaum et al. 1983). Hatchlings from a single clutch measured by Stebbins (1949c) reached 13–14 mm SVL and 19–20 mm TL.

SYSTEMATICS AND GEOGRAPHIC VARIATION. *Batrachoseps wrighti* is a well-differentiated species of *Batrachoseps* based on coloration and morphology. Populations are nearly uniform in color throughout the range except in the Columbia River Gorge, where individuals with gold or greenish gold dorsal stripes can be found. The average number of maxillary teeth tends to decline markedly from north to south (Brame 1964; Nussbaum et al. 1983).

DISTRIBUTION AND ADULT HABITAT. The Oregon slender salamander is restricted to the Cascade Mountains and Columbia River Gorge of western Oregon, where it inhabits Douglas-fir and subalpine forests up to 1340 m in elevation. Populations are primarily restricted to the west slopes of the Cascades, although seven east-slope populations have been found in Wasco Co., Oregon (Kirk 1991; Kirk and Forbes 1991). Oregon slender salamanders are most frequently found within large, well-decayed logs in mature and old-growth forests, but they have also been found on recent lava flows (Nussbaum et al. 1983).

BREEDING AND COURTSHIP. Females have been collected in April, May, and early June with spermatophore caps in their cloacae, suggesting that courtship occurs during the spring and early summer (Nussbaum et al. 1983). Courtship behavior has not been described.

REPRODUCTIVE STRATEGY. Researchers have rarely found clutches of eggs, suggesting that most females oviposit either underground or deep within large, decaying logs. Most females oviposit from April to June. Freshly laid eggs are cream colored and average about 4 mm in diameter (Stebbins 1949c). Each egg is surrounded by two jelly capsules, and adjoining eggs are held together in beadlike fashion by

Range of *Batrachoseps wrighti*

jelly strands about 20 mm long. Tanner (1953) found eight eggs and an attending female on 19 June beneath the bark of a large fir tree. The number of ovarian eggs in 38 females varied from 3 to 11 and averaged 6.3 (Nussbaum et al. 1983).

Each embryo develops a pair of forked, lobed gills that are lost just prior to hatching. Embryos incubated at 12°C hatch after about 4.5 months (Stebbins 1949c).

TERRESTRIAL ECOLOGY. Very little information is available on most aspects of the natural history of the juvenile and adult stages. Individuals are active on the ground surface after snowmelt in the spring, but reside below ground or deep within large logs during the late spring and summer months when the ground surface is dry. Males and females mature at about 33 and 35 mm SVL, respectively. The diet of Lane Co., Oregon, populations includes collembolans, pseudoscorpions, mites, dipterans, spiders, snails, beetles, centipedes, and earthworms (Nussbaum et al. 1983).

PREDATORS AND DEFENSE. Natural predators of this species have not been documented. Adults often remain immobile or coil into a spiral during simulated predator attacks (Brodie 1977). One individual uncoiled violently in watch-spring fashion and flipped away when handled (Stebbins and Lowe 1949).

COMMUNITY ECOLOGY. Nothing is known about ecological interactions of *B. wrighti* with other community members.

CONSERVATION BIOLOGY. Several studies suggest that the Oregon slender salamander is strongly affected by clear-cutting and that mature and old-growth forests are essential for maintaining high local densities (Bury and Corn 1988a; Gilbert and Allwine 1991; Nussbaum et al. 1983). Because of the strong affinity of *B. wrighti* for large, rotting logs, ecologists fear that the conversion of old-growth forests to young stands that are managed on short-rotation cycles will result in the long-term decline of this Oregon endemic.

Hydromantes brunus Gorman
Limestone Salamander
PLATE 67

Fig. 148. *Hydromantes brunus;* adult; Mariposa Co., California (E. D. Brodie, Jr.).

IDENTIFICATION. The limestone salamander is a medium-sized plethodontid with webbed feet, a short tail, and prominent eyes. The tongue is highly protrusible and mushroom shaped. Adults are uniformly brown above and the venter is pale. The underside of the tail is yellowish and there are 13 costal grooves. Adults measure 7–11 cm TL. Sexually active males have enlarged premaxillary teeth that protrude from the mouth. Young juveniles change from vivid yellow-green to dull yellow and then brown as they age (Gorman 1954, 1964). This species often coexists with *Aneides lugubris* and *Batrachoseps* species, and can be readily distinguished by its body shape, webbed feet, and uniformly brown dorsum.

DISTRIBUTION AND ADULT HABITAT. *Hydromantes brunus* is a rare species known from only a few localities in the Lower Merced River drainage in Mariposa Co., California. Populations occur at elevations of 365–760 m and inhabit limestone cliffs, ledges, and talus overlain with mosses (Gorman 1954, 1964). Associated plants include digger pine (*Pinus sabiniana*), California laurel (*Umbellularia californica*), manzanita (*Arctostaphylos*), and chamise (*Adenostoma fasciculatum*).

Range of *Hydromantes brunus*

COMMENTS. Most aspects of the natural history of this rare species are unknown, and a detailed natural history study is needed. Stebbins (1985) notes that individuals are active during the fall, winter, and early spring rainy season and that individuals coil their bodies when molested. Populations near Briceburg are protected in an ecology preserve purchased by the state of California. Gorman (1954) found seven large ova in the type specimen that measured 4.6 mm in diameter on average.

Hydromantes platycephalus (Camp)
Mt. Lyell Salamander
PLATE 68

Fig. 149. *Hydromantes platycephalus*; adult; Tuolumne Co., California (E. D. Brodie, Jr.).

IDENTIFICATION. The Mt. Lyell salamander is a medium-sized plethodontid with webbed feet, a mushroom-shaped tongue, short legs, and a short tail. The body and head are flattened and the eyes are large relative to head size. The dorsal ground color is brownish, and it is overlain with greenish yellow to grayish gold, lichenlike patches (Adams 1942; Stebbins 1951). The venter is light brown. Males have enlarged maxillary and premaxillary teeth, swollen cloacae, mental glands, and proportionately wider heads than females (Adams 1942). Adults measure 7–11.5 cm TL and there are 13 costal grooves. The young juveniles have a dark brown or blackish ground color with varying amounts of greenish gold frosting.

SYSTEMATICS AND GEOGRAPHIC VARIATION. Geographic variation in coloration occurs among populations and tends to correlate with the color of local bedrock; however, geographic trends are erratic and no subspecies are currently recognized (Adams 1942; Stebbins 1951).

DISTRIBUTION AND ADULT HABITAT. The Mt. Lyell salamander primarily inhabits granite domes, rock outcrops, and talus at elevations of 1200–3660 m in the Sierra Nevada of central California (Gorman 1964). Habitats vary from sparsely vegetated rock outcrops at high elevations to Douglas-fir–yellow pine forests at lower elevations. Specimens have most often been found under stones on north-facing slopes at the edges of snowfields in moist or wet soils (Adams 1942; Danforth 1950; Stebbins 1951, 1985).

BREEDING AND COURTSHIP. The breeding season is not known and detailed accounts of courtship are not available. This species presumably engages in a tail-straddle walk like other plethodontids. The maxillary and premaxillary teeth of males protrude from the mouth and are used to abrade the female's skin and introduce mental gland secretions during courtship (Arnold 1977; Noble 1931).

REPRODUCTIVE STRATEGY. The exact time of ovipositing is not known. Females presumably nest in deep fissures or underground cavities since the nests have not been discovered in surface searches. Females collected in late June and early August contain 6–14 mature ova, and the maximum size of ovarian eggs is 3.8 mm in diameter (Stebbins 1951). Young of the year have been found in June and early July.

TERRESTRIAL ECOLOGY. Adams (1942) found three distinct size groups based on head widths, which correspond to young of the year, yearlings, and adults. He estimates that individuals become sexually mature about 2.5 years after hatching.

The juveniles and adults are nocturnally active and are excellent climbers (Adams 1942). The short, blunt tail tip is used for bracing and the webbed feet for traction when climbing. Individuals remain beneath stones during the day and emerge at night to forage.

Range of *Hydromantes platycephalus*

The Mt. Lyell salamander is active on the ground surface from May to August (Adams 1942). Most individuals retreat into deep fissures and cracks with the arrival of dry summer weather following snowmelt (Stebbins 1947).

Hydromantes has a highly protrusible tongue that can be extended nearly half the length of the body to secure prey (Stebbins 1951). Dietary items in four specimens consisted of a variety of ground-dwelling invertebrates, including centipedes, spiders, a termite, beetles, flies, and fungus gnats (Adams 1942).

PREDATORS AND DEFENSE. When disturbed, adults often elevate the head and tail while arching the body downward (Hansen 1990; Stebbins 1951). This species also exhibits an unusual rolling escape behavior in which individuals coil the body and tail to form a spheroid that facilitates rolling downhill (Garcia-Paris and Deban 1995). Simulated predator attacks suggest that if a predator such as a bird drops a coiled *Hydromantes*, the prey may be able to escape by rolling downhill out of harm's way. The skin secretions appear to be quite toxic; a field researcher was temporarily blinded for 30 hours after he extensively handled a *Hydromantes* and then rubbed his eyes (Hansen 1990).

COMMUNITY ECOLOGY. No data are available on the community ecology of the Mt. Lyell salamander.

CONSERVATION BIOLOGY. Populations are found in national parklands and are protected. The greatest threat to the species may be overcollecting by herpetologists.

Fig. 150. *Hydromantes platycephalus;* intense coiling as seen in this individual may facilitate escape from predators by allowing individuals to roll downhill; Tuolumne Co., California (M. Garcia-Paris).

Hydromantes shastae Gorman and Camp
Shasta Salamander
PLATE 69

Fig. 151. *Hydromantes shastae;* adult; Shasta Co., California (W. P. Leonard).

IDENTIFICATION. The Shasta salamander is a small plethodontid with a short tail, webbed feet, and a mushroom-shaped tongue that can extend 4–6 cm beyond the reach of the mouth to capture prey (Gorman and Camp 1953). The dorsal ground color is dark reddish brown and is strongly mottled with grayish green to tan specks and patches. The venter is grayish and the lower sides and portions of the limbs have silvery flecks. Adults measure 7.5–11 cm TL and have 13 costal grooves. Hatchlings reach 15–17 mm SVL and 22–24 mm TL, and young juveniles resemble miniature adults (Gorman 1956).

SYSTEMATICS AND GEOGRAPHIC VARIATION. Specimens appear to be relatively uniform in morphology and coloration within the very restricted range of this species.

DISTRIBUTION AND ADULT HABITAT. The Shasta salamander occurs as isolated populations in moist, cool microhabitats in caves and cliff faces in forested limestone gorges and ravines at elevations of 300–910 m in the vicinity of Shasta Lake, Shasta Co., north-central California (Bury et al. 1980). During the wet season *H. shastae* has been collected in humus under stones, logs, and litter in the immediate vicinity of cave or rock wall habitats.

Range of *Hydromantes shastae*

BREEDING AND COURTSHIP. No information is available on the mating season or courtship behavior.

REPRODUCTIVE STRATEGY. The only nesting record is for two brooding females, each with a clutch of nine eggs, that were found on 23 September and 1 November (Gorman 1956). Nests at this site have been found in small tunnels about 15 cm long in the mud walls of a cave. Embryos have large yolk sacs and gills in late September and reach the hatching stages by November. The eggs are attached by one or two outer capsular strands to a central mass of intertwined strands to form a grapelike cluster. The strands are 8–11 mm long, and when two strands are present they occur at opposite ends of the egg. The eggs are roughly spherical and the outer capsules are 7–9 mm in diameter. The embryos have bilobed, allantoic gills.

TERRESTRIAL ECOLOGY. Little is known about the juvenile and adult stages. Adults probably forage at night outside their daytime retreats (Gorman and Camp 1953). This species is a good climber and uses its webbed feet to climb over slippery rocks. Individuals elevate and undulate the tail at about a 45° angle when disturbed but do not lash the tail (Brodie 1977).

COMMENTS. *Hydromantes shastae* was originally discovered in the early 1900s by Eustace Furlong but was not described because the preserved specimens were in such poor shape. Joseph Gorman rediscovered the species 40 years later near Shasta Reservoir in northern California. Most aspects of the Shasta salamander's natural history are undocumented.

PLATE 1. Ringed salamander, *Ambystoma annulatum*, adult.

PLATE 2. Streamside salamander, *Ambystoma barbouri*, adult.

PLATE 3. California tiger salamander, *Ambystoma californiense*, adult.

PLATE 4. Flatwoods salamander, *Ambystoma cingulatum*, adult.

PLATE 5. Flatwoods salamander, *Ambystoma cingulatum*, larva.

PLATE 6. Brown salamander, *Ambystoma g. gracile*, adult.

PLATE 7. Brown salamander, *Ambystoma g. gracile*, larva.

PLATE 8. Jefferson salamander, *Ambystoma jeffersonianum*, adults.

PLATE 9. Blue-spotted salamander, *Ambystoma laterale*, adult.

PLATE 10. Mabee's salamander, *Ambystoma mabeei*, adult.

PLATE 11. Mabee's salamander, *Ambystoma mabeei*, larva.

PLATE 12. Long-toed salamander, *Ambystoma macrodactylum*, adult.

PLATE 13. Western long-toed salamander, *Ambystoma m. columbianum*, adult.

PLATE 14. Spotted salamander, *Ambystoma maculatum*, adult.

PLATE 15. Spotted salamander, *Ambystoma maculatum*, larva.

PLATE 16. Marbled salamander, *Ambystoma opacum*, adult female and metamorph.

PLATE 17. Mole salamander, *Ambystoma talpoideum*, adult.

PLATE 18. Mole salamander, *Ambystoma talpoideum*, larva.

PLATE 19. Small-mouthed salamander, *Ambystoma texanum*, adult.

PLATE 20. Barred tiger salamander, *Ambystoma tigrinum mavortium*, adult.

PLATE 21. Arizona tiger salamander, *Ambystoma tigrinum nebulosum,* adult.

PLATE 22. Blotched tiger salamander, *Ambystoma tigrinum melanostictum,* adult.

PLATE 23. Eastern tiger salamander, *Ambystoma t. tigrinum,* adult.

PLATE 24. Barred tiger salamander, *Ambystoma tigrinum mavortium,* larva.

PLATE 25. JJL unisexual, *Ambystoma platineum,* adult.

PLATE 26. JLL unisexual, *Ambystoma tremblayi,* adult.

PLATE 27. Two-toed amphiuma, *Amphiuma means,* adult.

PLATE 28. One-toed amphiuma, *Amphiuma pholeter,* adult.

PLATE 29. Three-toed amphiuma, *Amphiuma tridactylum,* adult.

PLATE 30. Eastern hellbender, *Cryptobranchus a. alleganiensis,* adult.

PLATE 31. Idaho giant salamander, *Dicamptodon aterrimus*, gilled adult.

PLATE 32. Cope's giant salamander, *Dicamptodon copei*.

PLATE 33. Cope's giant salamander, *Dicamptodon copei*.

PLATE 34. California giant salamander, *Dicamptodon ensatus,* adult.

PLATE 35. Pacific giant salamander, *Dicamptodon tenebrosus,* adult.

PLATE 36. Pacific giant salamander, *Dicamptodon tenebrosus*, adult and juvenile.

PLATE 37. Pacific giant salamander, *Dicamptodon tenebrosus*, larva.

PLATE 38. Seepage salamander, *Desmognathus aeneus*, female and eggs.

PLATE 39. Apalachicola dusky salamander, *Desmognathus apalachicolae*, adult.

PLATE 40. Southern dusky salamander, *Desmognathus auriculatus*, adult.

PLATE 41. Southern dusky salamander, *Desmognathus auriculatus*, larva.

PLATE 42. Ouachita dusky salamander, *Desmognathus brimleyorum*, adult.

PLATE 43. Carolina mountain dusky salamander, *Desmognathus carolinensis*, adult.

PLATE 44. Dusky salamander, *Desmognathus fuscus*, adult.

PLATE 45. Dusky salamander, *Desmognathus fuscus*, larva.

PLATE 46. Imitator salamander, *Desmognathus imitator*, adult.

PLATE 47. Shovel-nosed salamander, *Desmognathus marmoratus,* adult.

PLATE 48. Shovel-nosed salamander, *Desmognathus marmoratus,* larva.

PLATE 49. Seal salamander, *Desmognathus monticola*, adult.

PLATE 50. Seal salamander, *Desmognathus monticola*, larva.

PLATE 51. Allegheny mountain dusky salamander, *Desmognathus ochrophaeus*, adult.

PLATE 52. Ocoee salamander, *Desmognathus ocoee*, adult mimic of *P. jordani*.

PLATE 53. Ocoee salamander, *Desmognathus ocoee*, adult.

PLATE 54. Blue Ridge dusky salamander, *Desmognathus orestes*, adult.

PLATE 55. Black-bellied salamander, *Desmognathus quadramaculatus,* adult.

PLATE 56. Black Mountain dusky salamander, *Desmognathus welteri,* female and eggs.

PLATE 57. Pygmy salamander, *Desmognathus wrighti,* adult.

PLATE 58. Red Hills salamander, *Phaeognathus hubrichti,* adult.

PLATE 59. Desert slender salamander, *Batrachoseps aridus*, adult.

PLATE 60. California slender salamander, *Batrachoseps attenuatus*, adult.

PLATE 61. Inyo Mountains salamander, *Batrachoseps campi*, adult (view sideways).

PLATE 62. Black-bellied slender salamander, *Batrachoseps nigriventris*, female and eggs.

PLATE 63. Garden slender salamander, *Batrachoseps pacificus major*, adult.

PLATE 64. Kern Canyon slender salamander, *Batrachoseps simatus*, adult.

PLATE 65. Tehachapi slender salamander, *Batrachoseps stebbinsi,* adult.

PLATE 66. Oregon slender salamander, *Batrachoseps wrighti,* adult.

PLATE 67. Limestone salamander, *Hydromantes brunus,* adult.

PLATE 68. Mount Lyell salamander, *Hydromantes platycephalus,* adult.

PLATE 69. Shasta salamander, *Hydromantes shastae,* female and eggs.

PLATE 70. Southern two-lined salamander, *Eurycea bislineata cirrigera,* female and eggs.

PLATE 71. Southern two-lined salamander, *Eurycea bislineata cirrigera*, larva.

PLATE 72. Junaluska salamander, *Eurycea junaluska*, adult.

PLATE 73. Junaluska salamander, *Eurycea junaluska*, larva.

PLATE 74. Three-lined salamander, *Eurycea guttolineata*, adult.

PLATE 75. Dark-sided salamander, *Eurycea longicauda melanopleura*, adult.

PLATE 76. Long-tailed salamander, *Eurycea l. longicauda*, adult.

PLATE 77. Dark-sided salamander, *Eurycea longicauda melanopleura,* larva.

PLATE 78. Cave salamander, *Eurycea lucifuga,* adult.

PLATE 79. Many-ribbed salamander, *Eurycea m. multiplicata*, adult.

PLATE 80. San Marcos salamander, *Eurycea nana*.

PLATE 81. Texas salamander, *Eurycea neotenes*.

PLATE 82. Dwarf salamander, *Eurycea quadridigitata*, adult.

PLATE 83. Texas blind salamander, *Eurycea rathbuni*, adult.

PLATE 84. Barton Springs salamander, *Eurycea sosorum*.

PLATE 85. Comal blind salamander, *Eurycea tridentifera*.

PLATE 86. Oklahoma salamander, *Eurycea tynerensis*.

PLATE 87. Big Mouth Cave salamander, *Gyrinophilus palleucus necturoides.*

PLATE 88. Blue Ridge spring salamander, *Gyrinophilus porphyriticus danielsi,* adult.

PLATE 89. Kentucky spring salamander, *Gyrinophilus porphyriticus duryi,* adult.

PLATE 90. West Virginia spring salamander, *Gyrinophilus subterraneus,* transformed adult.

PLATE 91. West Virginia spring salamander, *Gyrinophilus subterraneus,* mature larvae.

PLATE 92. Georgia blind salamander, *Haideotriton wallacei*, juvenile.

PLATE 93. Four-toed salamander, *Hemidactylium scutatum*, nesting females, with eggs.

PLATE 94. Midland mud salamander, *Pseudotriton montanus diastictus*, adult.

PLATE 95. Rusty mud salamander, *Pseudotriton montanus floridanus*, adult.

PLATE 96. Eastern mud salamander, *Pseudotriton m. montanus*, larva.

PLATE 97. Northern red salamander, *Pseudotriton r. ruber*, young adult.

PLATE 98. Northern red salamander, *Pseudotriton r. ruber,* old adult.

PLATE 99. Blue Ridge red salamander, *Pseudotriton ruber nitidus*, larva.

PLATE 100. Many-lined salamander, *Stereochilus marginatus*, adult.

PLATE 101. Many-lined salamander, *Stereochilus marginatus*, larva.

PLATE 102. Grotto salamander, *Typhlotriton spelaeus*, adult.

PLATE 103. Grotto salamander, *Typhlotriton spelaeus*, larva.

PLATE 104. Green salamander, *Aneides aeneus*, adult.

PLATE 105. Clouded salamander, *Aneides ferreus*, juvenile.

PLATE 106. Clouded salamander, *Aneides ferreus*, adult.

PLATE 107. Black salamander, *Aneides flavipunctatus*, female with eggs.

PLATE 108. Black salamander, *Aneides flavipunctatus*, adult.

PLATE 109. Sacramento Mountain salamander, *Aneides hardii*, adults.

PLATE 110. Arboreal salamander, *Aneides lugubris*, adult.

PLATE 111. Arboreal salamander, *Aneides lugubris*, adult.

PLATE 112. Large-blotched ensatina, *Ensatina eschscholtzii klauberi*, adult.

PLATE 113. Sierra Nevada ensatina, *Ensatina eschscholtzii platensis*, juvenile.

PLATE 114. Painted ensatina, *Ensatina eschscholtzii picta,* adult.

PLATE 115. Tellico salamander, *Plethodon aureolus.*

PLATE 116. Caddo Mountain salamander, *Plethodon caddoensis.*

PLATE 117. Red-backed salamander, *Plethodon cinereus,* adults (striped and unstriped morphs).

PLATE 118. Zigzag salamander, *Plethodon dorsalis,* adult.

PLATE 119. Dunn's salamander, *Plethodon dunni,* adult.

PLATE 120. Del Norte salamander, *Plethodon elongatus,* adult.

PLATE 121. Slimy salamander, *Plethodon glutinosus* (=*P. cylindraceus* sensu Highton), adult.

PLATE 122. Valley and Ridge salamander, *Plethodon hoffmani*, adult.

PLATE 123. Peaks of Otter salamander, *Plethodon hubrichti*, adult.

PLATE 124. Coeur d'Alene salamander, *Plethodon idahoensis*, adult and juvenile.

PLATE 125. Jordan's salamander, *Plethodon jordani*, adults ("red-legged" race).

PLATE 126. Jordan's salamander, *Plethodon jordani*, adult ("clemsonae" race).

PLATE 127. Jordan's salamander, *Plethodon jordani*, adult ("red-cheeked" race).

PLATE 128. Cumberland Plateau salamander, *Plethodon kentucki*, adult.

PLATE 129. Larch Mountain salamander, *Plethodon larselli*, adult.

PLATE 130. Jemez Mountains salamander, *Plethodon neomexicanus*, adult.

PLATE 131. Cheat Mountain salamander, *Plethodon nettingi*, adult.

PLATE 132. Southern Appalachian salamander, *Plethodon oconaluftee,* adult.

PLATE 133. Rich Mountain salamander, *Plethodon ouachitae,* adult.

PLATE 134. Pigeon Mountain salamander, *Plethodon petraeus,* adult.

PLATE 135. White-spotted salamander, *Plethodon punctatus,* adult.

PLATE 136. Ravine salamander, *Plethodon richmondi,* adult.

PLATE 137. Southern red-backed salamander, *Plethodon serratus,* adult.

PLATE 138. Shenandoah salamander, *Plethodon shenandoah*, adult.

PLATE 139. Siskiyou Mountains salamander, *Plethodon stormi*, adult.

PLATE 140. Van Dyke's salamander, *Plethodon vandykei*, female guarding eggs.

PLATE 141. Western red-backed salamander, *Plethodon vehiculum*, adult.

PLATE 142. Southern zigzag salamander, *Plethodon websteri*, striped morph.

PLATE 143. Wehrle's salamander, *Plethodon wehrlei*, adult.

PLATE 144. Weller's salamander, *Plethodon welleri*, adult with hatchlings.

PLATE 145. Yonahlossee salamander, *Plethodon yonahlossee*, adult.

PLATE 146. Alabama waterdog, *Necturus alabamensis*, adult.

PLATE 147. Gulf Coast waterdog, *Necturus beyeri*, adult.

PLATE 148. Neuse River waterdog, *Necturus lewisi*, adult.

PLATE 149. Mudpuppy, *Necturus m. maculosus*, adult.

PLATE 150. Mudpuppy, *Necturus m. maculosus*, larva.

PLATE 151. Dwarf waterdog, *Necturus punctatus,* adult.

PLATE 152. Cascade torrent salamander, *Rhyacotriton cascadae,* adult.

PLATE 153. Cascade torrent salamander, *Rhyacotriton cascadae,* larva.

PLATE 154. Columbia torrent salamander, *Rhyacotriton kezeri,* adult.

PLATE 155. Olympic torrent salamander, *Rhyacotriton olympicus,* adult.

PLATE 156. Southern torrent salamander, *Rhyacotriton variegatus,* adult.

PLATE 157. Black-spotted newt, *Notophthalmus m. meridionalis,* adult.

PLATE 158. Striped newt, *Notophthalmus perstriatus,* adult.

PLATE 159. Broken-striped newt, *Notophthalmus viridescens dorsalis,* adult.

PLATE 160. Red-spotted newt, *Notophthalmus v. viridescens,* breeding adult.

PLATE 161. Red-spotted newt, *Notophthalmus v. viridescens,* red eft.

PLATE 162. Broken-striped newt, *Notophthalmus viridescens dorsalis,* mature larva.

PLATE 163. Northern rough-skinned newt, *Taricha g. granulosa*, adult.

PLATE 164. Northern rough-skinned newt, *Taricha g. granulosa*, larva.

PLATE 165. Red-bellied newt, *Taricha rivularis*, adult (defensively posturing).

PLATE 166. Coast Range newt, *Taricha t. torosa*, adult.

PLATE 167. Coast Range newt, *Taricha t. torosa*, larva.

PLATE 168. Narrow-striped dwarf siren, *Pseudobranchus a. axanthus*, adult.

PLATE 169. Slender dwarf siren, *Pseudobranchus striatus spheniscus*, adult.

PLATE 170. Rio Grande siren, *Siren intermedia texana*, adult.

PLATE 171. Eastern lesser siren, *Siren i. intermedia*, larva.

PLATE 172. Greater siren, *Siren lacertina*, adult.

Tribe Hemidactyliini

Eurycea bislineata (Green)
Two-lined Salamander
PLATES 70, 71

Fig. 152. *Eurycea b. cirrigera*; adult; Powell Co., Kentucky (J. W. Petranka).

Fig. 153. *Eurycea b. wilderae*; adult male; Swain Co., North Carolina (J. W. Petranka).

IDENTIFICATION. The two-lined salamander is a small, slender plethodontid that has a broad, greenish yellow or orangish dorsum that is bordered on either side by a dark brown or black stripe that extends onto the tail. The dorsum usually contains numerous, scattered black spots or blotches. The tail is laterally compressed and the venter is yellow. Depending on subspecies, sexually active males may develop elongated cirri, swollen jaw musculature, conspicuous mental glands, caudal hedonic glands, and/or long, monocuspid premaxillary teeth during the mating season (Noble 1929a; Sever 1989a; Stewart 1958). Males also have papillose cloacal lips compared with the smooth cloacal lips of females. Average adult size is similar between the sexes; females in a large series of specimens from Ohio and vicinity measured on average only 2.5% larger than males (Guttman 1989a). Albino larvae and adults occur in a small percentage of populations (Bartley 1959; Dyrkacz 1981; Hensley 1959; Rubin 1963). Adults measure 6.5–12 cm TL and there are typically 13–16 costal grooves.

Hatchlings measure 7–9 mm SVL and 11–14 mm TL on average (Bahret 1996; Bishop 1941a; Bruce 1982a; Wilder 1924). Hatchlings and older larvae have stream-type morphology and are dusky colored above with six to nine pairs of light dorsolateral spots on the body. The body is streamlined and the tail fin stops near the insertion of the rear limbs. The venter is normally light colored with numerous iridophores, but large larvae in the extreme northern portion of the range sometimes have darkly pigmented throats and bellies (Trapido and Clausen 1940). Hatchlings superficially resemble *Desmognathus* larvae but have reddish gills with longer and more slender rami, and squared rather than rounded snouts (Eaton 1956).

SYSTEMATICS AND GEOGRAPHIC VARIATION. Several systematic problems have arisen in recent years concerning morphs or populations referable to *E. bislineata*. Sever (1979) described a morph of *E. bislineata* ("morph A") from the southern Appalachians that differs from *E. b. wilderae* by having strongly developed temporal musculature and by lacking cirri, hedonic glands, and enlarged maxillary teeth. Rose and Bush (1963) described specimens from central Alabama as a new species, the dark-sided salamander, *E. aquatica*. Relative to typical *E. bislineata*, this form has fewer vomerine teeth, is stockier, has a shorter tail, has fused nasal processes of the premaxilla, and has prominent prootic-squamosal crests. Mount (1975) and his colleagues later discovered several populations containing specimens referable to *E. aquatica* in the Ridge and Valley Province of Alabama, but they found intermediates between this form and *E. bislineata* in nearly every case. Other populations resembling *E. aquatica* have been found in northwestern Georgia and Davidson Co. in central Tennessee (e.g., Redmond and Scott 1996). A detailed analysis of morphological traits of populations in Alabama and northwestern Georgia confirmed Mount's original work (Jones 1980, cited in Sever 1989b). Both authors conclude that the evidence to date does not support the recognition of *E. aquatica*. A second stocky morph known as the "Cole Springs phenotype" occurs in north-central Alabama and is currently considered to be conspecific with *E. bislineata* (Mount 1975).

Jacobs (1987) conducted a broadscale survey of allozyme patterns of populations across the eastern United States and found several genetically differentiated groups. He suggested that populations of *E. bislineata* recognized as subspecies should be raised to the species level based on genetic distances between groups (Nei's distances [*D*] 0.38–0.40 among subspecies). Because he used a broadscale approach, Jacobs (1987) was unable to examine contact zones between subspecies or to determine the extent to which these forms interbreed. Noble and Brady (1930) report that *E. b. cirrigera* from North Carolina will readily court *E. b. bislineata* from Long Island, New York. Mittleman (1966) notes that *E. b. bislineata* and *E. b. cirrigera* intergrade in several counties in Virginia, whereas Howell and Switzer (1953) report a similar broad zone of intergradation in five counties in Georgia. Here I have not adopted the classification of Jacobs (1987). Instead, I recommend that subspecies continue to be recognized until contact zones are examined and the degree of gene exchange occurring between genetic subgroups is quantified.

In the same analysis, Jacobs (1987) concluded that morph A of Sever (1979) is conspecific with *E. b. wilderae* since genetic distances between these types are very low. His data further show an enigmatic pattern in which *E. junaluska* and *E. aquatica* are genetically very similar to each other and to populations of *E. bislineata* in south-central Indiana. In addition, *E. aquatica* is genetically distinct from nearby populations of *E. bislineata* in Alabama and Mississippi, whereas *E. junaluska* is very distinct from syntopic populations of *E. b. wilderae*. Until the systematic status of *E. aquatica* is better resolved through more detailed studies of genetic variation in populations in Alabama, Mississippi, and northwestern Georgia, I follow Mount's (1975) recommendation of considering this taxon to be conspecific with *E. bislineata*.

The status of morph A has yet to be fully resolved. The hypothesis that this form is simply a morph of *E. b. wilderae* that does not develop many of the secondary sexual characteristics of typical males is supported by Jacobs's analysis and by courtship trials, which show that female *E. b. wilderae* will readily court males of either morph (Reagan 1984). However, specimens referable to morph A have also been collected in the Piedmont and Coastal Plain of North Carolina, where *E. b. wilderae* is not found. This geographic pattern is not entirely consistent with the preceding interpretation. Alternative explanations are that the male polymorphism found in *E. b. wilderae* also occurs in *E. b. cirrigera*, or that morph A is a separate, widely distributed species (Sever 1989b).

Another issue concerns the systematic status of two largely parapatric groups of populations in Ohio and Indiana (Guttman and Karlin 1986). These groups show moderately strong genetic differentiation. A narrow hybrid zone occurs between the groups in north-central Ohio, where occasional hybrids and backcross progeny occur. Guttman and Karlin (1986) interpret the groups to be sibling species because of their genetic dissimilarities and the lack of a smooth zone of intergradation among forms. They also note that additional information on rate of gene exchange from other areas of their range is needed to clarify the evolutionary relationships of these groups. In his treatment of *E. bislineata* in Ohio, Guttman (1989a) questioned Jacobs's (1987) interpretation of the systematic affinities of southern Ohio two-lined salamanders and recommended that the nomenclatural status of the populations in Ohio remain unresolved until further data become available.

Guttman and Karlin's data provide the strongest evidence to date that *E. bislineata* is a semispecies or superspecies complex similar to the *Plethodon glutinosus* complex. Their data indicate that *E. bislineata* consists of two sibling species, whereas Jacobs's (1987) data indicate that as many as four species or semispecies may occur in this group. Here I recognize all populations referable to *E. bislineata* as a single geographically variable species until more refined data on gene exchange between subgroups become available.

Two-lined salamanders show marked geographic variation in morphology, coloration, and life history traits. Most treatments recognize three subspecies (Conant 1975; Mittleman 1966; Sever 1972). The northern two-lined salamander (*E. b. bislineata*) has a yellowish green or yellowish brown dorsum and dorsolateral stripes that extend midway to three-fourths the length of the tail. There are 14–16 costal grooves and cirri are usually absent on sexually active males. The Blue Ridge two-lined salamander (*E. b. wilderae*) has a bright yellowish orange dorsum and two black dorsolateral stripes that break up about midway along the tail. Adult males (except those of morph A) have prominent cirri during the mating season, and there are 13–14 costal grooves. The southern two-lined salamander (*E. b. cirrigera*) is similar to the northern two-lined salamander but has dorsolateral stripes that extend to near the tip of the tail and a row of light spots along each side of the body. There are usually 13–14 costal grooves, and males have prominent cirri during the breeding season. Geographic variation in life history traits is discussed in detail later in this account.

DISTRIBUTION AND ADULT HABITAT. The two-lined salamander is a wide-ranging species that occurs from southeastern Canada to the Gulf Coast and westward to Louisiana, Arkansas, and Illinois. Populations occur from near sea level to about 2000 m in elevation. Adults reside beneath rocks and logs along the margins of small, rocky streams or seeps, but also occur in forest floor habitats far from running water. During the breeding season adults can be found beneath submerged rocks and debris in streams. Coastal Plain populations typically inhabit bottomland forests and take shelter under flood debris and rotting logs. Although the adults typically oviposit in running water, on rare occasions they may use lakes in New York (Bahret 1996).

BREEDING AND COURTSHIP. The adults breed from September through May, based on the seasonal development of male secondary sexual characteristics and the presence of sperm caps in the cloacae of females (Noble and Weber 1929; Sever 1979; Stewart 1958; Weichert 1945). Although poorly documented, patterns in local and regional populations undoubtedly vary depending on the severity of winter weather and other environmental factors. Published observations of courtship in nature are not available, and the extent to

Fig. 154. *Eurycea b. wilderae*; adult male with nasal cirri; Swain Co., North Carolina (J. W. Petranka).

which adults court on land versus in water is unknown. In many populations, males and females move from hillsides into streams 1–3 months prior to the initiation of oviposition (Bishop 1941a; Wood 1953a). This movement is presumably tied to mating activity; however, adults in many populations undoubtedly mate on land during the late summer and autumn months prior to migrating toward streams.

Noble (1929a) observed partial courtship in several adult *E. b. bislineata* maintained in laboratory aquaria between January and April. The following is primarily a summary of his observations.

A male initially become restless and nudges other animals in his vicinity. Upon encountering a female, the male may lift the female up by pushing his snout under her cloaca or chest. The male assumes a distinctive pose with a sharp bending of his head around the snout of the female. Animals may remain posed like this for an hour or more. Eventually the female slips behind the male and straddles his tail with her forelimbs. The female presses her chin against the base of the male's tail, and he walks forward, undulating his tail in exaggerated fashion from left to right. With each bend of the tail, the female moves her head in the opposite direction. This tail-straddle walk may continue for 1 hour or more. The male presumably deposits a spermatophore and leads the female forward to pick the sperm mass up, as in other plethodontids. The spermatophore is conical, about 2.5 mm high, and slightly compressed and has a colorless stalk that tapers toward the tip (Organ and Lowenthal 1963).

Male and female *E. bislineata* exhibit sexual dimorphism in the shape of the premaxillary teeth. The teeth of breeding males are unicuspid and elongated relative to the bicuspid teeth of females. The premaxillary

Range of *Eurycea bislineata*

teeth, which often pierce the male's lip, are used to abrade the female's skin and introduce mental gland secretions into the female's circulatory system (Arnold 1977). Males replace their unicuspid teeth with typical bicuspid teeth after the breeding season ends (Arnold 1977; Stewart 1958). Courting males are aggressive and will attempt to drive off rival males that intrude (Arnold 1977).

REPRODUCTIVE STRATEGY. Females typically attach eggs singly to the undersides of submerged rocks in a tight monolayer (Baumann and Huels 1982; Dunn 1920; Wilder 1899; Wood 1949). In many cases females nest in submerged cavities on the undersides of rocks that are embedded in the stream bottom. Nesting usually occurs in running water, but Wood (1953a) found eggs in a stagnant drainage ditch in Virginia and Bahret (1996) described a population breeding in Lake Minnewaska, New York. At the latter site, adults remain year round in the lake and are found as deep as 18 m below the water's surface. Females lay eggs from late May to mid-July at depths of 9–13.5 m. In contrast to most populations, the females scatter the eggs loosely among the leaves of *Sphagnum* mosses in noncryptic sites and do not exhibit parental care.

In southeastern Ohio, the eggs are attached to rocks embedded in the substrate, and the number of clutches per rock is positively correlated with rock size (Baumann and Huels 1982). The eggs are deposited in tight clusters that average 50×32 mm (16 cm^2) and are usually attended by one or more adults. Two or three brooding females are often found under the same rock, a finding that suggests that ovipositional sites are in short supply. Joint nesting on the same substrate has been documented by others (Bishop 1941a; Weber 1928; Wood 1953a; Wood and McCutcheon 1954), and oviposition sites may possibly limit population size in some New York populations (Stewart 1968).

In Coastal Plain streams in Virginia that lack rocks, the eggs are attached to root fibers, leaves, watercress, logs, and planks (Richmond 1945; Wood 1953a). Eggs placed on the undersides of leaves, planks, or logs are usually in fairly tight clusters, whereas those placed on rootlets or watercresses are sometimes scattered over a relatively large area. Noble and Richards (1932) report similar modes of egg deposition in a rockless stream in North Carolina. When gravid females from this population are presented a choice of water weeds or rocks, they invariably oviposit on rocks. At a spring in Mississippi some females attach the eggs to the undersides of logs in fast-flowing water, whereas others bury the eggs in detrital sediments with little water flow and low oxygen levels (Marshall 1996).

When depositing eggs, a female flips on her back and arches the tail and back in order to make cloacal contact with the substrate. About 3 minutes are required to lay an egg, and one to several hours to deposit the entire clutch (Noble and Richards 1932). In some instances a female may actually excavate a cavity for oviposition (Wood 1953a). Freshly laid eggs are white to pale yellow, about 2.5–3.0 mm in diameter, and surrounded by two jelly membranes (Baumann and Huels 1982; Bishop 1941a; Noble and Richards 1932). Each egg is suspended by a short, broad stalk from a support structure. With time, the stalk and egg membranes become flimsy and the eggs dangle freely in the water. Females normally remain with their eggs through hatching and probably defend them from predators such as small crayfish and invertebrates.

The time of egg laying varies depending on latitude. Females in most northern and central populations begin laying with the arrival of spring weather, with peak activity occurring in April and May. Females oviposit in Massachusetts from late April through the first week of June (Wilder 1924), in New York from April through June (Bishop 1941a), and in West Virginia from mid-March through April (Green and Pauley 1987). Elsewhere, eggs in early to intermediate developmental stages have been found in western North Carolina from February to May (Bruce 1982a, 1988b; Sever 1983b), in Ohio and Indiana in May (Baumann and Huels 1982; Minton 1972; Wood 1949), and in eastern Tennessee in April (King 1939).

Populations in nonmountainous areas in the South oviposit during the winter and early spring. Freshly laid eggs can be found in central North Carolina from December through March (Brimley 1896) and in southeastern Virginia from late January to mid-April (Wood 1953a). Eggs collected on 4 April in Tennessee hatched 8 days later, suggesting that oviposition occurs in March in nonmountainous regions of the state (Dunn 1926). Eggs have been found in April and May in north-central Mississippi, but developmental stages were not reported (Marshall 1996).

Mean clutch size is geographically variable, but the pattern appears to be chaotic. Estimates of mean clutch size based on counts of eggs in nests include 18 for Massachusetts populations (Wilder 1924), 53 for a Mississippi population (Marshall 1996), 15 (Bahret 1996) and 30 (Bishop 1941a) for New York populations, and 39 (range = 15–110) for an Ohio population (Baumann and Huels 1982). Number of eggs in Virginia nests varies from 18 to 96 and averages 49–52, whereas number of mature ova in females from this same area varies from 29 to 115 and averages 71 (Wood 1953a; Wood and McCutcheon 1954). Estimates of mean clutch size based on ovarian egg counts are 50 (range = 15–114) for Alabama populations (Mount 1975) and 28 (range = 19–42; Bahret 1996) and 46 (range = 19–86; Stewart 1968) for New York specimens. Number of ovarian eggs is positively correlated with female SVL (Wood and Duellman 1951; Wood and McCutcheon 1954).

The incubation period lasts 4–10 weeks depending on local stream temperatures (Bishop 1941a; Wilder 1924). Hatching dates are also variable depending on latitude. Hatching occurs in late May and June in central Kentucky (Petranka 1984d), in June and early July in Massachusetts (Wilder 1924), in June and August in New York (Bishop 1941a), and from May to August in western North Carolina (Bruce 1982a,b, 1985b; Dunn 1920). Embryos in southern populations presumably hatch as early as January or February, but detailed observations are lacking.

AQUATIC ECOLOGY. The embryos hatch with conspicuous yolk reserves and probably do not begin feeding until most of the yolk is resorbed. The larvae live in slow-moving pools in streams and are rarely found in fast current except when drifting. They are primarily benthic feeders and prowl slowly about stream bottoms or over rocks in search of small prey.

Fig. 155. *Eurycea b. cirrigera*; larva; Bladen Co., North Carolina (R. W. Van Devender).

Average larval densities in a New Hampshire stream are 0.5–0.8 larva/m^2 of stream bottom (Burton and Likens 1975) compared with 10.5 larvae/m^2 of stream bottom in some New York populations (Stewart 1968). In Lake Minnewaska in New York, larvae occur to a depth of 19.5 m and larger larvae prefer lake bottom habitats. Larvae at this site attain exceptional sizes and may reach 46 mm SVL and 92 mm TL before transforming (Bahret 1996).

Larvae forage on stream bottoms and probably use chemical (Petranka et al. 1987b), tactile, and visual cues to locate prey. Major prey in the stomachs of 121 larvae from Indiana were chironomid larvae, copepods, fly pupae, and stonefly nymphs (Caldwell and Houtcooper 1973). Chironomids are most important in warm weather and copepods in cool weather. Midges, stoneflies, cladocerans, and copepods are the most numerically important prey in larvae from a New Hampshire stream (Burton 1976).

I examined diet and ontogenetic shifts in larval feeding behavior in a central Kentucky population (Petranka 1984d). Larvae at this site remain beneath cover objects during the day but emerge from cover at night and forage on the stream bottom. Larvae of all sizes feed continuously over a 24-hour period and have similar diurnal patterns of movement. As larvae grow, they incorporate larger prey into their diet, but continue to take large numbers of very small prey. Major prey items are isopods, amphipods, chironomid larvae, and zooplankton. Large *Eurycea* larvae prey rather heavily on hatchling *Ambystoma barbouri* for a short period of time in mid-April.

Although larvae are primarily benthic, they often drift downstream. Hatchling *E. bislineata* have been captured in drift nets in a Massachusetts stream from late June through mid-July (Johnson and Goldberg 1975). Captures are highest at night during periods of reduced water flow, but it is uncertain whether these captures reflect active or passive drift on the part of larvae. Drift nets placed in riffle sections of a South Carolina creek yield 1- and 2-year-old larvae in April and hatchlings in May and June (Stoneburner 1978). Larvae are not collected in drift samples taken in the remaining months of the year, even though bottom samples indicate that larvae and/or adults are present at these times. Other studies show that first-year larvae are more likely to move downstream than upstream, and that small larvae drift downstream disproportionately more than large larvae and adults (Bruce 1986). In contrast, upstream and downstream movements of older larvae and adults are about equal. Collectively, these studies suggest that hatchlings are more susceptible to downstream displacement than older larvae. It is uncertain if drift is active or passive; however, it appears to be an important dispersal mechanism that increases gene flow between local populations.

Length of the larval period tends to be longer in northern populations than in southern populations. Stewart (1968) reports a 2-year larval period for New York populations, with mean size at metamorphosis around 30 mm SVL and 58 mm TL. Most larvae in a southern Ohio population of *E. b. bislineata* transform when about 2 years old and measuring 23–32 mm SVL; however, a small percentage of larvae overwinter and transform when 3 years old (Duellman and Wood 1954). This same pattern occurs in Massachusetts (Wilder 1924) and Pennsylvania (Hudson 1955) populations. Animals judged to be in their third year of growth by Hudson (1955) measure 30 mm SVL and 58 mm TL on average. Larvae in a Quebec, Canada, population often reach >70 mm TL before metamorphosing, suggesting that many larvae require 3 years to transform (Trapido and Clausen 1940).

Length of the larval period, growth rate, and size at metamorphosis are highly variable in populations of *E. b. wilderae* in the southern Appalachians (Bruce 1982a,b, 1985b). The larval period lasts 1–2 years, but in any population either a 1- or a 2-year larval period predominates. Overall, there is little correspondence between growth rate and length of larval period among populations. Average size at metamorphosis varies from 18–19 mm SVL in 1-year-old larvae at one site (Bruce 1982a) to 32 mm SVL in 2-year-old larvae from a second site (Bruce 1982b). Metamorphosis occurs from April to July but in any given population is restricted to a 2- or 3-month period.

Eurycea b. wilderae in first-order streams in the southern Appalachians tend to have larval periods of about 1 year, whereas those in downstream sections of the same drainages have mixed larval periods of about 1 or 2 years depending on whether the larvae overwinter (Voss 1993b). The mean size of metamorphs tends to decrease seasonally from May through August, and metamorphs measure 18–25 mm SVL on average. Metamorphosis in New York populations occurs from April to September, with a peak in late June and early July (Stewart 1968).

TERRESTRIAL ECOLOGY. Bruce (1988b) estimates that juvenile *E. b. wilderae* that transform after 1–2 years become sexually mature after 3–4 years. The juvenile stage lasts about 2 years and the mean generation time is 4.4 years.

Juveniles and adults live along stream margins and in surrounding forests. In many populations adults undergo seasonal migrations to and from breeding streams, but patterns vary geographically and the proportion of adults that migrate is unknown. *Eurycea b. bislineata* adults in New York remain in the general vicinity of streams year round (Bishop 1941a), but *E. b. wilderae* often lives on the forest floor far from running water during the warm months. Adults in southern Ohio congregate in streams between late March and mid-April, but at other times of the year live in moist habitats that are often considerable distances from running water (Weichert 1945). Most radioactively tagged adults studied by Ashton and Ashton (1978) in southern Ohio remain in the vicinity of streams during the warm months and move <10 m between monitoring periods.

MacCulloch and Bider (1975) conducted an intensive, 9-year study of summer movements in one population in Quebec, Canada, using fine-grained sand transects. The salamanders have two bursts of uphill movements away from streams. A postbreeding migration of adults occurs in June, and a postmetamorphic migration composed predominantly of juveniles occurs in August. June migrants move >100 m from the stream, but August migrants remain relatively close to the stream. Seventy-five percent of the animals that migrate >100 m from the stream do not return, and most presumably die in the forest. During May animals make nightly treks to and from the stream, but there is no net movement away from the stream. Animals show a net movement back toward the stream in October with the arrival of fall weather. Movements are greatest immediately after dark and gradually slow during the night. Virtually all movements are at night during wet weather.

Adults in southern Ohio move to relatively warm streambank retreats, 8–82 cm beneath the soil surface, during periods of cold weather (Ashton and Ashton 1978), whereas adults placed in soil enclosures move >28 cm beneath the soil surface by late February (Vernberg 1953). Juveniles and adults feed year round, although food intake is reduced during the winter months (Ashton and Ashton 1978; Hamilton 1932; Weichert 1945). Four of 14 adults excavated from underground winter retreats by Ashton and Ashton (1978) had food in their stomachs. Weichert (1945) found food in the stomachs of metamorphosed animals in a southwestern Ohio population at virtually all times of the year. Food items include wood roaches, spiders, ticks, earthworms, isopods, millipedes, beetles, snails, springtails, flies, and hymenopterans.

During the warmer months of the year the juveniles and adults emerge at night and forage for small invertebrates on the forest floor or along streambanks. Adult surface activity is greatest during the first hour after dark in an Ohio population and correlates with a peak in the activity of potential prey (Holomuzki 1980). Food items in 47 New York specimens are beetles, spiders, sowbugs, mayflies, dipterans, annelids, stonefly nymphs, and thrips (Hamilton 1932). Burton (1976) lists a wide variety of invertebrates in adult specimens collected from woodlands, seepages, and streamside habitats. Individuals primarily feed on terrestrial prey, although stonefly nymphs, caddisfly larvae, and midges are occasionally consumed.

Adults appear to be territorial and will actively defend home shelters by aggressively posturing or biting (Grant 1955). Local densities as high as 11 adults/m^2 have been documented in some New York populations (Stewart 1968). However, adult densities in a New Hampshire watershed are only 0.02–0.04 individual/m^2 (Burton and Likens 1975).

PREDATORS AND DEFENSE. Two-lined salamanders are eaten by a variety of predators. Eastern screech owls (*Otus asio*), common garter snakes (*Thamnophis sirtalis*), and ringneck snakes (*Diadophis punctatus*) occasionally feed on juveniles and adults (Huheey and Stupka 1967; Rising and Schueler 1980; Uhler et al. 1939). In addition, spring and black-bellied salamanders are major predators on larvae and adults in some portions of the range (Beachy 1993a; Bruce 1979). Rainbow trout and two-lined salamanders are mutual predators. Trout feed on *E. bislineata* and adult salamanders feed on trout fry and embryos (Mathews 1982). Two-lined salamanders are distasteful to shrews (*Blarina brevicauda*), even though salamanders may be killed after prolonged attacks (Brodie et al. 1979). Individuals posture defensively by coiling the body with the head tucked beneath the vent and raising and undulating the tail (Brodie 1977).

Several researchers have examined the responses of two-lined salamanders to garter snakes, which are important natural predators. In staged encounters between *E. bislineata* and the common garter snake (*T. sirtalis*), most two-lined salamanders remain immobile when contacted by the head or body of a snake, but engage in protean flipping and flight when contacted by the snake's tongue (Ducey and Brodie 1983). Individuals that deviate from these rather stereotypical responses suffer greater mortality than those that do not. Specimens with intact tails are also more successful in avoiding capture than specimens with autotomized tails, in part because tail autotomy increases the probability of escape from snakes.

The specific responses of *E. b. bislineata* to tongue-flicking by *T. ordinoides* varies both among individuals and among geographic groups (Dowdey and Brodie 1989). Some individuals tend to remain immobile whereas others flee. Animals that flee tend to be better runners than those that do not. Thus the type of defense correlates with the physical abilities of the animals.

Eurycea b. bislineata is far more likely to autotomize the tail when attacked by garter snakes than is *Desmognathus ochrophaeus* or *D. fuscus* (Whiteman and Wissinger 1991). Tail autotomy increases the probability of an animal escaping. About 32% of animals collected from the field by Wake and Dresner (1967) had autotomized tails compared with 8% of museum specimens.

COMMUNITY ECOLOGY. Two-lined salamander larvae share breeding streams with other salamanders and with predatory fish, and several researchers have examined ecological interactions between these groups. *Eurycea bislineata* larvae that my colleagues and I studied in central Kentucky are primarily active outside cover at night and use chemical cues to monitor the presence of fish (Kats et al. 1988; Petranka et al. 1987b). Larvae respond to fish chemicals by seeking refuge beneath cover objects. Field experiments show that *Eurycea b. cirrigera* larvae will reduce their surface activity when confined with predatory brook trout and spring salamander larvae (Resetarits 1991). *Gyrinophilus* does not significantly affect the survival of *Eurycea* but can reduce larval growth rates.

Gustafson (1994) examined predator-prey interactions between *G. porphyriticus* larvae and *E. b. cirrigera* larvae in experimental streams and stream pools. Mortality rates of *Eurycea* larvae are size dependent and increase with the size of *Gyrinophilus* up to 30 mm SVL. *Eurycea* larvae also restrict their nighttime surface activity in the presence of *Gyrinophilus* and are more likely to remain in cover in the presence of large *Gyrinophilus* larvae.

A field experiment indicates that *E. b. wilderae* larvae are vulnerable to predation by large *D. quadramaculatus* larvae; however, there is no evidence of intra- or interspecific competition (Beachy 1993a). A related experiment shows that growth and survival of *Eurycea* in stream enclosures is independent of density, although *Eurycea* has lower survival in the presence of predatory *G. porphyriticus* and *D. quadramaculatus* larvae. Of the two predators compared, *Gyrinophilus* is the more effective predator (Beachy 1994).

CONSERVATION BIOLOGY. The two-lined salamander is a relatively common species and is in minimal need of protection. Like many other salamanders, this species is sensitive to intensive timbering, land clearing, stream pollution, and stream siltation, and it is often absent from urban areas or highly disturbed landscapes. Two-lined salamanders in New York are rarely found on soils with low pH (Wyman and Jancola 1992). Because sodium balance is disrupted at low pH (Frisbie and Wyman 1991), this and other species may be at risk if acid precipitation reduces or eliminates the acid-neutralizing capacity of soils.

Eurycea guttolineata (Holbrook)
Three-lined Salamander
PLATE 72

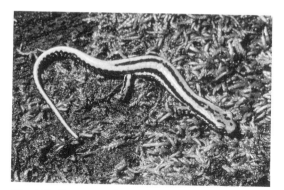

Fig. 156. *Eurycea guttolineata;* adult; Aiken Co., South Carolina (R. W. Van Devender).

Fig. 157. *Eurycea guttolineata;* ventral view of adult; Pitt Co., North Carolina (J. W. Petranka).

IDENTIFICATION. The three-lined salamander is a large *Eurycea* species with a dark middorsal stripe that bisects a yellowish orange to yellowish brown dorsum. A conspicuous dark brown or black stripe occurs along each side of the body. In many individuals widely separated yellow or whitish spots occur within the lateral stripes. Vertical bands on the tail are often partially fused to form a dark, wavy stripe. The belly and tail are conspicuously mottled with numerous greenish gray to black and white blotches. The tail becomes proportionately longer relative to body length as juveniles grow and may constitute 60-65% of the total length of adults.

Adults measure 10-18 cm TL and there are 13-14 costal grooves. Sexually active males have prominent cirri, papillose cloacal lips (smooth in females), and elongated maxillary teeth (Noble 1927a). Females also have cirri, but they are far less developed than those of males.

Larvae are the stream type with streamlined bodies and a dorsal fin that does not extend forward beyond the rear limbs. The gills are of medium length and are often more elongated than those of *E. bislineata*. Hatchlings measure 11-12.5 mm SVL on average (Bruce 1970). The dorsum of hatchlings is cream colored and uniformly stippled with melanophores; the belly is immaculate. The underside of the throat often has patches of melanophores. Within 1-2 months after hatching, larvae develop a dark, broad band on either side of the body and a more narrow and less conspicuous middorsal stripe. Recently transformed animals have black sides, a dark chin, and an olive-gray dorsum. Small larvae can be distinguished from *E. bislineata* larvae by the absence of paired light spots along the dorsum.

SYSTEMATICS AND GEOGRAPHIC VARIATION. *E. guttolineata* traditionally has been treated as a subspecies of *E. longicauda* (Ireland 1979); however, biochemical and morphological analyses suggest that these forms are distinct species (Carlin 1997). Details of the

Fig. 158. *Eurycea guttolineata;* mature larva; Watauga Co., North Carolina (R. W. Van Devender).

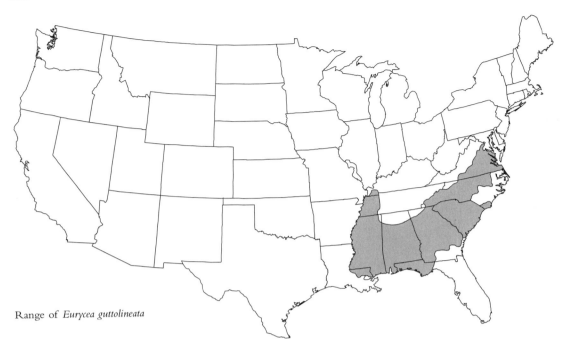

Range of *Eurycea guttolineata*

systematics of this group are presented under the account of *E. longicauda*.

DISTRIBUTION AND ADULT HABITAT. The three-lined salamander occurs from Virginia southwestward to northwestern Florida, then westward and northwestward to Louisiana, Mississippi, western Tennessee, and extreme southwestern Kentucky. Three-lined salamanders are primarily found below 800 m in elevation, although populations occur sporadically in the southern Appalachians up to 1000 m. Individuals can be found under logs and other surface cover near the margins of shaded seepages, springs, streams, bogs, and vernal ponds. In the Coastal Plain the preferred habitat is floodplain forests with logs or piles of flood debris (Gordon 1953; Mount 1975). Outside the Coastal Plain individuals are often found beneath rocks and logs along the margins of streams or in the vicinity of ditches, vernal ponds, and bogs that are fed by seepages or springs.

BREEDING AND COURTSHIP. Mating presumably occurs in the autumn and early winter as in *E. longicauda*, but detailed information on the mating season is not available. Courtship behavior has not been described. Reported sex ratios of males to females are about 1:1 (Gordon 1953) and 0.7:1 (Tinkle 1952) for Florida populations.

REPRODUCTIVE STRATEGY. Females move to breeding sites in the late autumn or early winter after activity on the ground surface ceases. Breeding sites include sluggish streams, seepages, cypress bays, vernal bogs, and bogs. Standing water habits used as breeding sites often are spring or seepage fed. Egg clutches have only rarely been collected, suggesting that females typically oviposit in subsurface retreats in or near streams or seepages. Mature ova are 2.5–3.0 mm in diameter, yellowish, and unpigmented, and are surrounded by two jelly envelopes.

Mount (1975) discovered several adults and eggs in early December in a covered concrete reservoir in a shallow spring in Alabama. The eggs were attached singly in groups of 8–14 to the sides of the concrete cylinder and were 5–25 cm below the water surface. Developmental stages varied from early to about midway to hatching. Bruce (1970) found late-term eggs and hatchlings in a western North Carolina in early to mid-March. The eggs were scattered on the bottom of a cistern. Observations from other regions suggest that females in many populations oviposit in December (Gordon 1953).

AQUATIC ECOLOGY. The larvae live in sluggish streams, ditches, and spring-fed vernal ponds, where they feed on small invertebrates. The larval period typically lasts <1 year, although a small percentage of animals, particularly in mountainous regions, may overwinter and transform the following summer. Bruce (1970, 1982a) collected metamorphosing larvae from June through August at sites in western North Carolina. Length of the larval period in most populations is 3.5–5.5 months, and average size at transformation is 22–27 mm SVL. A few larvae in one population overwinter and transform the following June, 15 months after hatching, when they measure 30–32 mm SVL. Martof (1955) found metamorphosing larvae in northern Georgia from the middle of May to the end of August.

TERRESTRIAL ECOLOGY. The smallest sexually mature males and females of *E. guttolineata* from Florida measure 43 and 46 mm SVL, respectively (Gordon 1953). Like most plethodontids, three-lined salamanders hide beneath litter or logs during the day but emerge on humid or rainy nights and search for prey. Adults are most active during the first few hours after dark and feed on a wide variety of invertebrates. This is one of the few plethodontid salamanders tested to date that does not appear to be territorial (Jaeger 1988).

Arachnids, beetles, flies, hymenopterans (mostly ants), and orthopterans are the major prey of Florida specimens (Tinkle 1952). Other prey include hemipterans, homopterans, lepidopterans, neuropterans, odonates, collembolans, mayflies, millipedes, isopods, nematodes, a *Gordius* worm, snails, and snail eggs.

PREDATORS AND DEFENSE. Three-lined salamanders exhibit a defensive posture that is similar to that of *E. bislineata* and *E. lucifuga*. When attacked, an individual will often coil its body, tuck its head beneath the tail, and raise and undulate its tail (Brodie 1977).

COMMUNITY ECOLOGY. In the southern Appalachians *E. guttolineata* is associated with bogs and sluggish streams and is often spatially segregated from *E. bislineata*, which inhabits higher-gradient streams. In other areas with lower-gradient streams, these two species often share breeding habitats. The extent to which larvae and adults of these species compete has not been examined.

CONSERVATION BIOLOGY. The widespread loss of bottomland hardwood forests throughout the Southeast has undoubtedly resulted in the loss of many local populations of the three-lined salamander. Leaving hardwood forest buffers along streams and associated wetlands would enhance habitats for this and many other species of wildlife.

Eurycea junaluska Sever, Dundee, and Sullivan
Junaluska Salamander
PLATES 73, 74

IDENTIFICATION. The Junaluska salamander superficially resembles the two-lined salamander but has a nondescript color pattern, proportionately longer forelimbs, and a proportionately shorter tail. In *E. junaluska* there are fewer than two costal folds between the toe tips of adpressed limbs and the tail is shorter than the body (typically 55–60% of body length in *E. bislineata*). The dorsum has brown mottling and lacks the dark dorsolateral stripes characteristic of *E. bislineata* (Sever 1983a; Sever et al. 1976). The sides often have dark blotches or wavy lines that may form a vague, broken dorsolateral line. The venter is light colored and lacks mottling. Sexually active males have very short cirri (0.4 mm in length), swollen jaw musculature, enlarged premaxillary teeth, and swollen cloacal and mental glands (Sever 1979). Adults reach 7.5–10 cm TL and usually have 14 costal grooves.

Hatchlings measure 7–9 mm SVL and 11–13 mm TL on average. Larvae closely resemble those of *E. bislineata* and in some cases cannot be reliably distinguished. Small larvae are best differentiated from those of *E. bislineata* using a combination of characters. In general *E. junaluska* larvae are darker and stockier than *E. bislineata* of comparable lengths (Bruce 1982b). The dorsal ground color is deep olive green to brown, compared with pale yellow or yellowish green in *E.*

Fig. 159. *Eurycea junaluska;* adult; Graham Co., North Carolina (R. W. Van Devender).

bislineata. The margin of dark pigmentation on the venter is more or less straight rather than wavy as in *E. bislineata*. Finally, *E. junaluska* larvae have dense, well-defined cheek patches and no iridophores on the venter, whereas *E. bislineata* has diffuse cheek patches and abundant iridophores on the venter (T. J. Ryan, pers. comm.). Large larvae have a pattern similar to metamorphosed animals except that larvae have more dorsolateral mottling on the tail (Sever 1983a). Larvae often have a row of six to seven large melanophores dorsal to the lateral mottling on the trunk. Specimens measuring >34 mm SVL are in all likelihood *E. junaluska,* since *E. bislineata* larvae almost never exceed this length in areas of sympatry.

SYSTEMATICS AND GEOGRAPHIC VARIATION. *E. junaluska* was originally described by Sever et al. (1976), and Jacobs (1987) has since shown it to be genetically distinct from syntopic populations of *E. b. wilderae*. Comparisons of *E. junaluska* and *E. bislineata* at a Tululah Creek site in Graham Co., North Carolina, indicate fixed differences at six gene loci. Jacobs's genetic data suggest that *E. junaluska* is most similar to *Eurycea* populations in Alabama and south-central Indiana. Sever (1976) compared specimens from the Great Smoky Mountains with those of the type series and found no evidence of morphological differences. The Smoky Mountains specimens were originally identified as *E. b. bislineata* × *E. b. cirrigera* (King 1939).

DISTRIBUTION AND ADULT HABITAT. The Junaluska salamander is a relatively rare species that is patchily distributed in extreme western North Carolina and adjoining areas in Tennessee. Specimens are known from the Cheoah River drainage in Graham Co., North Carolina; along Fighting Creek in Sevier Co., Tennessee; and along the Tellico River in Monroe

Range of *Eurycea junaluska*

Co., Tennessee, at elevations below 730 m (Sever 1983a). Adults reside beneath rocks and logs in and about large creeks, and they can be found on roads at night during spring and summer rains. The habitat requirements of this species appear to be generally similar to those of *E. bislineata*.

BREEDING AND COURTSHIP. Males collected between October and March contain spermatozoa in the vasa deferentia, a finding that suggests an extensive mating season from autumn through spring (Sever 1979). Very limited evidence based on observations of animals crossing roads suggests that adults migrate from surrounding forests to the breeding streams in late winter and early spring. Sever (1983b) collected five gravid females crossing a road on 21 March in Monroe Co., Tennessee. Courtship behavior has not been described.

REPRODUCTIVE STRATEGY. Females attach their eggs singly in small clusters to the undersides of rocks in stream channels in a manner identical to that of *E. bislineata*. Bruce (1982b) found four clutches with attending females during mid-May. Clutch size varied from 30 to 49 eggs, and the embryos were at intermediate developmental stages. Three of the four clutches were in compact groups, whereas the fourth was spread out over a 20 × 5 cm area on the underside of a rock. In all four instances, a female (40-42 mm SVL) was with the eggs. Embryos from three clutches that were transferred to the laboratory hatched in early June.

Number of mature ova in 10 females examined by Sever (1983b) varied from 41 to 68, averaged 51, and was positively correlated with SVL. The largest mature ova were 3.0 mm in diameter.

AQUATIC ECOLOGY. Most larvae in Santeetlah Creek in North Carolina transform in July, about 2 years after hatching (Bruce 1982b). A small percentage may overwinter and transform the following summer. Size at metamorphosis at this site varies from 35 to 42 mm SVL and averages 39 mm. In the Cheoh River some larvae may transform when only 1 year old. Metamorphs collected at this site measure 34-41 mm SVL, but larvae often exceed 42 mm SVL (T. J. Ryan, pers. comm.). Metamorphosis occurs from mid-May to August. Sever (1983b) collected larvae measuring 38-40 mm SVL.

Fig. 160. *Eurycea junaluska*; larva; Graham Co., North Carolina (R. W. Van Devender).

TERRESTRIAL ECOLOGY. Very little is known about many aspects of the natural history of this species. Metamorphs probably disperse during the summer from breeding streams into the surrounding forest, but the extent to which the terrestrial phase utilizes forest habitats is poorly documented. Adults can be found at streamside from the autumn through spring months.

PREDATORS AND DEFENSE. Natural predators have not been documented. Sever (1983b) surmised that predation from stocked trout may be one explanation for the rarity of this species.

COMMUNITY ECOLOGY. *Eurycea junaluska* is remarkably similar to *E. bislineata* in terms of its breeding biology and larval ecology. The extent to which these species interact negatively in stream and streamside communities is not known, but Sever (1983b) suggested that larval competition with *E. bislineata* may explain why *E. junaluska* is uncommon at most sites.

CONSERVATION BIOLOGY. The Junaluska salamander is a very rare species that appears to be uncommon throughout its small range. Because of its restricted range and general scarcity, *E. junaluska* should receive high priority for protection in all management decisions.

Eurycea longicauda (Green)
Long-tailed Salamander
PLATES 75, 76, 77

Fig. 161. *Eurycea l. longicauda;* adult; Bell Co., Kentucky (J. W. Petranka).

IDENTIFICATION. The long-tailed salamander is a large species of *Eurycea* with a conspicuously long tail. The tail becomes proportionately longer relative to body length as juveniles grow and often composes 60–65% of the total length of adults. The dorsum has a yellowish orange to yellowish brown ground color and numerous black spots that are often arranged in irregular or discontinuous lines (Ireland 1979). The venter is pale yellow to cream, and it may be immaculate or faintly mottled with light brown or gray spots. Adults measure 10–20 cm TL and there are typically 13–14 costal grooves. Sexually active males have prominent cirri, papillose cloacal lips (smooth in females), hedonic glands on the tail, and elongated maxillary teeth (Noble 1927a; Trauth et al. 1993b). Females also have cirri, but they are far less conspicuous than those of the males. Females from Ohio, Virginia, and West Virginia average 5–9% larger in SVL than males (Guttman 1989b; Hutchison 1956).

Larvae are the stream type, with streamlined bodies, short but well-formed gills (often more elongated in specimens collected from sluggish or standing water), and a dorsal fin that does not extend forward beyond the rear limbs. Hatchlings reach 10–12 mm SVL and 17–19 mm TL on average (Anderson and Martino 1966; Ireland 1974; Mohr 1943). The dorsum of hatchlings is cream colored and uniformly marked with melanophores, which contrast sharply with the immaculate belly (Anderson and Martino 1966; Bishop 1943; Franz 1967). The underside of the throat often has patches of melanophores. Within 1–2 months after hatching, larvae develop heavy, dark mottling on either side of the body. A weakly differentiated middorsal stripe may also develop. As larvae mature they gradually develop a color pattern that tends toward those of the adults of the respective subspecies (Banta and McAtee 1906). Metamorphs have black sides, a dark chin, and an olive-gray dorsum, and juveniles assume the adult color patterning within a few months after metamorphosing. The larvae lack conspicuous paired light spots on the dorsum, as seen in *E. bislineata* larvae.

SYSTEMATICS AND GEOGRAPHIC VARIATION. *Eurycea longicauda* has traditionally been treated as containing three subspecies. The taxonomic level of two of these subspecies, *E. l. guttolineata* and *E. l. longicauda*, has been debated because these forms coexist in close geographic proximity in scattered populations in the southern Appalachians with little evidence of interbreeding (Ireland 1979). The high peaks of the Appalachian Mountains form a natural barrier between *E. l. guttolineata* and *E. l. longicauda*. Near the southern extremity of the Appalachians this barrier disappears, and specimens showing evidence of intermediate color patterning have been reported from northwestern Georgia, northern Alabama, northeastern Mississippi, and along the Tennessee River in western Tennessee (Mount 1975; Redmond and Scott 1996; Valentine 1962).

Recent studies of protein variation along a north-south transect in the eastern region of this zone indicate marked genetic differences between forms (Nei's distance $[D] = 0.37$), with fixed differences occurring at 5 of 21 loci (Carlin 1997). A sixth fixed difference occurs between the southernmost sample of *E. l. longicauda* and the northernmost sample of *E. l. guttolineata*, and there is no evidence of interbreeding between the forms. Comparisons of morphological variation and color patterning in a broader sample of specimens does not provide evidence of a hybrid zone in northern Alabama and adjoining areas. Based on this evidence, I follow Carlin (1977) in recognizing these forms as separate species.

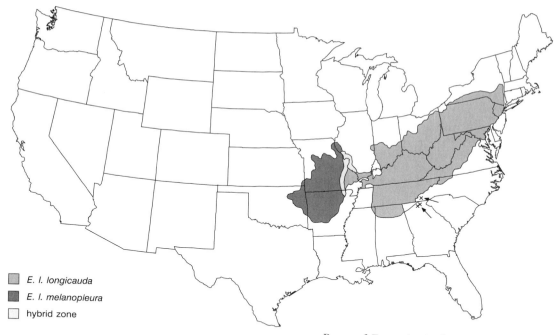

Range of *Eurycea longicauda*

Two subspecies of *E. longicauda* are currently recognized (Ireland 1979). The long-tailed salamander, *E. l. longicauda*, has black or brown spots that tend to coalesce on the sides to form lateral bands. The dorsum is orange to yellowish brown and has small, irregular black or brown blotches. The sides of the tail have elongated black spots that tend to form vertical bands, and the belly is cream to yellow. The dark-sided salamander, *E. l. melanopleura*, has a yellow to yellowish brown dorsum that is marked with numerous, irregular black marks. A broad, dark stripe that is marked with numerous light flecks occurs along each side. The belly is pale yellow and marked with indistinct light brown to gray spots. Specimens showing intermediate patterning occur in southwestern Illinois, eastern Missouri, and northeastern Arkansas and are presumed to be intergrades (Johnson 1977; Smith 1961).

DISTRIBUTION AND ADULT HABITAT. Long-tailed salamanders occur in the eastern United States from New York southwestward to northern Alabama, and westward to the Ozark and Boston mountains in Missouri, Arkansas, Oklahoma, and Kansas. Disjuncts are found in extreme western North Carolina and in northeastern Georgia. Juveniles and adults are most frequently found under rocks, logs, and other surface cover near the margins of shaded seepages, springs, or streams, but individuals are occasionally found far from running water in forest habitats (Ireland 1979; Minton 1972; Mount 1975; Smith 1961). This species is often locally abundant in wet shale banks. Long-tailed salamanders in the Ozark Uplift and in the Appalachians are often found in or near mines and caves (Bishop 1941a; Hutchison 1958; Holsinger 1982; Myers 1958a; Mohr 1944) and in some areas are abundant along pond margins (Anderson and Martino 1966, 1967).

BREEDING AND COURTSHIP. Data from several sources indicate that mating occurs primarily in the autumn and early winter. Mating has been observed in the field on 18 October (Cooper 1960) and in the laboratory on 18 November (Bishop 1943). The mating season of *E. l. melanopleura* in Arkansas lasts from November through February, based on analyses of male and female reproductive tracts (Ireland 1974). Females in this region breed annually. Detailed descriptions of courtship behavior are not available. On 18 October Cooper (1960) observed partial courtship of a pair of long-tailed salamanders in a mine that entailed the male chasing a female and rubbing his head around her cloaca and snout.

REPRODUCTIVE STRATEGY. Females move to breeding sites in seepages, streams, or stream-fed ponds in the late autumn after activity on the ground surface ceases. Franz and Harris (1965) found larvae in a reservoir in Maryland, an unusual habitat for this species. Egg clutches have only rarely been collected, and most have been found in caves, mine shafts, or cisterns. These observations suggest that females typically oviposit in dark, subsurface streams or seepages.

Data on seasonal changes in ovarian egg size (Guttman 1989b; Hutchison 1956; Minton 1972) and observations of eggs in the field indicate that egg laying occurs in late autumn through very early spring. Franz (1964) found a female *E. l. longicauda* on 23 November coiled about five eggs suspended just above a pool in a cave in Maryland. The eggs were in early developmental stages and were attached singly to the cave roof. Mohr (1943) discovered eggs of *E. l. longicauda* in early stages of development in a mine shaft in Pennsylvania on 2 January. The eggs were attached singly in running water to the tops and sides of rocks, and to the edges of submerged boards. Anderson and Martino (1966) believed that oviposition in New Jersey populations of *E. l. longicauda* occurs in underground passages sometime in January. Hatchlings in southern Illinois have been found from November through March, suggesting an extended egg-laying period in this region (Rossman 1960).

Based on examination of female reproductive tracts, Ireland (1974) concluded that *E. l. melanopleura* oviposits from December through March in Arkansas. Two clutches of eggs found in December were in a single row on the undersurface of a rock. Embryos in early cleavage discovered on 6 December hatched about 1 month later. In other areas, hatchling *E. l. melanopleura* have been found in springs in winter and early spring (Rudolph 1978).

Estimates of clutch size based on ovarian egg counts include a mean of 91 (range = 61-106) for seven *E. l. longicauda* (Hutchison 1956) and about 100 eggs for two other females (Minton 1972). The discrepancy between ovarian counts and the number of eggs observed in nature suggests that females deposit small groups of eggs in several localized areas in underground passages, a feat that would make guarding difficult. Although females and males have sometimes been found in the general vicinity of eggs, Franz (1964) is the only one to have observed a female tightly coiled about her eggs, and she may have been in the process of laying.

Mature ova are 2.5-3.0 mm in diameter, yellowish, and surrounded by two jelly envelopes (Franz 1964; Ireland 1974; Mohr 1943). The incubation period varies from 4 to 12 weeks depending on water temperature (Ireland 1974; Mohr 1943). Very small larvae or hatchlings have been found from November through March (Ireland 1974; Rossman 1960; Rudolph 1978).

AQUATIC ECOLOGY. The larvae live in both spring-fed caves and surface waters, where they feed on small invertebrates. Larval *E. l. melanopleura* inhabiting springs in Oklahoma consume a variety of aquatic invertebrates. The most numerically important prey are ostracods, copepods, snails, fly larvae, isopods, beetles, and mayfly nymphs (Rudolph 1978).

Data from several studies indicate that the larval period typically lasts <1 year, although a small percentage of animals in some populations may overwinter and transform the following summer. Length of the larval period and size at metamorphosis are similar for both subspecies, with transformation occurring primarily in June and July (Anderson and Martino 1966; Franz and Harris 1965; Huheey and Stupka 1967).

Anderson and Martino (1966) studied the larval ecology of *E. l. longicauda* populations in limestone sinkholes in New Jersey. Hatchlings first appear in the ponds in March and grow 4-5 mm SVL per month. Larvae metamorphose 65-70 days later in mid-June, when their average measurements are 20-21 mm SVL, 41 mm TL, and 24-25 mg. A second estimate for size at metamorphosis of *E. l. longicauda* is 18-21 mm SVL for metamorphosing larvae in Maryland (Franz and Harris 1965). Larvae occasionally overwinter and may reach 51 mm TL by mid-March (Franz 1967).

Growth rates of *E. l. melanopleura* in Arkansas increase from 1-2 mm SVL per month between January and March to 4-6 mm SVL per month in April and May (Ireland 1974). Larvae transform in June and July when they measure 23-28 mm SVL and are 4-7 months old. Most *E. l. melanopleura* larvae studied by Rudolph (1978) transformed in late summer and autumn, but some overwintered and transformed the following year when >1 year old.

Very few data are available on the behavior of larvae. Larvae hide in leaf litter or emergent vegetation during the day (Anderson and Martino 1966). I have observed them at night foraging on benthic invertebrates outside cover.

TERRESTRIAL ECOLOGY. Most juvenile *E. l. longicauda* studied by Anderson and Martino (1966) sexually mature about 2 years after transforming. Males become sexually mature when they reach >43 mm SVL, and females when they reach >46 mm SVL. In Arkansas *E. l. melanopleura* become sexually mature 1–2 years after transforming, when males and females measure 31–43 mm and 33–43 mm SVL, respectively (Ireland 1974).

Adults and juveniles often show marked seasonal shifts in microhabitat use. Individuals may migrate to slopes adjoining streams during periods of heavy spring rains (Duellman 1954b). Recently metamorphosed animals studied by Anderson and Martino (1966) in New Jersey disperse from pond margins within a few weeks after metamorphosing. Juveniles and adults remain in the vicinity of ponds throughout the summer, disperse to underground winter retreats in October, and emerge from the retreats in late April and early May. Adults overwinter in forests within 30 m of water and move to the margins of ponds within a few weeks of emergence. In Indiana adults are active on the ground surface from April through October (Minton 1972).

Large numbers of adults in a Pennsylvania population congregate in an old mine shaft for about 8 months out of the year (Mohr 1944). Adults move from the mine to the surrounding forest in late April and early May, then return about 4 months later and remain until the following spring. Similar seasonal migrations occur into and out of two Maryland caves (Franz 1967).

Adults are sometimes found in large aggregates in mines or beneath logs, suggesting that individuals are not territorial. Examples include 23 adults found beneath a 4-m-long log and 80 individuals found underneath a limestone slab in Ohio (Guttman 1989b). Over 300 adults were observed in a single day near the rear of a mine shaft (Mohr 1944).

Like most plethodontids, long-tailed salamanders hide beneath stones, litter, or logs during the day but emerge on humid or rainy nights and search for prey (Hutchison 1958; Smith 1961). Adults are most active during the first few hours after dark and feed on a wide variety of invertebrates. New Jersey long-tailed salamanders eat annelids, isopods, millipedes, centipedes, pseudoscorpions, phalangids, mites, ticks, spiders, homopterans, beetles, flies, ants, lepidopteran larvae, crickets, and thysanurans (Anderson and Martino 1967). Field observations of available prey suggest that millipedes, centipedes, earthworms, and isopods are less preferred than other prey types. Major prey of Virginia cave residents are flies, orthopterans, and beetles (Hutchison 1958). Miscellaneous prey include pseudoscorpions, spiders, and isopods.

PREDATORS AND DEFENSE. Little information is available on natural predators or defensive mechanisms of this species.

COMMUNITY ECOLOGY. Rudolph (1978) found that larvae of *E. longicauda, E. lucifuga, E. multiplicata, E. tynerensis,* and *Typhlotriton spelaeus* are spatially segregated to varying degrees along spring-fed headwater tributaries in Oklahoma. Larvae of all species are most abundant within the first 50 m or so of springheads. Fish predation and flooding are major factors limiting the downstream distribution of these species. Near spring sources, adult habitat requirements and competitive interactions among larvae may play a role in determining the local distribution of these species within streams.

CONSERVATION BIOLOGY. Long-tailed salamanders are widely distributed in forested habitats in the eastern United States and are in minimal need of protection. Strip mining and acid drainage from coal mining operations have undoubtedly affected many populations in the eastern coal fields, but few detailed studies of the impact of mining on this and other salamanders have been conducted.

Eurycea lucifuga Rafinesque
Cave Salamander
PLATE 78

Fig. 162. *Eurycea lucifuga;* adult; Fayette Co., Kentucky (J. W. Petranka).

IDENTIFICATION. The cave salamander is a large, slender species of *Eurycea* with a prehensile tail that makes up about 60–65% of the total length (Hutchison 1966). Adults have bright orange or reddish orange dorsums that are heavily marked with irregularly spaced black spots. Heavily melanized individuals are occasionally found that have the dorsal spots fused into large, irregular blotches. The limbs are long and the toe tips of adpressed limbs overlap. The head is broad and flattened dorsally, and the venter is whitish to yellowish and unmarked. Adults range from 10 to 20 cm in TL and have 14–15 costal grooves.

Both males and females have cirri, but those of sexually active males are longer. Sexually active males develop elongated, unicuspid maxillary teeth during the breeding season that are used in courtship (Noble 1927a). Males also have a round mental gland and papillae on the lining of the vent. Sexual dimorphism in body size varies geographically. Males in Indiana populations were on average 1% larger in SVL than females (Minton 1972), whereas males in an Illinois population were on average 10% larger (Williams 1980). In a large sample from throughout the range of the species, Hutchison (1966) found that females are 3% larger on average than males.

Hatchlings are whitish in coloration with inconspicuous pigmentation. Reported sizes are 9–11 mm SVL (McDowell 1988), 11–12 mm TL (Myers 1958b), and 17.5 mm TL (Banta and McAtee 1906). Larvae have stream-type morphology and hatch with functional limbs. Young larvae are sparsely and rather uniformly pigmented above and have three longitudinal series of light spots along each side (Hutchison 1966). During the last few months of the larval stage, individuals begin to develop the black spotting typical of metamorphosed animals. Metamorphs have a yellowish dorsal ground color and superficially resemble young *E. longicauda*. The dorsum turns reddish orange and resembles that of the adults within a few months after metamorphosis (Banta and McAtee 1906).

This species lacks traits, such as reduced eyes and reduced pigmentation, that are associated with obligate cave dwellers. The depressed body form, long legs, and prehensile tail are adaptations for living in cracks and climbing on rockfaces.

SYSTEMATICS AND GEOGRAPHIC VARIATION. The cave salamander is relatively uniform in morphology and coloration and no subspecies are recognized. Heavily melanized individuals with irregular dorsal blotches are more common in the western portion of the range but occur sporadically in populations elsewhere (Banta and McAtee 1906; Eigenmann and Kennedy 1903; Grobman 1943; Minckley 1959; Myers 1958a; Reese and Smith 1951).

Merkle and Guttman (1977) studied genetic variation in populations throughout the range and found very little genetic differentiation among populations. Average heterozygosity varies from 0.002 to 0.04 among regional groups, and genetic similarity values are ≥0.90.

DISTRIBUTION AND ADULT HABITAT. Cave salamanders occur in limestone regions from Indiana southward to northern Alabama and Mississippi, and from western Virginia to eastern Oklahoma. On rare occasions local populations are found on noncalcareous bedrock (Hutchison 1958). Adults are most commonly encountered on the floor or along rockfaces in the twilight zones of caves (Banta and McAtee 1906; Green et al. 1967; Hutchison 1956, 1958; Peck and Richardson 1976). However, they are regularly found in the dark zone hundreds of meters from cave en-

Range of *Eurycea lucifuga*

trances (Peck and Richardson 1976) and have sometimes been found in large numbers beneath stones in cave streams (Williams 1980).

The extent to which *E. lucifuga* is restricted to cave habitats has been exaggerated by some. Adults are often abundant in forested habitats in or near rock walls or ledges with springs at their bases and extensive cracks that allow access to underground passages (Blatchley 1897; Mount 1975; Sinclair 1950). Specimens have been collected under stones on dry hillsides, from a rubbish heap in a field, from a suburban yard, and from beneath logs on the forest floor. In some instances collection sites are >1 km from the nearest known caves or springs (Banta and McAtee 1906; Green et al. 1967; Minton 1972). Adults in southern Illinois are common around the margins of cypress swamps bordered by rock bluffs (Smith 1961). The restriction of this species to limestone regions is presumably more closely related to the microhabitats associated with caves and other solution features than to a direct dependence on high-pH substrates (Hutchison 1958).

BREEDING AND COURTSHIP. The mating season is not well documented but probably occurs during the summer and early autumn prior to the initiation of egg laying. Specimens that Organ (1968a) held in the laboratory courted in July. Males tend to outnumber females slightly in populations; reported sex ratios of males to females are 1.5:1 (Hutchison 1958) and 1.1:1 (Williams 1980).

Organ (1968a) observed incomplete courtship sequences in captive specimens. Courtship is somewhat similar to that of *E. bislineata*. The following is a proposed sequence based on Organ's fragmented observations. More detailed observations are needed to verify the sequence.

A male appears to initiate courtship by nudging and rubbing his chin and cheeks across the head or snout of the female. The male often poses with his cheeks pressed against the nasolabial grooves of the female. The male eventually moves forward while keeping his dorsum in contact with the female's chin. As the basal portion of his tail passes beneath the female, he pauses and begins push-ups with his hindlimbs while forcing his pelvic area and tail base up against the female. She straddles his tail and the two engage in a tail-straddle walk in which the male strongly arches his body forward and undulates the base of the tail in a circular motion. Spermatophore deposition and sperm pickup presumably ensue, but have not been witnessed. The spermatophore consists of a colorless, transparent stalk about 4 mm high with an oval base that is 6 mm long and 3.5 mm wide. The sperm cap is white and amorphous.

REPRODUCTIVE STRATEGY. Oviposition is staggered over a rather long period from September through February. In many populations a peak in oviposition occurs between October and January. Eggs have been found in West Virginia caves between 24 September and 5 November (Green et al. 1967), and in a Missouri cave on 2 January (Myers 1958b). Hatchlings have been found in February and March in Indiana (Banta and McAtee 1906), in winter and early spring (months not specified) in Oklahoma streams (Rudolph 1978), in January in Tennessee (Barr 1949), and from mid-November through January in West Virginia (Green et al. 1967). In Tennessee, larvae as large as 23 mm TL occur in December, whereas larvae as small as 14 mm TL have been found in early February (Sinclair 1950).

Biologists have rarely discovered the eggs, suggesting that females move into deep recesses within caves and springs to oviposit. Eggs in West Virginia are laid singly and either attached by a stalk to the sides of small rimstone pools or laid unattached on silt deposits in the bottoms of pools (Green et al. 1967). Eggs in a Missouri cave are attached singly to the bottoms and sides of submerged rocks in a stream (Myers 1958b). The most likely oviposition sites are underground springs, rimstone pools, and streams in caves. Banta and McAtee (1906) note that adults move away from cave mouths and into deeper areas of caves to oviposit. This behavior may be adaptive since larvae tend to wash downstream as they age (Green et al. 1967).

Freshly laid eggs are white, about 2.5–3.2 mm in diameter, and surrounded by two jelly envelopes (Barden and Kezer 1944; Green et al. 1967; Myers 1958b). The outer egg capsule is 4–5 mm in diameter. The number of large ova in 17 Virginia specimens was 49–87 and averaged 68 (Hutchison 1956); that in 11 Arkansas females was 60–120 and averaged 78 (Trauth et al. 1990).

AQUATIC ECOLOGY. Larvae live in both subsurface and surface streams, and in shallow pools and seepages formed by water dripping from rock formations (Rudolph 1978; Sinclair 1950). The presence of larvae in surface streams may reflect washout from subsurface nesting sites during periods of heavy rain. Larvae are benthic feeders and forage by moving

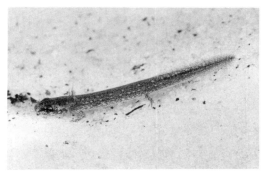

Fig. 163. *Eurycea lucifuga;* larva; Pulaski Co., Kentucky (R. W. Van Devender).

slowly over the substrate in search of small prey such as chironomid larvae. Dietary items listed by Rudolph (1978) are snails, a spider, ostracods, copepods, isopods, mayflies, stoneflies, beetles, and flies. Ostracods, dipteran larvae, and snails are the most numerically important prey.

Young larvae observed by Sinclair (1950) in stream pools outside Tennessee caves actively crawl about during the day and at night. Older larvae tend to be secretive during daylight hours and active at night. However, all larval stages in an Indiana population are photophobic (Banta and McAtee 1906).

The larval period lasts 6–18 months. Most larvae in Oklahoma streams transform during late summer or autumn of the year of hatching, but some overwinter and transform the following year (Rudolph 1978). Larvae in an Indiana cave transform in the autumn, winter, and spring (Banta and McAtee 1906). Most individuals transform 12–15 months after hatching, although a small percentage have longer larval periods. Larval samples showing two distinct year classes have been collected in April in Indiana (Minton 1972), in March in Missouri (Myers 1958b), and in February in Tennessee (Sinclair 1950). These findings suggest a larval period of perhaps 14–18 months in these populations.

The largest larva collected by Williams (1980) in southern Illinois measured 33 mm SVL and 70 mm TL, whereas the smallest metamorph measured 31 mm SVL and 68 mm TL. Larvae in Tennessee transform from June through August (Sinclair 1950). Two metamorphs collected in June had reached 59 and 60 mm TL; the smallest metamorph collected by Minton in Indiana measured 58 mm TL. Banta and McAtee (1906) raised two laboratory specimens that metamorphosed when they reached 56.5 and 57 mm TL. In Arkansas, larvae average 32 mm SVL in late October and 37 mm SVL in January (Trauth et al. 1990). Transforming larvae collected in late May average 36 mm SVL.

TERRESTRIAL ECOLOGY. Although the terrestrial ecology of juveniles is poorly documented, metamorphs appear to reach sexual maturity within 2 years after transforming. In a Virginia population, males and females become sexually mature when they measure about 46 mm SVL and 48–49 mm SVL, respectively (Hutchison 1956). Similarly, females in southern Illinois begin maturing sexually when they measure >49 mm SVL (Williams 1980).

Adults and juveniles often show seasonal movements in caves. Most adults in Virginia caves retreat to deep passages from September through March (Hutchison 1958). The number of visible animals increases in April and peaks in May and June. Seasonal changes in the visible population correlate with changes in moisture levels in the twilight zone. Cave salamanders in a southern Illinois cave tend to retreat deeper into a cave during the summer months when conditions become drier (Williams 1980). Seasonal changes in the number of visible individuals are similar to those reported by Hutchison (1958). Both authors note that human disturbance strongly reduces the number of individuals observed on subsequent trips, so seasonal patterns may be confounded by disturbance associated with sampling.

Cave salamanders have prehensile tails that facilitate climbing along rockfaces. Individuals that I have observed are surprisingly adept at maneuvering over wet rocks and ledges and use the tail for leverage and bracing. The adults are often found high up on ledges in caves, where they forage for invertebrates. Individuals inhabiting twilight zones tend to be nocturnally active; they remain in cracks or under bark and debris during the day and move actively about the cave at night (Green et al. 1967). Individuals living on the forest floor are secretive during the day but are active on the forest floor at night (Sinclair 1950). Adults show some tendencies to wander on rainy nights and are occasionally collected on wet roads.

The dietary items from 13 adults from caves examined by Hutchison (1958) were flies, orthopterans, beetles, thysanopterans, lepidopterans, collembolans, ants, phalangids, pseudoscorpions, mites, ticks, and isopods. Items from 112 adults from several caves in the eastern United States included 73 species of prey, among them earthworms, snails, crayfish, amphipods, millipedes, centipedes, pseudoscorpions, spiders, phalangids, and many types of insects (Peck 1974). Beetles and flies were particularly well represented. Prey in 213 specimens from central Tennessee and northern Alabama included over 101 prey taxa (Peck and Richardson 1976). Dietary items included three juvenile *Plethodon glutinosus* and relatively large numbers of caddisflies, dipterans, and orthopterans. Salamanders from the twilight zone contain a greater number and volume of prey than salamanders from either the cave entrance or the dark zone. Thus adults appear to congregate in areas that maximize feeding opportunities.

PREDATORS AND DEFENSE. Cave salamanders produce noxious tail secretions. When attacked, individuals coil their bodies, tuck their heads beneath their vents, and raise and undulate their tails (Brodie 1977). The percentage of animals with broken or regenerating tails was 28% and 4% in samples examined by Williams (1980) and Hutchison (1956).

COMMUNITY ECOLOGY. Because of their preferences for caves and rock ledges, adult cave salamanders appear to interact minimally with salamanders that inhabit the forest floor. Hutchison (1958) found a negative correlation between the numbers of *E. lucifuga* and *E. longicauda* in Virginia caves, which suggests that these species may compete. Cave salamanders appear to function as top predators in many cave communities, but their importance in structuring cave communities has not been investigated.

CONSERVATION BIOLOGY. Data are currently unavailable on the status of populations of cave salamanders throughout their range. This species would be a suitable candidate for inclusion in long-term monitoring projects on amphibians.

Eurycea multiplicata (Cope)
Many-ribbed Salamander
PLATE 79

Fig. 164. *Eurycea m. multiplicata;* adult; Polk Co., Arkansas (R. W. Van Devender).

IDENTIFICATION. The many-ribbed salamander is an elongated species of *Eurycea* with 19–20 costal grooves. The dorsum is brown to chocolate brown and is often bordered on either side by a dark dorsolateral line (Dundee 1965a; Loomis and Webb 1951). The sides have silvery white flecks, and the belly is yellow to pale gray. The undersurface of the tail is lemon yellow. Sexually active males have papillose cloacal lips and a swollen gland at the dorsal base of the tail, but are otherwise similar in size and form to the females (Noble 1931). The inner cloacal lips of females are smooth and folded. Adults measure 6–10.5 cm TL.

Hatchlings average about 10 mm SVL and have stream-type morphology. The gills of hatchlings and larvae have long rami and numerous filaments (Bishop 1943; Moore and Hughes 1941). Young larvae are yellowish above with varying amounts of brown flecking. They appear uniformly pigmented above and are immaculate below except for scattered melanophores that often occur on the chest, chin, and sides of the throat. Small larvae (<26 mm SVL) closely resemble those of *E. tynerensis* and in many cases cannot be accurately discriminated (Tumlison et al. 1990a). Color patterns of larvae nearing metamorphosis approximate those of the adults. Gilled adults have been collected in a small number of populations inhabiting permanent, spring-fed streams along the southwest edge of the Ozark Plateau (Brandon 1971a; Trauth et al. 1990).

SYSTEMATICS AND GEOGRAPHIC VARIATION. Two subspecies are currently recognized (Dundee 1965a). The many-ribbed salamander, *E. m. multiplicata,* may or may not have indistinct dorsolateral stripes bordering the dorsal stripe. Relatively few silvery white flecks occur on the sides, and the venter is yellow and lacks melanophores. This is the smaller subspecies. Adults reach a maximum of 43 mm SVL and gilled adults are not known.

The gray-bellied salamander, *E. m. griseogaster,* has conspicuous dorsolateral stripes and more prominent silvery white flecking along the sides. The venter is pale gray to lemon yellow and has concentrations of melanophores that produce either stippling or distinct patches of dark pigmentation. Gilled adults are found occasionally, particularly in cave populations. Adults of this subspecies grow to 54 mm SVL and 106 mm TL. Dundee (1965a) denotes a broad zone of intergradation in the southeastern Ozark Mountains, but genetic analyses have not been conducted to document the extent of gene exchange occurring among subspecies.

DISTRIBUTION AND ADULT HABITAT. *Eurycea multiplicata* occurs at elevations of 107–763 m above sea level in the Ozark and Ouachita mountains and adjoining rocky lowland formations in south-central and southwestern Missouri, eastern Oklahoma, and Arkansas (Dundee 1965a). Adult and larval habitats include caves, small springs and spring-fed ponds, and large, rocky creeks (Dundee 1965a; Ireland 1976; Loomis and Webb 1951; Moore and Hughes 1941). Adults are semiaquatic and are most frequently found beneath rocks and debris near streamside, or under submerged rocks in the stream proper (Loomis and Webb 1951; Moore and Hughes 1941). Populations utilize both permanent and seasonally ephemeral streams, and in some areas the adults frequent the twilight zones of caves.

BREEDING AND COURTSHIP. The mating season varies substantially among regional populations of *E. m. griseogaster* in Arkansas. Populations in springs or spring-fed habitats that are thermally stable tend to have relatively prolonged mating seasons relative to

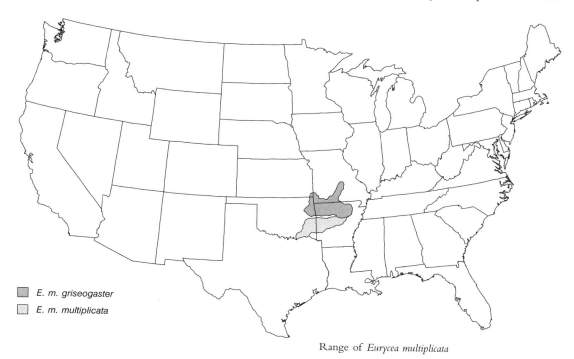

Range of *Eurycea multiplicata*

those in surface streams with widely fluctuating temperatures. Ireland (1976) reports that females have spermatozoa in their reproductive tracts from August through May at one site, and from July through February at a second site with spring-fed waters. Mating activity peaks between September and December. At a third site with widely fluctuating seasonal temperatures, females with spermatozoa are present from December to March. Mating activity in this population peaks in January. Courtship behavior has not been described.

REPRODUCTIVE STRATEGY. Females lay their eggs in springs or spring-fed pools and ponds from autumn through early spring. Gray-bellied salamanders studied by Ireland (1976) oviposit from August through March or April, with peak egg laying occurring from November through January. Nesting records include eggs in early developmental stages on 7 October attached to the underside of a rock in a pool near a cave (Spotila and Ireland 1970), and four clutches in early developmental stages found during December and February in spring-fed ponds used for trout production (Ireland 1976). At the latter site, the eggs are laid in single rows on the undersides of submerged rocks buried beneath 5–15 cm of gravel and are not attended by females. The incubation period for two clutches was 27 and 36 days. Trauth et al. (1990) report clutches of 8, 10, and 14 eggs laid by gilled adult *E. m. griseogaster* beneath several layers of small rocks at the mouth of a spring, and two gravid females with 8 and 13 mature ova. Mean number of mature ova in 112 females varied from 3 to 21, averaged 13, and was positively correlated with SVL (Ireland 1976).

Freshly laid eggs are creamy white, 2.0–2.6 mm in diameter, and surrounded by two envelopes (Spotila and Ireland 1970; Trauth et al. 1990). The outer envelope is about 5–6 mm in diameter.

AQUATIC ECOLOGY. Very little is known about diurnal activity patterns, foraging, and other aspects of larval behavior. The larvae are benthic feeders and are most active at night outside cover, where they feed on small benthic invertebrates and zooplankton. Ostracods and isopods are the most numerically important prey of larvae in an Oklahoma stream (Rudolph 1978). Other prey include snails, mites, copepods, amphipods, hymenopterans, dipterans, and nymphs of mayflies, caddisflies, and stoneflies.

Larvae grow rapidly and transform 5–8 months after hatching when they are approaching sexual maturity. In some populations gilled adults are common. All large larvae in a small stream in Cherokee Co., Oklahoma, are sexually mature (Dundee 1947). Most larvae

Fig. 165. *Eurycea m. griseogaster;* larva; Stone Co., Missouri (R. W. Van Devender).

studied by Ireland (1976) transform when they measure between 38 and 42 mm SVL. Length of the larval period is 5–6 months in one population and 7–8 months in two others. Transforming larvae measure 33–48 mm SVL and all larvae >42 mm SVL show evidence of metamorphosing.

TERRESTRIAL ECOLOGY. Ireland (1976) found that males become sexually mature during or shortly after transforming. All completely transformed males have spermatozoa, and 16% of male larvae in the process of transforming contain spermatozoa in the seminiferous tubules. Transformed females >36 mm SVL are sexually mature, and most females begin producing eggs soon after transforming. Individuals do not disperse far from streams and remain in cool, moist microhabitats in or near water.

Little is known about the seasonal activity patterns of adults. Adults are often beneath rocks, logs, and clumps of moss near or at the edges of streams during February, an observation that suggests that animals stay active on the surface except during periods of severe winter weather (Dundee 1947).

PREDATORS AND DEFENSE. Few data are available on natural predators or defensive mechanisms. Natural predators probably include crayfishes and raccoons. Fishes are major predators on the larvae and appear to exclude them from downstream sections of streams.

COMMUNITY ECOLOGY. Spring-fed headwater tributaries in eastern Oklahoma often contain mixtures of three to five species of salamander larvae, including *E. longicauda, E. lucifuga, E. multiplicata, E. tynerensis,* and *Typhlotriton spelaeus.* Spatial segregation of these species occurs with respect to the distance from springheads and may be due in part to competition (Rudolph 1978). Fish predation and flooding are major factors limiting the downstream distribution of larvae of all species.

CONSERVATION BIOLOGY. Because juveniles and adults do not stray far from water, *E. multiplicata* is less affected by land clearing and deforestation than many other salamanders. Nonetheless, spring-fed streams with protected watersheds appear to offer the best habitats for local populations.

Eurycea nana Bishop
San Marcos Salamander
PLATE 80

IDENTIFICATION. *Eurycea nana* is a diminutive, permanently gilled salamander that reaches 41–56 mm TL (Bishop 1941b; Brown 1967b; Tupa and Davis 1976). Larvae are plain brown above with a row of seven to nine yellowish flecks or paired pale spots down the back. The venter is whitish with a yellowish cast beneath the tail. Each eye has a dark ring around the lens. There are 16–17 costal grooves and six to seven costal grooves between the toes of adpressed limbs (Bishop 1941b, 1943; Brown 1967b). The gills are well developed and pigmented, and males have a

Fig. 166. *Eurycea nana;* adult; Hays Co., Texas (R. W. Van Devender).

Range of *Eurycea nana*

slightly larger cloacal region than females. The vent of males is lined with papillae, whereas that of females has smooth folds. Males have poorly defined mental and caudal hedonic glands that occur deep within the dermis (Sever 1985).

Tupa and Davis (1976) collected a hatchling that measured 8.2 mm TL. The specimen had limb buds, a gray dorsum, a whitish venter, and darkly pigmented eyes.

SYSTEMATICS AND GEOGRAPHIC VARIATION. This is one of several permanently gilled *Eurycea* species that live in the springs and underground water systems of the Edwards Plateau and Balcones Escarpment region of Texas. *Eurycea nana* is a well-differentiated species that most herpetologists have recognized as a valid species since its description in 1941. Tupa and Davis (1976) are the only researchers who have published detailed information on the life history of this species. The life history account that follows is based on their research.

DISTRIBUTION AND ADULT HABITATS. *Eurycea nana* is known only from the immediate vicinity of San Marco, Hays Co., Texas. This species is found primarily along the northern bank of Spring Lake, a large spring-fed body of water that forms the headwaters of the San Marcos River. Individuals live in dense mats of a blue-green alga (*Lyngbya*) that covers much of the substratum, in aquatic macrophyte beds, and beneath rocks and gravel. The highest densities are reached in algae mats and macrophyte beds. Although temperatures in these habitats are nearly constant year round (21–21.5°C), the mean critical thermal maxima of juveniles and adults is 36–37°C and similar to that of congeners that inhabit surface streams with much more variable seasonal temperatures (Berkhouse and Fries 1995). Specimens occur in the San Marcos River as far as 150 m downstream from the Spring Lake dam (Berkhouse and Fries 1995).

BREEDING AND COURTSHIP. Published data on courtship are not available. Breeding presumably can occur at any time of the year since oviposition appears to be strongly acyclical.

REPRODUCTIVE STRATEGY. The eggs have never been found in nature. Females oviposit throughout the year, and there is no well-defined peak in breeding activity. Hatchlings with yolk still evident have been collected in February, May, and June. Gravid females and very small larvae have been collected with about the same frequency during all months of the year. The maximum diameter of mature ovarian eggs is 2.0 mm and the maximum clutch size is about 20 eggs.

AQUATIC ECOLOGY. The San Marcos salamander reaches its greatest densities in algae mats and beds of aquatic macrophytes that provide refuge from predatory fishes and an abundance of small invertebrates for food. The estimated average density in optimal habitats is 116 individuals/m^2 of substrate, and the entire population in Spring Lake is estimated to be about 21,000 salamanders. Individuals feed mostly on amphipods and the larvae and pupae of tendipid midges, although other aquatic insects and snails are occasionally eaten. Males become sexually mature when they reach 19–23.5 mm SVL and females when they measure >21 mm SVL.

PREDATORS AND DEFENSE. Spring Lake contains several species of predatory fishes, including sunfishes (*Lepomis* spp.), bullheads (*Ictalurus melas*), and largemouth bass (*Micropterus salmoides*). Sunfishes appear to be important predators and are presumably the primary reason why *E. nana* is largely restricted to dense mats of algae and macrophytes that function as refuges. Crayfish are also present at the site and probably prey on *E. nana*.

COMMUNITY ECOLOGY. *Eurycea nana* coexists locally with the fountain darter (*Etheostoma fonticola*), which also feeds heavily on amphipods and lives in similar microhabitats. The extent to which these species compete for food resources has not been investigated.

CONSERVATION BIOLOGY. The San Marcos salamander is a federally threatened species that is protected by law.

Eurycea neotenes Bishop and Wright
Texas Salamander
PLATE 81

Fig. 167. *Eurycea neotenes*; Travis Co., Texas (R. W. Van Devender).

IDENTIFICATION. The Texas salamander consists mostly of gilled adults, although metamorphosed individuals occur in a small percentage of populations in mesic canyons of the Texas hill country (Sweet 1977b). Larvae in surface populations are finely mottled with yellowish brown above and have 10–12 inconspicuous pairs of dorsolateral light spots that extend from the head onto the tail. The throat, belly, and underside of the tail are whitish and translucent. A fairly prominent dark bar extends from each eye to the nostril (Brown 1967a) and the eyes are of nearly normal size (Bishop 1943). Populations restricted to caves have less pigmentation on the gills and body surface and have eyes that are reduced to varying degrees. Sexually mature gilled males can be distinguished from females by the presence of a swollen vent in males (Brown 1967a). Males have both mental and caudal hedonic glands, but these are found deep within the dermis and are not useful in sexing animals in the field (Sever 1985). Females often have eggs visible through the body wall.

Transformed adults have a uniformly dark dorsum and lack spots, bands, or other conspicuous markings. Scattered silvery specks occur on the dorsum and on the sides of the body and tail. The sides are grayish and finely mottled. The venter is whitish and translucent with scattered melanophores (Sweet 1977b). The smallest larva reported by Bishop (1943) measured 14 mm TL. Maximum size of adults is variable geographically, but in most populations is 35–50 mm SVL and 50–100 mm TL (Bruce 1976; Sweet 1977b, 1984). There are 15–17 costal grooves. Isolated cave populations of *E. neotenes* often exhibit local differentiation that may reflect genetic drift or introgressive hybridization with other species (Sweet 1982, 1984).

SYSTEMATICS AND GEOGRAPHIC VARIATION. The systematic status of many *Eurycea* populations in the eastern portion of the Edwards Plateau of central Texas has been a source of controversy over the years. Several isolated cave populations showing local differentiation have been described, including *E. latitans* (Smith and Potter 1946), *E. nana* (Bishop 1941b), *E. pterophila* (Burger et al. 1950), and *E. troglodytes* (Baker 1957). In his checklist of North American amphibians and reptiles, Schmidt (1953) considered all of the Edwards Plateau *Eurycea* described prior to 1953 to be subspecies of *E. neotenes*. Baker (1957, 1961), however, considered these to be valid species. Most workers have subsequently treated *E. pterophila* as an invalid taxon (Sweet 1978).

Sweet (1984) argued that isolated cave populations previously recognized as *E. troglodytes* and *E. latitans* are of hybrid origin between *E. neotenes* and *E. tridentifera*. *Eurycea tridentifera* is seemingly absent from the Valdina Farm Sinkhole, the type locality for *E. troglodytes*. However, specimens that are apparent hybrids between *E. tridentifera* and *E. neotenes* are present, as are individuals that approximate *E. neotenes*. In the Cascade Caverns system, the type locality of *E. latitans,* many specimens approximate *E. neotenes* but show evidence of minor influence from *E. tridentifera*. Both parental species and hybrids occur in Honey Creek Cave. Sweet (1984) considered *E. troglodytes* to represent a hybrid swarm and *E. latitans* to be a troglobitic population of *E. neotenes* that episodically incorporates individuals of *E. tridentifera*.

Sweet (1984) and Mitchell and Smith (1972) note that several additional forms from cave systems in the Edwards Plateau may be undescribed species, and Chippindale et al. (1993) described one as a new species, *E. sosorum*. Recent molecular studies of the *E. neotenes* complex have revealed substantial genetic diversity within the group, and a major taxonomic revision is currently being prepared (Chippindale 1995). At least three new species will be described soon, and some of the former names may be resurrected (P. T. Chippindale, pers. comm.). Here, I consider *E. latitans, E. pterophila,* and *E. troglodytes* to be junior synonyms of *E. neotenes* and recognize only *E. nana, E. neotenes, E. sosorum,* and *E. tridentifera* as valid taxa. *Eurycea neotenes* is clearly a species complex, and in the future the name *neotenes* will probably only apply to populations at or near the type locality at Helotes Creek Spring, Bexar Co., Texas.

DISTRIBUTION AND ADULT HABITAT. The Texas salamander occurs in the southeastern region of the Edwards Plateau in central Texas. Populations are restricted to cave waters or the immediate vicinity of springs that flow from limestone formations (Sweet 1982). Larvae and metamorphosed adults live in interstitial areas in gravel and beneath leaf litter, sticks, and stones in both fast-flowing water and pools (Brown 1942; Bruce 1976; Sweet 1977b).

Larvae are restricted to the immediate environs of springs because of the thermal, temporal, and physical stability of these habitats (Sweet 1982). Larvae are uncommon or absent in relatively warm sections of streams, where temperatures sometimes approach 30°C. Populations are most commonly found in springs or spring-fed streams that have temporally reliable surface flow. Surface populations of *E. neotenes* are rarely found in springs with erratic fluctuations in surface flow, or with surface flow that occurs for only a small portion of the year. However, these streams may have troglobitic populations (Sweet 1982).

BREEDING AND COURTSHIP. No data are available on most aspects of the breeding biology of this species in natural populations. Published descriptions of courtship are not available.

REPRODUCTIVE STRATEGY. No information is available on the mode of egg deposition or the seasonal time of ovipositing in natural populations. Two groups of researchers have induced females to lay in the laboratory (Barden and Kezer 1944; Roberts et al. 1995). The eggs are laid singly and are unpigmented, about 2 mm in diameter, and surrounded by three egg membranes. The embryonic period lasts 2-4 weeks in the laboratory and clutch size varies from 19 to 50 eggs. In the latter study females oviposited on 18-19 February, 13 March, 28 March (estimated based on the developmental stage when eggs were first discovered), and 10 May.

AQUATIC ECOLOGY. Larvae apparently spend much of their time burrowed in gravel beds or resting beneath cover during the day (Bishop 1943; Sweet 1982). They retreat to subsurface locations whenever

Range of *Eurycea neotenes* — type locality

surface flow stops but move back to the surface with the seasonal return of surface waters. Hatchlings maintained in a zoo aquarium grow as large as 60 mm TL within 6 months, at which time they are approaching sexual maturity (Roberts et al. 1995). In natural populations, males and females sexually mature during their second year, and females probably oviposit for the first time when 2 years old (Bruce 1976). Females begin maturing sexually when they reach >25 mm SVL and produce their first clutch of eggs when slightly larger. Males also mature when they measure >25 mm SVL. Annual survivorship of larvae is about 10% during the first year but increases thereafter.

The smallest sexually mature individual examined by Sweet (1977b) measured 25 mm SVL, but size at sexual maturity varies among populations. In one population males and females do not become sexually mature until they measure >30 and 33 mm SVL, respectively. However, mature females as small as 28–30 mm SVL are present in a second population (Sweet 1977b). Two size classes are evident in one population, suggesting a larval period of 1 year or more in individuals that transform.

TERRESTRIAL ECOLOGY. Most populations lack postmetamorphic stages. In the few populations that are not strict perennibranchs, transformed animals begin to mature sexually during or shortly after transformation (Sweet 1977b). Sexually mature larvae rarely metamorphose; those that do transform when they reach 30–36 mm SVL. Transformed animals live in water near spring sources; however, the presence of terrestrial isopods and collembolans in their stomachs suggests that adults may move short distances onto land at night to feed.

COMMENTS. Little has been published on many aspects of the life history of *E. neotenes*, even though specimens can often be collected in large numbers. Data are needed on ontogenetic and geographic variation in larval coloration and size, clutch and egg size, mode of egg deposition, breeding season, courtship, diet, larval behavior, and other life history components. Hunsaker and Potter (1960) observed massive die-offs of *E. neotenes* caused by "red leg" disease in one population. Other than disease, the greatest potential threat to this species is perhaps the long-term lowering of water tables from human use.

Eurycea quadridigitata (Holbrook)
Dwarf Salamander
PLATE 82

Fig. 168. *Eurycea quadridigitata*; adult; Macon Co., Alabama (J. W. Petranka).

IDENTIFICATION. The dwarf salamander is a small, slender *Eurycea* that somewhat resembles a miniature two-lined salamander. This is the only *Eurycea* that has four toes on each hindfoot. The dorsum has a broad bronze to yellowish brown stripe that is bordered by a narrower brown or black dorsolateral stripe on either side. The dorsolateral stripes are less conspicuous than those in *E. bislineata* and extend from the head to the tip of the tail (Mittleman 1967). Individuals sometimes have a middorsal row of small black spots that extends down the back. The venter is yellow in light-colored specimens. A dark morph that is dark brown above and silvery gray below occurs in many populations. The tail is keeled and makes up about 50–60% of the total length of adults. Sexually active males have cirri and elongated, monocuspid teeth (Noble 1927a). The average SVL is approximately the same for adult males and females (Semlitsch and McMillan 1980). Adults measure 5.5–9.0 cm TL and have 14–17 costal grooves.

Hatchlings show traits intermediate between typical pond- and stream-type morphology and measure 6.5–7 mm SVL and 7–10 mm TL on average (Goin 1951; Harrison 1973; Semlitsch 1980a). The hindlimbs are not fully formed at hatching, and the dorsal fin extends anteriorly beyond the insertion of the hindlimbs to the middle of the back. Hatchlings are uniformly grayish brown above and immaculate below (Goin 1951). The color pattern of older larvae is similar to that of the adults. An albino specimen is reported by Dyrkacz (1981).

DISTRIBUTION AND ADULT HABITAT. The dwarf salamander occurs primarily in Coastal Plain habitats from North Carolina to eastern Texas and southern Arkansas. Populations also occur in the upper Piedmont of South Carolina, and an isolate is found in west-central Florida. Adults have been found beneath logs and other surface objects around pond and swamp margins, in seepages, beneath pine straw and other litter around pine savanna ponds, and in leaf-filled springs. Carr (1940) notes that the adults in Florida are often found some distance from water during the summer and fall, but occur next to the water's edge from January to April.

SYSTEMATICS AND GEOGRAPHIC VARIATION. This species was once considered to be polytypic (Mittleman 1966), but no subspecies are currently recognized.

BREEDING AND COURTSHIP. Adults living away from pond margins migrate in late summer and the fall to the breeding ponds and are more likely to move during the day than other species, such as mole salamanders (*Ambystoma talpoideum*; Semlitsch and Pechmann 1985). In South Carolina migrations to breeding ponds occur from around mid-July through October following the onset of cooler autumn weather (McMillan and Semlitsch 1980; Semlitsch and McMillan 1980). In central Alabama the adults migrate in late autumn through early winter (Trauth 1983).

Mating occurs from late August or early September through February. Harrison (1973) collected females with spermatophores in their vents or observed courtship in the laboratory from September to early November in South Carolina populations. Males in breeding condition can be found from August through February in southern Louisiana (Sever 1975) and from October through February in Alabama (Trauth 1983). Courtship behavior has not been described but presumably involves a tail-straddle walk as in other

270 Plethodontidae: Plethodontinae: Hemidactyliini

Range of *Eurycea quadridigitata*

plethodontids. Sex ratios in South Carolina populations are about 1:1 (Harrison 1973; Semlitsch 1980a; Semlitsch and McMillan 1980), and local breeding populations vary from 10 to 10,000 animals (Gibbons and Semlitsch 1991).

REPRODUCTIVE STRATEGY. Females oviposit in midautumn through midwinter in woodland pools, seepages, roadside ditches, Carolina bays, and other standing bodies of water. Within a given population, oviposition appears to be restricted to a 1- to 2-month period. Females in a South Carolina population oviposit in October and November (Harrison 1973), whereas those in Alabama oviposit in February (Trauth 1983). Clutches have been found in Florida from 22 November to 15 January (Carr 1940; Goin 1951) and in central North Carolina on 20 December and 1 February (Brimley 1923).

Females attach their eggs singly to vegetation mats, leaves, pine straw, and other substrates in ponds or seepages. In Alabama the eggs of females have been found attached singly to rootlets, twigs, and debris on the underside of *Sphagnum* moss mats that are bathed by a steady flow of seepage water (Trauth 1983). Eggs at a North Carolina site are attached singly to leaves in seepages (Brimley 1923), whereas females in northern Florida sometimes attach the eggs in a monolayer to the undersides of submerged logs in a manner similar to that of *E. bislineata* (Carr 1940). Females maintained by Harrison (1973) in the laboratory laid eggs singly in loose groups attached to pine needles and other debris.

In some instances, females may oviposit in unflooded pond depressions. Females at a South Carolina site lay eggs in the autumn when the breeding site is dry (Taylor et al. 1988). The eggs apparently develop normally to advanced stages and hatch shortly after the site floods. Terrestrial nests have also been found in Florida in shallow depressions beneath wet logs along pond margins (Goin 1951).

Fig. 169. *Eurycea quadridigitata;* adult; Scotland Co., North Carolina (R. W. Van Devender).

As with *E. bislineata*, the nature of the substrate strongly influences whether eggs are laid in tight monolayers or scattered widely over the substrate. Trauth (1983) found females in close association with eggs, but the majority of evidence suggests that females abandon their eggs after ovipositing. Freshly laid eggs are slightly less than 2 mm in diameter, creamy white, and surrounded by two envelopes. The outer envelope is about 3-4 mm in diameter (Bishop 1943; Brimley 1923; Harrison 1973).

Marked differences in clutch parameters that are related to differences in average adult size sometimes occur between local populations (Semlitsch and McMillan 1980). Number of ovarian eggs in 31 females from one South Carolina population varies from 18 to 48 and averages 33, whereas that in 24 females in a second population varies from 7 to 42 and averages 21. Number of mature ova in 15 Alabama specimens varies from 14 to 59 and averages 34 (Trauth 1983), whereas that in South Carolina specimens varies from 13 to 36 and averages 22 (Harrison 1973). Clutch size of *E. quadridigitata* is positively correlated with female SVL (Harrison 1973; Semlitsch and McMillan 1980; Trauth 1983).

Length of the embryonic period is poorly documented in nature. Eggs maintained in the laboratory require 30-40 days to develop to the hatching stage (Harrison 1973). Hatchlings have been found in late January and early February in South Carolina (Harrison 1973; Semlitsch 1980a), and in March in North Carolina (Brimley 1923). Small larvae appear in woodland pools and flooded roadside ditches in Alabama in March (Mount 1975). Eggs from Florida populations that were maintained in the laboratory hatched 7 December-13 February (Goin 1951).

AQUATIC ECOLOGY. The larvae are primarily benthic feeders, and peak seasonal densities may sometimes exceed 10 larvae/m^2. Larvae in South Carolina feed rather heavily on small zooplankton. Ostracods, cladocerans, and chironomid larvae are the major dietary items, and larvae selectively feed on the larger individuals of cladocerans (Taylor et al. 1988).

The dwarf salamander has a brief larval period compared with most other species of *Eurycea*, with most larvae transforming 2-6 months after hatching. The larval period lasts only 2-3 months in certain North Carolina populations (Brimley 1923). Hatchlings in two South Carolina populations transform

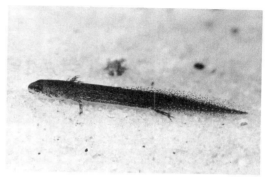

Fig. 170. *Eurycea quadridigitata;* larva; Robeson Co., North Carolina (R. W. Van Devender).

after 5-6.5 months of growth when they reach 21-26 mm SVL (Semlitsch 1980a), whereas those in other populations transform after 3-6 months of growth when they reach 17-20 mm SVL (Harrison 1973). In these studies, larvae transformed primarily between April and June (Harrison 1973) and from mid-June through early July (Semlitsch 1980a). Larvae in another South Carolina study measured about 22 mm SVL on average in mid-May and most transformed before June (Taylor et al. 1988).

TERRESTRIAL ECOLOGY. Individuals in South Carolina become sexually mature during the autumn of their first year of life. Harrison (1973) found that juveniles become sexually mature when they reach about 24 mm SVL. The smallest mature females collected by Semlitsch and McMillan (1980) and Trauth (1983) measured 23 and 22 mm SVL, respectively. Mature males in Alabama populations measure 23-31 mm SVL. In a South Carolina population, males mature sexually in September and October, 8-9 months after hatching (Semlitsch 1980a). Females also mature during their first year of life, but probably require an additional year before eggs are sufficiently large to oviposit.

The adults appear to be generalist feeders. McMillan and Semlitsch (1980) found a variety of small invertebrates in the stomachs of 124 adults from South Carolina. Dietary items include flies, beetles, hemipterans, homopterans, ants, wasps, collembolans, spiders, pseudoscorpions, mites, ticks, and millipedes. Ants, mites, and collembolans are the most common prey of this population. Prey in 25 Georgia specimens include ants and other hymenopterans, beetles, collembolans, mites, spiders, and insect larvae (Powders and Cate 1980), whereas prey in Florida specimens consists of amphipods, adult and larval beetles, spiders, and earthworms (Carr 1940).

The only study of population dynamics is that of Semlitsch et al. (1996), who conducted a 16-year study at Rainbow Bay at the Savannah River Site in South Carolina. The size of the breeding adult population was low at the initiation of the study, increased dramatically during year 7, then dropped sharply and remained consistently low during the last 6 years of monitoring (1989–1994). The size of the annual breeding population is not correlated with the amount of rainfall that occurs during the breeding season, but it is positively correlated with the number of metamorphs produced in subsequent years. The number of juveniles produced per female is independent of the initial number of larvae in the pond, and there is little evidence that larval production is negatively affected by the density of other salamander larvae in the pond.

PREDATORS AND DEFENSE. Although predation on this species is poorly documented, dwarf salamanders are probably eaten by numerous predators, including birds, small snakes, and large invertebrates such as crayfish and spiders. Lamb (1984) found four dwarf salamanders in the stomachs of pig frogs (*Rana grylio*) collected from southwest Georgia.

COMMUNITY ECOLOGY. Taylor et al. (1988) compared the diet of *E. quadridigitata* larvae with that of larval *Ambystoma talpoideum* and *Notophthalmus viridescens* that share a Carolina bay in South Carolina. Dietary overlap between *A. talpoideum* and *E. quadridigitata* is much lower than that between *A. talpoideum* and *N. viridescens*. These differences are probably due to size differences between larvae and the fact that *A. talpoideum* feeds in the water column at night whereas *E. quadridigitata* does not. All three species feed rather heavily on small zooplankton when younger but incorporate larger prey such as chironomid larvae into the diet as they grow.

CONSERVATION BIOLOGY. The dwarf salamander is moderately abundant throughout its range even though loss of wetlands has resulted in the elimination of many local populations. This species appears to be in minimal need of protection.

Eurycea rathbuni Stejneger
Texas Blind Salamander
PLATE 83

Fig. 171. *Eurycea rathbuni*; adult; Hays Co., Texas (R. W. Van Devender).

IDENTIFICATION. The Texas blind salamander is a troglobitic species with long, toothpick-thin legs; whitish, translucent skin; and a highly flattened snout with tiny, nonfunctional eyes (Emerson 1905; Norman 1900). There are 12 costal grooves between the limbs. The aquatic adults are gilled and measure 9–13.5 cm TL. Juveniles have proportionately larger eyes than the adults. Transformed specimens have never been collected, and it is unlikely that metamorphosis occurs in nature since the adults lack a functional thyroid gland (Dundee 1957; Dunn 1926).

SYSTEMATICS AND GEOGRAPHIC VARIATION. Species of *Eurycea* inhabiting the Edwards Plateau show varying degrees of specialization for living in caves. These involve changes in pigmentation, eye size and structure, relative limb length, head size and shape, and number of trunk vertebrae and teeth (Potter and Sweet 1981). Because of their geologic isolation, many cave populations are probably independent or semi-independent lineages that have undergone paral-

Range of *Eurycea rathbuni*

lel or convergent evolution, a phenomenon that further complicates resolution of the evolutionary histories of this group using morphological data.

Considerable controversy concerning the generic allocation of species has arisen over the years (Potter and Sweet 1981). Some authors feel that two species previously assigned to the genus *Typhlomolge* (i.e., *rathbuni, robusta*) are best treated as species of *Eurycea* that exhibit extreme troglobitic morphologies (Mitchell and Reddell 1965; Mitchell and Smith 1972); others feel that members of *Typhlomolge* are sufficiently distinct from Edwards Plateau *Eurycea* (sensu stricto) to warrant recognition of the genus. Recent analyses of skull characteristics by Potter and Sweet (1981) and secondary reproductive traits by Sever (1985) support the taxonomic validity of *Typhlomolge*. However, molecular data suggest that the recognition of *Typhlomolge* is not warranted (Chippindale 1995). Here I follow the latter interpretation.

DISTRIBUTION AND ADULT HABITAT. Specimens have been collected at seven localities in the Purgatory Creek system and along the San Marcos Springs Fault near San Marcos, Hays Co., Texas (Uhlenhuth 1921). Adults and immature larvae are well adapted for living in underground streams in caves, and many probably inhabit deep recesses that are not accessible to collectors. Specimens have been taken in deep pools with minimal current and nearly constant 21–22°C temperatures. The first specimens of this species were collected in 1895 from a newly constructed well that drew water from 58 m below the ground surface (Longley 1978).

BREEDING AND COURTSHIP. The time of breeding is poorly documented. Dunn (1926) notes that a specimen maintained in the laboratory laid a few eggs on 15 March and that a specimen collected in early fall had the spermatheca packed with spermatozoa. Very small juveniles have been found throughout the year, suggesting an aseasonal breeding pattern (Longley 1978).

Belcher (1988) observed one complete and two partial courtship bouts in captive specimens in which the female initiated courtship and the male remained passive initially. Courtship begins when the female approaches the male and rubs her chin on his dorsum. The female may also rub her cloaca on nearby rocks while rocking to and fro. If the male does not respond, the female may nip the male along his sides or engage in kicking behavior in which gravel is scratched with the hindlimbs. The female eventually straddles the tail of the male and rubs her snout above the tail base. The male responds by arching his pelvic region and fanning his tail between her legs. The female then rubs her snout more rapidly over the base of the tail. The male may lead the female forward and repeat the same cycle while slowly vibrating the anterior third of the tail. The male eventually bends the body laterally and moves the tail laterally at a right angle to the body while the female continues rubbing the base of the tail. The male then leads the female forward, bends his body into an S-shaped pattern, and deposits a spermatophore on the substrate. He next leads the female forward with the tail extended laterally until she picks up the spermatophore cap with her cloacal lips. The spermatophore consists of a crescent-shaped white sperm cap over a clear, gelatinous base that is about four times longer than it is wide.

REPRODUCTIVE STRATEGY. Almost nothing is known about the reproductive strategy of this species. One gravid female contained 39 mature ova (Longley 1978).

AQUATIC ECOLOGY. Texas blind salamanders are rather active animals that prowl about submerged rocks and prod their snouts into crevices, presumably in search of prey. The elongated limbs facilitate climbing over rocks. Individuals do not respond to light, are positively rheotactic, and will snap at meat held in forceps (Norman 1900). Known dietary items include a blind shrimp (*Palaemonetes antrorum*), snails, and amphipods (Longley 1978). In the laboratory larvae often feed by moving the head laterally and snapping at prey that they contact. These observations suggest that larvae locate prey using both chemical and tactile cues. One specimen was observed in nature swimming back and forth and skimming the water surface, apparently in search of insects that had landed on the water (P. T. Chippindale, pers. comm.). Most females become sexually mature when they reach 40-50 mm SVL (Brandon 1971a). Males mature at about 40 mm SVL. The size of breeding populations is not known. Uhlenhuth (1921) notes that about 100 *E. rathbuni* were collected annually during the first few years after a well was constructed. However, catches in the well declined to just a few animals per year thereafter.

PREDATORS AND DEFENSE. Natural predators are not known.

COMMUNITY ECOLOGY. The Texas blind salamander is a top predator in cave systems, but the extent to which it plays a role in organizing invertebrate communities is not known.

CONSERVATION BIOLOGY. *Eurycea rathbuni* is federally endangered and is protected by law.

Eurycea robusta Potter and Sweet
Blanco Blind Salamander

IDENTIFICATION. This extremely rare species was discovered in 1951 by a work crew excavating a spring in a dried bed of the Blanco River northeast of San Marcos, Hays Co., Texas. Four specimens were found, but only one preserved specimen remains. The site was subsequently covered by the Blanco River and filled with gravel and silt, and no additional specimens have been discovered to date (Potter and Sweet 1981). The following species description is based on a single preserved specimen redescribed by Potter and Sweet (1981). The specimen is a stout-bodied, depigmented, gilled salamander with very reduced eyes, 13 trunk vertebrae, robust limbs, and a thick tail with moderately high fins. The species somewhat resembles *E. rathbuni* but is stouter bodied and has a longer trunk and slightly shorter limbs. The specimen measures 57 mm SVL and 101 mm TL. This species is currently the subject of petitioning to be listed as federally endangered.

Range of *Eurycea robusta*

Eurycea sosorum Chippindale, Price, and Hillis
Barton Springs Salamander
PLATE 84

Fig. 172. *Eurycea sosorum;* Travis Co., Texas (W. Meinzer; courtesy of P. T. Chippindale).

IDENTIFICATION. The Barton Springs salamander was described by Chippindale et al. (1993) based on morphological and allozyme comparisons with other perennibranch *Eurycea* species found in the Edwards Plateau. Sweet (1984) had studied this population earlier and recommended that it be formally recognized as a new species. Because *E. sosorum* is restricted to Barton Springs and is the only species of *Eurycea* found there, identification based on locality alone is feasible.

Eurycea sosorum does not transform and is permanently aquatic. Individuals have varying degrees of dorsal blotching and mottling owing to an irregular mixture of melanophores, pigment gaps, and iridophores. The iridophores are often highly concentrated in patches and may partially obscure the darker pigments in the skin (Chippindale et al. 1993). The general dorsal color varies from dark gray to yellowish cream. The head has a slight shovel-nose appearance, and the eyes are more reduced than those in other surface-dwelling *Eurycea* populations of central Texas. The maximum size of sexually mature larvae measured by Chippindale et al. (1993) is 37 mm SVL and 63 mm TL, although larger larvae have been observed in the field.

DISTRIBUTION AND ADULT HABITAT. The Barton Springs salamander lives in a group of hydrologically connected springs (Barton, Parthenia, Eliza, and Sunken Garden springs) in the Barton Creek

Range of *Eurycea sosorum*

drainage within the city of Austin, Travis Co., Texas. Most specimens have been collected from a wading pool and from the outflow of a larger spring that fills Barton Springs pool in Zilker Park. This species is primarily a surface-dweller, although individuals appear to be well suited to living underground (Chippindale et al. 1993). The population lives at depths of up to 4-5 m in water that is almost constantly 20°C. Specimens have been found under rocks, in gravel, and in aquatic vegetation or algae mats, where they forage on small invertebrates.

COMMENTS. Few data are available on the natural history of this species. Hatchlings have been found in November, March, and April, and females with well-developed eggs in September through January. Most individuals measuring >22.5 mm are sexually mature (Chippindale et al. 1993). The diet consists mostly of amphipods.

The Barton Springs population was depleted significantly because of maintenance procedures that involved the removal of aquatic plants from Barton Springs pool and the use of chlorine for pool cleaning. New procedures that are more environmentally sound are being used to help speed the recovery of the population. *Eurycea sosorum* is currently being considered for listing as an endangered species. The greatest future threat to this species may be deterioration of the surface waters that supply the springs as a result of pollution and siltation associated with development of the watershed for housing.

Eurycea tridentifera Mitchell and Reddell
Comal Blind Salamander
PLATE 85

Fig. 173. *Eurycea tridentifera*; Comal Co., Texas (P. T. Chippindale).

IDENTIFICATION. *Eurycea tridentifera* is a permanently gilled, troglobitic species that has a large head, short trunk, thin elongated limbs, and reduced eyes that typically lack lenses (Mitchell and Reddell 1965; Sweet 1977a, 1984). The head is large and about 40% wider than the body. The eyes are minute and tend to be irregularly shaped, but are not buried deeply in the head. The snout is depressed abruptly at the level of the eyes and is truncated between the external nares (Mitchell and Reddell 1965). There are 11-12 costal grooves, and the toes of adpressed limbs either overlap or are separated by one or two costal folds. The maximum size of specimens examined by Mitchell and Reddell (1965) is 37 mm SVL and 74 mm TL, but specimens may reach 46 mm SVL and 85 mm TL.

Adults have cream to pale yellow dorsal coloration, which is overlain with diffuse brown or gray mottling. Four to as many as 15 pairs of dorsolateral light spots occur on the body and base of the tail. The venter is unpigmented and translucent. Larvae tend to be darker than adults and are generally less yellow (Sweet 1977a). Sexually active males have swollen cloacal lips (Brandon 1971a) and the adults measure 5-8.5 cm TL. Sexual dimorphism in SVL is slight; males measured 4% longer than females on average in a series of 15 animals with SVL >22 mm examined by Mitchell and Reddell (1965).

SYSTEMATICS AND GEOGRAPHIC VARIATION. *Eurycea tridentifera* is sometimes placed in the genus *Typhlomolge* following Wake (1966). However, Mitchell and Reddell (1965) and Mitchell and Smith (1972) provide evidence that this species is more closely affiliated with *Eurycea*. Populations vary in morphology among cave systems, and some populations are considered by Mitchell and Reddell (1965) and Mitchell and Smith (1972) to warrant recognition as separate species (Sweet 1977a).

Range of *Eurycea tridentifera*

DISTRIBUTION AND ADULT HABITAT. This species was originally described from Honey Creek Cave in Comal Co., Texas, at the southeastern margin of the Edwards Plateau. It has since been found in several caves within a few kilometers of this site and in Elm Springs Cave and Genesis Cave in Bexar Co. (Sweet 1977a; P. T. Chippindale, pers. comm.).

COMMENTS. Most aspects of the life history and ecology are not known. This species inhabits underground passageways and is difficult to collect in large numbers. Females produce from 7 to 18 mature ova per clutch that measure 3.5 mm in diameter on average (Sweet 1977a). Embryos hatch when they reach about 7 mm SVL and 13 mm TL. Sweet (1977a) reports that males mature at 25–27 mm SVL and females at 28–32 mm SVL, whereas Brandon (1971a) concludes that males and females mature when they reach 22–25 mm SVL and 25 mm SVL, respectively. The stomachs are often filled with detritus, suggesting that individuals may graze the substrate for tiny invertebrate prey (P. T. Chippindale, pers. comm.).

Eurycea tynerensis Moore and Hughes
Oklahoma Salamander
PLATE 86

IDENTIFICATION. The Oklahoma salamander is a permanently gilled species, although specimens treated with thyroxin will transform (Kezer 1952b). The body is elongated with 19–20 costal grooves and 7–11 costal grooves between the toe tips of adpressed limbs. The dorsal ground color is cream to gray and is finely mottled with black spots or streaks. The dorsum of the tail has a broad brownish stripe, and a series of light spots is usually present along each side of the body. The venter

Fig. 174. *Eurycea tynerensis;* Oklahoma (R. W. Van Devender).

Range of *Eurycea tynerensis*

is pale and unpigmented except for a few melanophores on the chin and lower jaw. Adults measure 51–81 mm TL (Dundee 1965b). Hatchlings measure 9–13 mm TL and have stream-type morphology. Tumlison et al. (1990a) provide information on traits that are useful in separating *E. tynerensis* from *E. multiplicata*.

SYSTEMATICS AND GEOGRAPHIC VARIATION. Detailed data on morphological and genetic variation between populations are not available. *Eurycea tynerensis* is closely related to *E. multiplicata*, but phylogenetic relationships of this and other *Eurycea* species have yet to be delineated.

DISTRIBUTION AND ADULT HABITAT. The Oklahoma salamander occurs in deep gravel beds of small spring-fed streams in eastern Oklahoma, southwestern Missouri, and northwestern Arkansas (Dundee 1947, 1958; Moore and Hughes 1939; Rudolph 1978). This species prefers shallow, slow-flowing areas near stream edges that have medium-sized rocks with moderate degrees of embeddedness and high invertebrate densities (Tumlison et al. 1990c).

COMMENTS. Most aspects of the life history and ecology of this species have not been studied. Courtship precedes oviposition by several months, and females can store viable sperm in the spermathecae for many months (Dundee 1958). Larvae collected by Rudolph (1978) from two sites measured 10–39 mm SVL, but distinct year classes are not evident. Larvae mature sexually after 2–3 years of growth when they reach 26–27 mm SVL (Dundee 1958, 1965b; Tumlison et al. 1990b). Estimates of clutch size are based on a small number of dissected animals and include one specimen with 12 large ova (Moore and Hughes 1939) and three Arkansas specimens with 1–11 ova (Trauth et al. 1990). A single egg measured in the former study was 1.8×1.5 mm in diameter.

Ostracods, isopods, and mayfly nymphs are the major prey items in Oklahoma specimens (Rudolph 1978). Other prey include snails, copepods, amphipods, crayfish, stonefly nymphs, homopterans, beetles, fly larvae, and caddisflies. Chironomids, mayflies, and isopods are the most important prey in specimens from several streams in the western Ozarks (Tumlison et al. 1990b). Other prey are ephydrid dipterans, stoneflies, beetles, caddisflies, hymenopterans, thysanopterans, ostracods, amphipods, crayfishes, mites, snails, and leech cocoons. Juveniles consume more chironomids and fewer amphipods than adults. Differences in diet also occur between sexes and local stream populations.

Gyrinophilus palleucus McCrady
Tennessee Cave Salamander
PLATE 87

Fig. 175. *Gyrinophilus palleucus;* Mud Flats Cave, Knox Co., Tennessee (R. W. Van Devender).

IDENTIFICATION. The Tennessee cave salamander is a stout-bodied, troglobitic species with greatly reduced eyes, a broad head, and a truncated snout. The gills are long and bright red, and the tail is flattened laterally and oarlike. The skin is not heavily pigmented and appears pinkish to flesh colored from the underlying blood capillaries. Larvae in most populations rarely exceed 100 mm SVL (Dent and Kirby-Smith 1963; McCrady 1954), although three males collected by Brandon (1965a) measured 120–124 mm SVL. Adults reach 10–23 cm TL and there are 17–19 costal grooves.

Specimens rarely transform in nature, although they will transform in the laboratory when treated with metamorphic agents (Blair 1961; Brandon 1971a; Dent and Kirby-Smith 1963). Two metamorphosed specimens are known from Tennessee caves (Simmons 1976; Yeatman and Miller 1985). The smallest hatchlings that have been collected measured 10 mm SVL (Simmons 1975).

SYSTEMATICS AND GEOGRAPHIC VARIATION. Cave populations are isolated from each other to varying degrees and have undergone regional differentiation in morphology and coloration. Three subspecies are currently recognized that differ in body pigmentation, head width, leg length, eye size, and modal number of trunk vertebrae (Brandon 1966b, 1967a); however, certain populations are difficult to assign to any of the recognized subspecies. Adults of the Sinking Cove Cave salamander (*Gyrinophilus p. palleucus*) lack the dark, spotted dorsal pigmentation found in the other subspecies. This subspecies occurs in several caves in the Crow Creek drainage in Franklin Co., Tennessee, southward into northern Alabama and extreme northeastern Georgia. Adults of the Big Mouth Cave salamander (*Gyrinophilus p. necturoides*) have 19 trunk vertebrae and prominent, dark dorsal spots. The spots often have faint, poorly defined edges and may appear more as smudges. This subspecies is known from caves in the Elk River drainage in Grundy Co., Tennessee. The Berry Cave salamander (*Gyrinophilus p. gulolineatus*) has only 18 trunk vertebrae and tends to be more heavily pigmented than other forms. This species has a dark stripe on the throat, a conspicuously large head, and dorsal spotting on a medium to dark brown dorsum. It is known from localities in the Ridge and Valley Province in Knox, McMinn, and Roane counties, Tennessee. Small larvae of the latter two subspecies are uniformly darker than those of *G. p. palleucus*. Specimens that may be intergrades between *G. p. palleucus* and *G. p. necturoides* occur in northeastern Alabama (Brandon 1967c). In addition, several populations that were not assigned to subspecies have been reported from northwestern Georgia and northern Alabama (Brandon 1966b, 1971a; Cooper and Cooper 1968).

Populations in the Central Basin in Rutherford and Wilson counties, Tennessee, are light brown with relatively large chocolate brown spots on the back. These forms differ in coloration from the three described subspecies, inhabit a different physiographic unit, and are not easily assigned to any of the recognized subspecies. Genetic variation is currently being examined in some populations; Redmond and Scott (1996) note that one form in Tennessee may be sufficiently distinct based on biochemical variation to recognize as a separate species.

DISTRIBUTION AND ADULT HABITAT. The Tennessee cave salamander inhabits cave systems in central to eastern Tennessee, northwestern Georgia, and northern Alabama (Brandon 1967a; Cooper 1968; Cooper and Cooper 1968; Mount 1975; Redmond

Range of *Gyrinophilus palleucus*

and Scott 1996). Specimens have been found under rocks in shallow pools and in both rocky and sandy substrates in quiet pools (McCrady 1954; Simmons 1975). Local populations in Tennessee are usually found in sinkhole-type caves or in phreatic cave systems that are in the vicinity of sinkholes. The association with sinkholes is tied to the relatively high nutrient input that these formations receive and the large invertebrate populations that sustain salamander populations (R. S. Caldwell, pers. comm.).

BREEDING AND COURTSHIP. The mating season is not known, but one male was collected on 24 August with a spermatophore protruding from its cloaca (Lazell and Brandon 1962). Females appear to have very irregular annual breeding patterns and may skip one or more years before breeding again (Simmons 1975).

REPRODUCTIVE STRATEGY. The eggs have never been discovered and little is known about the nesting biology. Simmons (1975) found small hatchlings in two caves on 21 December and 15 February, suggesting that females oviposit in the autumn or early winter.

AQUATIC ECOLOGY. Food items of 19 specimens from several caves included oligochaetes, amphipods, isopods, crayfish, cladocerans, beetles, stoneflies, mayflies, dipterans, caddisflies, thrips, and two conspecifics (Simmons 1975). Amphipods are the most frequent prey and constitute 21% of the diet. Specimens from other caves feed predominantly on isopods but also eat chironomid larvae and earthworms (Brandon 1967b).

Growth rates of larvae are presumably slow relative to those of *Gyrinophilus* that live in surface streams, and larvae probably require many years to reach sexual maturity. Males as small as 66 mm SVL may be sexually mature (Simmons 1975), and all specimens of *G. p. palleucus* measuring >70 mm SVL are sexually mature. Ovarian eggs in one female averaged 3.2 mm in diameter (Dent and Kirby-Smith 1963).

Surveys of populations rarely find more than 10–20 animals per cave visit; this observation suggests that local populations are very small. Cave Cove Cave has perhaps the largest population in Tennessee; a recent estimate of the population size in 834 m of passageway using mark-recapture was 48 animals (R. S. Caldwell, pers. comm.). Three other caves contain local populations consisting of an estimated 25, 32, and 88 individuals, with densities of 0.11, 0.17, and 0.15 animal/m^2 of stream (Simmons 1975).

Individuals are often highly sedentary and rarely

move more than 3-4 m between repeated surveys. Many animals have been found in exactly the same place in which they were observed weeks or months earlier (Simmons 1975). During winter and spring floods, specimens may wash downstream and individuals may move into cover to avoid floodwaters. Cooper and Cooper (1968) describe the invertebrate faunas associated with known localities of this species.

PREDATORS AND DEFENSE. Simmons (1975) documents cannibalism and Lazell and Brandon (1962) report the remains of a urodele that is probably a conspecific in the stomach of one specimen. Lee (1969b) found a large adult in the stomach of a cave-dwelling bullfrog. These are the only documented predators, although crayfish undoubtedly feed on small larvae.

COMMUNITY ECOLOGY. The range of *G. palleucus* overlaps that of *G. porphyriticus*. Several researchers have reported that these two species do not occur syntopically in the same cave (Brandon 1966b, 1971a; Cooper and Cooper 1968); however, there are now at least two confirmed cases of microsympatry in caves (Simmons 1975; R. S. Caldwell, pers. comm.). The extent to which competitive exclusion or competition occurs between these species is not known.

CONSERVATION BIOLOGY. Some populations of *G. palleucus* appear to be declining as a result of disturbances and alterations of watersheds that drain into sinkholes that support local populations. Deforestation and urbanization have increased water flow and silt loads in some caves and have adversely affected local populations (Simmons 1975; R. S. Caldwell, pers. comm.). Because of its restricted range and association with caves, *G. palleucus* should receive maximum protection in land management decisions. Local populations are best maintained by protecting watersheds that drain into sinkhole systems.

Gyrinophilus porphyriticus (Green) Spring Salamander
PLATES 88, 89

IDENTIFICATION. The spring salamander is a large, stout-bodied plethodontid with a salmon to pinkish orange ground color. The ground color is overlain with diffuse black streaks or spots that sometimes form a vague reticulate pattern. A light line extends from each eye to the tip of the snout along a raised ridge—the canthus rostralis—and is often paralleled by a faint gray to bold black line below. The chin and sides of the lower jaw of southern Appalachian races are often brightly mottled with black and white. Brandon and Rutherford (1967) report that 2-3% of individuals in a West Virginia cave are albinos. There is no conspicuous sexual dimorphism and males lack a well-defined mental gland. Adults measure 11-21 cm TL and there are 17-19 costal grooves.

Hatchlings have stream-type morphology with dorsal fins that terminate near the rear limbs. Hatchlings range in size from 18-22 mm TL in the southern Appalachians (Bruce 1978a) to 26 mm TL in New York (Bishop 1924). Larvae have long, truncated snouts and small eyes relative to head size. The dorsal ground color varies from light yellowish brown to light gray or lavender and frequently has fine reticulations or flecking that is most conspicuous on older animals. Larvae lack the conspicuous dorsal light spots seen in many *Eurycea* and *Desmognathus* species and the dorsal dark dots that are typical of *Pseudotriton* species. Juveniles generally resemble the adults, but are more brightly colored.

Fig. 176. *Gyrinophilus p. danielsi;* Avery Co., North Carolina (R. W. Barbour).

SYSTEMATICS AND GEOGRAPHIC VARIATION. Genetic variation has not been examined, but researchers will probably find significant regional differentiation based on the extent to which populations differ in coloration and life history parameters. A small and a large race of *G. porphyriticus* from the Cowee Mountains, western North Carolina, are sexually incompatible because of ethological barriers (Beachy 1996). Paired mating trials suggest that these parapatric forms may be genetically isolated to a large degree.

Four subspecies are currently recognized, and intergrades that are difficult to assign to subspecies occur over broad geographic areas (Brandon 1966b, 1967c). The northern spring salamander (*G. p. porphyriticus*) has dark pigments above that form a mottled or reticulate pattern and lacks distinct spots or flecks. The Kentucky spring salamander (*G. p. duryi*) is salmon colored and has small black spots and flecks on the back and upper sides that are concentrated dorsolaterally. The light stripe from the eye to the nostril is indistinct and is not bordered by black. The Carolina spring salamander (*G. p. dunni*) has profuse flecking on the back and sides and a pronounced light stripe that is bordered by black. The Blue Ridge spring salamander (*G. p. danielsi*) is similar to *G. p. dunni* but has distinct spots rather than flecking on the dorsum. Specimens from high elevations usually have conspicuous black and white mottling along the lower jaw. Brandon (1966b) provides a comprehensive analysis of geographic variation in this species.

DISTRIBUTION AND ADULT HABITAT. Spring salamanders occur from 100 to 2000 m in elevation from southern Quebec and southern Maine to central Alabama in formations associated with the Appalachian uplift. This species is most abundant in springs, seepages, caves, and headwater sections of small tributaries that lack predatory fish (Bishop 1941a; Bruce 1972a; Wright and Haber 1922). Populations in the Piedmont of South Carolina are restricted to springs and small streams in deep ravines with mature hardwood forests (Bruce 1972a). In more mountainous regions in western North Carolina specimens can be found in or near springs, seepages, small streams, and wet roadside ditches. Larvae are uncommon in caves in Alabama but frequently inhabit caves in Kentucky, Virginia, and West Virginia (Cooper and Cooper 1968; Green and Brant 1966).

BREEDING AND COURTSHIP. Males examined by Bruce (1972a, 1978a) from several populations in North Carolina contain sperm in their vasa deferentia from October through June, a finding that suggests autumn and spring mating seasons. Bishop (1941a) believed that mating in New York populations occurs primarily in the autumn, but Dieckmann (1927) collected females in New York with moderate to abundant sperm in their spermathecae in June, July, August, and November. Sperm is absent from the vasa deferentia of males from June through mid-September. Most females in the southern Appalachians and vicinity oviposit annually, and sex ratios are near 1:1 (Bruce 1969, 1972a).

Bishop (1941a) observed a male and female in a spring in New York that were engaged in an aggressive encounter in which both were biting and wrestling each other. It is uncertain if this reflected courtship behavior, aggressive displays, or an attempt to cannibalize. Individuals engage in a tail-straddle walk (Beachy 1996), but detailed descriptions of courtship behavior are not available.

REPRODUCTIVE STRATEGY. Most females probably lay their eggs in deep underground recesses in streams or seeps as evidenced by the fact that so few nests have been found. Females attach their eggs in monolayers to the undersides of rocks or other objects. When depositing eggs, a female turns upside down and uses her arched body and tail to brace against the substrate (Noble and Richards 1932). As each egg is extruded, the female pushes her cloaca to the substrate and undulates her tail. She then shifts her position slightly and lays another. Mature ova are light yellow,

Fig. 177. *Gyrinophilus p. duryi;* northeastern Kentucky (R. W. Barbour).

Range of *Gyrinophilus porphyriticus*

- G. p. danielsi
- G. p. dunni
- G. p. duryi
- G. p. porphyriticus

about 3.5–4.0 mm in diameter, and surrounded by three jelly coats (Bishop 1941a; Bruce 1969, 1972a).

Most females oviposit during the summer and the embryos hatch in late summer or autumn (Bruce 1972a, 1978a, 1980; Green 1925; Organ 1961c). Females in New York populations begin ovipositing as early as May (Bishop 1941a), but most nest records are for the summer months. In Pennsylvania, 15 eggs and an attending adult were found on 8 August (Green 1925). The eggs were in early developmental stages and were attached singly to the underside of a submerged stone in a spring. Organ (1961c) discovered two egg masses and attending females in a small spring in a spruce-fir forest in southwestern Virginia. One clutch, found on 14 July, contained 41 eggs in intermediate stages of development, whereas the second, found on 6 August, contained 66 eggs in advanced developmental stages. In both instances the eggs were attached singly in a monolayer to the underside of a large rock that was partially embedded in the streambank. The first clutch covered a circular area about 18 cm in diameter, and the second an oblong area that measured 8 × 20 cm.

In North Carolina 24 eggs at the tail bud stage were found with an attending female in a small rivulet on 13 July (Bruce 1978a). The eggs were attached singly to the underside of a rock buried about 25 cm below the stream surface. A second clutch along with a female was found near Mt. Mitchell on 22 October (Bishop 1924). The eggs were near hatching and were attached singly to the underside of a large rock in relatively still water.

Populations in the Piedmont and Blue Ridge Province of South and North Carolina differ markedly in clutch size, which is related in part to differences in female body size (Bruce 1969, 1972a). Mean numbers of mature ova in populations in western North Carolina are 39-63, and individual clutches contain 16-106 ova. Clutch size is positively correlated with female SVL. At a common SVL, females in high-elevation populations tend to have lower fecundities than those at lower elevations. Regional populations in the Cowee Mountains, North Carolina, also differ in adult body size, age at maturity, and clutch size (Bruce 1978a).

Hatching occurs in the late summer or autumn. Larvae as small as 23-24 mm TL have been found in mid-April and early May in Virginia, but these presumably hatched the previous fall (Organ 1961c).

AQUATIC ECOLOGY. Larvae are most often collected in gravel beds and beneath stones and logs in springs, seepages, or spring-fed creeks, but they have occasionally been taken from lakes in New York (Bishop 1941a). Larvae are secretive during the day and reside in subterranean haunts far below the surface of streambeds (Bruce 1980). At night individuals often emerge from daytime cover and forage on the streambed (Resetarits 1991).

Stomach contents of larvae from a North Carolina population include oligochaetes, spiders, isopods, crayfish, centipedes, odonates, mayflies, stoneflies, caddisflies, flies, salamander eggs, adult and larval *Eurycea bislineata,* and adult *Desmognathus ocoee* (Bruce 1979). *Gyrinophilus porphyriticus* larvae in New Hampshire populations feed on two-lined salamander larvae and a variety of aquatic invertebrates, including beetles, caddisflies, and dipterans (Burton 1976).

Analyses of feeding behavior suggest that larvae use mechanoreception to locate prey (Culver 1973). Larvae do not respond to dead prey, but rise up on their limbs in a feeding posture when live prey are placed in tanks. Isopods respond to *G. porphyriticus* larvae by remaining motionless, a defense that reduces their susceptibility to predation.

In North and South Carolina there is no relationship between the seasonal surface activity of larvae and the elevation of local populations (Bruce 1972a). Specimens can be found in aquatic sites during every month of the year. Although larvae are generally far less abundant than those of other plethodontids such as *Desmognathus* and *Eurycea* species, densities in some

Fig. 178. *Gyrinophilus p. porphyriticus;* mature larva; Grayson Co., Virginia (J. W. Petranka).

Virginia streams may reach 5-10 larvae/m^2 of streambed (Resetarits 1991, 1995).

Estimating length of the larval period is difficult for spring salamanders because larvae have variable growth rates and are difficult to collect in large numbers. The best attempt to date is that of Bruce (1980), who estimates a modal length of around 4 years for a population in western North Carolina. Bishop (1941a) and Weber (1928) conclude that the larval period lasts 3 years in New York populations; however, their analyses are based on small data sets and may not be reliable. Bruce (1972a, 1978a, 1980) found that most larvae transform when they reach 55-65 mm SVL in low-elevation (<1200 m) populations and 61-70 mm SVL in montane (>1200 m) populations. Transformation occurs from late June through late August. Larvae in New York populations transform from March through October (Bishop 1941a).

TERRESTRIAL ECOLOGY. Adults are most frequently found beneath surface or subsurface objects in or near springs or seepages, or on roads during rainy nights. Adults move to subsurface retreats in or near springs and seepages during the winter as well as in summer, when surface flow is greatly reduced (Bishop 1941a). During the warmer months, adults in the southern Appalachians are occasionally found on the forest floor at night far from running water or seepages.

The seasonal duration and extent of surface activity of metamorphosed animals in North and South Carolina increase with elevation (Bruce 1972a). Proportionately few juveniles and adults occur in lowland samples from the Piedmont, and individuals are active on the ground surface from about mid-October to late March. At intermediate elevations in the Blue Ridge Embayment, metamorphosed individuals compose a greater proportion of samples, and juveniles and adults are active on or near the ground surface between mid-September and late June. In high-elevation populations in the Nantahala Mountains of western North Carolina, metamorphosed salamanders tend to be active near the surface during all months of the year.

Males at low to intermediate elevations in the southern Appalachians become sexually mature immediately after metamorphosing, whereas those at higher elevations require as much as a year longer to mature after metamorphosing (Bruce 1972a). The minimum size at maturity varies substantially among populations. In a group of lowland populations, all males >55 mm are sexually mature. In a second group from higher elevations, some juveniles may not mature until they exceed 81 mm SVL.

Females exhibit similar trends and manifest delayed development in high-elevation populations. Some females in low-elevation populations mature shortly after transforming, but those in high-elevation populations may not mature for 1 year or more after metamorphosing. Depending on the locality, females may mature when as small as 61 mm SVL or as large as 82 mm SVL. Bruce (1979) hypothesized that size at metamorphosis is influenced evolutionarily by the availability and size spectrum of food resources available to the larval and adult stages.

Gyrinophilus porphyriticus is notorious for its habit of eating other salamanders (Bishop 1941a; Wright and Haber 1922), but the extent to which adults feed on salamanders varies geographically. Salamanders are an important component of the diet in the southern Appalachians, where the surface density of salamanders is often very high (Bruce 1972a, 1979; Huheey and Stupka 1967; Martof 1955). About half of the spring salamanders with identifiable remains from montane populations in North Carolina contained salamanders, including *E. bislineata, D. ocoee, Plethodon jordani, P. serratus, Pseudotriton ruber,* and a conspecific (Bruce 1972a). The remainder of the diet consists of invertebrates, including large earthworms, beetle larvae, slugs, and snails. In a related study adults were found to feed on oligochaetes, isopods, spiders, centipedes, insects, adult and larval *E. bislineata,* and adult *D. ocoee* and *Plethodon* species. Other salamanders taken include *P. oconaluftee* and *D. wrighti* (Huheey and Stupka 1967; King 1939). Salamanders compose a larger component of the diet in metamorphosed *G. porphyriticus* than in larvae (Bruce 1979).

In northern populations the adults feed primarily on invertebrates, including both aquatic and terrestrial taxa. Prey items in 26 Pennsylvania specimens included flies, beetles, earthworms, snails, and myriopods (Surface 1913), whereas those in 80 New York specimens included mayflies, caddisflies, stoneflies, beetles, flies, hymenopterans, hemipterans, earthworms, spiders, centipedes, millipedes, snails, and mites (Bishop 1941a). Two specimens of *G. porphyriticus* contained *D. fuscus* and *G. porphyriticus* larvae. Hamilton (1932) lists similar invertebrate prey in 26 specimens from New York, as well as an occasional specimen of *Desmognathus,* but Burton (1976) lists only invertebrates for 25 New Hampshire specimens.

Gyrinophilus porphyriticus is a gape-limited predator. Feeding trials indicate that adults can handle small to medium-sized *D. ocoee* but cannot consume large adults (Formanowicz and Brodie 1993). *Gyrinophilus porphyriticus* is more efficient than *D. quadramaculatus* at capturing *D. ocoee.*

PREDATORS AND DEFENSE. Spring salamanders are occasionally eaten by northern water snakes (*Nerodia sipedon*) and common garter snakes (*Thamnophis sirtalis;* Uhler et al. 1939). When attacked, individuals often posture with the head tucked beneath the body and the tail raised and undulated. Adults produce noxious skin secretions that repel shrews (Brodie et al. 1979). The spring salamander is brightly colored and noxious, and may be part of a Müllerian mimicry complex involving species of *Pseudotriton* and

Fig. 179. *Gyrinophilus p. porphyriticus;* defensive posturing during shrew attack; Catskill Mountains, New York (E. D. Brodie, Jr.).

Notophthalmus viridescens. Details of this mimicry complex are presented under the account of *P. ruber*.

COMMUNITY ECOLOGY. Although *G. porphyriticus* larvae reach their highest densities in stream sections without fish, they often coexist with predatory fish. Resetarits (1991) examined interactions of spring salamander larvae with brook trout and two-lined salamanders in artificial streams. Under these conditions, *G. porphyriticus* has lower survival and growth in the presence of adult trout than when growing alone. Larvae do not respond to trout by moving into refuges and so are vulnerable to predation. The presence of *G. porphyriticus* does not significantly reduce *E. bislineata* survival; however, *E. bislineata* larvae have slower growth because they reduce their surface activity at night. In a related study, Resetarits (1995) reports similar trends for growth and survival when using fingerling trout in artificial stream experiments, and demonstrates that *G. porphyriticus* larvae use shallower microhabitats in the presence of trout.

Gustafson (1994) examined ecological interactions between *E. bislineata* larvae and small and large *G. porphyriticus* larvae in experimental streams and pools. Both sizes of *G. porphyriticus* larvae prey upon *E. bislineata*, and *E. bislineata* is less active on the surface when large *G. porphyriticus* are present compared with smaller ones. Large *G. porphyriticus* reduce the growth of smaller conspecifics, but the reverse does not occur. Beachy (1994) found that *E. bislineata* has lower survival in the presence of predatory *G. porphyriticus* and *D. quadramaculatus* larvae; however, *G. porphyriticus* is the more effective predator. Although both species share a common resource, there is no evidence that *G. porphyriticus* and *D. quadramaculatus* compete for food.

CONSERVATION BIOLOGY. Like many eastern species with mountain affinities, *G. porphyriticus* reaches its highest densities in protected watersheds with hardwood forests. Deforestation is perhaps the single most important factor in eliminating local populations.

Gyrinophilus subterraneus Besharse and Holsinger
West Virginia Spring Salamander
PLATES 90, 91

IDENTIFICATION. *Gyrinophilus subterraneus* is a large, troglobitic species of *Gyrinophilus* that is known from a single cave system in southeastern West Virginia. It coexists microsympatrically with *G. porphyriticus* at the type locality. The two species are morphologically similar, but *G. subterraneus* differs from *G. porphyriticus* is several ways. Relative to those of the latter species, larvae of *G. subterraneus* reach a larger body size and have paler, more fleshy-colored skin with dark reticulations. Larvae of *G. subterraneus* also have smaller eyes, a wider head, and more premaxillary and prevomerine teeth. Larvae may reach 112 mm SVL, and large larvae usually have two or three irregular rows of pale yellow spots laterally. The adults retain the pale reticulate pattern and reduced eyes of the larvae and have an indistinct canthus rostralis (Besharse and Holsinger 1977). Adults measure 10–18 cm TL.

SYSTEMATICS AND GEOGRAPHIC VARIATION. This species was described in 1977 (Besharse and Holsinger 1977), and its taxonomic validity was immediately questioned by Blaney and Blaney (1978), who argued that *G. subterraneus* is an extreme variant of *G. porphyriticus*. Very limited electrophoretic data (Green and Pauley 1987) suggest that *G. subterraneus* is distinct from *G. porphyriticus*.

Fig. 180. *Gyrinophilus subterraneus*; transformed adult; Greenbrier Co., West Virginia (R. W. Van Devender).

Range of *Gyrinophilus subterraneus*

DISTRIBUTION AND ADULT HABITAT. This species is restricted to General Davis Cave, Greenbrier Co., West Virginia. Besharse and Holsinger (1977) collected adults along muddy banks near the streamside and larvae from the stream proper.

COMMENTS. Most aspects of the life history of this very rare species are not documented. Larvae transform when they reach >95 mm SVL (Besharse and Holsinger 1977). The largest larvae in samples are sexually mature, or nearly so, but it is uncertain if individuals reproduce as gilled adults.

Fig. 181. *Gyrinophilus subterraneus;* mature larva; Greenbrier Co., West Virginia (R. W. Van Devender).

Haideotriton wallacei (Carr)
Georgia Blind Salamander
PLATE 92

Fig. 182. *Haideotriton wallacei;* juvenile; Jackson Co., Florida (R. W. Van Devender).

IDENTIFICATION. This small, permanently gilled troglobitic species is readily distinguished from all other urodele species east of the Mississippi River by its pale body, slender legs and greatly reduced eyes. The body is slender, pinkish white, and covered by scattered, inconspicuous melanophores dorsally and laterally (Brandon 1967d; Carr 1939; Means 1992b). The head is broad, but not flattened, and the eyes appear as minute dark dots. Adults measure 51–76 mm TL and there are 12–13 costal grooves. Specimens collected by Lee (1969a) and Peck (1973) measured 14–30 mm SVL. Transformed specimens have never been collected and larvae show little response to metamorphic agents (Dundee 1962).

DISTRIBUTION AND ADULT HABITAT. The Georgia blind salamander inhabits subterranean streams and aquifers in limestone formations in the Dougherty Plain region of southwestern Georgia and adjoining portions of Florida. The holotype was collected from an artesian well in 1939, and specimens have since been collected from streams or isolated pools in caves in Decatur Co., Georgia, and Jackson Co., Florida.

COMMENTS. So few specimens of this rare species have been collected that little is known about its natural history. Individuals typically rest on bottom sediments in underground streams or pools, but occasion-

Range of *Haideotriton wallacei*

Fig. 183. *Haideotriton wallacei*; Jackson Co., Florida (R. W. Van Devender).

ally climb on limestone sidewalls, where they move over ledges or vertical faces. Pylka and Warren (1958) observed specimens walking slowly about the bottoms of small, silt-covered pools in a Florida cave. Individuals show no response to light, suggesting that they are blind. Gravid females have been collected in May and November, and breeding may be aseasonal (Means 1992b).

Brandon (1971a) examined 14 specimens ranging from 13 to 28 mm SVL and all were sexually immature; specimens as large as 43 mm TL collected by Pylka and Warren (1958) were also sexually immature. The holotype described by Carr (1939) is a sexually mature female that measures 44 mm SVL and 77.5 mm TL and contains eggs 2.0–2.2 mm in diameter (Brandon 1971a).

Haideotriton wallacei probably relies on tactile or olfactory cues when feeding. Larvae will not attempt to capture prey until the prey come into very close contact (Pylka and Warren 1958). Ostracods and amphipods were the most numerically important prey in 32 specimens examined by Lee (1969a) from a Florida cave. Other prey include isopods, copepods, a mite, and beetles. A small number of amphipods, copepods, and ostracods are listed as dietary items from eight other specimens (Peck 1973).

Haideotriton wallacei is currently protected by state law and cannot be collected without special permits.

Hemidactylium scutatum (Temminck and Schlegel)
Four-toed Salamander
PLATE 93

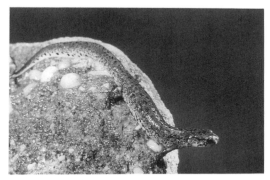

Fig. 184. *Hemidactylium scutatum*; adult; eastern Kentucky (R. W. Barbour).

IDENTIFICATION. The four-toed salamander is a small plethodontid that is readily identified by the presence of only four toes on each hindfoot, a constriction around the base of the tail, and a white venter with bold black spots or blotches. The dorsum is rusty brown above and the sides of the body are grayish.

The tail of adults makes up about 57% of the total length on average and is proportionately shorter in females (Berger-Bishop and Harris 1996; Blanchard and Blanchard 1931). Sexually active males have truncated snouts (rounded in females) and enlarged premaxillary teeth that are visible when the mouth is shut (Bishop 1941a). Females measure about 15% greater in SVL on average than males (Bishop 1941a; Blanchard and Blanchard 1931). Adults measure 5–10 cm TL and there are 13–14 costal grooves.

Hatchlings reach 11–15 mm TL and have front limbs with well-formed toes and rear limbs with toe buds (Bishop 1919, 1941a). In some instances hatchlings may not emerge from the egg capsules until the rear toes are fully formed (R. N. Harris, pers. comm.). The larvae have pond-type morphology and are rather nondescript, with a mottled yellowish brown dorsum, a slender build, prominent eyes, and a dorsal fin that extends forward nearly to the head (Bishop 1919). Small juveniles generally resemble the adults but have

Range of *Hemidactylium scutatum*

proportionately shorter tails (Berger-Bishop and Harris 1996).

SYSTEMATICS AND GEOGRAPHIC VARIATION. Detailed analyses of geographic variation in genetic and phenotypic characteristics have not been conducted, and the extent to which geographic isolates differ genetically from contiguous populations is not known.

DISTRIBUTION AND ADULT HABITAT. The four-toed salamander ranges from Nova Scotia southward to the Gulf of Mexico and westward to Oklahoma, Missouri, and Wisconsin. Populations are discontinuous in many areas in the southern and southwestern portions of the range. Juveniles and adults inhabit forests surrounding swamps, bogs, marshes, vernal ponds, and other fish-free habitats that serve as breeding sites. Adults in northern locales are often collected outside the breeding season beneath cover objects on the forest floor, but those in the southern portion of the range tend to be more fossorial. During the breeding season, females are most easily obtained by searching moss mats or turning over cover objects along the margins of ponds and bogs.

BREEDING AND COURTSHIP. Females mate in the autumn and possibly the early winter months prior to migrating to breeding ponds. In Michigan females can be found in late October and November with spermatozoa in their cloacae, and adults collected in September and October will court soon after being returned to the laboratory (Blanchard 1933a). Specimens collected in Michigan on 2 November deposited over 120 spermatophores through 2 December (Branin 1935). Sex ratios of a large series of juveniles collected in Michigan are close to 1:1, with females slightly outnumbering males (Blanchard 1935). Data on the mating seasons of southern populations are not available.

Four-toed salamanders engage in a tail-straddle walk as do other plethodontids. The following account of courtship in laboratory specimens is from Branin (1935). A male initiates courtship by showing heightened activity characterized by active nosing of conspecifics. A male that encounters a female during this stage will often circle about with the tail bent sharply at a right angle to the body. The female eventually straddles the male's tail and presses her chin against the dorsal surface of the base of his tail. The male then moves forward and undulates his tail from side to side. The female follows while maintaining the same relative position and tracks the movement of the tail with her chin. The tail-straddle walk may last for 20 minutes. Branin (1935) observed females pausing over spermatophores for several minutes, but did not witness spermatophore pickup. The spermatophore consists of an expanded, gelatinous disk about 2–2.3 mm wide that tapers upward into a 1-mm-diameter stalk that supports a pale yellowish sperm cap. The entire spermatophore is around 2.5 mm tall (Blanchard 1933a).

REPRODUCTIVE STRATEGY. Data from drift fence captures indicate that females in most populations migrate from their winter retreats to nesting sites shortly before ovipositing in the spring. Peak migrations of females to a Michigan breeding site occur during the last week of March and the second week of April, but small numbers of gravid females arrive at drift fences through the second week of May (Breitenbach 1982). Migrating females have been observed during the second week of April in New York (Bishop 1941a) and arrive at drift fences in Virginia ponds in early April (Harris and Gill 1980). In Arkansas, females have been found moving toward ponds in October, November, February, and March, suggesting that autumnal migrations may occur in some southern populations (Saugey and Trauth 1991).

Field observations indicate a seasonal progression of egg laying from southern to more northern latitudes or from coastal to cooler, inland areas. Nesting records include a single nest in Alabama on 24 February (Mount 1975), eggs as early as 24 March in Maryland (Cooper 1955), eggs in Missouri in early April (Easterla 1971), and eggs in Ohio from 6 April to 22 May (Daniel 1989). Oviposition in southern Michigan occurs from mid- to late April (Blanchard 1934b) and in New York from mid-April through mid-May (Bishop 1919, 1941a). In Virginia eggs have been found from mid-February through the second week of April in the Coastal Plain and Piedmont (Goodwin and Wood 1953; Wood 1953b) and from April through mid-May in mountainous regions to the west (Harris and Gill 1980).

The egg-laying period in any population lasts 2–6 weeks, but most oviposition is confined to a 2- to 3-week period (Blanchard 1923; Harris and Gill 1980; Wood 1955). Females typically nest slightly above the water line in clumps of mosses at the margins of swamps, bogs, marshes, vernal ponds, or sluggish streams (Bishop 1919, 1941a; Blanchard 1923; Gilbert

Fig. 186. *Hemidactylium scutatum;* typical nesting site (this moss clump contained a joint nest with 13 females in the process of ovipositing); Bell Co., Kentucky (J. W. Petranka).

Fig. 185. *Hemidactylium scutatum;* nesting female with eggs; Bell Co., Kentucky (J. W. Petranka).

1941; Wood 1955). Raised moss mats on islets within swamps or bogs are common nesting sites (Breitenbach 1982; Wood 1955), as are steep, moss-covered banks next to the water's edge. *Sphagnum* mosses seem to offer optimal microhabitats for egg deposition (Bishop 1941a; Breitenbach 1982; Wallace 1984), but females readily nest in other mosses, such as *Thuidium, Mnium,* and *Climacium,* that form thick mats or carpets (Gilbert 1941; Wood 1953b, 1955). Many nesting sites that I have observed in Alabama, eastern Kentucky, North Carolina, and eastern Tennessee lack *Sphagnum,* and females nest in moss clumps of other species.

Although moss clumps are preferred nesting habitats, other microhabitats or nesting substrates are sometimes used. Somewhat unusual nesting sites include a moss mat along a wet cliff face 1.5 m above the floor of a gorge in Kentucky (Green 1941) and moss mats along the margin of a small woodland stream in Missouri (Easterla 1971). Small, fishless forest streams and springs with moss mats appear to be the preferred habitat of this species in Arkansas (Saugey and Trauth 1991). Females will occasionally nest in microhabitats other than moss mats. Eggs have been found under loose bark and in leaf litter, rotted wood, clumps of grasses and rushes, and mounds of pine needles (Blanchard 1922; Breitenbach 1982; Harris and Gill 1980; Wood 1955). The preferred nesting site in a New York swamp is loose moss mats that cover fallen logs in the water (Gilbert 1941).

Females construct crude cavities or use existing cavities either below or several centimeters within moss mats (Bishop 1941a; Gilbert 1941; Wood 1955). When ovipositing, a female usually turns upside down and deposits her eggs attached singly to rootlets, moss strands, or other substrates in a loose cluster within the cavity. Several minutes are required to deposit a single egg (Bishop 1941a), and 12-72 hours are required to deposit an entire clutch (Harris et al. 1995). Freshly laid eggs are 2.5-3.0 mm in diameter, very lightly pigmented above, and surrounded by two envelopes (Bishop 1919, 1941a; Wood 1953b). The eggs have sticky outer coats and readily adhere to debris or other eggs within the nest.

Females frequently share moss clumps and lay eggs in joint nests (Bishop 1919; Blanchard 1934a; Breitenbach 1982; Goodwin and Wood 1953; Harris and Gill 1980; Wood 1953b). Joint nests in Michigan may contain as many as 1110 eggs from 30-35 females (Blanchard 1934a). The majority of females seem to arrive at joint nests fairly synchronously (Blanchard 1934a), although sequential arrivals of females do occur (Harris and Gill 1980). Estimates of the percentage of joint nests in different populations are 12% (Breitenbach 1982) and 61% (Blanchard 1934a) in Michigan, 6% in New York (Gilbert 1941), and 41% (Harris and Gill 1980) and 40% (Wood 1953b) in Virginia. (The last estimate is conservative and is based on the assumption that nests with >80 eggs are joint.)

Joint nesting decreases the reliability of clutch size estimates based on counts of eggs in nests. In New York, estimates are 22-64 (mean = 50; Bishop 1941a) and 12-65 (mean = 24; Gilbert 1941) eggs per nest. Estimates of mean clutch size based on ovarian egg counts are 42 (range = 27-57) for 31 Arkansas females (Trauth et al. 1990), 29 (4-46) for 217 Michigan females (Blanchard 1936), and 47 (range = 29-80) for 32 Virginia females (Wood 1953b). In all three studies clutch size is positively correlated with female SVL.

One unusual feature of the brooding behavior of *H. scutatum* is that only one or two females typically remain with the eggs in a joint nest during the first few weeks after oviposition (Blanchard 1934a; Breitenbach 1982; Harris and Gill 1980; Wood 1953b). In detailed studies of two Virginia populations, Harris and Gill (1980) and Harris et al. (1995) found that brooding enhances embryonic survival, but survival does not differ among solitary versus joint nests that are brooded. There is no relationship between brood size and embryonic survivorship, even though nests are rarely guarded by more than one female. Females will desert nests if they are flooded following heavy rains. Two estimates of average survivorship to hatching after flooding are 9% and 21%.

Harris et al. (1995) evaluated several hypotheses concerning nesting behavior in *H. scutatum*. Females at their Virginia site either lay eggs in solitary nests and brood them, lay eggs in joint nests and brood them, or lay eggs in joint nests and leave. Most joint nests have one brooder, which most frequently is the first female that oviposits. The number of days that a female broods is positively correlated with loss of body mass, suggesting that brooding females incur an energetic cost. Both field data and the results of a seminatural experiment indicate that the frequency of joint nesting does not increase with population density, and that joint nesting is not due to a shortage of nesting sites. However, the frequency of females that lay and leave joint nests increases with population density. Nesting females are not aggressive to one another, and there is little evidence of oophagy. This observation suggests that brooding of joint nests by a single female is not due to brood parasitism or aggressive usurpation. A more likely explanation is that joint nesting and brooding by individual females are the result of kin selection.

In some cases, all females will desert a nest site. In Michigan 0–44% of the nests in early developmental stages are unattended when initially uncovered, and all are abandoned prior to hatching (Breitenbach 1982). Most females in Virginia populations also desert their nests before the embryos hatch (Wood 1953b). Others (Bishop 1941a; Blanchard 1934a) report that females remain with the eggs through hatching. Brooding females sometimes eat eggs, but it is uncertain if the embryos are dead or alive (Harris and Gill 1980). The stomachs of 243 brooding females from Virginia were empty even though there were many dead eggs in the nests (Wood 1953b).

The incubation period averages 38 days in Michigan populations (Blanchard 1923). Other estimates include 52–62 days for clutches in New York (Bishop 1941a) and 61 days for a clutch in Virginia (Wood 1955). The time of hatching varies depending on breeding dates and local site conditions. Hatchlings or late-term embryos have been found in late May in Michigan (Blanchard 1923) and between 19 May and 2 June in New York (Gilbert 1941).

AQUATIC ECOLOGY. Surprisingly few data are available on the ecology and natural history of the larvae. Shortly after hatching, the larvae wiggle through moss mats or other surface cover and enter the ponds (Blanchard 1923). They feed on small zooplankton and other invertebrates in the ponds and have very brief larval periods. The larval period in Michigan populations lasts about 6 weeks, with transformation occurring in July when larvae reach 18–24 mm TL (Blanchard 1923). Other estimates are 39 days (Bishop 1941a) and 21 days (Berger-Bishop and Harris 1996) for natural populations.

Berger-Bishop and Harris (1996) examined the effects of larval density and food level on relative tail length and larval growth parameters. Their data indicate that relative tail length varies with the developmental stage of larvae, but food level and larval density have little effect on relative tail length at any developmental stage. Larvae established at low and high experimental densities do not differ in relative tail length, SVL at metamorphosis, mass at metamorphosis, or length of the larval period. As in natural populations, the larval period is very brief (means of 23–27 days and 38–39 days for two experiments) and larvae transform when they measure only 11–14 mm SVL and 17–25 mm TL on average.

Based on the frequency of submissive and aggressive behaviors, behavioral trials suggest that larvae are not capable of recognizing kin or familiar conspecifics (Carreno et al. 1996). Larvae are not strongly aggressive and rarely bite or lunge at one another.

TERRESTRIAL ECOLOGY. Juveniles disperse away from breeding ponds within a few weeks after transforming and live in the surrounding forests until sexually mature. In Michigan, both sexes become sexually mature about 28 months after hatching, when males reach 49–57 mm TL and females reach 62–68 mm TL (Blanchard and Blanchard 1931). In Virginia populations, individuals become sexually mature roughly 2 years after hatching (Wood 1953b). The smallest brooding females measure 29 mm SVL.

Adults and juveniles may congregate with other amphibians in late autumn in or near overwintering sites. On separate occasions in November, Blanchard (1933b) found 18 *H. scutatum* in or adjacent to a rotten log, and over 200 in cavities or depressions under leaf litter in an area of <2 m^2. Another aggregate consisting of 53 juveniles and 2 adults was found on 16 November while snow was falling. Specimens are rarely found during the colder months after mid-November and presumably spend the winter in subsurface retreats.

Adults maintained in the laboratory are weakly aggressive toward conspecifics. This observation suggests that individuals may defend burrows or cover objects during the warmer months of the year (Grant 1955).

The only dietary study is that of Bishop (1919), who found beetles, lepidopterans, spiders, mites, and bristletails in transformed New York specimens.

PREDATORS AND DEFENSE. Although predation is poorly documented, four-toed salamanders are probably preyed upon by numerous vertebrates, such as shrews, small woodland snakes, and birds. Trauth and Cochran (1991) found a female *H. scutatum* that had been bitten by a pygmy rattlesnake (*Sistrurus milarius*).

When attacked or molested, *H. scutatum* often coils tightly, tucks the head beneath the tail, and remains immobile (Bishop 1919; Brodie 1977; Brodie et al.

1974b). An individual also may raise and undulate its tail while exuding noxious secretions and, in extreme cases, autotomize the tail at the constriction near the base. The tail wiggles after being autotomized and distracts would-be predators from attacking the animal (Bishop 1941a). Four-toed salamanders can voluntarily autotomize their tails. In contrast, the tails of almost all other salamanders must be grasped by an attacker before they will break off. Most specimens that Green and Pauley (1987) inadvertently left in a collecting bag in the sun autotomized their tails. Bishop (1919) induced an animal to autotomize its tail by applying acetic acid.

COMMUNITY ECOLOGY. Four-toed salamanders often share breeding habitats with ambystomatid salamanders such as *Ambystoma maculatum* and *A. opacum* and may coexist with terrestrial species such as *Plethodon cinereus* and *P. glutinosus* outside the breeding season. Ecological interactions between *H. scutatum* and these species have not been investigated. My colleagues and I found that the larvae are palatable to fish and lack any obvious defenses against predatory fish (Kats et al. 1988). This finding may explain why four-toed salamanders almost never use breeding habitats that are populated by fish.

CONSERVATION BIOLOGY. *Hemidactylium scutatum* has rather specialized breeding requirements that make it vulnerable to habitat disturbance. Populations thrive in areas with mature hardwood forests containing suitable breeding sites. Although the phenomenon is poorly documented, this species has undoubtedly declined markedly throughout its range because of the loss of vernal ponds, bogs, and other wetlands associated with land clearing and development (e.g., Daniel 1989).

A canopy of mature deciduous trees helps retain moisture on the forest floor and encourages the growth of moss mats around pond margins. Mature forests also have large amounts of fallen woody debris and organic soils that offer optimal habitats for adults outside the breeding season. Land managers or landowners interested in constructing vernal ponds to protect local populations of *H. scutatum* should consider adding raised hummocks of earth in ponds or raised hills along pond margins to encourage the growth of moss mats and sedges.

Pseudotriton montanus Baird
Mud Salamander
PLATES 94, 95, 96

Fig. 187. *Pseudotriton m. montanus;* adult; James City Co., Virginia (R. W. Van Devender).

IDENTIFICATION. The mud salamander is a stout-bodied, short-tailed salamander that has an orangish brown to bright crimson dorsum that is usually marked with widely scattered black or brown spots. Populations in Florida and southern Georgia normally are not spotted. The eyes are brown and the tail makes up about 40% of the TL. The venter is light orange to pinkish orange and may or may not have black spotting depending on the subspecies. Adults darken with age, and the dorsal spots are less conspicuous in old animals. Geographic variation in color patterns and body size is described in detail in the subspecies accounts. Adult females measured by Bruce (1975) averaged 19% greater in SVL than adult males. Adults measure 7.5-19.5 cm TL and the mean number of costal grooves is 16-17.

Hatchling *P. m. floridanus* reach 7.5-9.0 mm SVL and 12-13.5 mm TL (Goin 1947), whereas those of *P. m. montanus* reach 10-13 mm SVL (Bruce 1974, 1978b). Hatchlings are the stream type and are light brown above and immaculate below (Goin 1947;

Martof 1975b). Older larvae are light brown above and often have dark, widely scattered spots similar to those of young adults. Populations in the upper Piedmont of North and South Carolina may have a streaked or reticulate pattern, with streaking particularly evident along the sides. Those in the lower Piedmont and Coastal Plain have little or no streaking.

SYSTEMATICS AND GEOGRAPHIC VARIATION. *Pseudotriton montanus* shows marked geographic variation in color pattern and body size. Genetic variation has not been documented but will no doubt show substantial regional differentiation given the low dispersal rates and regional variation that are evident in this species.

Four subspecies are currently recognized (Conant 1975). The eastern mud salamander (*P. m. montanus*) is a large race that reaches a maximum TL of around 21 cm. Juveniles and young adults have a clouded, orangish red to red dorsum with scattered, round black spots. Old adults are usually dark reddish brown and have more numerous but less conspicuous spots. The venter is lighter than the dorsum and is often spotted or flecked with brown. The midland mud salamander (*P. m. diasticus*) is the most brightly colored subspecies. Adults are usually bright crimson and have fewer dorsal spots than those of *P. m. montanus*. The venter is light red and unspotted. The Gulf coast mud salamander (*P. m. flavissimus*) is a diminutive race that is smaller and more slender than the northern subspecies. Adults reach a maximum TL of about 12 cm and have a light orangish brown or reddish brown dorsum with scattered dark spots or flecks. The ground color darkens with age and the spots become obscure. The venter is lighter colored than the dorsum and lacks dark spots.

Fig. 188. *Pseudotriton m. diasticus*; adult; eastern Kentucky (R. W. Barbour).

The rusty mud salamander (*P. m. floridanus*) is also a diminutive, slender, dark-colored race that reaches about 12 cm TL. The dorsum is dark yellowish or orangish brown and lacks dark spots on the body. Scattered dark spots are sometimes present on the tail. The sides of the head and body are mottled, blotched, or streaked with dull yellow coloration. The venter is buff colored and marked with small, irregular dark spots. Zones of intergradation among subspecies are not well documented. Neill (1948) noted that *P. m. flavissimus* and *P. m. montanus* intergrade along the Fall Line in eastern Georgia and western South Carolina.

DISTRIBUTION AND ADULT HABITAT. *Pseudotriton montanus* is absent from most of the higher elevations of the Appalachian Mountains (Huheey and Stupka 1967), and populations occur as two geographic isolates. The western group inhabits the Interior Lowlands and portions of the Appalachian Plateau from southwestern West Virginia and southern Ohio westward to western Kentucky and southward to southeastern Tennessee. The eastern group occurs predominantly in the Coastal Plain and Piedmont from extreme southern Pennsylvania (McCoy 1992) and New Jersey southwestward to Alabama, Mississippi, and Louisiana. Populations of the eastern group are more or less confined to the Coastal Plain in the north, whereas farther south they are common in the Piedmont. Populations extend westward in North and South Carolina to the lower elevations of the southern Blue Ridge (Bruce 1978b; Conant 1957) and northward in Alabama beyond the Coastal Plain provinces.

The mud salamander inhabits muddy or mucky microhabitats in or along the margins of swamps, bogs, springs, floodplain forests, and small headwater tributaries. Populations are primarily found at elevations of <700 m. In central Kentucky populations are often found in rocky, spring-fed headwater tributaries or in floodplains of larger streams that lack muddy substrates (Hirschfeld and Collins 1963; Kats 1986).

Mud salamanders show less of a tendency to disperse away from aquatic breeding and wintering sites than *P. ruber*. However, specimens have occasionally been taken far from breeding sites (e.g., Barbour 1953). Most Alabama specimens have been found beneath logs in low, wooded floodplains (Mount 1975). Specimens in the Coastal Plain of North Carolina are fossorial and are rarely found beneath cover objects in the forest floor (Funderburg 1955).

Range of *Pseudotriton montanus*

BREEDING AND COURTSHIP. Adults mate primarily during the late summer and autumn. Males in South Carolina have vasa deferentia packed with sperm from mid-August through November (Bruce 1975), and adults in breeding condition have been collected in central Kentucky in late September (Robinson and Reichard 1965). Adults in most populations appear to breed annually. Courtship behavior has not been described.

REPRODUCTIVE STRATEGY. Females oviposit during the autumn or early winter and the embryos hatch in winter. The eggs of *P. montanus* have rarely been found, suggesting that females normally nest in cryptic, underground sites in or near aquatic habitats. Measurements of large ovarian eggs suggest that freshly laid eggs are around 3.5 mm in diameter (Bruce 1975). The eggs lack dark pigmentation and are attached to the substrate by a gelatinous stalk about 4 mm long (Brimley 1923; Fowler 1946; Goin 1947).

A group of about 25 eggs of *P. m. montanus* was discovered on 27 December in a small cavity in a seep at the base of a hillside in Maryland (Fowler 1946). The eggs were attached either singly or in groups of as many as six to rootlets that formed the wet walls of the cavity. Developmental stages of the embryos varied from being freshly laid to near hatching, and the eggs were the product of at least two females. Other eggs of *P. m. montanus* were found on 23 December and 1 February in central North Carolina (Brimley 1923). The eggs were bunched together and attached to dead leaves in a spring-fed ditch. Bruce (1974) collected hatchling *P. m. montanus* in the Piedmont of South Carolina in March that measured about 13 mm SVL on average and were estimated to be several weeks old.

A clutch of 27 eggs of *P. m. floridanus* at the hatching stage was found on 14 January in northern Florida in a boggy, *Sphagnum*-filled seepage area (Goin 1947). The eggs were suspended singly by short stalks from rootlets that hung down into the water from the edge of an undercut bank. The eggs were clustered in groups of two to six and were accompanied by a spent female. Netting and Goin (1942) collected a *P. m. floridanus* on 10 April that had numerous large ovarian eggs and surmised (perhaps erroneously) that oviposition occurs in April.

The mud salamander has one of the highest average clutch sizes of North American plethodontids. The mean number of ovarian eggs in 30 females examined by Bruce (1975) varied from 77 to 192, averaged 126, and was positively correlated with female SVL. The only other record is 66 ova in a Kentucky specimen (Robinson and Reichard 1965).

Fig. 189. *Pseudotriton m. montanus;* larva (note fine streaking along side); Watauga Co., North Carolina (R. W. Van Devender).

AQUATIC ECOLOGY. Larvae reside beneath rocks, leaf litter, and aquatic vegetation in seepages, ditches, muddy streams, or floodplain ponds, and they prefer microhabitats with sluggish flow. Hatchlings have conspicuous yolk sacs and probably rely on yolk as a primary energy source for 1 month or more after hatching. Older larvae feed primarily on aquatic invertebrates.

Bruce (1974, 1978b) compared growth and developmental rates of Piedmont and Blue Ridge populations and documented geographic variation in both parameters. Larvae in Piedmont populations metamorphose from mid-May through early September. Most individuals metamorphose 15-17 months after hatching when they measure 35-44 mm SVL; however, some larvae grow an additional year before transforming. Individuals in one population grow 2 mm SVL/month during the first 5 months after hatching (Bruce 1974). Individuals in a nearby Blue Ridge population grow more slowly and most metamorphose after 29-30 months when about the same size as metamorphosing Piedmont animals. Differences in length of the larval period between populations appear to be due primarily to temperature differences among sites. Larvae of *P. m. floridanus* may grow to 38-42.5 mm SVL before transforming (Netting and Goin 1942).

TERRESTRIAL ECOLOGY. The juveniles and adults typically live in muddy habitats next to springs, streams, and swampy sloughs and ponds in bottomland forests. Bruce (1975) observed animals in vertical burrows in soft, damp mud. Many specimens at this site live within 1 m of the water's edge, but some can be found 15-20 m away in the surrounding forest soil. The burrows lead downward into a complex series of water-filled channels that provide access to the surrounding watercourses. Individuals often pose at the entrances to burrows hidden by leaf litter and will rapidly withdraw their heads when exposed. Young juveniles live beneath leaves and debris at the margins of watercourses.

South Carolina males studied by Bruce (1975) become sexually mature within 1 year after transforming and reproduce annually thereafter. Age at first reproduction varies from 20 to 32 months depending on the length of the larval period. Most females first oviposit when about 4 years old. Subsequent reproduction is irregular, with females usually laying annually but occasionally skipping years.

Dietary studies of this species are scant. Dunn (1926) notes that adults may prey on smaller salamanders, including *Eurycea bislineata*.

PREDATORS AND DEFENSE. Mud salamanders are regularly eaten by common garter snakes (*Thamnophis sirtalis*) and water snakes (*Nerodia* spp.; Brown 1979; Carr 1940; Kats 1986). The juveniles and adults have a defensive posture that is somewhat similar to that of *P. ruber* (Brandon et al. 1979b). When molested, individuals assume a coiled position by tucking their snouts against or beneath the body. They may also extend the rear limbs and curl the tail above or around the head. Mud salamanders have been hypothesized to be part of a Müllerian mimicry complex involving the red eft stage of eastern newts, spring salamanders, and *Pseudotriton*. Details of the complex are presented under the discussion of *P. ruber*.

COMMUNITY ECOLOGY. The mud salamander occasionally preys on other salamanders, but the ecological importance of this species in bottomland forest communities is unknown.

CONSERVATION BIOLOGY. Because of its highly fossorial nature and use of silted, muddy habitats, the mud salamander can probably tolerate habitat disturbance better than many eastern salamanders. The status of populations is generally unknown, and long-term monitoring programs are needed to track these and other amphibians in eastern North America.

Pseudotriton ruber (Latreille)
Red Salamander
PLATES 97, 98, 99

Fig. 190. *Pseudotriton r. schencki;* adult; Macon Co., North Carolina (J. Bernardo).

IDENTIFICATION. The red salamander is a relatively large, stout, reddish salamander with a short tail and short legs. The tail of adults averages about 38% of the TL, and the pupils of the eyes are yellow (Martof 1975a). The dorsum and sides of the body vary from purplish brown to bright crimson and are heavily marked with irregularly rounded black spots. The belly varies from pinkish to red and contains scattered black spots in adults. The ventral black spots are sometimes absent in juveniles and young adults. In most areas small juveniles are bright crimson and boldly marked with black spots. Individuals darken with age and old adults are often orangish to purplish brown with fused black spots that are less distinct. In general, individuals cannot be readily sexed using external morphology (Bishop 1941a). Adults measure 9.5–18 cm TL and there are 16–17 costal grooves. Females in southern Blue Ridge populations are on average about 10% larger in SVL than males (Bruce 1978c).

Hatchlings reach 11–14 mm SVL (Bruce 1978c; Pfingsten 1989b; Semlitsch 1983b) and larvae have stream-type morphology. Hatchlings and small larvae are rather uniformly light brown above and dull white below. As larvae mature, the dorsum often becomes weakly mottled or streaked. Some local populations throughout the range and in the upper Piedmont of North and South Carolina are usually spotted and lack streaking. Older larvae retain the light brown dorsal color, whereas those nearing transformation may acquire reddish coloration (Bishop 1941a). Older larvae usually do not develop black dorsal spots or black flecking on the chin as seen in adults, and juveniles acquire the adult color pattern within a few months after metamorphosing.

SYSTEMATICS AND GEOGRAPHIC VARIATION. The red salamander is geographically variable in coloration and color patterning, and four subspecies are recognized. In addition, juveniles and young adults in the Mississippi Embayment are less brightly colored than those in more northern or eastern populations. Martof (1975a) recommended that subspecies not be recognized until a detailed analysis of geographic variation could be conducted; however, most workers continue to recognize the following subspecies.

The northern red salamander (*P. r. ruber*) is red or reddish orange above with black spots. Individuals often have black flecking on the margin of the chin. This is the largest subspecies, reaching a maximum TL of around 18 cm. The Blue Ridge red salamander (*P. r. nitidus*) is similar to *P. r. ruber*, but has little or no spotting on the top of the posterior half of the tail and little or no black flecking on the chin, and reaches a maximum size of only 12 cm TL. The black-chinned red salamander (*P. r. schencki*) generally resembles *P. r. ruber*, but has heavy black flecking under the chin and reaches a maximum TL of about 15 cm. The southern red salamander (*P. r. vioscai*) is generally more dully colored that the other subspecies. The ground color is purplish brown, and tiny white flecks are concentrated on the snout and sides of the head. The dark spots on the dorsum often fuse and tend to form a herringbone pattern. Zones of intergradation are not well documented for most subspecies. *Pseudotriton r. ruber* and *P. r. vioscai* intergrade over a broad region in Alabama and the Piedmont of Georgia (Mount 1975; Neill 1948).

DISTRIBUTION AND ADULT HABITAT. Red salamanders occur from southern New York southwestward to Indiana and southward to the Gulf coast. This species is absent from much of the Atlantic Coastal Plain from Virginia southward to eastern Florida. Pop-

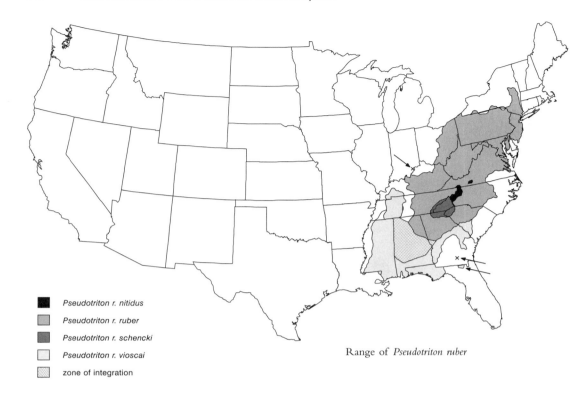

- ■ Pseudotriton r. nitidus
- ▨ Pseudotriton r. ruber
- ▨ Pseudotriton r. schencki
- □ Pseudotriton r. vioscai
- ▨ zone of integration

Range of *Pseudotriton ruber*

ulations occur at elevations ranging from near sea level to greater than 1500 m, although specimens are relatively scarce above 1200 m. Adult *P. ruber* are found in both terrestrial and aquatic habitats in or near small headwater streams, seepages, and spring-fed bogs that serve as breeding sites and overwintering areas. The species is rarely associated with large, swiftly flowing streams. Adults and juveniles are often locally abundant in or near thick accumulations of leaf litter in streams. They also can be found beneath fallen bark, logs,

Fig. 191. *Pseudotriton r. ruber;* adult; Wolfe Co., Kentucky (J. W. Petranka).

rocks, and other surface cover in or near aquatic sites and in adjoining forests, meadows, and pastureland (Bishop 1941a; Bruce 1978c; Mount 1975).

BREEDING AND COURTSHIP. The breeding season is geographically variable and may occur at almost any time except during the coldest months of the year. Adults in southern Blue Ridge populations mate primarily during the summer, and males have sperm-filled vasa deferentia from mid-June through September (Bruce 1978c). Populations in Virginia and Maryland court primarily during the autumn and spring months. Males with cloacal spermatophores have been collected in October, May, and June (Organ and Organ 1968). Most adults breed annually (Bruce 1978c). The ratio of males to females in samples taken by Bruce (1978c) is 1.89, which suggests that males may potentially compete for mates.

Organ and Organ (1968) observed various aspects of courtship behavior in *P. r. nitidus* from Virginia and *P. r. ruber* from New Jersey. Two nearly complete courtship sequences were observed on 22–23 June between *P. r. nitidus* in a terrarium. The following account is based on these observations.

A male initially approaches a female and rubs his snout across her snout, cheeks, and chin. He next moves his head and body beneath her chin and begins to undulate his tail. The female places her chin against the base of the male's tail and straddles his tail. The pair then engages in a tail-straddle walk that lasts for <2 minutes and terminates with the male depositing a spermatophore. During spermatophore deposition the male presses his cloaca to the substrate and rhythmically contracts his cloacal region while violently undulating the tail. The male then arches his tail, flexes it strongly to one side, and leads the female forward over the spermatophore. The female picks up the sperm cap as she moves under it, and the couple separates soon thereafter. Rival males do not appear to be aggressive, and a male normally deposits a maximum of two spermatophores per night.

The spermatophore is about 3.5 mm high, greatly compressed, and wedge shaped. The base is 5.5 × 2.5 mm in diameter and tapers into a colorless, transparent stalk. The white, adhesive sperm cap rests on the top of a stalk that is 3.0 mm long and 1.5 mm wide.

Organ and Organ (1968) note that courtship in *P. ruber* is less complex than in other plethodontids, a finding that further supports the view that *Pseudotriton* is one of the most primitive plethodontids. Males occasionally will court other males and dupe them into depositing spermatophores (Organ and Organ 1968). This behavior presumably reflects sexual interference rather than an inability to recognize members of the opposite sex (Arnold 1977).

REPRODUCTIVE STRATEGY. Females may not lay eggs for several months after courting and are capable of long-term sperm storage. Females oviposit during the autumn or early winter in springs, headwater streams, seepage-fed mountain bogs, and other suitable breeding sites (Bishop 1941a; Bruce 1972b, 1978c; Semlitsch 1983b). Nests have rarely been found, suggesting that most females oviposit in cryptic sites within seepages, springs, or streambanks. The eggs are attached singly by gelatinous stalks to the underside of a rock or other support and are often submerged in water. Each embryo is surrounded by a vitelline membrane and two jelly envelopes, and freshly laid ova are about 4 mm in diameter (Bishop 1925, 1941a). The largest ovarian eggs in females examined by Bruce (1969) are 4.0 mm in diameter.

In New York clutches have been found from October through early February (Bishop 1925, 1941a). One clutch with 72 late-term eggs was on the underside of a rock that was buried about 15 cm below the water surface; 290 additional eggs were discovered in November on the undersides of stones in springs. Embryos in some of the eggs were at the hatching stage. In one instance 156 eggs were attached singly to the underside of a large rock in small clusters containing 14–48 eggs. This was presumably a communal nest involving two females. Fowler (1962) found a clutch of eggs with hatchlings on 11 November in Virginia.

The mean number of mature ovarian eggs in a Piedmont population (89 ova) is significantly greater than that in a Blue Ridge population (80 ova; Bruce 1969). The number of mature ova in 29 gravid females examined by Bruce (1978c) varied from 29 to 130, averaged 70, and was positively correlated with female SVL.

Females most likely brood their eggs until hatching since females disappear from surface sites in the autumn and do not reappear until shortly after hatching begins in late autumn or winter. The incubation period in southern Blue Ridge populations lasts about 3 months, with hatching occurring from about mid-December through mid-February (Bruce 1978c). Similarly, the incubation period in northern populations lasts 2–3 months (Bishop (1941a). Hatchlings or late-term embryos have been discovered in November in New York and Virginia (Bishop 1941a; Fowler 1962), in January in Coastal Plain streams in South Carolina (Semlitsch 1983b), and in March in Ohio (Pfingsten 1989b) and the Carolina Piedmont (Bruce 1974). Specimens in the latter study are larger than hatchlings from the Blue Ridge and probably hatch in late February.

AQUATIC ECOLOGY. Larvae aggregate in slow-moving sections of springs, seepages, and stream pools where decaying leaves, bottom debris, and aquatic plants are abundant. These microhabitats not only provide cover but also harbor relatively high densities of invertebrate prey. Larvae feed on worms, insect larvae, and other invertebrates (Bishop 1941a).

Annual growth rates vary depending on seasonal water temperatures and length of the growing season. Larvae in Coastal Plain populations grow 1.2–2.0 mm/month during the first 6 months after hatching (Semlitsch 1983b), whereas those in a Piedmont population grow 1.2 mm/month during the first 5–6 months after hatching (Bruce 1974). Larvae in two

Fig. 192. *Pseudotriton r. nitidus*; larva; Watauga Co., North Carolina (R. W. Van Devender).

populations grow 10–12 and 10–11 mm SVL during their first and second years of life (Bruce 1972b).

The larval period is prolonged and may last 1.5–3.5 years. Transformation occurs from April to November, but most larvae transform in late spring through midsummer (Bishop 1941a; Bruce 1972b; Dunn 1915; Semlitsch 1983b). Northern populations tend to have longer larval periods than southern or Coastal Plain populations. The larval period lasts 3.5 years in one New York population (Bishop 1941a), and very limited data for Pennsylvania specimens suggest that most larvae transform 2.5–3.5 years after hatching (Bell 1956). Data provided by Pfingsten (1989b) suggest a 27- to 31-month larval period for an Ohio population, with larvae transforming at 44–52 mm SVL.

In western North Carolina most larvae metamorphose between May and July after a larval period of 27–31 months (Bruce 1972b; Gordon 1966). Size at metamorphosis in two studies varies from 34 to 46 mm SVL (Bruce 1972b) and from 62 to 86 mm TL (Gordon 1966). In two Coastal Plain populations in South Carolina, the larval period lasts only 18–23 months, and larvae metamorphose from June to November when they average 47 mm SVL (Semlitsch 1983b). Mean size at metamorphosis is 43 and 39 mm SVL in Piedmont and Blue Ridge populations in the Carolinas (Bruce 1972b, 1974), respectively.

Bruce (1972b) estimates annual survivorship at one site to be about 50% per year. Larvae tend toward type II survivorship, in which mortality rates are independent of larval age and size.

TERRESTRIAL ECOLOGY. Because larvae transform at a relatively large size, the juvenile stage is rather abbreviated. Most males studied by Bruce (1978c) matured within 1 year after metamorphosing and bred during their fourth summer when nearly 4 years old. A few males required an additional year before mating for the first time. Males mature when they reach 53–63 mm SVL, and females when they reach 55–68 mm SVL. Most females oviposit for the first time when about 5 years old or older, then reproduce annually thereafter. The sex ratio of juveniles is nearly 1:1, but the number of mature males collected in all seasons is greater than the number of mature females. Gordon (1966) reports that males mature when they measure 93 mm TL, compared with 96 mm TL for females. Sexual maturity is reached 43–48 months posthatching.

Red salamanders overwinter in aquatic sites and often concentrate in springs or streams during the late fall (Bishop 1941a; Bruce 1978c; Pfingsten 1989b). They are often seen on roads on rainy nights during the fall and spring months as they disperse to and from overwintering sites. Juveniles and adults in the southern Blue Ridge disperse from springs and streams into surrounding habitats in early spring (Bruce 1978c). They remain in terrestrial sites until early summer, then gradually return to aquatic sites over the summer and autumn. Individuals in a local population in New York rarely disperse >30 m from their overwintering site (Axtell and Axtell 1948). Red salamanders can be found during the spring months in Alabama in relatively dry habitats far from aquatic sites (Mount 1975). This observation suggests that individuals in southern populations leave overwintering sites in late winter or very early spring.

Adults often live in burrows in soft sediments in streamside habitats (Bishop 1941a), but also reside beneath logs and other cover on the forest floor. Individuals are secretive during the day and are generally more active on the ground surface at night, where they feed on invertebrates and small amphibians (Bishop 1941a; Organ and Organ 1968).

The adults are opportunistic foragers and regularly eat other salamanders in addition to a wide variety of invertebrates. Records of food items include a specimen in New York that ate a *Plethodon cinereus* (Bishop 1941a), two Ohio juveniles that each ate a *P. cinereus* (Bock and Fauth 1992), and 59 Pennsylvania specimens that contained frog skin, a *Plethodon cinereus,* four "undeveloped" salamanders, earthworms, snails, slugs, spiders, millipedes, and numerous types of insects

Fig. 193. *Pseudotriton r. schencki;* defensive posturing; Macon Co., North Carolina (E. D. Brodie, Jr.).

(Surface 1913). Prey in 10 specimens from northern Florida included larval and adult dytiscids, haliplid larvae, dipteran larvae, and hellgrammites (Carr 1940).

PREDATORS AND DEFENSE. Natural predators are poorly documented but probably include woodland birds, shrews, raccoons, and skunks. Uhler et al. (1939) found a copperhead (*Agkistrodon contortrix*) that had consumed a *P. ruber.*

When pinched or prodded, red salamanders usually assume a defensive posture that involves curling the body, elevating and extending the rear limbs, and placing the head in a protective position under the tail. While the animal is in this position, the tail is usually elevated and undulated slowly from side to side (Brandon et al. 1979b; Brodie and Howard 1972).

The terrestrial red eft stage of the eastern newt (*Notophthalmus viridescens*) produces a powerful neurotoxin in the skin that is an effective chemical defense against predators. In many areas in the Appalachian Mountains, efts are active on the ground surface during the day, particularly following periods of sustained rain. The bright crimson color of the efts is aposematic and facilitates a predator's ability to learn to avoid efts after an initial attack.

Howard and Brodie (1971, 1973) hypothesized that *Pseudotriton* is a palatable Batesian mimic of *Notophthalmus,* and that avian predators are the primary agents selecting for the mimicry complex. Although *Pseudotriton* is nocturnally active, on rare occasions individuals are active outside cover during the day, when red efts and birds forage (Brandon and Huheey 1975).

A series of studies by these and other authors in the early 1970s provided strong support for the Batesian mimicry hypothesis. Both wild and domestic birds, for example, could be conditioned to avoid red salamanders after being exposed to red efts (Howard and Brodie 1971, 1973). In addition, red salamanders appeared to be palatable to predators (Brodie and Howard 1972) and the defensive posture of red salamanders was found to resemble that of red efts (Brodie 1977; Brodie and Howard 1972). The mud salamander (*P. montanus*) was later hypothesized to be a Batesian mimic of red efts for similar reasons (Pough 1974).

Huheey and Brandon (1974) and Pough (1974) noted that geographic variation in skin coloration covaries among *Notophthalmus* and *P. ruber.* Both the red efts of *Notophthalmus* and *P. ruber* are brightly colored in montane areas where efts are abundant and diurnally active, but dully colored in nonmountainous areas where efts are uncommon and rarely active during the day. These authors hypothesized that this pattern reflects selection for the mimicry complex in mountainous areas and selection for cryptic coloration in regions where efts are uncommon and secretive during the day. Huheey and Brandon (1974) assumed that brown efts in nonmountainous areas were less toxic than red efts because diurnally active red efts have more encounters with predators. However, subsequent studies revealed that both red and brown efts are equally toxic. Furthermore, chickens learned to avoid brown efts nearly as readily as red efts (Brandon et al. 1979a). These findings caused some to question the cryptic coloration hypothesis.

Pseudotriton adults greatly exceed red efts in size, and critics of the Batesian mimicry hypothesis argued that birds may have little trouble distinguishing between efts and large *Pseudotriton.* This rationale was used to explain ontogenetic shifts in color patterns of *Pseudotriton* from being bright red as small juveniles to dull brown as large adults. Pough (1974) suggested that ontogenetic shifts in color pattern reflect a switch in selection pressures from mimetic to cryptic coloration as *Pseudotriton* grows. Brodie (1976), however, argued that size differences between efts and brightly colored *Pseudotriton* might actually enhance selection for the mimicry complex because larger *Pseudotriton* would elicit a greater avoidance response from predators.

Several researchers challenged the Batesian mimicry hypothesis after discovering that red and mud salamanders are intermediate in palatability between red efts and palatable mountain dusky salamanders (Brandon et al. 1977, 1979a,b; Brandon and Huheey 1981; Huheey and Brandon 1977). Research also revealed that *Pseudotriton* has defensive postures that are more similar to those of closely related plethodontids than to those of newts. In addition, both *P. ruber* and *P. montanus* produce highly toxic skin secretions that are concentrated on the back skin. These researchers concluded that *P. ruber* and *P. montanus* are not Batesian mimics of red efts since they have reduced palatabilities of their own that are reinforced by skin color and defensive posturing. Instead, *Notophthalmus, Gyrinophilus,* and *Pseudotriton* are now considered to be part of a Müllerian mimicry complex in which all species are unpalatable and benefit by evolving similar aposematic coloration. This hypothesis is more consistent with patterns of geographic variation in body size, coloration, and toxicity.

COMMUNITY ECOLOGY. The red salamander often feeds on other salamanders and could potentially be an important species structuring certain streamside salamander communities. We currently have few data on the ecological importance of this species in natural communities.

CONSERVATION BIOLOGY. Populations of the red salamander thrive in mature, deciduous forests with clean, unsilted streams. Deforestation, acid drainage from coal mines, and stream siltation and pollution have undoubtedly resulted in the loss of many local populations throughout the range of *P. ruber.*

Stereochilus marginatus Cope
Many-lined Salamander
PLATES 100, 101

Fig. 194. *Stereochilus marginatus;* adult; Bladen Co., North Carolina (R. W. Van Devender).

IDENTIFICATION. The many-lined salamander is a small, slender, short-tailed salamander that is brownish to dull yellow above with an indistinct, dark dorsolateral band that runs through the eye onto the tail. A series of fine, parallel, dark lines or streaks usually occurs along the sides of the body. In many cases these are indistinct or reduced to a series of spots. The venter is yellowish and is scattered with brown or black flecks (Rabb 1966). The tail is laterally compressed toward the tip and has a prominent dorsal keel. Adults measure 6.5–11.5 mm TL and there is no sexual dimorphism in body size (Bruce 1971).

The smallest hatchlings collected by Bruce (1971) measured 8 mm SVL. Hatchlings are strongly bicolored and are dark brown above and yellow below. Small white to yellowish spots occur on the top and sides of the head and body. Unlike many pond-dwelling larvae, the hatchlings have functional limbs at hatching. The dorsal fin extends anteriorly nearly to the head; however, the fin becomes reduced during the early larval period and is confined to the tail in larvae measuring >15 mm SVL (Bruce 1971; Schwartz and Etheridge 1954). Older larvae generally resemble the adults in color and general form, but often have mottling on the dorsum (Bishop 1943).

SYSTEMATICS AND GEOGRAPHIC VARIATION. Patterns of geographic variation in morphology, coloration, or genetic characteristics have not been analyzed in detail. No subspecies are recognized.

DISTRIBUTION AND ADULT HABITAT. *Stereochilus marginatus* inhabits gum and cypress swamps,

Range of *Stereochilus marginatus*

woodland ponds, borrow pits, drainage ditches, canals, sluggish streams, and other permanent aquatic habitats in the Coastal Plain from extreme southeastern Virginia to extreme northern Florida (Christman and Kochman 1975). Individuals tend to aggregate in leaf litter, sphagnum mats, and other aquatic microhabitats that provide cover. Although the species is mostly aquatic, it can be collected beneath logs, in wet *Sphagnum* mats, and in other semiaquatic microhabitats along the margins of ponds and sluggish streams. Foard and Auth (1990) found specimens during a drought on land beneath mats of exposed vegetation.

BREEDING AND COURTSHIP. The many-lined salamander is an autumn breeder and both sexes reproduce annually. In North Carolina breeding peaks in November, when most males have swollen vents and vasa deferentia packed with sperm (Bruce 1971). The ratio of males to females is about 1:1 in this population, compared with 2.5:1 in a Georgia population (Foard and Auth 1990).

Stereochilus marginatus engages in the typical plethodontid courtship pattern, characterized by a preliminary period of nosing and rubbing of the female's snout by the male, followed by a tail-straddle walk (Noble and Brady 1930). Females often turn their tails forward while performing the tail-straddle walk. The description of courtship behavior is sketchy and a detailed analysis is needed.

REPRODUCTIVE STRATEGY. Females oviposit during the winter months in microhabitats in or near aquatic sites. The eggs are attached singly by short pedicels to rootlets, pine needles, plant stems, moss fibers, and the undersides of logs. Wood and Rageot (1963) noted that females in southeastern Virginia do not nest in *Sphagnum* mats even though these are abundant in many aquatic sites.

When ovipositing, a female flips on her back and presses her cloaca to the substrate while bracing herself with the arched back or tail (Noble and Richards 1932). The mode of egg deposition varies depending on the substrate and whether or not the eggs are submerged. When ovipositing on land, females tend to clump their eggs relatively tightly and brood the eggs in about 30% of cases (Rabb 1956; Schwartz and Etheridge 1954). When ovipositing in water, females scatter the eggs more widely and almost never brood.

Nesting records include a terrestrial nest found inside a decaying gum log near the edge of a pond (Rabb 1956), and a terrestrial nest and brooding female found beneath a small log lying near the margin of a flooded borrow pit (Schwartz and Etheridge 1954). The most detailed data on the nesting ecology are those for 43 clutches in southeastern Virginia that were laid both in water and on land (Wood and Rageot 1963). In this area most nests can be found 8-15 cm below the water surface in quiet pools. The eggs are typically attached in loose, irregular groupings to an aquatic moss (*Fontinalis*). They are positioned either singly or in compact, adherent clusters of three to six eggs. In a few instances, eggs are attached to the undersides of logs or other detritus. The terrestrial nests are placed in *Fontinalis* clumps around the margins of ponds.

The ova are surrounded by a vitelline membrane and two jelly capsules. The outer capsule is adhesive, and the egg is attached by a short pedicel to the substrate (Noble and Richards 1932; Wood and Rageot 1963). Diameters of the outer capsules of eggs collected by Rabb (1956) were 2.5-3.4 mm and averaged 3.0 mm. Information on the size of recently laid eggs in natural habitats is lacking. Females induced to ovulate lay eggs that are 2.0-2.5 mm in diameter (Noble and Richards 1932).

Field observations suggest a weak seasonal progression in the time of egg laying from southern to more northern localities. Oviposition occurs in January and February in the Carolinas (Bruce 1971). Nesting records include a female and 62 freshly laid eggs found on 27 January in South Carolina (Rabb 1956), and a female brooding 15 eggs near hatching found on 27 March in North Carolina (Schwartz and Etheridge 1954). Eggs in early developmental stages can be found in southeastern Virginia from late February through early April (Wood and Rageot 1963).

Estimates of clutch sizes are based primarily on counts of eggs in the field and are similar to those of other small plethodontids such as *Eurycea bislineata*. Estimates are 6-92 eggs (mean = 37) in 35 aquatic nests in Virginia (Wood and Rageot 1963), 9-45 eggs (mean = 22) in 8 terrestrial nests in Virginia (Wood and Rageot 1963), 22-29 mature ova in three North Carolina females (Bruce 1971), and 16-121 eggs (mean = 70) for 19 females induced to ovulate (Noble and Richards 1932). This latter average is much higher than those reported for natural populations and may be an artifact since females were maintained in the laboratory for nearly a year prior to experimentation. The incubation period lasts 1-2 months and hatching occurs in late March and April in the Carolinas (Bruce 1971).

AQUATIC ECOLOGY. Many aspects of the larval ecology are poorly documented. Larvae tend to aggregate in bottom debris and *Sphagnum* mats, where they feed on small invertebrates. Larvae in a North Carolina population grow an average of 15-16 mm SVL during the first year after hatching and 9-10 mm during their second year of life (Bruce 1971). Most larvae metamorphose during late spring and summer, 25-28 months after hatching, although some individuals transform after only 13-16 months of growth. Size at metamorphosis typically is 31-40 mm SVL, but in extreme cases individuals may transform when as small as 27 mm SVL. Larvae in Georgia populations transform when they reach 30-35 mm SVL (Foard and Auth 1990).

The juveniles and adults live in aquatic or semiaquatic habitats. Bruce (1971) found that most males mature sexually within a few months after transforming and reproduce the following autumn when about 33 months old. Sexual maturation occurs at 33-40 mm SVL. The small percentage of males that transform 13-16 months after hatching first reproduce when 21 months old. Most females mate for the first time when 4 years old. Unlike males, females remain as juveniles for over 1 year after transforming and mature when they reach 37-45 mm SVL.

Foard and Auth (1990) report a wide variety of prey in Georgia specimens. Amphipods, isopods, chironomids, and ostracods are the most common prey, and the diets of larvae and adults differ only slightly. There is a weak tendency for prey size to increase with salamander SVL.

PREDATORS AND DEFENSE. Natural predators are unknown but probably include aquatic insects, such as the larvae of dragonfly and dytiscid beetles, along with aquatic snakes, fishes, and wading birds.

COMMUNITY ECOLOGY. This species often shares habitats with a variety of amphibians and reptiles, but no data are available on ecological interactions of *S. marginatus* with other community members.

CONSERVATION BIOLOGY. The many-lined salamander occurs in scattered populations throughout its range, but nothing is known about the conservation status of this species. The draining of wetlands has probably caused more losses of local populations than any other form of human disturbance.

Typhlotriton spelaeus Stejneger
Grotto Salamander
PLATES 102, 103

Fig. 195. *Typhlotriton spelaeus;* adult; Christian Co., Missouri (R. W. Van Devender).

IDENTIFICATION. The grotto salamander is a cave form with atrophied eyes and a white, pinkish white, or light brown dorsum and venter. The eyes of adults appear as small black spots and the eyelids are partially fused (Brandon 1965c, 1970; Mittleman 1950; Smith 1960). In sexually mature males and females, the lip near the nasolabial groove is slightly swollen and a short cirrus extends downward from each nasolabial groove. Sexually active males have a circular mental gland at the tip of the lower jaw, and longer, more swollen, cirri than females (Brandon 1971b). Adults measure 7.5–13.5 cm TL and there are 16–19 (usually 17) costal grooves.

Hatchlings measure 13 mm SVL and 17 mm TL on average and are the stream form (Brandon 1970). Larvae have a tan dorsum and sides and are often stippled or weakly mottled with slightly darker pigment. There are 16–19 costal grooves and the eyes are smaller than in *Eurycea* larvae of similar size. Larvae collected from caves tend to be noticeably lighter than those from surface streams. Small and intermediate-sized larvae have functional eyes. Fusion of the lids and atrophy of the eyes begin near the end of the larval stage and continue into the adult stage (Brandon 1970; Smith 1960; Stone 1964).

SYSTEMATICS AND GEOGRAPHIC VARIATION. Genetic variation has not been examined. Western populations tend to have a higher modal number of trunk vertebrae than eastern populations (Brandon 1970), but subspecies are not recognized.

DISTRIBUTION AND ADULT HABITAT. The grotto salamander is found in the Salem and Springfield plateaus of the Ozark Uplift in southern Missouri, extreme southeastern Kansas, and adjacent areas in Arkansas and Oklahoma. The adults inhabit stream-fed caves or springs, but the larvae are commonly found in streams and brooks outside caves (Brandon 1971b; Hendricks and Kezer 1958). Larvae are most common in springs with clear water and sandy or gravelly substrates with little silt (Smith 1960). Both adults and larvae have been found beneath stones and in pools deep within caves (Dunn 1926).

BREEDING AND COURTSHIP. Adults mate during the summer months and females appear to oviposit within 1–4 months after mating. Changes in seasonal development of the cirri and mental glands suggest that mating occurs from May through August (Brandon 1971b). Gravid females have been collected in May and June, but they disappear from the surface from July through December, when oviposition presumably occurs (Smith 1960). No information is available on courtship behavior.

REPRODUCTIVE STRATEGY. Females oviposit in cryptic sites within caves and springs. The eggs are unpigmented and are surrounded by three envelopes. A female that received pituitary implants laid 13 eggs that were 2.0–2.2 mm in diameter (Barden and Kezer 1944; Kezer 1952a). The eggs are laid singly or in lines of two or three eggs. The largest ova are 2.7–3.0 mm in diameter (Brandon 1966c; Trauth et al. 1990).

Smith (1960) found eggs with embryos near hatching on 29 and 31 December in an Arkansas cave. The eggs—which were presumed to be those of *Typhlotriton*—were stirred up from gravel beds and were attached singly to rocks. *Eurycea lucifuga* and *E. longicauda* are also present at this site and deposit their eggs in a similar manner. Thus additional verification of this observation is needed.

Range of *Typhlotriton spelaeus*

AQUATIC ECOLOGY. Hatchlings have been found from 15 December to 1 January in one Arkansas cave (Smith 1960). Although hatchlings are often found within caves, many drift from caves into surface streams. Larvae are often abundant in streams outside caves, and larval drift may be important in dispersing individuals from one cave system to another.

Because the adults are almost always found in caves, some researchers have suggested that larvae migrate from surface streams back into caves shortly before reaching sexual maturity. Brandon (1971b) noted that there is no empirical evidence that surface-dwelling larvae return to caves and surmised that most adults are derived from larvae within caves.

The larvae are secretive and remain in recesses in gravel or underneath large, flat rocks, where they forage for small invertebrates (Smith 1960). Larvae are largely sit-and-wait predators that do not show strong tendencies to search actively for prey. Individuals often posture with the anterior portion of the body raised well above the substrate, a behavior that may enhance prey detection (Dodd 1980). An isopod (*Lirceus happinae*) is the major prey item of larvae in some populations (Smith 1948a,b). *Lirceus* probably serves as a secondary host of *Ophiotaenia cestodes*, a parasite that infects many of the larvae. Brandon (1971b) found only dipteran larvae and snails in larvae.

Length of the larval period is poorly documented. Larvae have been collected from cave streams and springs in Kansas throughout the year (Collins 1993), and very limited data suggest that the larval period lasts 2–3 years (Brandon 1966c; Hendricks and Kezer 1958; Trauth et al. 1990). Larvae metamorphose when they measure 36–56 mm SVL.

Although Bishop (1944), Mohr (1950), and Smith (1960) report that gilled adults occur in nature, Brandon (1966c) found no evidence to support this view. His observations suggest that gilled adults either do not occur or are very rare in local populations. Larvae as

Fig. 196. *Typhlotriton spelaeus;* larva; Christian Co., Missouri (R. W. Van Devender).

large as 50–60 mm SVL and 90–120 mm TL occur at some sites (Smith 1960).

TERRESTRIAL ECOLOGY. *Typhlotriton spelaeus* larvae metamorphose shortly before reaching sexual maturity, and the juvenile stage is extremely abbreviated. Almost all of 220 transformed animals that Brandon (1971b) collected from a Missouri cave were sexually mature, with sex ratios approximating 1:1. Although adults are often found in water, most reside on moist cave walls beyond the twilight zone (Brandon 1971b; Hendricks and Kezer 1958). Animals are normally most active during the spring and summer months when rock walls are wet, food is abundant, and the adults are courting. The adults feed on both aquatic and terrestrial prey. Food items include heleomyzid flies, mosquito larvae, and beetles in 122 postmetamorphic animals (Brandon 1971b) and an isopod (*Lirceus*) in specimens examined by Smith (1948a,b). Bat guano is a major source of nutrients that supports insect prey for the larvae, juveniles, and adults.

PREDATORS AND DEFENSE. Natural predators are not known. The larvae are undoubtedly eaten by crayfish and other stream invertebrates; the adults presumably have few natural predators.

COMMUNITY ECOLOGY. The role of this and other species of salamanders in structuring cave communities is largely unknown. The grotto salamander functions as a top predator in many cave communities, and adult populations may be limited by the food supply.

CONSERVATION BIOLOGY. Vertebrates that inhabit caves are often vulnerable to human disturbance and overcollecting. Systematic surveys of caves and a long-term monitoring program of selected populations of this and other cave-dwelling salamanders in the Ozarks are needed to assess the long-term stability of these populations.

Tribe Plethodontini

Aneides aeneus Cope and Packard
Green Salamander
PLATE 104

Fig. 197. *Aneides aeneus;* adult; eastern Kentucky (R. W. Barbour).

IDENTIFICATION. The green salamander is readily identifiable by its greenish color, flattened body, long legs, and expanded, squared toe tips that are adaptations for living in rock crevices. The dorsum is black and is overlain with distinctive yellowish green, lichenlike patches. The venter is light colored and unmarked, and the base of each leg is often faintly washed in yellow. The adults measure 8–14 cm TL and there are 14–15 costal grooves.

Sexually active males differ from females by having a mental gland that is often yellowish orange, papillae on the anterior wall of the cloaca, enlarged jaw muscles, and elongated premaxillary and maxillary teeth that penetrate the upper lip (Gordon 1967; Wake 1963). Mature females also have somewhat enlarged jaw muscles and protruding teeth, so these traits are not always reliable for sexing individuals in the field (Juterbock 1989). Hatchlings are similar in color to adults and reach 18.5–23 mm TL (Gordon 1952).

SYSTEMATICS AND GEOGRAPHIC VARIATION. Sessions and Kezer (1987) recognize two chromosomally distinct groups. One occurs in southern Pennsylvania, West Virginia, and southern Kentucky, with a disjunct in southwestern North Carolina. The second occurs in southern Tennessee and northern Alabama. A third chromosomally distinct form that was probably collected in eastern Tennessee is described by Morescalchi (1975). Local populations often vary in the degree of green mottling and darkness of the dorsum, but no subspecies are recognized.

DISTRIBUTION AND ADULT HABITAT. Green salamanders inhabit cliffs and rockface habitats at elevations <1340 m in mountain formations from extreme southwestern Pennsylvania to northern Alabama and extreme northeastern Mississippi. Populations are largely confined to the Appalachian Plateau and the Blue Ridge Province, although some occur in the Interior Low Plateau and Ridge and Valley provinces of Tennessee. A large disjunct from the main range occurs in southwestern North Carolina and adjoining areas in South Carolina and Georgia. Weller (1931) reported a disjunct population along the eastern slope of Mt. LeCount in the Great Smoky Mountains National Park, but this population has not been located since.

Populations are most frequently encountered in sandstone, granite, and schist formations with deep, shaded crevices that are moist, but not dripping wet (Bruce 1968; Gordon and Smith 1949; Mount 1975; Netting and Richmond 1932; Schwartz 1954). Populations have also been found on quartzite and limestone cliffs (Walker and Goodpaster 1941). Specimens in eastern Kentucky and adjacent portions of Tennessee and southwestern Virginia are occasionally collected beneath the loose bark of fallen trees (Barbour 1953; Fowler 1947; Gordon 1952; Pope 1928; Welter and Barbour 1940). Green salamanders are weakly arboreal (Bishop 1928), but this habit is not nearly as developed as in some western *Aneides* species (Barbour 1953; Pope 1928; Schwartz 1954).

BREEDING AND COURTSHIP. Males and females often pair together in rock crevices during the breeding season. Breeding can occur at almost any time during

Range of *Aneides aeneus*

the warmer months of the year, but the majority of adults mate in May and June. A secondary bout often occurs in September and October. Courtship in eastern Kentucky populations occurs between late March and early November, but most individuals mate during the spring and fall months (Cupp 1971, 1991). Mating in West Virginia populations occurs primarily in late May and early June, but spermatogenic data and observations of paired adults indicate that a secondary breeding period occurs in September and October (Canterbury and Pauley 1994).

Observations from other sites suggest that mating is confined almost entirely to May and June. In northeastern Mississippi, pairing and mating occur in late May and June (Cupp 1991). Breeding is apparently restricted to late May through mid-June in Blue Ridge populations since this is the only time of the year when males possess swollen vents and adults have been observed courting (Brooks 1948a; Gordon 1952). Annual breeding patterns are not fully documented, but the available data suggest that males breed annually and females biennially (Canterbury and Pauley 1994).

Cupp (1971) observed courtship on 17 October between a pair of green salamanders in a sandstone crevice about 1 m from the ground. The following sequence is based on his observations. After encountering each other, the male and female first move in a tight circle with the chin of each resting on the tail of the other. As they circle, the male occasionally bites and nudges the female. At times, both individuals crawl over each other and mutually rub their snouts on each other while circling. During most of the courtship, the female straddles the male's tail. Periodically, the male stops and undulates the base of his tail laterally a few times. After again engaging in the tail-straddle walk, the male undulates the base of his tail, then arches the tail and deposits a spermatophore. The pair then moves forward and the female picks up the spermatophore while undulating the base of her tail. After the female picks up the spermatophore, the male exhibits jerking movements of the body and laterally undulates the tail. The pair soon parts and courtship activity ceases.

REPRODUCTIVE STRATEGY. Females deposit their eggs in secluded, damp rock crevices and attach the egg masses to the roofs of horizontal crevices or to the sides of vertical crevices. On rare occasions individuals may nest beneath bark on large logs. Females lay eggs in small clusters that are suspended by several mucus strands and remain with their eggs through hatching. The eggs are whitish yellow and average 4.5 mm in diameter, including the outer membranes (Gordon 1952; Gordon and Smith 1949).

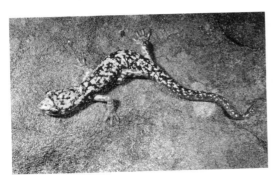

Fig. 198. *Aneides aeneus*; adult; Bell Co., Kentucky (J. W. Petranka).

Fig. 199. *Aneides aeneus*; female nesting in rock crevice (P. Cupp).

When ovipositing, a female lies on her back and then secretes a glob of adhesive material on the ceiling before laying the first egg (Cupp 1991; Gordon 1952). This material later elongates to form one of the strands or cords that support the egg mass. As many as 12 eggs may be laid in a 14-hour period (Cupp 1991). Females require 1–2 days to lay an entire complement of eggs and may use the same nesting site from year to year (Cupp 1991; Gordon 1952). Eggs occasionally fall from their supports and are lost. Females that are experimentally displaced 0.4–9.0 m from their breeding crevices return within 24 hours (Gordon 1961).

Females lay their eggs in June and July shortly after mating. Eggs have been found in June or July in Alabama (Mount 1975), Kentucky (Pope 1928; Cupp 1991), North Carolina (Eaton and Eaton 1956; Gordon 1952; Gordon and Smith 1949), and West Virginia (Canterbury and Pauley 1994; Green and Pauley 1987). All of the oviposition sites were rock crevices except for a fallen tree, beneath the bark of which 14 eggs were found (Pope 1928).

Detailed studies suggest that the egg laying period lasts 2–4 weeks in any local population. Females in eastern Kentucky populations deposit eggs from 25 June to 25 July, although most eggs are laid from 1 to 16 July (Cupp 1991). Most females at a central West Virginia site oviposit during the first 2 weeks of June (Canterbury and Pauley 1994), whereas most eggs at a North Carolina site are deposited in early to mid-June (Gordon 1952; Gordon and Smith 1949).

Clutch size for 32 North Carolina nests varied from 10 to 26 and averaged 17–19 (Gordon 1952; Gordon and Smith 1949), whereas that for 14 West Virginia nests varied from 12 to 27 and averaged 18 (Canterbury and Pauley 1994). Seventeen females collected from the same site contained 12–30 ovarian eggs (mean = 22). The number of eggs in 21 Kentucky nests varied from 14 to 20 (Cupp 1991), whereas ovarian complements of five West Virginia females ranged from 20 to 32 ova (Lee and Norden 1973).

Females brood their eggs and hatchlings for several months and have little opportunity to feed during this period. This behavior may explain why most females breed biennially. Estimates of the duration of the incubation period are 67–82 days (mean = 73 days) in Kentucky (Cupp 1991), 82 days in Mississippi (Woods 1969), 84–91 days in North Carolina (Gordon 1952), and 82–90 days in West Virginia (Canterbury and Pauley 1994). Hatching has been documented from 14 September to 6 October in Kentucky (Cupp 1991), from 7 to 20 October in Mississippi (Woods 1969), in early September in North Carolina (Gordon 1952), in late September in Ohio (Walker and Goodpaster 1941), and from 31 August to 9 September in West Virginia (Canterbury and Pauley 1994).

Nest failure is common, particularly if females desert their eggs. Snyder (1973) found that orphaned clutches invariably fail to survive. Brooding females regularly eat dead eggs in clutches and probably enhance embryonic survivorship by protecting clutches from pathogens and predators. The rate of nest failure is 20–40% and increases during seasonal droughts. Embryonic survivorship was <70% in Gordon's (1952) study, and survival of unguarded eggs was near zero.

TERRESTRIAL ECOLOGY. Females may brood their young for 3–5 weeks after they hatch (Cupp 1991), but the juveniles disperse from natal areas within a few months of hatching and take up residence in cracks and crevices (Gordon 1952). Data on growth rates and age at first reproduction are scant. Lee and Norden (1973) document three distinct size classes in

Fig. 200. *Aneides aeneus*; territorial dispute (note individual that has been lifted off the ground by a rival) (P. Cupp).

West Virginia specimens, suggesting that juveniles mature sexually 2–3 years after hatching. Both sexes reach sexual maturity at about 45 mm SVL (Juterbock 1989).

Male green salamanders are territorial and aggressively guard home territories in rock crevices (Canterbury and Pauley 1991; Cupp 1980). Resident males engage intruders by either biting them or vigorously pressing their snouts along their backs, and they are usually successful in defending crevices. In laboratory trials, male *A. aeneus*, *Plethodon glutinosus*, and *P. kentucki* are all mutually aggressive; however, *A. aeneus* cannot successfully defend its space from the *Plethodon* species (Canterbury and Pauley 1991). Females are not aggressive toward conspecifics.

During the warmer months of the year, individuals emerge from rock crevices at night to forage on rockfaces, particularly during periods of rain (Brandon and Huheey 1975; Gordon 1952; Netting and Richmond 1932; Richmond 1952; Synder 1991). Gordon (1952) studied the movements of 26 marked adults and found that most moved <3.7 m from the point of original capture. In extreme cases, individuals may move 76–91 m from their release point. Subadults may move as far as 31 m (Woods 1969).

Adults reach their highest densities on rockfaces from late October to mid-December after females leave their hatchlings (Cupp 1991; Gordon 1952). Both juveniles and adults retreat to deep crevices in winter and are rarely encountered during the winter months (Cupp 1991; Gordon 1952; Woods 1969). Individuals monitored by Cupp (1991) in Kentucky begin to reappear in rock crevices in early March and increase in number through late April. Gravid females emerge later than other individuals and eventually move to crevices occupied by territorial males.

The extent to which individuals disperse from one isolated rock outcrop to another is not known. Several researchers (Cupp 1991; Williams and Gordon 1961) have found animals crossing roads during April–June. These may reflect a postemergent dispersal phase from winter hibernacula to surrounding rock outcrops that are used as summer retreats. Animals are rarely taken in adjoining forest floor habitats even when collecting in the immediate vicinity of rock outcrops with large populations (Synder 1991). All of the records of green salamanders being found beneath the bark of fallen trees are from the 1920s to the early 1950s, when substantial amounts of old-growth forest covered eastern Kentucky and Tennessee (Barbour 1953; Fowler 1947; Gordon 1952; Pope 1928; Welter and Barbour 1940). The large, thick slabs of bark found on large old-growth logs presumably provided better microhabitats for foraging and nesting than the small logs that are now present in secondary forests that are regenerating after timbering.

Very few data are available on diet and all are for West Virginia specimens. Hymenopterans (mostly ants), beetles, and mites are the most numerically important prey in West Virginia specimens, and mites are more important in the diet of very small juveniles (Canterbury and Pauley 1990). Other prey include orthopterans, spiders, dipterans, hemipterans, and lepidopterans. Lee and Norden (1973) list beetles, mosquitoes, ants, pseudoscorpions, mites, spiders, snails, shed skin, and detritus for other West Virginia specimens.

PREDATORS AND DEFENSE. Although hatchling green salamanders typically remain immobile when disturbed (Brodie et al. 1974b), adults in rock crevices may turn the tail toward the opening and undulate it from side to side if poked with a twig (Brodie 1977). Animals that I have attempted to drive out of cracks by prodding with twigs respond by moving deeper into cracks. Thus retreating into deep recesses may be the most common defensive mechanism.

COMMUNITY ECOLOGY. Because of their microhabitat preferences, green salamanders probably do not compete with salamanders that restrict their activity to the forest floor. However, slimy salamanders often frequent rock outcrops and may potentially compete with *A. aeneus*. Green salamanders and slimy salamanders sometimes stratify vertically (Cliburn and Porter 1986, 1987). Green salamanders typically select cracks that are higher on rockfaces or in experimental chambers than those preferred by *Plethodon* species. This microhabitat segregation could potentially reduce competition for food between the species.

CONSERVATION BIOLOGY. Because of its unusual habitat requirements, the green salamander is patchily distributed and generally uncommon throughout most of its range. Populations in the Blue Ridge Escarpment experienced a dramatic decline during the late 1970s in association with a prolonged drought, but appear to be recovering (Synder 1991). Timbering in the immediate vicinity of rock outcrops dries crevices used for foraging and nesting and can lead to the extinction of local populations. Whenever feasible, forest managers should leave forest buffers of 100 m or more around rock outcrops with green salamander colonies. The extent to which timbering of old-growth forest has reduced gene flow between rock outcrop populations is not known but should be investigated.

Aneides ferreus Cope
Clouded Salamander
PLATES 105, 106

Fig. 201. *Aneides ferreus;* adult; Humboldt Co., California (R. W. Van Devender).

Fig. 202. *Aneides ferreus;* profile of toes; Benton Co., Oregon (W. P. Leonard).

IDENTIFICATION. The clouded salamander is a medium-sized salamander with long limbs, squarish toe tips, enlarged jaw musculature, and a triangularly shaped head. The rounded tail is prehensile and aids in climbing. The toes on the hindlimbs are exceptionally long, and the toe tips of adpressed limbs either overlap or are separated by a maximum of 1.5 costal folds. The length of the outer toe relative to toes 2–4 is longer in this species than in western *Plethodon* species. The dorsal ground color of adults is brownish and is overlain with greenish gray, pale gold, or reddish blotches scattered with brassy flecks that become less pronounced with age (McKenzie and Storm 1970; Nussbaum et al. 1983). The venter is dusky gray with light flecking. Sexually active males have swollen jaw musculature and more triangularly shaped heads than females. Mature females average slightly larger than males. Nussbaum et al. (1983) report the maximum size of males and females as 63 and 65 mm SVL, respectively.

Light and dark color phases are recognized by some authors (e.g., Wake 1965a). Dark-phase animals are almost uniformly dark brown above, whereas light-phase animals are heavily suffused with ashy gray, greenish gray, or other light colors that mask the underlying brown ground color. These color phases are best viewed as two extremes of a continuum, since variation in dorsal color appears to be continuous

Fig. 203. *Aneides ferreus;* adult; Benton Co., Oregon (W. P. Leonard).

rather than discrete. Two albinos are known from Humboldt Co., California (Houck 1969). Adults measure 7.5–13 cm TL and there are 16–17 costal grooves.

Hatchlings measure 25–26 mm TL on average (McKenzie 1970; Storm 1947) and have a deep chocolate brown dorsum with a brassy dorsal stripe that runs from the neck to the tip of the tail. A brassy triangle occurs between the eyes, and the venter is dark (McKenzie and Storm 1970; Storm 1947). Juveniles slowly lose the dorsal brassy stripe of hatchlings and acquire the adult color patterns as they age (Nussbaum et al. 1983; Stebbins 1954).

SYSTEMATICS AND GEOGRAPHIC VARIATION. Kezer and Sessions (1979) and Sessions and Kezer (1987) document geographic variation in chromosome characteristics and delineate two well-differentiated genetic forms within this species. One consists of two allopatric groups located on Vancouver Island and adjacent islands in British Columbia, and in extreme northern California southward to the southern limit of the range of the species. The second genetic form occurs in Oregon and extreme northern California. The chromosomal evidence suggests that *A. ferreus* may contain two distinct species that contact each other in extreme northern California. Addition studies in the northern California contact zone are currently under way that will clarify the specific status of these forms (D. B. Wake, pers. comm.). No subspecies are recognized.

DISTRIBUTION AND ADULT HABITAT. Clouded salamanders live in forests on Vancouver Island and many surrounding islands in British Columbia, and in coastal forests in Oregon and northern California. Populations are conspicuously absent from mainland Washington and occur from near sea level to around 1700 m in elevation. This species is characteristically associated with large logs and talus throughout its range (Corn and Bury 1991; Whitaker et al. 1986) and is one of the most arboreal salamanders in the United States and Canada. Specimens have been observed 7–40 m above ground (Leonard et al. 1993; Van Denburgh 1916). The adults are frequently found beneath loose bark on recently fallen trees, and in cracks in lightly decayed logs and stumps (Bury and Corn 1988a; Fitch 1936; Stebbins 1951; Stelmock and Harestad 1979). In California, specimens have been found under objects on the ground surface in redwood forests, beneath mats of vegetation on the tops of sawed stumps, and beneath loose bark on sawed stumps of Port Orford cedar and Douglas-fir (*Pseudotsuga menziesii;* Bury and Martin 1973; Stebbins 1951). Clouded salamanders are often abundant where rockfaces or talus provide deep cracks for foraging, resting, and nesting.

Although *A. ferreus* prefers conditions associated with old-growth forests (Corn and Bury 1991), it is often abundant in areas that have recently burned or have been logged and have numerous stumps or large amounts of coarse, woody debris on the forest floor (Bury and Corn 1988a; McKenzie and Storm 1970; Welsh and Lind 1988, 1991).

BREEDING AND COURTSHIP. McKenzie (1970) found little seasonal variation in the appearance and size of testes. Sperm can be found in vasa deferentia in every month except September, suggesting that mating is aseasonal. Additional studies are needed to confirm this unusual pattern. The sex ratio of specimens from Vancouver Island is about 1:1 (Stelmock and Harestad 1979). Courtship behavior has not been described.

REPRODUCTIVE STRATEGY. The few observations of nesting to date suggest that females lay eggs in late June and July and hatchlings emerge in late August through September. Nesting records include two egg masses in early developmental stages found on 7 July in Oregon within a rotting Douglas-fir log and a third mass of late-term embryos found within the same log on 20 August (Storm 1947). In northern California records include a clutch of late-term eggs

Range of *Aneides ferreus*

discovered on 16 August under the bark of a fallen Douglas-fir (Dunn 1942) and a clutch found in August in a fern clump at the base of a tree limb located 30–40 m above ground (Welsh and Wilson 1995).

Females suspend the eggs from a common base by gelatinous strands about 2 cm long that are twisted about each other. Freshly laid eggs are cream colored, average 4–5 mm in diameter, and are surrounded by two jelly capsules (Dunn 1942; Storm 1947). The egg capsules of late-term eggs are about 6 mm in diameter, and the advanced embryos have a prominent, leaflike gill on each side of the head. Nests have been found with only a female guarding the eggs, with both a male and a female guarding the eggs, and with no adults guarding the eggs (Dunn 1942; Storm 1947; Welsh and Wilson 1995). One clutch without adults was in a recently felled tree, and the adults may have abandoned the eggs because of disturbance.

Reported clutch sizes are 6 and 9 eggs in northern California nests (Dunn 1942; Welsh and Wilson 1995); 8, 9, and 17 eggs in Oregon nests; and 13 eggs laid by a captive Oregon female (Storm 1947). Estimates of clutch size from counts of ovarian eggs include 12 ova for a single female from Oregon (Fitch 1936) and an average of 18 ova (range = 14–26) for seven females from Vancouver Island (Stelmock and Harestad 1979). The maximum diameter of ova is 4 mm.

TERRESTRIAL ECOLOGY. Juveniles and adults appear to be generalist feeders and tend to take fewer but larger prey on average as they grow (Harestad and Stelmock 1983). Individuals often forage as sit-and-wait predators from beneath bark or in the immediate vicinity of downed logs and feed on small invertebrate prey associated with these habitats. Invertebrates in the stomachs of 63 specimens (presumably from Oregon)

Fig. 204. *Aneides ferreus;* adult; Oregon (J. W. Petranka).

included isopods, members of several families of insects, spiders, pseudoscorpions, and mites (Storm and Aller 1947). Other dietary reports include mites, beetles, and a weevil in five specimens from Oregon (Fitch 1936) and snails, arachnids, isopods, millipedes, centipedes, and insects in Vancouver Island specimens (Stelmock and Harestad 1979). Similar prey are reported for animals from northern California (Bury and Martin 1973).

Whitaker et al. (1986) conducted a detailed dietary analysis using 650 specimens collected from a cleared area with large Douglas-fir logs. Males and females at this site eat similar prey but shift their diets seasonally. Hatchlings eat very small prey, particularly mites, springtails, flies, and small beetles. Juveniles and adults feed on a wide variety of larger prey, with flies, sowbugs, beetles, ants, and earwigs being the most volumetrically important. Other dietary items include spiders, millipedes, centipedes, homopterans, neuropterans, hemipterans, lepidopterans, orthopterans, phalangids, and shed skin.

Data on growth rates are not available. Males apparently mature during their second year of growth when they reach >36 mm SVL, whereas females first oviposit during their third year when they reach >55 mm SVL (McKenzie 1970).

The adults are aggressive: in one study, 32% of field-collected specimens had scars that were presumably the result of conspecific attacks (Staub 1993). Males have a higher percentage of scars than females, suggesting that competition for mates or space may occur at certain times of the year. Unlike some plethodontid salamanders, clouded salamanders do not appear to use chemical signals from feces to delineate the territories of either conspecifics or heterospecifics (Ovaska and Davis 1992).

Clouded salamanders are rarely collected in pitfall traps even though large numbers of individuals can be collected by searching in logs or beneath bark (Bury and Corn 1988a). This observation suggests either that individuals remain within their territories and rarely move long distances or that they are good at avoiding traps. On Vancouver Island average surface densities are estimated to be 0.008–0.016 salamander/m^2 (Stelmock and Harestad 1979).

PREDATORS AND DEFENSE. Natural predators are poorly documented but no doubt include snakes, woodland birds, and mammals. When attacked, individuals may posture, remain immobile, or flip at random and then remain immobile. When posturing, individuals raise their bodies high above the ground and raise and undulate the tails (Brodie 1977).

COMMUNITY ECOLOGY. *Aneides ferreus* often coexists with other forest salamanders in the Pacific Northwest, but competitive interactions with other community members have not been examined experimentally.

CONSERVATION BIOLOGY. The clouded salamander may thrive for a decade or so following clearcutting. However, local densities eventually decline as stumps and logs decay and critical microhabitats for nesting and foraging are eliminated. Corn and Bury (1991) estimate that several hundred years may be required before populations fully recover from logging. They question whether this species could survive in intensively managed forests maintained on short rotation cycles. Mature and old-growth forests provide the best habitats for assuring the long-term survival of local populations of this and many other North American salamanders.

Aneides flavipunctatus (Strauch)
Black Salamander
PLATES 107, 108

Fig. 205. *Aneides flavipunctatus;* adult; Humboldt Co., California (R. W. Van Devender).

IDENTIFICATION. *Aneides flavipunctatus* is a medium-sized salamander that superficially resembles the slimy salamander, *Plethodon glutinosus* (Lynch 1974, 1981). The tail is prehensile and the jaw musculature is greatly hypertrophied, giving the head a triangular appearance. The toes are expanded, but not flattened. The dorsum is dark gray to black with varying amounts of brassy pigmentation and white or cream-colored spots. The venter is grayish black. Adults measure 10–17 cm TL and the modal number of costal grooves is 16–17. Adult males have a heart-shaped mental gland and small gray glands on the belly, and are about the same size as adult females (Lynch 1981).

Brassy pigmentation is most evident in hatchlings and small juveniles and is usually absent in adults except in the northwestern portion of the range (Lynch 1981). Juveniles have yellow pigmentation at the limb bases, as well as proportionately larger heads, longer limbs, and shorter tails than adults. Black salamanders in northern California and Oregon are often frosted with green coloration that is particularly conspicuous in hatchlings and young juveniles (Nussbaum et al. 1983). A partial albino was reported from California (Seeliger 1945).

SYSTEMATICS AND GEOGRAPHIC VARIATION. California populations show a high level of genetic differentiation (Larson 1980). In particular, the Shasta and Santa Cruz isolates are well differentiated (Nei's distances [D] = 0.17 and 0.19) from the main part of the range of the species. There is no evidence of marked geographic variation in chromosomal morphology or banding patterns (Sessions and Kezer 1987).

Myers and Maslin (1948) recognize two subspecies (*niger* and *flavipunctatus*) that differ in coloration around the cloacal region, number of costal grooves, and degree of pigmentation. Lynch (1981) provides a detailed analysis of geographic variation in coloration and notes that some of the features used by Myers and Maslin (1948) to separate subspecies are ontogenetic rather than racial differences. In addition, geographic variation in many traits is clinal. In general, adults in northern populations grow larger and have proportionately longer legs, smaller white spots, more brassy pigmentation, and more vertebrae than southern populations. Here I follow Lynch's (1981) recommendations and do not recognize subspecies.

DISTRIBUTION AND ADULT HABITAT. *Aneides flavipunctatus* occurs in lowland forests from the Applegate Valley in extreme southern Oregon to the Santa Cruz Mountains in central California in areas that receive >75 cm of annual precipitation. Populations occur as high as 1700 m, but most are at elevations <600 m in mesic forests that do not experience sustained freezing temperatures (Lynch 1981; Nussbaum et al. 1983). Specimens are often found in seeps in talus slopes and in wet soil beneath logs or rocks along streams. They have also been collected from a mine shaft, from beneath rocks and logs in fields and old pastures, and from beneath debris in recently burned areas (Staub 1993; Stebbins 1951; Wood 1936).

This species occurs as two or three geographic isolates (Larson 1980; Lynch 1981). The Santa Cruz isolate is clearly disjunct from the others. A second group of populations south of Mt. Shasta and east of the Trinity Mountains may be isolated from populations to their immediate west.

BREEDING AND COURTSHIP. No published data are available on the breeding season or courtship behavior.

Range of *Aneides flavipunctatus*

REPRODUCTIVE STRATEGY. The few observations of nesting suggest that most females lay eggs in July or early August in cavities below ground. The eggs are 5.9 to 6.4 mm in diameter and are surrounded by two tough, thin, gelatinous sheaths. Each egg is suspended by a short stalk about 5 mm long that is attached to a common base (Stebbins 1951; Storer 1925). Single nests have been found 23 and 38 cm below ground (Storer 1925; Van Denburgh 1895). Clutch size estimates based on ovarian complements are 25 ova for a single specimen (Van Denburgh 1895), 8–24 ova for nine specimens (Stebbins (1951), and 5–25 ova for 112 specimens (Lynch 1981). The mean for the latter estimate (estimated from figure in text) is 12, and clutch size is positively correlated with female SVL.

TERRESTRIAL ECOLOGY. Black salamanders may be active on the ground surface year round in streamside habitats in the southern portion of the range, but move underground during the summer in the northern portion of the range. Adults forage for small invertebrates on the ground surface during wet weather. Specimens from northern California contain a wide

Fig. 206. *Aneides flavipunctatus;* adult; Humboldt Co., California (W. P. Leonard).

Fig. 207. *Aneides flavipunctatus;* eggs; Humboldt Co., California (W. P. Leonard).

variety of prey, but millipedes, beetles, termites, hymenopterans, flies, and collembolans make up most of the diet (Lynch 1985). Total volume of food in the stomach tends to increase with body size to about 50 mm SVL, then remain steady. The mean number of prey consumed declines with body size, suggesting that larger individuals tend to ignore small prey.

Adults are territorial and will aggressively defend familiar areas in captivity. In staged laboratory trials, residents respond to nonsubmissive intruders with aggressive posturing and threat displays (Staub 1993). These are often followed by more aggressive acts that include chasing, biting, or bite-holding, in which the resident bites and then pins the intruder to the substrate while using the trunk and tail as leverage. Unlike many plethodontid salamanders, intruders in this species often fight back, and the skin of both contestants is frequently scarred or lacerated. Field-caught animals also have scars, and males have a higher frequency of scars than females.

PREDATORS AND DEFENSE. Black salamanders have several antipredator behaviors, including biting, fleeing, remaining immobile, defensively posturing, and producing copious amounts of gluelike skin secretions. Large adults will often bite western terrestrial garter snakes (*Thamnophis elegans*), and in some cases the snakes suffer serious injuries (Lynch 1981). Adults often flee when uncovered in the field, but juveniles normally remain immobile, a response that makes them less conspicuous to predators (Jones 1984). On rare occasions, a juvenile may defensively posture in a manner similar to that used during agonistic displays toward conspecifics. This posture involves extending the legs outward, arching the back, moving the body from side to side, and undulating the distal third of the tail.

COMMUNITY ECOLOGY. Other salamander species that share habitats with *A. flavipunctatus* have similar diets, although some species-specific differences are evident (Lynch 1985). The extent to which *A. flavipunctatus* competes interspecifically for food resources with other community members is not known.

CONSERVATION BIOLOGY. The black salamander was formerly common in many areas of its range but has become uncommon or rare in recent years (D. B. Wake, pers. comm.). A long-term monitoring program for this and many other California salamanders is needed to quantify long-term population trends.

Aneides hardii (Taylor)
Sacramento Mountain Salamander
PLATE 109

Fig. 208. *Aneides hardii*; adult; Otero Co., New Mexico (R. W. Van Devender).

IDENTIFICATION. *Aneides hardii* is a small, short-legged species with a dark brown dorsum that is mottled with greenish gray or goldish bronze. The toe tips are rounded and the upper teeth project slightly beyond the lip. The venter is light brown to cream colored and the throat is usually lighter colored than the belly. Adults measure 7.5–11.5 cm TL and there are 14–15 costal grooves. Males tend to have a more triangularly shaped head than females owing to hypertrophied jaw musculature (Wake 1965b). Males in populations studied by Schad et al. (1959) are 5–9% longer in SVL than females. Juveniles have a brown or bronze dorsal stripe and a flesh-colored or light gray throat.

Range of *Aneides hardii*

SYSTEMATICS AND GEOGRAPHIC VARIATION. Populations in the Capitan, Sacramento, and White mountains are geographically isolated from each other, but have undergone little morphological, molecular, or chromosomal differentiation (Pope and Highton 1980; Schad et al. 1959; Sessions and Kezer 1987). Genetic divergence is low (Nei's $D = 0.03$), and the three groups probably shared a common ancestor during the late Pleistocene (Pope and Highton 1980). No subspecies are recognized.

DISTRIBUTION AND ADULT HABITAT. *Aneides hardii* occurs in Douglas-fir forests, Engelmann spruce–white fir forests, and alpine tundra at elevations of 2400–3570 m in the Capitan, Sacramento, and White mountains of New Mexico (Scott and Ramotnik 1992). Specimens live in well-rotted logs and beneath bark, logs, rocks, and rubble (Johnston and Schad 1959; Lowe 1950; Schad et al. 1959; Stebbins 1951). Populations primarily inhabit moist north- or east-facing slopes and are associated with large, downed logs (Schad et al. 1959; Scott and Ramotnik 1992). Although specimens are most often encountered in coniferous forests, they are occasionally found beneath rocks and in moss mats in alpine tundra above the timberline (Moir and Smith 1970).

BREEDING AND COURTSHIP. An analysis of male reproductive tracts suggests that mating occurs sometime after the brooding season when adults are in subterranean retreats (Williams 1978). Because of the very short growing season, females probably breed every second or third year after becoming sexually mature. The overall sex ratio for several populations is near 1:1 (Johnston and Schad 1959). Courtship behavior has not been described.

REPRODUCTIVE STRATEGY. Adults spend 7–8 months below ground and are only active on the surface during the warm summer months. Individuals emerge from underground retreats in June or early July and females oviposit shortly thereafter (Johnston and Schad 1959). Hatching occurs in August or early September just before the population retreats to underground winter quarters. Adults probably nest in both underground retreats and large, rotting logs. However, virtually all nests found to date have been in decaying logs.

Nests have been found in July and August in hollow chambers in rotting logs. A female lays a small cluster of eggs and remains with the young through hatching. Each egg is suspended by a pedicel that is attached to a common base, and the late-term embryos have large, flat, trilobed allantoic gills that are characteristic of the genus. The outer capsules are 7–9 mm in diameter. Nesting records include a cluster of three eggs in late developmental stages within a large, decaying Douglas-fir log in August (Lowe 1950), a female with 10 late-term embryos in a Douglas-fir log on 12 August (Schwartz 1955), and clutches of one, four, four, and six eggs inside chambers within large, rotten fir logs in July (Johnston and Schad 1959).

Estimates of clutch size based on counts of mature ova are 9–11 ova for 3 females (Schwartz 1955) and an average of 8 ova for 16 females (Williams 1978). The maximum size of mature ova is 4.3 mm in diameter. The discrepancy between clutch sizes observed in nature and number of mature ovarian eggs in females indicates either that embryonic mortality rates are high or that females do not lay their entire complement of eggs. Williams (1978) suggested that only about half of the mature ovarian eggs are deposited during a given season.

TERRESTRIAL ECOLOGY. Individuals are active on the ground surface in summer when seasonal conditions are warm and wet. Growth rates are undoubtedly slow in juveniles because of the short growing season, but age at first reproduction is unknown. Males and females sexually mature when they reach about 40 mm SVL (Johnston and Schad 1959; Schad et al. 1959; Williams 1978). Two instances of agonistic behavior have been observed in laboratory animals (Staub 1993).

Scar frequencies of field-caught specimens suggest that this species is less aggressive than other *Aneides* species.

Johnston and Schad (1959) found snails, spiders, mites, orthopterans, hemipterans, carabid and buprestid beetles, ants, and wasps in 16 specimens. Most animals contain nematodes, which are probably parasites, in their guts. Adults appear to be sight-oriented predators that actively hunt prey that are within a short distance.

PREDATORS AND DEFENSE. Natural predators are not known but presumably include birds and small mammals.

COMMUNITY ECOLOGY. Because of its geographic isolation from other salamanders, *A. hardii* lives in amphibian communities with little or no diversity. Studies of the competitive ability of this species relative to species that inhabit more complex amphibian communities might provide important insights into the evolution of competitive ability in vertebrates.

CONSERVATION BIOLOGY. The extent to which this species is dependent on old-growth forests with large decaying logs is not known. Scott and Ramotnik (1992) surveyed sites 0.5–3.5 years after an old-growth forest was logged and found *A. hardii* to be present. However, juveniles compose a smaller portion of the population on logged sites compared with unlogged sites, suggesting that reproductive rates decline following timber removal.

Aneides lugubris (Hallowell)
Arboreal Salamander
PLATES 110, 111

IDENTIFICATION. The arboreal salamander is a large species of *Aneides* that has a dark brown dorsal ground color with small cream to sulfur-yellow spots. Spots occur on the head, trunk, tail, and limbs, but these vary in intensity and distribution. In some populations they are nearly absent, and in many individuals the spots are more concentrated on the sides of the body. The venter is creamy white and the undersides of the tail and feet are dull yellow. The head is widest behind the eyes and is very large and triangularly shaped owing to hypertrophied jaw musculature. The

Fig. 209. *Aneides lugubris;* adult; California (R. W. Van Devender).

toe tips are expanded and squarish, and the tail is prehensile (Ritter and Miller 1899). The front teeth are enlarged and project beyond the lips.

This is the largest species of *Aneides*, with adults reaching 11-18 cm TL. Sexually active males have a heart-shaped mental gland. Hatchlings and young juveniles are frosted with silvery color, and lack or show reduced yellow spotting. Hatchlings measure 26-32 mm TL (Stebbins 1951; Storer 1925).

SYSTEMATICS AND GEOGRAPHIC VARIATION. The most conspicuous geographic variation is in spotting patterns (Morafka and Banta 1976). The population on South Farallon Island has more prominent spotting and a smaller adult body size than most other populations. This form was recognized as a separate subspecies, *A. l. farallonensis*, by Van Denburgh (1905). Most recent treatments do not recognize the South Farallon Island population as a valid subspecies because similarly spotted specimens occur in the Gabilon Range about 150 km southeast of the Farallon Islands; these similarities may reflect an early Pliocene land connection to the islands (Morafka and Banta 1976). Spotting in Sierra Nevada populations is somewhat reduced compared with that in populations elsewhere.

Aneides lugubris consists of two chromosomally differentiated groups that intergrade in south and east-central Medocino Co. about 90 km north of the San Francisco Bay region (Sessions and Kezer 1987).

DISTRIBUTION AND ADULT HABITAT. The arboreal salamander occurs in yellow pine and black oak forests in the Sierra Nevada, and in coastal oak forests from northern California to Baja California. Geographic isolates from the main range occur in the foothills of the Sierra Nevada and on South Farallon, Santa Catalina, Los Coronados, and Año Nuevo islands. This species is largely absent from areas receiving <25 cm of precipitation per year. Specimens have been collected in logged clearings in redwood forests (Bury and Martin 1973), from under rocks and woody surface cover, and in stumps, logs, wood rat houses, mine shafts, rodent burrows, and stone walls (Rosenthal 1957; Stebbins 1951; Storer 1925). Climbing is facilitated by the expanded digits and prehensile tail, and individuals are sometimes found 9-18 m above ground in trees (Ritter 1903; Stebbins 1951). A 5-million-year-old fossil of this species was discovered in the Sierra Nevada foothills (Clark 1985).

Fig. 210. *Aneides lugubris*; adult; Humboldt Co., California (W. P. Leonard).

BREEDING AND COURTSHIP. Data on the breeding season are not available and courtship behavior has not been described.

REPRODUCTIVE STRATEGY. Most females oviposit in June and July during the dry season and guard their eggs through hatching. Eggs have been found under surface objects, in underground cavities, and in tree holes from July through September. Freshly laid eggs are about 7 mm in diameter, whitish to cream, and surrounded by two jelly envelopes. The outer capsule is 7-9.5 mm in diameter, and each egg is supported by an 8- to 20-mm-long pedicel attached to an overhead support. The eggs are often attached to a common base and become wrapped around each other to form a cluster, although single eggs may be deposited within a few millimeters of the main cluster (Kessel and Kessel 1942; Stebbins 1951; Storer 1925). The embryos have allantoic, trilobed gills that are lost immediately after hatching.

Nesting records include a female found guarding 19 late-term eggs in late July below ground at the base of a palm tree (Ritter and Miller 1899), and a cluster of 17 late-term eggs found on 19 September attached to the underside of a large rock buried 60 cm below ground (Kessel and Kessel 1942). However, most nests have been found in tree holes, including 12 egg masses and over 100 salamanders in oak trees on the University of California campus at Berkeley (Ritter 1903; Stebbins 1951). The eggs are suspended from overhangs within tree cavities and the females coil about them. Males are often present in the nesting cavities.

The number of eggs per clutch in the 12 masses found at Berkeley ranged from 12 to 18 (Ritter 1903)

Range of *Aneides lugubris*

compared with 14, 16, 17, and 24 eggs for other clutches (Stebbins 1951; Storer 1925). The number of mature ova in specimens examined by P. K. Anderson (1960) varied from 5 to 26 and was positively correlated with SVL. Individuals on islands generally produce smaller clutch sizes than those on the mainland.

The embryos hatch in the fall after a 3- to 4-month incubation period. The timing of breeding in this species permits hatchlings to disperse from arboreal nests after the arrival of fall rains.

TERRESTRIAL ECOLOGY. Arboreal salamanders are active on the ground surface during and shortly following the rainy season when soil moisture is high. Like many other salamanders, this species is a sit-and-wait predator. At night individuals emerge from cover and wait for passing prey. As animals grow, increased gape width allows them to incorporate larger prey into their diet. Adults feed proportionately less on small prey than juveniles, but still consume substantial numbers of small prey (Lynch 1985). The spatial distribution of individuals on the ground surface is best explained by the distribution of cover objects rather than the distribution of food resources (Maiorana 1978a).

Arboreal salamanders occasionally prey upon *Batrachoseps* species (Miller 1944), but the primary food is invertebrates. Zweifel (1949) lists beetles, caterpillars, sow bugs, centipedes, and ants in the stomachs of 13 animals from the Santa Monica Mountains in southern California. The stomachs of 157 individuals from northwestern California contained millipedes, beetles, termites, hymenopterans, flies, and collembolans (Lynch 1985). The prey of other specimens from northern California include insects, millipedes, centipedes, isopods, worms, and snails (Bury and Martin 1973).

Juvenile growth rates and age at first reproduction are poorly documented. Size histograms of animals collected at several sites in California suggest that three or more years are required to reach sexual maturity (P. K. Anderson 1960).

The arboreal salamander is more tolerant of dry conditions than more aquatic salamanders and is often the last to go beneath ground as the summer dry season begins (Cohen 1952; Ray 1958). Individuals often move into tree holes when conditions become dry. Adults are active on the ground from November through early May in natural habitats near Berkeley, California (Rosenthal 1957).

Adults appear to be territorial during certain times of the year and use agonistic displays and biting to contest resources (Staub 1993). About 15% of field-caught specimens in this study had scars that were presumably the result of aggressive encounters between conspecifics. Males and females do not differ in the proportion with scars. Densities of *A. lugubris* on offshore islands are generally higher than those on the mainland. Local densities in some areas may exceed 4900 salamanders/ha (P. K. Anderson 1960).

PREDATORS AND DEFENSE. Individuals often produce a high-pitched squeak when disturbed or molested (Ritter and Miller 1899). The adults will sometimes bite humans attempting to handle them and use biting as an antipredator defense against natural predators (Storer 1925). The adults will viciously bite predatory western terrestrial garter snakes (*Thamnophis elegans*), and in rare instances a snake may die from injuries suffered during an *Aneides* defensive strike (Lynch 1981).

COMMUNITY ECOLOGY. Ecological interactions between *A. lugubris* and other salamanders are poorly understood. Significant dietary differences occur between syntopic populations of *A. lugubris* and *Batrachoseps attenuatus* in northern California (Lynch 1985). However, broad dietary overlap may occur between individuals of similar size in habitats with few large prey (Maiorana 1978a). When arboreal salamanders feed only on small prey, large individuals tend to consume less food per unit body weight than smaller conspecifics. The extent to which these species compete for food is still not fully known.

CONSERVATION BIOLOGY. The arboreal salamander is common in many areas; however, populations appear to have declined during the last 20 years in some areas of the range (D. B. Wake, pers. comm.). Large oaks that are used for nesting and aestivation are critical microhabitats and should be preserved whenever possible.

Ensatina eschscholtzii Gray
Ensatina
PLATES 112, 113, 114

IDENTIFICATION. The ensatina is a medium-sized plethodontid with a relatively short body and long legs. The tail is rounded dorsally and has a prominent constriction that encircles its base. This species is polytypic and shows marked geographic variation in color patterns. Individuals in most coastal populations tend to be uniformly colored, whereas those in the Sierra Nevada and surrounding inland regions are strongly blotched. Detailed descriptions of the seven subspecies are given under "Systematics and Geographic Variation." Relative to females, males have a more truncated snout, swollen cloacal glands, and a more enlarged upper lip. In addition, sexual dimorphism in tail

Fig. 211. *Ensatina e. platensis;* adult; California (R. W. Van Devender).

length and morphology is extreme. Males have long, slender tails and females have short, stout tails.

Detailed studies of geographic variation in sexual dimorphism in body size have not been conducted, but size differences between the sexes appear to be minimal. Adult females in one population are 6% larger in SVL than adult males (Gnaedinger and Reed 1948). Adults measure 7.5–15.5 mm TL and there are 12–13 costal grooves.

Hatchlings measured by Stebbins (1954) averaged 20 mm SVL, but smaller individuals undoubtedly occur in natural populations. Hatchling color patterns have not been described for most subspecies. Hatchling *E. e. oregonensis* have a black dorsum overlain with purplish brown blotches. The dorsum of the tail is rust to orangish red and the groin area is vivid orange (Norman and Norman 1980). Hatchlings of the blotched subspecies also have very dark dorsums and orangish red limb bases (D. B. Wake, pers. comm.).

SYSTEMATICS AND GEOGRAPHIC VARIATION. *Ensatina* is a geographically and genetically variable taxon that traditionally has been treated as a single species with seven recognized subspecies. These include both blotched and unblotched color forms. *Ensatina* has also traditionally been treated as a "ring" species whose subspecies form a ring-shaped distribution around the Central Valley of California and do not interbreed where the ends of the ring overlap in southern California (Brown 1974; Jackman and Wake 1994; Stebbins 1949a; Wake and Yanev 1986; however, see other evidence discussed below in this section). One unusual element of the ring pattern is *E. e. xanthoptica*, which occurs on both the east and west sides of the Central Valley. The Sierra Nevada populations of *E. e. xanthoptica* probably reflect invasion of coastal populations across the Central Valley when past climatic conditions were cooler and moister.

Early interpretations of systematic relationships among members of the ring species are based primarily on color patterns (Brown 1974; Stebbins 1949a). These analyses suggest that hybrid zones between subspecies range from very narrow to very broad, although recent evidence based on molecular data indicates that contact zones are very narrow (Wake et al. 1989). Brown (1974) found that Sierra Nevada populations of *E. e. platensis* and *E. e. xanthoptica* hybridize along a narrow zone, but a clear zone of intergradation is not readily evident. Thus limited evidence from analyses of external traits suggests that many of the members of the ring are at levels between subspecies and full species.

Detailed studies of protein and mitochondrial DNA variation have helped resolve evolutionary relationships between members of the ring complex (Jackman and Wake 1994; Moritz et al. 1992; Wake and Yanev 1986; Wake et al. 1986, 1989). These studies indicate that geographic patterns of genetic variation are complex and probably reflect repeated extinctions and invasions of populations within the range of the species. Genetic distances between local populations both within and among subspecies are, in many cases, higher than those that occur among coexisting plethodontid species elsewhere (Nei's distances [D] often >0.40–0.50). These data suggest that gene exchange is occurring at low rates both within and among many of the recognized subspecies, and that some groups are at or near the species level of evolutionary differentiation.

Gene flow between groups of the complex is very restricted or absent in two general regions. The first is in the Sierra Nevada foothills, where gene flow is very limited between populations of *E. e. platensis* and *E. e. xanthoptica* (Wake et al. 1989). The second is in southern California, where Brown (1974) and Wake et al. (1986) found evidence of slight introgression and hybridization of *E. e. eschscholtzii* and *E. e. klauberi* at three sites. At a fourth site the two forms coexist microsympatrically without hybridizing.

One interpretation of these patterns is that the complex is a double ring species, i.e., there are two zones where overlapping subspecies show little or no gene exchange (Jackman and Wake 1994). A second interpretation is that the group consists of two or more species that form a superspecies complex. Frost and Hillis (1990) recommend recognizing *E. e. klauberi* as a separate species because it is disjunct from other blotched forms to the north and occurs sympatrically with *E. e. eschscholtzii* with little or no hybridization. Using similar reasoning (i.e., ignoring the ring species nature of the complex), one could argue that the Sierra Nevada isolate of *E. e. xanthoptica* also merits recognition as a separate species. Jackman and Wake (1994) argue against splitting the complex into several species because the genetic evidence is consistent with the ring species concept. Either of these views is valid depending on one's taxonomic philosophy.

Ensatina is perhaps best viewed as being both a

Range of *Ensatina eschscholtzii*

semispecies complex and a ring species. *Ensatina* has characteristics of both a single species and a species complex, and its members do not readily conform to any of the discrete taxonomic categories used to classify organisms. Nomenclatural changes will not rectify the problem, but will almost certainly lead to nomenclatural instability. As such, I follow Wake and his colleagues' recommendation of recognizing the *Ensatina* complex as a single polytypic species that is a ring species.

Seven subspecies are recognized, and the following descriptions are condensed from detailed information presented by Brown (1974) and Stebbins (1949a, 1951). Adult Oregon ensatinas (*E. e. oregonensis*) are light to dark brown above, often with minute pale yellow or orange flecks. The sides are pale orange to yellowish orange and are usually mottled with light-colored markings concentrated in the costal grooves. The eyelids are dark. The upper surface of the proximate portion of the limbs and the lower surface of the limbs and tail are light orange or yellowish. The remaining ventral surfaces are whitish, or are marked to varying degrees with pale orange or yellow spots or blotches and fine black stippling. Juveniles generally resemble adults of *E. e. picta* in having dark blotching on the body and tail.

The yellow-eyed ensatina (*E. e. xanthoptica*) has orangish brown coloration above that extends on the sides downward to the region of the upper surface of the limbs. The lower limit of dark coloration on the sides is irregular, and the eye has a bright yellow patch. The eyelids, venter, and sides of the head, tail, and body are bright reddish orange.

The Monterey ensatina (*E. e. eschscholtzii*) is uniformly reddish brown to pale brown above. The upper eyelids are usually pinkish to brown and lighter than the dorsal ground color, and the eyes are black. The tail often has vague orange blotching, and the ventral

Fig. 212. *Ensatina e. oregonensis*; adult; Clark Co., Washington (W. P. Leonard).

surfaces of the head, body, and tail are whitish. The undersurfaces of the limbs and tail are pinkish white to pale orangish red. The proximal segments of the limbs are orange or reddish orange above.

The painted ensatina (*E. e. picta*) has a dark tan to brown ground color with blackish brown dorsolateral blotches. The sides of the body carry light orange to yellow mottling, and the ventral surface is light orange with a pinkish cast.

The Sierra Nevada ensatina (*E. e. platensis*) has reddish orange spots on the dorsum and sides of the head, body, and tail. The spots have irregular borders and vary from flecks to spots 3 mm in diameter. The dorsal ground color is dark brown, and the ventral surface is pale gray to whitish. The upper eyelids are spotted and yellow to orange. Juveniles are darker and have less spotting above.

The yellow-blotched ensatina (*E. e. croceater*) has a black dorsal ground color overlain with irregular lemon yellow to yellowish cream spots up to 7 mm in diameter. The venter is pale gray.

The large-blotched ensatina (*E. e. klauberi*) has a black dorsal ground color overlain with large orange to pale, flesh-colored blotches often 5 mm or more in maximum diameter. The blotches are often connected to form diagonal or transverse bands and are regular in outline. The venter is deep gray, often with a purplish tinge.

Geographic variation in color patterns in this species may reflect cryptic coloration related to soil color, or possibly mimicry of *Taricha* species (Stebbins 1949a; Wake et al. 1989). The coastal forms appear to be *Taricha* mimics, whereas the blotched forms may reflect selection for cryptic coloration. A general cline from relatively small to large body size occurs from north to south. *Ensatina e. picta* deviates from this pattern in being exceptionally small.

DISTRIBUTION AND ADULT HABITAT. Ensatinas occur from southwestern British Columbia and Vancouver Island southward through the Coastal and Sierra Nevada ranges to Baja California (Mahrdt 1975). This relatively common species occurs in a wide variety of terrestrial habitats in hilly or mountainous terrain from near sea level to slightly over 3350 m. Habitats include chaparral, mesic coastal forests, coastal live oak woodlands, pine-oak woodlands, coastal sage scrub, and mixed conifer-hardwood forests (Stebbins 1949a). Ensatinas normally inhabit mesic microhabitats with moist, rocky soils and thick leaf mulch or litter. They are often found beneath decaying logs and woody debris (Corn and Bury 1991; Stebbins 1949a, 1951). Populations in arid regions of southern California primarily occur on north-facing slopes of deep canyons or in other microhabitats that provide cool, moist conditions. In southern California ensatinas are often found in the vicinity of streams where soils are relatively moist, or in shaded, moist habitats where there is good canopy cover. Farther north, they occupy a broader variety of habitats (Stebbins 1951). In Oregon ensatinas are sometimes less abundant in streamside habitats than on adjoining slopes (McComb et al. 1993a,b).

Specimens live beneath moss mats, rocks, boards, logs, and other surface cover, and in rotten logs, leaf litter, and rodent burrows (Aubry et al. 1988; Bury and Corn 1988a; Gnaedinger and Reed 1948; Stebbins 1949a, 1951). This species reaches its highest local densities where large amounts of coarse, woody debris are present on the forest floor.

BREEDING AND COURTSHIP. Female *E. e. xanthoptica* have sperm caps protruding from the vents from November to March, whereas males have spermatozoa in the sperm ducts from autumn through March (Stebbins 1951, 1954). Males collected in April usually lack viable sperm in the ducts. These observations indicate that breeding occurs from the time of autumnal emergence until adults begin to disappear from the surface in late March. *Ensatina e. platensis* is sexually active in late April and early May, suggesting that interior populations have later breeding seasons than coastal ones. Mating in many populations appears to be most intense in late winter and early spring, a time when adults are often found in pairs beneath cover objects (Gnaedinger and Reed 1948; Stebbins 1949a, 1954).

Courtship of *E. e. xanthoptica* is described in detail by Stebbins (1949b, 1951, 1954). The following is a summary based on his descriptions. The male initially creeps to the side of the female with his body carried close to the ground. He noses the side of the female's neck, head, throat, and face with the sides of his head and neck. He then slides past her, moving beneath her gular region until his sacral region is against her throat. He massages her throat by rotating his hindquarters. If sufficiently stimulated, the female will position her throat against the male's back and follow him as he slowly moves forward with the back arched sharply upward. The couple engages in a tail-straddle walk with the female straddling the male's tail. This stage may last for several hours, and in some cases as long as 6 hours (Arnold 1977).

The male finally stops, extends his legs laterally, and places his vent against the substratum. He then rocks laterally and deposits a spermatophore as the female strokes his sacral region with her throat in a direction counter to his. After stroking him for a minute or so, she lifts her head and the two advance until the female is above the spermatophore. The female squats and picks up the spermatophore while the male lurches backward and throws his tail over her back. The distal portion of the tail writhes violently over the back and tail base of the female. The pair then separates, but may court again within a few moments. Male homosexual behavior often occurs that is presumed to be a form of sexual interference (Arnold 1977). Sex ratios are typically near 1:1 (Staub et al. 1995).

REPRODUCTIVE STRATEGY. Females oviposit in late spring in central and southern coastal populations, and in early summer in northern coastal areas (e.g., Norman 1986) and higher-elevation sites in the Sierra Nevada (Stebbins 1949a, 1951; D. B. Wake, pers. comm.). Each female lays a single cluster of eggs in an underground passage, beneath bark, or in or beneath logs. Eggs of ensatinas have also been found in the nest chamber of a mountain beaver (Stebbins 1954). Oviposition requires about 1 day, and the female may lie on her side or back while laying eggs (Jones and Aubry 1985; Stebbins 1954). Freshly laid ova measure 4.5–6 mm in diameter on average, are whitish to cream colored, and are surrounded by two jelly envelopes. They are deposited in grapelike clusters but are not connected by jelly cords (Stebbins 1951, 1954). Two clusters described by Norman (1986) were 11–15 mm wide and 16–19 mm long.

Diameters of ova and the surrounding capsules of Washington specimens are 3.5–4.5 mm and 5–7 mm, respectively (Norman 1986; Norman and Norman 1980). Two clutches from northern California have ova and egg capsules that are 5.0–5.5 mm and 7.6–7.8 mm in diameter, respectively (Storer 1925). The late-term embryos have large, highly vascularized, leaflike gills, and females brood the embryos through hatching.

Nesting females have been discovered in numerous locations, and the descriptions of the nests and eggs are similar for all. Stebbins (1951) discusses several unpublished accounts of eggs collected in California. Nesting records for *E. e. picta* include clutches of 19 and 25 late-term eggs found on 9 August within a large, rotted spruce log in northern California, a cluster of 13 midterm eggs found on 4 June nearly 60 cm below ground in a rodent burrow in west-central California, and two clutches of 16 eggs at the toe bud stage found on 26 July in damp soil beneath slabs of redwood in northern California (Storer 1925). In all instances, females were guarding the eggs.

Nesting records for *E. e. platensis* include clutches of 8 and 9 eggs at the hatching stage found on 9 September under a rotten log in east-central California, 12 eggs in early developmental stages found on 21 May beneath the bark of a fallen Douglas-fir in the Sierra Nevada, and 11 eggs with embryos in late developmental stages found on 8 August beneath bark on the same log (Howard 1950; Stebbins 1951). The outer capsules of eggs found earlier in the season are 6–7 mm in outer diameter, whereas those found later are 10–11.5 mm in diameter. Nineteen females that oviposited in the laboratory after receiving hormonal injections laid an average of 12 eggs (range = 9–16; Collazo and Marks 1994).

Norman and Norman (1980) report finding clusters of 8 and 10 eggs with brooding female *E. e. oregonensis* in large, rotting logs in Washington. The two females at this site selected small, rotten cavities about 13×18 mm in size for nesting and did not attach the eggs to the substrate. Other observations in Washington include a female ovipositing 10 eggs in a cavity in a large, rotting log (Jones and Aubry 1985) and brooding females with clutches of 8 and 9 eggs in cavities in rotting logs (Norman 1986).

The number of large ovarian eggs in five *E. e. oregonensis* varied from 14 to 16 (Gnaedinger and Reed 1948). The mean number and range of large ovarian eggs reported by Stebbins (1954) for different subspecies are as follows: *E. e. croceater*, 16 (only one specimen examined); *E. e. eschscholtzii*, 13 (8–17); *E. e. klauberi*, 11 (8–14); *E. e. oregonensis*, 11 (7–14); *E. e. picta*, 15 (10–21); *E. e. xanthoptica*, 11 (8–17). Clutch size is positively correlated with female SVL. The incubation period lasts about 4–5 months, and the embryos hatch in late summer or early autumn when they have reached 20–26 mm SVL (Stebbins 1951, 1954).

TERRESTRIAL ECOLOGY. Ensatinas are euryphagic predators that show little tendency to specialize on specific prey. Juvenile ensatinas consume a wider size range of prey than juveniles of other sympatric plethodontids, a pattern that reflects the relatively large head of this species (Lynch 1985). Food items in 21 Oregon specimens included sowbugs, mites, spiders, millipedes, centipedes, collembolans, and beetles (Gnaedinger and Reed 1948). Other Oregon specimens contained snails, annelids, and a variety of arthropods, particularly mites, spiders, collembolans, hemipterans, flies, and hymenopterans (Altig and Brodie 1971). A diverse array of dietary items occurred in 37 specimens from redwood forests (Bury and Martin 1973), including collembolans, beetles, spiders, millipedes, isopods, and snails. The principal prey of 45 specimens from southern California were isopods, centipedes, spiders, collembolans, and beetles (Zweifel 1949). Prey in specimens examined by Stebbins (1954) were similar to those reported from other California studies.

Growth is rapid for the first 2–3 years of life, then slows dramatically as individuals approach sexual maturity. In a detailed study by Stebbins (1954), male and female *E. e. xanthoptica* became sexually mature when they reached 48–55 and >60 mm SVL, respectively. Juveniles grow 9–14 mm SVL per year and mature sexually about 3–4 years after hatching. The largest adults are estimated to be at least 8.5 years old. Growth rates for adult *E. e. platensis* are 1.8 and 3.4 mm/year for males and females, respectively (Staub et al. 1995). Adults in this population are estimated to be 5.5–14.9 years old.

Juveniles and adult ensatinas are most active on the surface during wet weather when temperatures are moderate (Stebbins 1949a; Storer 1925). Ensatinas remain beneath ground throughout the dry summer in most areas of their range and can tolerant substantial dehydration (Stebbins 1945). They emerge with the first rains of autumn and are active on the surface through spring except in areas where severe winter weather occurs. Surface activity is most intense immediately following rains and decreases whenever the interval between rains is sufficient to dry the soil (Stebbins 1954).

Adults tend to be more diurnally active than other western plethodontids (Stebbins 1949a). Surface activity of adults diminishes after the onset of the egg laying period, but juveniles remain on the surface until the arrival of dry summer weather. Individuals cannot burrow through firmly packed soil. Instead, they frequent microhabitats with rodent burrows, root channels, and other passages that allow them to move beneath ground during dry weather. Adults may live 60–90 cm below the soil surface in summer, and rarely have large amounts of food in their stomachs when unearthed (Stebbins 1954). They have occasionally been taken within rotten logs or in moist duff in the summer months.

Laboratory observations of staged encounters between residents and intruders from four California populations representing three subspecies (*oregonensis, platensis, xanthoptica*) suggest that ensatinas are territorial outside the breeding season (Wiltenmuth 1996). When same-sex, similar-sized intruders from the same population are introduced into resident containers, residents are more overtly aggressive than intruders, and resident males are more than four times more aggressive than resident females. Males generally exhibit more agonistic behaviors than females, whereas females rely more on sensory behaviors such as nose-tapping to mediate interactions. Residents presumably mark their territories with fecal material and/or cloacal secretions, and intruders orient to these and the resident's presence by nose-tapping frequently. The frequency of most behaviors is independent of the size of the paired contestants, although avoidance behavior tends to increase with the size of the male intruder. The relative size of contestants may significantly influence the frequency of several behaviors associated with avoidance and passive aggression, and these vary between the sexes.

Stebbins (1954) tracked the movements of marked individuals over a 4-year period and found that adult males tend to move about twice as far between captures as adult females. Young *E. e. xanthoptica* show a greater tendency to remain beneath the same cover object between captures than adults. The average distance moved between captures (for animals that moved) is 5.5 m for mature females and 10 m for mature males. Young animals move an average of 6.1 m between captures. Frequency distributions of movements for young, subadult, and adult salamanders are skewed to the right, and in a few instances animals move >30 m between captures. The average maximum width of the home range of eight females was 10 m, and that of six males was 19.5 m. There is no evidence of territorial behavior in either laboratory or field animals.

Staub et al. (1995) conducted a 5.3-year study of movements of *E. e. platensis* that involved monitoring relatively large plots. Males at this site are generally more active than females and tend to engage in long-range movements more than females. The maximum distances moved by males and females are 120 and 61 m, respectively. In some cases individuals move relatively long distances but eventually return to their original site of capture. Estimated densities for two populations are 0.17 ensatina/m^2 in favorable habitats in California (Brown 1976) and 1.5 ensatinas/m^2 in Oregon (Gnaedinger and Reed 1948).

PREDATORS AND DEFENSE. Garter snakes (*Thamnophis*) and Steller's jay (*Cyanocitta cristata*) prey upon ensatinas (Beneski 1989; Fitch 1940; Stebbins 1954). Snakes often gape repeatedly after attacking ensatinas, a behavior that suggests that the tail secretions serve to repel potential predators (Hubbard 1903). When attacked, an individual will often stand stiff-legged on its toes with the body arched downward, the tail elevated and arched upward, and the head held horizontally or directed somewhat downward. The tail and its milky, astringent secretions are often lashed at the attacker. Individuals may also weakly vocalize by making hissing or squeaking noises when attacked by predators or conspecifics (Brodie 1978; Wiltenmuth 1996).

If the tail is grasped by a predator, an individual may autotomize the tail at its constricted base (Hubbard 1903; Nussbaum et al. 1983; Stebbins 1951, 1954;

Fig. 213. *Ensatina e. klauberi;* defensive posturing; San Diego Co., California (E. D. Brodie, Jr.).

Storer 1925). The wiggling tail distracts predators such as garter snakes from the fleeing animal. The tail is an important storage organ, and an individual that loses its tail may compromise future growth or reproduction. Individuals will almost never autotomize their tails unless the tail is grabbed and autotomy is the only means of escape (Beneski 1989). Ensatinas require about 2 years to regenerate autotomized tails (Staub et al. 1995; Stebbins 1954). Typically, 8–13% of animals in natural populations have regenerating tails (Beneski 1989; Gnaedinger and Reed 1948; Staub et al. 1995; Stebbins 1954; Wake and Dresner 1967).

COMMUNITY ECOLOGY. Ensatinas are abundant in many forest communities, but the role that this and other western salamanders play as forest-floor predators of invertebrates has not been examined experimentally.

CONSERVATION BIOLOGY. *Ensatina e. eschscholtzii* is a common species and seems to tolerate intensive timbering better than many western salamanders. Ensatinas are common in stands of all ages in Oregon and Washington (Aubry et al. 1988; Bury and Corn 1988a; Corn and Bury 1991). However, they are more abundant in old-growth forests in northern California than in regenerating forests that have been logged (Welsh and Lind 1988, 1991). This difference presumably reflects the generally moister conditions in Oregon and Washington forests.

Plethodon aureolus Highton
Tellico Salamander

PLATE 115

Fig. 214. *Plethodon aureolus;* adult; Graham Co., North Carolina (R. W. Van Devender).

IDENTIFICATION. *Plethodon aureolus* is a small member of the *P. glutinosus* group that was described by Highton (1983) based on electrophoretic evidence. The Tellico salamander resembles the slimy salamander and has a grayish black or black dorsum with abundant brassy spotting (Highton 1986a). The chin is light colored and the sides of the body have more concentrated white or yellow spotting. This and other *Plethodon* species have tails that are rounded in cross section. Sexually active males have circular mental glands just behind the chin, and the average number of costal grooves is 16. The maximum size of adults is 72 mm SVL and 151 mm TL. This species coexists locally with *P. oconaluftee* and occurs allopatrically with *P. glutinosus,* which is located immediately to its west, north, and south.

The Tellico salamander is best identified by its small size, coloration, and range. Coloration and size can be used to separate this species from *P. oconaluftee* within areas of sympatry. Sympatric *P. oconaluftee* have white spotting on the dorsum and may reach 90 mm SVL. There are no known gene loci that will completely differentiate *P. aureolus* from sympatric *P. oconaluftee.* However, local populations of the two species sometimes show fixed differences in allozymes in areas of microsympatry. Because some allopatric populations of *P. glutinosus* are indistinguishable from those of *P. aureolus,* these species are best identified by range. In general, the smaller size of *P. aureolus* distinguishes it from most nearby populations of *P. glutinosus.* Highton (1983) found only one locality where all three species coexist locally and there is no evidence of interbreeding.

SYSTEMATICS AND GEOGRAPHIC VARIATION. *Plethodon aureolus* is uniform in coloration throughout most of its small range. The brassy dorsal spotting may be reduced or absent in some individuals found at high elevations in the northeastern portion of the range (Highton 1983). This species hybridizes with *P. jordani* on Sassafras Ridge in the Unicoi Mountains, but in general it appears to be genetically distinct from *P. glutinosus, P. jordani,* and *P. oconaluftee.*

DISTRIBUTION AND ADULT HABITAT. *P. aureolus* has a very restricted range and is currently known only from both mountainous and lowland habitats in Monroe and Polk counties in southeastern Tennessee, and Cherokee and Graham counties in extreme southwestern North Carolina (Highton 1983).

BREEDING AND COURTSHIP. Courtship behavior is indistinguishable from that of *P. glutinosus* (Dawley 1986a,b). The male initially places his nasolabial grooves and mental gland in contact with the female's head, body, or tail. The male then engages in a "foot dance" in which he raises and lowers his rear limbs either alternately or simultaneously. The male eventually proceeds forward to the female's head, then pushes his head under the female's chin and passes beneath. The male undulates his tail as it passes under the female's chin, then stops his forward progress. The female straddles the tail and the pair moves forward and engages in a tail-straddle walk. During the tail-straddle walk the male undulates his tail and often flexes his body laterally to slap the snout of the female with his mental gland. The male eventually deposits a spermatophore, then raises the vent and flexes his tail to one side. The couple then moves forward, and the female picks up the sperm cap with her cloacal lips. Greater detail of this general sequence is provided in the account of *P. glutinosus.*

Chemical cues appear to be important in species

Range of *Plethodon aureolus*

recognition and in preventing interbreeding of *P. aureolus* with syntopic *P. oconaluftee*. Male *P. aureolus* show strong preferences for odors of female conspecifics over odors of sympatric *P. oconaluftee* (Dawley 1984b, 1986a). However, females tend to show the reverse preferences when allowed to choose between male conspecifics and heterospecifics.

COMMENTS. Currently, there is almost no published information on the life history, ecology, and behavior of *P. aureolus*.

Plethodon caddoensis Pope and Pope
Caddo Mountain Salamander
PLATE 116

IDENTIFICATION. This member of the *P. ouachitae* complex is restricted to the Caddo Mountains of west-central Arkansas. Adults are smaller than *P. ouachitae* and reach a maximum size of 52 mm SVL and 11 cm TL. The back and sides of the body are black and profusely marked with small white spots and brassy flecking (Pope 1964). Individuals with small amounts of red pigmentation on the dorsum are occasionally collected, but most lack this coloring. The undersides are dark gray to black, except for the throat and portions of the chest, which are white. Males have conspicu-

Fig. 215. *Plethodon caddoensis;* adult; Arkansas (R. W. Van Devender).

Plethodontidae: Plethodontinae: Plethodontini

Range of *Plethodon caddoensis*

ous, rounded mental glands. Adults reach 9–11 cm TL and there are 16 costal grooves.

SYSTEMATICS AND GEOGRAPHIC VARIATION. Duncan and Highton (1979) studied electrophoretic protein variation in this and other members of the *P. ouachitae-caddoensis* complex and found that *P. caddoensis* is moderately distinctive from the other forms (Nei's $D = 0.30$). Details of genetic variation in this complex are discussed in the account of *P. ouachitae*.

DISTRIBUTION AND ADULT HABITAT. The Caddo Mountain salamander is endemic to forested habitats in the Caddo Mountains of west-central Arkansas and is locally abundant in or near talus slopes or other rocky sites, particularly on north-facing slopes that support mature, mesic forests. Deep talus in heavily shaded forest provides important microhabitats for foraging as well as passageways for moving into protective underground retreats during hot, dry weather.

BREEDING AND COURTSHIP. Breeding probably occurs in the autumn and spring (Taylor et al. 1990), but detailed information on the seasonal time of mating is not available. Courtship behavior has not been described, but males are aggressive toward one another during the breeding season (Arnold 1977).

REPRODUCTIVE STRATEGY. Females presumably oviposit in deep underground retreats in early summer. Heath et al. (1986) report finding egg clusters in a mine shaft for two consecutive years, but details about the mode of egg deposition and egg structure are not provided. Number of ovarian eggs in 22 specimens averaged 11 and was not correlated with female SVL (Taylor et al. 1990). Embryos hatch in late summer and autumn, and the juveniles reach sexual maturity when they measure about 40 mm SVL (Pope 1964).

COMMENTS. Many aspects of the natural history and behavior of this species have not been studied. Individuals are active on the ground surface during the cooler months of spring and autumn, particularly on rainy nights (Spotila 1972). During the summer months animals move below ground into deep talus or abandoned mine shafts. Adults are common in abandoned mines between June and September, particularly on rock walls near pools of water (Saugey et al. 1985).

The adults will aggressively defend areas from conspecifics by biting, and they are rarely found in groups beneath the same cover object (Anthony 1993; Thurow 1976). In laboratory trials, adult males are attracted to the odors and fecal pellets of other males (Anthony 1993). This result is surprising given that fecal pellets are often used to advertise territories.

Plethodon cinereus (Green)
Red-backed Salamander
PLATE 117

Fig. 216. *Plethodon cinereus*; adult; Washtenaw Co., Michigan (R. W. Van Devender).

Fig. 217. *Plethodon cinereus*; ventral view of adult; Shenandoah National Park, Virginia (J. W. Petranka).

IDENTIFICATION. The red-backed salamander is a small eastern species of *Plethodon* that has short legs relative to body size and an average of 18–20 costal grooves. Two color morphs occur in most populations. The striped or red-backed morph has a broad, straight-edged, orangish red or red (rarely light tan) dorsal stripe that extends from the head onto the tail. The sides of the body are dark, and the venter is strongly mottled with black and white. The unstriped or lead-backed morph is similar but lacks the reddish dorsal stripe. The entire dorsum is dark above, and the animals are unicolored. In addition to the striped and unstriped morphs, albinos (Dyrkacz 1981; Hensley 1959) and an erythristic morph that has bright crimson on the back, sides, and legs occur in some populations. As in other *Plethodon* species, the tail is rounded in cross section.

The phenotypic expression of the genes affecting color morphs appears to be influenced by epistatic interaction of two or more gene loci, and individuals that are intermediate between the striped and unstriped morphs are occasionally encountered (Highton 1959, 1975). Color variants include individuals with continuous stripes that occur only on the back or tail and individuals with stripes that occur in disconnected sections along the body. All morphs often have small white spots and brassy flecks on the back, head, and sides of the body (Bishop 1941a; Highton 1962a).

External features that are unique to sexually active males include greatly swollen nasolabial glands, hedonic glands on the tail, a crescent-shaped mental gland near the apex of the lower jaw, and elongated premaxillary teeth (Dawley and Crowder 1995; Noble 1927a; Smith 1963). Males of *P. cinereus* and other members of the *P. cinereus* group have strongly curved cusps on the premaxillary teeth that are not present in other *Plethodon* species (Highton 1962a). Sexual dimorphism in body size is minimal; females average from 0% to 4% larger than males (Blanchard 1928; Nagel 1977; Pfingsten 1989a; Sayler 1966). Hatchlings measure 19–25 mm TL and there is no larval stage. Hatchlings and small juveniles resemble miniature adults but have conspicuously shorter tails relative to body size (Bishop 1941a; Cockran 1911; Piersol 1910b) and proportionately broader heads (Maglia 1996). The adults measure 6.5–12.5 cm TL.

Range of *Plethodon cinereus*

SYSTEMATICS AND GEOGRAPHIC VARIATION. Populations that occur southwest and west of the French Broad River were previously referred to as *P. cinereus* but are now considered to be *P. serratus*. Details of an electrophoretic analysis of protein variation in this species pair are given in the account of *P. serratus*. Populations from formerly glaciated regions of the range are almost identical genetically, whereas populations to the south are more variable (Highton and Webster 1976). These genetically uniform populations are presumably derivations from northern ancestral stock that repopulated the region following the last glacial retreat 10,000–15,000 years ago.

Geographic variation in the proportion of color morphs in local populations has been documented by various workers, and several hypotheses have been advanced to explain the patterns (Brown 1965; Lotter and Scott 1977; Pfingsten and Walker 1978; Test 1952; Thurow 1961; Williams et al. 1968). Most populations contain both the striped and unstriped morphs, but their relative proportions vary markedly. Populations consisting entirely of striped morphs are common in the northernmost areas of the range, whereas those consisting entirely of unstriped morphs are uncommon and occur sporadically throughout the range of *P. cinereus* (Highton 1962a).

In many areas of the eastern United States, the proportion of striped and unstriped morphs in local populations changes predictably over local or regional zones. Angleberger and Chinnici (1975), for example, document a clinal change in the frequency of unstriped morphs from 0.13 to 0.00 over a distance of 8 km in Virginia. No obvious environmental factor correlates with the change in morph frequency.

Most populations in the Upper Peninsula of Michigan contain only striped morphs (Test 1952). The unstriped morph occurs in low proportions in populations in the extreme southern Upper Peninsula and northern Upper Peninsula, but progressively increases in frequency toward southern Michigan. The unstriped morph prevails in populations in north-central Ohio, particularly on islands in Lake Erie, but progressively declines in populations to the south, east, and southwest (Pfingsten and Walker 1978). The proportion of morphs in some populations appears to vary little with time; however, wide temporal swings occur in the relative proportions of the morphs in some disturbed sites. In New York, the proportions of morphs remained at about the same level in one population over a 10-year period (Brown 1965).

The striped morph predominates in New England populations and generally increases in abundance from northern to southern New England. This pattern correlates with changes in average temperatures (Lotter and Scott 1977). In general, unstriped morphs are more prevalent throughout the range of *P. cinereus* in regions with relatively warm climates (Greer 1973). However, numerous exceptions occur at local and regional levels, and marked variation is sometimes evident among local populations (Test 1955).

Striped morphs are less active than unstriped morphs on the ground surface during relatively warm periods (Moreno 1989). In addition, surface activity is less energetically costly for unstriped morphs during warm weather because they have lower metabolic demands. Despite this advantage, unstriped morphs incur a cost because of increased predation from predators such as garter snakes that are more active during warm weather. Moreno (1989) hypothesized that dorsal color is pleiotropically linked to metabolic rates, and that temperature-related selection indirectly favors different frequencies of the morphs in local demes.

Highton (1977) examined microgeographic variation in morph frequency, trunk vertebrae number, and isozymes in the Delmarva Peninsula. Significant changes in morph frequencies in this region often occur over distances of 10 km or less. Morph frequency and trunk vertebrae number tend to vary discordantly, but variation in isozymes is highly concordant. Thus electrophoretic data are more useful than morphological data in determining the genetic relationships between populations.

Populations on the western half of Long Island, New York, usually contain only striped morphs, whereas most populations on the eastern half contain only unstriped morphs (Williams et al. 1968). The change from striped to unstriped populations is abrupt, and only a few populations along the northern side of the island have both morphs. Geographic shifts in color morphs on the island correlate with changes in plant communities and mean number of trunk vertebrae. This pattern contrasts with that in mainland populations, which usually have both morphs and lack strong concordance between color and trunk vertebrae number.

The erythristic morph occurs sporadically in populations throughout much of New England westward to Ohio (deMaynadier 1995; Lotter and Scott 1977; Mueller and Himchak 1983; Tilley et al. 1982; Thurow 1961). In the majority of populations it occurs at very low frequencies, but in some areas it may compose 15–20% of local populations (Lotter and Scott 1977; Mathews 1952; Pfingsten 1969; Tilley et al. 1982). The erythristic morph is most common in populations with low proportions of the unstriped morph (Lotter and Scott 1977). Most of the known populations with the erythristic morph occur near the limits of the Cary substage of the Wisconsin glaciation, but some individuals have been collected as far north as Canada (Rosen 1971; Westell and Ross 1974). One explanation for this pattern is that the erythristic morph increased in semi-isolated, small populations at the glacial boundary via genetic drift, then spread regionally (Thurow 1961). Stochastic events unrelated to glaciation could explain the presence of erythristic morphs in local populations north of the glacial boundary.

Lotter and Scott (1977) hypothesize that the erythristic morph is a Batesian mimic of red efts. Experiments with blue jays in the laboratory (Tilley et al. 1982) and robins and other avian predators in semi-natural settings (Brodie and Brodie 1980) show that birds exposed frequently to red efts feed proportionately less on erythristic morphs than striped or unstriped morphs of *P. cinereus*. One interesting finding of Brodie and Brodie's study is that birds do not avoid erythristic morphs when first exposed to them, even though red efts occur naturally at the study site. This observation suggests that birds do not encounter red efts frequently enough in their natural environment to learn to associate the erythristic morphs with the models. Although the mimicry hypothesis is supported to some degree by experimentation, it cannot explain the rarity or absence of the morph in many areas where efts are exceedingly common and diurnally active (Tilley et al. 1982).

Regional variation occurs in the number of costal

grooves. The most conspicuous pattern is between Appalachian Plateau populations, which have a modal number of 18 costal grooves, and Valley and Ridge populations, which mostly have a modal number of 19 costal grooves (Highton 1972).

DISTRIBUTION AND ADULT HABITAT. The red-backed salamander ranges from southern Quebec and the Maritime Provinces southward to western and southeastern North Carolina and westward to western Minnesota, where a geographic isolate occurs. Adults inhabit forest litter habitats in deciduous, northern conifer, and mixed deciduous-conifer forests. Populations reach their greatest densities in well-drained, forested habitats. Mature forests with deep soils and scattered logs or rocks provide optimal habitats (Burger 1935). Populations are usually absent or occur at low densities in highly acidic soils, in soils that are perennially wet, and in shallow, rocky soils.

BREEDING AND COURTSHIP. Red-backed salamanders have a prolonged mating season that lasts from autumn through early spring. Mating in a New York population begins during the second week of October (Hood 1934), and most mating occurs during the autumn in southern Michigan populations. Females in these populations have sperm in their cloacae from late October through late April, and females have fresh spermatophores between 31 October and 9 December (Blanchard 1928). In a Tennessee population, females have spermatophores in their cloacae from December through March (Nagel 1977). In a Maryland population, the vasa deferentia are packed with sperm from September to May and females with spermatophores are present from 28 October to 18 April (Sayler 1966).

Males breed annually whereas females breed annually or biennially depending on geographic locality and age. Sayler (1966) found that most females oviposit every other year and only breed during the second year of the egg maturation cycle, when ovarian eggs are >1.3 mm in diameter. Data on populations in Michigan (Test and Bingham 1948), New York (Bishop 1941a), Ohio (Pfingsten 1989a), and Wisconsin (Vogt 1981) suggest that most females oviposit biennially in northern populations. In these locales the growing season is relatively short and females apparently cannot obtain sufficient energy in 1 year to yolk a clutch.

In a Connecticut population females breed every other year when young, but shift to annual breeding after exceeding 44 mm SVL (Lotter 1978). In contrast, mature females of all size classes reproduce annually in certain populations in Michigan (Werner 1971) and eastern Tennessee (Nagel 1977). Sex ratios of specimens collected by Test (1955) in Michigan, Burger (1935) and Hood (1934) in New York, and Mathis (1991b) in Virginia are near 1:1 except during the nesting season, when many females are underground and males outnumber females on the ground surface.

Courtship of *P. cinereus* involves a tail-straddle walk that is characteristic of plethodontids. Males have enlarged premaxillary teeth with recurved anterior cusps that function in abrading the skin and presumably passing mental gland secretions into the female's circulatory system during courtship. The teeth are either pulled across the skin or used to pierce the skin when using a snapping motion of the body (Arnold 1977).

Gergits and Jaeger (1990b) observed courtship behavior on 15 October in a natural population of *P. cinereus* in Virginia. Animals at this site aggregate in patches that may contain as many as 7–10 individuals/m^2. Males appear to locate gravid females by following pheromone trails. After tracking and approaching a female, a male moves in front of her and sometimes nose-taps the female as he moves forward. The male then arches his tail upward just distal to the vent and, with the tip on the ground, undulates the tail from side to side. If the female moves away, the male reorients in front of her and undulates the tail again. The male then rubs his mental gland back and forth on the female's back while moving rostrally along the female's body.

The male next moves forward and aligns himself along the female's body and undulates his tail. The female approaches and places her chin on his dorsum just above the vent. The male's tail remains curled and is undulated if the female's head slips from its position. The couple then performs a tail-straddle walk until the male deposits a spermatophore. The pair next moves forward, and the female picks up the spermatophore after positioning herself above it. The pair separates shortly thereafter. Courting males sometimes break from females and bite approaching males. An intruder male may also bite a courting male and disrupted courtship.

Males prefer to eat soft-bodied termites over hard-bodied ants and are more efficient foragers on termites (Jaeger et al. 1995a). Termites are high-quality food that is digested faster and with higher efficiency than ants (Gabor and Jaeger 1995). In laboratory trials, gravid females are attracted to male fecal pellets composed of high-quality termites more often than those composed of lower-quality ants (Walls et al. 1989). Gravid females often squash pellets with their snouts, which may be a way of assessing chemical signals. Field censuses indicate that males with termites in their diets are more likely to be next to females than are males with ants in their diet. These data suggest that courting females profit in fitness by selecting males with high-quality food in their territories. Jaeger and Wise (1991) provide additional support for this hypothesis by demonstrating that, in general, only gravid females squash male pellets.

REPRODUCTIVE STRATEGY. Females deposit their eggs in grapelike clusters within natural cavities or crevices. The eggs are usually suspended from the roof of the cavity by a short pedicel, and the female remains coiled about her clutch through hatching. The pedicel is a compound structure formed from mucus strings of individual eggs that are attached to a common point and intertwined (Piersol 1910b, 1914). Occasionally, females lay their eggs in a compact cluster directly on the substrate and do not suspend them by a pedicel (Bishop 1941a). Freshly laid ova are pale yellow to yellowish white, 3.0–4.0 mm in diameter, and surrounded by two jelly envelopes. The entire structure is 3.5–5.0 mm in diameter (Bishop 1941a; Cockran 1911; Piersol 1910b; Sayler 1966).

In southern populations, females oviposit in subsurface retreats (e.g., Nagel 1977). In northern or mountainous populations, females often oviposit in crevices within decaying logs, or in cavities beneath logs or rocks embedded in the soil. The eggs are usually attached directly to a rock or log, which forms the roof of the cavity (Cockran 1911; Friet 1995; Piersol 1910b, 1914). Nests in northern Michigan are often found in decaying conifer logs containing a network of cracks, but are occasionally found in mammal burrows or in cavities at the mineral soil–leaf litter interface (Test and Heatwole 1962). The use of cavities below matted leaf litter may be common in some northern populations, particularly where logging has reduced the density of large, decaying logs on the forest floor. Nests are rarely encountered in southern Michigan and are typically found in subsurface burrows of vertebrates or invertebrates.

Red-backed salamanders lay a few, large eggs, and the average clutch size in local populations varies from 6 to 9 eggs. Number of eggs in 23 Massachusetts nests varies from 1 to 14 and averages 8 (Lynn and Dent 1941), whereas that in 20 New York nests varies from 1 to 11 and averages 6 (Bishop 1941a). These values are similar to clutch size estimates based on ovarian egg counts: number of mature ova in 40 Connecticut females ranges from 3 to 12 and averages 7 (Lotter 1978), whereas that in 91 Michigan females ranges from 5 to 13 and averages 9 (Blanchard 1928). Females in a Tennessee population average 8 eggs (Nagel 1977), whereas the number of hatchlings from 35 Virginia clutches varies from 3 to 11 and averages 7 (Angleberger and Chinnici 1975).

Number of mature ova is positively correlated with female SVL in two populations (Lotter 1978; Nagel 1977), and with body mass (but not SVL) in a third (Fraser 1980). The results of the latter study suggest that the amount of food that a captive animal receives does not influence the number of mature ova produced in subsequent months. However, the data do suggest that low food levels can reduce clutch size during the first year of a biennial reproductive cycle.

In New Brunswick, Canada, a pair of adults is usually found with each egg mass, suggesting that both parents brood (Friet 1995). However, in most populations only females guard the eggs. In Virginia 95% of the nests ($n = 117$) are attended by brooding females (Highton and Savage 1961). This energetically expensive behavior protects eggs from predators and may minimize egg dehydration. There is no evidence that skin secretions from brooding females have antibiotic properties (Vial and Preib 1966).

Brooding females will aggressively defend their eggs from conspecific females. In the laboratory, intruders occasionally show a strong interest in the eggs of brooding females and may chase resident females from their nests (Bachmann 1984). In most instances, however, resident females drive off intruders by biting, snapping, or lunging at them. Responses to males and juveniles are similar, but females show less aggression toward hatchlings. Females that encounter ringneck snakes (*Diadophis punctatus*) desert their eggs and flee in 80% of trials. On rare occasions, a female may pick up her eggs and move them if the nest is uncovered (Watermolen 1996).

Females will cannibalize eggs both in the laboratory and in the field. Females that are induced to ovulate in the laboratory often prey upon their own eggs, and residents that drive females from their nests may eat the eggs (Highton and Savage 1961; Piersol 1914). Survivorship of embryos to hatching is very low if females are removed from their clutches. There is one record of a wild-caught female with three eggs in her stomach (Burger 1935), but almost no information on the extent to which oophagy and nest piracy occur in nature.

Females remain with their clutches as the embryos develop and have limited opportunities to feed when brooding. Ng and Wilbur (1995) examined the cost of brooding by placing gravid females in cages in the forest floor and adding either supplemental food or another female to the cages. Gravid females that lay eggs and brood grow significantly less over a 10-week period than intruder females. In addition, the presence of a second female reduces the growth of resident animals. These results suggest that brooding females incur a major energetic cost when brooding.

Most females oviposit in late spring or early summer, but on rare occasions a female may oviposit in late summer or early autumn (Test and Bingham 1948). Oviposition typically occurs in June in populations in Maryland (Sayler 1966), New Jersey (Burger 1935), New York (Bishop 1941a), and Pennsylvania. Females in a Massachusetts population oviposit from 18 June to 20 July (Lynn and Dent 1941), whereas those in a population near Toronto, Canada, oviposit from 16 June to 3 July (Piersol 1910b). Unusual nesting records include a clutch of eggs found in New York in late October (Sherwood 1895) and observations of a female ovipositing in northern Michigan on 2 August when most other clutches found in the field were near hatching (Davidson and Heatwole 1960).

The average incubation period is about 6 weeks (Burger 1935; Davidson and Heatwole 1960). Mature embryos have three gills on either side of the head that are lost immediately before or shortly after hatching (Piersol 1910b). Hatching has been observed in August and/or September in southern Canada (Piersol 1910b), Maryland (Sayler 1966), Massachusetts (Lynn and Dent 1941), Michigan (Davidson and Heatwole 1960), New York (Bishop 1941a), and Virginia (Highton 1959).

TERRESTRIAL ECOLOGY. *Plethodon cinereus* is a euryphagic predator that will apparently eat any palatable prey that it can capture. The adults are opportunistic cannibals and will eat conspecific eggs or juveniles in both the laboratory (Highton and Savage 1961; Piersol 1914) and the field (Burger 1935; Burton 1976; Heatwole and Test 1961; Surface 1913). Cannibalism in the field appears to be rather rare: in one study there were 2 cases of cannibalism in 317 specimens examined from northern Michigan.

Small invertebrates are the dietary mainstay. Major prey of 149 New York specimens examined by Jameson (1944) were beetles, flies, ants and other hymenopterans, earthworms, and spiders. The remaining prey were lepidopterans, thysanopterans, snails, slugs, spiders, mites, centipedes, and millipedes. Ants were the primary prey in 86 adults from West Virginia (Pauley 1978b), whereas mites, spiders, snails, and representatives of numerous insect families were reported for 200 New Hampshire specimens (Burton 1976). Southern Michigan specimens eat their shed skins along with beetles, ants, spiders, snails, millipedes, bugs, mites, springtails, pseudoscorpions, and miscellaneous invertebrates (Blanchard 1928). Similar prey are reported in other dietary accounts (Cockran 1911; Hamilton 1932; Jaeger 1972; Maglia 1996; Mitchell and Woolcott 1985).

In three populations in northeastern Tennessee, individuals do not shift their diet markedly as they grow. Nonetheless, small prey such as mites and collembolans are more prevalent in juveniles, whereas larger prey such as beetles and hymenopterans occur more frequently in adults (Maglia 1996). Prey size is independent of head size in adults but increases with head size in smaller juveniles. Juveniles have proportionately broader heads than adults, which may allow juveniles to handle a relatively wide range of prey despite their small body size.

Studies of specimens dug out of abandoned ant mounds in winter show that individuals may feed while in deep underground retreats, even though feeding rates are greatly curtailed during the coldest months. Caldwell (1975) and Caldwell and Jones (1973) list insects, worms, spiders, and millipedes in specimens from ant mounds.

Juveniles often remain in the nests with the females for 1–3 weeks after hatching, then disperse away from the nest site (Burger 1935; Highton 1959; Piersol 1910b; Test 1955). The extent to which they remain in their parents' territories is poorly documented.

Data from studies in Maryland (Saylor 1966), Michigan (Blanchard 1928; Werner 1971), and Tennessee (Nagel 1977) indicate that juveniles reach sexual maturity about 2 years after hatching. Hatchlings in a Maryland population grow rapidly (about 15 mm SVL) during their first year of life (Sayler 1966). Females have biennial breeding cycles and probably first oviposit 3.5 years after hatching, when they measure >34–39 mm SVL. Males breed annually and mature sexually when they reach 32–37 mm SVL. Connecticut females reach sexual maturity when they measure 34–38 mm SVL, and the unstriped color morph matures at a larger size than the striped morph (Lotter 1978).

Juveniles in an eastern Tennessee population grow an average of 15 mm SVL during their first year of life, but only 8 mm SVL during the second (Nagel 1977). Surprisingly, growth rates during the winter months are as high as those in the spring and summer months. Females reproduce annually, in contrast to the biennial pattern seen in many northern populations.

Several studies suggest that a large reservoir of *P. cinereus* is underground and that individuals regularly move vertically between the soil and soil surface (Fraser 1976a; Taub 1961; Test and Bingham 1948). The extent to which vertical patterns of movement vary geographically is poorly documented. Taub (1961) found that when animals in field enclosures are provided underground passageways, many remain underground and never move to the surface. Thus both surface counts and mark-recapture techniques may underestimate true population size.

When all individuals found beneath surface cover in Michigan are removed during four consecutive searches conducted about 1 week apart, the number of individuals collected in successive samples shows little tendency to decline with time (Test and Bingham 1948). The results of this and additional studies in New Jersey (Taub 1961) and Virginia (Fraser 1976a) have led many to conclude that most individuals are below ground on any sample day. Nonetheless, individuals often surface following rains to forage on leaf litter invertebrates. Soil moisture and temperature appear to be the primary factors affecting the vertical distribution of individuals in the soil (Taub 1961).

Juveniles and adults are often very abundant in well-drained forest sites. Mark-recapture estimates of densities are an average of 0.89 salamander/m^2 of forest floor in Michigan during peak summer activity (Heatwole 1962), 0.25 salamander/m^2 in a New Hampshire population (Burton and Likens 1975), 0.21 salamander/m^2 for a Pennsylvania population (Klein 1960), and 2.2 and 2.8 salamanders/m^2 for Virginia populations (Jaeger 1980a; Mathis 1991b). A 14-year study by Jaeger (1980b) on Hawksbill Mountain in Virginia suggests that populations are stable and presumably at carrying capacity.

Surface-active animals remain under cover during the day but emerge at night to forage and mate when weather conditions permit. Individuals forage directly on the forest floor or climb on vegetation at night (Burton and Likens 1975; Cockran 1911). In a Virginia population, individuals that are on plants have a greater volume of food in their stomachs than individuals that are on the forest floor (Jaeger 1978). This finding suggests that climbers have greater foraging opportunities than nonclimbers.

Both stomach analyses and field observations show that individuals feed continuously throughout the day and night (Burger 1935; Jaeger 1978). Adults use both visual and olfactory cues to detect prey (David and Jaeger 1981). Olfaction may be important for locating nonmobile prey such as insect pupae, or when feeding on mobile prey in dark, subsurface retreats. Individuals use encounter rates and visual information to assess prey density (Jaeger et al. 1982a), but do not appear to be capable of remembering where high concentrations of food are located or of estimating the profitability of prey based on prey size alone (Jaeger and Rubin 1982). The former characteristic may reflect the fact that patchy food resources on the forest floor are highly ephemeral and selection pressures for optimal patch foraging behaviors are weak (Hill et al. 1982). Nonetheless, individuals do forage optimally by selecting more profitable prey as the density of prey increases (Jaeger and Barnard 1981; Jaeger et al. 1982a) and by learning through foraging experiences to assess the energetic rewards associated with prey types (Jaeger and Rubin 1982). Territorial animals that are aggressive tend to shift from being specialists on profitable prey to being generalists, and the extent to which an animal shifts to a more generalized diet is positively correlated with its level of aggression (Jaeger et al. 1983).

Adults are most active on the ground surface during the spring and autumn months except in some montane regions, where they are also active during the summer. They move into deep subsurface retreats with the arrival of freezing weather and emerge in late winter or spring with the arrival of warmer weather (Buhlmann et al. 1988; Cockran 1911; Highton 1972; Lotter 1978; Sayler 1966; Vernberg 1953). In winter, specimens have been found nearly 1 m deep in the soil in February (Grizzell 1949), beneath stumps of recently cut white oaks (Hoff 1977), and beneath stones in small streams in Maryland (Cooper 1956). In the latter case animals probably migrated to the dried beds of streams that later flooded. In the Atlantic coastal states, individuals are often active in winter during prolonged periods of warm weather (Highton 1962a).

In southwestern Indiana, red-backed salamanders aggregate in winter in deserted ant mounds (Caldwell and Jones 1973). Individuals are more likely to have food in their guts in December (100%) than in February (21%). Adults have difficulty burrowing through many substrates but can readily enlarge existing crevices, holes, and burrows in the soil (Heatwole 1960).

In areas with hot summer weather, individuals reduce their surface activity during the hottest months (Blanchard 1928; Fraser 1976a; Highton 1972; Maglia 1996; Nagel 1977; Taub 1961; Test 1955). The number of animals beneath logs, stones, and bark in southern Michigan is highest in the cool spring months and declines during the summer and autumn (Test 1955). Although many animals move underground during the summer, those that remain on the surface tend to congregate beneath large logs where moisture content is high. Groups of as many as 30 adults may congregate beneath logs or in other moist microhabitats. In Virginia groups of four to seven salamanders are occasionally encountered under the same cover object during the spring, but not in the summer, when adults establish territories (Jaeger 1979). These trends suggest that levels of agonistic behavior and the tendency to establish summer territories vary markedly depending on geographic locality.

Red-backed salamanders in Michigan are active in deciduous leaf litter during wet weather (Heatwole 1962). During dry weather they retreat both vertically into the humus layer and mineral soil and horizontally into logs or other moist microhabitats, where they may reach local densities of almost 2 individuals/m^2 of cover. Individuals are rarely found in conifer litter or lichen mats, where summer temperatures sometimes exceed the critical thermal maximum of the species. Temperature and substrate moisture appear to be the primary factors affecting local distributions and movements.

In a Virginia population surface densities may not change appreciably over a period of 22 days and individuals tend to move horizontally to logs when conditions become dry (Jaeger 1979, 1980a). Similar patterns occur in Michigan (Test 1955). Collectively, these studies suggest that as the forest floor dries, individuals first move to logs, depressions, or other microhabitats with high moisture content, then into subsurface retreats.

Both male and female red-backed salamanders are territorial and aggressively defend moist microhabitats such as decaying logs (Mathis 1989). In addition, a significant percentage of the population may consist of floaters, which are typically smaller animals that do not hold territories. About 49% of the animals in a Virginia population consist of floaters (Mathis 1991b). Territory holders have a greater number of broken tails than floaters but also have relatively longer tails, suggesting that territories are high-quality foraging sites. The home areas of males, females, and juveniles do not differ significantly in size and average 0.16–0.33 m^2. Surprisingly, the size of a home area is inversely proportional to the resident's size.

In studies of Virginia populations, individuals of the same sex tend to space out their territories, whereas those of different sexes often have overlapping territories (Jaeger et al. 1995b; Mathis 1991b). Juveniles often move into adult territories during dry weather and move out of territories into the leaf litter during wet weather (Jaeger et al. 1995b). Although adults generally consume larger prey than juveniles, overlap in diet is high, and competition for scarce food resources probably occurs during dry weather. Laboratory trials indicate that juveniles are attracted to adult territories and that adult males are less aggressive toward juveniles than toward other adult males. In addition, adult males are more likely to tolerate territorial intrusion from familiar versus unfamiliar juveniles. The authors hypothesize that this behavior may reflect kin recognition.

Individuals in a Virginia population often utilize the same surface sites throughout the year and do not shift their territories seasonally (Gergits and Jaeger 1990a). Radioactively tagged specimens in a Michigan population move the greatest distances after periods of rain (Kleeberger and Werner 1982). Daily movements average 0.43 m, but individuals often move >1 m when rainfall exceeds 1 cm. About 39% of the specimens are below ground; the remainder occur in surface microhabitats. Home ranges monitored over an average of 53 days averaged 13 m^2 for males and juveniles and 24 m^2 for females. Individuals displaced as far as

90 m sometimes successfully return to their home ranges, whereas those displaced 30 m almost always return home. The mechanism used in homing is not known; however, studies show that experimentally displaced animals do not use their own pheromones as directional cues when returning to their homes (Jaeger et al. 1993).

The exact reason for territoriality in *P. cinereus* is not fully understood, although defense of feeding areas and mates seems most important (Jaeger et al. 1982b). In a detailed study of competition in a Virginia population, salamanders preferred large cover objects over small ones in laboratory trials (Mathis 1990a). Furthermore, the body size of animals beneath surface objects in natural habitats was positively correlated with the resident's size. This observation suggests that large animals are effective in defending high-quality resources from smaller animals. In natural habitats, logs from which resident animals are removed are invaded significantly more often than logs from which the residents are not removed. Furthermore, animals that invade logs are significantly smaller than the residents that are removed. Collectively, these data indicate that adults compete for high-quality cover objects, and that larger individuals are more successful competitors for food than smaller animals (Mathis 1990a).

A related study also suggests that larger individuals hold higher-quality territories since the number (but not volume) of prey in a territory is positively correlated with the resident's SVL (Gabor 1995). Cover object size is not significantly correlated with the resident's SVL or the number of dipteran or mite prey. This finding suggests that cover objects are defended because they are refuges from extreme heat and dryness, rather than areas of high prey density.

Other studies show that larger males occupy higher-quality territories (Mathis 1990a) and are more likely to attract females than smaller males (Mathis 1991a). Once a female moves into a large male's territory, the resident is more likely to defend his territory from smaller males (Jaeger et al. 1995b). In addition, both territorial residents and intruders are more likely to contest space when the resident is fed high-quality food versus low-quality food (Gabor and Jaeger 1995). Floaters are presumably at a disadvantage since individuals that do not possess territories are less likely to forage in an optimal manner (Jaeger et al. 1981).

Jaeger (1972, 1980c) found that foraging success is highest during or immediately following periods of rainy weather, when the salamanders can move away from surface cover and forage in the leaf litter. During dry periods, individuals are forced to remain under cover with moist substrates. Calculations based on information on assimilation efficiencies, caloric content of food in the gut, and metabolic demands indicate that many individuals do not maintain positive energy budgets during the warm summer months. However, indirect evidence from other studies suggests that individuals in many populations do maintain positive energy budgets throughout the summer. For example, Maryland juveniles and subadults grow as rapidly during the summer months as during the cooler months of spring (Sayler 1966), and Tennessee females show constant rates of increase in follicle size throughout the year (Nagel 1977).

One behavioral conflict with regard to aggression concerns a male's response to a female. Several studies show that males are generally more aggressive toward other males than toward females (Jaeger 1981, 1984; Thomas et al. 1989). Males that are aggressive to females may drive off competitors for food or cover, but lose potential mates. Thomas et al. (1989) tested the response of males to gravid versus nongravid females and found that males are equally aggressive toward both. Males may tolerate nongravid females because such behavior increases the probability that these females will be in close proximity the following year when they are gravid and willing to mate.

Red-backed salamanders have an array of mechanisms for marking territories and mediating territorial disputes (Jaeger 1986; Jaeger and Forester 1993). Both males and females use glandular secretions and feces to mark and recognize home territories (Horne and Jaeger 1988; Jaeger 1981, 1986; Jaeger and Gergits 1979; Jaeger et al. 1986; Simon and Madison 1984). Chemicals from the shoulder, a postcloacal gland on the tail, the urinary ducts, and the feces trigger exploratory behavior in other males and may be involved in territorial marking (Jaeger and Gabor 1993; Simons and Felgenhauer 1992; Simons et al. 1994, 1995). These chemicals may function in mate selection and to reduce the cost of territorial defense by allowing the recognition of neighbors. Jaeger (1981), for example, found that individuals tend to be less aggressive and more submissive toward familiar neighbors than unfamiliar conspecifics.

Fecal pellets appear to be one of the most important substances for marking territories. Males that encounter another male's fecal pellets often respond by acting submissive and moving away from the pellets into their own burrows (Jaeger et al. 1986). Females that encounter another female's pellets are less likely to vacate an area and are more likely to assume threatening postures (Horne and Jaeger 1988). Females not only tap the fecal pellets of female conspecifics but also squash them with their snouts, a behavior that may help them obtain detailed information about other individuals.

Information concerning the body size and gender of conspecifics can be gathered from chemicals in fecal pellets, and individuals may alter either the rate of fecal pellet production or the size of fecal pellets when exposed to pheromones from conspecifics (Mathis 1990b). The results of this study further suggest that females use fecal pellets primarily to advertise territories, whereas males use them primarily to attract mates.

Red-backed salamanders often exhibit nose-tapping behavior in which the cirri and nasolabial grooves on the upper lip are briefly touched to the substrate or to fecal pellets (Horne and Jaeger 1988; Nunes 1988). Tapping facilitates chemoreception by transporting water-borne chemicals on the substrate via capillary action to internal receptors in the nares (Brown 1968; Dawley and Bass 1989; Graves 1994; Jaeger 1981).

In laboratory studies, resident males increase their nose-tapping and exploratory behavior when presented with cotton swabs containing pheromones of conspecifics, but do not re-mark their territories more frequently. Large males often tend to become more aggressive when exposed to swabs with pheromones, whereas small males tend to become more submissive (Mathis and Simons 1994). Tapping rate and duration are significantly higher when individuals are handled then returned to their home dishes, versus being placed in dishes that previously held conspecifics (Tristram 1977). Responses of individuals when exposed to odors from conspecifics are generally no different from those when placed in control dishes that are not chemically conditioned. Individuals also tap more frequently when exposed to chemical cues of familiar versus unfamiliar conspecifics (McGavin 1978).

Chemical cues are also important in species recognition. Male and female *P. shenandoah* and *P. cinereus* in general show a preference for substrates marked with their own chemical markings versus those of heterospecifics (Jaeger and Gergits 1979). The one exception is that female *P. cinereus* do not respond differentially to their own versus heterospecific markings. This preference for an individual's own pheromonal markings is due to the fact that adults are both attracted to their own markings and repulsed by those of unfamiliar individuals (Jaeger et al. 1986).

Territorial disputes that are not resolved through chemical signaling may escalate into direct confrontations in which individuals assume threatening postures and/or attack one another (Jaeger 1984; Jaeger et al. 1982b, 1983). Because combatants may be injured while fighting, and larger individuals usually win aggressive encounters (R. G. Jaeger, pers. comm.), aggressive and submissive posturing is a means of communicating the extent to which individuals are likely to fight. *Plethodon cinereus* exhibits agonistic posturing that involves raising the trunk and tail. This is a graded threat that can range from weak (head and body raised slightly) to strong (head, trunk, and tail held high off the substrate with the trunk arched). The stronger the threat posture, the longer intruders respond with submissive behaviors (Jaeger and Schwarz 1991). In general, the longer a male salamander occupies a territory, the more aggressive and less submissive it becomes to intruders (Nunes and Jaeger 1989).

Both gravid and nongravid females are territorial during the breeding season (Horne 1988). Nongravid females are more submissive to male intruders than gravid females, perhaps because gravid females are less likely to be attacked by sexually active males. In addition, gravid females are more submissive toward nongravid intruders than are nongravid female residents. Male intruders spend more time in burrows and less time escaping from gravid females than from nongravid females. Salamanders that are maintained on low food levels tend to avoid aggressive encounters either by spending more time inside burrows or by escaping from potential aggressors (Nunes 1988).

Observations of numerous instances of biting in the field verify that laboratory behaviors of *P. cinereus* are similar to those that occur under natural conditions (Gergits and Jaeger 1990b). During aggressive encounters, individuals often bite the nasolabial grooves of conspecifics, an action that lessens their ability to find nonmobile prey (Jaeger 1981). Thus, biting of the nasolabial grooves can ultimately lower fitness by reducing the foraging rates of injured animals. About 12% of adults in a natural population have scars on the snouts that are presumed to be the result of intraspecific aggression (Jaeger et al. 1982b).

PREDATORS AND DEFENSE. Even though red-backed salamanders possess noxious skin secretions, they are eaten by woodland snakes (Arnold 1982; Cockran 1911; Uhler et al. 1939) and birds that forage in the leaf litter (Brodie et al. 1979; Fenster and Fenster 1996; Lotter and Scott 1977). There is even evidence that spiders prey on *P. cinereus* (Lotter 1978). In the laboratory, the size of an individual, antipredator behaviors, and the presence of a tail reduce the rate at which garter snakes (*Thamnophis elegans*) consume *P. cinereus* (Arnold 1982). When individuals use two or more defenses, this behavior has a multiplicative rather than an additive effect in reducing predation rate.

Studies with the ringneck snake (*Diadophis punctatus*) indicate that these predators can distinguish *P. cinereus* from nonedible prey and can distinguish substances derived from the tail from those derived from other portions of the body (Lancaster and Wise 1996). Secretions from postcloacal glands on the ventral portion of the tail are probably used for chemical discrimination. Although these snakes can discriminate between body secretions, they cannot strongly differentiate between substrates marked in a more natural manner by tailed versus tailless *P. cinereus*. Additional studies are needed to determine the ecological significance of this phenomenon.

Red-backed salamanders often assume a coiled, motionless position when uncovered in the field and may autotomize their tails when handled roughly (Cockran 1911; Piersol 1910b). Individuals that remain immobile when first uncovered are presumably less conspicuous to visually oriented predators such as birds (Dodd 1990c). In two samples about 10% (Piersol 1910b) and 13% (Wake and Dresner 1967) of specimens have broken or regenerating tails.

COMMUNITY ECOLOGY. Highton (1972) summarizes much of the distributional data on *Plethodon* in the eastern United States and emphasizes that several pairs of ecologically similar species tend to be spatially segregated geographically or altitudinally. The importance of competition between *P. cinereus* and other small *Plethodon* species has been addressed by Fraser (1976a) and in a series of studies by Jaeger (1970, 1971a,b, 1972, 1974) and Lancaster and Jaeger (1995).

Plethodon shenandoah is restricted to talus on all three mountains it occupies in the Blue Ridge of north-central Virginia (Highton and Worthington 1967). On one mountain *P. cinereus* coexists with *P. shenandoah* near talus, whereas on the remaining mountains it is strongly segregated microspatially from *P. shenandoah* (Jaeger 1970). One explanation for the pattern at the latter sites is that *P. cinereus* competitively excludes *P. shenandoah* from areas outside talus, but is unable to live in talus because of susceptibility to desiccation (Jaeger 1971b). To test this hypothesis, Jaeger (1971a) placed *P. shenandoah* and *P. cinereus* in either mixed or pure groups in small enclosures buried in microhabitats varying from bare rock to deep soil outside talus. In shallow and deep soils, *P. shenandoah* had significantly lower survivorship in the presence of *P. cinereus* than when in isolation. No *P. cinereus* survived for the duration of the experiment in enclosures with bare rock or rock with minimal soil development.

Although Jaeger's field experiment supports the competitive exclusion hypothesis, the results are difficult to interpret because the experiment was not replicated and salamanders were established at artificially high densities. Consequently, it is uncertain if competition occurs at natural densities, or if the experimental results could be duplicated with any degree of certainty.

Subsequent laboratory studies reveal that *P. shenandoah* is as good or better at obtaining prey than *P. cinereus* when the two species are placed together, and that *P. shenandoah* tends to displace *P. cinereus* from artificial burrows when both species are maintained at high densities (Jaeger 1972, 1974; Kaplan 1977). When both species are paired, the more aggressive of the two tends to win contests. However, neither species wins more often than the other (Wrobel et al. 1980). In addition, both species appear to be equally efficient at assimilating energy from food (Bobka et al. 1981). Because of the sometimes conflicting results from laboratory and field studies, it is still uncertain which interaction between the two species is responsible for the microspatial segregation that occurs in the field. However, the only *P. shenandoah* that are found on the forest floor more than a few meters from the talus edge are large adults. This finding suggests that competitive exclusion of juvenile *P. shenandoah* by adult *P. cinereus* is responsible for the restriction of *P. shenandoah* to talus (Jaeger 1972).

Fraser (1976a) designed a series of experiments to test for competitive interactions between *P. cinereus* and *P. hoffmani*. These species are parapatrically distributed in most areas and are rarely microsympatric. His data suggest that food may periodically become limiting during dry weather, when the salamanders cannot forage widely on the forest floor. However, interspecific competition for food during dry weather would only occur if salamanders shared the same surface retreat. Competition might also occur if animals simultaneously surfaced following rains to feed on temporarily scarce surface prey. Fraser (1976a) found that neither of these phenomena occurs and concluded that intense interspecific competition for food is unlikely. Pauley (1978b) found high dietary overlap between co-occurring *P. cinereus* and *P. wehrlei,* but it is not known if these species compete for food.

Other researchers have examined interactions of *P. cinereus* with more distantly related salamanders. Adult *P. cinereus* are as aggressive to juvenile *P. glutinosus* as they are to adult conspecifics and appear to defend territories against the latter (Lancaster and Jaeger 1995). *Desmognathus ochrophaeus* will aggressively defend cover objects from *P. cinereus* and can drive *P. cinereus* from occupied sites (Smith and Pough 1994). In staged encounters in the laboratory, *Ambystoma maculatum* prey on *P. cinereus* in about 9% of trials (Ducey et al. 1994). The ecological consequences of these behaviors in natural populations are not known.

CONSERVATION BIOLOGY. Red-backed salamanders are rare or absent in highly acidic soils with pH ≤3.7. Experimental evidence suggests that extremely acidic soils may disrupt sodium balance and be potentially lethal to these and other terrestrial salamanders (Frisbie and Wyman 1991, 1992; Wyman 1988; Wyman and Hawksley-Lescault 1987; Wyman and Jancola 1992). These findings raise concerns that continued acidification of soils via acid precipitation could eventually harm terrestrial salamanders in regions with low soil buffering capacity.

Like many other woodland salamanders, *P. cinereus* is sensitive to clear-cutting and other intensive harvesting practices. Studies suggest that local populations may return to relatively high levels within 30–60 years after timber removal (deMaynadier and Hunter 1995; Pough et al. 1987).

COMMENT. *Plethodon cinereus* has proven to be an excellent model for examining the ecological and evolutionary significance of aggression and territoriality in amphibians. An excellent review of these topics is provided in Mathis et al. (1995).

Plethodon dorsalis Cope
Zigzag Salamander
PLATE 118

IDENTIFICATION. The zigzag salamander is a small *Plethodon* species that is about the same size and shape as *P. cinereus.* The tail of adults makes up about half of the body length and there are usually 18 costal grooves. Two color morphs occur in many populations. The striped morph has a wavy yellowish brown to orangish red dorsal stripe that extends from the head to the tail tip. In almost all populations east of the Mississippi River, the dorsal stripe is wavy anteriorly but more straight-edged toward the tail. In the Ozark Mountains and in extreme southwestern Illinois, the dorsal stripe is nearly straight-edged, and specimens superficially resemble *P. cinereus* (Thurow 1956a, 1957b). The unstriped or dark morph lacks the dorsal stripe, but often has a suffusion of orange or red near the insertion of the limbs and on the venter. Interme-

Fig. 218. *Plethodon dorsalis;* striped morph; Jessamine Co., Kentucky (J. W. Petranka).

Fig. 219. *Plethodon dorsalis;* unstriped (lead) morph; Edmonson Co., Kentucky (J. W. Petranka).

diates between these two color phases are common in many populations, and partial albinos are known from Indiana (Thurow 1955).

Both the striped and unstriped morphs have numerous, tiny, silvery white spots and brassy flecks dorsally that often produce a frosted appearance. The belly is mottled with fine black, white, and orange coloration (Highton 1962a). Sexually active males have a conspicuous, oval mental gland, nasolabial swellings, and a pair of lips at the posterior end of the vent. Adults measure 6.5–11 cm TL.

SYSTEMATICS AND GEOGRAPHIC VARIATION. Populations in central and southwestern Alabama, western Georgia, southeastern Louisiana, central Mississippi, and western South Carolina that were previously referred to as *P. dorsalis* are a genetically distinct sibling species, *P. websteri* (Highton 1986e). A summary of electrophoretic studies of protein variation is given in the account of *P. websteri*.

Two subspecies of *P. dorsalis* are currently recognized (Thurow 1966). The eastern zigzag salamander (*P. d. dorsalis*) has a broad, wavy dorsal stripe. In contrast, the Ozark zigzag salamander (*P. d. angusticlavius*) has a narrower and more straight-edged dorsal stripe. Detailed studies of geographic variation in the frequencies of color morphs have not been conducted. In most areas the striped morph outnumbers the unstriped morph (Highton 1962a; Minton 1972). However, the striped morph is absent in southeastern Kentucky and in areas of sympatry with *P. websteri* in northern Alabama and *P. serratus* in eastern Tennessee (Highton 1962a, 1979).

Larson and Highton (1978) and Highton (1997) examined genetic variation in populations currently assigned to *P. d. dorsalis* and documented northern and southern groups that differ by an average genetic distance (D) of 0.18. At a contact zone in Lincoln Co., Kentucky, the two forms interbreed to produce a 13-km-wide hybrid zone. The distribution of gene loci within the hybrid zone does not differ significantly from Hardy-Weinberg equilibrium. Highton (1997) elected to recognize these forms as separate species (northern form: *P. dorsalis;* southern form: *P. ventralis*) and to raise *P. d. angusticlavius* to the level of full species. Here I consider the northern and southern forms to be conspecific and closer to the level of subspecies or semispecies. Because the northern and southern forms are genetically distinct yet appear to interbreed freely, I recommend that populations in the Ozark Mountains continue to be recognized as a subspecies (*P. d. angusticlavius*) until more compelling evidence is produced that eastern and western forms are separate biological species.

DISTRIBUTION AND ADULT HABITAT. The zigzag salamander occurs from west-central Indiana and southern Illinois southward to northern Alabama and westward to the Ozark Mountains of southwestern Missouri, northwestern Arkansas, and northeastern Oklahoma. Populations occur in mesic forests at up to 610 m and usually reach their greatest abundance in or about caves, talus slopes, rocky hillsides, and other habitats with rocky substrates that allow access to deep underground passages (Minton 1972; Mount 1975; Reinbold 1979; Smith 1961; Thurow 1957b). Specimens have been found in caves as deep as 2.4–7.6 m below the ground surface (Thurow 1963).

BREEDING AND COURTSHIP. Mating occurs from autumn through midspring, but patterns in local populations vary depending in part on the severity of winter weather. In Indiana males or females have spermatophores in their vents in November, March, and April (Minton 1972; Sever 1978). Females in some Arkansas populations have spermatozoa in their spermathecae from November to April, but not in September and October (Wilkinson et al. 1993). In one Arkansas population the mating season lasts from January through April based on the seasonal development of the vasa deferentia (Meshaka and Trauth 1995). Adults in this population appear to migrate 100–150 m from the base of a rocky bluff to a nearby cedar

Range of *Plethodon dorsalis*

glade to mate. There are no published accounts of courtship behavior. Adults presumably engage in a tail-straddle walk similar to that of *P. cinereus* and other *Plethodon* species.

REPRODUCTIVE STRATEGY. Females oviposit shortly after moving to underground retreats in late spring or early summer. Females in Indiana have ovarian eggs as large as 4.0 mm in diameter in April, but females collected in June are spent (Sever 1978). Females in Arkansas and Kentucky disappear from the ground surface during the late spring or early summer and do not resurface until September or October (Meshaka and Trauth 1995; Petranka 1979; Wilkinson et al. 1993). Hatchlings that measure 11-16 mm SVL also appear on the ground surface beginning in late September. Collectively, these observations indicate that females begin ovipositing soon after moving below ground in late spring or early summer.

The only published record of nesting is that of Mohr (1952), who observed 12 adults brooding eggs in intermediate developmental stages in a cave in Kentucky in late June and early July. The nests at this site are in small cavities in fluted cave formations and contain two to five eggs. The eggs lack stalks and are attached to the substrate by their sticky outer membranes. The ova are 4.0-4.5 mm in diameter and are surrounded by the vitelline membrane and two envelopes. Late-term embryos are present by 21 August, suggesting an incubation period of about 3 months. Estimates of average clutch size for two samples of Arkansas females are 5 ova (ranges = 1-10 and 3-9) for both groups based on ovarian egg counts (Wilkinson et al. 1993). In both populations, number of ova is independent of female SVL and adults reproduce annually.

TERRESTRIAL ECOLOGY. Very few data are available on the natural history of the juveniles. Hatchlings first appear on the ground surface in autumn, and the juveniles use habitats similar to those of the adults. The diet of Indiana specimens mostly consists of spiders along with smaller numbers of beetles, mites, miscellaneous arthropods, and slugs (Holman 1955).

Individuals in an Arkansas population become sexually mature when 3 years old (Wilkinson et al. 1993). Adult males and females reach 30-42 mm SVL and 32-45 mm SVL, respectively, and the largest immature specimen measured 33 mm SVL. In a second Arkansas population individuals reach sexual maturity when they are 2 years old and measure >34 mm SVL (Meshaka and Trauth 1995). Growth rates of juveniles average 9.5 mm SVL per year.

Zigzag salamanders are active on the ground surface

during the cooler months of spring and autumn and during bouts of warm weather during the winter months. Individuals move below ground with the onset of hot summer weather and are rarely found between May and September (Meshaka and Trauth 1995; Minton 1972; Smith 1961; Wilkinson et al. 1993). At a southern Illinois site, *P. d. angusticlavius* is inactive during the colder months of the year except for a few individuals that reside in rock rubble and leaf debris along a spring run (Rossman 1960). Individuals have occasionally been found submerged beneath running water, as has been reported for *P. cinereus*.

Populations are strongly associated with rocky habitats and are abundant in limestone regions where caves, talus, or rock crevices provide easy access to deep, underground passageways. Underground passages are important refuges from severe winter weather. I was unsuccessful in finding *P. dorsalis* at many sites in central Kentucky where individuals were abundant in the autumn prior to the most severe winter on record (Petranka 1979). Populations inhabiting talus slopes were not affected, but those on wooded hillsides with scattered rocks appeared to have suffered heavy mortality.

Unlike many *Plethodon* species, *P. dorsalis* is rarely aggressive toward conspecifics (Thurow 1976). Two or more adults are often present beneath the same cover object, a finding that suggests that adults do not defend cover sites.

PREDATORS AND DEFENSE. Little information is available on natural predators, but screech owls are known to prey on this and other small salamanders (Huheey and Stupka 1967). Zigzag salamanders are also undoubtedly eaten by other woodland birds, small snakes, shrews, and other small predators. Adults will avoid substrates marked with scent from ringneck snakes (*Diadophis punctatus*), which are major predators of this and other small salamanders (Cupp 1994). When uncovered beneath surface objects, individuals typically remain immobile (Brodie 1977).

COMMUNITY ECOLOGY. Very little is known about the community ecology of *P. dorsalis*. Reinbold (1979) compared microhabitats of *P. cinereus* and *P. dorsalis* in Indiana at a site where the two occur sympatrically and found that *P. dorsalis* is more frequently encountered beneath rocks than is *P. cinereus*. Adults can reach relatively high densities in optimal habitats, but their influence in shaping soil invertebrate communities is not known.

CONSERVATION BIOLOGY. Deforestation and the conversion of forest lands into agricultural and urban areas have eliminated many populations throughout the range of the zigzag salamander. Nonetheless, populations are still common in areas supporting mesic hardwood forests with rocky substrates.

Plethodon dunni Bishop
Dunn's Salamander
PLATE 119

IDENTIFICATION. Dunn's salamander is a relatively large western *Plethodon* species that has a modal number of 15 costal grooves and a broad, uneven, yellowish to olive green dorsal stripe that extends from the head almost to the tail tip. The sides of the body are black and are overlain with blotches similar in color to the dorsal stripe. The venter is slate gray with pale white or yellowish flecks, and there are two phalanges on the outermost hind toe (Nussbaum et al. 1983; Stebbins 1951; Storm and Brodie 1970a).

Adult males are slightly smaller on average than females and have proportionately longer tails and

Fig. 220. *Plethodon dunni*; adult; southwestern Oregon (J. W. Petranka).

Range of *Plethodon dunni*

wider heads. Males also have a large, concave mental gland, vent flaps, and papillose vent linings. Females lack mental glands and vent flaps and have pleated vent linings (Brodie 1968b, 1970; Stebbins 1951). Adults measure 10–15.5 cm TL.

Hatchlings reach 13–16 mm SVL (Dumas 1956). The dorsal stripe of small juveniles is brighter and more straight-edged than that of adults and in some individuals may extend to the tail tip.

SYSTEMATICS AND GEOGRAPHIC VARIATION. No subspecies are currently recognized. Brodie (1970) provides detailed accounts of local and regional variation in tooth number and external morphology. He also describes an unstriped *Plethodon* species that occurs in Benton, Lincoln, and Lane counties, Oregon, as *P. gordoni*. This form (the Mary's Peak salamander) differs from sympatric *P. dunni* in body form, tooth number, number of costal grooves, and relative length of the limbs. The Mary's Peak salamander closely resembles *P. dunni* and was previously treated as a melanistic color morph of *P. dunni* (Stebbins 1951). Evidence based on protein variation strongly suggests that *P. gordoni* is conspecific with *P. dunni* (Feder et al. 1978). Nei's genetic distance (D) for these two forms is only 0.001, and there are no fixed allelic differences. Here I treat *P. gordoni* as a junior synonym of *P. dunni* based on data provided by Feder et al. (1978).

Fig. 221. *Plethodon dunni*; Multnomah Co., Oregon (W. P. Leonard).

DISTRIBUTION AND ADULT HABITAT. Dunn's salamander inhabits rocky forest habitats from sea level to about 1000 m from southwestern Washington to extreme northwestern California. Populations are absent from most of the Willamette Valley in Oregon. This species frequents semiaquatic habitats and is often locally abundant in mesic, heavily shaded forests within the immediate vicinity of seepages, streams, waterfalls, and other aquatic sites with rocky substrates (Bishop 1943; Nussbaum et al. 1983; Slater 1933;

Storm 1955). Populations are also locally abundant on or near moist talus slopes (Brodie 1970; Dumas 1956; Fitch 1936).

BREEDING AND COURTSHIP. The breeding season is not precisely known. Females have been collected in October and April with either sperm or spermatophores in their cloacae; this finding suggests that mating occurs from autumn through spring (Dumas 1956; Nussbaum et al. 1983). In the former study females outnumbered males by more than 2:1 in samples from talus slopes. Courtship behavior has not been described.

REPRODUCTIVE STRATEGY. Data on the nesting biology of *P. dunni* are scant. Females presumably oviposit in underground retreats in rocky habitats during the spring or early summer and brood their eggs. Mature ova are 4.5–5.5 mm in diameter, pigmentless, and surrounded by two jelly envelopes (Dumas 1955; Nussbaum et al. 1983; Stebbins 1951). Clutch size varies from 4 to 18 and averages 9.4 eggs (Nussbaum et al. 1983; Slater 1939). A female that received a pituitary implant oviposited 13 eggs in the laboratory between 25 and 26 May (Stebbins 1951).

The only nesting record is that of Dumas (1955), who found a grapelike cluster of nine eggs in early developmental stages on 6 July in western Oregon. The eggs were attached by a 4-mm stalk to a slab of slate in a crevice of a heavily shaded rock outcrop within 2 m of a stream. The clutch was 38 cm deep within the crevice and was attended by a female. The outer jelly envelope was 4.8–5.3 mm in diameter (average = 5.1 mm). Embryos maintained at field temperatures reached the hatching stages by mid-September, and the incubation period was estimated to be 70 days.

TERRESTRIAL ECOLOGY. The juveniles and adults are active on wet nights on the ground surface, where they forage for invertebrates. Food items of 36 Oregon specimens consisted of snails, annelids, isopods, centipedes, millipedes, scorpions, pseudoscorpions, mites, phalangids, and numerous types of insects, particularly collembolans, beetles, and dipterans (Dumas 1956). Mites, collembolans, and dipterans were the major prey of both small and large specimens examined by Altig and Brodie (1971). Other prey included snails, spiders, hemipterans, beetles, hymenopterans, and noninsect arthropods. Large individuals occasionally cannibalize juveniles (Altig and Brodie 1971; Riesecrer et al. 1996).

Most juveniles become sexually mature when they reach 50–55 mm SVL (Brodie 1970; Nussbaum et al. 1983). Juveniles probably require 2–4 years to reach these sizes, but specific information on growth rate and length of the juvenile stage is not available.

Seasonal activity patterns vary geographically depending on the local climate (Nussbaum et al. 1983). In warm coastal habitats Dunn's salamander may be active on the ground surface during any month of the year in periods of wet weather. In many areas surface activity is restricted because of cold winter temperatures or summer droughts. Adults are agile and rapidly run for cover when disturbed.

Dumas (1956) found that individuals remain on or near talus slopes during the winter months, but move to rocky streamsides during the warmer spring months. During the dry summer months they move deep within talus, into deep cracks in rocks, or beneath stones and rocks in moist, cool microhabitats along streams. Individuals emerge from summer retreats with the return of autumn rains and are again active on the ground surface. Surface activity is greatest during April, and autumnal emergence usually begins in October.

Ovaska and Davis (1992) found that adults respond to chemical markers in conspecific fecal pellets and also avoid burrows with fecal pellets. Thus fecal pellets may be used as territorial markers. Individuals can distinguish between fecal pellets of conspecific males and females, and between fecal pellets of *P. vehiculum* and conspecifics. However, they are unable to distinguish pellets of *P. vandykei* from those of conspecifics. In natural habitats individuals are usually found singly beneath cover, but in a few instances as many as three individuals have been found under the same cover object (Dumas 1956).

PREDATORS AND DEFENSE. Steller's jay (*Cyanocitta stelleri*) and the Northwestern garter snake (*Thamnophis ordinoides*) are known predators (Dumas 1956; Nussbaum et al. 1983). Individuals primarily rely on immobility or rapid fleeing to avoid predators (Brodie 1977).

COMMUNITY ECOLOGY. Dumas (1956) compared microsympatric populations of *P. dunni* and *P. vehiculum* in Oregon and found that the two species differ substantially in diet and microhabitat preferences. In general, *P. dunni* eats a more diverse array of prey than *P. vehiculum* and prefers wetter substrates. The author believed that competition for food rarely occurs because prey are very abundant in talus slope habitats and adults can survive without food for over 5 months. Competition for moist microhabitats may occasionally occur during periods of dry weather.

Plethodon vehiculum and *P. dunni* from Washington can distinguish between each other's fecal pellets, suggesting that interspecific competition via territoriality could potentially occur (Ovaska and Davis 1992).

However, neither species avoids burrows marked with the other's fecal pellets, and it is thus uncertain how chemical marking with fecal pellets would result in the spacing out of congeners.

CONSERVATION BIOLOGY. This species does not show strong affinities for old-growth forests and is often locally abundant in forest stands of all ages (Corn and Bury 1991). Nonetheless, populations are more likely to be present in logged stands when mature timber is present upstream than when stands upstream have been cut (Corn and Bury 1989).

Plethodon elongatus Van Denburgh
Del Norte Salamander
PLATE 120

Fig. 222. *Plethodon elongatus*; adult; Siskiyou Co., California (R. W. Van Devender).

IDENTIFICATION. *Plethodon elongatus* is a slender, elongated species with 6.5–7.5 intercostal folds between adpressed limbs and a modal number of 18 costal grooves. Adults have a dark brown to black dorsal color that is overlain with a straight-edged, reddish or reddish brown dorsal stripe that usually extends from the head to the tail tip. The stripe decreases in intensity with age and in many adults may be obscure or missing (Brodie 1970). The venter is dark gray, and the throat is light gray and mottled. Sexually mature males are on average slightly larger than females, have posteriorly concave mental glands, and have slightly swollen vent lobes. Adults reach 11–15 cm TL.

Hatchlings measure about 18 mm SVL on average.

Hatchlings and small juveniles generally resemble the adults but have more conspicuous dorsal stripes (Livezey 1959; Stebbins and Reynolds 1947).

SYSTEMATICS AND GEOGRAPHIC VARIATION. Brodie (1970), Nussbaum et al. (1983), and Stebbins (1951) summarize information on geographic variation in morphology and coloration. No subspecies are currently recognized; however, inland populations in California are on average larger, are lighter colored, and have more conspicuous dorsal stripes than populations found elsewhere. In addition, inland specimens tend to have scattered white flecking along the sides.

DISTRIBUTION AND ADULT HABITAT. This species is confined to terrestrial habitats in humid coastal forests in southwestern Oregon and northwestern California and is characteristically associated with moist rock rubble and talus throughout its range (Diller and Wallace 1994; Herrington 1988). Although strongly associated with talus, *P. elongatus* also lives beneath bark, logs, and other surface cover on the forest floor of coastal forests (Brodie 1970; Wood 1934). In redwood and Douglas-fir forests that receive relatively high precipitation, *P. elongatus* is most common on north-facing slopes in talus that is covered with organic debris (Diller and Wallace 1994). Population density is influenced by stand age, and moderately high densities of *P. elongatus* sometimes occur in talus along road cuts in young-growth forests.

Range of *Plethodon elongatus*

BREEDING AND COURTSHIP. Mating can apparently occur anytime from the fall through the spring months, but timing varies among regional populations. Mature males have spermatozoa in their vasa deferentia during April and May (Nussbaum et al. 1983). Two California females with spermatophores were collected during the fall months, but none during the spring months; this finding suggests a fall and winter mating season in northern California (Welsh and Lind 1992). Sex ratios are 1:1, and about half of all females from this region have mature or maturing ovarian eggs and a biennial reproductive cycle. In one intensively studied population, mature ovarian eggs that are visible through the body wall are present in only 22% of females. Descriptions of courtship behavior are not available.

REPRODUCTIVE STRATEGY. Most females oviposit during the spring or very early summer and brood their clutches in underground nests during the summer months. Mature ova are white and average 5.1 mm in diameter (Nussbaum et al. 1983). Each embryo is surrounded by the vitelline membrane and two jelly envelopes, and the mature embryos have trilobed gills.

The only nest record is for a brooding female and a grapelike cluster of 10 eggs in intermediate developmental stages found on 27 July in a small cavity under a redwood post in northwestern California (Livezey 1959). The eggs are attached by short pedicels to a central peduncle that is attached to an overhanging support. The outer jelly capsule averages slightly more than 10 mm in diameter.

Estimates of clutch size come mostly from ovarian egg counts of California specimens. Welsh and Lind (1992) estimate the average clutch size to be 7–8 eggs based on the number of ova visible through the body wall of live specimens. The number of mature ova in 18 females from northwestern California varied from 3 to 11 and averaged 8 (Nussbaum et al. 1983), whereas two other females contained 10 and 11 ova (Stebbins 1951).

TERRESTRIAL ECOLOGY. Hatching occurs in late summer or early autumn. Most young of the year remain in underground retreats in the fall and do not emerge on the ground surface until the following spring. Annual growth in adult females and males is 1.1 and 2.4 mm/year in a northern California population (Welsh and Lind 1992). Individuals mature sexually when they reach about 55 mm SVL (Brodie and Storm 1971).

Individuals in a northern California population move very little between recaptures, and most remain within a 7.5 × 7.5-m area within a single year (Welsh and Lind 1992). On rare occasions a male may move as far as 36 m over a 6-month period. Estimated density is 0.3–0.9 salamander/m^2 depending on the sampling method used. Adults captured in the spring are heavier than those captured during the fall, suggesting that individuals do not feed much when in underground retreats in the summer. Diller and Wallace (1994) found 0.03–0.83 animal/m^2 by turning rocks and digging through debris in talus. Immature individuals composed about 50% of their sample compared with only 20% of specimens collected by Welsh and Lind (1992).

The diet consists of small invertebrates. Prey in 24 adults from northwestern California included mites, spiders, centipedes, isopods, hemipterans, homopterans, beetles, lepidopterans, collembolans, and hymenopterans (Bury and Johnson 1965). Beetles and centipedes are the most important prey volumetrically.

PREDATORS AND DEFENSE. Natural predators are not well documented but probably include small snakes, small mammals such as shrews, and woodland birds. Individuals often remain immobile when uncovered or attacked (Brodie 1977).

COMMUNITY ECOLOGY. *Plethodon elongatus* and *P. vehiculum* are allopatrically distributed, suggesting that competitive exclusion may occur between these two ecologically similar species (Stebbins 1951). Interactions of *P. elongatus* with this and other salamander species have yet to be examined.

CONSERVATION BIOLOGY. Populations of *P. elongatus* in moist coastal forests do not appear to be strongly affected by timbering (Diller and Wallace 1994). However, populations in areas that receive less fog and annual precipitation are strongly affiliated with old-growth forests, and their numbers often decline sharply following the intensive harvesting of timber (Raphael 1988; Welsh 1990; Welsh and Lind 1988, 1991).

Plethodon glutinosus (Green)
Slimy Salamander
PLATE 121

Fig. 223. *Plethodon glutinosus;* adult; Pike Co., Ohio (R. W. Van Devender).

IDENTIFICATION. The slimy salamander is a large *Plethodon* species that is aptly named because individuals produce copious amounts of adhesive skin secretions from the tail when handled roughly. The secretions function to deter predators (Brodie et al. 1979) and when dry are difficult to remove from human skin. Adults have a black to dark bluish black ground color that is overlain with small, scattered silvery white or metallic gold spots. Larger white, gray, or yellow spots occur along the sides of the body, and, in extreme cases, may fuse to form a broad band. The venter is grayish black and is slightly lighter colored than the dorsum. As in other *Plethodon* species, the tail is rounded in cross section.

Sexually mature males have prominent circular mental glands and small, round yellow or orange glands on the belly (Bishop 1941a; Highton 1956). Like many other salamanders, males of this species have papillose cloacal linings. Sexual dimorphism in SVL is slight. Females typically average 0–6% larger in SVL than males (Highton 1956; Pfingsten 1989a; Pope and Pope 1949; Semlitsch 1980b). Albino specimens have been found in Florida and Indiana (Hensley 1959; Highton 1956; Piatt 1931). Adults measure 11.5–20.5 cm TL and the average number of costal grooves is 16.

Hatchlings lack pigmentation on the undersides of the body and are uniformly gray to black above with pigment-free areas that appear as scattered light spots (Wells and Gordon 1958). The length of hatchlings varies among populations. Reported values are 12–15 mm SVL and 20–26 mm TL (Highton 1956), 18 mm SVL and 31 mm TL (Wells and Gordon 1958), and 18–26 mm TL (Minton 1972).

SYSTEMATICS AND GEOGRAPHIC VARIATION. Populations that traditionally were recognized as *P. glutinosus* represent a complex of geographically and genetically variable groups that occur throughout the eastern United States (Highton 1962a,b; Highton et al. 1989). Ancestral forms of this group apparently were fragmented and then reunited one or more times during their history. Divergence in some groups has reached the species level; however, other groups show evidence of gene exchange that ranges from slight to moderate. Two groups that have reached the species level are *P. aureolus* and *P. kentucki*. Both of these species coexist with other members of the complex and rarely interbreed. A third species (*P. oconaluftee*; formerly *P. teyahalee*) is found in the southern Appalachians and interbreeds to varying degrees with other forms traditionally referred to as *P. glutinosus* (sensu Highton 1962a). In Polk Co. in southeastern Tennessee the two forms do not hybridize at one site. However, these forms do hybridize parapatrically in two areas to the northeast, and they hybridize extensively with *P. glutinosus*-like forms in northeastern Georgia (Highton 1983, 1995). Highton (1995) does not provide quantitative information on the extent of hybridization or width of hybrid zones in Tennessee and Georgia. Although *P. oconaluftee* has many characteristics of a semispecies, my view is that the nomenclatural status of this form should not be altered until detailed data are published that will allow a critical evaluation.

Highton et al. (1989) describe geographic variation in isozymes and serum albumins in the remaining groups within the complex. Microcomplement fixation using three antisera is not effective in resolving taxonomic groupings. However, many populations show moderate genetic differentiation based on isozyme data and can be arranged in subgroups that reflect these differences. The authors formally recognize 13 subgroups as different species (*P. albagula, P. chattahoochee, P. chlorobryonis, P. cylindraceus, P. glutinosus* [sensu stricto], *P. grobmani, P. kiamichi, P. kisatchie, P. mississippi, P. ocmulgee, P. savannah, P. sequoyah,* and *P. variolatus*) after electing a posteriori to use a genetic

Fig. 224. *Plethodon glutinosus;* adult; central Pennsylvania (J. W. Petranka).

distance or *D* value of 0.15 (with two minor exceptions) to define species. Their rationale is that the *D* >0.15 criterion defines groups that are geographically continuous and often morphologically different. In addition they argue that genetic divergence at this level is typical of vertebrate species that are generally considered taxonomically valid.

Many researchers have argued against using genetic distance as the primary criterion for recognizing allopatric or parapatric species because this approach assumes that genetic distance tightly correlates with the development of reproductive isolating mechanisms. Tilley et al. (1990) provide data that suggest that this assumption may not hold in plethodontid salamanders. Frost and Hillis (1990) object to splitting *P. glutinosus* into 13 species based solely on arbitrarily selected genetic distances and cite a variety of perceived problems, including several biases in estimating genetic distances.

Detailed information on the extent to which gene exchange is occurring among the 13 parapatric groups recognized by Highton et al. (1989) is not available. Limited data suggest that some parapatric groups appear to be at or near the species level with only slight gene exchange, whereas others hybridize extensively and may be closer to the level of subspecies or semispecies. The contact zones for many pairs have yet to be examined in detail, and there is no published information on the degree of gene exchange occurring between most parapatric forms. Overall, the *P. glutinosus* complex appears to include parapatric groups that have differentiated to levels ranging from subspecies through semispecies to full species. My view is that formal taxonomic recognition of genetic subgroups of *P. glutinosus* is premature and should be deferred until detailed studies of contact zones provide evidence that will lead to a stable nomenclature. Here I do not formally recognize these 13 groups as species, and I refer to all as *P. glutinosus*.

Plethodon glutinosus shows marked geographic variation in body size and color patterns (Carr 1996; Highton 1962a,b, 1970, 1983; Highton et al. 1989). Specimens from inland populations usually are larger on average than those from the Coastal Plain. Carr (1996) conducted a detailed morphological analysis of members of the *Plethodon glutinosus* complex and found that 61% of individuals can be accurately assigned to the proper species (sensu Highton et al. 1989) using discriminant analysis. The size of adults is the most useful character in sorting out groups and differentiating the smaller Coastal Plain groups from larger inland groups. Variation among populations within species is about as great as that between different species, and populations of the same species often do not cluster with each other.

Geographic variation in color patterning and coloration is more evident than morphological variation other than adult size. Specimens with small, brassy spots and grayish sides are most common in Florida populations (Allen and Neill 1949). Individuals in portions of the Blue Ridge and Piedmont provinces of North Carolina, South Carolina, and Virginia have light-colored chins, white spots on the dorsum, and little or no brassy spotting. In populations in northeastern Georgia, many individuals lack dorsal spotting altogether, a trait that may in part reflect hybridization with *P. jordani*. Individuals in the Coastal Plain of North Carolina, South Carolina, and Virginia are relatively small and have small dorsal spots and yellow spotting along the sides. More detailed descriptions of geographic variation in *P. glutinosus* are given in Highton (1962a) and Highton et al. (1989).

Several subspecies have been described in the past (Highton 1962a), but most recent accounts recognize only two subspecies: the white-throated slimy salamander (*P. g. albagula*), which is found in central Texas, and the slimy salamander (*P. g. glutinosus*), which is found elsewhere. Here subspecies are not recognized because white-throated forms of *P. glutinosus* occur in some populations outside Texas and dark-throated forms occur in some central Texas populations (Highton et al. 1989).

DISTRIBUTION AND ADULT HABITAT. The slimy salamander inhabits forests from central New York to central Florida and from central Missouri

Fig. 225. *Plethodon glutinosus*; adult (note conspicuous mental gland of male); locality unknown (R. W. Van Devender).

southward to central Texas. Geographic isolates occur in New Hampshire and in portions of Louisiana, Arkansas, and eastern Texas (Highton 1962a,b; Warner 1971). This species is absent from most areas of southwestern North Carolina where *P. oconaluftee* occurs. *Plethodon glutinosus* is most characteristically associated with eastern deciduous forests and lives in moist forest floor microhabitats at elevations ranging from near sea level to about 1500 m. This species is also locally abundant in many areas in the Coastal Plain from Virginia through Texas, where populations inhabit bottomland hardwoods, swamp forests, and wet pinewoods. Adults are most easily collected by turning rocks or logs on the ground surface during the day and by searching the forest floor at night. Individuals also frequent caves (Noble and Marshall 1929; Peck 1974; Wells and Gordon 1958). Populations previously referred to as *Plethodon g. albagula* occur in wooded canyons and ravines in karst regions of the Edwards Plateau (Grobman 1944).

BREEDING AND COURTSHIP. Time of mating varies geographically depending on latitude. Males in northern Florida have vasa deferentia packed with sperm from mid-February through early August (Highton 1956). Mating begins in March in Alabama populations (Trauth 1984) and peaks in September and October in northern populations (Bishop 1941a; Highton 1962b). A small percentage of adults in northern populations may also breed in the spring. Captive specimens from New Jersey and Virginia court in the laboratory during September and October (Organ 1960a, 1968b), and a pair of adults courted on 19 August in the mountains of Virginia (Pope 1950).

Range of *Plethodon glutinosus*

Yearly breeding patterns vary with latitude. Both males and females reproduce annually in Florida (Highton 1956, 1962b). In contrast, females in northern populations reproduce biennially, whereas male reproduce annually (Highton 1962b; Pfingsten 1989a; Semlitsch 1980b). These differences reflect the amount of time that is available annually for females to forage on the ground surface and store energy to produce eggs.

Organ (1960a) provides a detailed summary of courtship behavior. The following is a summary of his observations of Virginia animals.

When a male encounters a female, he places his nasolabial grooves and mental gland in contact with her head, body, or tail. The male then engages in a "foot dance" in which he raises and lowers his rear limbs either alternately or simultaneously. When raising the limbs simultaneously, the tail is used as a brace to support the body. As the dance proceeds, all four limbs are raised and lowered as if the male is marking time. The male proceeds toward the female's head. While moving forward, the male repeatedly rubs his nasolabial grooves on the female, then gently grasps her body or tail with his mouth. His grip is maintained for a short period of time, then released.

The male eventually reaches the head of the female and moves his mental gland over her head and nasolabial grooves. He then pushes his head under her chin and passes beneath while maintaining contact with the chin with his body and tail. Foot dancing continues throughout these early stages of courtship. The male undulates his tail as it passes under the female's chin, then stops his forward progress. The female straddles the tail and the pair moves forward and engages in a tail-straddle walk. During the tail-straddle walk the male undulates his tail and often flexes his body laterally to slap the snout of the female with his mental gland. The tail-straddle walk is prolonged and may cover a linear distance of >5 m.

Prior to spermatophore deposition the male stops moving forward and begins rocking his sacral region laterally. The female moves her head laterally in synchrony, but counter to the direction of movement of the male. The male lowers his vent to the substrate and deposits a spermatophore. He then raises the vent and flexes his tail to one side. The couple then moves forward with the female keeping her chin against the tail base of the male. As the female passes over the spermatophore she picks up the sperm cap with her cloacal lips. The couple usually breaks up shortly thereafter.

The spermatophore is mushroom shaped, bilaterally

symmetric, and laterally compressed. The stalk is colorless and has a spike that projects upward into the center of the sperm cap. In addition, a mound of jelly occurs behind the posterior border of the cap. The sperm cap is pale yellow to white and is adhesive. The entire spermatophore measures on average about 4 mm tall and 3.5 mm wide at the base.

Patrolling males do not discriminate between sexes until they make physical contact (Organ 1960a). Males often viciously bite other males attempting to approach and court them. In addition, courting males often bite passive males they approach and attempt to court. Males occasionally trick other males into courting by mimicking female behavior. This is a form of sexual competition that results in rivals wasting gametes (Arnold 1977).

Dawley (1984b) found little evidence that individuals can identify sex via odors deposited on substrates. However, both males and females prefer female airborne odors over those of males. Other mating trials show that male *P. glutinosus* prefer odors of female conspecifics over those of sympatric female *P. aureolus, P. jordani,* and *P. kentucki* during the breeding season (Dawley 1986a,b).

Female *P. glutinosus* show no preference for conspecific males when tested against *P. aureolus* males, but do show preference for conspecific males when tested against male *P. jordani* and *P. kentucki*. In breeding experiments in which males are confined with conspecifics and heterospecifics, male *P. glutinosus* always initiate courtship with conspecifics first. In rare instances they may court and inseminate female *P. kentucki* if prolonged attempts to court conspecifics are unsuccessful.

REPRODUCTIVE STRATEGY. Each female lays a globular cluster of eggs that is often suspended from the ceiling of a natural cavity. Freshly laid eggs are 3.5–5.5 mm in diameter on average, creamy white, and surrounded by two jelly envelopes (Noble and Marshall 1929; Wood and Rageot 1955). The outer envelopes of adjoining eggs tend to stick together and may sometimes fuse partially after oviposition. The outer capsule is 5–8 mm in diameter (Brode and Gunter 1958; Noble and Marshall 1929; Wells and Gordon 1958; Wood and Rageot 1955). Females guard their eggs and often coil about them. The mature embryos have trilobed, antlerlike gills that disappear within a few days after hatching (Highton 1962b; Noble and Marshall 1929). Limited observations indicate that females may remain with their hatchlings for 1–2 weeks before the young disperse from the nesting cavities (Brode and Gunter 1958).

Most females presumably oviposit in cavities beneath the ground since relatively few surface nests have been discovered. Eggs have been found beneath or in decaying logs in Florida (Highton 1956, 1962b), in a pile of discarded bags in Virginia (Wood and Rageot 1955), in a decaying stump in West Virginia (Fowler 1940), and in rock crevices in caves or mine shafts in Alabama, Arkansas, Mississippi, and Missouri (Brode and Gunter 1958; Heath et al. 1986; Highton 1962b; Noble and Marshall 1929; Wells and Gordon 1958). One clutch was in a cave 60 m from the entrance.

Seasonal patterns of vitellogenesis and nesting records indicate that females in most southern populations oviposit during late summer (Highton 1956, 1962b). Females in eastern Kentucky, Maryland, Pennsylvania, and Virginia and other northern populations oviposit in late spring or early summer (Bush 1959; Highton 1962b). Cave records include over a dozen clutches between 15 and 28 September in a cave in northern Alabama (Highton 1962b), clutches of 18 and 10 eggs on 17 August and 3 September in an Arkansas cave (Noble and Marshall 1929), a clutch of 17 eggs in a Mississippi cave on 14 August (Brode and Gunter 1958), and two clutches of 8 and 11 eggs with full-term embryos on 27 October in a Missouri cave (Wells and Gordon 1958). Records of surface nests include six clutches (mean = 8; range = 5–11 eggs) with brooding females found in northern Florida between late August and mid-October (Highton 1956, 1962b), a report of nests being found in early July in Pennsylvania (Hudson 1954), a clutch (size not reported) discovered on 30 May in the Coastal Plain of Virginia (Wood and Rageot 1955), and a clutch of 15 eggs in a rotting stump in West Virginia on 3 June (Fowler 1940). Two females that Highton (1962b) placed in jars in the field in Virginia and West Virginia oviposited in late May or June based on the developmental stages of embryos examined in late July and August.

Estimates of clutch size based on counts of large ovarian eggs are generally higher than counts of eggs in nests. Reported values for ovarian complements include means of 16 and 17 (range = 10–22) for two Florida populations (Highton 1962b), a mean of 25 for southeastern Kentucky specimens (Bush 1959), a mean of 17 (range = 13–25) for Pennsylvania specimens (Highton 1962b), a mean of 26 (range = 16–34) for Maryland specimens (Highton 1962b), a range of 17–38 for New York specimens (Bishop 1941a), and a mean of 23 (range = 17–33) for Virginia specimens (Pope and Pope 1949). Clutch size is positively correlated with female SVL in Maryland and Pennsylvania populations (Semlitsch 1980b).

The discrepancy between number of mature ovarian eggs and clutch sizes observed in the field reflects embryonic mortality and the fact that females do not always deposit their full complement of eggs (Highton 1956). Differences in ovarian egg counts among populations are primarily due to differences in average adult body size.

The embryonic period lasts 2–3 months and hatching dates vary depending on latitude (Highton 1956, 1962b). Hatchlings have been found from 3 November to 10 December in a northern Alabama cave (Highton 1962b), during the second week of October in Indiana (Minton 1972), in early August through mid-September in a Mississippi cave (Brode and Gunter 1958), and in late October in a Missouri cave (Wells and Gordon 1958).

TERRESTRIAL ECOLOGY. Marked differences in life history patterns exist between Florida populations and northern populations in Maryland, Pennsylvania, and Virginia (Highton 1956, 1962b). In Florida, hatchlings first emerge on the ground surface about 2 months after hatching and grow to 25–26 mm SVL during their first year of life. Most females become sexually mature when they are 2 years old and measure 46–56 mm SVL, but do not oviposit until the following year when nearly 3 years old. Some males have enlarged testes when 1 year old, and breed the following year when nearly 2 years old. The remainder presumably breed when nearly 3 years old. Males become sexually mature when they reach 40–53 mm SVL, and both males and females reproduce annually.

Individuals in northern populations take longer to mature than Coastal Plain populations and reproduce biennially rather than annually (Highton 1962b; Pfingsten 1989a; Semlitsch 1980b). Hatchlings first emerge on the ground surface during the spring, roughly 7 months after hatching. Growth rates of Pennsylvania juveniles are similar to those in Florida populations during the first 7 months of surface activity following hatching (Highton 1962b). However, annual growth in Florida specimens is greater because specimens are active year round.

Analyses of size frequencies indicate that most individuals in northern populations begin maturing sexually 4 years after hatching and reproduce for the first time the following year. Males mature sexually at 45–52 mm SVL in a western Virginia population, and females between 58 and 65 mm SVL. Individuals in Maryland and Pennsylvania populations mature at 53–70 mm SVL, with males reaching sexual maturity at smaller sizes than females. Individuals in southern Illinois appear to mature after 3 years when they measure 47–58 mm SVL (Highton 1962b). The smallest mature male collected by Wells and Wells (1976) from the Piedmont of North Carolina measured 50 mm SVL.

Seasonal patterns of surface activity vary depending on climate. In general, slimy salamanders tend to be active on the ground surface throughout the year except during droughts and during periods of extreme heat or cold. Individuals near southern coastal regions are active on the ground surface throughout the winter except during periods of freezing weather (e.g., Eaton 1953). Slimy salamanders in northern Florida are much less active on the ground surface during the warmer summer months than during the cooler months of the year (Highton 1956). Individuals in northern populations show the reverse trend, with minimal seasonal activity and little growth occurring during the cold winter months (Highton 1962b, 1972; Semlitsch 1980b). Slimy salamanders are active on the ground surface in southern Indiana from late March through early November (Minton 1972), whereas individuals in the southern Appalachians are active from May to October. In West Virginia individuals at low elevations are inactive during the summer, whereas those at high elevations are active throughout the summer (Green and Pauley 1987).

In Florida individuals are very sedentary and are rarely recaptured more than 60 cm from their original sites of capture (Highton 1956). In a North Carolina population juveniles do not move from their home logs during the first year of life (Wells and Wells 1976). Adults move significantly greater distances between captures than juveniles, but adult females and males do not differ significantly in the average distance moved. The maximum distance moved between captures is 92 m; however, most adults move <9 m. The most distant moves are made by juveniles shortly after becoming sexually mature. Older adults are relatively sedentary.

Slimy salamanders often frequent caves and have prehensile tails that aid in climbing cliff faces and rock formations (Brode and Gunter 1958). In many populations the use of caves is seasonal. In a Georgia population individuals migrate from a ravine into a nearby cave during droughts (Humphries 1956); in an Alabama population gravid females move into caves to oviposit (Highton 1962b). I observed *P. glutinosus* fanning out of a cave in eastern Kentucky during the onset of a heavy rain. Individuals apparently forage on the adjoining forest floor in wet weather and return to the cave at other times.

Slimy salamanders will aggressively defend territories in laboratory terraria from both conspecifics and other large *Plethodon* species (Thurow 1976). The extent to which this behavior results in animals spacing out in nature is not well known. Wells (1980b) examined how groups of *P. glutinosus* found beneath logs in east-central North Carolina vary with respect to sex and state of maturity. In 83% of the cases a single salamander was found under a log on a given day. However, captures of 417 different individuals under 78 logs over a 3-year period indicated that some logs are used by up to several dozen salamanders. When two or more individuals are found beneath the same log, males are associated with other males less frequently than would be expected by chance. Males and females are together more frequently than would be expected by chance, but females do not appear to avoid other females. Juveniles associate with males, females, and other juveniles equally. These data suggest that females and juveniles show little tendency to avoid logs with occupants. The underrepresentation of male-male groupings, and overrepresentation of male-female groupings, may reflect the fact that males defend home logs from other males, or that males actively search logs until they find females. Groups of 10–25 animals can be found under large logs during summer droughts (Wells and Wells 1976). In some cases the animals pile on top of one another, a behavior that presumably functions to reduce dehydration.

Adult body size and density of salamanders sometimes vary locally in populations only a few hundred meters apart (Semlitsch 1980b). Average densities of Maryland and Pennsylvania populations in optimal habitats are 0.52–0.81 salamander/m^2 based on mark-recapture estimates.

Like many other salamanders, juveniles and adults of this species emerge from cover and forage on the ground surface at night if conditions are sufficiently moist. Individuals in a southern Appalachian population are most active during the first few hours after dark (Gordon et al. 1962). Slimy salamanders are generalist feeders and eat a wide variety of invertebrate prey. Food items in 58 adults from western Virginia included collembolans, homopterans, hemipterans, lepidopterans, dipterans, coleopterans, hymenopterans, phalangids, pseudoscorpions, spiders, mites, millipedes, centipedes, earthworms, and snails (Pope 1950), whereas those in 170 Pennsylvania specimens consisted of earthworms, snails, spiders, millipedes, and insects (Surface 1913). The bulk of the diet of 64 New York specimens was insects—particularly ants and beetles—followed by centipedes, spiders, slugs, and snails (Hamilton 1932). A similarly wide array of prey was described for 33 Texas adults, with ants, beetles, and sowbugs composing much of the diet (Oliver 1967). Ants and beetles constituted most of the bulk of the diet of 100 Virginia specimens (Davidson 1956). Other prey included snails, isopods, earthworms, centipedes, millipedes, arachnids, phalangids, pseudoscorpions, and miscellaneous insects.

Slimy salamanders occasionally cannibalize and may consume smaller salamanders of other species. Powders (1973) records an instance of a large adult consuming a smaller conspecific, and Powders and Tietjen (1974) found what appeared to be the bones of salamanders in another specimen. These studies probably include mixed samples of *P. glutinosus* and *P. oconaluftee*. Oliver (1967) documents cannibalism in *P. glutinosus* from Texas.

The diets of slimy and Red Hills salamanders are similar in areas of microsympatry in southern Alabama (Brandon 1965b). Slimy salamanders consume millipedes, spiders, phalangids, beetles, ants and other hymenopterans, and miscellaneous insect larvae. The diets of *P. glutinosus* and *Eurycea lucifuga* inhabiting caves in the eastern United States differ in some respects. In general *P. glutinosus* eats fewer and less diverse prey than *E. lucifuga*; 60% of 108 specimens of *P. glutinosus* had empty digestive tracts compared with only 10% of 112 *E. lucifuga* (Peck 1974).

PREDATORS AND DEFENSE. Despite their slimy tails, slimy salamanders are occasionally eaten by predators such as copperheads (*Agkistrodon contortrix*) and garter snakes (*Thamnophis*; Uhler et al. 1939). Cannibalism occurs occasionally, and cave salamanders (*E. lucifuga*) sometimes prey on small *P. glutinosus* (Peck and Richardson 1976). Individuals often engage in body-flipping and tail-lashing when attacked by shrews (Brodie et al. 1979). On rare occasions, slimy salamanders may emit a squeak when roughly handled (Highton 1956; Mansueti 1941; Neill 1952).

COMMUNITY ECOLOGY. Slimy salamanders are common members of many forest communities and probably play important roles in structuring invertebrate communities inhabiting the forest floor. Despite numerous dietary studies showing that these and other salamanders prey heavily on forest floor invertebrates, we still have almost no understanding of the ecological importance of salamanders in controlling these invertebrates.

CONSERVATION BIOLOGY. This and other species of large eastern *Plethodon* reach their highest densities in mature hardwood forests. Large fallen logs are important microhabitats that are used as surface cover and foraging sites. Deforestation and the conversion of hardwoods to pine monocultures have reduced or eliminated many *Plethodon* populations in eastern North America. Bennett et al. (1980) found that *P. glutinosus* in South Carolina is 8–12 times more abundant in oak-hickory forest than in young pine monocultures, whereas Grant et al. (1994) found *P. glutinosus* to be more common in older pine stands than in pine stands <3 years old. *Plethodon glutinosus* and many other salamanders in the southern Appalachians are far more abundant in mature hardwood forests than on recent clear-cuts (Petranka et al. 1993, 1994).

Plethodon hoffmani Highton
Valley and Ridge Salamander
PLATE 122

Fig. 226. *Plethodon hoffmani;* adult; Monroe Co., West Virginia (R. W. Van Devender).

IDENTIFICATION. *Plethodon hoffmani* is an elongated, short-legged, small eastern species with a tail that makes up 50% or more of the total length. The dorsal ground color is brown to brownish black with small scattered white spots and abundant brassy flecking (Highton 1972, 1986b). Small amounts of red pigment are sometimes present on the dorsum. The throat is light colored, and the belly is dark and mottled with brown and white.

A rare striped morph with a narrow red dorsal stripe on the body occurs in populations from the vicinity of Reddish Knob and Shenandoah Mountain in Virginia and West Virginia (Highton 1972, 1986b; Highton and Jones 1965). Males have a crescent-shaped mental gland near the apex of the lower jaw (Dodd and Brodie 1976). Adults measure 8–13.5 cm and have an average of 20–21 costal grooves. Hatchlings have not been described, but juveniles resemble the adults in coloration.

SYSTEMATICS AND GEOGRAPHIC VARIATION. The most conspicuous geographic variation concerns the presence of a striped morph in populations from the vicinity of Reddish Knob and Shenandoah Mountain in Virginia and West Virginia. Some populations on Shenandoah Mountain, Virginia, have a modal number of 20 costal grooves compared with 21 in populations elsewhere (Highton 1972). No subspecies are recognized.

DISTRIBUTION AND ADULT HABITAT. *Plethodon hoffmani* is found in the Valley and Ridge Physiographic Province from central Pennsylvania southwestward to the New River in Virginia and West Virginia.

Range of *Plethodon hoffmani*

Populations also occur in adjoining portions of the Appalachian Plateau and Blue Ridge provinces in Virginia and West Virginia and have been collected as high as 1402 m above sea level (Pauley 1980). Specimens can be found by turning rocks, logs, and other surface cover along ravines or mountain slopes in deciduous forest. This species sometimes inhabits the twilight zones of West Virginia caves (Cooper 1961).

BREEDING AND COURTSHIP. Adults in Maryland and Pennsylvania populations mate primarily in the spring, but have a secondary breeding bout in the autumn. Males have convoluted, sperm-packed vasa deferentia from late September to May, but females with spermatozoa in their spermathecae have only rarely been collected during the autumn (Angle 1969). Females have biennial reproductive cycles and mate every other year shortly before ovipositing. Courtship behavior has not been described, but presumably involves a tail-straddle walk similar to that of other *Plethodon* species.

REPRODUCTIVE STRATEGY. Females lay eggs in underground retreats. Although nesting females have never been found, data on the seasonal development of ovarian eggs indicate that females oviposit in late May or early June. The maximum size of ova is 4.0 mm in diameter. Number of mature ovarian eggs in 39 females varied from 3 to 8 and averaged 4.7 (Angle 1969). Hatching probably occurs about 2 months after the eggs are deposited.

TERRESTRIAL ECOLOGY. Embryos in Pennsylvania and Maryland populations hatch in late August or September, but the young do not emerge on the ground surface until March (Angle 1969). Individuals grow 10–12 mm SVL during the first year after hatching. Most juveniles mature sexually the following autumn about 2 years after hatching. Males mature sexually when they reach 38–44 mm SVL. Females mature when they reach 39–47 mm SVL and reproduce biennially. Most females initially oviposit the year after becoming mature, but a small percentage may require an additional year before ovipositing for the first time.

Plethodon hoffmani is most active on the ground surface during the spring and autumn months and lives underground during the hottest and coolest months of the year (Angle 1969; Fraser 1976b; Highton 1962a). The large tail serves as a fat storage organ and provides energy when animals are underground. Only a small proportion of the population is above ground at any time, even when conditions are ideal for surface foraging (Fraser 1976b).

Adults reside beneath rocks and logs on the forest floor during the day and actively forage for small invertebrates outside cover at night. Food items in Virginia specimens consist of snails, beetles, oligochaetes, dipterans, spiders, millipedes, centipedes, collembolans, and hymenopterans (Fraser 1976a,b). Ants, insect larvae, oligochaetes, millipedes, and centipedes constitute most of the volume of prey, and the maximum size of prey increases with salamander head width.

PREDATORS AND DEFENSE. Data on natural predators and defensive mechanisms are not available.

COMMUNITY ECOLOGY. *Plethodon hoffmani* and *P. richmondi* are parapatric in western Virginia, where they are separated by the New River. Because rivers do not serve as complete barriers to dispersal in *Plethodon* species, mutual exclusion of some sort is likely to be occurring (Highton 1972). The distributions of *P. hoffmani* and *P. cinereus* also tend to be parapatric, although in several regions *P. cinereus* is sympatric with *P. hoffmani*. Fraser (1976a) concludes that strong competition for food is unlikely between these species because they differ in microhabitat preferences and nocturnal foraging schedules (details of this study are presented in the account of *P. cinereus*).

In a related study Fraser (1976b) found that diet and surface cover are similar for adult *P. hoffmani* and juvenile *P. punctatus* at a site where the two co-occur microsympatrically in Virginia. Differences in the ecological requirements of the adults of both species are greater and are related to differences in adult size. Staggered feeding schedules and differences in adult requirements reduce inter- and intraspecific competition. However, strong overlap between juvenile *P. punctatus* and adult *P. hoffmani* could serve as an ecological bottleneck that results in strong competition. Competition has never been demonstrated, and it is uncertain if these species compete in nature.

CONSERVATION BIOLOGY. Mature hardwood forests on well-drained soils with scattered rocks provide optimal habitats for *P. hoffmani*. Like those of all eastern *Plethodon* species, populations are rapidly eliminated when forests are cleared and converted into agricultural lands or urban areas.

Plethodon hubrichti Thurow
Peaks of Otter Salamander
PLATE 123

Fig. 227. *Plethodon hubrichti*; adult; Bedford Co., Virginia (R. W. Van Devender).

IDENTIFICATION. This small eastern *Plethodon* is similar to the Cheat Mountain salamander (*P. nettingi*), but it has heavy brassy flecking on the dorsum and a modal number of 19 costal grooves (Highton 1986d; Thurow 1957a,b). The dorsal ground color is brown and is overlain with brassy flecks. Inconspicuous white spotting occurs on the dorsum and along the sides of the body. The venter is gray to grayish black and males have small mental glands immediately behind the chin. Adults measure 8–13 cm TL and have 18–20 costal grooves. Hatchlings have a distinct dorsal stripe consisting of reddish spots.

SYSTEMATICS AND GEOGRAPHIC VARIATION. *Plethodon hubrichti*, *P. nettingi*, and *P. shenandoah* are a closely related group of species that were previously treated as subspecies of *P. nettingi*. These forms are in all likelihood remnants of a more widely distributed ancestral form whose populations became fragmented and isolated on mountain ranges in Virginia and West Virginia. Highton and Larson (1979) raised these forms to the status of full species after comparing protein similarities. Their data indicated moderate lev-

Range of *Plethodon hubrichti*

els of genetic divergence, with Nei's genetic distances (*D*) varying from 0.22 to 0.35 for different pairs of species (genetic distances based on five specimens of each species). Researchers have debated whether these forms merit recognition as full species. Courtship trials or transplant experiments would help clarify whether these forms are reproductively isolated, but such tests have not been conducted to date. Here I follow Highton and Larson's scheme and treat previously recognized subspecies of *P. nettingi* as full species.

DISTRIBUTION AND ADULT HABITAT. *Plethodon hubrichti* occupies mesic forest floor habitats in deciduous forests along a 19-km length of the Blue Ridge Mountains in Bedford, Botetourt, and Rockbridge counties in the Peaks of Otter region of west-central Virginia. This species is generally found at elevations above 845 m.

BREEDING AND COURTSHIP. No data are currently available on the mating season and courtship behavior of this species.

REPRODUCTIVE STRATEGY. Females presumably nest underground during the summer months, but nests have not been discovered and the seasonal time of oviposition has not been clearly delineated.

TERRESTRIAL ECOLOGY. Kramer et al. (1993) conducted an intensive study of *P. hubrichti* in a 10 × 10-m plot on a forested mountainside. Densities average 4.5 salamanders/m^2 and individuals have a slightly clumped distribution. Individuals grow 0.8–0.11 mm SVL per day between April and October and large adults grow less than smaller individuals.

Surface activity is greatest following periods of rain, and individuals often climb ferns to forage at night. On a given night typically only 3–5% of the population can be observed on the ground surface. Most animals are active between 2100 and 2400 h. Median home ranges average only 0.6 m^2, and the median distance between movements for marked animals is about 1 m. Young of the year, which presumably hatch in late summer, become active on the surface in August and September.

Plethodon hubrichti is aggressive and will actively defend cover objects from conspecifics (Thurow 1976). Whether this behavior reflects competition for food, mates, or other resources is not known.

PREDATORS AND DEFENSE. The adults produce noxious skin secretions that act to deter predators. Dodd et al. (1974) found that the adults are noxious to avian predators and hypothesized that *D. ochrophaeus* is a Batesian mimic of *P. hubrichti*.

COMMUNITY ECOLOGY. Very little is known about the functional role that this and other small *Plethodon* species play in structuring forest floor communities. The high local densities reached by *P. hubrichti* suggest that this species could play an important role in regulating populations of forest floor invertebrates.

CONSERVATION BIOLOGY. Some populations of *P. hubrichti* are fully protected. Because of its very low dispersal rates, some researchers are concerned that intensive timbering and habitat fragmentation could be highly detrimental to this species (Kramer et al. 1993).

Plethodon idahoensis Slater
Coeur d'Alene Salamander
PLATE 124

Fig. 228. *Plethodon idahoensis;* adult; Clearwater Co., Idaho (R. W. Van Devender).

Fig. 229. *Plethodon idahoensis;* adult; Kootenai Co., Idaho (W. P. Leonard).

IDENTIFICATION. *Plethodon idahoensis* is closely allied with *P. vandykei* and is treated as a subspecies of the latter by many authors. This species has a modal number of 14 costal grooves and indistinct parotoid glands. The legs are relatively long and there are only 0.5–3 intercostal folds between the adpressed limbs. The dorsum is dark brown to black with a yellow, green, orange, or red stripe that does not extend to the tail tip. The dorsal stripe is slightly wavy and narrow, and the limb bases are dark with light flecking. The undersides are dark except for a pale yellow throat. Mature females measure 47–58 mm SVL but males are slightly smaller (44–54 mm SVL; Nussbaum et al. 1983).

Females have slightly smaller vent lobes and lack nasolabial cirri that are present on mature males. Hatchlings reach 15–18 mm SVL. Hatchlings and young juveniles generally resemble the adults in coloration and color patterning.

SYSTEMATICS AND GEOGRAPHIC VARIATION. *Plethodon idahoensis* is closely allied with *P. vandykei* and has been treated as a subspecies of *P. vandykei* by many authorities. A summary of genetic studies of this species is given in the account of *P. vandykei*. Geographic variation in color patterning and morphology has not been described. Five populations examined by Howard et al. (1993) from throughout the range of *P. idahoensis* are only slightly different genetically ($D < 0.08$).

DISTRIBUTION AND ADULT HABITAT. *Plethodon idahoensis* occurs in the northern Rocky Mountains in northern Idaho, northwestern Montana, and southeastern British Columbia at elevations ranging from 488 to 1524 m (Brodie 1970; Groves et al. 1996). The southern limit of its range occurs in the Selway River drainage in Idaho and Sweathouse Creek in the Bitterroot River drainage in Montana. The northern limit occurs along Copper Creek on the Moyie River drainage in Idaho and the South Fork of the Yaak

Range of *Plethodon idahoensis*

River in Montana. Six populations also occur along the southeastern corner of Kootenay Lake, in British Columbia (Groves et al. 1996).

This is one of the most aquatic *Plethodon* species in North America. Individuals inhabit seepages, the splash zones of waterfalls, rocky habitats along streams, moist talus in moist conifer forests, and, on rare occasions, abandoned mine shafts (Brodie 1970; Slater and Slipp 1940; Teberg 1963). Most populations are found in areas having steep terrain (mean grade = 62%; range = 10–90%), angular rocks, and conifer forests with >25% canopy cover (Groves et al. 1996). During wet weather individuals are occasionally found beneath bark or logs in forest floor habitats. The wet, rocky habitats that this species occupies allow animals to move to subsurface refugia during both dry summer weather and cold winter weather.

BREEDING AND COURTSHIP. Adults have a prolonged breeding season and mate from late summer through the spring months. Females in Idaho have spermatophores in their cloacae shortly before and after winter hibernation (Nussbaum et al. 1983), and courtship has been observed in the field at three localities during August, September, and October (Lynch and Wallace 1987). These observations suggest that most mating occurs in late summer and autumn. The following is a summary of the courtship sequence based on the latter authors' observations.

Initially, a male approaches a female and engages in snout-to-snout contact. During snout contact, the animals move their heads from side to side in opposite directions. If the female is receptive, snout-to-snout contact progresses to snout-to-shoulder contact. The male then moves his snout along the female's flank. The female does the same, moving her snout down the flank of the male. When the female's snout reaches the male's tail base, he turns, she steps over his tail, and the pair begins a tail-straddle walk that can last >1 hour and cover a distance of >1 m. During the walk, the male twitches his tail and makes repeated snout-to-groin and snout-to-tail base contact with the female. The male eventually stops, makes exaggerated contortions of the tail, and squats and deposits a spermatophore. The male leads the female forward until her cloaca is directly above the sperm mass, then curls his tail about the shoulder and face of the female. The female then picks up the spermatophore cap with her expanded cloacal lips. The spermatophore is about 5.5 mm high and has a base about 3.6 mm wide. Females may store sperm in their cloacae for as long as 9 months before fertilizing the eggs.

REPRODUCTIVE STRATEGY. No data are available on the nesting biology of *P. idahoensis*. Females appear to oviposit during April and May in underground cavities or rock crevices, but nests have not been found in the wild (Groves et al. 1996). The number of mature ova in females examined from northern Idaho varies from 4 to 12 and averages 6.7 (Nussbaum et al. 1983).

TERRESTRIAL ECOLOGY. Juveniles grow rather slowly, and males and females appear to reach sexual maturity at approximately 3.5 and 4.5 years of age, respectively (Groves et al. 1996). Males presumably mate annually, whereas females oviposit every other year.

Individuals inhabiting forests and talus slopes in northern Idaho emerge from winter hibernation in late March and remain active near the ground surface until late May (Nussbaum et al. 1983). From late May to mid-September, they move beneath ground or to seepages and other aquatic habitats. Individuals are again active on the surface from mid-September through early November, at which time they move underground to hibernate. Populations in seepages are often active on the ground surface throughout the summer months.

Wilson and Larsen (1988) conducted a detailed study of a population inhabiting a series of cold seepages in Montana. Individuals forage for insects and other prey on wet rockfaces and rubble. The diet consists mostly of aquatic and semiaquatic insects. Dipteran larvae are numerically and volumetrically the most important prey in the diet. Other items include collembolans, mayflies, stoneflies, psocopterans, hemipterans, bugs, beetles, caddisflies, lepidopterans, hymenopterans, millipedes, mites, spiders, phalangids, snails, and oligochaetes.

Surface activity occurs almost entirely at night and is negatively correlated with the time since the last rain. The relative number of salamanders observed is generally constant, except on cold nights when animals move below ground within a few hours after dark as temperatures drop below 4°C. Individuals remain in close proximity to the splash zone, except on wet nights when they sometimes move as far as 10 m from seepages.

PREDATORS AND DEFENSE. The American robin and the red-sided garter snake (*Thamnophis s. parietalis*) are the only documented predators (Wilson and Simon 1985; Wilson and Wilson 1996). Defensive mechanisms have not been described.

COMMUNITY ECOLOGY. The community ecology of this species has not been investigated and nothing is known about its interactions with other members of seepage and headwater stream communities.

CONSERVATION BIOLOGY. The Coeur d'Alene salamander is uncommon and patchily distributed. Local populations appear to consist of relatively small isolated demes that are vulnerable to local extinction and should receive high priority for protection. Groves et al. (1996) provide detailed information on management requirements for this species.

Plethodon jordani Blatchley
Jordan's Salamander
PLATES 125, 126, 127

IDENTIFICATION. Jordan's salamander is a large species of *Plethodon* that shows marked geographic variation in color patterning. The dorsal ground color varies from slate gray to bluish black and the back usually lacks both red pigment and white spots. Depending on geographic location, the adults may be unmarked or have red cheeks, red legs, or brassy frosting on the back. The venter varies from light gray to grayish black depending on latitude. The tail of this and other large eastern *Plethodon* species produces

Fig. 230. *Plethodon jordani*; adult; Swain Co., North Carolina (J. W. Petranka).

slimy secretions that are a deterrent to certain predators, particularly birds. Sexually active males have conspicuous, rounded mental glands. Adults measure 8.5–18.5 cm TL and there are usually 16 costal grooves.

Hatchlings and small juveniles in some populations are light brown and sometimes have inconspicuous brassy iridophores on the head and eyelids (Gordon 1960). Young juveniles of red-cheeked and red-legged forms may have pairs of red spots down the back (Wood 1947a,b).

SYSTEMATICS AND GEOGRAPHIC VARIATION. *Plethodon jordani* shows marked geographic variation in color patterns and average adult size. Hairston (1950) recognized seven subspecies of *P. jordani*. Many of these were considered to be separate species before intensive collecting in the southern Appalachians revealed zones of intergradation between adjoining races (Bruce 1966; Hairston 1950). In his revision of the genus *Plethodon,* Highton (1962a) did not recognize subspecies because some characters do not vary concordantly among the recognized taxa. In later papers, Highton (1970, 1972) recognized 12 geographic isolates in southern populations of *P. jordani* (22 isolates for the entire species) and argued that recognizing subspecies would serve no useful purpose. Here I follow this recommendation and simply describe patterns of geographic variation that reflect divergence among geographic isolates, as well as hybridization with other *Plethodon* species.

Individuals in the Great Smoky Mountains have bright red or yellowish red cheek patches on an otherwise gray to bluish black dorsum. Populations in the Nantahala and Tusquitee mountains and on Cheoah Mountain, North Carolina, have bright red or yellowish red coloration on the dorsal surfaces of the legs. Individuals with conspicuous brassy frosting on the dorsum and spotting along the sides occur in extreme northwestern South Carolina and on and near Fishhawk Mountain in North Carolina (Bruce 1967; Highton 1970). Specimens from most remaining areas tend to be uniformly dark gray to bluish black above and have a gray belly. Individuals in the northern portion of the range tend to have light bellies and are smaller on average than individuals in the southern portion of the range. Hairston (1950) notes that the light- and dark-bellied forms meet near the top of the Balsam Mountains in North Carolina.

Populations in the southwestern portion of the species' range often have either irregular or rounded white spotting along the sides of the body and head that may reflect hybridization with *P. glutinosus* or *P. oconaluftee* (Bruce 1967; Hairston and Pope 1948; Highton 1962a, 1970; Howell and Hawkins 1954). Populations in mountainous regions of northeastern Georgia once considered to be *P. j. rabunensis* have pronounced white spotting along the sides of the body and head, and lack conspicuous brassy markings on the dorsum. These populations may reflect swamping of ancestral *P. jordani* stock by *P. glutinosus* (Highton 1970).

Plethodon oconaluftee and *P. jordani* are more closely related genetically than any other species of *Plethodon* (Highton 1983). Northern populations of *P. jordani* are reproductively isolated from microsympatric populations of members of the *P. glutinosus* complex, but populations in the southern Appalachians hybridize extensively with *P. oconaluftee* in areas south of the Great Smoky Mountains (Highton 1970, 1983; Highton and Henry 1970). Specimens from hybrid zones have varying amounts of red on the legs and white

Fig. 231. *Plethodon jordani;* adult; Macon Co., North Carolina (J. W. Petranka).

Fig. 232. *Plethodon jordani;* adult; Grayson Co., Virginia (J. W. Petranka).

Range of *Plethodon jordani*

spotting on the body. *Plethodon jordani* also hybridizes with *P. aureolus* in extreme southwestern North Carolina and southeastern Tennessee (Highton 1983).

Hairston et al. (1992) examined changes in the width of two hybrid zones over time. In the first, involving intergradation of red-cheeked and gray-cheeked forms of *P. jordani* in the Great Smoky Mountains, the width of the zone of intergradation between two geographic races did not change over an 18-year period. The stability of this zone is interpreted as being due to random diffusion of a neutral trait. At a second site in the Nantahala Mountains where *P. jordani* and *P. oconaluftee* hybridize, the hybrid zone spread upward during a 20-year span. The authors estimate that hybridization began about 60–65 years ago in association with intense timbering in the area, and that selection for *oconaluftee* traits has since resulted in a rapid widening of the zone with time.

DISTRIBUTION AND ADULT HABITAT. *Plethodon jordani* is restricted to cool, mesic forests in mountainous terrain from southwestern Virginia to extreme northeastern Georgia. Specimens have been collected from elevations of 213 to over 1951 m (Grobman 1944), but populations are generally restricted to elevations above 600–925 m throughout most of the range (Hairston 1949; Hairston and Pope 1948). Populations occur at lowest elevations in the southern portion of the range, where deep gorges and high annual precipitation provide suitable conditions for survival (Bruce 1967). This species can be readily collected by turning logs or rocks during the day or by searching the forest floor at night after juveniles and adults have emerged to forage.

BREEDING AND COURTSHIP. Courtship has been observed in the field from mid-July through early October (Arnold 1976; Hairston 1983a; MacMahon 1964; Organ 1958). Females apparently mate every other year, a pattern that correlates with their biennial reproductive cycle. Males will not aggressively court sexually mature females with small ovarian eggs, but will actively court females with large ovarian eggs (Arnold 1976). Laboratory trials show that males prefer female airborne odors over those of other males (Dawley 1984b). In addition, both sexes of *P. jordani* prefer the odors of conspecifics of the opposite sex to those of *P. oconaluftee* of the opposite sex during the breeding season (Dawley 1986a,b). When offered a choice between female *P. jordani* and *P. oconaluftee*, males only court and inseminate conspecifics. These results apply to populations that rarely hybridize. In

Fig. 233. *Plethodon jordani;* courting adults in a tail-straddle walk (the male is facing posteriorly; from this position he may slap the female across her nasolabial region with his mental gland); Grayson Co., Virginia (S. J. Arnold).

the Nantahala Mountains, where hybrids are common, interspecific matings may occur as frequently as intraspecific matings in paired laboratory trials (N. G. Hairston, pers. comm.).

Courtship has been described by several workers based on observations made in both the laboratory and the field (Arnold 1976, 1977; MacMahon 1964; Organ 1958). The following account is mostly based on Arnold's and Organ's descriptions.

A male approaches a female and begins nudging, nosing, or tapping her with his snout. He then places his mental gland and nasolabial grooves in contact with the back, sides, or tail of the female and engages in a "foot dance" in which the limbs are raised and lowered off the substrate one at a time. The male eventually moves forward toward the female's head. After reaching the head of the female, the male presses his mental gland along the side of her head and over her nasolabial grooves. The male then turns his head under the female's chin and lifts. Next, the male circles under the female's chin and laterally undulates his tail as he passes. If the female is responsive, she places her chin on his tail and moves forward to the base of the tail. The pair then engages in a tail-straddle walk that may last for 1 hour. During the walk the male may turn and slap the female across her nasolabial region with his mental gland. Green and Richmond (1944) report that one pair covered an area about 3 m in diameter during a 25-minute interval.

The male eventually stops moving and begins a series of lateral rocking movements of his sacrum. The female begins a series of synchronous lateral head movements counter to the lateral movements of the sacral region of the male. The male then presses his vent to the substratum and deposits a spermatophore. He next flexes the tail to one side and leads the female forward. She stops when her vent is over the spermatophore, then lowers her sacrum and picks up the sperm cap. During this process the male arches the sacral region and does a series of push-up motions with the rear limbs. The pair usually splits up and terminates courtship shortly after spermatophore deposition, even if the female is unsuccessful in picking up the spermatophore. Courtship takes about 1 hour in laboratory animals and is identical for four geographic isolates (Arnold 1976).

The spermatophore has a flattened base that tapers to a spiked stalk. The spike functions to secure the sperm cap, which rests on top. The entire structure is laterally compressed and the spike normally points anteriorly toward the male (Arnold 1976; Organ 1958). A male produces one spermatophore per courtship (very rarely two) and only deposits spermatophores every 7-8 days on average (Arnold 1976, 1977). Females may court repeatedly and be inseminated by two or more males over the course of a season.

Males will sometimes court other males and mimic the behavior of females (Arnold 1976; Organ 1958). This is a form of sexual competition in which males cause other males to waste gametes. Aggressive encounters among males are common, have been observed in the field (Hutchison 1959), and often involve one male pursuing and biting a second.

Madison (1975) found that, in some instances, individuals prefer the odors of neighboring versus non-neighboring conspecifics of the same sex when presented simple choices in the laboratory. Before the breeding season, individuals generally prefer the odor of neighbors. However, neither sex shows a significant preference during the breeding season.

REPRODUCTIVE STRATEGY. Nests of *P. jordani* have never been found. Gravid females move underground in late spring or early summer and presumably oviposit in deep underground recesses. Females oviposit in May in the southern Appalachians based on the presence or absence of gravid females in samples (Hairston 1983a). Specific data on ovarian complements have not been published; however, Hairston (1983a) provides a regression equation for estimating ovarian complements. Hatching probably occurs in late summer or early autumn, about 2–3 months after the eggs are deposited.

TERRESTRIAL ECOLOGY. Newborns in western North Carolina remain below ground for 10–12 months after hatching (Hairston 1983a). Small individuals judged to be nearly 1 year old first appear on the surface in May or June. Individuals increase by an average of 12 mm SVL between May and October of their second year of life and 9 mm SVL during their third year of life. Males mature sexually about 3 years after hatching, when they reach >43 mm SVL. An estimated 25% of females oviposit for the first time 4 years after hatching, whereas another 25% oviposit for the first time 1 year later. The remaining animals first oviposit 6 years or more after hatching, then every other year for the remainder of their lives. The smallest gravid females measure 46 mm SVL. Males in a population studied by Howell and Hawkins (1954) mature sexually at 50 mm SVL.

Based on surface densities of juveniles collected from plots, Hairston (1983a) estimates annual survivorship to be 36% and 48% during the second and third years of life, respectively. Annual adult survival is 81% and the mean generation time is nearly 10 years.

Plethodon jordani is active on the ground surface from mid-May to early October in the mountains of western North Carolina. Individuals retreat to underground wintering quarters with the onset of freezing weather in the autumn and remain underground even during bouts of warm weather. At low elevations, individuals often remain active on the ground surface throughout the winter except during periods of freezing weather (Schwartz 1957).

During the warmer months of the year individuals live in burrows and other underground passageways or beneath rocks, logs, or other surface objects (Brooks 1946). Juveniles and adults emerge at dusk from their daytime haunts and forage on the forest floor. Individuals will sometimes climb a short distance up trees or vegetation to forage (Hairston 1987). Individuals may climb 61–122 cm above ground on spruce and hemlock trunks in western Virginia (Green 1939) and 1 m or so above ground in western North Carolina (Gordon et al. 1962). I have occasionally observed animals over 2 m high on tree trunks in western North Carolina.

Individuals are most active during periods of rainy weather, but moderate numbers can be seen on most summer nights. On dry nights individuals forage from the entrances of burrows, whereas on wet nights they actively roam the forest floor in search of prey. At sites where herbaceous plants are common, the majority of individuals may forage from vegetation at night. Individuals are most active during the first 2–4 hours after dusk (Gordon et al. 1962; Hairston 1987; Schwartz 1957) and do not emerge synchronously to forage at night (Merchant 1972). Males and females eat the same sizes of prey, and the average prey size and diversity of dietary items increase with SVL (Mitchell and Taylor 1986; Whitaker and Rubin 1971).

Food items in 204 specimens from the Great Smoky Mountains included annelids, snails, millipedes, centipedes, isopods, phalangids, pseudoscorpions, mites, spiders, and a variety of insects (Powders and Tietjen 1974). Millipedes, annelids, beetles, and insect larvae are the most volumetrically important prey. Millipedes are more important during the spring, and insect larvae more important during the autumn months. In addition, collembolans and annelids tend to increase in importance with altitude.

The most comprehensive dietary analysis is that of Whitaker and Rubin (1971), who examined prey in the stomachs of 821 *P. jordani* from North Carolina. The 10 most important prey by volume are ants, spiders, lepidopteran larvae, beetle larvae, collembolans, millipedes, centipedes, mites, snails, and dipteran larvae. Dietary items include over 60 families of insects along with shed skin and the fecal pellets of rodents. Wood roaches and carabid beetles are uncommon in the diet even though both of these groups are abundant on the forest floor or beneath logs. Small prey such as ants, mites, and collembolans tend to decrease in importance with size, whereas large prey such as earthworms, millipedes, and craneflies show the reverse trends.

The adults often live in extensive burrow systems. Individuals can enlarge existing burrows and may actively construct openings to the surface. The burrows at one western North Carolina site average 2.5 cm in diameter and run 3–8 cm below the ground surface

(Chadwick 1940). Occasional vertical branches lead deep underground or upward to the surface, where they usually open beneath rocks or logs. At another site the burrows are at least 46 cm deep and are 1.3 cm wide at the entrances (Brooks 1946).

Individuals tend to remain in localized areas on the forest floor and rarely move large distances. Madison (1969) found that the maximum distance that individuals move from their initial points of capture over a 10-week period varies from <4 m for juveniles to <11 m for males. Displaced animals successfully return to their home ranges after being moved as far as 305 m from the initial site of capture; however, the proportion successfully homing generally decreases with displacement distance. Juveniles are less successful than adults in homing. Vision is not necessary for homing, but olfaction may be important in orienting to home areas (Madison 1969, 1972). Experimentally displaced animals often climb trees immediately after being released (Madison and Shoop 1970). This behavior may enhance detection of airborne chemical cues used in homing.

Nishikawa (1990) found that *P. jordani* occupies fixed home ranges with little overlap between individuals of the same age or sex. Younger animals are more likely to share their home ranges with adults than with other juveniles. The mean distances moved between captures (mean interval between captures = 17 days) for 2-year-olds, 3-year-olds, adult males, and adult females are 1.92, 2.47, 2.59, and 1.69 m, respectively, and do not differ significantly between ontogenetic groups or sexes. The respective mean sizes of home ranges for the groups are 1.52, 2.98, 5.04, and 1.87 m^2. Merchant (1972) estimates average home ranges for males, females, and juveniles as 11.4, 2.8, and 1.7 m^2, respectively. Estimates of densities are 0.18 animal/m^2 at a site in western North Carolina (Ash 1988) and 0.86 animal/m^2 in a Great Smoky Mountains population.

PREDATORS AND DEFENSE. *Plethodon jordani* produces slimy tail secretions that are noxious to many predators (Brodie et al. 1979). Nonetheless, some predators such as spring salamanders, black-bellied salamanders, and common garter snakes (*Thamnophis sirtalis*) prey regularly upon *P. jordani* (Bruce 1972a; Feder and Arnold 1982; Huheey and Brandon 1961; Huheey and Stupka 1967). Hatchlings often remain immobile when prodded or attacked, but adults actively attempt to escape (Brodie et al. 1974b).

If attacked by a garter snake, *P. jordani* may wrap its tail around the snake's head while releasing tail secretions, autotomize the tail, bite the snake, or thrash wildly while attempting to escape (Feder and Arnold 1982). In laboratory trials *P. jordani* responds to the flicking of a garter snake's tongue with protean flipping behavior (Ducey and Brodie 1983). When attacked by shrews (*Blarina*), individuals often flip about and position the tail toward the mouth of the shrew (Brodie et al. 1979). The tail is frequently arched and undulated or lashed toward the predator, particularly after an individual is bitten. Specimens often have missing or damaged tails; in one study 28% of 455 specimens had broken tails (Wake and Dresner 1967).

COMMUNITY ECOLOGY. N. G. Hairston and his associates conducted a series of studies examining interspecific competition between *P. jordani* and *P. oconaluftee* in the southern Appalachians. These species tend to replace each other altitudinally in the southwestern part of the range of *P. jordani*. However, in other areas *P. jordani* often broadly overlaps elevationally with either *P. oconaluftee* or other members of the *P. glutinosus* complex (Bruce 1967; Hairston 1949, 1951; Highton 1970, 1972). To examine competitive interactions between *P. jordani* and *P. oconaluftee*, Hairston (1980a, 1983b) contrasted the responses of each species to the experimental removal of the other. Experiments were conducted in the Great Smoky Mountains, where the elevational overlap is slight, and in the Balsam Mountains, where the overlap is much more extensive.

In both cases the removal of *P. jordani* increases the surface densities of *P. oconaluftee*. Removal of *P. oconaluftee* does not affect the densities of *P. jordani*, but it does increase the proportion of juveniles in the population. In both instances the response is stronger or occurs sooner in the Great Smoky Mountains than in the Balsam Mountains. Hairston concluded that competition occurs between the two species, but was unable to pinpoint any resource that was in short supply.

Subsequent studies (Hairston 1987; Hairston et al. 1987) show that interspecific differences in diet and foraging sites are not greater in the Balsam Mountains than in the Great Smoky Mountains, as might be predicted if exploitative competition for food was occurring. Based on this relationship, competition for food seems unlikely. Other species in the community (*Desmognathus imitator; D. ocoee; D. wrighti; Eurycea bislineata; P. serratus*) are not affected by the removal of either large *Plethodon* species (Hairston 1981).

In a related study, Nishikawa (1985) found that *P. oconaluftee* from a population in the Great Smoky Mountains is more aggressive toward both conspecifics and heterospecifics than is *P. oconaluftee* from a population in the Balsam Mountains. *Plethodon jordani* in the Balsam and Great Smoky mountains do not differ significantly in their overall aggressiveness toward either conspecifics or heterospecifics. Because the experiment was restricted to one population from each region, it is uncertain if these patterns would hold for Balsam and Great Smoky mountains populations in general. Nonetheless, the results are consistent with those from reciprocal transplant experiments with *P. jordani* from the Balsam and Great Smoky mountains (Hairston 1980b) and suggest that interference competition for space is the primary factor mediating interactions between these species.

My colleagues and I find that levels of intraspecific aggression do not vary markedly among populations from throughout the range of *P. jordani* (Selby et al. 1996). Levels of aggression in an intensively studied population are independent of sex and are positively correlated with the SVL of resident animals. Overall, adults are about four times more aggressive than juveniles.

Powders and Tietjen (1974) report that sympatric *P. jordani* and *P. oconaluftee* in the Great Smoky Mountains have high dietary overlap, although *P. oconaluftee* tends to take larger prey on average than *P. jordani* during the spring months. The authors surmise that competition for food could occur during those times of the year when dietary overlap is highest.

CONSERVATION BIOLOGY. Local populations of *P. jordani* and other salamanders are often absent or greatly reduced in number on recent clear-cuts in western North Carolina (Ash 1988; Petranka et al. 1993, 1994). Less intensive harvesting practices that leave the basic structure of the forest intact would benefit this and other salamander species in southern Appalachian forests. Hairston and Wiley (1993) monitored *P. jordani* populations in mature forests in western North Carolina and found no evidence of long-term population declines over a 15-year period.

COMMENTS. Since the late 1920s herpetologists have debated the hypothesized existence of a Batesian mimicry complex between *Plethodon jordani* and certain members of the *D. ochrophaeus* complex (i.e., *D. ocoee* and *D. imitator*). Highton and Henry (1970) recognized 12 geographic isolates of *P. jordani* south of the French Broad River, including several isolates with black dorsums and either red cheeks or red legs. The two forms of *P. jordani* that serve as presumed models are a red-cheeked group restricted to the Great Smoky Mountains and vicinity, and a red-legged group restricted to the Nantahala and Tusquitee mountains and Cheoah Mountain to the south. Feeding trials demonstrate that *P. jordani* with red legs or red cheeks have slimy tails that are noxious to potential avian predators (Brodie and Howard 1973; Hensel and Brodie 1976; Huheey 1960). These races appear to be more noxious than races of *P. jordani* lacking red legs or red cheeks. Thus the red cheeks or red legs of *P. jordani* may have an aposematic function.

Desmognathus ocoee, one of the presumed mimics, is polymorphic in the southern Appalachians and has morphs with red cheeks, red legs, and both red cheeks and red legs (colors on the cheeks and legs may actually vary from yellow or orange to bright red; for simplicity, they are referred to here as being red). A second member of the *D. ochrophaeus* complex, *D. imitator,* has a red-cheeked morph that is also thought to be a mimic. Feeding trials show that all color morphs of the *D. ochrophaeus* complex are palatable to a wide variety of predators, and that birds will avoid red-cheeked specimens after several encounters with noxious *P. jordani* (Brodie and Howard 1973; Brodie et al. 1979).

Despite a substantial body of evidence that supports the mimicry hypothesis, several lines of evidence complicate the problem. The geographic distribution of populations of members of the *D. ochrophaeus* complex that contain individuals with red cheeks, red legs, or both red cheeks and red legs is well documented (Bishop 1947; Brimley 1928; Hairston 1949; Huheey 1966a,b; Huheey and Brandon 1961; Labanick 1988; Martof and Rose 1963; Tilley et al. 1978). Tilley et al. (1978) discovered that all red-cheeked individuals in the Great Smoky Mountains and vicinity are *D. imitator,* and that all *D. ocoee* where red-cheeked *P. jordani* occur lack red cheeks. In addition, numerous populations of red-cheeked *D. ocoee* occur far beyond the range of red-cheeked *P. jordani*. The red-cheeked morph in these populations may compose as much as 25% of the population. Populations with relatively high frequencies of animals with both red legs and red cheeks also occur outside the range of red-legged *P. jordani* (Labanick 1988). This finding presents an unusual situation in *D. ocoee*: presumed red-cheeked mimics occur at higher frequencies outside than within the range of the model. This revelation by Tilley et al. (1978) makes the hypothesis of mimicry between red-cheeked *P. jordani* and red-cheeked *D. ocoee* untenable. The possibility that red-legged *D. ocoee* and red-cheeked *D. imitator* are mimics of *P. jordani* is plausible, given that the presumed mimics are sympatric with their respective models.

Another problem is that researchers have not been able to demonstrate clearly an aposematic function of red coloration in *P. jordani* (Hensel and Brodie 1976; Huheey 1960; Orr 1967). Previous workers who had demonstrated the avoidance of *P. jordani* after conditioning assumed that the predator is warned by the red coloration of the model. However, Hensel and Brodie (1976) found that predators appear to use gestalt perception to recognize models instead of red coloration. The obliteration of the red coloration on models, or the addition of red coloration to mimics by the application of black or red acrylic paint, does not influence the probability of an individual being killed. Some of the most recent work investigating the possibility of a mimicry complex between red-legged *D. ocoee* and *P. jordani* (Labanick and Brandon 1981) provides weak support for the mimicry hypothesis.

Several alternative explanations have been advanced to explain the geographic distribution of the red-legged and red-cheeked morphs of *P. jordani* and members of the *D. ochrophaeus* complex, but none is supported by experimental evidence. Noble (1931) felt that red coloration has little survival value and is pleiotropically linked to other traits that are adaptive. Brimley (1928) suggested that common environmental factors or common ancestry might explain the patterns, whereas Cody (1969) proposed an "aggression-associated convergence" hypothesis to explain the phenomenon. According to this theory, the convergence of closely competing species promotes interspecific aggression and simultaneously reduces competition as territorial tendencies increase. This hypothesis was discredited by Brodie and Howard (1973) because of a lack of evidence of interspecific aggression between *P. jordani* and members of the *D. ochrophaeus* complex.

Members of the *D. ochrophaeus* complex have been implicated as being Batesian mimics of several other species of salamanders, including *E. bislineata*, *P. cinereus*, *P. hubrichti*, *P. nettingi*, and *P. welleri* (Brodie 1981; Brodie and Howard 1973; Dodd et al. 1974). Virtually all of the common color morphs of members of the *D. ochrophaeus* complex have at one time or another been interpreted as mimics of other salamanders. Additional work is clearly needed to test these hypotheses.

Plethodon kentucki Mittleman
Cumberland Plateau Salamander
PLATE 128

IDENTIFICATION. *Plethodon kentucki* is a large *Plethodon* species that has a black ground color with small dorsal white spots and larger lateral spots. This species closely resembles *P. glutinosus* but is smaller on average and has a lighter chin and throat, fewer and smaller dorsal spots, and a larger mental gland (Highton 1986d). These traits are not completely diagnostic since the range of variation in traits between the two species overlaps widely in some populations. Identification is most reliable using electrophoretic comparisons of proteins. The average number of costal grooves is 16.

Hatchlings from southeastern Kentucky reach 14–15 mm SVL and weigh 0.08–0.10 g (Marvin 1996).

Fig. 234. *Plethodon kentucki*; adult; Washington Co., Virginia (R. W. Van Devender).

Range of *Plethodon kentucki*

Males have conspicuous mental glands and the adults measure 9.5–17 cm TL. Females in a population studied by Marvin (1996) average about 5% larger than males. The sex ratio varies among habitats and years, but the average is about 1:1.

SYSTEMATICS AND GEOGRAPHIC VARIATION. *Plethodon kentucki* is a sibling species of *P. glutinosus* that was described by Mittleman (1951). Later workers were unable to find traits that are diagnostic for *P. kentucki* and considered *P. kentucki* to be a junior synonym of *P. glutinosus* (Clay et al. 1955). Highton and MacGregor (1983) resurrected *P. kentucki* after finding it to be electrophoretically distinct from sympatric *P. glutinosus* (Nei's genetic distance $[D] = 0.45$). Immunological evidence (Maha et al. 1983) and the demonstration of ethological barriers (Dawley 1986b) further verify the distinctiveness of *P. kentucki*.

Marked geographic variation in isozymes occurs in *P. kentucki* (Highton and MacGregor 1983). In particular, populations in West Virginia are very distinct from those in eastern Kentucky and western Virginia; several pairs of populations have $D > 0.40$.

DISTRIBUTION AND ADULT HABITAT. *Plethodon kentucki* occurs in forests in the Cumberland Plateau of eastern Kentucky, western West Virginia, and southwestern Virginia to the New and Kanawha rivers. The southernmost record is from Scott Co., Tennessee, adjoining the Kentucky state line. The microhabitats used by *P. kentucki* are similar to those of slimy salamanders in the same vicinity. Individuals can be collected by turning logs and other cover, or by searching the leaf litter on wet nights when adults forage for invertebrates outside cover. Individuals are occasionally found in rock crevices and in sandstone and shale rock outcrops. In West Virginia specimens are most frequently found beneath rocks, and populations reach their greatest densities on west-facing slopes (Bailey and Pauley 1993).

BREEDING AND COURTSHIP. Data provided by Marvin (1996) indicate that males reproduce annually. Females oviposit every 2 years (sometimes less frequently) and presumably only mate during the summer or fall preceding egg laying. The adults appear to mate in late July or early August through October, when the mental glands of males are conspicuously swollen. Males often outnumber females during the mating season, when a subset of the female population is brooding beneath ground.

The following general description of courtship is based on Marvin and Hutchison's (1996) detailed laboratory studies of individuals from Big Black Mountain, Harlan Co., Kentucky. The courtship sequence is generally similar to that of other large eastern *Plethodon* and involves a rather prolonged period of persuasion (range = 5–320 minutes), followed by a tail-straddle walk (range = 1–73 minutes) and spermatophore deposition and pickup.

When a male initially approaches a female, he may nose-tap his partner and engage in a foot dance in which the fore- and hindlimbs are raised one at a time while the male remains stationary. Females may initially bite approaching males (or rarely vice versa), but both partners reduce their aggressive behavior soon after an encounter. After locating a female, the male may nudge the female with his snout, tap the female's nearly vertically raised snout with his mental gland, or slide his head and mental gland over the female's tail before progressing along the body toward the head. Individuals often foot dance when sliding the head over the tail and other body regions. If sufficiently aroused, the female typically initiates the tail-straddle walk by placing her chin on the dorsal base of the male's tail. The pair then periodically walks forward in tandem. During this time, the male undulates the base of his tail laterally against the female's chin and keeps his tail base and head elevated above the substrate. He may also turn and slap the snout of the female with his mental gland.

Near the termination of the tail-straddle walk, the male lowers his vent to the substrate, begins to undulate his tail about twice as rapidly as before, and deposits a spermatophore. He then stops undulating the tail, raises the vent, and swings the tail laterally while moving forward to lead the female over the spermatophore. The female stops and lowers her vent to pick up the sperm cap. During this period the male flexes his hindlimbs to raise the vent, and the female maintains chin contact with the dorsal base of the tail. If spermatophore pickup is successful, the female usually abandons the male and the male may eat the remaining part of the spermatophore. Otherwise the pair may continue with the tail-straddle walk and the male may eventually deposit a second spermatophore. On rare occasions as many as three spermatophores may be deposited in a single courtship bout. The spermatophore is laterally compressed, and measures about 2.7 × 3.3 mm at the base, and is 3.5 mm high. The sperm cap is about 1.6 mm high. Unusual aspects of courtship in this species relative to that of other large *Plethodon* species include a high frequency of mental gland tapping prior to the tail-straddle walk and the initiation of the walk by the female rather than by the male.

In laboratory trials male and female *P. kentucki* prefer the odors of conspecifics over those of *P. glutinosus* of the opposite sex (Dawley 1986a). When presented with choices between female conspecifics and heterospecifics in breeding experiments, male *P. kentucki* only inseminate conspecifics. However, in 23% of trials, male *P. glutinosus* inseminate *P. kentucki* females.

REPRODUCTIVE STRATEGY. The only comprehensive study of the natural history of *P. kentucki* is that of Marvin (1996), who studied a population on Big Black Mountain, Harlan Co., Kentucky. Females in this population oviposit in underground retreats in July and the embryos hatch in October. Observations of five captive females that oviposited in jars either maintained in the laboratory or buried in soil in the field indicate that females lay their eggs in grapelike clusters within brooding chambers. Females typically position themselves immediately below the eggs. They guard the developing embryos through hatching and may remain with the hatchlings for several weeks posthatching. The wet mass of females that oviposit and brood through hatching declines an average of 37% (range = 24–50%).

Freshly laid eggs have ova that are 4.0–4.5 mm in diameter, but eggs with mature embryos may reach 7 mm in diameter. Captive females that were maintained in the laboratory laid clutches of 9, 9, and 12 eggs, and embryos from eggs buried in the field hatched from 5 to 23 October.

TERRESTRIAL ECOLOGY. The hatchlings remain in close proximity to the females for several weeks until the yolk reserves are expended, then presumably disperse into underground burrows and passageways (Marvin 1996). Hatchlings begin emerging on the ground surface the following June or July when 8–9 months old. Juveniles grow an average of 15 mm SVL during their first year on the ground surface, then 10 and 6 mm SVL in subsequent years. Growth rates may be reduced significantly in years when summer rainfall is low and opportunities to forage on the ground surface are reduced. Males reach sexual maturity 3–4 years after hatching when they measure about 47–48 mm SVL. Females first oviposit 4–5 years after hatching, after exceeding 52 mm SVL.

Marvin (1996) estimates annual survivorship of 2–

and 3-year-olds as being at least 48% and 68%, respectively, based on recapture data. Annual survivorship of adults is high and ranges from 72% to 91% based on Jolly-Seber mark-recapture estimates. The lower estimated survival for juveniles could be biased if many juveniles were floaters that dispersed off study plots. The densities of 2-year-olds, 3-year-olds, and adult males and females at this site vary among years and plots, but are <0.20 animal/m^2 for any group.

Plethodon kentucki is active on the ground surface on moist nights, when individuals either actively move about or perch in burrow entrances in wait for invertebrate prey. The adults are presumably territorial, but detailed studies of aggression and territoriality are not available. Individuals in West Virginia populations show strong seasonal activity patterns and are most active on the ground surface in March and April (Bailey and Pauley 1993). Individuals are most active on the ground surface within 2–3 hours after dusk. Two marked animals moved 1.3 and 1.8 m between captures. Individuals in a Kentucky population appear to be active throughout the summer during periods of rainy weather (Marvin 1996). During summer droughts individuals move underground, where there are few opportunities to feed. Small juveniles in this population appear to grow more slowly and to suffer higher mortality during summers with reduced rainfall.

PREDATORS AND DEFENSE. No data are available on predators or defense.

COMMUNITY ECOLOGY. *Plethodon kentucki* and *P. glutinosus* occur sympatrically in many areas and appear to have similar life-styles and habitat requirements. Detailed studies of competition between these species have not been published.

CONSERVATION BIOLOGY. Mature hardwood forests offer optimal habitats for this species and almost all other woodland salamanders in the eastern United States.

Plethodon larselli Burns
Larch Mountain Salamander
PLATE 129

Fig. 235. *Plethodon larselli*; adult; Skamania Co., Washington (R. W. Van Devender).

IDENTIFICATION. The Larch Mountain salamander is a small western species of *Plethodon* that has dark brown to black sides and a broad dorsal stripe that extends from the head to the tail tip. The dorsal stripe varies from red to chestnut. The venter is pinkish to red with scattered patches of melanophores, and the sides are flecked with gold and silvery white pigment (Brodie 1970; Burns 1962; Nussbaum et al. 1983). The outermost toe is very short and the modal number of costal grooves is 15. Males lack mental glands, have poorly developed vent lobes, and have swollen areas around the nasolabial grooves. Adult males and females are about the same size on average. Adult females have an average SVL of 50 mm compared with 45 mm for males (Herrington and Larsen 1987). Adults measure 7.5–10.5 cm TL and hatchlings reach 17–18 mm SVL.

SYSTEMATICS AND GEOGRAPHIC VARIATION. Washington adults are smaller on average and darker than Oregon adults (Brodie 1970), but no subspecies are recognized. Howard et al. (1983) compared the genetic similarities of paired populations of *P. larselli* that occur on opposite sides of the Columbia River Gorge. All four populations are extremely similar, although each has one variant allele that is missing from all other populations. Mean heterozygosity is very low (range = 0.002–0.01) and only 5 of 30 loci are poly-

Range of *Plethodon larselli*

morphic. The extremely high similarity of populations on either side of the gorge ($D < 0.01$) is surprising and suggests that the Columbia River has not been an effective barrier to gene flow.

DISTRIBUTION AND ADULT HABITAT. The Larch Mountain salamander is restricted to a very small area in the Columbia River Gorge in southern Washington and northern Oregon. Isolated populations also occur in the central Cascades of Washington (Aubry et al. 1987). Populations in the Columbia River Gorge are found almost exclusively on talus slopes formed from adesitic or basaltic lava flows (Brodie 1970; Herrington and Larsen 1987).

Preferred microhabitats include stones with considerable moss and humus in dense stands of Douglas-fir (Burns 1954, 1964) and areas where accumulations of gravel 1–6 cm in diameter allow animals to move easily within talus (Herrington and Larsen 1985, 1987). Most populations are in mixed forest consisting of Douglas-fir and mixed hardwoods. Individuals use microhabitats that have relatively large amounts of decaying leaves from the surrounding canopy, but little soil or herbaceous growth. This species typically occurs below 600 m but is found as high as 1190 m on Larch Mountain (Burns 1962). Populations are more patchily distributed in Washington than in Oregon, where talus habitats are more abundant. One specimen was collected at 1100 m in Lewis Co., Washington, in an old-growth forest with a rocky forest floor and scattered rock outcrops. Three additional specimens were collected from a lava tube entrance and cave in Skamania Co., Washington, at 470 m (Aubry et al. 1987).

Fig. 236. *Plethodon larselli*; ventral view of adult; Kittitas Co., Washington (W. P. Leonard).

BREEDING AND COURTSHIP. Mating occurs during the autumn and spring months (Herrington and Larsen 1987; Nussbaum et al. 1983). Males at this time have spermatozoa in the vasa deferentia, whereas females often have spermatophore caps in their cloa-

Fig. 237. *Plethodon larselli;* ventral view of adults; Skamania Co., Washington (R. W. Van Devender).

cae. Females require 2–3 years to store enough energy reserves to produce a clutch (Herrington and Larsen 1987). The sex ratio does not differ significantly from 1:1. There are no published accounts of courtship behavior.

REPRODUCTIVE STRATEGY. Nesting females have never been found. Females presumably oviposit deep underground within talus during the late spring or early summer. The embryos hatch in late summer or early autumn. The number of mature ova in 43 Oregon and Washington specimens varies from 2 to 12, averages 7.3, and is not correlated with female SVL (Herrington and Larsen 1987). Mature ova in 105 females from Oregon vary from 3 to 11, average 7, and range from 4.0 to 4.3 mm in diameter (Nussbaum et al. 1983).

TERRESTRIAL ECOLOGY. Very young juveniles appear on the ground surface in late October–November and measure 20–22 mm SVL (Herrington and Larsen 1987). These probably hatch 1–2 months earlier and remain underground until fall rains arrive. Juveniles reach a modal SVL of 36 mm within 30 months of hatching. Growth is 6–8 mm SVL per year for juveniles and 2–4 mm SVL for adults. Males reach sexual maturity when they are 3–3.5 years old and measure 39–42 mm SVL. Females first oviposit when they are 4–4.5 years old and measure >43 mm SVL. Washington and Oregon specimens sexually mature at slightly below and slightly above 40 mm SVL, respectively (Brodie 1970).

Individuals may be active on the ground surface during all months of the year, but are most active during the spring and autumn (Burns 1962; Nussbaum et al. 1983). Surface activity is reduced during the dry summer months and from late November through early January during periods of freezing weather (Herrington and Larsen 1987).

Mites and collembolans are the dietary mainstay of *P. larselli* from a site in Oregon, and the diversity of prey consumed increases with salamander size (Altig and Brodie 1971). Other prey include snails, annelids, spiders, hemipterans, beetles, lepidopterans, and hymenopterans. Adults often eat earthworms, snails, and other large prey that are rarely eaten by juveniles (Nussbaum et al. 1983).

PREDATORS AND DEFENSE. The natural predators of *P. larselli* are not known, but probably include small snakes, woodland birds, and small, carnivorous mammals. Individuals coil and uncoil rapidly when molested, a behavior that may serve to fling the animals out of reach of a potential predator (Nussbaum et al. 1983).

COMMUNITY ECOLOGY. No information is available on the community ecology of this species.

CONSERVATION BIOLOGY. Many populations of *P. larselli* are being threatened by clear-cutting of old-growth forest and the use of talus slope material for road construction (Herrington and Larsen 1985). Old-growth forest buffers should be left around talus slopes used by *P. larselli* because thick leaf litter from overhanging trees provides critical foraging microhabitats for the juveniles and adults. The population in Lewis Co., Washington, is highly susceptible to extinction from habitat alteration and should be protected from timbering or other forms of disturbance (Aubry et al. 1987).

Plethodon neomexicanus Stebbins and Riemer
Jemez Mountains Salamander

PLATE 130

Fig. 238. *Plethodon neomexicanus;* adult; Sandoval Co., New Mexico (R. W. Van Devender).

IDENTIFICATION. The Jemez Mountains salamander is a slender, elongated, short-legged *Plethodon* species with a brown dorsal ground color that is moderately flecked with brassy pigment. The venter is nearly pigmentless and appears somewhat translucent (Stebbins and Riemer 1950; Williams 1973). The outermost toe on each hindlimb is shorter than that in most *Plethodon* species (Brodie and Altig 1967; Stebbins 1951). Adults measure 9.5–14.3 cm TL and there are usually 18–19 costal grooves. Females average about 2% larger in SVL than males. Hatchlings often have a brassy stripe and reach 17 mm SVL on average.

SYSTEMATICS AND GEOGRAPHIC VARIATION. No conspicuous geographic variation is evident and no subspecies are recognized. Highton and Larson (1979) found that this species is most closely related to *P. larselli* (Nei's $D = 1.1$) and is very distantly related to eastern *Plethodon* species.

DISTRIBUTION AND ADULT HABITAT. *Plethodon neomexicanus* is found on north-facing slopes and in steep canyons in montane forests of the Jemez Mountains in north-central New Mexico. A few specimens have been found in meadows adjoining forests; however, most specimens are found beneath rocks and in and under conifer logs near talus slopes in mixed conifer forests (Reagan 1972; Stebbins 1951; Stebbins and Riemer 1950). Populations occur at elevations of 2190–2800 m (Williams 1973).

Plethodon neomexicanus is geographically isolated from other *Plethodon* species in the United States. Blair (1958) hypothesized that ancestral forms of *P. neomexicanus* colonized the southern Rockies through Oklahoma during the Wisconsin glacial period. However, electrophoretic comparisons indicate that this species is more closely related to western than to eastern *Plethodon* species and that eastern and western species have been separated for an estimated 40 million years (Highton and Larson 1979).

BREEDING AND COURTSHIP. Mating occurs in July and August, and females mate every other year shortly before depositing eggs on a biennial schedule. Spermatozoa enter the vasa deferentia of males in late autumn or early winter. Males collected during the summer months have vasa packed with sperm produced the previous year. Females have sperm in their cloacae in July and August (Reagan 1972; Williams 1978). Courtship behavior has not been described.

REPRODUCTIVE STRATEGY. Females oviposit sometime between mid-August and the following spring (Reagan 1972; Williams 1978). The nests have never been found despite extensive searching. Two females maintained in the laboratory laid clusters containing an average of seven eggs (Williams 1973, 1978). Diameter of the outer envelopes is 6.8–7.3 mm and the eggs are attached by a pedicel to a support structure. Females in natural populations presumably oviposit deep underground on talus slopes and brood their eggs through hatching. The number of mature ova in 45 females was 5–12 and averaged 7.7 (Reagan 1972).

TERRESTRIAL ECOLOGY. Individuals emerge from underground retreats during late spring, are active on the ground surface during the wet summer months, then retreat to subsurface retreats in early autumn (Reagan 1972). This species is most active on the ground surface following summer rains, when surface temperatures are 10–13°C. Females often outnumber males in counts of surface-active individuals.

Range of *Plethodon neomexicanus*

Individuals live in extensive burrow systems in talus that often extend >1 m below ground. Juveniles and adults emerge from their burrows or from surface cover at night to forage on the ground surface. Food items in 39 specimens from a talus slope included earthworms, spiders, pseudoscorpions, beetles, flies, ants, termites, lepidopterans, and snails (Reagan 1972). Ants, lepidopteran larvae, and beetle larvae are the most common prey.

Males and females reach sexually maturity 2–3 and 3 years after hatching, respectively (Reagan 1972; Williams 1973, 1978). The smallest mature males and females measure 49–51 mm and 52–56 mm SVL, respectively.

PREDATORS AND DEFENSE. No data are available on natural predators or defensive mechanisms.

COMMUNITY ECOLOGY. *Plethodon neomexicanus* lives in geographic isolation from other salamanders, and its ecological role in high-elevation forest communities in the Jemez Mountains is not known.

CONSERVATION BIOLOGY. No data are available on long-term population trends or on whether *P. neomexicanus* is adversely affected by timbering or other forms of disturbance.

Plethodon nettingi Green
Cheat Mountain Salamander
PLATE 131

IDENTIFICATION. The Cheat Mountain salamander is a small eastern species of *Plethodon* that has a black or dark brown dorsal color overlain with moderate amounts of brassy flecking that tends to be concentrated on the head. Small white spots may occur on the dorsum. The venter is gray to grayish black. Adults reach 8.0–11 cm TL and there are usually 18 costal grooves (Green and Pauley 1987; Highton

Fig. 239. *Plethodon nettingi;* adult; Randolph Co., West Virginia (R. W. Van Devender).

1986c). The hatchlings have red dorsal pigmentation, lack dorsal stripes, and reach 17–18 mm TL (Green and Pauley 1987).

SYSTEMATICS AND GEOGRAPHIC VARIATION. *Plethodon hubrichti, P. nettingi,* and *P. shenandoah* are closely related species that were previously treated as subspecies of *P. nettingi.* Highton and Larson (1979) raised these to the status of full species after comparing their protein similarities. Their data show moderate levels of genetic divergence, with genetic distances (*D*) varying from 0.22 to 0.35 for different species pairs.

DISTRIBUTION AND ADULT HABITAT. *Plethodon nettingi* is confined to the Cheat Mountains in Grant, Pendleton, Pocahontas, Randolph, and Tucker counties in eastern West Virginia at elevations of 908–1463 m (Pauley 1993). This species consists of 68 known populations that are geographically isolated from one another. Pauley (1980, 1981) collected specimens at elevations of 1052–1378 m, but the vertical distribution at any site is rarely greater than 67 m. *Plethodon nettingi* is largely restricted to red spruce–yellow birch forests but has occasionally been collected in mixed deciduous hardwoods (Clovis 1979; Green and Pauley 1987). Brooks (1948b) reports that specimens are more abundant in young-growth spruce forests as opposed to mature stands. Surface-active animals can be found under rocks, within and beneath logs, and on the ground surface at night.

BREEDING AND COURTSHIP. No data are available on the breeding season or courtship behavior.

REPRODUCTIVE STRATEGY. The most extensive nesting records are those of Brooks (1948b), who found 29 egg masses of *P. nettingi* between 28 April and 25 August. Eggs in surface nests are laid in cavities within well-rotted spruce logs and are similar in size and shape to those of *P. cinereus.* Each cluster is sus-

Range of *Plethodon nettingi*

pended from a natural cavity by a short pedicel, and one or two adults are often with the eggs. Number of eggs per clutch varies from 4 to 17, and most logs contain only one clutch. Other records include clutches of 9, 10, and 8 eggs with attending females found on 3 June, 23 July, and 4 September, respectively (Bishop 1943; Green and Pauley 1987). Embryos in one clutch hatched in the laboratory during the third week of August. Freshly laid eggs are pale yellow and about 4 mm in diameter.

TERRESTRIAL ECOLOGY. Individuals become active on the ground surface immediately following snowmelt in April. They remain active on the ground surface throughout the summer unless dry, hot weather drives them underground. Surface activity begins to decline in early to mid-October and most animals move beneath ground by November (T. K. Pauley, pers. comm.). The juveniles and adults remain in leaf litter or beneath limbs, logs, and rocks during the day but emerge at night and crawl about the forest floor in search of invertebrate prey. Individuals may be active during both wet and dry periods and will climb tree trunks and limbs on damp nights to heights of nearly 2 m above ground (Brooks 1945, 1948b; Green and Pauley 1987). The diet of 42 animals consisted of mites, springtails, beetles, flies, ants, and miscellaneous prey (Green and Pauley 1987).

PREDATORS AND DEFENSE. Natural predators are undocumented but no doubt include shrews and woodland birds.

COMMUNITY ECOLOGY. No data are available on the community ecology of *P. nettingi*.

CONSERVATION BIOLOGY. *Plethodon nettingi* is a federally threatened species that is protected by law.

Plethodon oconaluftee (Hairston)
Southern Appalachian Salamander
PLATE 132

Fig. 240. *Plethodon oconaluftee*; adult; Macon Co., North Carolina (J. W. Petranka).

IDENTIFICATION. *Plethodon oconaluftee* is a large, light-chinned species of the *P. glutinosus* group that superficially resembles the slimy salamander. Adults have a grayish black dorsum with small white spots. Larger spots may occur along the sides and small red spots are sometimes present on the legs (Highton 1983). The belly is uniformly gray and the throat is lighter than the belly. Like that of the slimy salamander, the tail produces sticky secretions that function in defense against predators. Sexually active males have conspicuous, rounded mental glands. Adults measure 12–21 cm TL and there are usually 16 costal grooves. Juveniles resemble the adults in coloration and spotting patterns.

SYSTEMATICS AND GEOGRAPHIC VARIATION. The nomenclatural status of populations that were previously referred to as *P. teyahalee* is being debated. The holotype for *P. teyahalee* is from a hybrid zone between pure *P. jordani* and pure *P. teyahalee*. Hairston (1993) considered the name *teyahalee* to be invalid because the type specimen is a hybrid. He redescribed this species as *P. oconaluftee* using specimens from an area outside the hybrid zone. R. C. Highton (pers. comm.) considers *P. oconaluftee* to be a junior synonym of *P. teyahalee* because the rules of nomenclature for hybrids only apply to F_1 hybrids. Although Highton's interpretation may be correct, it is uncertain what percentage of the genome of the type specimen reflects *jordani* versus *teyahalee* genes. Individuals at the

Range of *Plethodon oconaluftee*

type locality contain on average about two-thirds *teyahalee* genes and one-third *jordani* genes (R. C. Highton, pers. comm.). However, the exact genetic complement of the type specimen cannot be determined. Because of chance, assortative mating, or other mechanisms, the majority of the genes of the holotype could conceivably be those of *P. jordani*. If so, *teyahalee* would be invalid. An official ruling will be required to resolve the issue. Here I follow Hairston's (1993) nomenclature because the holotype for *P. oconaluftee* is the only one that unequivocally does not contain a majority of *P. jordani* genes. This specimen best characterizes the species in terms of coloration, color patterning, and molecular and morphological characteristics.

Plethodon oconaluftee is rather uniform in coloration throughout its range. This species hybridizes to varying degrees with other large *Plethodon* species, and hybrids often deviate in coloration from pure *P. oconaluftee*. Highton et al. (1989) surmised that *P. oconaluftee* may be of hybrid origin between *P. glutinosus* and *P. jordani*. Evidence to support this hypothesis includes the presence of red pigmentation on the dorsum of the legs and the tendency for *P. oconaluftee* to hybridize with *P. jordani* in the southern portion of the range of both species.

Plethodon jordani and *P. oconaluftee* hybridize extensively in major mountain ranges immediately south of the Great Smoky Mountains (Bishop 1941b; Highton 1970, 1983; Highton and Henry 1970). Hybrids typically have less white spotting on the dorsum than pure *P. oconaluftee* and less red coloration on the legs than pure *P. jordani*. *Plethodon oconaluftee* also hybridizes with red-cheeked *P. jordani* in the northeastern portion of the latter's range, but the two species co-occur elsewhere without interbreeding. In areas of sympatry with gray-cheeked *P. jordani*, *P. oconaluftee* coexists locally without hybridizing. *Plethodon glutinosus* and *P. oconaluftee* are parapatrically distributed in most areas, but co-occur microsympatrically at two sites in the Unicoi Mountains in Tennessee (Highton 1983). At one site they hybridize, but at the second they do not. These two species hybridize in northeastern Georgia and vicinity, and in areas northeast of Polk Co., Tennessee (Highton 1995). However, the extent of hybridization and widths of hybrid zones have yet to be reported in detail.

DISTRIBUTION AND ADULT HABITAT. *Plethodon oconaluftee* occurs west of the French Broad River in the Blue Ridge Physiographic Province of southwestern North Carolina and adjoining portions of Tennessee.

Scattered populations also occur in Rabun Co., Georgia, and in Abbeville, Anderson, Oconee, and Pickens counties, South Carolina (Highton 1983). Adults are typically found in deciduous forests up to 1550 m in elevation. Specimens can be found by turning large logs or rocks or by patrolling the forest floor at night when individuals are feeding on the ground surface.

BREEDING AND COURTSHIP. Although poorly documented, breeding presumably occurs from July through October since this is the time of year when *P. jordani* mates, and hybrids between the two species are common in some areas. Courtship has not been described in detail, but it is undoubtedly similar to that of *P. glutinosus* and *P. jordani*.

Laboratory trials indicate that ethological barriers contribute to reproductive isolation between *P. oconaluftee* and other congeners. Male *P. oconaluftee* show strong preferences for odors of female conspecifics over odors of sympatric *P. aureolus* (Dawley 1984b). Male *P. oconaluftee* and *P. jordani* from an area in the Great Smoky Mountains where these two species do not hybridize prefer chemical cues from conspecific females to those from heterospecific females (Dawley 1987). A similar pattern occurs for male *P. oconaluftee* collected next to a hybrid zone in the Nantahala Mountains. However, male *P. jordani* from the same locality show no preference for conspecifics over heterospecifics. Female *P. oconaluftee* collected next to the hybrid zone prefer male heterospecific odors, whereas none of the other females show preferences when tested in matchups analogous to those used to test males. These results suggest that male *P. jordani* near the hybrid zone sometimes make mistakes and court *P. oconaluftee*. When hybrids from the Nantahala Mountains are used in courtship trials, *P. jordani* does not discriminate between conspecifics and hybrid mates (N. G. Hairston, pers. comm.).

REPRODUCTIVE STRATEGY. Females nest underground, but the eggs have never been discovered. The nesting season is similar to that of *P. jordani*, with females moving below ground to nest in late spring or early summer. Hatching occurs 2–3 months after the eggs are laid.

TERRESTRIAL ECOLOGY. Many aspects of the life history of this species are poorly documented. Juveniles and adults feed on forest floor invertebrates and are most active on rainy nights. No published data are available on growth rates, fecundity, and longevity. Females first lays eggs when 5 years old or older (Hairston 1983a).

The density of a Great Smoky Mountains population based on mark-recapture estimates is 0.23 salamander/m^2 (Merchant 1972). Estimates of average home ranges for males, females, and juveniles at this site are 14.3, 6.5, and 7.5 m^2, respectively. In a second population in the Great Smoky Mountains, the respective mean distances that 2-year-olds, 3-year-olds, adult males, and adult females move are 0.81, 0.51, 0.67, and 1.44 m over an average of 14 days (Nishikawa 1990). The respective home ranges are 0.37, 0.06, 0.49, and 1.03 m^2. Overlap in home ranges is slight and least evident among individuals of the same sex or age. Individuals <3 years old are found in retreat holes significantly less often than adults and 3-year-olds. Home ranges are centered near retreat holes, suggesting that this resource is defended by older animals.

Volumetrically important prey of five specimens of *P. "glutinosus"* (presumably *P. oconaluftee*) examined by Rubin (1969) from western North Carolina included ants, lepidopteran larvae, spiders, camel crickets, slugs, millipedes, snout beetles, and earthworms. Powders and Tietjen (1974) examined slimy salamanders from several sites in or near the Great Smoky Mountains in North Carolina and Tennessee that may include *P. glutinosus* and *P. oconaluftee*, although most specimens are presumably *P. oconaluftee*. Dietary items of these specimens included annelids, gastropods, millipedes, centipedes, isopods, phalangids, pseudoscorpions, spiders, mites, and representatives of over 20 families of insects. Millipedes and insects are the dietary staples in all populations. Large adults on rare occasions contain the remains of conspecifics (Powders 1973) or unidentified salamander bones (Powders and Tietjen 1974).

PREDATORS AND DEFENSE. No data are available on natural predators, but spring and black-bellied salamanders, small snakes, shrews, and woodland birds undoubtedly prey upon *P. oconaluftee*. When attacked or molested, this species produces slimy tail secretions that deter predators.

COMMUNITY ECOLOGY. *Plethodon jordani* and *P. oconaluftee* replace each other altitudinally in many areas, a finding that suggests that interspecific competition may influence altitudinal distributions. Details of competition between these closely related species are provided in the account of *P. jordani*.

CONSERVATION BIOLOGY. *Plethodon oconaluftee* occurs primarily on federal lands and many populations are receiving some degree of protection. This species reaches its highest densities in mature, mesic, hardwood forests. Hairston and Wiley (1993) monitored *P. oconaluftee* populations in mature forests in western North Carolina for nearly two decades and found no evidence of long-term population declines.

Plethodon ouachitae Dunn and Heinze
Rich Mountain Salamander
PLATE 133

Fig. 241. *Plethodon ouachitae;* adult; Polk Co., Arkansas (J. W. Petranka).

Fig. 242. *Plethodon ouachitae;* adult; LeFlore Co., Oklahoma (R. W. Van Devender).

IDENTIFICATION. The Rich Mountain salamander is a large species of *Plethodon* that consists of four geographic variants (Buck Knob, Kiamichi Mountain, Rich Mountain, and Winding Stair Mountain) that differ in coloration and maximum adult size (Blair and Lindsay 1965; Duncan and Highton 1979; Pope and Pope 1951). Individuals from all populations have a dark dorsum (sometimes overlain with red) with small white spots similar to those of *P. glutinosus*. White spotting is often abundant along the sides of the body and in some cases may fuse to form a more or less continuous band. The Buck Knob variant (sensu Blair and Lindsay 1965) has two longitudinal rows of large white spots or blotches along the back.

The Rich Mountain and Winding Stair variants have small white spots, and the Rich Mountain variant has brassy iridophores that produce a frosted appearance. Red pigmentation occurs on the dorsum of many specimens, but the proportion of the dorsum covered is highly variable both within and between populations. The degree of frosting varies from being nearly absent to covering most of the dorsum. Red dorsal coloration is most prevalent in the Rich Mountain variant and in some cases may cover most of the dorsum. In other populations, most individuals lack red pigmentation or have only a small proportion of the dorsum covered with red. Red pigmentation is absent in the Buck Knob variant and rare in the Kiamichi Mountain variant. The throat is light colored in all variants, but the remainder of the undersides is black. Sexually active males have circular mental glands and swollen nasolabial regions. Females in samples analyzed by Pope and Pope (1951) were on average 3% larger in SVL than males. Adults measure 10–17 cm TL and the average number of costal grooves is 16 (Blair 1967b).

SYSTEMATICS AND GEOGRAPHIC VARIATION. Duncan and Highton (1979) studied electrophoretic and color variation in members of the *Plethodon ouachitae-caddoensis* group in the Ouachita Mountains of Arkansas and Oklahoma. These forms compose five geographic groups that differ in body size and color patterns (Blair 1967b; Blair and Lindsay 1965; Duncan and Highton 1979). The groups occupy adjacent mountain ranges and are named the Rich Mountain, Winding Stair Mountain, Kiamichi Mountain, Buck Knob, and Caddo Mountain variants. Previously the last variant was considered to constitute one species, *P. caddoensis*, and the remaining four geographic isolates a second species, *P. ouachitae*. Electrophoretic analyses of protein variants show that three major subgroups occur among the five: the Caddo Mountain variant, the Buck Knob variant, and the remaining three variants. Within the last subgroup, the Kiamichi Mountain variant is almost as differentiated from the Winding Stair Mountain variant as some subgroups are from each other. In addition, the Buck Knob variant hybridizes freely over an approximately 5-km-wide zone with members of the Rich Mountain variant. Nei's genetic distances (D) between the three subgroups vary from 0.28 to 0.30.

Duncan and Highton (1979) discuss the difficulties of interpreting the systematic status of the variants and suggest several alternative classification schemes. They ultimately recommend that the Buck Knob isolate be recognized as a separate species, *P. fourchensis* (Highton 1986c). Here I follow Blair and Lindsay's original interpretation and consider *P. fourchensis* to be conspecific with *P. ouachitae* because the two forms intergrade freely in the zone of contact. Although the primary zone of intergradation is narrow (about 5 km), data presented by the authors indicate that the influence of the two variants on each other is not confined entirely to the primary zone of intergradation. I retain *P. caddoensis* as a separate species because electrophoretic and morphological evidence indicates that it is one of the most differentiated members of the complex.

DISTRIBUTION AND ADULT HABITAT. *Plethodon ouachitae* is endemic to forest communities in the Ouachita Mountains in a five-county region in western Arkansas and southeastern Oklahoma (Blair 1967b; Duncan and Highton 1979). Specimens have been collected from forested habitats at 320-869 m, but populations are rarely encountered below 457 m. In the Winding Stair Range, specimens are most abundant on northwest-facing talus slopes (Black 1974). *Plethodon ouachitae* lives beneath logs, bark, boards, rocks, and other debris on the ground surface and is often seen in caves (Black 1974). In one instance 117 specimens were taken from a pile of shingles about 4.6 m in diameter (Pope and Pope 1951).

BREEDING AND COURTSHIP. The mating season of *P. ouachitae* is poorly documented. Testes of males collected during the spring do not contain mature spermatozoa, suggesting that mating occurs during the autumn or winter months (Pope and Pope 1951). Adults presumably engage in a tail-straddle walk, but this has not been verified. Rival males are often aggressive toward each other, and males sometimes court rival males and induce them to waste spermatophores (Arnold 1977).

REPRODUCTIVE STRATEGY. Females undoubtedly oviposit shortly after moving underground in early summer, but nests have not been discovered in the field. Oviposition occurs in May and June and females reproduce biennially (Pope and Pope 1951; Taylor et al. 1990). Estimates of mean clutch size based on number of mature ovarian eggs are 17 (range = 13-23) for 22 females (Pope and Pope 1951), and 14 for the Buck Knob variant compared with 15 for individuals from other areas (Taylor et al. 1990). Clutch size is not correlated with SVL in either of the latter two groups.

TERRESTRIAL ECOLOGY. Rich Mountain males and females mature sexually when they reach 47-49 mm SVL and 52-53 mm SVL, respectively (Pope and Pope 1951). Animals <1 year old have rarely been collected, suggesting that hatchlings remain underground for 1 year or so before moving to the ground surface. Individuals mature sexually about 3 years after hatching, and sex ratios of surface populations are near 1:1 (Pope and Pope 1951).

Individuals are often abundant on the ground surface after heavy rains during the spring and autumn, but are uncommon during the summer (Black 1974; Blair 1967b; Spotila 1972). Talus areas in the Ouachita Mountains provide critical microhabitats that allow *P. ouachitae* to escape to physically benign subsurface re-

Range of *Plethodon ouachitae*

treats during the hottest and coldest months of the year. Food items in 12 animals included annelids, centipedes, mites, bugs, orthopterans, adult and larval beetles, and hymenopterans (Black 1974).

The adults are aggressive and will defend territories in laboratory aquaria against conspecifics and *P. caddoensis*. Residents will attack and drive off larger intruders and in rare instances may bite off and eat their tails (Thurow 1976). In laboratory trials the adults are far more likely to bite and chase other adults than juveniles. Adults are extremely aggressive toward each other, with levels of aggression being about 18 times greater than those documented for other *Plethodon* species (Anthony and Wicknick 1993). Adults will often bite opponents, then engage in lateral rolling of the body that effectively repels their challengers.

Juveniles do not avoid substrates marked with adult scent even though adults often attack juveniles in encounters. Males are attracted to the odors and fecal pellets of other males, a phenomenon that contradicts the behaviors of many other plethodontid salamanders (Anthony 1993). In natural populations adults are almost always found alone beneath cover objects, and males are never found together.

PREDATORS AND DEFENSE. There are no descriptions of natural predators or defensive mechanisms. Woodland birds, small snakes, and shrews are probably the major predators. About 20% of a sample of museum specimens had broken tails (Wake and Dresner 1967), but it is uncertain to what extent this observation reflects attempted predation versus attacks from other salamanders.

COMMUNITY ECOLOGY. *Plethodon ouachitae* coexists locally with *P. glutinosus* and the two occupy similar habitats. *Plethodon ouachitae* is aggressive toward *P. glutinosus* and can successfully defend and usurp cover objects from *P. glutinosus* (C. D. Anthony, pers. comm.).

CONSERVATION BIOLOGY. Populations of *P. ouachitae* can reach high densities in optimal habitats, which are northwest-facing talus slopes in mature hardwood forests. Hardwood buffers left around the margins of talus would help maintain viable populations of both *P. caddoensis* and *P. ouachitae* by providing leaf litter and shading.

COMMENTS. Anthony et al. (1994) and Duncan and Highton (1979) note that this species is often more infested with mites than *P. caddoensis* and *P. glutinosus*, and that males have more mites than females. Typically, 5–10% of individuals in samples have mites clogging the nasolabial grooves (Anthony et al. 1994). The extent to which this condition compromises feeding or mating success is not known.

Plethodon petraeus Wynn, Highton, and Jacobs
Pigeon Mountain Salamander
PLATE 134

Fig. 243. *Plethodon petraeus*; adult; Walker Co., Georgia (R. W. Van Devender).

Fig. 244. *Plethodon petraeus*; adult; Walker Co., Georgia (R. W. Van Devender).

IDENTIFICATION. The Pigeon Mountain salamander is a large species of *Plethodon* that is most easily recognized by its unique toes and dorsal coloration. The fourth toe on the forefeet and fifth toe on the hindfeet are disproportionately long and extend beyond the second joint of the adjacent digit when adpressed against it. The toes have blunt, expanded tips that aid in climbing rocks. Individuals have a black ground color that is overlain by a reddish brown dorsal color that often extends onto the head and tail. Small white iridophore spots with brassy flecks occur throughout the reddish brown dorsum, and yellow iridophore spots without brassy flecks form a band along the sides of the body. Young animals often have red spots on the dorsum that fuse with age to form first a zigzag pattern, then a complete dorsal band. The maximum sizes of females and males are around 84 mm and 73 mm SVL, respectively. The 10 largest females in one sample averaged 12% larger in SVL than the 10 largest males. Adults measure 11.5–18 cm TL and there are usually 16 costal grooves.

COMMENTS. The Pigeon Mountain salamander was described by Wynn et al. (1988) and is presently known from only two localities in northwest Georgia that are 7.5 km apart. Both are in limestone outcroppings on the eastern slope of Pigeon Mountain. *Plethodon petraeus* is largely restricted to cave mouths, rock outcrops, and detritus-filled crevices between rocks. The truncated, expanded toe tips are adaptations for climbing in rockface habitats. This species is locally abundant, although its range is very restricted.

390 Plethodontidae: Plethodontinae: Plethodontini

Range of *Plethodon petraeus*

Plethodon punctatus Highton
White-spotted Salamander
PLATE 135

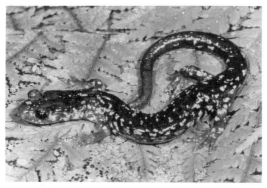

Fig. 245. *Plethodon punctatus;* adult; Pendleton Co., West Virginia (R. W. Van Devender).

IDENTIFICATION. The white-spotted salamander is a large eastern *Plethodon* species that is closely allied to *P. wehrlei*, but superficially resembles *P. glutinosus*. Individuals have a grayish black dorsum overlain with conspicuous white or yellow spots. The throat is light colored and the belly dark and often marked with yellow spots. The modal number of costal grooves is usually 17 or 18, and adults measure 10–17 cm TL.

This species is similar to *P. wehrlei* in having strongly webbed feet and a modal number of costal grooves >16. It differs from *P. wehrlei* by lacking reddish dorsal spots on juveniles and by having white or yellow spots on the dorsum similar to those of *P. glutinosus*. In addition, brassy flecking on *P. punctatus* is less prevalent than that in most *P. wehrlei* from similar latitudes. Hatchlings and juveniles have not been described.

SYSTEMATICS AND GEOGRAPHIC VARIATION. Biogeographic and genetic evidence suggests that *P. punctatus* was derived from ancestral *P. wehrlei* stock that became geographically isolated from the main population (Highton 1972). This species was thought to be separated from *P. wehrlei* to its west by 20–40 km

Range of *Plethodon punctatus*

(Highton 1972, 1987), but populations have been found within 2 km of each other in West Virginia (T. K. Pauley, pers. comm.).

Electrophoretic comparisons of New York *P. wehrlei* with Virginia *P. punctatus* (Highton and Larson 1979) indicate that these species are only moderately differentiated genetically ($D = 0.18$). Highton (1972) questioned whether *P. punctatus* should be treated as a separate species because it is closely related to *P. wehrlei*, yet geographically isolated, so that the standard test of reproductive isolation cannot be directly applied. Although electrophoretic data do not strongly support the recognition of *P. punctatus*, I recommend that it continue to be recognized until more detailed data on genetic variation in Virginia and West Virginia populations become available.

DISTRIBUTION AND ADULT HABITAT. *Plethodon punctatus* occurs on the Great North and Shenandoah mountains in north-central Virginia and adjoining portions of northeastern West Virginia. This species inhabits mixed deciduous forest with numerous rock outcrops. Populations are not known from below 810 m (Green and Pauley 1987), and most occur above 960 m on north-facing slopes (Buhlmann et al. 1988). Adults are most frequently collected beneath rocks.

BREEDING AND COURTSHIP. No information is available on the breeding season or courtship behavior. Mating presumably occurs in the autumn and/or spring months when the adults are most active on the ground surface.

REPRODUCTIVE STRATEGY. Females presumably oviposit after moving to underground retreats in the summer, but nests have not been discovered and detailed information on ovarian cycles is not available.

TERRESTRIAL ECOLOGY. The adults sometimes climb on tree trunks or rocks at night and are most active seasonally in the late spring and autumn (Buhlmann et al. 1988). Individuals emerge from beneath rocks and logs on wet nights to forage on the forest floor, but are inactive on the ground surface during prolonged periods of dry weather (Fraser 1976b). Individuals tend to be clumped with respect to cover objects, and larger animals generally tend to occupy the largest cover objects. Individuals emerge from underground winter retreats in April and retreat to deep burrows by late October. Seasonal surface activity at this site peaks during the early summer and progressively declines until the animals move to winter retreats. Individuals reach sexual maturity about 3 years after hatching.

The white-spotted salamander feeds on invertebrates, including hymenopterans, collembolans, millipedes, centipedes, spiders, insect larvae, flies, oligochaetes, beetles, and snails (Fraser 1976a). The size of prey eaten is positively correlated with the size of the salamander.

PREDATORS AND DEFENSE. Natural predators have not been documented but probably include shrews, small birds, woodland snakes, opossums, skunks, and other predators.

COMMUNITY ECOLOGY. Fraser (1976a,b) examined competitive interactions between juvenile *P. hoffmani* and *P. punctatus* and found little evidence of strong interspecific competition. Details of this work are presented in the account of *P. hoffmani*.

CONSERVATION BIOLOGY. This species appears to be sensitive to clear-cutting and other intensive timbering practices. Buhlmann et al. (1988) provide evidence that suggests that *P. punctatus* reaches its greatest abundance on rocky sites with virgin forest and is least abundant on recent clear-cuts and in white pine monocultures.

Plethodon richmondi Netting and Mittleman
Ravine Salamander
PLATE 136

Fig. 246. *Plethodon richmondi*; adult; Jessamine Co., Kentucky (J. W. Petranka).

IDENTIFICATION. The ravine salamander is an elongated, short-legged, slender salamander that has a modal number of 20-23 costal grooves. The dorsum is dark brown to black above and is flecked with varying amounts of silvery white and brassy specks (Netting and Mittleman 1938). The venter is uniformly dark except for moderate light flecking on the throat. The tail of adults makes up about 50% of the total length. Males have a crescent-shaped mental gland near the apex of the lower jaw and papillae on the inside of the vent. Mature females in populations in northern Kentucky and northeastern Tennessee average 5% larger in SVL than mature males (Nagel 1979; Wallace 1969). Adults measure 7.5-14.5 cm TL.

Hatchlings are light gray above, have an immaculate belly, and reach an average of 14-15 mm SVL (Wallace and Barbour 1957). Juveniles resemble adults in coloration but have proportionately shorter tails and longer legs (Duellman 1954b). This species superficially resembles *P. hoffmani* and the unstriped phase of *P. cinereus*. Geographic range, throat color, and degree of mottling on the belly are useful in separating these species.

SYSTEMATICS AND GEOGRAPHIC VARIATION. There are no conspicuous patterns of geographic variation in coloration or patterning. The modal number of costal grooves varies from 20 to 23 in regional populations. Major trends are as follows: 20 costal grooves for a population on Whitetop Mountain, Virginia; 21 costal grooves for populations in southeastern Kentucky, northwestern North Carolina, eastern Ohio, northeastern Tennessee, southwestern Virginia, southern West Virginia, and Paddy and Reddish knobs on the Virginia–West Virginia state line; 22 costal grooves for populations in southeastern Indiana, northeastern Kentucky, western Ohio, the Appalachian Plateau of western Pennsylvania, the vicinity of Huntington, West Virginia, and the Valley and Ridge Province of Pennsylvania, Virginia, and West Virginia; and 23 costal grooves for populations in the Appalachian Plateau of West Virginia (Highton 1962a).

Range of *Plethodon richmondi*

DISTRIBUTION AND ADULT HABITAT. The ravine salamander lives in ravines and on wooded hillsides in deciduous forests from western Pennsylvania and north-central Ohio southward to northeastern Tennessee and northwestern North Carolina. This species is strongly associated with rocky habitats and is often locally abundant on or about talus slopes and rock outcroppings. It is also abundant on wooded slopes with friable soil and large, flat rocks. Individuals show a preference for rocky cover even when logs and other surface cover are available (Duellman 1954b), and specimens are rarely found in floodplains or on dry ridgetops (Netting 1939).

BREEDING AND COURTSHIP. The mating season extends from autumn through early spring. Mating in an eastern Tennessee population occurs from November through March, a period when males have sperm-filled vasa deferentia and females have spermatophores in their cloacae (Nagel 1979). In West Virginia the adults mate in April and May and do not have a fall breeding season (Jewell and Pauley 1995).

Little information is available on courtship in *P. richmondi*. Adults engage in a tail-straddle walk like other plethodontids. Body snapping and pulling are used to abrade the skin of females and to introduce mental gland secretions of the male into the circulatory system (Arnold 1977).

REPRODUCTIVE STRATEGY. Most females oviposit in deep underground passageways. I was unable to locate 22 radioactively tagged females during the summer nesting season, presumably because individuals move deep within talus to oviposit (Petranka 1979). In all likelihood females brood their eggs through hatching like other small eastern *Plethodon* species.

In Ohio, females with large ovarian eggs occur in surface populations in March and April (Duellman 1954b; Wood 1945). Based on seasonal patterns of movement below ground, most females oviposit from late April through May (Pfingsten 1989a; Wood 1945). Egg laying in West Virginia also occurs in late April or May and females oviposit biennially (Jewell and Pauley 1995). Females in a Tennessee population appear to oviposit in late May through June, a period that corresponds to the time when females move from the ground surface into underground summer retreats (Nagel 1979). Ovarian eggs reach their maximum size in May, when all gravid females have mature ova >3.0 mm in diameter. Females in northern Kentucky develop ova 3.0 mm in diameter in March and April (Wallace 1969).

Estimates of clutch size are based almost entirely on counts of ovarian complements. The average number of mature ova in 94 Kentucky females was 8.5 (Wallace 1969) compared with 11.5 (range = 9–15) in six Ohio specimens (Wood 1945). The number of mature ova in 50 Tennessee females varied from 5 to 11, averaged 8.3, and was independent of female SVL (Nagel 1979).

Only two nests have been found. Duellman (1954b) discovered 12 eggs in intermediate stages of development beneath a limestone slab on a talus slope in southern Ohio on 14 July. The eggs were in a depression beneath the rock and the outer capsule averaged 5 mm in diameter. The ova averaged 3.5 mm in diameter and each was surrounded by a vitelline membrane and two clear jelly envelopes. Wallace and Barbour (1957) found two adults, two hatchlings, and two eggs with embryos at the hatching stage on 23 August in northern Kentucky. The nest was 25 cm beneath the ground under a flat limestone rock. Netting and Mittleman (1938) collected a hatchling in October.

TERRESTRIAL ECOLOGY. Hatching occurs in late summer or early fall; however, most hatchlings do not move to the ground surface until the following spring (Nagel 1979; Pfingsten 1989a). A cluster of 12 overwintering animals, including 9 small juveniles, was found buried in clay near a rock outcrop on 5 March in Ohio (Duellman 1954b). The small animals may have been hatchlings that remained at the nest site through the winter.

The diet of Indiana specimens consists of insect larvae, ants, beetles, spiders, earthworms, mites, and larval ticks (Minton 1972). Ants and sowbugs are the most volumetrically important prey in Ohio specimens (Duellman 1954b); however, numerous other prey are eaten, including earthworms, snails, centipedes, flatworms, collembolans, roaches, hemipterans, beetles, flies, spiders, mites, and pseudoscorpions. Specimens feed throughout the winter, although food intake is reduced during this period. Three other Ohio specimens contain mostly sowbugs, along with spiders and rove beetles (Seibert and Brandon 1960).

Nagel (1979) did not directly report growth rates, but data provided in histograms indicate that hatchlings grow about 14–15 mm SVL during their first year of life. These estimates assume that embryos hatch during September–October when they reach 14–15 mm SVL. Data provided by Pfingsten (1989a) on Ohio populations are in close agreement.

Males in a Tennessee population mature sexually slightly more than 2 years after hatching, when they reach 38–42 mm SVL (Nagel 1979). Some females begin maturing when 2 years old and probably oviposit the following spring or summer; however, most appear to require an additional year before ovipositing for the first time. The smallest female with maturing eggs in Kentucky specimens measured 35 mm SVL (Wallace 1969). In Ohio the smallest mature males and females measured 38 and 40 mm SVL, respectively (Pfingsten 1989a). Females in all populations reproduce biennially (Nagel 1979; Pfingsten 1989a).

Ravine salamanders are most active on the ground surface during the spring and autumn. During the hottest summer months individuals retire to underground retreats (Duellman 1954b; Nagel 1979; Pfingsten 1989a), where they either oviposit or aestivate. Specimens have been unearthed as deep as 1.2 m below the ground surface (Green and Pauley 1987). The tail is used as a fat storage organ, and individuals that emerge in the fall after summer aestivation have tails that are noticeably smaller in diameter than those of animals collected in late spring (Duellman 1954b; Green and Pauley 1987; Netting 1939). This finding suggests that individuals feed very little while underground.

Ravine salamanders are sometimes active on the surface during periods of warm winter weather. Large numbers of individuals can be found in Tennessee during January and February (Nagel 1979), and lethargic specimens can been found in Ohio when air temperatures are below freezing and snow is on the ground (Duellman 1954b). Individuals are more abundant during winter thaws and are found at shallow depths beneath rocks near talus slopes. Following the most severe winter on record for Kentucky, I found very few ravine salamanders at collection sites in central Kentucky that lacked talus. Individuals had been common at these sites the previous autumn and may have frozen to death (Petranka 1979).

Adult *P. richmondi* show little evidence of being territorial. This species is much less aggressive than many eastern *Plethodon* species (Thurow 1976), and as many as four to eight individuals have been found under the same rock (Duellman 1954b; Minton 1972).

PREDATORS AND DEFENSE. When uncovered in the field, individuals may assume a coiled, motionless position that makes them inconspicuous to predators (Dodd 1990c). Individuals will avoid substrates marked with odors of the ringneck snake (*Diadophis punctatus*), which is a major predator of this and other salamanders (Cupp 1994).

COMMUNITY ECOLOGY. No data are available on the community ecology of *P. richmondi*.

CONSERVATION BIOLOGY. The ravine salamander often reaches high local densities in mesic hardwood forests with rocky substrates. Deforestation and urbanization are the primary factors that have eliminated local populations of this species throughout its range.

Plethodon serratus Grobman
Southern Red-backed Salamander
PLATE 137

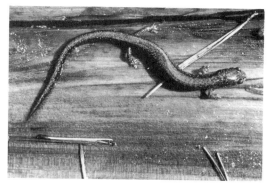

Fig. 247. *Plethodon serratus*; adult; Macon Co., North Carolina (J. W. Petranka).

IDENTIFICATION. The southern red-backed salamander is closely allied with *P. cinereus* and at one time was considered conspecific with it. This is a small salamander with a rounded tail that composes about 50% of the total length. Both striped and unstriped color morphs occur in some populations. The striped morph has a dark brown dorsal ground color with a red or orangish red stripe that extends from the head to the tail tip. The dorsal stripe is serrated along the back in populations in the Ouachita Mountains and central Louisiana, but tends to be straight-edged or weakly serrate in populations elsewhere. The unstriped morph is brown above and usually has small amounts of red spotting on the back, sides, and belly.

The sides of both morphs are brown and have varying amounts of white spotting. The belly is mottled black and white, and sometimes has small amounts of red pigment, especially near the limb insertions (Highton 1986f). The number of costal grooves ranges from 18 to 21 but varies regionally. Adults measure 6.5–10.5 cm TL, and females in Georgia, North Carolina, and Tennessee are on average 5–7% larger in SVL than males (Camp 1988). Mature males have swollen nasolabial glands and a crescent-shaped mental gland near the apex of the lower jaw. Hatchlings have proportionately shorter tails than older animals and an average SVL of 12.5 mm (Camp 1988).

SYSTEMATICS AND GEOGRAPHIC VARIATION. *Plethodon serratus* was formerly considered to comprise two southern subspecies of *P. cinereus* (*P. c. serratus* and *P. c. polycentratus*). Highton and Larson (1979) Highton and Webster (1976) analyzed protein variation and found that these forms differ moderately from *P. c. cinereus* (Nei's $D = 0.32$) and have fixed differences for five loci. They recommended that the southern populations be treated as a separate species, *P. serratus*.

Color patterning, modal number of costal grooves, and frequency of striped versus unstriped morphs vary geographically in *P. serratus*. The modal number of costal grooves is 18 for populations in Missouri; 19 for populations in Louisiana, the southern Appalachian Mountains, and the Ouachita Mountains of Arkansas and Oklahoma; and 20 or 21 for populations in Alabama and Georgia (Highton 1986f). Both striped and unstriped color morphs occur in some populations, but the unstriped morph is rare except in Alabama and

Range of *Plethodon serratus*

Georgia (Highton 1962a; Highton and Webster 1976). Although the extent to which the dorsal stripe is serrate varies somewhat predictably between eastern and western groups (less serrate to the west), *P. serratus* is currently considered to be monotypic.

DISTRIBUTION AND ADULT HABITAT. *Plethodon serratus* exists as four widely separated geographic isolates. Three major groups of populations occur in the Salem Plateau in southeastern Missouri, in the Ouachita Mountains of west-central Arkansas and southeastern Oklahoma, and in the Piedmont and Blue Ridge provinces from northwestern Georgia and northeastern Alabama to eastern Tennessee and western North Carolina. Isolated populations also occur in central Louisiana.

Populations occupy mesic to moderately dry hardwood forests with abundant rocks or logs that serve as surface cover. In Alabama and in the southern Appalachians *P. serratus* inhabits mesic forests, whereas in west-central Louisiana populations are found on rocky hillsides in longleaf pine forests (Keiser and Conzelmann 1969). In the Ouachita Mountains in Arkansas, specimens live beneath logs and in leaf litter, and have been collected in moderately dry habitats (Spotila 1972; Thurow 1957b). Specimens have been found at elevations as high as 1690 m (Huheey and Stupka 1967).

BREEDING AND COURTSHIP. Information on the mating season is only available for eastern populations. Females have spermatophores in their cloacae from February to March in the Georgia Piedmont, and during December in the southern Blue Ridge (Camp 1988). The sex ratio in these populations does not differ significantly from 1:1. Courtship behavior has not been described.

REPRODUCTIVE STRATEGY. Females presumably lay their eggs in underground burrows and brood their young, but nests have not been discovered in the wild. Indirect data from three studies suggest that females oviposit during June and July (Camp 1988; Taylor et al. 1990; Trauth et al. 1990). Mean number of mature ova in females from Piedmont and Blue Ridge populations are both 5.5, and clutch size is independent of female SVL (Camp 1988). The mean number of mature ova in Arkansas specimens is 5.9 (Taylor et al. 1990) and 7.0 (Trauth et al. 1990) and is positively correlated with female SVL. Females that Camp (1988) buried in boxes laid two clutches containing five eggs each between 10 and 30 July. The egg capsules are 4.5 mm in diameter.

TERRESTRIAL ECOLOGY. The embryos hatch in late summer, but almost no data are available on the natural history of the juveniles. Sexually mature males and females from the Georgia Piedmont measure 33–45 and 33–47 mm SVL, respectively, whereas those from the Blue Ridge of North Carolina and Tennessee measure 33–39 and 35–46 mm SVL (Camp 1988). The length of the juvenile stage is not known, but it probably lasts about 2 years based on data for other small *Plethodon* species.

Southern red-backed salamanders are generally inactive on the ground surface during the summer months, but remain active throughout the winter in areas where winter weather is not severe. In eastern populations, juveniles first appear on the ground surface in late October, and adults surface in early November (Camp 1988). Individuals are common in leaf litter or under cover objects from November through March. Adults become increasingly difficult to find in April and juveniles disappear beneath ground by mid-May.

Camp and Bozeman (1981) list a variety of invertebrates in 55 Georgia specimens. Ants and beetles are the most volumetrically important prey. Other food items include snails, annelids, mites, spiders, pseudoscorpions, millipedes, centipedes, isopods, and members of several orders of insects.

PREDATORS AND DEFENSE. No data are available on predators or defensive mechanisms. The primary predators are probably small woodland snakes, shrews, birds, and small mammals such as skunks.

COMMUNITY ECOLOGY. Hairston (1981) experimentally reduced the densities of *P. jordani* and *P. oconaluftee* in plots in the southern Appalachians and found that this manipulation had no detectable effect on surface numbers of *P. serratus*. His data suggest that competition between *P. serratus* and large *Plethodon* species is minimal.

CONSERVATION BIOLOGY. The conversion of hardwoods to intensively managed pine forests appears to be depleting many populations of this species in Georgia (Camp 1986). Old hardwood forests on moderate to steep relief provide optimal habitats for this species in Georgia.

Plethodon shenandoah Highton and Worthington
Shenandoah Salamander
PLATE 138

Fig. 248. *Plethodon shenandoah;* adult; Hawksbill Mountain, Virginia (J. W. Petranka).

IDENTIFICATION. The Shenandoah salamander is a small eastern species of *Plethodon* that is closely allied to *P. hubrichti* and *P. nettingi*. This species has a modal number of 18 costal grooves and has two color morphs. The unstriped morph has a black ground color overlain with small amounts of brassy flecking. Very small red dots are often evident in the middorsal region. The striped morph has a narrow red to yellow middorsal stripe that extends from the head onto the tail, and a uniformly dark venter. The belly is dark with only a few scattered light markings (Highton and Worthington 1967). Adults reach 7–10 cm TL. The Shenandoah salamander closely resembles *P. cinereus* but has a narrower dorsal stripe and uniformly dark belly.

SYSTEMATICS AND GEOGRAPHIC VARIATION. *Plethodon hubrichti, P. nettingi,* and *P. shenandoah* are closely related species that were previously treated as subspecies of *P. nettingi*. Highton and Larson (1979) raised these to the status of full species after comparing their protein similarities. Their data indicate moderate levels of genetic divergence, with genetic distances (D) varying from 0.22 to 0.35 for different pairs of species. The extent to which reproductive isolating mecha-

Range of *Plethodon shenandoah*

nisms have developed in association with genetic divergence is not known.

DISTRIBUTION AND ADULT HABITAT. *Plethodon shenandoah* is restricted to talus slopes in deciduous forests on north or northwest slopes of The Pinnacles, Stony Man Mountain, and Hawksbill Mountain in the Blue Ridge of north-central Virginia (Highton 1972; Highton and Worthington 1967). Populations occur at elevations between 914 and 1143 m, and individuals are most common along the edges of talus where the soil is shallow (Jaeger 1970).

COMMENTS. Most aspects of the ecology and natural history of this species have not been studied.

Females appear to reach sexual maturity during their third year of life. Clutch size averages 13 eggs and females oviposit every other year (Jaeger 1980b). A long-term study of population trends suggests that populations inhabiting talus with little or no soil may be locally extirpated during severe droughts. Recolonization of these drier sites occurs by dispersal of individuals from adjoining areas with deeper soils, and full recovery may require 8 years or more.

Plethodon cinereus and *P. shenandoah* replace each

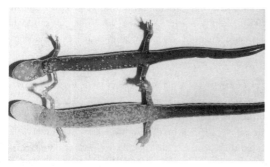

Fig. 249. Comparison of ventral markings on *Plethodon shenandoah* (upper) and *P. cinereus* (lower); Hawksbill Mountain, Virginia (J. W. Petranka).

Fig. 250. Aspect of talus habitats occupied by *Plethodon shenandoah*; Hawksbill Mountain, Virginia (J. W. Petranka).

other microgeographically around most talus slope habitats (Jaeger 1970). Competitive interactions between these two species are discussed in detail in the account of *P. cinereus*.

Plethodon shenandoah will establish and actively defend territories when introduced into terraria and is one of the most aggressive species of small eastern *Plethodon* (Thurow 1976). Individuals remain immobile when uncovered for a significantly shorter period of time (mean time = 6 seconds) than other *Plethodon* species (Dodd 1989a).

Plethodon shenandoah is found entirely within the Shenandoah National Park and is a federally endangered species.

Plethodon stormi Highton and Brame
Siskiyou Mountains Salamander
PLATE 139

Fig. 251. *Plethodon stormi*; adult; Siskiyou Co., California (W. P. Leonard).

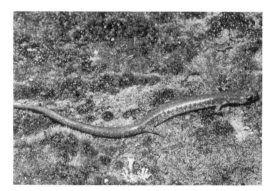

Fig. 252. *Plethodon stormi*; adult; Siskiyou Co., California (W. P. Leonard).

IDENTIFICATION. The Siskiyou Mountains salamander is an elongated, short-limbed species of *Plethodon* that closely resembles *P. elongatus*. It differs from the latter species in having a modal number of 17 costal grooves, 4–5.5 intercostal folds between adpressed limbs, and a less conspicuous dorsal stripe. Adults have a light brown to purplish brown dorsal ground color that is overlain by an inconspicuous light brown stripe that extends from the head to about half the length of the tail. The venter is lavender to purplish in adults, and the throat is lighter colored than the remaining undersides (Brodie 1971b). White to yellowish flecks are scattered on most parts of the body and are concentrated on the limbs and sides of the body. Adults measure 9.8–14 cm TL and males are on average slightly smaller than females.

Hatchlings reach 17–18 mm SVL on average. Juveniles have a darker ground color than the adults and an olive green dorsal stripe that turns light brown with age (Nussbaum et al. 1983).

SYSTEMATICS AND GEOGRAPHIC VARIATION. *Plethodon stormi* is so similar to *P. elongatus* that some treat it as a subspecies of *P. elongatus*. These two species occur within 1 km of each other in Oregon without showing clear evidence of interbreeding. However, one population in northern California is intermediate between the two species (Brodie 1970). *Plethodon elongatus* and *P. stormi* are as closely related biochemically as any other pair of western *Plethodon* species (Highton and Larson 1979). Nei's distance (D) between the two species is 0.33, a value that does not provide strong evidence either for or against recognizing *P. stormi* as a separate species. Here I follow the recommendation of Nussbaum et al. (1983) in continuing to recognize *P. stormi* until further data are available on its genetic relationship with *P. elongatus*.

Brodie (1970) and Nussbaum et al. (1983) describe patterns of geographic variation. Individuals from southern Oregon and northern California are virtually identical in coloration and patterning. However, local

Range of *Plethodon stormi*

populations sometimes vary in the modal number of costal grooves.

DISTRIBUTION AND ADULT HABITAT. *Plethodon stormi* is restricted to the Siskiyou Mountains of southern Oregon and northern California at elevations of 488–1078 m (Nussbaum et al. 1983). Individuals are almost invariably found in the vicinity of talus slopes or rock crevices, but during the wettest months may move into the surrounding forest and reside beneath bark, logs, and other forest debris. Heavily forested, north-facing slopes with talus harbor the densest populations.

BREEDING AND COURTSHIP. Courtship behavior has not been described. The seasonal presence of sperm in the vasa deferentia of males suggests that mating occurs during the autumn and spring, but this hypothesis is poorly substantiated (Nussbaum et al. 1983). Females appear to breed biennially.

REPRODUCTIVE STRATEGY. Nests have never been found. Indirect evidence suggests that females oviposit during the spring and brood their eggs during the summer months in subsurface nesting sites deep within talus. Mature ova are 4.2–5.2 mm in diameter and white. The number of mature ova in 37 females examined by Nussbaum et al. (1983) varied from 2 to 18, averaged 9, and was not correlated with female size.

TERRESTRIAL ECOLOGY. Hatching probably occurs in early autumn, but the hatchlings do not emerge from underground nest sites until the following spring. Juvenile males and females grow 6.5 and 8.0 mm SVL per year, respectively, and mature when 5–6 years old (Nussbaum et al. 1983). Individuals are most active on the ground surface between March and April, and from September to early November. During the summer individuals remain below ground during the day but move to surface sites at night to feed on invertebrates. During dry weather individuals may pose near entrances to underground retreats and rapidly dart forward to capture passing prey. During wet weather individuals actively prowl about talus slopes at night in search of prey. Spiders, pseudoscorpions, mites, ants, collembolans, and beetles are the major prey of *P. stormi* (Nussbaum et al. 1983).

PREDATORS AND DEFENSE. Natural predators and defensive mechanisms have not been described.

COMMUNITY ECOLOGY. *Plethodon stormi* is one of many western salamanders that show strong affinities for talus. Ecological interactions within talus slope communities have received little attention from ecologists. The roles of this and other salamanders in structuring invertebrate communities that inhabit talus have not been explored.

CONSERVATION BIOLOGY. *Plethodon stormi* has a very limited geographic range and rather specialized habitat requirements that make it sensitive to environmental disturbance. Timbering around talus slopes with local populations of *P. stormi* would in all likelihood be highly detrimental to the species; current management plans recommend full protection of existing populations.

Plethodon vandykei Van Denburgh
Van Dyke's Salamander
PLATE 140

Fig. 253. *Plethodon vandykei*; adult; Lewis Co., Washington (W. P. Leonard).

Fig. 254. *Plethodon vandykei*; adult (note light throat region); Lewis Co., Washington (W. P. Leonard).

IDENTIFICATION. *Plethodon vandykei* is a small western salamander that has a modal number of 14 costal grooves and indistinct parotoid glands. The legs are relatively long, and there are 0.5–3 intercostal folds between the adpressed limbs. Three color morphs occur (Brodie 1970; Nussbaum et al. 1983). The dark morph has a dark brown to black dorsal ground color with a straight dorsal stripe. The stripe may be yellow, green, orange, or red, and it extends to the tail tip. The undersides are dark except for a pale yellow throat. The yellow morph has a tan or yellowish ground color above and below that is often indistinguishable from the dorsal stripe. The venter is yellow and the throat pale yellow. The rose morph is similar to the yellow morph except that the dorsal coloration is pinkish to salmon. Females are slightly larger than males on average, have slightly smaller vent flaps, and lack the nasolabial cirri that are present on mature males. Adults reach 9.5–10.5 cm TL and the hatchlings resemble miniature dark-morph adults.

SYSTEMATICS AND GEOGRAPHIC VARIATION. Populations that have been referred to as *P. vandykei* occur as four disjunct groups in the northern Rocky Mountains in northern Idaho, northwestern Montana, and western Washington (Brodie 1970). Populations in Idaho and Montana were originally described as a separate species, *P. idahoensis* (Slater and Slipp 1940). This form was later treated as a subspecies of *P. vandykei* (Lowe 1950), but Brodie (1970) found discordant variation in certain traits and recommended that subspecies not be recognized.

Highton and Larson (1979) examined protein variation in a small sample of each form and recommended splitting the group into two species after finding moderate levels of genetic differentiation. Howard et al. (1993) completed a more extensive genetic analysis of populations from all four disjuncts and verified that Idaho and Montana populations are moderately differentiated from Washington populations. Nei's genetic distances (*D*) between these groups av-

Fig. 255. *Plethodon vandykei;* profile of feet; Lewis Co., Washington (W. P. Leonard).

erages 0.33. They recommend recognizing *P. idahoensis* because Rocky Mountain populations have been isolated from Washington populations for millions of years and have diverged both genetically and phenotypically from the latter. Here I follow this recommendation and treat *P. idahoensis* as a separate species. Studies of mating compatibility would help clarify whether these forms have reached the level of biological species.

Brodie (1970) summarizes geographic variation in number of costal grooves, vomerine teeth, and coloration. Local populations often have two, or rarely all three, of the color morphs described previously. Some traits, such as number of vomerine teeth, do not vary concordantly with color patterns. Populations in the Willapa Hills and Olympic Mountains are virtually identical genetically, whereas those in the Cascades are moderately differentiated ($D = 0.14$) from the coastal groups (Howard et al. 1993).

DISTRIBUTION AND ADULT HABITAT. Van Dyke's salamander occurs from sea level to 1560 m in the Willapa Hills and Olympic and southern Cascade mountains of western Washington (Brodie 1970). Populations are patchily distributed throughout these areas and often occur at low densities. Populations are found in regions receiving >150 cm average annual precipitation, and they have an upper limit at the boundary between temperate mesophytic and subalpine forests (Wilson et al. 1995). Disjunct groups are separated by glacial and alluvial deposits, which appear to limit the distribution of local populations.

This is perhaps the most aquatic *Plethodon* species in North America. Individuals live in seepages, in splash zones of waterfalls, and in rocky habitats along streams in moist conifer forests (Brodie 1970). In interior regions *P. vandykei* is associated with moist talus slopes in well-shaded, north-facing slopes, but it may occasionally be found beneath bark or logs in forest floor habitats during wet weather (Jones 1989;

Range of *Plethodon vandykei*

Nussbaum et al. 1983; Stebbins 1951). In coastal regions, populations are not as strongly associated with rocky habitats and frequently reside beneath woody debris on the forest floor (Wilson et al. 1995).

BREEDING AND COURTSHIP. Information on the mating season is largely lacking. Partial courtship has been observed during the spring, but it is uncertain if mating also occurs in the autumn (Nussbaum et al. 1983). Courtship has not been described in detail but presumably is similar to that of *P. idahoensis* (Lynch and Wallace 1987).

REPRODUCTIVE STRATEGY. Most females oviposit during spring, and embryos hatch in late summer or early autumn when they reach 15–18 mm SVL (Nussbaum et al. 1983). Nesting records include a grapelike cluster of eggs (date unspecified) attached by a gelatinous pedicel to the underside of a moss-covered stone (Noble 1925), and a female and cluster of seven eggs in early developmental stages found on 21 May (Jones 1989). The latter eggs were in a partially rotted log about 2 m from a headwater stream. Freshly laid eggs are cream colored, average 5.3 mm in diameter, and are surrounded by a thin jelly coat. Two gravid females from Washington contained 11 and 14 mature ova that were 4.3–5.0 mm in diameter (Stebbins 1951).

TERRESTRIAL ECOLOGY. Most aspects of the natural history of the juvenile and adult stages are not documented. Adults are most active on the ground surface during the spring and fall months, and males and females mature when they reach >44 and 47 mm SVL, respectively.

Ovaska and Davis (1992) found that *P. vandykei* does not respond to chemical odors from the fecal pellets of either conspecific males or *P. dunni*. Individuals do respond to odors from *P. vehiculum* but do not avoid burrows containing their fecal pellets. Thus the ecological significance of chemically mediated behavioral interactions between these species is uncertain.

PREDATORS AND DEFENSE. No data are available on natural predators or defensive mechanisms.

COMMUNITY ECOLOGY. *Plethodon vandykei* is generally uncommon and its interactions with other community members are unknown. It is uncertain if the spottiness of local populations and general scarcity of this species reflect narrow habitat tolerances or negative interactions with competitors or predators in local communities.

CONSERVATION BIOLOGY. Van Dyke's salamander is patchily distributed and uncommon throughout its range and should receive high priority for protection by resource managers. This species is currently protected in the state of Washington and many populations are on lands protected by the National Park Service. Local populations of this species are at risk of extirpation in intensively managed forests (Welsh 1990). Leaving forest buffers around headwater streams and talus areas is perhaps the single most important action that will protect local populations in managed forests.

Plethodon vehiculum (Cooper)
Western Red-backed Salamander
PLATE 141

IDENTIFICATION. The western red-backed salamander is a small species of *Plethodon* that has a modal number of 16 costal grooves and an even-edged dorsal stripe that extends from the snout to the tail tip. The stripe varies from yellow or red to olive green or tan; however, the red variant is most common. A fine line of dark pigment often occurs along the midline of the back. Unstriped individuals occur in many populations. In addition, weakly melanized specimens whose entire body is the color of the dorsal stripe occur in many populations (Brodie 1970; Stebbins 1951; Storm and Brodie 1970b). The proximal segments of the limbs are colored the same as the dorsal stripe, and the vent has numerous scattered light marks on a dark ground color that produces a salt-and-pepper effect. An albino specimen is reported by Hensley (1959).

Mature males have squared snouts; protruding, recurved premaxillary teeth; and small vent flaps that

Fig. 256. *Plethodon vehiculum;* adult; Jefferson Co., Washington (W. P. Leonard).

Fig. 257. *Plethodon vehiculum;* adult; Skamania Co., Washington (R. W. Van Devender).

extend slightly beyond the posterior margin of the vent. Females lack the vent flaps and enlarged teeth, and have more rounded snouts (Stebbins 1951). Adult females in a large sample of Oregon specimens were on average 7% larger in SVL than adult males. Adults measure 7–11.5 cm TL.

SYSTEMATICS AND GEOGRAPHIC VARIATION. No subspecies are currently recognized. Brodie (1970) and Nussbaum et al. (1983) present detailed accounts of regional and local variation. Frequencies of color morphs vary considerably among populations. Usually 50–60% of the individuals in a population are red striped; however, in many island populations almost all individuals are red striped. Vancouver Island specimens tend to be larger on average, to have proportionately longer legs, and to have a much higher percentage of red-striped morphs than mainland populations. Melanistic individuals and individuals with reduced melanism are both common in coastal populations.

DISTRIBUTION AND ADULT HABITAT. This relatively common species occurs from sea level to 1250 m in moist coniferous forests from Vancouver Island and adjoining portions of mainland British Columbia southward to southwestern Oregon (Nussbaum et al. 1983). It occurs from the crest of the Cascades to the western coast, but only occasionally at elevations above 760 m. Populations also occur on several islands off the Washington mainland (Brown and Slater 1939; Slater and Brown 1941). The western red-backed salamander is common on and about talus slopes but also inhabits forest floor habitats, where individuals live beneath leaf litter, bark, logs, and other cover (Brodie 1970; Corn and Bury 1991; Dumas 1956; Nussbaum et al. 1983). This species is most commonly found in Washington in moderately decaying logs with diameters of 10–30 cm (Aubry et al. 1988). In the northern Puget Trough and foothills of the Cascade Range in Washington, populations are often found in rocky seeps, springs, and small streams (Leonard 1996). *Plethodon dunni* and *Rhyacotriton* species do not occur in this region, an observation that suggests that they may competitively exclude *P. vehiculum* from wet microhabitats elsewhere.

BREEDING AND COURTSHIP. The mating season is geographically variable and extends from September through January depending on the locale. Peak breeding activity occurs from November through January in Oregon populations, but mating may begin as early as September (Dumas 1956). Females collected November–January often have spermatophores in their vents, whereas males have highly enlarged vasa deferentia (Peacock and Nussbaum 1973). Individuals on Vancouver Island breed from September to November, with a peak in October and November (Ovaska and Gregory 1989).

Females in many populations have biennial reproductive cycles and mate every other year shortly before ovipositing (Peacock and Nussbaum 1973). Females on Vancouver Island oviposit every 3 years because of reduced surface feeding opportunities resulting from cold winters and dry summers (Ovaska and Gregory 1989). Sex ratios do not differ significantly from 1:1, although sexually active males often outnumber sexually active females.

Range of *Plethodon vehiculum*

Adults are aggressive during the mating season, particularly rival males that are competing for females (Ovaska 1987). Males engage in a variety of threat displays and also bite and grip their opponents during physical encounters. In some cases, individuals emit high-pitched clicking sounds during encounters.

Courtship behavior has not been described. Adults presumably engage in a tail-straddle walk similar to that of other *Plethodon* species. Males have large, recurved premaxillary teeth that are used to lacerate the skin of females during courtship.

REPRODUCTIVE STRATEGY. Females oviposit during the spring or early summer (Peacock and Nussbaum 1973), but the nests of *P. vehiculum* have rarely been found, suggesting that most females nest in subsurface retreats. The eggs are laid in grapelike clusters and are guarded by the female through hatching. On rare occasions, males have been collected with the females (e.g., Norman and Swartwood 1991). The eggs are yellowish white and are surrounded by two jelly envelopes. Females attach the eggs by broad, flaring, gelatinous bases to the sides and roofs of moist cavities in logs, or to the undersides of rocks or bark (Carl 1943; Hanlin et al. 1979). Nesting records include clutches of eight and nine recently laid eggs on 12 May in western Oregon with eggs that averaged 4.7 and 5.0 mm in diameter (Hanlin et al. 1979), and a clutch in early developmental stages on 13 May in Washington with ova that averaged 4.4 mm in diameter (Norman and Swartwood 1991). A Washington female that Stebbins (1951) implanted with pituitary tissue laid a clutch of nine eggs in the laboratory with ova and outer capsules 4.5 mm and 5.8–6.0 mm in diameter, respectively.

Clutch size estimates based on counts of mature

ovarian eggs are 6–19 ova (mean = 10) for 65 females from west-central Oregon, 4–18 (mean = 10) for females from west-central Washington, and 11–18 (mean = 14) for females from Destruction Island, northwestern Washington (Nussbaum et al. 1983; Peacock and Nussbaum 1973). At the Oregon site clutch size is positively correlated with female SVL, whereas at the mainland Washington site it is not. The largest ovarian eggs are 4.3 mm in diameter.

TERRESTRIAL ECOLOGY. Individuals in an Oregon population hatch in August or September when they reach 13–15 mm SVL (Peacock and Nussbaum 1973). Hatchlings are abundant on the ground surface during the late autumn and early winter, when the soil is saturated with water. Individuals grow about 10 mm SVL annually during their first 3 years of life. Both males and females mature between their second and third years of life when they reach 42 mm SVL and 44 mm SVL, respectively. Females reproduce biennially, whereas males breed annually. Size at sexual maturity varies geographically from 40 to 50 mm SVL (Brodie 1970).

Hatchlings in a Vancouver Island population first appear on the ground in the autumn but are more numerous in the spring (Ovaska and Gregory 1989). Individuals grow 10.5 mm SVL during their first year. Growth rates decline thereafter, and little growth occurs after juveniles reach sexual maturity. Males and females in this population reach sexual maturity when 3 and 4 years old, respectively. The smallest sexually mature males and females measure 38 and 42 mm SVL. Adult males measure 47 mm SVL on average compared with 51 mm SVL for females.

Western red-backed salamanders in a population studied by Dumas (1956) move deep within talus during the winter months. With the arrival of spring weather, individuals wander extensively from the talus. During the dry summer months, individuals are most abundant in talus, in rock crevices, and along rocky stream margins with moist microhabitats.

Microhabitat use varies among size classes of animals on Vancouver Island (Ovaska and Gregory 1989). Animals measuring ≤30 mm SVL tend to use small rocks and leaf litter more than larger animals, which prefer large rocks and other cover objects. Overall, surface densities are 0.30–1.16 salamanders/m^2, and they are highest in March–June and September–October. Juveniles are more active during the dry summer months than are adults. The adults retreat underground except when they briefly resurface following occasional summer rains.

Western red-backed salamanders are capable of detecting chemical signals from the fecal pellets and bodies of conspecifics; however, the function of these signals is not as apparent as in eastern species such as *P. cinereus*. Both males and females will avoid airborne odors of conspecific males, but not those of females (Ovaska 1988a). This finding suggests that chemical signals may be used to space out males, perhaps as they compete for females during the mating season. Although chemical signals could be used to mark feeding territories, there is little evidence that they function in this regard.

Individuals show a high degree of site specificity in the field, and adults maintain small home ranges of only a few square meters (Ovaska 1988b). The fact that adults frequently share cover objects and are rarely aggressive suggests that this species is not territorial. Adults can recognize the fecal odors of conspecifics, but do not avoid burrows with fecal pellets (Ovaska and Davis 1992). Thus the adaptive value of fecal recognition is not readily evident.

Ovaska (1989) found that the nose-tapping rates of both males and females from Vancouver Island are greater when individuals are exposed to chemical odors from conspecifics from the mainland, compared with those from a second population on the island. It is not known whether this observation reflects pheromonal divergence or simply attraction to the odors of strangers.

The diet consists of small invertebrates. The most numerically important prey in a sample of Oregon specimens were collembolans, mites, and other tiny invertebrates (Dumas 1956). Other prey include earthworms, snails, isopods, millipedes, centipedes, pseudoscorpions, spiders, phalangids, beetles, scorpionflies, dipterans, lepidopterans, and ants and other hymenopterans.

PREDATORS AND DEFENSE. Natural predators are poorly documented. Dumas (1956) found a dead *P. vehiculum* in a tunnel made by a shrew-mole (*Neurotrichus gibbsi*). He also observed an American dipper (*Cinclus mexicanus*) capture either a *P. vehiculum* or a *P. dunni* from among rocks adjoining a creek. Carabid beetles prey upon small juveniles in the laboratory (Ovaska and Smith 1988). The beetles are equally likely to attack salamanders of all sizes, and salamanders rarely posture defensively when approached by beetles. Field observations suggest that beetle predation on juveniles also occurs in nature.

COMMUNITY ECOLOGY. Dumas (1956) compared microsympatric populations of *P. dunni* and *P. vehiculum* in Oregon and found that the two species differ substantially in diet and microhabitat preferences. Competition for food seems unlikely, although competition for moist microhabitats might occasionally occur during periods of dry weather. Western red-backed salamanders do not respond to the fecal pellets of *P. vandykei* (Ovaska and Davis 1992). This finding suggests that these species do not compete for space via interference competition.

CONSERVATION BIOLOGY. *Plethodon vehiculum* is well suited for inhabiting a variety of forest habitats and is often common in forest stands of all ages (Aubry and Hall 1991; Aubry et al. 1988; Corn and Bury 1991). This is one of the few species of salamanders in the Pacific Northwest that appear to thrive in young forests.

Plethodon websteri Highton
Southern Zigzag Salamander
PLATE 142

Fig. 258. *Plethodon websteri*; striped morph; Lee Co., Alabama (R. W. Van Devender).

IDENTIFICATION. *Plethodon websteri* is a sibling species of *P. dorsalis*, and the description of the latter species generally applies to *P. websteri*. Both striped and unstriped morphs occur in many populations. The striped morph has a wavy yellowish brown to orangish red dorsal stripe that extends from the head to the tail tip. The unstriped or dark morph lacks the dorsal stripe but may have scattered red pigmentation on the dorsum. Intermediates between these two extremes are common in many populations. All color morphs have tiny silvery white spots and brassy flecks dorsally that often produce a frosted appearance. The belly is mottled with fine black, white, and orange coloration. The modal number of costal grooves is normally 18, but it is 19 in South Carolina populations (Highton 1986e).

Sexually active males have nasolabial swellings and small, round to oval mental glands. Adults measure 69–82 mm TL, and males and females have about the same average SVL (Semlitsch and West 1983). Hatchlings reach 12–13 mm SVL (Semlitsch and West 1983). Hatchlings and small juveniles generally resemble the adults, but have proportionately smaller tails.

This species can be easily confused with *P. dorsalis*. In *P. websteri* the dorsal stripe tends to be wavy only on the anterior half of the body. Unfortunately, some *P. dorsalis* also show this pattern, so the trait is not diagnostic. Lazell and Mann (1991) note that Alabama and Mississippi specimens of *P. websteri* have paler sides and bellies and more boldly ringed fingers and toes than those of *P. dorsalis*. Despite some slight differences in external traits, *P. websteri* is best distinguished from *P. d. dorsalis* by its geographic distribution and electrophoretic characteristics.

SYSTEMATICS AND GEOGRAPHIC VARIATION. Highton (1979) and Larson and Highton (1978) found that populations of *P. dorsalis* in the southern portion of this species' range are highly differentiated genetically from populations to the immediate north (estimates of Nei's $D = 1.5$–2.0). These southern populations are now recognized as *P. websteri*. Although remarkably dissimilar genetically, *P. websteri* and *P. dorsalis* are virtually identical in their morphology, behavior, and ecological requirements.

The proportion of striped and unstriped morphs

Range of *Plethodon websteri*

varies among local populations, particularly in the zone of contact with *P. dorsalis*. The ranges of *P. dorsalis* and *P. websteri* are allopatric except for one contact zone in Jefferson Co., Alabama, where the two occur microsympatrically over a distance of about 1 km (Highton 1979, 1985). In the zone of contact, character displacement appears to have occurred in color morphs; most *P. dorsalis* are unstriped, whereas most *P. websteri* are striped. Elsewhere, both species tend to have similar degrees of polymorphism within local populations. No subspecies are recognized.

DISTRIBUTION AND ADULT HABITAT. Populations of *P. websteri* are more or less continuously distributed in central Alabama and western Georgia and occur as geographic isolates in central Mississippi, southeastern Louisiana, southwestern Alabama, and western South Carolina (Highton 1986e). This species inhabits mesic deciduous forests and can be collected beneath rocks, logs, and other surface cover during the autumn, winter, and spring months (Blaney and Relyea 1967; Ferguson and Rhodes 1958).

BREEDING AND COURTSHIP. *Plethodon websteri* breeds primarily in winter. In a South Carolina population the adults breed from January through March, a period when the vasa deferentia are packed with sperm and mental glands reach their maximum seasonal enlargement (Semlitsch and West 1983). Both sexes reproduce annually and sex ratios are near 1:1. Courtship behavior has not been described, but it presumably involves a tail-straddle walk similar to that of other *Plethodon* species.

REPRODUCTIVE STRATEGY. South Carolina females oviposit in June or July shortly after moving to subsurface summer retreats (Semlitsch and West 1983). Ovarian follicles are smallest in October and largest in May and June, when average diameters reach 2.6 mm. Brooding and oviposition presumably occur between June and September since hatchlings and spent females are present in October. The number of mature ova varies from 3 to 8, averages 5.8, and is positively correlated with female SVL.

TERRESTRIAL ECOLOGY. Semlitsch and West (1983) have provided the only comprehensive life history study of *P. websteri* to date. At their site in South Carolina adults are active on the ground surface from October through May and move to underground retreats during the warmest months of the year. Growth rates during the active season average 1.3, 1.1, and 0.33 mm SVL/month for 1-year-old, 2-year-old, and adult age classes, respectively. Growth rates during the first year of life average 10 mm SVL and there is little or no growth over the summer months.

Males and females become sexually mature during the summer, about 21–26 months after oviposition. Males and females first court the following winter or spring, and females oviposit for the first time several months later. Males become sexually mature when they reach 29–30 mm SVL and females when they reach 28–31 mm SVL.

Prey in 34 specimens from Georgia and Alabama included snails, annelids, mites, spiders, pseudoscorpions, centipedes, isopods, and a variety of insects (Camp and Bozeman 1981). Mites and other small prey make up a greater percentage of the diet in smaller individuals than in larger ones.

PREDATORS AND DEFENSE. The natural predators of *P. websteri* are undocumented but probably include skunks, opossums, small woodland snakes, birds, and shrews.

COMMUNITY ECOLOGY. *Plethodon websteri* and *P. dorsalis* have parapatric distributions except for a very narrow zone of overlap in Alabama (Highton 1985). This pattern suggests that one species may be excluding the other via competition or other mechanisms.

CONSERVATION BIOLOGY. The optimal habitat for this species is mature hardwood forest with rocky substrates and large amounts of coarse, woody debris. Deforestation and the conversion of deciduous forests to pine monocultures maintained on short harvesting cycles have adversely affected many populations of this and other small woodland salamanders.

Plethodon wehrlei (Fowler and Dunn) Wehrle's Salamander
PLATE 143

Fig. 259. *Plethodon wehrlei;* Randolph Co., West Virginia (R. W. Van Devender).

IDENTIFICATION. Wehrle's salamander is a large *Plethodon* species that has a modal number of 17 costal grooves and hindfeet that are more strongly webbed than those of other large eastern *Plethodon* species. The dorsal ground color is dark brown to brownish black and the sides of the body are heavily marked with bluish white to yellow spots that may fuse to form blotches or irregular bands. Small, widely scattered white spots occur on the dorsum. In southern populations, hatchlings and small juveniles often have paired red spots on the dorsum compared with numerous brassy flecks on the adults (Grobman 1944; Highton 1962a, 1987). In many instances the spots are retained by the adults (Brooks 1945; Newman 1954), or adults may have paired yellow spots on the back. The venter is uniformly gray, except for the throat and upper chest, which often are blotched with white or yellowish pigments. Adult males have cloacal papillae and enlarged mental glands, and are slightly smaller on average than adult females. Adults measure 10–17 cm TL.

SYSTEMATICS AND GEOGRAPHIC VARIATION. *Plethodon wehrlei* is a monotypic species, but several unusual geographic variants have been described. Local populations in southwestern Virginia were described as *P. dixi* (Pope and Fowler 1949) and *P. jacksoni* (Newman 1954). However, Highton (1962a) believed both to be geographic variants of *P. wehrlei* and synonymized them with the latter species. The *dixi* type is found in Dixie Caverns and Blankenship Cave in Roanoke Co. and has a dorsum that is heavily frosted with light flecks and brassy markings. Adult *jacksoni*

Range of *Plethodon wehrlei*

types described by Newman (1954) differ from typical specimens in having red spotting on the dorsum and a more prominent mental gland. An isolated rockface population of *P. wehrlei* that has paired yellow spots on the dorsum occurs in Letcher Co., Kentucky (Cupp and Towles 1983). This form has also been collected in Summers Co., West Virginia, and Campbell Co., Tennessee (Redmond and Jones 1985).

DISTRIBUTION AND ADULT HABITAT. Wehrle's salamander is largely confined to forested hillsides in mountainous or hilly terrain in the Appalachian Plateau from southwestern New York southward to western Virginia and northwestern North Carolina. Geographic isolates consisting of yellow-spotted individuals occur in southeastern Kentucky and northeastern Tennessee (Cupp and Towles 1983; Redmond and Jones 1985). Ohio records for this species may not be valid (Pfingsten 1989a).

Populations occur along forested slopes in both red spruce–yellow birch and mixed deciduous forests. Specimens occur in West Virginia at elevations of 183–1463 m (Green and Pauley 1987; Pauley and England 1969). Forests with rock ledges or abundant, flat rocks are ideal habitats for this species (Bishop 1941a; Brooks 1945; Hall and Stafford 1972; Hassler 1932). Specimens have occasionally been taken in caves (e.g., Cooper 1961; Holsinger 1982; Netting et al. 1946; Pope and Fowler 1949; Reese 1933) and can be an important cave element in some areas in West Virginia (Green and Pauley 1987).

Wehrle's salamander is often abundant on dry slopes that are generally unsuitable for other *Plethodon* species. Pauley (1978a) found that *P. wehrlei* is more abundant than *P. cinereus* in relatively dry microhabitats at a West Virginia site.

BREEDING AND COURTSHIP. Mating occurs from autumn through spring depending on the geographic location. New York and Pennsylvania populations breed in September and October (Hall and Stafford 1972), but in West Virginia most adults breed in March and April (Pauley and England 1969). Courtship behavior has not been described.

REPRODUCTIVE STRATEGY. Females presumably oviposit in deep underground cavities since eggs have never been found on the ground surface (Green and Pauley 1987). The only nesting record is for a female found brooding a clutch of six eggs on 19 August in a small cavity in a Virginia cave (Fowler 1952). The ova average 5.0 mm in diameter, are creamy white and unpigmented, and are surrounded by the vitelline membrane and two jelly envelopes.

The eggs are laid in grapelike clusters and adjoining eggs may be loosely fused together by tubular extensions of the outer envelope. The outer capsule averages 6.0 mm in diameter.

The average size of maturing ova in New York and Pennsylvania specimens increases between April and May (Hall and Stafford 1972). Some specimens in November have ova as large as 4.0 mm in diameter, whereas females collected in April and May are spent or have small ova. These data suggest that females oviposit sometime in midwinter through early spring in northern populations. Similar analyses of ovarian egg sizes indicate that West Virginia females oviposit in March or April (Pauley and England 1969).

Number of mature ova varies from 7 to 24 and is positively correlated with female SVL (Hall and Stafford 1972). Respective mean ovarian complements in New York and Pennsylvania populations are 11 and 16 eggs. This difference is significant and reflects a difference in average female size. Most females in New York reproduce biennially, but many Pennsylvania females reproduce annually. Males breed annually in all populations.

TERRESTRIAL ECOLOGY. In New York and Pennsylvania populations, individuals grow about 20 mm SVL during the first year of life (Hall and Stafford 1972). During their second, third, and fourth years, males grow 11, 10, and 9 mm SVL, respectively. Females grow 9–10 mm SVL during their second year of life. Males begin maturing sexually during their third year, and first breed during their fourth year when >61 mm SVL. Females probably do not breed until their fourth or fifth year of life.

Like most other *Plethodon* species, Wehrle's salamander remains beneath cover during the day and actively forages on the ground surface at night. In West Virginia, individuals occasionally climb tree trunks, and seasonal surface activity varies with elevation (Green and Pauley 1987). In high-elevation spruce forests, adults do not emerge from hibernation until late April and are active on the surface until mid-October. At lower elevations, individuals emerge from hibernation in March and are active on the surface until early June. Individuals remain beneath ground during the summer months and resume surface activity from late September until the arrival of cold weather.

The estimated density in favorable habitats at one Pennsylvania site is 1000 salamanders/ha or 0.1 salamander/m^2 (Hall and Stafford 1972). In some samples, males outnumber females, a fact that may reflect sexual differences in surface activity during the brooding season. Individuals will actively defend territories from other large *Plethodon* species and in some instances may bite the tails off smaller *Plethodon* species (Thurow 1976).

Ants were the dietary mainstay of 83 juveniles and adults from West Virginia (Pauley 1978b). Other prey included mites, spiders, beetles, and collembolans. The gut contents of 191 Pennsylvania specimens included annelids, gastropods, millipedes, centipedes, isopods, phalangids, spiders, mites, collembolans, orthopterans, homopterans, hemipterans, beetles, flies, and unidentified insect larvae (Hall 1976). Insect larvae, centipedes, spiders, weevils, and orthopterans make up the bulk of summer prey, and the relative importance of prey groups varies both seasonally and ontogenetically. Bishop (1943) lists lepidopteran larvae, ants, a cricket, beetles, craneflies, hymenopterans, aphids, and mites in New York specimens.

PREDATORS AND DEFENSE. Very few data are available on natural predators. The ringneck snake (*Diadophis punctatus*) is the only known predator (Hall and Stafford 1972).

COMMUNITY ECOLOGY. Pauley (1978b) examined the diets of *P. cinereus* and *P. wehrlei* at a West Virginia site where the two species occur microsympatrically. The diets of the two species overlap widely, with ants being the major shared resource. Although these species coexist locally, they tend to sort out microspatially, with *P. wehrlei* preferring drier microhabitats. Thus competition between these species appears to be minimal.

CONSERVATION BIOLOGY. The systematic affinities of yellow-spotted isolates in Kentucky, Tennessee, and West Virginia have not been determined. These populations should receive high priority for protection by conservation biologists and land managers.

Plethodon welleri Walker
Weller's Salamander
PLATE 144

Fig. 260. *Plethodon welleri*; Watauga Co., North Carolina (R. W. Van Devender).

IDENTIFICATION. Weller's salamander is a small eastern species of *Plethodon* with a modal number of 16 costal grooves and a black dorsum that is heavily flecked or blotched with brassy pigmentation. The belly is dark and finely spotted with white, except for populations on Grandfather Mountain, North Carolina, which tend to have uniformly dark bellies (Thurow 1956b; Walker 1934). Males develop small, slightly oval mental glands during the breeding season and have papillose vent linings (smooth in females). Adults measure 64–92 mm TL.

Hatchlings lack the heavy brassy patterning of the adults, but develop pairs of brassy spots along the anterior portion of the back within a few days after hatching (Organ 1960b). Hatchlings reach 12.5–15.0 mm SVL and 18.5–23.5 mm TL (Organ 1960b).

SYSTEMATICS AND GEOGRAPHIC VARIATION. Individuals on Grandfather Mountain have less brassy pigmentation on the dorsum and a less spotted belly than those from elsewhere in the range. Thurow (1956b) described the Grandfather Mountain population as a subspecies (*P. w. welleri*) separate from the remaining populations (*P. w. ventromaculatus*). Larson and Highton (1978) found that genetic similarities among subspecies are greater than those between two populations of *P. w. ventromaculatus*. Here I do not recognize subspecies because of discordance between color patterning and underlying genetic variation.

DISTRIBUTION AND ADULT HABITAT. Weller's salamander inhabits mesic deciduous and spruce-fir forests from the Mt. Rogers region of southwestern Virginia southward to portions of western North Carolina and eastern Tennessee (Hoffman 1953; Thurow 1956b). Most populations appear to be isolated mountaintop relicts of an ancestral form that was more widespread (Thurow 1956b). This species is strongly associated with high-elevation spruce-fir forests that occur above 1500 m, but occasionally is found in rich hardwood forests as low as 701 m (Thurow 1963). Populations are often found in or near talus and have been collected beneath small logs, pieces of bark, and rocks.

BREEDING AND COURTSHIP. Specimens from southwestern Virginia will court in the laboratory only during October (Organ 1960b). Males collected in the autumn have well-developed, circular mental glands, but those collected in other months do not. These observations suggest that Virginia populations court in the autumn shortly before moving to underground winter hibernacula. Thurow (1963) found males with their vasa deferentia swollen with sperm in both April and September, suggesting that a secondary breeding bout may occur during the spring in more southern populations.

The only published description of courtship is that of Organ (1960b). The following is a summary of his account.

Males become restless shortly before courting and begin actively moving about. When a male contacts a female he immediately begins to undulate his tail back and forth about 1.5 times per second. The male then places his nasolabial projections and mental gland on the female and slowly advances toward her head while maintaining contact with her skin. After reaching her head, the male breaks contact and circles under the female's chin. As he passes under her, he holds his back and undulating tail in contact with her chin. As his tail passes under her chin, the male arches the base of his tail and begins a series of vertical undulations that alternate with horizontal undulations. The female then straddles the male's tail and

Range of *Plethodon welleri*

presses her chin against the base of his undulating tail. An abbreviated tail-straddle walk occurs that rarely covers a distance of more than 8 cm.

Just prior to spermatophore deposition the male begins a series of lateral sacral rocking movements. The female responds by moving her head counter to the direction of the male's movement. The male then lowers his vent to the substrate and deposits a spermatophore. He raises his vent, arches the base of his tail upward, and flexes the tail sharply to one side of the female. The female keeps her chin pressed against the bent tail of the male, and the two move forward rapidly until the vent of the female is positioned over the spermatophore. The female then picks up the sperm cap with her cloacal lips, leaving the spermatophore stalk behind.

Males will attempt to court both sexes. Females that are approached by courting males either flee or remain passive. Males approached by courting males will often turn and viciously bite approaching males on the legs or tail. Males appear to use both olfactory and visual cues to locate other salamanders (Organ 1960b).

The spermatophore is mushroom shaped and consists of a pale yellow sperm cap and a colorless, jellylike stalk. The stalk has a spike at the top that supports the cap. The cap is laterally compressed and appears pinched inward on the anterior end. The base of the stalk is about 2 mm in diameter and the entire spermatophore is 4 mm high.

REPRODUCTIVE STRATEGY. Nests have rarely been found and the limited evidence to date suggests that females oviposit in May or June. Organ (1960b) found three nests on 16 August in southwestern Virginia with late-term embryos, and five additional nests between 18 August and 5 September with hatchlings and empty egg capsules. The surface nests at this site are located in well-rotted conifer logs between the upper surface of the log and moss mats. Nests may be placed within 30 cm of each other but are usually found in separate logs. Females coil about the eggs or young and apparently remain with their young until they disperse from the nests. The range of eggs in three nests was 5–10, whereas the range of hatchlings in five nests was 4–11.

Eggs are deposited in a tight, grapelike cluster and are usually suspended by a tough, whitish stalk. Individual eggs are attached either directly to the stalk or to other eggs. The outer capsule of individual eggs is 5.5–6.5 mm in diameter and the embryos are surrounded by the vitelline membrane and two jelly capsules. The outer capsule is much thicker than the inner capsule and its outside surface is adhesive. Late-term embryos have three pairs of antlerlike gills that are lost at or near hatching.

TERRESTRIAL ECOLOGY. Most aspects of the natural history of the juveniles and adults are undocumented. Males reach sexual maturity when they measure >31 mm SVL and females when they measure >35 mm SVL (Highton 1962a; Thurow 1963). Most individuals mature when about 3 years old.

During wet or humid nights individuals forage on the ground surface for small invertebrates. Dietary items include pseudoscorpions, mites, spiders, beetles, lepidopteran larvae, hemipterans, flies, and shed skin (Thurow 1963). The adults do not appear to be aggressive outside the breeding season (Thurow 1976).

PREDATORS AND DEFENSE. *Plethodon welleri* produces noxious skin secretions and has been hypothesized to be part of a Batesian mimicry complex involving members of the *D. ochrophaeus* complex (Brodie 1981; Brodie and Howard 1973). Details are provided in the account of *P. jordani*.

COMMUNITY ECOLOGY. No data are available on the community ecology of *P. welleri*.

CONSERVATION BIOLOGY. Because of the species' affinity for high elevations, most populations of *P. welleri* are protected to some degree. The die-off of spruce-fir forests in the southern Appalachians may constitute the greatest long-term environmental threat to this species.

Plethodon yonahlossee Dunn
Yonahlossee Salamander
PLATE 145

Fig. 261. *Plethodon yonahlossee;* Buncombe Co., North Carolina (J. W. Petranka).

IDENTIFICATION. *Plethodon yonahlossee* is the largest and most distinctive of the large eastern series of *Plethodon*. The dorsum ground color is black and the dorsum is covered with blotches of light reddish brown to red pigment. The blotches usually fuse in older animals to form a more or less continuous wide band from the neck to the base of the tail. In some individuals, particularly from the southern portion of the range, the red may be reduced to scattered flecks or blotches. The sides of the head, body, and tail are strongly marked with gray to silvery white blotches that often fuse to form a band along the sides of the body. The head, tail, and limbs are black above with lesser amounts of white or gray flecking. The tails of this and other large eastern *Plethodon* species produce slimy secretions that deter certain predators.

The average SVL of females is 7% greater than that of males, and sexually active males have a snout that is more swollen around the nasolabial grooves than that of females (Pope 1950). Adults measure 11–22 cm TL and have a modal number of 16 costal grooves. Young juveniles have four to six pairs of red spots on a dark ground color and a light belly. The red spots eventually fuse so that older juveniles resemble the adults in general coloration.

SYSTEMATICS AND GEOGRAPHIC VARIATION. A long-legged, crevice-dwelling form of *P. yonahlossee* found at low elevations in western North Carolina was described by Adler and Dennis (1962) as a separate species, *P. longicrus*. Morphological and electrophoretic comparisons of *P. longicrus* and *P. yonahlossee* by Guttman et al. (1978) suggest that *P. longicrus* is only a geographic variant of *P. yonahlossee;* however, others have continued to recognize *P. longicrus* as a valid spe-

Range of *Plethodon yonahlossee*

cies. Nei's genetic distances (*D*) between regional populations of this group that range from Grandfather Mountain, North Carolina, and Unaka Mountain, Tennessee, southward to the vicinity of Bat Cave, North Carolina, vary from 0.001 to 0.08 and are within the range seen within a single species. Here I follow the interpretation of Guttman et al. (1978) and treat *P. longicrus* as being conspecific with *P. yonahlossee*.

DISTRIBUTION AND ADULT HABITAT. The yonahlossee salamander inhabits deciduous forests in mountainous terrain at elevations of 436–1737 m (Guttman et al. 1978). Individuals are active on the ground surface at night and can be found by turning stones and large logs on the forest floor. Martof et al. (1980) note that this species is often locally abundant on hillsides where rockslides are carpeted with mosses and ferns. Hairston (1949) reports that *P. yonahlossee* in the Black Mountains of western North Carolina is generally found within 30 m of streams and is most common in virgin forests. I have collected this species in a variety of forested habitats that are far removed from streams and find that the preferred habitats are similar to those of *P. glutinosus* and *P. jordani*.

BREEDING AND COURTSHIP. Little information is available on the time and duration of the mating season. Pairs of males and females in breeding condition have been found in August on White Top Mountain, Virginia (Pope 1950). Detailed descriptions of courtship behavior have not been published. Presumably courtship is very similar to that of other large eastern *Plethodon* species, such as *P. glutinosus* and *P. jordani*, and involves a tail-straddle walk followed by spermatophore deposition and pickup.

REPRODUCTIVE STRATEGY. Most aspects of the nesting ecology are poorly documented. Females probably deposit their eggs underground during late spring or early summer and the embryos hatch in 2–3 months. Ovarian complements for three females examined by Pope (1950) were 19, 24, and 27 mature ova.

TERRESTRIAL ECOLOGY. The few data that are available suggest that juveniles become sexually mature when about 3 years old and that males mature when they reach 56 mm SVL (Pope 1950). Juveniles and adults emerge from cover at night and forage on the forest floor. Animals may either pose at burrow entrances with their heads exposed or move about at the bases of trees or along fallen logs and stumps in search of prey (Gordon et al. 1962). Young animals are most abundant on the surface for about 1 hour after sunset; adult activity peaks 2–3 hours after sunset. Individuals seem to prefer large logs with a thick layer of litter near the ground. Juveniles and adults will aggressively defend territories that they establish in terraria (Thurow 1976), but it is uncertain if they maintain territories in natural communities.

Food items of individuals from two sites in North Carolina include spiders, mites, pseudoscorpions, millipedes, centipedes, earthworms, gastropods, nematodes, and a wide variety of insects (Rubin 1969). This study suggests that crevice-dwelling populations consume a less diverse assemblage of prey than forest-floor populations. Dietary items in Virginia specimens include collembolans, orthopterans, termites, homopterans, hemipterans, lepidopterans, dipterans, coleopterans, hymenopterans, pseudoscorpions, spiders, mites, millipedes, centipedes, isopods, and snails (Pope 1950). Comparisons of dietary items with those of syntopic *P. glutinosus* indicate that the diets of these two species are very similar. One animal in a terrarium ate a smaller *P. jordani* (Thurow 1976).

PREDATORS AND DEFENSE. Predators and defensive mechanisms have not been documented. *Plethodon yonahlossee* produces slimy tail secretions that are noxious to birds and other predators.

COMMUNITY ECOLOGY. *Plethodon yonahlossee* often coexists locally with other large *Plethodon* species (*P. glutinosus* and *P. jordani*) as well as with other salamander species. The large *Plethodon* species appear to have similar life-styles and diets, but none seems capable of competitively excluding *P. yonahlossee*. An experimental analysis of interactions between these species would provide insights into mechanisms that allow coexistence in the diverse salamander communities found in the southern Appalachians.

CONSERVATION BIOLOGY. The yonahlossee salamander is locally abundant in many areas of its range, but it is generally less abundant than other large species of *Plethodon*. Clear-cutting can eliminate or greatly reduce local populations of this and other *Plethodon* species in the southern Appalachians (Petranka et al. 1993, 1994). Less intensive harvesting practices that leave the basic structure of the forest intact would benefit this and many other vertebrates in southern Appalachian forests.

Family Proteidae
Waterdogs and Mudpuppy

The Proteidae is a small family of perennibranch salamanders of which fossils have been found extending back to the Paleocene. There are two extant genera (*Necturus* and *Proteus*) consisting of six species. *Necturus* contains five species and is restricted to eastern North America. Fossil *Necturus* are known from the Paleocene of Canada and the Pleistocene of Florida (Hecht 1958). *Proteus* is a cave form that occurs in southern Europe; fossils of this form have been found from the Pleistocene of Germany (Duellman and Trueb 1986).

The proteids are relatively large salamanders that may reach 30–40 cm TL. Members of this family lack maxillae and have only two gill slits. The prevomerine teeth are in lateral rows and parallel the premaxillary teeth. *Necturus* inhabits lakes and large streams. The adults are larviform with caudal tail fins and three pairs of conspicuously bushy gills. Adults also possess lungs and have only four toes on each rear foot.

Necturus alabamensis (Viosca)
Alabama Waterdog
PLATE 146

Fig. 262. *Necturus alabamensis*; adult; Wakulla Co., Florida (R. W. Van Devender).

IDENTIFICATION. The Alabama waterdog is a medium-sized species of *Necturus* that usually lacks spotting on the venter. The dorsum varies from reddish brown to nearly black. In some populations spotting is present on the dorsum, particularly on young animals. There are four toes on each hindlimb, the tail is compressed laterally, and the gills are red and bushy. Sexually active males have swollen cloacae and two enlarged cloacal papillae that project posteriorly. Adults measure 15–22 cm TL.

Hatchlings are mottled above and have a few light dorsal spots. Juveniles lack the prominent stripes characteristic of *N. maculosus*, except for populations in the West Sipsey Fork of the Black Warrior River that have striped larvae. The Alabama waterdog is geographically variable in degree of spotting and dorsal coloration and is best identified by its geographic range.

SYSTEMATICS AND GEOGRAPHIC VARIATION. The genus *Necturus* has proven to be a challenge for systematists because all members are permanently gilled and have rather generalized larval morphology with few derived characters. *Necturus lewisi*, *N. maculosus*, and *N. punctatus* are well differentiated based on electrophoretic, karyological, and immunological data, and are clearly valid species (Ashton et al. 1980; Guttman et al. 1990; Maxson et al. 1988; Sessions and Wiley 1985). The taxonomic status of *Necturus* in Alabama, western Florida, western Georgia, Louisiana, Mississippi, and eastern Texas has been addressed by several researchers (see historical reviews by Bart et al. 1997; Gunter and Brode 1964; Hecht 1958; Maxson et al. 1988; and Mount 1975). Earlier field guides (Behler and King 1979; Conant 1975) followed the interpretations of some authors (Neill 1963; Viosca 1937) in recognizing the western spot-bellied and eastern immaculate-bellied forms in these states as two species, *N. alabamensis* and *N. beyeri*. However, Gunter and Brode (1964) and Hecht (1958) found evidence of hybridization between these forms in Mississippi and recognized them as subspecies. Mount (1975) found that none of the morphological criteria useful in recognizing these forms applied well to Alabama specimens as a whole. Of special interest are populations in the West Sipsey Fork of the Black Warrior River that have striped larvae and adult patterning that is more characteristic of *N. maculosus*.

Maxson et al. (1988) found that populations referable to *N. alabamensis* in Alabama and adjoining states can be divided into two groups based on microcomplement fixation data. One is indistinguishable from *N. maculosus* (including the Sipsey Fork population from the Black Warrior River system), whereas the second is clearly distinct. However, the phylogenetic relationships between *N. alabamensis*, *N. beyeri*, and *N. maculosus* cannot be resolved using this technique.

The most recent study using electrophoretic evidence suggests that *N. alabamensis* and *N. beyeri* are distinct species and that the Sipsey Fork population is most closely allied with *N. alabamensis* (Guttman et al. 1990). Studies under way suggest that the striped form restricted to the upper Black Warrior River system may be sufficiently distinct to recognize as a separate species (Bart et al. 1997). Because the type specimen of *N. alabamensis* appears to be this striped form, populations currently referable to *N. alabamensis* in the lower Coastal Plain of southwestern Georgia, western Florida, southern Alabama, and southeastern Mississippi may eventually be assigned a new name. Here I follow the interpretations of Guttman et al (1990) and recognize five species of *Necturus*. The striped form restricted to the upper Black Warrior River system is

Range of *Necturus alabamensis*

tentatively assigned to *N. alabamensis* until sufficient data become available to assess its taxonomic status definitively. Genetic evidence suggests that populations currently assigned to *N. beyeri* may contain more than one species (Bart et al. 1997), so future revisions of this taxon are also likely.

DISTRIBUTION AND ADULT HABITAT. The Alabama waterdog is most common in medium to large streams in eastern Mississippi, north-central to southern Alabama, western Georgia, and the panhandle of Florida. Neill (1963) notes that this species seems to prefer rivers and streams with logjams or other debris that provide cover and oviposition sites for the adults.

COMMENTS. Virtually no published data are available on the life history of *N. alabamensis*, and a detailed life history study is needed. Adults in breeding condition have been collected from December to February, and individuals are most active during the winter months (Neill 1963).

Necturus beyeri Viosca
Gulf Coast Waterdog
PLATE 147

IDENTIFICATION. The Gulf Coast waterdog is a medium-sized *Necturus* species that is brownish above and light brown below. Numerous conspicuous dark brown to black spots occur on the back, sides, and undersides of the body. Spots on the venter are smaller than those on the dorsum (Viosca 1937). There are four toes on each hindlimb, the tail is compressed laterally, and the gills are red and bushy. Sexually active

Fig. 263. *Necturus beyeri;* adult; Nacogdoches Co., Texas (R. W. Van Devender).

Range of *Necturus beyeri*

males have swollen cloacae and two enlarged cloacal papillae that project posteriorly. Adults measure 16–22 cm TL. Hatchlings are mottled above and have a few dorsal light spots. Juveniles are spotted and lack the prominent stripes characteristic of *N. maculosus*.

SYSTEMATICS AND GEOGRAPHIC VARIATION. The taxonomic status of *N. alabamensis* and *N. beyeri* has been a source of controversy for many years, but recent electrophoretic evidence suggests that these two forms are distinct species (Guttman et al. 1990). A detailed summary of research on the systematics of *Necturus* is given in the account of *N. alabamensis*.

DISTRIBUTION AND ADULT HABITAT. The Gulf Coast waterdog occurs as two isolated groups. One is found from eastern Texas to central Louisiana and the second from southeastern Louisiana to central Mississippi. This species is most common in medium to large streams. Logjams and leaf beds are important habitats for the adults.

BREEDING AND COURTSHIP. Most of the information on the life history of *N. beyeri* comes from a detailed study of populations in southern Louisiana (Shoop 1965b). Males in this region have greatly swollen cloacae and sperm in the vasa deferentia in December and January and only slightly swollen cloacae in April. In May males are no longer in breeding condition. The only record of a female containing a spermatophore is for a specimen collected on 28 December. These observations suggest that mating occurs primarily in late autumn through early winter. All females collected from December to May from this site have sperm in their spermathecae, and individuals are capable of storing viable sperm for at least 6 months after mating (Sever and Bart 1996). Published accounts of courtship behavior are not available. Sex ratios of juveniles and adults do not differ significantly from 1:1 (Shoop 1965b).

REPRODUCTIVE STRATEGY. Most information of the reproductive biology of *N. beyeri* comes from studies of a population in Talisheek Creek in St. Tammany Parish, Louisiana. Females at this site attach their eggs singly to the undersides of large boards, railroad ties, and pine logs embedded in sandy sections of the stream (Shoop 1965b). Freshly deposited eggs are yellowish to yellowish green and are surrounded by three jelly membranes. The eggs form oblong monolayers that measure 10 × 27 cm on average, and the nests can be found 15–65 cm below the water's surface. Ten embryos in early neural fold stages measured 5.6–5.8 mm in diameter, and the attached egg capsules were 12–19 mm long and 7–9 mm wide. Embryos maintained in the laboratory hatch after slightly more than 2 months when they reach 13–16 mm SVL.

Females in Talisheek Creek oviposit in late April and May (Sever and Bart 1996; Shoop 1965b). Nesting records include five nests without females discovered on 4–6 May (Shoop 1965b) and four nests with brooding females discovered on 22 May (Sever and Bart 1996). The number of eggs in nests without females was 4–40, compared with 26–37 in nests with females (mean = 32). The number of mature ova in 13 females ranged from 28 to 76 and averaged 47.5 (Sever and Bart 1996). The number of mature ovarian eggs in 25 specimens from the Pearl River system varied from 37 to 67, averaged 57, and was independent of female SVL (Shoop 1965b). The largest ovarian eggs were 4.6–6 mm in diameter.

AQUATIC ECOLOGY. Males and females in southern Louisiana become sexually mature when they reach 112–123 mm SVL and 115–135 mm SVL, respectively (Shoop 1965b). The time required to reach sexual maturity is not known, but probably is on the order of 4–6 years based on studies of other *Necturus* species (e.g., Cooper and Ashton 1985). Adults live at least 6–7 years in nature (Bart and Holzenthal 1985).

Hatchlings and young juveniles frequent bottom debris and leaf litter where the current is slow and food is concentrated. The adults use both leaf litter and logjams as cover. At a southern Louisiana site, *N. beyeri* almost exclusively uses leaf bed habitats in sluggish sections of streams (Bart and Holzenthal 1985). Catch per unit of effort is highest during the colder months of the year and decreases progressively with the arrival of warm summer weather. The authors surmise that *N. beyeri* aestivates in burrows during the summer and autumn, when invertebrate prey are scarce in leaf litter beds.

At another Louisiana site, most adults are found beneath logs, flood debris, and other obstructions in slow-moving sections of the stream (Shoop and Gunning 1967). Some animals live in burrows in the streambanks. During the colder months a few animals venture forth at night into shallow open waters, presumably to feed. Most individuals remain within a restricted portion of the stream over long periods of time. Mean movements between captures are 10 m for females, 16 m for males, and 25 m for juveniles. All recaptures are within 64 m of the initial release point, even after as many as 724 days.

Fig. 264. *Necturus beyeri;* larva; St. Tammany Parrish, Louisiana (R. W. Van Devender).

Crayfish are the dietary staple of one Louisiana population (Shoop and Gunning 1967), whereas prey in 90 specimens from a southern Louisiana population included isopods, amphipods, mayflies, caddisflies, dragonflies, dytiscid beetles, midges, and sphaeriid clams (Bart and Holzenthal 1985). Individuals at the latter site eat a greater number and diversity of prey during the coldest months of the year, and progressively fewer prey with the onset of warmer weather. Many specimens collected between July and October have empty guts. These trends parallel a decrease in the number of invertebrates in leaf litter beds where the animals feed. Juveniles feed on isopods, amphipods, chironomids, and a burrowing mayfly. As juveniles grow, they incorporate larger and more diverse prey into their diets. Dietary analyses suggest that adults often forage away from leaf litter beds along logs and in sandy substrates.

PREDATORS AND DEFENSE. Fish and crayfish are probably the major predators, but natural predators have not been identified.

COMMUNITY ECOLOGY. Species of *Necturus* inhabiting the Coastal Plain feed far less during the warmer months of the year (Bart and Holzenthal 1985; Braswell and Ashton 1985; Neill 1963; Shoop and Gunning 1967). Because predatory fish are inactive during winter months when *N. beyeri* actively feeds, Shoop and Gunning (1967) hypothesized that reduced feeding during the warm months is an adaptation to minimize predation risk from fishes. Bart and Holzenthal (1985), however, found that invertebrate prey reach a seasonal low during the summer months, which may also explain the scarcity of prey in *N. beyeri* collected during this time. In all likelihood, both the high risk of predation and low food rewards favor summer inactivity in *N. beyeri*.

CONSERVATION BIOLOGY. Many local populations of our native fishes have been extirpated or severely depleted by pollution and siltation of streams and rivers in the United States and Canada. The extent to which *N. beyeri* has been affected by stream siltation and pollution is not known, but other riverine species such as *Cryptobranchus* have been adversely affected.

Necturus lewisi (Brimley)
Neuse River Waterdog
PLATE 148

Fig. 265. *Necturus lewisi*; adult; Wake Co., North Carolina (R. W. Van Devender).

IDENTIFICATION. The Neuse River waterdog is a medium-sized species of *Necturus* with a rusty brown dorsum that is marked with numerous, large, bluish black spots or blotches. The venter is dull brown to gray and is also spotted. The spots on this species are larger, but less numerous, than those on *N. maculosus*. The snout is dorsally compressed and truncated, and the tail is laterally compressed and keeled. There are four toes on each hindlimb, and the gills are dull red and bushy (Bishop 1943; Martof et al. 1980). Sexually active males have swollen cloacae and two enlarged cloacal papillae that point to the rear. The sexes are similar in size. Adults have 14 costal grooves and measure 16.5–28 cm TL.

Hatchlings reach 15–16 mm SVL and 22–24 mm TL, and are uniformly dark brown above with a white spot behind each eye (Ashton and Braswell 1979). The undersides are immaculate and contrast sharply with the pigmented lower sides of the body. Individuals measuring 21–41 mm SVL have a broad, light tan dorsal stripe that extends from the head to the tail. The light dorsal stripe is usually flanked on either side by a dark lateral stripe that extends from the snout to the tail (Ashton and Braswell 1979). The lateral stripes gradually disappear as individuals mature, and are sometimes poorly defined in small larvae. The venter either is white or has a faint, reticulate pattern.

SYSTEMATICS AND GEOGRAPHIC VARIATION. *Necturus lewisi* was originally described by Brimley (1924) as a subspecies of *N. maculosus*, but it has been recognized since 1937 as a valid species. Recent electrophoretic and chromosomal comparisons of *Necturus* species by Ashton et al. (1980) and Sessions and Wiley (1985) indicate that *N. lewisi*, *N. maculosus*, and *N. punctatus* are distinct species. Cooper and Ashton (1985) review the taxonomic history of this species.

Detailed studies of geographic variation in morphology and color patterns are not available. However, no conspicuous geographic trends are evident and no subspecies are recognized.

DISTRIBUTION AND ADULT HABITAT. *Necturus lewisi* is endemic to the Neuse and Tar-Pamlico river basins in the eastern Piedmont and Coastal Plain of North Carolina. Animals are most abundant in stream sections where the main channel flow is >10 cm/second and the stream is >15 m wide and 1 m deep (Braswell and Ashton 1985). Capture success is highest in stream sections with clay or hard soil substrate and lower in leaf beds, even though some (Brimley 1924; Martof et al. 1980) consider leaf beds to be the preferred habitat.

BREEDING AND COURTSHIP. Males have sperm in the vasa deferentia from November through May, and females have sperm in their spermathecae from

Range of *Necturus lewisi*

December through May (Cooper and Ashton 1985). Animals maintained in the laboratory courted on 8 March (Ashton 1985). These observations indicate that mating occurs from December at least through March, and perhaps through April or May.

The only description of courtship is that of Ashton

Fig. 266. Typical habitat of *Necturus lewisi*; Wake Co., North Carolina (J. W. Petranka).

(1985), who observed partial courtship between a pair of adults in an aquarium. The pair was first observed crawling slowly with a male following 2–4 cm behind the female's tail. When the female stopped, the male moved forward and positioned his snout just behind the rear legs of the female. During this period, the female's gills were distinctly flared, but the male's were held close to his neck. The male then moved across her body at the base of her tail. Once their bodies were parallel, the male began to stroke and rub the female with his chin. Stroking began on the top of the head, then moved posteriorly along the female's neck and middorsum, then moved back to the head. The female raised her head at a 30° angle to the substrate each time the male's chin contacted her neck or head. This sequence was repeated 12 times during a 5-minute interval. The entire chin rubbing phase lasted 18 minutes. During the second phase the male slowly circled the female in a clockwise direction while maintaining contact with her. Three complete circles were made in 6 minutes, at which point the male moved parallel to the female and placed his limbs over her dorsum. The pair remained in this position for 30 minutes, then moved to a retreat. Further observations of courtship were not made.

REPRODUCTIVE STRATEGY. Females move to nesting sites beneath large rocks in fast current during the spring. Oviposition occurs in April and May, which corresponds to the time when ovarian eggs reach their maximum seasonal size (Ashton 1985; Cooper and Ashton 1985). The only known nest was found on 2 July beneath a flat rock and was attended by a male (Ashton and Braswell 1979). The nest contained 35 egg capsules attached singly by a blunt stalk to a 60-cm^2 area on the underside of a flat granite rock. The nest was in midstream, about 2 m from shore and 1.2 m below the water surface. All but three of the embryos had hatched, and the capsules that remained were 8–9 mm in diameter. Guarding of nests by males has not been reported in other *Necturus* species, so it is questionable whether the male discovered by Ashton and Braswell (1979) was protecting the eggs.

Fig. 267. *Necturus lewisi;* larva; Wake Co., North Carolina (R. W. Van Devender).

AQUATIC ECOLOGY. Very little is known about the ecology of small larvae. They appear to seek out quiet waters and are often found in leaf beds, which provide both cover and foraging sites (Braswell and Ashton 1985). Males and females become sexually mature after reaching 102 mm SVL and 100 mm SVL, respectively (Cooper and Ashton 1985). The estimated age at maturity is 5.5 years for males and 6.5 years for females.

Home ranges of two adult females monitored by Ashton (1985) were 16–19 m^2, whereas those of three males were 49–90 m^2. Males move greater distances between captures than females. Movements of both sexes increase following moderate rains but decrease after heavy rains of 40 mm or more. Movements also increase whenever barometric pressure falls or remains low, or when the moon phase changes from full to dark. The level of movements is highest during the spring and fall months.

During the winter the adults mostly reside in burrows in the banks or under granite rocks. During the spring the adults move to large bedrock outcrops or beneath large boulders in relatively fast, well-oxygenated current, where nesting presumably occurs. Summer habitats are similar. Home ranges contain bank areas with animal burrows or rock overhangs; large, flat rocks on sand-gravel substrates; and slack water areas where leaves and detritus form mats during the fall and winter. Juveniles mostly remain beneath granite boulders except in early spring, when some move to leaf beds.

Adults in laboratory aquaria construct retreats by shoveling sand and gravel with their snouts to form a cavity beneath a cover object (Ashton 1985). In some instances the animals pick up gravel in their mouths and transport it to the periphery of the retreat. In the field, adults construct entrances to their retreats on the downstream sides of rocks. Animals measuring <47 mm TL do not construct retreats.

Females, and to a lesser extent males, actively defend their retreats from intruders. Residents exhibit threat displays in which they flare and pulsate the gills and curl the upper lip. Intruders that fail to retreat are often attacked and bitten at the base of the tail or, less frequently, on the snout. In laboratory aquaria *N. lewisi* will viciously bit the head and gills of *N. punctatus*. Subsequent infections sometimes result in the death of the intruder.

Adults either feed from the mouths of their retreats or move about in search of prey at night. Both olfaction and sight appear to be important in locating food. Animals are active away from cover at night (Braswell and Ashton 1985). Individuals are also diurnally active during periods of rising water levels and increased turbidity. Individuals are inactive during warmer weather, when minimum stream temperatures exceed 18°C.

The most numerically important prey in 36 larvae from numerous sites in North Carolina were ostracods, copepods, mayflies, true flies, and beetles (Braswell and Ashton 1985). Earthworms, cladocerans, isopods, amphipods, collembolans, odonates, stoneflies, and unidentified insects composed the remainder of the prey. The major prey of 118 adults were snails, earthworms, isopods, amphipods, mayflies, stoneflies, caddisflies, true flies, and fish. Other prey included slugs, leeches, spiders, crayfish, centipedes, millipedes, odonates, hellgrammites, beetles, caterpillars, salamanders, and an adult worm snake.

PREDATORS AND DEFENSE. The inactivity of this and other Coastal Plain *Necturus* species during the warmer months is thought to be in part a behavioral response to minimize predation from fish (Braswell and Ashton 1985; Neill 1963; Shoop and Gunning 1967). The extent to which fish prey on *N. lewisi* is unknown. Brandon and Huheey (1985) found that skin extracts from this species cause distress when injected into mice; this finding suggests that *N. lewisi* possesses chemical defenses against predatory fish. Direct feeding trials would verify whether *N. lewisi* is distasteful.

COMMUNITY ECOLOGY. This and other species of *Necturus* probably compete with certain fishes for food, but ecological interactions between these groups have not been studied. Braswell and Ashton (1985) note that strong dietary overlap occurs between microsyntopic *N. lewisi* and *N. punctatus* and surmise that these species may compete for food.

CONSERVATION BIOLOGY. Recent surveys indicate that populations of *N. lewisi* occur throughout the range of the species, although streams with potentially significant pollution problems yielded few or no specimens (Braswell and Ashton, 1985).

Necturus maculosus (Rafinesque) Mudpuppy
PLATES 149, 150

Fig. 268. *Necturus m. maculosus*; adult; Tazewell Co., Virginia (R. W. Van Devender).

IDENTIFICATION. The mudpuppy is a large, permanently gilled salamander with four toes on each hindlimb and dark red, bushy gills. A dark bar extends through the eyes to the gills and the snout is strongly truncated. The dorsum varies from rusty brown to gray or black and is marked with scattered, bluish black spots or blotches. In rare instances these markings may be either absent or fused to form dorsolateral stripes. The venter is whitish to grayish and is sometimes spotted with bluish black markings. Albino adults are rare but have been reported from Arkansas (McAllister et al. 1981).

Males in breeding condition have swollen cloacae. The male's cloaca has two prominent papillae directed backward, whereas the female's cloaca is slitlike and often surrounded by a light-colored area. Adults measure 20-49 cm TL and most individuals reach an average TL of around 30 cm (Bishop 1941a).

Hatchlings have prominent yolk sacs and measure 14-15 mm SVL and 21-25 mm TL (Bishop 1941a; Shoop 1965b). A dark dorsal band extends down the midline of the back and is bordered on both sides by a light yellow stripe. A broad dark band occurs below the yellow stripes along the sides of the body (Bishop 1926; Shoop 1965b; Smith 1911b). Juveniles have the same general color pattern as hatchlings but are more conspicuously marked with yellow and black stripes along the body. Rare albino morphs as well as melanistic morphs that lack the yellow stripes are known (Cahn and Shumway 1925; Hensley 1959). The juvenile pattern begins to approximate that of the adults after larvae exceed 13-15 cm TL (Bishop 1941a).

SYSTEMATICS AND GEOGRAPHIC VARIATION. Two subspecies are generally recognized. The mudpuppy (*N. m. maculosus*) has a rusty brown to gray dorsum with conspicuous bluish black spots and a gray venter that varies from being unspotted to heavily spotted. The Louisiana waterdog (*N. m. louisianensis*) has a light yellowish brown to tan dorsum. Many individuals show evidence of a dark dorsal stripe that is bordered on either side by a faint light stripe. The dorsum and sides of the belly are marked with large

Fig. 269. *Necturus m. maculosus;* adult; central Kentucky (R. W. Barbour).

spots and blotches, but the center of the belly is light colored and immaculate.

Mudpuppies in northeastern Wisconsin and the adjacent upper peninsula of Michigan have a dark gray or black dorsum with scattered, small black spots. The dark ground color is usually suffused with tan pigment. In some individuals, large dark blotches may occur on the back, sides, and venter. The venter is dark brown to gray, and in some individuals a lighter area occurs along the midline. The line through the eye is inconspicuous or lacking in large specimens. These populations are treated by Hecht (1958) as a poorly defined subspecies (*N. m. stictus*) that often resembles *N. m. maculosus* from elsewhere. Here I treat both groups as members of the nominate subspecies.

A broadscale analysis of genetic variation in populations referable to *N. maculosus* has not been conducted, but it would be valuable in determining the similarities of regional populations that differ in morphology and coloration. Preliminary data provided by Ashton et al. (1980) for populations in Massachusetts, Minnesota, and North Carolina suggest almost no genetic variation among populations (Nei's $D < 0.001$).

DISTRIBUTION AND ADULT HABITAT. Mudpuppies inhabit a variety of permanently aquatic habitats from southeastern Manitoba and southern Quebec to northern Georgia, Alabama, Mississippi, and Louisiana. Populations occur as far west as eastern Oklahoma, eastern North Dakota, and adjoining areas in Manitoba. In the northern part of the range they occur in muddy canals and weed-choked bays, in large streams with fast-flowing water and rocky substrates, and in clear, cool lakes (Bishop 1926). Farther south, mudpuppies often inhabit reservoirs and sluggish, turbid streams. Specimens reside beneath rocks, logs, and other cover during the day and live as deep as 27 m below the surface of lakes (Reigle 1967).

BREEDING AND COURTSHIP. Adults in northern populations mate primarily in the autumn, but may mate sporadically through April (Bishop 1941a; Harris 1961). Mating in Wisconsin occurs during late October and November (Vogt 1981), and females retain sperm in their reproductive tracts for 6 months or more before ovipositing in late May or early June. In New York adults gather in groups of two to eight in depressions beneath logs and rocks in late September and October. Females have spermatophores in their vents during October, but can also be found in groups with males in breeding condition during February and April (Bishop 1926, 1932, 1941a). This suggests that a secondary breeding bout occurs in late winter and early spring in some populations.

A study of *N. m. louisianensis* in Louisiana suggests that southern populations breed primarily in winter (Shoop 1965b). Males have fully swollen cloacae between mid-December and February, and females have spermatophores in their vents in January. Vasa deferentia contain sperm during December and January, and spermatophore deposition has been observed as late as March.

Detailed accounts of courtship are not available. Bishop (1926) gave an account of partial courtship in which the male swam and crawled over the tail or between the legs of a stationary female. Spermatophore deposition did not occur. The spermatophore consists of a gelatinous base that supports a milky white sperm mass (Bishop 1932). The entire spermatophore is surrounded by a thin layer of jelly and is 6-8 mm in diameter and 10-12 mm high. The extent to which males compete for females is not known. Reported sex ratios are approximately 1:1 (Cagle 1954; Gibbons and Nelson 1968; Shoop 1965b).

REPRODUCTIVE STRATEGY. Males abandon the nest sites before females begin laying eggs (Bishop 1941a). Females construct nests by excavating depressions beneath rocks, logs, boards, and other cover. Common nesting sites are inside submerged rotten logs and beneath boards, rocks, and other structures that are partially embedded in bottom sediments (Bishop 1941a; Eycleshymer 1906; Fitch 1959; Pfingsten and White 1989; Smith 1911b). Females at-

Range of *Necturus maculosus*

tach their eggs singly to the undersides of a support structure in a monolayer that is typically 15–30 cm in diameter (Bishop 1926, 1941a; Smith 1912a). During oviposition, the female turns upside down and shifts a short distance to the side after depositing an egg. A female can lay at least 12 eggs in 1 day (Eycleshymer 1906).

Each egg is attached by a disklike expansion of the outer envelope and has a short stalk. Freshly laid eggs are cream to light yellow, 5–6.5 mm in diameter, and surrounded by three jelly envelopes (Bishop 1941a; Shoop 1965b; Smith 1912a). The entire structure is 8–11 mm in diameter and 12–21 mm long (Bishop 1941a; Shoop 1965b).

Females typically construct nests in water >0.5 m deep, but occasionally nest as little as 10 cm below the water surface. Nests in New York streams are constructed beneath rocks at depths of 10–60 cm (Bishop 1941a). The nests are near riffles, but not in the fastest-flowing water, and the entrances to the cavities face downstream. Eggs in a Wisconsin lake are attached to the undersides of large, flat rocks in water 20–46 cm deep (Smith 1911b).

Females in western New York and western Pennsylvania usually begin ovipositing in late May or early June (Bishop 1926, 1941a), but may begin as early as the first or second week of May during warm springs. Females in a Louisiana population also oviposit in May (Shoop 1965b), whereas females in southern Michigan oviposit as early as late April (Fitch 1959). Oviposition in local populations is synchronized and typically lasts 1 week or less (Bishop 1941a; Fitch 1959; Smith 1911b). Females guard the eggs through hatching from predators such as crayfishes and fishes (Bishop 1941a).

Estimates of clutch size based on counts of eggs in nests are generally lower than those based on dissections of gravid females. This finding may reflect predation, cannibalism by large adults, or the fact that females do not lay their entire clutch of ovarian eggs. Fitch (1959) reports a mean of 60 eggs (range = 28-101) in 11 Michigan nests of *N. m. maculosus;* however, three gravid females from the same population contained 105-140 ova. Similarly, the number of mature ova in 25 Michigan females examined by Lagler and Goellner (1941) varied from 75 to 193 and averaged 122. Other nest records include three Pennsylvania nests with 87-140 eggs (mean = 107; Bishop 1941a), five Wisconsin nests with 18-87 eggs (mean = 66; Smith 1911b), and a nest in Wisconsin with 62 eggs (Eycleshymer 1906).

The number of mature ova in 48 *N. m. louisianensis* varied from 32 to 91, averaged 54, and was positively correlated with female SVL (Shoop 1965b). Arkansas specimens had 48-174 ova and the average count was 106 (Trauth et al. 1990).

Length of the embryonic stage varies depending on water temperature, but hatching in northern populations occurs from July to August (Bishop 1941a; Smith 1911b). Estimated incubation periods are 38-63 days in natural populations of *N. m. maculosus* and 69-70 days for eggs of *N. m. louisianensis* incubated at 18°C in the laboratory (Shoop 1965b).

AQUATIC ECOLOGY. The mudpuppy is a generalist predator that will consume prey ranging in size from *Daphnia* ephippia to fish. Cochran and Lyons (1985) report chironomid larvae, diptera pupae, hydropsychid larvae, mayfly naiads, and mites in the stomachs of 16 larvae estimated to be about 1 month posthatching. Larger juveniles and adults consume a much larger size range of prey. Gut contents include plant material and small pebbles; however, it is uncertain whether these are intentionally ingested (Harris 1959).

The most important prey by bulk in 340 New York specimens were crustaceans (33%), insects (30%), fish (13%), annelids (11%), mollusks (5%), and amphibians (4%) (Hamilton 1932). Crayfish, snails, amphipods, mayflies, caddisflies, chironomid larvae, and beetle larvae are important invertebrate prey, whereas sculpins are the most important fish in the diet. Amphibians consumed include *Desmognathus* species, *Notophthalmus viridescens, Eurycea bislineata,* and *Necturus* eggs.

Food items in 18 Pennsylvania specimens included earthworms, isopods, spiders, crayfish, and insects (Surface 1913), whereas those in 33 adults from Wisconsin lakes included crayfish, insect larvae, fish, and snails (Pearse 1921). Lagler and Goellner (1941) report a variety of prey in 105 specimens from a Michigan lake, including several species of fishes, a frog, a stinkpot turtle, mayflies, odonates, crayfish, earthworms, and snails. Small numbers of representatives from miscellaneous families of insects are also present. In Louisiana, crayfish are the dietary staple of one population (Shoop and Gunning 1967), and crayfish, odonates, and a fish are reported for a second (Cagle 1954).

Harris (1959) reviews the literature on *Necturus* diet and provides a summary of prey items. Vertebrate prey include over 15 species of fish, 5 species of salamanders, tadpoles, and a turtle. Several earlier accounts indicate that *Necturus* feeds on fish eggs, and large adults will often eat the eggs and smaller larvae of conspecifics (Bishop 1926; Eycleshymer 1906; Harris 1959).

Most aspects of the natural history of juveniles are poorly documented. Hatchlings have conspicuous yolk reserves and probably do not feed for the first week or two after emerging from the egg capsules. Hatchlings and juveniles prefer microhabitats with relatively low current. They can be found in leaf litter accumulations in Louisiana (Cagle 1954; Shoop and Gunning 1967), or in the deeper waters of large streams in northern populations (Bishop 1926). Juveniles often share the same retreat, and this observation suggests that they are not territorial (Bishop 1926).

Hatchlings in populations studied by Bishop (1926) reach an average of 56 mm TL when 13 months old. One-year-olds subsequently grow an average of 26-39 mm TL per year to reach an average TL of about 187 mm when 5 years old. Sexual maturity is reached shortly thereafter when individuals exceed 200 mm TL. Male and female *N. m. louisianensis* studied by Shoop (1965b) become sexually mature at 130-137 mm SVL and 127-157 mm SVL, respectively.

Mudpuppies take shelter in burrows in streambanks, or beneath rocks, logs, or flood debris (Bishop 1926; Shoop and Gunning 1967). Individuals normally remain beneath cover during the day and emerge at night to forage (Cagle 1954; Harris 1959). Populations that inhabit vegetation-choked or muddy habitats are sometimes active during the day (Bishop 1941a). Individuals normally walk or crawl slowly over the bottom in search of food but are capable of rapid bursts of speed if disturbed. This species is abundant in northern lakes; Bishop (1926) recalls a catch of about 2000 mudpuppies in one haul of a commercial seine in Michigan.

Adult Louisiana waterdogs move an average of 81 m between captures in a Louisiana stream (Shoop and Gunning 1967). Thirty-three animals that were displaced about 128 m downstream began moving upstream, and four were recaptured near the site where originally collected. Animals released at the original site of capture immediately sought shelter and did not strongly orient upstream. These observations suggest that *N. maculosus* is capable of homing. *Necturus m. maculosus* in a southern Michigan lake migrate to shoreline areas in the spring, then retreat to deeper waters during the summer (Gibbons and Nelson 1968). The exact reason for these seasonal migrations is uncertain.

Adults in both northern and southern populations are active during the winter. Mudpuppies have been seen swimming beneath ice during the winter months (Morse 1904), and ice fishers frequently catch mudpuppies in Ohio (Pfingsten and White 1989). Individuals are also regularly caught in the spring months using baited hooks (Bishop 1926). *Necturus m. louisianensis* is most abundant during the winter months in sluggish water with rocks, logs, or other cover (Cagle 1954).

PREDATORS AND DEFENSE. Natural predators are poorly documented but undoubtedly include crayfishes, fishes, and turtles. Northern water snakes (*Nerodia sipedon*) feed on mudpuppies (Collins 1993), but humans appear to be one of the most important sources of mortality. Fishers regularly catch them and discard them on land. Bishop (1926) notes that around 500 *Necturus* were caught in Evanston, Illinois, on a single day.

COMMUNITY ECOLOGY. The ecological importance of this species in lake and stream systems is unknown. Several dietary accounts indicate that *N. maculosus* feeds on fish eggs (Harris 1959), and some have suggested that the mudpuppy could significantly affect some commercial species (e.g., Hacker 1956). Nonetheless, detailed dietary analyses indicate that the importance of egg predation has been greatly exaggerated.

CONSERVATION BIOLOGY. Despite the widespread pollution and siltation of streams in eastern North America, *N. maculosus* is still abundant in many northern lakes and rivers and appears to be in minimal need of protection.

Necturus punctatus (Gibbes)
Dwarf Waterdog
PLATE 151

IDENTIFICATION. The dwarf waterdog is the smallest species of *Necturus*, and like other members of the genus it is permanently gilled. The dwarf waterdog is a slender *Necturus* species that has a depressed, blunt snout. Adults in most populations are uniformly slate gray to brown above and usually lack spots. Populations in the Cape Fear and Lumber river systems often have a spotted morph with small dark spots that are most prevalent on the tail (Martof et al. 1980). The venter of both spotted and unspotted morphs is pale and lacks spots, and there are four toes on each hindlimb. Sexually active males have swollen cloacae and two enlarged cloacal papillae that point backward. Mature males and females are similar in average SVL, but females have proportionately longer tails than

Fig. 270. *Necturus punctatus;* adult; Moore Co., North Carolina (R. W. Van Devender).

Range of *Necturus punctatus*

males (Meffe and Sheldon 1987). The adults measure 11.5–19 cm TL.

Hatchlings and small larvae are uniformly brown above and lack the black and yellow striping typical of *N. maculosus* (Bishop 1943). The tail fin is slightly mottled, and the venter is bluish white. Coloration gradually shifts from brown to gray as larvae reach 40–50 mm SVL (Folkerts 1971).

SYSTEMATICS AND GEOGRAPHIC VARIATION. No subspecies are currently recognized. Ashton et al. (1980) found that *N. punctatus* is distinguishable electrophoretically from *N. maculosus* (six fixed loci differ-

Fig. 271. *Necturus punctatus;* adult; Moore Co., North Carolina (R. W. Van Devender).

ences) and *N. lewisi* (four fixed loci differences). The spotted morph of *N. punctatus* from the Pee Dee River drainage is indistinguishable from the unspotted morph from the Neuse River drainage and appears to be conspecific with other *N. punctatus* populations.

DISTRIBUTION AND ADULT HABITAT. Dwarf waterdogs inhabit small and medium-sized streams from the Chowan River in southeastern Virginia to the Ocmulgee-Altamaha River system in southern Georgia. This species occurs mostly in the Atlantic Coastal Plain, but sporadically in the lower Piedmont. Specimens have primarily been collected from deeper sections of streams that are sluggish and have accumulations of mud, silt, and leaves (Folkerts 1971; Meffe and Sheldon 1987). This species is rare or absent from the mainstream of the Neuse and Tar rivers (Braswell and Ashton 1985).

BREEDING AND COURTSHIP. Little is known about the courtship behavior and breeding biology of the dwarf waterdog. Mating probably occurs in winter, followed by oviposition from March to May (Meffe and Sheldon 1987). A gravid female collected on 12 April in South Carolina had 33 ovarian eggs with an average diameter of 4.2 mm (Folkerts 1971). Other females collected at the same time had already oviposited. Meffe and Sheldon (1987) collected a single female on 20 February with enlarged ovarian eggs that averaged 4 mm. Specimens taken during the autumn contain 15-55 mature ova, and the number of ova is positively correlated with female SVL. Most individuals contain 20-40 mature ova.

REPRODUCTIVE STRATEGY. Nests have not been discovered. Females presumably attach their eggs singly to the undersides of logs or other debris, as do females of other *Necturus* species.

AQUATIC ECOLOGY. Many aspects of the natural history of the juvenile and adult stages are not documented. Larvae live in both shallow and deep waters (Brimley 1924) and sometimes burrow in silt deposits (Martof et al. 1980). Adults are active during the winter and often congregate in leaf beds during this time (Brimley 1924; Martof et al. 1980).

Most individuals mature sexually when they reach 65-70 mm SVL (Hecht 1958). In a South Carolina study all specimens measuring <60 mm SVL are juveniles, whereas those measuring >75 mm SVL are adults (Meffe and Sheldon 1987). The smallest mature male and female in a sample taken by Folkerts (1971) measured 84 and 81 mm SVL, and a size-frequency analysis suggests that most individuals mature during their fifth year of life.

The diets of *N. lewisi* and *N. punctatus* are similar at sites where the two occur sympatrically (Braswell and Ashton 1985). Ostracods, blackflies, ceratopogonids, and dytiscid beetles were the most numerically important prey in 14 immature *N. punctatus*. Other prey include oligochaetes, cladocerans, copepods, isopods, amphipods, mayflies, caddisflies, and flies. Larger prey such as isopods and caddisflies are more important in the diets of adults. Additional prey include snails, clams, pseudoscorpions, amphipods, centipedes, mayflies, stoneflies, dobsonflies, true flies, hymenopterans, and lepidopteran larvae.

Food items of six immature specimens collected in March and April include millipedes, caddisfly larvae, fly larvae, amphipods, and oligochaetes (Folkerts 1971). Of these, earthworms are the most important food item. Nine of 20 adults that were examined had empty stomachs, suggesting that the adults feed little during the breeding season. In South Carolina 54% of specimens examined by Meffe and Sheldon (1987) also had empty digestive tracts. Prey of the remaining specimens included earthworms, mayflies, chironomids, salamanders, crayfish, and unidentified insects; plant material was also present.

PREDATORS AND DEFENSE. Natural predators have not been identified but undoubtedly include fish. Palatability studies have not been conducted, and it is not known if the dwarf waterdog produces toxic skin secretions.

COMMUNITY ECOLOGY. Strong dietary overlap occurs between microsyntopic *N. lewisi* and *N. punctatus*, which suggests that these species may compete for food (Braswell and Ashton 1985). However, studies of competitive interactions have not been reported.

CONSERVATION BIOLOGY. *punctatus* at 54 of 361 sites sampled in North Carolina and verified that viable populations occur in most areas of the range.

Family Rhyacotritonidae
Torrent Salamanders

The Rhyacotritonidae contains a single genus (*Rhyacotriton*) with four species in North America. This genus was previously included in the Ambystomatidae, but Good and Wake (1992) recommended placing *Rhyacotriton* in a separate family based on biochemical and morphological data. Fossil forms have yet to be found. Rhyacotritonids can be distinguished from all other salamander families by the presence of expanded, squarish glands that are lateral and posterior to the vent in adult males. Adults frequent cold, clear headwater streams and seepages. The eggs are unpigmented and adults abandon their eggs shortly after ovipositing. Larvae have stream-type morphology with short, stubby gills and a tail fin that does not extend onto the back.

Rhyacotriton cascadae Good and Wake
Cascade Torrent Salamander
PLATES 152, 153

Fig. 272. *Rhyacotriton cascadae;* adult; Skamania Co., Washington (W. P. Leonard).

IDENTIFICATION. The cascade torrent salamander is a small to medium-sized salamander with a small head, slim body, and short tail. This species is brown above and yellow below, with varying degrees of dorsal blotching (Good and Wake 1992). In most populations the adults are heavy blotched and spotted along the dark dorsal surfaces, particularly the sides of the body. The line of demarcation between the dorsal and ventral ground color is relatively distinct and straight. The venter is bright yellow and often has fine grayish flecking that is most abundant on the throat and chest. In some populations a dark band occurs across the venter and immediately behind the cloaca, whereas in others spotting is greatly reduced or absent. The eyes are prominent, and the snout is relatively short.

Males have swollen glands (vent flaps) on either side of the vent that produce a squared shape (Nussbaum and Tait 1977; Nussbaum et al. 1983). The largest 10% of females in museum specimens examined by Good and Wake (1992) were 2% larger in SVL on average than a comparable sample of males. Adults reach 7.5–11 cm TL and have 14–15 costal grooves.

The smallest hatchlings measure 15.8 mm SVL (Nussbaum and Tait 1977). Larvae are the mountain brook type with greatly reduced gills and tail fins (Valentine and Dennis 1964). The larvae have short, rounded snouts; tail fins that lack mottling; and prominent, dorsally positioned eyes (Myers 1943). The dorsum is light brown above and the venter in cream to yellowish. Small black dots usually are scattered on the dorsum and venter, particularly in older animals.

SYSTEMATICS AND GEOGRAPHIC VARIATION. *Rhyacotriton* is a somewhat enigmatic group whose systematic affinities have been debated for many years. This taxon is thought to be a primitive genus of salamanders that evolved many derived traits in response to living in fast-flowing streams (Nussbaum and Tait 1977; Worthington and Wake 1971). Among these are a streamlined body, functional limbs at hatching, and reduced gills, lungs, and fins.

Biochemical evidence suggests that *Rhyacotriton* is an ancient and distinct lineage composed of several genetically distinct subgroups (Good and Wake 1992; Larson 1991). Previous workers included *Rhyacotriton* either as a monotypic subfamily in the Ambystomatidae or as a member of the Dicamptodontidae, which includes *Dicamptodon* and *Rhyacotriton* (see Good and Wake 1992 for a detailed summary of past systematic treatments of *Rhyacotriton*). After reviewing morphological and biochemical evidence used in classifying *Dicamptodon, Rhyacotriton,* and ambystomatids, Good and Wake (1992) found no evidence for including these groups in a single monophyletic family and recommended that each be placed in a separate family. Here I follow this classification and recognize the Dicamptodontidae, Rhyacotritonidae, and Ambystomatidae.

Until recently, *Rhyacotriton* was considered to be a single species composed of two subspecies (Stebbins and Lowe 1951). However, studies of protein variation indicate that this taxon consists of four genetically cohesive subgroups with well-differentiated populations within each group (Good and Wake 1992; Good et al. 1987). Members of two coastal subgroups (*R. kezeri* and *R. variegatus*) have fixed differences at nine loci ($D = 0.55$) and do not interbreed in a zone of contact along the Little Nestucca River in Oregon; these are clearly distinct species. The remaining two subgroups are geographically isolated from other *Rhyacotriton* populations. Biochemical evidence indicates that each of the isolates is a genetically cohesive group that has diverged markedly from the remaining subgroups (typically nine or more fixed allele differences). Although

Range of *Rhyacotriton cascadae*

there is no direct evidence that the geographic isolates are reproductively isolated from other *Rhyacotriton* populations, analyses of protein similarity suggest that gene exchange has not occurred among groups for 3.2–9.5 million years. Based on this evidence, Good and Wake (1992) argue that the geographically isolated subgroups have in all likelihood reached the species level of evolution. Here I follow this interpretation and recognize four species of torrent salamanders: *R. cascadae, R. kezeri, R. olympicus,* and *R. variegatus.*

The four species of *Rhyacotriton* closely resemble each other in osteology and external morphology and often overlap in color pattern and degree of spotting. Metamorphosed individuals are brown above, yellow to orange below, and often covered with small black spots. White flecking usually occurs along the sides of the body and, to a lesser extent, on the belly. Because color patterning and morphology often cannot be used reliably to recognize species, torrent salamanders are best identified by range and collection locality. Local populations of *R. cascadae* vary markedly in the degree of spotting, but geographic patterns are rather chaotic, and no subspecies are recognized.

DISTRIBUTION AND ADULT HABITAT. The cascade torrent salamander is restricted to the western slopes of the Cascade Mountains from just north of Mt. Saint Helens, Washington, to Lane Co., Oregon. This and other species of *Rhyacotriton* inhabit cold, permanent, heavily shaded streams in humid forests from sea level to 1200 m (Bury 1983; Bury and Corn 1988a,b; Good and Wake 1992; Nussbaum et al. 1983; Welsh 1990; Welsh and Lind 1991). Preferred microhabitats are sites beneath moss-covered, water-washed rocks in streams, watercourses beneath rock rubble in streambanks, fissures in streamheads, and cracks in wet cliff faces (Nussbaum and Tait 1977; Stebbins 1951; Stebbins and Lowe 1951). *Rhyacotriton* is largely aquatic and is rarely found more than a few meters from aquatic habitats. In rare instances adults have been collected on land as far as 50 m from the nearest stream (Good and Wake 1992).

BREEDING AND COURTSHIP. Unlike that of most North American salamanders, mating of *R. cascadae* is not strongly seasonal and occurs throughout most of the warmer months of the year. Females have been collected with spermatophores in their cloacae in October, November, February, March, May, June, and July (Nussbaum and Tait 1977). Young females maturing their first clutch of eggs mate shortly before ovipositing. In contrast, older females that have oviposited at least once sometimes mate shortly after ovipositing when their remaining ovarian eggs are very small. Females

Fig. 273. *Rhyacotriton cascadae;* adult male (note conspicuous vent lobes); Multnomah Co., Oregon (W. P. Leonard).

Fig. 275. *Rhyacotriton cascadae;* larva; Skamania Co., Washington (W. P. Leonard).

probably mate two or more times throughout the year, and most females appear to oviposit annually. Courtship behavior has not been described.

REPRODUCTIVE STRATEGY. The eggs of *R. cascadae* have not been found, suggesting that females oviposit in cryptic recesses in a manner similar to *R. kezeri*. The eggs are presumably laid singly and scattered about in deep cracks and crevices. Most females oviposit during the warmest months of the year. Mean size of ovarian eggs in a Columbia River Gorge population reaches a maximum from May through July (Nussbaum and Tait 1977). Most females lay eggs during or shortly after this period, although some appear to oviposit in the fall and winter. The number of mature ovarian eggs varies from 2 to 13 and is positively correlated with female SVL. The mean clutch size of 135 females was 8 ova, which is identical to that reported for 7 females examined by Good and

Fig. 274. *Rhyacotriton cascadae;* adult female; Multnomah Co., Oregon (W. P. Leonard).

Wake (1992). In one population 18% of females contained one to five atretic eggs (Nussbaum and Tait 1977). Average ovum diameter is 3.4 mm, and maximum ovum diameter is 4.2 mm.

AQUATIC ECOLOGY. Larvae of this and other *Rhyacotriton* species live in loose gravel and beneath stones in springs, seepages, and small streams that are often fast flowing. Both the adults and larvae are rarely found submerged in more than a few millimeters of water, although they often use deep, rapidly flowing water as a refuge (Good and Wake 1992). The larvae feed on small invertebrates, and growth rates are slow relative to those of other stream-breeding salamanders that inhabit warmer streams.

Intermediate-sized larvae in a Columbia River Gorge population grow 0.7 mm/month during the summer and fall compared with 0.37 mm/month for older larvae nearing metamorphosis (Nussbaum and Tait 1977). The annual growth rate of older larvae over a 1-year period is only 0.30 mm/month. The larvae transform when they reach 37–45 mm SVL and they are 3–4 years old (Good and Wake 1992; Nussbaum and Tait 1977). Most larvae transform in the autumn, but a small percentage appear to transform during other times of the year.

Larvae can reach high densities and biomass in optimal microhabitats. Estimates based on mark-recapture methods are 28–41 larvae/m^2 and 27–34 g/m^2 in the Cascades (Nussbaum and Tait 1977). Marked larvae rarely move more than 2 m between captures, and the number of upstream movements is slightly higher than the number of downstream movements. In extreme cases larvae may move as far as 22 m from their original capture site over the summer.

TERRESTRIAL ECOLOGY. Most aspects of the natural history of the juveniles and adults are poorly documented. Juveniles mature 1–1.5 years after metamorphosing when 5.5–6.0 years of age (Nussbaum and Tait 1977). Females and males reach maturity when they measure 40–47 and 40–46 mm SVL, respectively. Information on diet is not available. The juveniles and adults probably take a mixture of aquatic and semiaquatic insects, along with a variety of other miscellaneous invertebrate prey.

PREDATORS AND DEFENSE. Natural predators are not known; however, *Dicamptodon* species and garter snakes (*Thamnophis*) are the most likely candidates. Large larvae and transformed adults respond to attacks by coiling, elevating, and undulating the tail. The skin secretions contain poisons and are effective in repelling shrews and other predators (Nussbaum et al. 1983).

COMMUNITY ECOLOGY. Very little is known about the community ecology of *R. cascadae*. The restriction of torrent salamanders to cold, rocky microhabitats with minimal siltation is presumably due more to physiological constraints than to interactions with other stream species.

CONSERVATION BIOLOGY. Old-growth forests provide optimal habitats, and populations in Oregon and Washington have been greatly reduced or eliminated by the clear-cutting of virgin forests (Bury 1983; Bury and Corn 1988a,b; Good and Wake 1992; Welsh 1990; Welsh and Lind 1991). Retention of old-growth buffers around headwater streams would help assure the long-term survival of local populations of this species.

Rhyacotriton kezeri Good and Wake
Columbia Torrent Salamander
PLATE 154

Fig. 276. *Rhyacotriton kezeri*; adult; Pacific Co., Washington (W. P. Leonard).

IDENTIFICATION. *Rhyacotriton kezeri* generally resembles other *Rhyacotriton* species in being a small to medium-sized salamander with a small head, slim body, and short tail. The males have distinctive swollen glands on either side of the vent that produce a squared shape. Both sexes lack dorsal spotting or blotching, except for populations near the contact zone with *R. variegatus* in Tillamook Co., Oregon (Good and Wake 1992). The line of demarcation between the dorsal and ventral ground color is relatively indistinct and straight. The venter is bright yellow or orangish and lacks spots in most specimens. Adults reach 7.5–11.5 cm TL and have 14–15 costal grooves. The largest 10% of males in museum specimens examined by Good and Wake (1992) were 6% larger in SVL on average than a comparable sample of females.

Hatchlings reach 22.5–25.8 mm TL and have two longitudinal rows of dark spots along the back (Nussbaum 1969a). Larvae are the stream (mountain brook) type, with greatly reduced gills and tail fins (Valentine and Dennis 1964). They have short, rounded snouts; tail fins that lack mottling; and prominent, dorsally positioned eyes (Myers 1943). The dorsum is light brown above and the venter is cream to yellowish.

SYSTEMATICS AND GEOGRAPHIC VARIATION. The most recent systematic treatments of *Rhyacotriton* are discussed in detail in the account of *R. cascadae*. This species closely resembles other *Rhyacotriton* spe-

Range of *Rhyacotriton kezeri*

cies and is best identified by collection locality and biochemical data.

DISTRIBUTION AND ADULT HABITAT. *Rhyacotriton kezeri* occurs in the Coast Ranges from the vicinity of the Chehalis River in Grays Harbor Co., Washington, southward to the zone of contact with *R. variegatus* along the Little Nestucca River and the Grande Ronde Valley in Polk, Tillamook, and Yamhill counties, Oregon. This and other species of *Rhyacotriton* inhabit cold, permanent, heavily shaded streams in humid forests from sea level to 1200 m. Preferred microhabitats include sites beneath water-washed rocks, wet rock rubble in streambanks, fissures in streamheads, and cracks in wet cliff faces.

BREEDING AND COURTSHIP. Data are not available on the mating season and courtship behavior. This species presumably has a prolonged mating season like *R. cascadae* and *R. variegatus*, but a detailed life history study is needed for verification.

REPRODUCTIVE STRATEGY. The seasonal time of egg laying is poorly documented. Nussbaum (1969a) excavated two clutches of eggs in late tail bud stages on 14 December from the mouth of a spring in Oregon. The eggs were laid singly in cold (8°C), flowing water in cracks within sandstone and consisted of two groups that each contained 16 eggs. The eggs lacked pedicels and were lying unattached to the substrate in cracks. Based on the number discovered, the eggs were the product of several females.

Groups of eggs that are incubated in the laboratory at 8°C hatch after an average of 210-290 days. The embryos continue to subsist on yolk for another 70-85 days before reaching the feeding stages. The average incubation period for this and other *Rhyacotriton* species in nature is not known, but it is probably the longest of any North American salamander. Females abandon the

Fig. 277. *Rhyacotriton kezeri*; adult male; Pacific Co., Washington (W. P. Leonard).

eggs after ovipositing, perhaps because the energetic cost of brooding over an extended incubation period outweighs the benefits of guarding the eggs.

COMMENTS. Very little is known about the natural history of this species. Larvae live in very shallow microhabitats such as wet rock rubble and feed on small invertebrates. Although not documented, larval growth rates are presumably very slow relative to those of other stream-breeding salamanders. Larvae in a Tillamook Co. population transform when they measure between 36 and 42 mm SVL; however, most transform when they reach >38 mm SVL (Good and Wake 1992). Based on studies of *R. cascadae* (Nussbaum and Tait 1977), the larval period of *R. kezeri* probably lasts 3-4 years.

Because of their specialized habitats, torrent salamanders probably have fewer predators than many other salamanders. Nussbaum (1969a) found two *R. kezeri* eggs in the stomach of a large *Dicamptodon tenebrosus* larva. As with other *Rhyacotriton* species, old-growth forests appear to provide optimal habitats.

Rhyacotriton olympicus (Gaige)
Olympic Torrent Salamander
PLATE 155

Fig. 278. *Rhyacotriton olympicus;* adult; Mason Co., Washington (W. P. Leonard).

IDENTIFICATION. *Rhyacotriton olympicus* is the largest of the *Rhyacotriton* species, with adults reaching a maximum SVL of about 60 mm. As in other *Rhyacotriton* species, the adults have relatively small heads and short tails. The Olympic torrent salamander has a dark, unspotted dorsum and a yellow to orange venter with well-defined black spots or blotches. The line of demarcation between the dorsal and ventral ground color is distinctly wavy and well defined. Individuals are best identified by collection locality since this species is geographically isolated from all other *Rhyacotriton* species. Males have conspicuous vent flaps. The SVL of the largest 10% of males in museum specimens examined by Good and Wake (1992) was on average 3% greater than that of a comparable sample of females.

Larvae are of the mountain brook type, with greatly reduced gills and tail fins. They have short, rounded snouts; tail fins that lack mottling; and prominent, dorsally positioned eyes (Myers 1943). The dorsum is light brown above and the venter is cream to yellowish.

SYSTEMATICS AND GEOGRAPHIC VARIATION. The genus *Rhyacotriton* consists of four closely related, allopatric or parapatric species that are best identified using geographic range. The most recent systematic treatments of *Rhyacotriton* are discussed in detail in the account of *R. cascadae*.

DISTRIBUTION AND ADULT HABITAT. *Rhyacotriton olympicus* is restricted to the Olympic Peninsula of Washington, where it inhabits cold, permanent, heavily shaded streams in humid forests (Good and Wake 1992; Nussbaum et al. 1983). Preferred microhabitats are sites beneath water-washed rocks in streams, watercourses beneath rock rubble in streambanks, fissures in streamheads, and cracks in wet cliff faces.

BREEDING AND COURTSHIP. The mating season is poorly documented, but in all likelihood it is as prolonged as in *R. cascadae* and *R. variegatus*. Pairs of males and females have been found in streams near Lake Cushman between 7 and 24 June (Noble and Richards 1932). Courtship behavior has not been described.

Range of *Rhyacotriton olympicus*

REPRODUCTIVE STRATEGY. Most females appear to oviposit during the spring and early summer. Eggs from Washington specimens (probably a mixture of *olympicus* and *kezeri*) are largest from April through early July, although a few individuals collected during the fall contain large eggs (Stebbins and Lowe 1951). Eleven females examined by Good and Wake (1992) contained an average of eight mature ova.

Females presumably scatter small numbers of eggs singly in cryptic sites in streams. Females that received pituitary implants laid from three to eight (mean = five) eggs singly on the sides and upper surfaces of rocks in laboratory tanks (Noble and Richards 1932). A single egg was found in nature attached to a tendril on the underside of a rock on 7 June, and a second egg was found on 6 July. Nussbaum (1969a) questioned the identity of these eggs and suggested that they were tailed frog (*Ascaphus truei*) eggs.

Freshly laid eggs are white and about 4.5 mm in diameter. The eggs are surrounded by six jelly capsules, but only three capsules are discernible with the naked eye. The largest mature ova in females examined by Stebbins and Lowe (1951) were 3.9 mm in diameter.

AQUATIC ECOLOGY. The larvae live in loose gravel or beneath stones in springs, seepages, and small streams, where they feed on small invertebrates. Growth rates, length of the larval period, and age at sexual maturity have not been reported.

COMMENTS. Many aspects of the life history of this species have not been studied. Adults emerge from daytime cover after dark and are often active along the margins of streams, where they forage for small invertebrates (Slater 1933).

Fig. 279. *Rhyacotriton olympicus;* larva; Jefferson Co., Washington (W. P. Leonard).

Rhyacotriton variegatus Stebbins and Lowe
Southern Torrent Salamander
PLATE 156

Fig. 280. *Rhyacotriton variegatus;* adult; Douglas Co., Oregon (R. W. Van Devender).

Fig. 281. *Rhyacotriton variegatus;* adult male; Oregon (W. P. Leonard).

IDENTIFICATION. The southern torrent salamander resembles other *Rhyacotriton* species in being a small to medium-sized salamander with a small head, slim body, and short tail. This species has evenly distributed black spots on both the dorsum and the venter, but the degree of spotting can vary from light to moderately heavy in local populations. The dorsal ground color is brown to greenish brown and the venter is bright yellow. Heavily spotted individuals usually have a strong preorbital stripe anterior to each eye. The line of demarcation between the dorsal and ventral ground color is relatively straight, and light flecks are often present on the head and body.

Males have conspicuously swollen glands on either side of the vent that produce a squared shape. Adults measure 7.5–11.5 cm TL and have 14–15 costal grooves. The largest 10% of females in three populations examined by Good and Wake (1992) were 1–5% larger in SVL on average than a comparable sample of males.

The smallest known hatchling measured 13.5 mm SVL (Nussbaum and Tait 1977). Larvae are of the mountain brook type and have short, rounded snouts; tail fins that lack mottling; and prominent, dorsally positioned eyes. The dorsum is light brown above and the venter in cream to yellowish. Scattered black dots usually occur on the dorsum and venter of older larvae.

SYSTEMATICS AND GEOGRAPHIC VARIATION. The most recent systematic treatments of *Rhyacotriton* are discussed in detail in the account of *R. cascadae*. This species closely resembles other *Rhyacotriton* species and is best identified by collection locality and biochemical data. Spotting is reduced in individuals near the northern limit of the species' range, where individuals closely resemble nearby populations of *R. kezeri*.

DISTRIBUTION AND ADULT HABITAT. *Rhyacotriton variegatus* occurs from southern Mendocino Co., California, northward to the Little Nestucca River and the Grande Ronde Valley in Polk, Tillamook, and Yamhill counties, Oregon. A geographic isolate occurs on the west slope of the Cascades in the vicinity of Steamboat Springs, Douglas Co., Oregon. Like other torrent salamanders, adults of this species inhabit cold mountain springs and seepages that have year-round flow. Relative to other salamander species, this and other *Rhyacotriton* species can tolerate very little water loss before becoming critically stressed (Ray 1958). Populations in the redwood forests of northwestern California are most often found in stream sections with relatively steep slopes, high-gradient riffles, and substrates composed mostly of cobble or gravel rather than sand or silt (Diller and Wallace 1996). Populations appear to reach their highest densities in mature, heavily shaded forests on landscapes having northerly aspects. Welsh and Lind (1996) conducted an extensive analysis of habitat characteristics in a broader region of northwestern California. They characterize the pre-

Range of *Rhyacotriton variegatus*

ferred habitat as being cold, clear, low-order streams with loose, coarse substrates. Humid forests with large conifers, abundant moss, and >80% canopy closure provide optimal habitats.

Populations of these and other *Rhyacotriton* species occur in isolated patches of suitable habitat throughout their range. Gene flow between demes is low, and this observation may explain the high degree of genetic differentiation that occurs between many regional populations.

BREEDING AND COURTSHIP. The mating period of *R. variegatus* is similar to that of *R. cascadae* and extends throughout much of the year. Females with spermatophores in their cloacae have been found during October, March, April, and May in Oregon (Nussbaum and Tait 1977) and in the autumn in northern California (Welsh and Lind 1992).

Mating reaches a seasonal low during the winter months and from May to August. Like those of *R. cascadae,* young females maturing their first clutch of eggs breed shortly before ovipositing. Older females that have oviposited at least once sometimes breed shortly after ovipositing, when their remaining ovarian eggs are very small. Females probably mate two or more times throughout the year and reproduce annually (Nussbaum and Tait 1977). The sex ratio in northern California populations averages about 1:1 (Diller and Wallace 1996). Courtship behavior has not been described.

REPRODUCTIVE STRATEGY. The eggs have not been found. As in other *Rhyacotriton* species, the females presumably attach their eggs singly in cryptic recesses during the warmer months of the year and do not brood. The mean size of ovarian eggs in an Oregon population reaches a maximum from about April through June (Nussbaum and Tait 1977). Most oviposition in Oregon populations occurs during this period, although the discovery of hatchlings throughout the year suggests that a small percentage of females oviposit at other times. Gravid females with 3-mm ova have been found in February in California, suggesting that southern populations oviposit earlier in the year than northern populations (Stebbins and Lowe 1951). Based on incubation periods reported for *R. kezeri,* hatching probably occurs 5–6 months after the eggs are laid.

Females produce smaller clutches than most stream-breeding salamanders of similar size. The number of mature ovarian eggs in females examined by Nussbaum and Tait (1977) varied from 4 to 16 and was positively correlated with female SVL. The mean clutch size for 40 females was 9.9 ova, compared with

Fig. 282. *Rhyacotriton variegatus;* larva; Oregon (R. W. Van Devender).

means of 8.4, 8.7, and 10.0 ova for females in three other populations (Good and Wake 1992). Average and maximum ovum diameters are 2.9 mm and 3.6 mm.

AQUATIC ECOLOGY. Larvae live in loose gravel or beneath stones in shallow sections of streams. The larval period in one population in the Coast Range is estimated to last 2-2.5 years (Nussbaum and Tait 1977). Larvae grow 0.50 and 0.42 mm SVL per month during the first and second years of the larval period, and metamorphs reach 30-39 mm SVL. Larvae in three populations in Oregon and California transform when they reach 31-40, 36-39, and 35-41 mm SVL (Good and Wake 1992). Comparisons with other populations suggest that mean size at metamorphosis differs significantly among local populations.

Larvae in a northern California population grow only 2.3 mm/year (Welsh and Lind 1992). Combined densities of larvae and adults based on mark-recapture estimates are 14-22 salamanders/m^2, and annual survival of all animals is 44%. This site was selected because of the dense population that is present; most populations sampled in northern California have much lower densities of larvae and adults. Diller and Wallace (1996) estimate 0.18-5.5 animals/m^2 (mean = 0.83 animal/m^2) based on surface searches of stream sections known to contain this species, whereas Welsh and Lind (1996) report an average of 0.68 animal/m^2 of substrate in other streams in northwestern California. Estimated density and biomass in Nussbaum and Tait's (1977) study are 13 larvae/m^2 and 6.8 g/m^2 wet mass. Corn and Bury (1989) provide conservative estimates of 0.29 salamander/m^2 and 0.23 g/m^2 of biomass based on plot searches.

Juveniles in a coastal population mature about 1-1.5 years after metamorphosing, when 4.5-5.0 years old (Nussbaum and Tait 1977). Females and males become sexually mature when they measure 38-44 and 36-42 mm SVL, respectively. Length of the juvenile stage is similar in a Cascade Mountain population, but individuals mature when 5.5-6.0 years of age.

The juveniles and adults feed mostly on aquatic and semiaquatic invertebrates. Amphipods are the most important prey based on frequency of occurrence in 42 metamorphosed specimens from California (Bury and Martin 1967). Other items include collembolans, stoneflies, mayflies, caddisflies, beetles, flies, hymenopterans, spiders, millipedes, snails, and oligochaetes.

PREDATORS AND DEFENSE. Natural predators have not been documented, but Pacific giant salamanders are probably among the most important. Large larvae and transformed animals of this and other *Rhyacotriton* species respond to attacks by coiling, elevating, and undulating the tail, which contains poison glands (Nussbaum et al. 1983).

COMMUNITY ECOLOGY. *Rhyacotriton variegatus* is parapatric with *R. kezeri*. Although local populations occur in close proximity to each other, they are not known to occur syntopically. Good and Wake (1992) suggest that competitive exclusion may be occurring between these two species. The diet of tailed frogs is very similar to that of *Rhyacotriton* (Bury 1970), but competition between these species has not been studied.

CONSERVATION BIOLOGY. Populations of this and other torrent salamanders have been greatly reduced or eliminated by the clear-cutting of old-growth forests (Bury 1983; Bury and Corn 1988a,b; Corn and Bury 1989; Good and Wake 1992; Welsh 1990; Welsh and Lind 1991). Studies in northwestern California indicate that populations are present in many generating forests in the redwood belt, where annual rainfall is relatively high (Diller and Wallace 1996). However, it is uncertain to what extent past timbering of old-growth forests affected *Rhyacotriton* species in this region (Welsh and Lind 1996). Maintaining forest buffers around headwater streams is the single most important forestry practice that will ensure the survival of local populations of torrent salamanders in managed forests.

Family Salamandridae
Newts

The salamandridae consists of 13 extant genera and about 53 species that are found primarily in Europe, Asia, and northern Africa. Eight extinct genera have been described, and fossil forms are well represented in Cenozoic deposits in Europe (Duellman and Trueb 1986). Phylogenetic relationships within the family are discussed by Wake and Özeti (1969). Two genera (*Notophthalmus* and *Taricha*), each with three species, occur in North America. *Notophthalmus* is restricted to the eastern United States and eastern Mexico, whereas *Taricha* is widespread in western North America. The earliest North American fossils are from the Upper Oligocene about 25 million years ago (Van Frank 1955). *Taricha* and *Notophthalmus* are well differentiated by the middle Miocene and apparently have been separated geographically since that period (Estes 1981; Tan and Wake 1995). Fossil *Notophthalmus* have been collected in Miocene and Pleistocene deposits only in the eastern United States (Estes 1963; Holman 1977; Mecham 1967a,b, 1968; Tihen 1974).

Salamandrids lack nasolabial grooves and distinct costal grooves, have two longitudinal rows of teeth on the palate that extend backward between the orbits, and have several technical features of the skeleton that distinguish them from other families of salamanders (Duellman and Trueb 1986). Mating and oviposition usually occur in ponds or streams, and sexual dimorphism in breeding coloration, size of the tail fin, and other traits is often conspicuously evident. All species appear to produce skin toxins, and many species have bright, aposematic coloration. The skin often appears grainy or warty, in contrast to the smooth skin of other salamanders.

Notophthalmus meridionalis (Cope)
Black-spotted Newt
PLATE 157

Fig. 283. *Notophthalmus meridionalis;* adult; Tamaulipas, Mexico (R. W. Van Devender).

Fig. 284. *Notophthalmus meridionalis;* courting pair; Tamaulipas, Mexico (R. W. Van Devender).

IDENTIFICATION. The black-spotted newt has an olive green dorsum with small yellow or gold flecks that are sometimes concentrated to form larger spots or vermiculations. Individuals from northern populations have a wavy, broken yellow dorsolateral line that extends from the base of the head onto the tail. Some specimens have a faint stripe down the midline of the back that varies from brown or russet to pale yellow. The venter is yellowish orange to orange, and the entire body is marked with conspicuous, scattered black spots. Males in breeding condition have hedonic pits in a line behind each eye and develop broad tail fins, swollen cloacae, and cornified toe tips. Males lack the cornified ridges on the underside of the thighs, and the yellowish spot on the posterior margin of the vent, that are evident in *N. perstriatus* and *N. viridescens* (Mecham 1968). Adults reach 7–11 cm TL and do not exhibit strong sexual dimorphism in size. Larvae have not been described in detail, but closely resemble those of *N. viridescens* (Strecker 1922).

SYSTEMATICS AND GEOGRAPHIC VARIATION. Reilly (1990) analyzed genetic variation in species of *Notophthalmus* and found that *N. meridionalis* is more similar to *N. perstriatus* (Nei's $D = 0.56$) than to *N. viridescens* ($D = 1.19$). Two subspecies are currently recognized, but only one occurs in the United States (Mecham 1968). The black-spotted newt (*N. m. meridionalis*) is a relatively stocky subspecies that has an olive green dorsum with conspicuous black spots. A broken, irregular line of yellow spots or vermiculations is present along either side of the back. The Mexican newt (*N. m. kallerti*) is slimmer and generally darker, and has a gray brown to dull brown dorsum with black spots that are inconspicuous because of the dark ground color. Diffuse yellow or gold flecks occur on the dorsum, but these are not organized into irregular lines along the sides of the back. Mecham (1968) collected intergrades near Acuña and Gonzales in Tamaulipas, Mexico.

DISTRIBUTION AND ADULT HABITAT. Black-spotted newts (*N. m. meridionalis*) occur from southeastern Texas to northern Mexico and inhabit both seasonally ephemeral and permanent habitats. Breeding adults have been collected in habitats that are >60 cm deep and heavily vegetated with submergent plants (Mecham 1968).

COMMENTS. Most aspects of the natural history of this species are undocumented. Mating can occur during any month of the year and is closely tied to rainfall patterns (Mecham 1968). Eggs have been found in

Range of *Notophthalmus meridionalis*

March and April attached singly to the leaves of submerged plants in small sloughs (Strecker 1922).

The black-spotted newt apparently lacks a well-defined eft stage. The juveniles remain in water until mature unless pond drying or high temperatures drive them onto land. Mecham (1968) collected juveniles and adults from beneath rocks near a recently dried pond, and this finding suggests that individuals do not move far from the ponds when forced to leave because of pond drying. He also collected a female with oviductal eggs and gill rudiments, suggesting that gilled adults may occur in some populations.

Notophthalmus perstriatus (Bishop)
Striped Newt
PLATE 158

Fig. 285. *Notophthalmus perstriatus;* young juvenile; Putnam Co., Florida (R. W. Van Devender).

IDENTIFICATION. The striped newt is a facultative perennibranch, and local populations often contain both gilled and transformed adults. Transformed adults are dark brown to olive green above. A red stripe bordered in black occurs on either side of the midline. The stripe extends nearly the full length of the body, but tends to break up on the head and tail. A series of spots or elongated red streaks bordered by black is often present along the sides of the body (Mecham 1967a). The dorsal midline is often slightly lighter than the adjoining ground color. The venter is yellow and marked with small black spots that are more widely spaced than those in *N. viridescens*. Gilled adults reach the same size as transformed adults (Bishop 1941b). The adults measure 5–10.5 cm TL and are smaller on average than those of other *Notophthalmus* species.

Males have an orange-colored gland cluster on the posterior margin of the vent that is present year round (Dodd 1993). Sexually active males have hedonic pits in a line behind each eye, fleshy tail fins, and horny, black ridges and pads on the inner surfaces of the thighs. Females in a population studied by Dodd (1993) are on average about 2% larger in SVL than males.

Hatchlings are the pond type; they have broad dorsal fins and balancers, and lack functional limbs. The hatchlings are similar in coloration to those of *N. v. viridescens* and measure about 8 mm TL on average (Mecham and Hellman 1952). Two broad, dark dorsolateral bands extend from the head to the tip of the tail, and a dark stripe extends from each eye to the gills. The dorsolateral bands disappear with age, but older larvae retain the dark marking on the head and develop a series of pale spots along the sides (Bishop 1941b). Older larvae measuring >30 mm SVL may develop red spots or red stripes along the sides, but in many cases larvae as large as 42 mm SVL may lack stripes (K. Dodd, pers. comm.). Gilled adults have a ground color similar to that of transformed adults and may have either red spots or red stripes along the sides. Efts resemble the adults but have a dull orange dorsal ground color, more roughened skin, and a less compressed tail.

SYSTEMATICS AND GEOGRAPHIC VARIATION. Reilly (1990) studied genetic variation in species of *Notophthalmus* and found that *N. perstriatus* is more similar to *N. meridionalis* (Nei's $D = 0.56$) than to *N. viridescens* ($D = 0.95$). Geographic variation in coloration and morphology has not been carefully analyzed. However, conspicuous geographic trends are not evident and no subspecies are recognized.

DISTRIBUTION AND ADULT HABITAT. Striped newts are found in southern Georgia and northern Florida. Records for disjunct populations in southern Florida and extreme western Florida are questionable and have not been confirmed by recent collecting. The adults occupy small ponds, drainage ditches, and other bodies of standing or sluggish water during the breeding season and live in surrounding forests at other times of the year. This species is not common and consists of widely scattered, local populations that occur sporadically throughout the range. Known, extant populations in Georgia occur at five widely separated sites and are found on deep, well-drained, sandy soils adjacent to rivers or large streams (Dodd and LaClaire 1995). These sites have breeding ponds with intact hardwood canopies and are surrounded by relatively mature pine forests that have not been altered by recent timbering operations. The adults breed in seasonally ephemeral habitats that usually lack fish and have a high diversity of other amphibians, particularly

Range of *Notophthalmus perstriatus*

anurans (Dodd 1992, 1993; Dodd and LaClaire 1995). Characteristic breeding habitats in Florida include flatwoods ponds in pine-palmetto habitats, sinkhole ponds in limestone regions, and ponds in scrub or sandhill regions (Christman and Means 1991).

BREEDING AND COURTSHIP. The adults live in uplands surrounding breeding ponds and migrate to ponds and other aquatic habitats shortly before breeding. Dodd (1996) trapped 12 striped newts an average of 225 m from the nearest water source (range = 42–709 m) at a Florida site, suggesting that terrestrial adults move several hundred meters from ponds after breeding. Although poorly documented, the breeding season lasts from midwinter through spring.

Adults in a seasonally ephemeral pond in Florida move into or away from the breeding site in response to pond filling and drying (Dodd 1993; Dodd and Charest 1988). A major movement into the pond occurs during periods of heavy rains from January to March, shortly before and during the breeding season. The sex ratio of males to females is on average 1:1.46. Elsewhere, adults in breeding condition have been found in ponds in February and June (Bishop 1941b; Mecham and Hellman 1952).

Courtship is very similar to that of *N. viridescens* (Mecham and Hellman 1952), but detailed descriptions of courtship behavior are not available. The spermatophore has a broad, flattened disk about 7.5 mm across, a stalk about 3 mm high, and a spherical sperm cap 1 mm in diameter.

REPRODUCTIVE STRATEGY. Most aspects of the reproductive ecology of the species are poorly documented. Females attach their eggs singly to aquatic plants in ponds in a manner similar to that of *N. viridescens*. Females in some Florida populations oviposit in March and hatchlings first appear in April (Christman and Means 1991), but in other populations mating appears to begin in late January or February (Dodd 1993; Dodd and Charest 1988). Oviposition in Georgia occurs from late January through April based on the size of larvae collected at different sites (Dodd and LaClaire 1995).

AQUATIC ECOLOGY. The larval stage lasts a minimum of 2–3 months, but in populations with gilled adults larvae may remain in ponds permanently. Larvae may transform as early as April in Georgia and some Florida populations, but metamorphosis may continue through December depending on local conditions (Christman and Means 1991; Dodd and LaClaire 1995). In wet years metamorphs leave a Florida site from late June to late August, but in years with sea-

Fig. 286. *Notophthalmus perstriatus;* larva; Putnam Co., Florida (R. W. Van Devender).

sonal droughts mortality is catastrophic and no juveniles are recruited into the terrestrial population (Dodd 1993).

Gilled adults are common in some populations, but the time required to reach sexual maturity is not precisely known (Mecham 1967a). Based on data for *N. viridescens,* the gilled adults probably reach sexual maturity within 8-24 months after hatching. Gilled larvae as large as 44 mm SVL that appear to be sexually mature can be found in a Florida pond in November and December (K. Dodd, pers. comm.).

TERRESTRIAL ECOLOGY. The efts disperse from breeding sites after transforming, but may return periodically before reaching sexual maturity (Dodd 1993). They are sometimes active after downpours in forest litter, where they feed on small invertebrates. A series of efts examined by Bishop (1941b) measured 43-51 mm TL; adults measured 52-79 mm TL. Dodd (1993) considers animals measuring <25 mm SVL to be sexually immature. The time required to reach sexual maturity is poorly documented, but it is undoubtedly shorter in individuals that become gilled adults than in those that transform into terrestrial efts.

Information on adult movements is based entirely on research by Dodd (1992, 1993) and Dodd and Charest (1988), who studied a population inhabiting a pond in Florida for over 5 years. A drought occurred during most of this period, and the breeding site dried and refilled repeatedly. Males and females at this site are active on the same days and do not differ in seasonal activity periods. Newts enter and exit the pond from November through March; however, most newts enter from January through March and leave from January through June. Adults leave the site if it dries or becomes very shallow, but return with the onset of heavy rains. During a 4-year drought, juveniles were not recruited into the population, the population declined markedly, and the size structure of the population shifted toward larger animals. Newts collected in July through September were extremely emaciated.

The striped newt appears to be a generalist feeder that will attempt to catch most prey it encounters. The stomachs of 59 Florida adults contained oligochaetes, snails, crustaceans, arachnids, aquatic and terrestrial insects, and frog eggs (Christman and Franz 1973). Prey types shift seasonally with availability. In January newts feed heavily on eggs of the spring peeper (*Pseudacris crucifer*) and on terrestrial prey that wash into ponds during heavy rains. At other times of the year individuals feed mostly on aquatic insects and crustaceans.

PREDATORS AND DEFENSE. No data are available on natural predators. Like other North American newts, this species produces highly toxic skin secretions that repel predators.

COMMUNITY ECOLOGY. Very little is known about ecological interactions of striped newts with other community members. This species sometimes occurs sympatrically with *N. viridescens* (K. Dodd, pers. comm.). Competitive interactions of *N. perstriatus* with *N. viridescens* have not been examined but may play a role in explaining the patchy distribution of *N. perstriatus.*

CONSERVATION BIOLOGY. The striped newt is a relatively rare salamander that is patchily distributed throughout its range. This species is currently a candidate for federal protection (Dodd 1993). A recent study in Georgia suggests that populations have been eliminated throughout the range because of the conversion of native longleaf pine forests into intensively managed slash pine plantations and the conversion of native forests into agricultural lands (Dodd and LaClaire 1995). Efforts are needed to locate existing populations and better characterize the aquatic and terrestrial habitats. Local populations should receive high priority for protection. Aquatic habitats should be protected and forest buffers that are important habitats for the efts and adults should be left around breeding sites (Dodd 1996).

Notophthalmus viridescens (Rafinesque)
Eastern Newt

PLATES 159, 160, 161, 162

Fig. 287. *Notophthalmus v. dorsalis;* aquatic adult; Moore Co., North Carolina (R. W. Barbour).

Fig. 288. *Notophthalmus v. viridescens;* terrestrial adult; Powell Co., Kentucky (J. W. Petranka).

IDENTIFICATION. The eastern newt has one of the most complex and variable life cycles of any North American salamander. Most populations have four distinct life history stages: the egg, aquatic larva, terrestrial red eft, and aquatic adult. Larvae transform into a terrestrial, juvenile stage termed the red eft. After several years on land, the efts migrate back to breeding sites and transform into aquatic adults. The aquatic adults have lungs and lack gills. They may remain in water for the remainder of their lives, but under certain conditions they may leave the water and live temporarily on land (Gill 1978a; Hurlbert 1969; Massey 1990; Noble 1926). Adults that leave ponds for prolonged periods undergo further morphological changes, including reduction in the dorsal tail fin and development of more granular skin.

Some populations contain gilled adults that are derived from larvae that undergo partial metamorphosis but retain the gills and associated structures (Brandon and Bremer 1966; Healy 1974a). Gilled adults or subadults may lose their gills and transform if breeding sites dry (Healy 1970; Noble 1929b). In other populations larvae metamorphose but the red eft stage is skipped. The transformed juveniles remain in the ponds and eventually mature sexually. Populations with gilled adults and aquatic, lunged juveniles are most common in sandy coastal habitats that lack suitable cover for the efts (Bishop 1941a; Brandon and Bremer 1966; Noble 1926, 1929b). In summary, several morphologically distinct stages may occur in the life cycle, including embryos, larvae, aquatic juveniles and adults with lungs, terrestrial juveniles (efts) and adults with lungs, and aquatic adults with gills.

The aquatic, lunged adults are light yellow below and olive green to yellowish brown above. The skin is slightly granular, and small black specks and blotches are scattered over the entire body. A series of conspicuous red spots bordered by black occurs on either side of the body. Depending on subspecies, these may be absent or coalesce to varying degrees to form broken lines down the back. The tail makes up about 50% of the total length and is narrowly keeled above and below. The skin is less granular than that of the efts. Lunged adults measure 31–54 mm SVL and 65–112 mm TL. Males and females generally tend to be similar in average length; however, significant differences in mean size occur in some local populations because of sexual differences in growth or survival (e.g., Caetano and LeClair 1996).

During the mating season the males develop broadly keeled tails; swollen vents; dark, cornified toe tips; and horny, black ridges and pads on the inner surfaces of the thighs (P. H. Pope 1924). Males also have hedonic pits in a line behind the eye that may be missing or reduced in females, and a yellowish, glandular spot on the posterior margin of the vent that is absent in females. Sexually inactive males have larger rear limbs than females, and the sexes are similar in average SVL.

Fig. 289. *Notophthalmus v. viridescens;* red eft; Bell Co., Kentucky (J. W. Petranka).

Fig. 290. *Notophthalmus v. viridescens;* defensive posturing by red eft; Macon Co., North Carolina (E. D. Brodie, Jr.).

Gilled adults have a yellowish brown or brownish green ground color and may retain the dark stripe through the eyes. Gilled males in breeding condition have swollen cloacae and develop the horny black ridges and pads typical of lunged adults. Terrestrial adults resemble the aquatic adults, but often lack keeled tails and have more granular skin. Terrestrial adults collected outside the breeding season are often difficult to distinguish from large, dark-colored efts (Hurlbert 1969).

Red efts vary from bright vermilion to dull red or greenish brown and have conspicuously granular, coarse skin. Small efts are typically bright vermilion, whereas large specimens that are approaching sexual maturity are duller. Red and black spotting is similar to that of the adults. Efts in coastal populations are less brightly colored than those in mountainous areas. Aquatic juveniles resemble miniature aquatic adults in having laterally compressed tails, olive coloration, and less granular skin than the efts and terrestrial adults.

Hatchlings have pond-type morphology, an average TL of 7–9 mm, and balancers on the sides of the head. The general ground color is yellowish green, and two broad, dark bands extend down either side of the back (Bishop 1941a). Mature larvae are light brown to yellowish brown above and have slender bodies and blunt snouts. A conspicuous dark stripe extends from the snout through the eyes. In some subspecies, individuals nearing metamorphosis develop reddish spots or lines on the dorsum. Individuals take on the bright red color of the efts within 1–2 weeks after metamorphosing (Chadwick 1950).

SYSTEMATICS AND GEOGRAPHIC VARIATION. Tabachnick (1977) and Reilly (1990) examined genetic variation in this species and found genetic distances between geographically distant populations to be surprisingly low (mean modified Nei's $D = 0.13$ in the latter study). Merritt et al. (1984) reached similar conclusions after completing a detailed study of populations in the Northeast ($D < 0.03$ among populations sampled). In contrast to general trends in salamanders, populations with gilled adults are almost as genetically variable (mean heterozygosity $[H] = 0.12$) as those without ($H = 0.16$).

Four subspecies are currently recognized (Mecham 1967b). The red-spotted newt (*N. v. viridescens*) has red dorsal spots encircled by black. The central newt (*N. v. louisianensis*) is smaller than the red-spotted newt and usually lacks red spotting on the back. The peninsula newt (*N. v. piaropicola*) is dark olive to black above, lacks red spotting, and has a venter that is heavily marked with black spots. The broken-striped newt (*N. v. dorsalis*) has broken red dorsolateral stripes that are bordered in black. All three of the Coastal Plain subspecies tend to be smaller and more slender than the red-spotted newt.

DISTRIBUTION AND ADULT HABITAT. *Notophthalmus viridescens* is the second most widely distributed salamander in North America. Populations occur throughout much of eastern North America from southern Florida westward to eastern Texas, and northward to southern Canada. Individuals can be found in communities ranging from northern boreal

Range of *Notophthalmus viridescens*

forests to coastal pine savannas and subtropical forests. The adults are found in all sorts of permanent or semipermanent bodies of water, including farm ponds, natural lakes, reservoirs, swamps, marshes, ditches, and sluggish streams and canals (Bishop 1941a; Gates and Thompson 1982). Sites with emergent or submergent macrophytes often harbor large populations of eastern newts. In western Maryland eastern newts prefer relatively deep ponds that contain aquatic vegetation near the shoreline. Local populations are common in both upland and bottomland habitats (Gates and Thompson 1982).

BREEDING AND COURTSHIP. The eastern newt breeds from autumn through very early summer, and in some populations the breeding season lasts for as long as six months. The duration and time of breeding vary depending on latitude and climate. In many northern populations adults court during the autumn and spring, but females do not lay eggs until the spring (Bishop 1941a; Gage 1891). At intermediate latitudes, adults mate primarily in the spring. In eastern Kansas *N. v. louisianensis* breeds from mid-March to late April (Ashton 1977), whereas in mountainous regions of Virginia *N. v. viridescens* breeds from March to June and the adults often overwinter on land (Gill 1978a; Massey 1988, 1990).

Adults in southern populations breed in late autumn or early winter through spring. Breeding records in the South include a single late-term egg of *N. v. louisianensis* in northern Florida on 8 January (Goin 1951), observations of courting pairs as early as 15 January in the Great Smoky Mountains National Park (King 1939), breeding in early March in Louisiana (Liner 1954), and breeding during late winter (months not specified) in a South Carolina bay (Taylor et al. 1988).

Data from Virginia populations indicate that most adults breed annually, although certain females may skip breeding for one or more years (Gill 1985; Gill et al. 1983). Sex ratios vary markedly between populations, but males often outnumber females in breeding sites and compete for females when courting (e.g., Chadwick 1944; Healy 1974b; Hurlbert 1969; Massey 1990; P. H. Pope 1924). The average ratio of males to females in 15 Pennsylvania ponds is 2.6:1 (Bellis 1968) compared to around 2:1 in Virginia ponds (Gill 1978a). The ratio of males to females in samples from four Quebec lakes varies from 0.7 to 1.7, with females outnumbering males at two of the sites (Caetano and LeClair 1996). Females outnumber males in samples of efts examined by Hurlbert (1969), but the reverse occurs in samples of adults from ponds. Similarly, sex ratios in a large collection of juveniles and adults made by Healy (1974b) are near 1:1, but males outnumber females in collections made during the breeding season. This discrepancy could reflect a sampling bias related to sexual differences in habitat preferences. Other reported sex ratios are about 1:1 in a population of *N. v. dorsalis* (Harris et al. 1988) and 1.8–2.5 males per female in a Virginia pond (Massey 1988).

Courtship behavior is described by Arnold (1977), Bishop (1941a), Gage (1891), Hardy and Dent (1988), Humphries (1955), Jordan (1891), P. H. Pope (1924), and Verrell (1982a, 1983, 1986, 1990a) and appears to be identical among subspecies (Verrell 1990a). The following is a synthesis based on the observations of these authors.

Males patrol slowly about ponds during the mating season. If a female stays close to and is responsive to a male when he first approaches, the male may perform a brief lateral display, or "hula," in front of her that involves undulating the body and tail. If the female nudges his tail with her snout, the male deposits one or more spermatophores, and the female picks these up with her expanded vent (Arnold 1977; Humphries 1955; Verrell 1982a, 1990a).

This very abbreviated courtship typically occurs in <30% of encounters in staged laboratory trials. If the female is unresponsive when a male first approaches, he rapidly darts above the female and grasps her with his large rear limbs. The pair may remain amplexed for several hours with the male typically grasping the female just in front of her front limbs. During this time the male periodically alternates between rubbing the snout of the female with the forelimbs and the sides of his head, where genial glands are located (Arnold 1977). At the same time, the male curls his tail forward and rapidly fans it back and forth for a few seconds; this action wafts cloacal secretions toward the female's nostrils.

Males frequently shift their head and tail positions from one side of the female to the other between fanning and rubbing bouts. After a prolonged period of fanning that may last for over an hour, courtship activity intensifies. The male increases his fanning activity and begins to make violent swimming contortions of his body, which cause the female to be dragged and jerked about. The cloaca of the male is strongly everted and pressed against the back of the female during this time. This aggressive activity intensifies over a period of as long as 10 minutes, at which point the male dismounts and moves just in front of the female. While moving forward, the male elevates his tail, strongly everts his cloaca, and undulates his body. The female follows the male and often presses her head lightly onto his tail or cloacal region. The male responds by depositing a spermatophore. If the female again presses against the tail, the male may move forward a few centimeters and deposit one or two additional spermatophores. The spermatophore has a broad, disk-shaped base and an abruptly narrowing, spinelike stalk with a sperm mass on top (Bishop 1941a). The entire structure is 4–7 mm tall.

After spermatophore deposition, the male moves forward about a body length and turns his body sideways to block the female's path. The female moves forward while keeping contact between her venter and the sperm cap. This tactile stimulus appears to be important in allowing the female to line up the sperm cap with her vent. As the female's cloaca approaches the sperm cap, she arches and positions her back so that the cap can be removed with the protruding cloacal lips.

In many cases the female may not respond to the male once he releases her. Instead, she swims away after being released. A female that is receptive will often raise her tail vertically at a 90° angle during amplexus. Small amounts of sperm can be found in females that courted 10 months earlier, but it is not known if females are able to fertilize eggs with sperm obtained from the previous year's mating (Massey 1990).

Males often compete for females and will try to dislodge rival males that are amplexing and courting females. Males will spend more time trying to dislodge other males if the female being contested for is large (fecund) or the male rival is relatively small (Verrell 1989b). In about 90% of encounters the amplexed male is not dislodged by its rival. Males prefer to court large, fecund females and use either chemical or visual cues to assess the size and fecundity of potential mates (Verrell 1982b, 1985a). Males are attracted to water conditioned with females (Dawley 1984a), suggesting that males use chemical cues to recognize females in ponds. Females court repeatedly during the mating season and most presumably mate with several males (Gill 1978b).

The hula courtship is much briefer and energetically less expensive, but results in a female being inseminated about half as often as the amplexus mode of courtship (Verrell 1982a, 1985c). Laboratory studies indicate that courting males are more likely to amplex (and not hula) if rival males are nearby (Verrell 1983). However, the number of females present does not influence the frequency of these behaviors. Although rival males are almost always unsuccessful in dislodging amplexed males, they may mimic female behavior during spermatophore deposition. This behavior typically entails the rival slipping between the male and female and inducing the male to deposit a spermatophore by nudging its tail. At the same time the rival male may deposit a spermatophore and inseminate the female that is nudging his tail. Males that engage in hula courtship are most vulnerable to this form of sexual competition.

Most of the information on courtship in *N. viridescens* has been derived from studies of laboratory populations. Massey (1988) reports detailed observations of courtship at natural breeding sites in New York and Virginia. Of 131 courtship bouts involving male-female amplectic pairs in Virginia, only 6% resulted in the insemination of the female by the courting male. In 61% of the unsuccessful courtships, females deserted males immediately after they dismounted, and in 35% of courtships they deserted after rival males slipped between the pair. In 6% of courtships a rival male deposited a spermatophore and the female was successfully inseminated. In 62.5% of instances involving male interference, rival males nudged the cloacae of courting males that had dismounted and induced them to deposit spermatophores. Similar behaviors occur in a New York population, and in both populations the insemination success of courting and interfering males is approximately the same. Rival males that attempt to break up amplexed male-female pairs are never successful. However, rival males that interfere with courtship often cause courting males to expend more time courting females.

Males that clasp other males often release their hold when the male assumes a head-down posture (Arnold 1977). In some instances males do not dismount and continue to court amplexed males. Duration of male-male pairs does not differ significantly from that of male-female pairs (Massey 1988). When a courting male dismounts, the courted partner nudges the male's cloaca, and by displaying this pseudofemale behavior induces the courting male to deposit spermatophores about 56% of the time. In a North Carolina population, males may eat the spermatophores of rival males.

The number of females present with a male can influence the male's behavior (Verrell 1985b). When males are experimentally presented with four females, each male usually completes only one or two courtships, but in rare cases may court as many as four consecutive times. When paired with individual females, males typically only court once. Males that engage in two courtships reduce the time of each courtship and produce fewer spermatophores per courtship relative to males that court once. Once a male has courted and inseminated a female, he is less likely to select the same female in subsequent courtships.

REPRODUCTIVE STRATEGY. Females attach their eggs singly to the leaves and stems of aquatic plants, decaying leaves, or other detritus. When depositing eggs, a female grasps the substrate with her rear legs, pushes her thighs together, and forces the eggs out one at a time (Bishop 1941a; P. H. Pope 1924). The eggs are often concealed by portions of vegetation wrapped about the sticky outer envelopes. Freshly laid eggs measure about 1.5 mm in diameter on average. They have light to dark brown animal poles and yellowish green vegetal poles, and are surrounded by three elliptical envelopes. Each female lays a few eggs per day and scatters the eggs widely in the breeding habitat. Thus oviposition by any individual takes several weeks to complete. Hurlbert (1970b) found that overwintering adults oviposit earlier in the season than efts that are returning to breed for the first time.

Females in southernmost populations probably begin ovipositing in early winter based on Goin's (1951) observation of a single egg of *N. v. louisianensis* at the hatching stage on 8 January. Oviposition in northern populations and in the southern Appalachians begins in late March through mid-April and continues until May, June, or early July depending on the locale (Bishop 1941a; Chadwick 1944; P. H. Pope 1924). In Kansas oviposition occurs from mid-March through late April (Ashton 1977), whereas in the Sandhills of North Carolina it occurs from April to June (Harris et al. 1988). Hatchlings can be found in mid-May and eggs as late as 28 June in a Maryland pond (Worthington 1969). This finding suggests an oviposition period from around April through June. The incubation period depends on water temperature and is 20–35 days (Bishop 1941a; Gage 1891).

Very few data are available on fecundity. Five females examined by Bishop (1941a) contained 232–376 mature ova; however, it is uncertain whether females deposit all of their eggs in a single breeding season. Morin (1983a) reports that fecundity is positively correlated with body size, but does not report clutch sizes.

AQUATIC ECOLOGY. All life history stages of *N. viridescens* appear to be generalist carnivores that feed on almost any prey that is palatable. Hatchlings feed on small invertebrates and can reach peak seasonal densities of 7–23 individuals/m^2 of pond bottom (Harris et al. 1988). Older larvae feed primarily on invertebrates and use both chemical and visual cues to locate food (P. H. Pope 1924). Large larvae will cannibalize smaller larvae in the laboratory (Walters 1975), but it is uncertain if this occurs in nature. The relative abundance of prey in 300 larvae from a New York pond is very similar to that in the environment, and larvae appear to be generalist feeders that feed on prey in direct proportion to their availability and accessibility (Hamilton 1940). Ostracods, copepods, chironomid larvae, snails, and finger clams are the most important prey by bulk. Miscellaneous items include beetle larvae, mosquitoes, and water mites.

Larvae in a New Hampshire lake feed on odonates, chironomids, protozoans, cladocerans, ostracods, amphipods, clams, and snails (Burton 1977). Ostracods and aquatic snails compose around 95% of the bulk of prey of larvae in a pond in southern Illinois (Brophy 1980). The remaining prey include turbellarians, copepods, water mites, aphids, dipterans, and sphaeriid clams. Larvae in a South Carolina bay feed rather heavily on cladocerans and incorporate larger prey such as chironomid larvae into their diets as they grow

Fig. 291. *Notophthalmus v. dorsalis;* larva; Scotland Co., North Carolina (R. W. Van Devender).

(Taylor et al. 1988). Copepods are underrepresented in the diet, probably because they are difficult to capture owing to their rapid, darting movements. Larvae also selectively feed on large cladocerans.

Length of the larval period, size at metamorphosis, and seasonal time of metamorphosis vary markedly depending on life history patterns and site characteristics. Size at metamorphosis is highly variable in populations with gilled adults compared with populations with typical life cycles in which all larvae transform into red efts. In the latter, most larvae transform when they reach 19–21 mm SVL and 35–38 mm TL, and the larval period lasts 2–5 months. Representative averages for size at metamorphosis based on measurements of transforming larvae or recent metamorphs are as follows: 19 mm SVL for a southern Illinois population (Brophy 1980), 20 mm SVL for a Maryland population (Worthington 1968), 20 mm SVL for Massachusetts populations (Healy 1973, 1974a), 22 mm SVL for a coastal North Carolina population (Taylor et al. 1988), and 21 mm SVL and 38 mm TL for a western North Carolina population (Chadwick 1950). In contrast, metamorphosing newts in a Massachusetts pond with gilled adults range from 46 to 75 mm TL (Noble 1929b).

Metamorphs or transforming larvae have been collected in New York from July through early November (Bishop 1941a; Hurlbert 1970b) and in Virginia from mid-August through the end of November (Gill 1978a). In the former studies a peak in metamorphosis occurs in August and September. Elsewhere, metamorphosis has been observed in September in Illinois (Brophy 1980), in late June in Maryland (Worthington 1968), in July and August in Massachusetts

(Noble 1929b; Smith 1920), in September in coastal North Carolina (Taylor et al. 1988), and in August and September in western North Carolina (Chadwick 1950).

The larval period lasts about 2 months in New York populations (Bishop 1941a) and slightly less than 2 months in a Maryland population (Worthington 1968). Metamorphs collected by Harris et al. (1988) in a Sandhills pond in North Carolina averaged 33 and 50 mm TL during two consecutive years. The larval period in this population typically lasts 4–5 months, although a very small percentage of larvae may overwinter. Growth appears to be density dependent and may average only 5 mm TL/month when larval density is high, compared with 11 mm/month when larval density is low. In populations with gilled adults, length of the larval period is much more variable since some larvae may not transform until nearly sexually mature (Noble 1929b).

Few data are available on larval survivorship to metamorphosis, but most accounts indicate very high mortality during the larval period. Total juvenile production in Virginia populations studied by Gill (1978a) is highly variable among ponds and years and is not correlated with pond age or the size of the breeding population. In most ponds studied over a 3-year period, juvenile recruitment is insufficient to replace adult losses. Larval survivorship to metamorphosis is presumably <1–2% in most ponds, and leech predation is believed to be the major cause of high larval mortality. Other estimates of premetamorphic survival are 0% in ponds studied by Massey (1990) for 2 years and <2–3% (estimate from data presented in the text) in a North Carolina site studied by Harris et al. (1988).

Very few observations have been made on larval behavior. Larvae appear to segregate by size in ponds, and this behavior may be a mechanism by which small larvae avoid being cannibalized by larger individuals (Harris et al. 1988). Larvae hide in bottom debris during the day, but actively forage in the water column at night (Morin 1983a).

The aquatic adults feed on worms, insects, amphibians, leeches, crustaceans, mollusks, and small fishes such as sticklebacks (Bishop 1941a). Predation on the eggs or larvae of spring-breeding frogs and ambystomatid salamanders is well documented (e.g., Bishop 1941a; Fauth 1990; Gill 1978a; Hamilton 1932; Walters 1975; Wilbur and Fauth 1990; Wood and Goodwin 1954), and in some instances may be sufficient to nearly eliminate certain species from ponds (Morin 1983a). Adults forage on yellow perch eggs at spawning sites (George et al. 1977) and cannibalize larvae (Burton 1977; Morgan and Grierson 1932).

Adult newts in a New Hampshire lake feed predominantly on odonate larvae and cladocerans between July and October (Burton 1977). Other prey include mayflies, stoneflies, lepidopterans, dipterans, amphipods, clams, and newt larvae. Cannibalism of larvae is widespread in July and August. Sphaeriid clams are common in the diets of adults from western North Carolina (Behre 1953), whereas mayflies, caddisflies, and dipterans are the most numerically important prey in a sample of 300 adults from a Pennsylvania pond (Ries and Bellis 1966). Other prey include snails and clams, oligochaetes, leeches, hemipterans, and beetles.

Newts are sluggish in Massachusetts during the coldest winter months, but about 20–30% of the individuals collected from December to February contain food in their stomachs (Morgan and Grierson 1932). Peak feeding activity occurs from July through October. Stomach items in 373 adults included crustaceans, mollusks, dipterans, odonates, beetles, hemipterans, caddisflies, mayflies, amphibian larvae, and shed skins. Kesler and Munns (1991) note that aquatic adults exhibit a diel feeding pattern in which most feeding occurs during the early morning. The presence of centrarchid fish does not influence the pattern.

Although adults do not have highly specialized diets, temperature and water clarity can affect feeding rates, and prey density can affect prey choice. In trials conducted by Attar and Maly (1980), individual newts tended to specialize on cladocerans or amphipods during short-term feeding bouts, then switch to an alternate prey. Adults appear to be stimulated more by visual cues than by chemical or tactile cues when feeding (Attar and Maly 1980; Martin et al. 1974). The time required for food to pass through the gastrointestinal tract varies from as little as 33 h at 25°C to >15 days at 5°C (Jiang and Claussen 1993).

Aquatic adults are active throughout the day and forage as they move slowly about pond bottoms or rest in beds of macrophytes. Although adults often tend to concentrate in vegetation near the shoreline, they occasionally forage at greater depths. Adults in a New York lake often forage at depths of 9–13 m in beds of the aquatic alga *Nitella flexilis* (George et al. 1977). Estimates of densities are 0.20–0.26 adult/ m^2 in lake habitats with rooted macrophytes (Burton 1977), 0.16-2.5 adult/m^2 in Sandhills ponds in North Carolina (Morin 1983a), 1 adult/m^2 in a Pennsylvania pond (Bellis 1968), and 4 and 0.8 adults/m^2 for maximum seasonal densities during two consecutive years in Sandhills ponds (Harris et al. 1988).

The extent to which adults establish home ranges varies among local populations. Aquatic adults in a Pennsylvania population remain in the same area of the pond for many weeks (Bellis 1968). Males move more than females, and individuals will return to their home ranges if the pond level drops and then returns to normal. In contrast, there is little evidence of intrapond homing behavior or the restriction of newts to limited home ranges in a small pond in Virginia (Harris 1981). Individuals can identify neighbors with which they are closely associated (Wise et al. 1993).

Pitkin and Tilley (1982) observed an unusual aggregate of adults in a Massachusetts pond in March. Large numbers of adults gathered together in ice-free areas of the pond where water temperatures were 5-6°C. One cluster contained at least 589 adults with both sexes equally represented. Morgan and Grierson (1932) observed smaller aggregates of 20-46 newts in aquatic habitats in Massachusetts in winter. The exact function of these aggregates is unknown.

Although adults have sometimes been stereotyped as being permanently aquatic, they often abandon breeding sites and reside on land (Gill 1978a; Gray 1941; Hurlbert 1969; Massey 1990). Adults will leave ponds in the summer if water levels drop and temperatures reach high levels (Gill 1978a; Hurlbert 1969), but may also abandon ponds to remove parasites. Adults in some Virginia populations move onto land to remove heavy infestations of leeches, which transmit the blood endoparasite *Trypanosoma diemyctyli* (Gill 1978a). Leeches appear to be the major source of adult mortality, and adults spend much time trying to bite or scratch them from their bodies.

In many populations some or all adults may vacate ponds in late summer or autumn and return the following spring to breed (Gill 1978a; Hurlbert 1969; Massey 1990). In the latter two studies, adults in Virginia populations leave ponds in August and September, hibernate on land, and return in March. The terrestrial adults show strong fidelity to the breeding ponds and consistently return to their home ponds after winter hibernation. If adults are transplanted to adjoining ponds, most will return to their native ponds (Gill 1979). Newts use olfaction and a light-dependent magnetic compass to locate home ponds (Hershey and Forester 1980; Phillips 1985a,b, 1986, 1987; Phillips and Borland 1992, 1994). Adults are capable of true navigation, and individuals that are displaced 10-50 km from their home ponds will orient toward home ponds even if deprived of directional information during experimental displacement (Phillips et al. 1995).

Adult densities of *N. v. dorsalis* in a North Carolina pond are highest in the winter (Harris et al. 1988). Densities progressively decline and reach their lowest levels during the summer. Most individuals presumably leave the pond during the summer and return in the fall and winter as the breeding season nears. Rather than leave ponds, adult *N. v. dorsalis* and *N. v. louisianensis* may hide in moist mud or beneath plant debris when ponds completely dry, and sometimes they become emaciated and severely dehydrated (Fauth and Resetarits 1991; Liner 1954; Morin 1983a).

Size at sexual maturity varies depending on life history, geographic race, and geographic locale. On average, adult *N. v. viridescens* in coastal populations are smaller than adults in inland populations. Adult males as small as 51 and 64 mm TL have been found in two northern coastal regions (Noble 1926, 1929b). In contrast, the smallest adult *N. v. viridescens* collected from a western North Carolina population measured 80 mm TL (Chadwick 1944). Similarly, the largest *N. v. viridescens* efts collected by Hurlbert (1969) measured 85-90 mm TL. Maturing efts of *N. v. dorsalis* collected by Harris et al. (1988) measured only 28 mm SVL on average, but this and other subspecies are smaller on average than *N. v. viridescens*.

Sex ratios of migrating efts in four Virginia populations are near 1:1 (Gill 1978a, 1985), but males are nearly twice as abundant as females in breeding cohorts. This disparity may in part reflect higher annual adult survival in males, which averages about 73% versus 68% in females. Annual survival of adults differs significantly among ponds, among years, and between sexes. Adults tend to have type II survivorship schedules, with age-specific mortality being roughly constant for all age groups. Based on survivorship data, the respective maximum longevities of males and females in these populations are around 15 and 12 years (Gill 1978a, 1985). Adults in a Maryland population are only 4-9 years old based on estimates using skeletochronology (Forester and Lykens 1991). Adults in Quebec lakes vary from 2 to 13 years of age, and the age structure differs markedly among local populations (Caetano and LeClair 1996). Most adults are 3–8 years old, and the maximum age varies from 9 to 13 years among lakes.

Data in Gill et al. (1983) show that population dynamics vary markedly among populations. In one pond the breeding population increased nearly 20-fold during a 5-year period, and in two others it roughly doubled or tripled. The size of two other populations remained relatively constant over the study period. In two ponds with high adult densities, survivorship and body mass declined when the adult density was experimentally increased (Gill 1979). Transplanted animals became emaciated, presumably because of food depletion in the pond rather than transplant stress.

In the Virginia studies discussed previously, larval survivorship in most populations is so low that juvenile recruitment is insufficient to replace adult losses in any given year. Immigration of efts from populations with high juvenile production appears to be the major factor maintaining many adult populations. Juvenile production in individual ponds is highly erratic from year to year. Juvenile production is low in some ponds for several years, then increases dramatically. The reverse occurs in other ponds. Juvenile production is low in drought years when ponds nearly dry and in ponds with large leech populations. Leeches either prey upon larvae or parasitize females and drain energy stores needed for egg production.

Factors that influence the probability of a larva developing into a gilled adult are not fully understood, but probably involve both genetic and environmental components. Populations with gilled adults occur sporadically throughout the range of *N. viridescens* and are common in some coastal regions. Gilled adults have been reported from Florida, Illinois, Indiana, Louisiana, Massachusetts, New Jersey, New York, North Carolina, and Tennessee (Brandon and Bremer 1966; Gage 1891; Healy 1970; Noble 1929b) and probably occur in many Coastal Plain populations elsewhere. The common occurrence of gilled adults in populations inhabiting unfavorable terrestrial habitats suggests a genetic component in the propensity to develop into gilled adults.

Harris (1987) conducted experiments that indicate that the tendency to metamorphose is influenced by environmental factors. When established at low experimental densities, larvae grow rapidly and are far more likely to become gilled adults than slow-growing larvae established at high densities. Individuals that become gilled adults reach sexual maturity in as few as 7 months. In two ponds in coastal Massachusetts the proportion of gilled adults gradually decreased over a 5-year span (Healy 1970). The decline was in part related to environmentally induced transformation associated with pond drying during a drought. In addition, the proportion of individuals transforming increased across years, even when water levels in ponds were normal.

TERRESTRIAL ECOLOGY. *Notophthalmus viridescens* is one of the best known salamanders to laymen because the efts are brightly colored and are more active during the day than most salamanders. The bright coloration of the efts functions as warning coloration, and the efts are thought to be part of a Müllerian mimicry complex involving several other salamander species. Details of this complex are described in the species account of *Pseudotriton ruber*. Red efts are exceedingly abundant in many areas in the Northeast, but are less frequently encountered in southern populations. Boy Scouts at a camp in New York, for example, collected 8650 efts over four summers (Evans 1947).

Red efts feed in the leaf litter, particularly following rains when invertebrates are active. Efts appear to be opportunistic feeders and often aggregate near rotting mushrooms or other food patches where invertebrates are concentrated. The diets of red efts and terrestrial adults are very similar except for a weak tendency for adults to take larger prey (MacNamara 1977). Individuals in this population eat a wide variety of prey, including species from 25 orders and 58 families of invertebrates. Springtails (collembolans), mites, fly larvae, and spiders are the most numerically abundant prey. In some instances an eft may contain over 2000 springtails in its stomach.

Efts are present in most populations of *N. viridescens* and require greater time to reach sexual maturity than individuals that skip the eft stage and develop into aquatic juveniles or gilled adults. The eft stage is estimated to last 2–3 years in New York populations (Bishop 1941a) and 4 years in mountain populations in western North Carolina (Chadwick 1944). These estimates are based on analyses of size distributions and may be conservative because age classes overlap in size and are difficult to distinguish in older age groups. The eft stage lasts 4–7 years in a Maryland population, based on skeletochronological aging (Forester and Lykens 1991). Aging of aquatic adults from five lakes in Quebec suggests that the eft stage lasts a minimum of 3–4 years, with females maturing slightly later than males (Caetano and LeClair 1996). In a Coastal Plain population in North Carolina, maturing efts are only slightly larger than metamorphosing larvae. This finding suggests that some individuals may reproduce when only 1 year old (Harris et al. 1988).

Healy (1973, 1974a) compared life history patterns in coastal populations with aquatic juveniles to those in an inland population with efts. Larvae in coastal populations in Massachusetts transform about 6 months after hatching when their average SVL is 20 mm. Juveniles, which resemble miniature aquatic adults, remain in the ponds and become sexually mature about 2 years after hatching. Larvae in the inland population also transform in the autumn when they reach about 20 mm SVL. The eft stage typically lasts 4–5 years, but in some cases as long as 7 years, and most individuals do not become sexually mature until 5–6 years old. Efts in the inland population grow an average of 5–6 mm SVL per year, whereas aquatic juveniles in ponds grow about twice as fast. Efts grow at slower rates than aquatic juveniles because their feeding is restricted during periods of dry or cool weather. Growth rates are highest in aquatic juveniles during the spring months and in efts during the summer. Aquatic males in some Massachusetts populations mature slightly over 1 year after hatching, although most aquatic juveniles require 2 years to reach sexual maturity (Noble 1929b). In artificial pond experiments, gilled adults may reach sexual maturity in as few as 7 months posthatching (Harris 1987). In Quebec lakes, the aquatic adults typically grow about 2–4 mm SVL per year. Growth rates often are slightly higher for females, may vary among lake populations, and decline only moderately with age (Caetano and LeClair 1996).

In populations with typical life cycles, metamorphs begin migrating away from their natal ponds shortly after transforming. Unlike that of ambystomatid salamanders, the trip to permanent forest residences is prolonged and may require as long as 1 year. Hurlbert (1970b) found that migrations occur in two or more waves that correspond to larval size classes in ponds. Migrants measure 28–47 mm TL, and most migrate at night during rainy weather.

Postlarval efts tracked by Healy (1974a, 1975) require about 1 year to migrate to woodlands 800 m from the natal pond. Efts reside under leaf litter during periods of dry weather and are most active on the ground surface during rainy weather when temperatures exceed 12°C. The average home range of 10 specimens collected six or more times during a 1-year period was about 270 m^2. Resident efts shift to different microhabitats seasonally. In spring they congregate close to the bases of trees and stumps, in June they are abundant in the beds of temporary pools, and in late August and September they often cluster about decaying mushrooms, where they feed on larval dipterans. Most of the local movements appear to occur in response to shifts in local food patches. The estimated density of efts in one Massachusetts population is 0.03 eft/m^2 of forest floor (Healy 1975).

Efts remain in forest-floor habitats until nearly mature, then migrate back to breeding habitats, where they transform into aquatic adults. Individuals typically reach sexual maturity shortly before or immediately after arriving at the breeding ponds. Patterns of seasonal migrations of efts back to breeding ponds vary locally. Depending on the population, migrations may occur in the autumn, in the spring, or during both seasons. In New York migrations have been observed in May, August, and September (Bishop 1941a). In western North Carolina efts emigrate during the autumn (Chadwick 1944), but in mountainous regions of Virginia most efts return to the ponds in the spring (Gill 1978a). In a Massachusetts population, large efts begin moving toward their natal ponds in July and first arrive at the ponds in August (Healy 1975).

At eight breeding ponds in New York the mature efts migrate to ponds during both spring and fall (Hurlbert 1969). Migrating efts tend to follow linear depressions such as streambeds or slope junctions while moving to ponds. Migrations are most intense during periods of rainy weather, particularly in the spring and late fall, and occur during the day and at night. During the spring, males tend to arrive at the ponds before females. Peak migrations occur in April, early May, August, and September, and new arrivals may make several excursions into and out of the pond before finally settling into an aquatic existence. A spectacular mass migration of efts that involved many thousands of animals occurred at a Massachusetts site. Efts were so abundant below the outflow of a dam that as many as 20 could be picked up with one scoop of the hand (Stein 1938). On 3 October over 1000 animals were collected during a half-hour period, and on 23 October an additional 1200 or so were collected from an area of <5 m².

Because adults show very strong fidelity to home ponds, the eft stage is the primary dispersal stage in the life cycle. Hurlbert (1969) observed thousands of adults breeding in seasonally ephemeral pools that normally dry before larvae can metamorphose. These were presumably colonized by efts that did not return to their natal ponds.

Factors influencing the population dynamics of adult populations are poorly understood. In a detailed 16-year study at the Savannah River Site in South Carolina, the population of breeding adults increased dramatically during the first 8 years of monitoring but subsequently declined to very low numbers (Semlitsch et al. 1996). The production of juveniles at this site is highly episodic. In 2 years >15,000 efts left the pond, but in 9 years no efts were collected in drift fences. The size of the breeding population is not correlated with the number of efts produced in subsequent years, and the number of efts produced per breeding female each year is independent of initial larval densities. Density of *Ambystoma* larvae in the pond appears to weakly negatively affect *N. viridescens* survival to metamorphosis.

PREDATORS AND DEFENSE. Larval, juvenile, and adult *N. viridescens* produce toxic skin secretions and are unpalatable to many predators (e.g., Brodie et al. 1974a; Hurlbert 1970a; Kats et al. 1988; Mosher et al. 1964; Wakeley et al. 1966). Eastern garter snakes (*Thamnophis sirtalis*), for example, will not attack efts even though they will attack other salamander species (Ducey and Brodie 1983).

Feeding trials indicate that newts are less acceptable to diurnally active, terrestrial predators than to nocturnally active or aquatic predators (Hurlbert 1970a). Bullfrogs (*Rana catesbeiana*), painted turtles (*Chrysemys picta*), and common snapping turtles (*Chelydra serpentina*) usually will eat efts and adults, but raccoons (*Procyon lotor*), American toads (*Bufo americanus*), garter snakes (*Thamnophis*), killdeers (*Charadrius vociferus*), and red-tailed hawks (*Buteo jamaicensis*) will not. Red efts are more than 10 times more toxic than the aquatic adults, and the skin on the back is more toxic than that on the belly (Brodie 1968a). Although garter snakes and northern water snakes are fairly resistant to newt toxins, many amphibians and reptiles have severe reactions to the toxins, including death.

Fig. 292. *Notophthalmus v. dorsalis;* defensive posturing by terrestrial adults; Scotland Co., North Carolina (J. W. Petranka).

Despite obvious chemical defenses, the larvae, efts, and adults are occasionally preyed upon. Lesser sirens feed on newt larvae in Coastal Plain ponds (Fauth and Resetarits 1991) and larvae of predaceous diving beetles (*Dytiscus*) largely avoid the toxic skin secretions by using piercing-sucking mouthparts (Brodie and Formanowicz 1981). There is one record of an eastern hognose snake (*Heterodon platyrhinos*) eating a newt (Uhler et al. 1939) and another of a Boy Scout eating several red efts wrapped in bread with no ill effects to the scout (Evans 1947). The latter account is surprising given that humans have died or become severely ill from eating western newts (*Taricha*) that contain similar skin toxins. Caetano and LeClair (1996) note that creek chubs (*Semotilus atromaculatus*) do not suffer any apparent ill effects after being fed red efts.

In western North Carolina 27 efts were found that had been partially eaten by an unknown predator (Shure et al. 1989). Many of the specimens were decapitated and had slits along the belly that allowed access to the guts. Comparisons of dead and live animals indicated that larger efts are more susceptible to the predator. Trout will die if force-fed newts, and fish will usually reject newts that are offered to them (Webster 1960). Adult newts are occasionally taken by smallmouth bass (*Micropterus dolomieui*) in Mirror Lake, New Hampshire (Burton 1977). However, an examination of several thousand predatory fishes from a New York lake containing newts produced no evidence of fish predation (George et al. 1977).

Efts and adults will often engage in defensive posturing if attacked. When terrestrial adult *N. v. dorsalis* and *N. v. louisianensis* are disturbed in the field, the animals exhibit an "unken" posture with the head and tail bent upward, the eyes closed and depressed, and the tail curled (Neill 1955; Petranka 1987). This defensive posture reveals the brightly colored venter, which is aposematic. The highly toxic red efts of *N. v. viridescens* are more likely to remain immobile and engage in exaggerated antipredator posturing than are the adults (Ducey and Dulkiewicz 1994). Only about 15% of adults assume weak or full unken postures when subjected to simulated predator attacks, compared with 36% of efts. Adults will avoid substrates with newt skin extracts, but do not avoid other test materials such as skin extracts from plethodontid salamanders (Marvin and Hutchison 1995).

COMMUNITY ECOLOGY. Several researchers have examined interactions between newts and other vertebrates in artificial pond experiments. Newts often function as keystone predators in artificial ponds, where they can alter the outcome of competition between anuran tadpoles (e.g., Alford 1989; Morin 1981, 1983b; Wilbur et al. 1983) and affect the abundances of zooplankton and phytoplankton species (Morin 1995; Morin et al. 1983). The extent to which newts organize anuran and plankton communities in more complex natural communities is not known.

Although newts are generally viewed as having a negative impact on anuran tadpoles, the reverse phenomenon may also occur. Gill et al. (1983) provide evidence of one indirect negative interaction of wood frogs (*Rana sylvatica*) on newts. These frogs are one of the first amphibians to breed in the spring. The tadpoles provide a blood meal for leeches that, in turn, provides energy for egg production. Small leeches that hatch from the eggs attack newt larvae and adults. This behavior ultimately causes low juvenile production.

When wood frogs are fenced from ponds, large numbers of efts may subsequently emerge.

Fauth and Resetarits (1991) examined interactions between lesser sirens (*Siren intermedia*), newts, and anuran tadpoles in artificial pond communities. The results show that *N. viridescens* does not affect the survival or growth of *S. intermedia* and that the effect of *S. intermedia* on *N. viridescens* is dependent on newt densities. At low newt densities, *S. intermedia* reduces newt reproductive success by preying on larvae, whereas at high newt densities *S. intermedia* enhances newt reproductive success by reducing the survival of adults that prey upon and compete with newt larvae. *Siren intermedia* does not alter the role of *N. viridescens* as a keystone predator. Bristow (1991) demonstrated that red-spotted newts and banded sunfish (*Enneacanthus*) can potentially compete, but it is uncertain if competition occurs in nature.

Newt and tiger salamander larvae that coexist in a pond in southern Illinois do not strongly partition food resources (Brophy 1980). A similar pattern is evident for *Ambystoma talpoideum* and *N. viridescens* larvae that share a South Carolina bay (Taylor et al. 1988). The extent to which *A. talpoideum* larvae compete for food resources with newt larvae is not known.

CONSERVATION BIOLOGY. The eastern newt is one of the few species of salamanders in North America that may have benefited from European colonization. This species readily colonizes weed-choked farm ponds and other habitats that contain predatory fish. Populations of eastern newts are common throughout much of eastern North America and appear to be increasing in number in association with the reintroduction of beavers into many areas of their former range.

Taricha granulosa (Skilton)
Rough-skinned Newt
PLATES 163, 164

IDENTIFICATION. The rough-skinned newt is a large North American salamandrid with a complex life cycle that typically includes a larval stage and a terrestrial juvenile stage. The terrestrial adults tend to shift seasonally from land to water, and in a few high-elevation populations larvae may develop into gilled adults. Terrestrial adults have a granular or warty dorsum that varies from light brown to blackish brown. The eyes have yellow irises and do not protrude beyond the profile of the head when viewed from above. The upper and lower eyelids are uniformly dark (Stebbins 1951). The palatine teeth form a V-shaped

Fig. 293. *Taricha granulosa;* adult; Curry Co., Oregon (J. W. Petranka).

Fig. 294. *Taricha granulosa;* ventral view of adult; Thurston Co., Washington (W. P. Leonard).

pattern as opposed to the Y-shaped pattern in *T. torosa*. The venter and lower sides of the body range from light yellow to orangish red. In some populations a small percentage of the individuals have varying amounts of dark mottling on the venter, although most individuals are uniformly light colored (Myers 1942). In a few restricted locales most individuals are heavily blotched on either the dorsum or the venter.

During the breeding season terrestrial males that enter aquatic sites develop soft, spongy, lighter-colored skin; swollen vents; prominent tail crests; and cornified nuptial pads on the toe tips (Oliver and McCurdy 1974). Aquatic females have tail crests and raised, conical cloacal apertures. In addition, their skin is lighter colored and smoother than that of terrestrial females. Development of smooth skin, tail crests, and lighter body color is less pronounced in aquatic females than in males. Adults measure 12.5–22 cm TL and males in local populations are on average 1–9% larger in SVL than females (Neish 1971; Taylor 1984).

Bishop (1943) and Stebbins (1951) reported average hatchling sizes of 12 mm TL, whereas specimens measured by Riemer (1958) measured 7.6–10.3 mm TL. Hatchlings are the pond type and have a weak, dark dorsal stripe on either side of the body that becomes diffuse and ill-defined within a few weeks posthatching (Twitty 1966). Older larvae have reticulate or diffuse dark markings on the body and lack well-defined stripes. A dark bar extends from each eye to the nostril, and the fin is usually mottled with dark pigment. Two rows of light spots are usually evident along the body, one near the dorsal fin and the other on the sides above the limb insertions. Isolated populations containing adults with gills or remnants of the gill arches and gill rakers are found in San Mateo Co., California; Latah Co., Idaho; and Crater Lake, Oregon (Bishop 1943; Farner and Kezer 1953; Nussbaum and Brodie 1971; Riemer 1958). Populations of largely gilled adults also occur in seven lakes in southern Oregon at elevations of 1743–1847 m (Marangio 1978).

Taricha granulosa and *T. torosa* are often difficult to distinguish in areas of sympatry. The relative size of the eyes is the most reliable trait for distinguishing the two. When viewed from above, the eyes of *T. granulosa* do not meet the margin of the head, whereas those of *T. torosa* do. A behavioral difference is also useful. When assuming an intense unken posture (one should tap repeatedly on the dorsum to induce this), *T. torosa* holds the tail tip out straight, whereas sympatric *T. granulosa* curls the tail tip into a single coil (Riemer 1958).

SYSTEMATICS AND GEOGRAPHIC VARIATION. Many authors previously considered *T. granulosa* and *T. torosa* to be conspecific (Nussbaum and Brodie 1971; Pimentel 1958, 1960), but behavioral, molecular, and chromosomal studies indicate that these species are reproductively isolated and genetically distinct (Davis and Twitty 1964; Hedgecock 1976; Hedgecock and Ayala 1974; Tan and Wake 1995). Although all three western species of *Taricha* are interfertile, they rarely hybridize in nature because of behavioral isolation (Ayala 1975; Twitty 1961b, 1964). Within *T. granulosa*, moderate genetic differences occur between Oregon and California populations ($D = 0.15$), and average heterozygosity is 0.10 (Hedgecock 1976).

Riemer (1958) provides a detailed summary of geographic variation in external morphology and body size in *T. granulosa*. Individuals in the southern portion of the range generally grow larger than those farther north. Individuals with dark mottling or blotching on

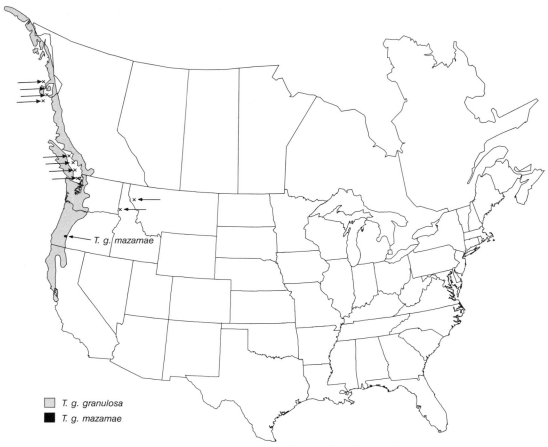

Range of *Taricha granulosa*

the venter occur at relatively high frequencies in populations on Gravina Island, Alaska, and in Crater Lake, Oregon. Specimens with dark blotching on the dorsum occur in Fay Lake, northwestern Oregon, and in 13 lakes in Del Norte Co., northwestern California (Garber and Garber 1978). All of these sites have harsh winter environments and occur at either high elevations or far northern latitudes.

As many as four subspecies (*granulosa, mazamae, similans,* and *twittyi*) have been concurrently recognized in the past based on differences in color pattern and body size (Myers 1942; Stebbins 1951). Most authors now recognize only two subspecies. The Crater Lake rough-skinned newt (*T. g. mazamae*) has heavy, dark blotching on the belly and is restricted to Crater Lake, Oregon, and its immediate environs (Farner and Kezer 1953; Myers 1942). Belly mottling in this subspecies is often absent in larvae and young juveniles.

The northern rough-skinned newt (*T. g. granulosa*) has little or no blotching on the belly and occupies the remainder of the species' range. Dorsally blotched populations are interspersed locally with unblotched populations (Garber and Garber 1978) and are best treated as local variants of the nominate subspecies.

DISTRIBUTION AND ADULT HABITAT. Rough-skinned newts range from southeastern Alaska to just south of San Francisco Bay in west-central California. Populations occur from near sea level to 2800 m. Isolated populations of *T. granulosa* that may be introductions occur in the Rocky Mountains in Latah Co., Idaho, and Saunders Co., Montana (Nussbaum and Brodie 1971; Nussbaum et al. 1983).

Adults have terrestrial and aquatic phases. Aquatic habitats include seasonally ephemeral ponds as well as permanent habitats, such as lakes, beaver ponds, ditches, and sluggish portions of streams (Evenden 1948; Garber and Garber 1978; Stebbins 1951). Although *T. granulosa* occupies streams of all sizes, it is most common in large streams with sluggish pools and is rarely found in fast-flowing water (Bury 1988). Concentrations of adults have been seen as deep as 12 m below the water surface in British Columbian lakes (Neish 1971). Terrestrial juveniles and adults are typically found beneath logs, bark, and other surface cover in or near mesophytic forest in mountainous or hilly terrain. Local populations occasionally inhabit grasslands, pastures, or other open areas (Riemer 1958).

BREEDING AND COURTSHIP. The seasonal time of breeding varies markedly with latitude and elevation and has been recorded in every month except November. The adults court in aquatic sites and usually begin breeding within one to several weeks after arriving at the breeding sites. Peak breeding activity at low to moderate elevations occurs from March through early May. Males in breeding condition have been collected in Oregon as early as February, but spermatophore deposition occurs primarily during the spring months (Fitch 1936; Oliver and McCurdy 1974). Oviposition may occur as early as January in California, but peak egg laying at low-elevation sites usually does not occur until early March through early April (Stebbins 1951; Twitty 1935).

Adults in many high-elevation lakes breed during the summer and early autumn. Gravid, amplexed, or spawning females have been observed in late October in high-elevation lakes in northern California (Garber and Garber 1978) and on 4 October and 10 August at high-elevation lakes in Oregon (Chandler 1918; Marangio 1978). *Taricha g. mazamae* also breed during the summer months in Crater Lake, Oregon (Farner and Kezer 1953).

Geographic variation in annual patterns of breeding is poorly documented. Most females return annually to breed in Lake Marion, British Columbia (Efford and Mathias 1969), and at least some females breed annually in an Oregon population (Pimentel 1960). Most adults in the latter population appear to breed biennially.

Patterns of seasonal migration to and from breeding sites vary markedly among local populations. In British Columbia, adults of both sexes enter Marion Lake around mid-April and leave before late September or early October to overwinter on land (Neish 1971). In

Fig. 295. *Taricha granulosa;* male and female in amplexus; Benton Co., Oregon (W. P. Leonard).

contrast, males and females in lakes on southern Vancouver Island exhibit marked differences in seasonal migration patterns (Oliver and McCurdy 1974). Females migrate to breeding sites between mid-March and mid-April, remain through the summer, then migrate back to land between mid-September and mid-November. Males, however, remain in aquatic sites throughout the year except for brief excursions onto land. Adults using streams in California usually remain in water year round unless forced out by severe winter flooding (Packer 1961; Twitty 1942).

At a low-elevation site in Oregon, adults emerge from terrestrial burrows and engage in sporadic surface movements prior to migrating to a temporary pond to breed (Pimentel 1960). Adults migrate to and from the pond primarily at night during rainy weather when temperatures are >5°C. A few individuals arrive at the ponds in November, but a peak in arrival occurs between late January and mid-March. Males begin arriving before females and show much less tendency to migrate en masse. When migrating to the pond, individuals tend to use inlet and outlet streams as migratory corridors.

Postmating migrations from this pond occur primarily in late April and early May and appear to be in response to either high water temperatures or low water levels. Individuals migrating to ponds have moderately to well-developed secondary sexual traits, whereas those leaving ponds to migrate to terrestrial sites do not. Most individuals migrate <400 m away from the pond margin. Field observations at several other localities indicate that males typically remain 7–10 months in ponds, whereas females remain 6–8 months.

During the breeding season many males and females make short excursions into and out of the pond. In some cases these last as long as 23–45 days, but most are much shorter. Newts exiting from the pond orient toward vegetation along an adjoining creek, and laboratory experiments suggest that the adults orient toward the dark horizon produced by the vegetation. The use of light and sun-compass orientation may complement orientation via olfaction (Landreth and Ferguson 1967).

Courtship of *T. granulosa* is very similar to that of *T. rivularis* and virtually identical to that of *T. t. torosa* (Davis and Twitty 1964). Most of the account that follows is from Propper's (1991) account of laboratory trials.

Courtship begins when a male dorsally amplexes an approaching female. The male places his forelimbs just behind the female's forelimbs and places his hindlimbs in front of the female's hindlimbs. The pair then engages in a prolonged courtship in which the male periodically rubs the female's snout with his chin using lateral head movements, and strokes the female's abdomen, shanks, and vent region with his hindlimbs. During this period the male presses his cloaca against the female's dorsum and the cloaca becomes distended as the rubbing phase progresses. This component of courtship lasts for several hours (range = 40 minutes to 2 days), and the intensity and frequency of limb stroking increase with time. Near the termination of this phase, the female becomes immobile and dorsally reflexes her head. This action presumably signals her sexual receptivity. Both adults expel air bubbles and the male soon dismounts to deposit a spermatophore.

After dismounting, the male curves his pelvic region slightly and moves it away from the female until his body is in front of and at a right angle to the female. The male deposits a spermatophore while the female maintains her snout near the male's cloaca. The male then pivots on one forelimb and rotates laterally while the female moves with the male and keeps her snout near the male's cloaca. This movement brings the female's cloaca into close proximity with the spermatophore. The female moves laterally back and forth until she aligns the spermatophore below her cloaca, then picks up the sperm cap. The male then usually circles behind the female and amplexes her again.

Pairs may remain in a quiescent state with no snout rubbing or limb stroking after amplexing a second time (Propper 1991). Pairs may remain amplexed for as long as 4 days; during this time, males engage in snout rubbing and limb stroking, which lead to additional spermatophore production in 37% of the cases. However, only 7% of females are successfully inseminated a second time because the original sperm cap blocks the uptake of a second. In general, mated females are unreceptive to courtship from additional males. Based on laboratory data, Propper (1991) surmised that females rarely mate with more than one male during the breeding season.

Janzen and Brodie (1989) observed numerous amplexed pairs as well as aggregates of two or more males with a female in an Oregon pond where males outnumber females by 3.8:1. Field observations indicate that an unpaired male typically remains stationary in the pond until a female approaches. The male then pursues and dorsally amplexes the female. During amplexus the male rubs his chin several times from side to side on the top of the female's head. If an unpaired male approaches an amplexed pair, the amplexed male may vigorously transport the female away with powerful strokes of his tail. If the amplexed pair does not escape, the unpaired male may try to dislodge the amplexed male. Unpaired stationary males are significantly smaller than paired males as well as males that attempt to dislodge amplexed pairs. Positive assortative pairing by body size does not occur. Males attempting to dislodge other males have proportionately longer tails than stationary unpaired males and amplexed males. Because tail size may influence the ability of amplexed males to escape intruder males and to surface with paired females to breathe, males with large tails may have an advantage in competing for females.

REPRODUCTIVE STRATEGY. Females begin ovipositing shortly after mating and attach their eggs singly to aquatic vegetation, rootlets, or detritus. Females observed on Vancouver Island lay eggs singly in the leaf nodes of *Ceratophyllum demersum* and often roll the leaves around the eggs in the process (Oliver and McCurdy 1974). Oviposition begins in late April and continues through July. Others have observed egg laying in late June and early July in British Columbia (Efford and Mathias 1969) and in late October in northern California (Garber and Garber 1978).

Freshly laid eggs are pigmented above and are surrounded by the perivitelline membrane and either two or three envelopes. Eggs measured by Twitty (1935) were on average 1.8 mm in diameter, whereas those measured by Bishop (1943) varied from 1.85 to 2.0

Fig. 296. *Taricha granulosa*; larva; Klickitat Co., Washington (W. P. Leonard).

mm in diameter. The outer membranes are 3–4 mm in diameter.

AQUATIC ECOLOGY. Larvae begin feeding on small invertebrates shortly after hatching. Larvae in one stream population are secretive during the day and hide beneath rocks or in vegetation (Licht and Brown 1967). Individuals congregate in warm microhabitats along the shallow margins of the stream during the morning, but disperse to cooler, deeper waters as temperatures rise to 30°C. Detailed descriptions of diurnal activity patterns and larval feeding behavior are not available.

Length of the larval period is variable depending on site conditions. Most larvae in lowland populations transform during late summer or early autumn after a 4- to 5-month larval period. In contrast, larvae in permanent, cold, high-elevation lakes overwinter and transform the following summer when about 1 year old. Larvae of *T. g. mazamae* hatch in August in Crater Lake and reach 65–95 mm TL by the following summer, when they transform (Farner and Kezer 1953). Larvae in west-central California streams commonly reach 75 mm TL before transforming in late summer or early autumn before the arrival of heavy autumnal rains (Twitty 1935). A small percentage of stream larvae may overwinter and transform the following spring or summer (Bishop 1943).

TERRESTRIAL ECOLOGY. Most aspects of the ecology of the juveniles are poorly documented. Juveniles presumably disperse away from breeding sites within a few weeks after transforming and do not return to water until sexually mature. About 300 metamorphs were observed dispersing from a lake on Vancouver Island between 28 September and 2 October (Oliver and McCurdy 1974). In California nearly equal numbers of adults and juveniles are collected in terrestrial traps, a finding that suggests that juveniles are as active on the ground surface as adults during the wetter portions of the year (Twitty et al. 1967a).

Chandler (1918) estimates that 4–5 years are required for individuals to become sexually mature. However, male and female *T. torosa* × *rivularis* hybrids initially breed 6–8 and 8 years, respectively, after marked metamorphs are released (Twitty 1961b).

Taricha granulosa is the most aquatic of the western *Taricha* species, and the adults often remain in permanent habitats for many months after breeding. Efford and Mathias (1969) estimate average adult density in Marion Lake, British Columbia, to be 0.03 newt/m^2, although local densities near the edge of the lake may be as high as 0.27 newt/m^2. Animals displaced as far as 550 m from their original point of capture almost always successfully return to their home ranges. One animal displaced 500 m swam home in 9 hours. Growth rates of males and females average 0.93 and 0.59 mm/year, respectively, and sex ratios are near 1:1.

Patterns of diurnal activity vary geographically. Individuals in an Oregon lake often hide among rocks on the lake bottom, but may occasionally rise to the water surface and float for a while before returning to the bottom (Evenden 1943). Individuals in Marion Lake are most active at night and are rarely seen or trapped during the day (Efford and Mathias 1969). In contrast, newts in ponds in Oregon tend to be equally active during the day and night (Pimentel 1960).

Large aggregates of *T. granulosa* have been documented in some populations. The most impressive was in a channel of a reservoir in Oregon, where an estimated 5000 animals were observed on 7 September in one aggregate composed of about equal numbers of males and females that covered about 18 m^2 (Coates et al. 1970). The number of animals in the aggregate slowly declined until only a few hundred newts were observed on 1 November. The newts were in postreproductive condition and appeared to be dispersing from the aggregate to terrestrial overwintering sites. On 6 September in Crater Lake, Farner and Kezer (1953) observed an aggregate of 259 newts that consisted of large larvae, metamorphs, and adults. They also collected groups of as many as 12–15 adults under the same cover object along the shores of Crater Lake.

Newts in Marion Lake, British Columbia, concentrate in areas containing thick clumps of vegetation or tree roots, and as many as 190 animals have been collected from a 15-m^2 area (Neish 1971). Areas of concentration tend to shift seasonally from shallow-water sites in the spring months to deep-water sites later in the season. Adults shift their home ranges periodically, although displaced animals readily home back to their place of capture.

Very little is known about the distribution and behavior of the terrestrial adults. Packer (1961) compared diets of adults collected on land with diets of those collected in water and found that all individuals have full stomachs. Thus adults that are forced to abandon streams during periods of high water quickly resume feeding on land.

Dietary studies have focused on the aquatic adults and no data are available on the diets of larvae or terrestrial juveniles. Larvae presumably feed on zooplankton, aquatic insects, and other aquatic organisms. The aquatic adults eat a broad array of prey, such as zooplankton, snails, aquatic insects, and amphibian eggs and larvae. Over 50 different prey items were recorded for 104 adults from five aquatic sites in Oregon, including cladocerans, snails, water boatmen, midges, phantom midges, and the eggs of conspecifics and hylid and ranid frogs (Evenden 1948). Mayflies are the most important prey overall. Specimens from Crater Lake contain amphipods, snails, aquatic insect larvae, and a few terrestrial arthropods (Farner 1947; Farner and Kezer 1953).

Prey in adults from Marion Lake include amphipods, terrestrial and aquatic insects, leeches, tadpoles, clams, cladocerans, ostracods, copepods, and oligochaetes (Efford and Tsumura 1973). Amphipods, terrestrial insects, and caddisflies compose most of the volume of food. The diet of *T. granulosa* is somewhat similar to that of other salamanders and fish in the lake, but the importance of competition between these species has not been examined. Adults in lowland populations also feed on *Ambystoma* and *Rana* eggs and eat the hatchlings of *Rana aurora* (Neish 1971; Nussbaum et al. 1983).

Dietary items from adults from a temporary and permanent pond near Corvallis, Oregon, include ostracods, dipterans, gastropods, nematodes, and numerous amphibian eggs (White 1977), whereas those from adults from four lake populations in Oregon include ostracods, copepods, amphipods, aquatic insects, clams, and spiders (Taylor 1984). The diet varies markedly among years and lakes. Within a given lake, the diet of *T. granulosa* overlaps significantly with that of microsympatric *Ambystoma gracile*.

PREDATORS AND DEFENSE. *T. granulosa* has skin glands that produce tetrodotoxin, a powerful neurotoxin that is a chemical defense against predators (Mosher et al. 1964; Twitty 1937; Wakeley et al. 1966). Experiments in which vertebrates either receive an injection of skin preparations or are force-fed newts indicate that potential predators are highly susceptible to the skin toxin (Brodie 1968c). Birds and mammals are extremely vulnerable; as little as 0.0002 cc of back skin can kill a white mouse in <10 minutes. Humans are also highly susceptible to tetrodotoxin, and consumption of newts can be fatal. A human died after ingesting a single *T. granulosa* on a dare, and a second individual became severely ill but survived after consuming five newts (Bradley and Klika 1981).

Snakes are generally less susceptible than mammals and birds. Amazingly, red-spotted garter snakes (*Thamnophis sirtalis concinnus*) eat adult *Taricha* without ill effects, although previous studies showed that some garter snakes will reject *Taricha* or suffer ill effects after eating them. Gregory (1978) compared the diets of three species of garter snakes on Vancouver Island and found that only *T. sirtalis* preys heavily on *T. granulosa* and other amphibians. Most *T. sirtalis* show no ill effects when force-fed whole newts, but *T. elegans* and *T. ordinoides* suffer loss of motor function (Macartney and Gregory 1981). Dietary analyses indicate that naive young of *T. elegans* and *T. sirtalis* are more similar in their preferences for amphibian odors and their tendency to attack live amphibians than are field-caught snakes. For adults, *T. sirtalis* is more likely than *T. elegans* to attack amphibians. In contrast, *T. ordinoides* shows a strong innate tendency to avoid amphibians. These data suggest that *T. elegans* may learn to avoid newts to some extent after one or more predatory encounters in the wild.

Taricha granulosa from Vancouver Island are far less toxic than those from mainland populations (Brodie and Brodie 1990, 1991). In addition, *T. sirtalis* from Vancouver Island and from areas outside the range of *T. granulosa* are less resistant to tetrodotoxin than snakes from areas where the newts are highly toxic.

Taricha granulosa is susceptible to its own toxin, but a large amount is necessary to cause detectable symptoms (Brodie 1968c). The eggs of *T. granulosa* are also toxic and contain tetrodotoxin (Brodie 1968c; Twitty and Johnson 1934). The skin of *T. granulosa* is several times more toxic than that of *T. rivularis* or *T. torosa* (Brodie et al. 1974a). However, the eggs of all three species had levels of toxins similar to that found in the skin of *T. torosa* and *T. rivularis*. Although newts cannibalize their own eggs, it is uncertain if poisoning results from oophagy.

Despite their well-documented toxicity, rough-skinned newts are preyed upon by a number of vertebrates. Besides *T. sirtalis*, which feeds on both the larvae and metamorphosed stages (Farner and Kezer 1953; Fitch 1936), a trout, a catfish, a bullfrog, and a mallard have been found that had recently eaten *Taricha* (Brodie 1968c; Farner and Kezer 1953). The catfish, bullfrog, and mallard were dead, but the trout appeared healthy. Fish occasionally eat newts (Chandler 1918; Vincent 1947), but strongly prefer more palatable salamanders such as *Ambystoma gracile* (Neish 1971).

When attacked, rough-skinned newts often assume an "unken" posture in which the back is depressed, the head is raised vertically, the eyes are closed, the tail is raised forward over the body, and the limbs are extended outward (Johnson and Brodie 1975; Riemer 1958; Stebbins 1951). Both the bright color of the venter and defensive posturing per se appear to have an aposematic function.

COMMUNITY ECOLOGY. In California the rough-skinned newt often coexists with *Taricha rivularis*, although the two forms tend to segregate to some extent, with *T. granulosa* preferring slower-flowing stream sections. Competitive interactions between larvae that share breeding streams have not been

Fig. 297. *Taricha granulosa;* defensive posturing; Curry Co., Oregon (J. W. Petranka).

examined, nor have the mechanisms that allow coexistence of these closely related species.

CONSERVATION BIOLOGY. Studies of forest stand characteristics suggest that populations of *T. granulosa* in the Cascades of southern Washington and the Coast Range of Washington reach their highest densities in mature and old-growth forests (Aubry and Hall 1991; Corn and Bury 1991). In contrast, populations in the Oregon Cascades reach relatively high densities in younger forest stands. Resource managers should consider these regional differences when designing management plans. Little information is available on the migratory corridors used by adults when moving to and from breeding sites. Forest buffers left around ponds and other breeding sites would provide optimal habitat for the juveniles and adults of these and many other amphibians that breed in standing-water habitats.

Taricha rivularis (Twitty)
Red-bellied Newt
PLATE 165

IDENTIFICATION. The red-bellied newt resembles other western *Taricha* species in being a stocky, medium to large salamander with grainy skin, indistinct costal grooves, and a dark-colored dorsum. This species differs from *T. granulosa* and *T. torosa* in having more prominent eyes, a brown iris, and a tomato red venter (Stebbins 1951). The dorsal ground color is brownish black in terrestrial adults and brownish in aquatic adults. Males in breeding condition have smooth skin, but the dorsal fin is less prominent and the cloacal lips are less enlarged than in other *Taricha* species (Twitty 1935). A broad dark band occurs across the cloaca of males that is usually absent in females. Adults reach 14–19.5 cm TL.

Range of *Taricha rivularis*

Hatchlings are the pond type, but the balancers are incompletely developed or absent and the dorsal fin is not as well developed as in other *Taricha* species. Hatchlings and older larvae have numerous, fine, evenly dispersed black spots along the sides and dorsum of the body.

SYSTEMATICS AND GEOGRAPHIC VARIATION. No subspecies are recognized. Davis and Twitty (1964) found that on rare occasions local populations contain hybrids between *T. granulosa* and *T. rivularis*.

DISTRIBUTION AND ADULT HABITAT. The red-bellied newt is found in Sonoma, Mendocino, and Humboldt counties in northwestern California. This species is found predominantly in redwood forests along the northern California coast. Adults breed in clean, rocky streams with moderate to fast flow. This species avoids ponds and other standing-water habitats that are used by other *Taricha* species.

BREEDING AND COURTSHIP. Almost all of our understanding of the natural history of this species is based on the elegant studies by Victor Twitty and his colleagues of the breeding ecology of red-bellied newts that inhabit Pepperwood Creek in Sonoma Co.,

California. Except where noted, the life history summary that follows is based on studies at this site.

Breeding of *T. rivularis* is closely tied to the time when streams begin to recede as heavy winter and spring rains taper off seasonally. The adults move randomly across the ground during the autumn months, but begin moving directionally toward breeding sites in late January (Packer 1960, 1961, 1963). Individuals primarily move toward the stream during the first 5 hours following sunset, and a peak in movement occurs in early March. Adults often migrate on days without rain, and heavy rains inhibit movements.

Fig. 298. *Taricha rivularis;* adult; Sonoma Co., California (E. D. Brodie, Jr.).

During days with no rainfall, migrations to the stream are generally greater on warm, humid nights.

Males begin arriving at the breeding streams 1–3 weeks prior to the first arrival of females, and most males are in the breeding streams when the arrival of females reaches its seasonal peak (Twitty 1942, 1961a). Newts often leave the water during periods of heavy rain, when there is a greater risk of being injured or killed. Breeding in one population studied by Twitty (1942) occurs primarily in April, whereas breeding at Pepperwood Creek usually begins in early March, peaks in mid- to late March, and ends by mid- to late April (Twitty 1961a). Females court within a few days after entering the streams, and both sexes abruptly leave after breeding. Data on recapture success of marked individuals indicate that most males breed annually, whereas most females skip one or more years between breeding bouts (Twitty 1961a).

After leaving the stream the adults fan out over a gradually widening area, but almost always remain in the same drainage (Twitty et al. 1967a). Only 1 of 22,000 marked animals moved from its home stream to a neighboring stream. Adults tend to move in an upstream direction after exiting streams (Twitty et al. 1967b).

Adults return to the same segment of Pepperwood Creek to breed year after year, and most animals return to within 15 m of their original capture point in subsequent breeding years (Packer 1963; Twitty 1959; Twitty et al. 1967a). Although migrations occur primarily on land, some individuals enter the water near the end of the migration and move to home sites. Within a given year most remain within a 15-m length of stream. Some individuals may move as far as 76–152 m from their initial point of capture. In extreme cases, males move over 274 m from their initial point of capture. These unusually long movements are associated with breeding activity, and are usually followed by a return to the home area. Animals displaced up to 150 m downstream from their original point of capture successfully return home (Packer 1962).

Through a series of displacement experiments, Twitty and his colleagues found that adults have remarkable homing abilities. Many adults can successfully return to their home stretch of stream when displaced 0.8 km downstream or 1.6 km upstream from their site of capture. Of about 1000 females displaced roughly 5 km from the streams in another watershed, 18 returned to the home stream over 2 years later, and many more in the years that followed (Twitty 1959; Twitty et al. 1967a). In another study, 77% of males displaced to an adjoining creek over 3 km from their home stream returned home within 3 years after their release. When displaced 4 km, 38% of 747 males returned to the home stream (Twitty et al. 1964).

Animals displaced distances less than 210 m orient strongly toward home, whereas those displaced greater distances tend to orient randomly when released. Blinded animals are equally effective at homing as sighted controls, and there is little evidence that the animals orient by memorizing topographic features in their environment (Twitty et al. 1967a). Individuals normally orient and begin moving toward their homes immediately after they are released (Twitty et al. 1967c).

Twitty and his associates surmised that olfaction is an important cue in orientation, and later studies support this hypothesis (Grant et al. 1968). Other studies, however, indicate that *T. granulosa* may orient using celestial cues in addition to olfaction (Landreth and Ferguson 1967). There is also evidence that *T. torosa* uses kinesthetic orientation, in which sensory data on body position in the immediate past are used to keep the individual on a straight path after maneuvering around obstacles (Endler 1970). All three factors presumably play a role in facilitating navigation by *Taricha* species. Celestial cues are probably important at the start of the journey in giving the proper initial orientation toward streams or ponds. Kinesthesia would be most important during the middle part of the journey, when newts often have to traverse rugged terrain and move around logs or other obstacles. Finally, olfaction may play an important role in locating home areas toward the end of a migration.

Davis and Twitty (1964) observed over 100 complete courtships of *T. rivularis*. The following account is based on their generalized observations, and, with minor exceptions, also applies to *T. granulosa* and *T. torosa*.

Upon recognizing a female, a male clasps some part of the female's body, then quickly positions himself over the back and clasps the pectoral region. The male's forelimbs are placed immediately behind those of the female, and the hindlimbs dangle free or are pressed against the female's hindquarters. The male usually hooks his chin over the female's snout, which he rubs vigorously by lateral movements of his head. At the same time, he presses his cloaca spasmodically against the female's dorsum and strokes the abdomen, shanks, and vent region with his hindlimbs. The male's vent becomes distended as the rubbing phase progresses. With time, chin rubbing increases and the male begins to undulate his tail back and forth. The amplexed pair usually engages in this phase of courtship for about 1 hour, with the female frequently moving about. As mating progresses, the female's vent also becomes distended.

Eventually, the female raises her snout slightly and the male rubs his throat and chin on it vigorously. He then releases a bubble of air, moves off to the left or right side of the female's head, and positions himself at an oblique angle with his flank about 3 cm from the female's snout. The male then spreads his hindlimbs, presses his cloacal lips over the substrate, and deposits a spermatophore. He then curves his pelvic region and moves his hips away from the female until his body is at a right angle to the female. The male next begins rocking laterally to and fro while spasmodically striking the substratum with his outstretched feet. This activity continues for 15–30 seconds.

The female sidesteps along the male's flank toward his curved pelvic region. This action brings the female's cloaca into close contact with the spermatophore, at which point she spreads her legs and picks it up. In some cases the female advances upon the male while he is depositing a spermatophore and either pushes parts of his flank with her front feet or walks over him. In this situation the distance the male moves his hips from the spermatophore is altered, and the female may fail to encounter the spermatophore with her cloaca as she moves sideways along the male's flank. In such cases the female usually realigns herself at various distances from the male and probes with her vent for the spermatophore. Persistent females often find spermatophores, but courtship success is diminished under these circumstances. The male may circle behind the female and amplex her a second time, or court other females. Females appear to court many times before ovipositing. Males at Pepperwood Creek outnumber females by as much as 12:1, but this unbalanced sex ratio may have been exaggerated by the removal of females from the population (Twitty 1961a).

The spermatophore has a wide, oval base that tapers into a narrow, question mark-shaped stalk with an open center. Anterior processes curve forward as free arms. The sperm cap weakly adheres to the apex of the central unit of the stalk. The oval, gelatinous base measures 10 × 16 mm on average, and the entire spermatophore stands about 10 mm high. The spermatophore of this species is more adhesive than that of other *Taricha* species, a trait that may be an adaptation to breeding in fast-flowing water.

Taricha granulosa and *T. rivularis* often coexist in streams, but rarely court each other. Crosses indicate that these species are fully interfertile even though hybridization is very rare in nature (Twitty 1961b). Chemical cues play a critical role in mediating courtship and preventing interbreeding. Male *T. rivularis* are attracted to sponges soaked with conspecific female skin secretions, but ignore control sponges with no secretions (Twitty 1955).

REPRODUCTIVE STRATEGY. Shortly after mating, females deposit their eggs in small masses on the undersurfaces or overhanging edges of stones, or attached to submerged rootlets (Mosher et al. 1964; Twitty 1935, 1942). The masses are laid in fast-flowing water, are usually only one egg layer thick, and appear flattened. The eggs are dark gray to grayish brown above and have an average diameter of 2.75 mm. As many as 70 egg masses may be attached to the underside of a small stone (Twitty 1935, 1942). The number of eggs in a sample of 20 masses varies from 5 to 15 and averages 9.

The incubation period in the laboratory is 16–20 days at 23°C and 30–34 days at 15°C (Licht and Brown 1967). Incubation periods in nature are intermediate between these values.

AQUATIC ECOLOGY. Most aspects of the larval stage have not been studied. Larvae observed by Licht and Brown (1967) hatch before late April and transform about 4 months later in late August. Larvae at this site remain in vegetation or under stones during the day. During the early morning hours, individuals concentrate in warm microhabitats near the effluent of hot springs and along the shallow edges of the stream until temperatures exceed 30°C. In laboratory studies larvae prefer temperatures of 20–26°C. The larvae transform when they reach 45–55 mm TL (Stebbins 1951).

TERRESTRIAL ECOLOGY. Juveniles are secretive, and many aspects of their natural history have not been studied. Metamorphs disperse into forests surrounding the breeding streams, but the time required to reach permanent terrestrial sites is not known. Twitty et al. (1967a) caught adults or subadults in terrestrial traps about 58 times more often than juveniles. This finding suggests that juveniles spend much of their time in subsurface retreats and are rarely active on the ground surface until they near sexual maturity. The juvenile stage lasts about 5 years (Licht and Brown 1967).

The adults begin emerging from underground retreats within a few weeks following the onset of fall rains (Packer 1960, 1961). Unlike some California salamanders, the adults of this species usually do not emerge in the fall until several bouts of rain have occurred. Emergence typically occurs non-synchronously over a period of several months. Adults are active outside cover primarily at night, but a small percentage of individuals are sometimes active during the day during periods of rainy weather or when in heavily shaded microhabitats.

Recapture data indicate that annual survivorship of adults is very high, and it is probably >90% in most years (Twitty 1961b). The long juvenile stage and high adult survivorship suggest that many adults live 12–15 years; an estimated 40% of marked animals survive for 11 years. Licht and Brown (1967) note that 48 adults died after accidentally entering hot springs adjoining a breeding stream.

Almost no data are available on the diet of *T. rivularis*. Insects and other small invertebrates presumably make up the bulk of the diet of both the larvae and the adults. The adults do not feed while in breeding streams (Packer 1961).

PREDATORS AND DEFENSE. When attacked, *T. rivularis* will often assume an "unken" posture with the tail and head elevated to expose the bright tomato red ventral surface, which serves as warning coloration (Brodie 1977). This and other *Taricha* species produce tetrodotoxin in their skin and are toxic to predators (Brodie et al. 1974a). Details of chemical defenses in this and other *Taricha* species are given in the account of *T. granulosa*.

COMMUNITY ECOLOGY. Breeding adults reach very high densities in many streams and deposit large numbers of eggs. The larvae are predators on stream invertebrates, but their role in structuring invertebrate communities has not been investigated.

CONSERVATION BIOLOGY. *Taricha rivularis* is common in many streams, but long-term population trends of this and other western salamanders are undocumented and in need of study.

COMMENTS. Twitty (1966) presents a delightful summary of much of his lifetime work with *T. rivularis* that is highly recommended reading. Although the life history studies conducted at Pepperwood Creek are very detailed, comparative studies of other populations are needed to determine the extent to which life history characteristics vary geographically.

Taricha torosa (Rathke)
California Newt
PLATES 166, 167

IDENTIFICATION. The California newt superficially resembles the rough-skinned newt, *T. granulosa*, but has larger eyes, vomerine teeth that form a Y-shaped pattern, and light-colored lower eyelids. Terrestrial adults are yellowish brown to dark brown above and pale yellow to orange below. Unlike those of *T. rivularis*, the iris is yellow and the belly is never tomato red. Individuals in most populations develop warty skin. During the breeding season males develop smooth skin, a lighter body color, swollen cloacal glands, enlarged mental glands, enlarged tail fins, and minute cornified papillae on the toe tips and bases of the hindlimbs (Storer 1925). Adults measure 12.5–20

Fig. 299. *Taricha torosa*; adult; California (R. W. Van Devender).

cm TL, and both albino larvae and adults have occasionally been found (Dyrkacz 1981; Riemer 1958; Wells 1963).

Hatchlings are the pond type and are light yellow above with two dark, narrow bands on the back. The dark bands are regular in *T. t. torosa* and irregular in *T. t. sierrae* (Twitty 1942). Dark spots or blotches occur along the sides of the body of *T. t. sierrae*, but are usually absent in *T. t. torosa*. Color patterns of older larvae are generally similar to those of hatchlings; however, the sides of the body of *T. t. sierrae* tend to have more prominent dark blotches. Hatchling *T. t. torosa* and *T. t. sierrae* reach 10-14 mm TL and 13-14 mm TL, respectively (Riemer 1958; Stebbins 1951).

Taricha granulosa and *T. torosa* are often difficult to distinguish in areas of sympatry in northern coastal areas and in the northern Sierra Nevada. Differences in relative eye size (smaller in *T. torosa*) and defensive posturing induced by tapping on the body (tail tip held straight out in *T. torosa*; tail tip curled in sympatric *T. granulosa*) are the most useful traits for identifying animals in the field.

SYSTEMATICS AND GEOGRAPHIC VARIATION. The California newt exhibits much regional differentiation in coloration, genetic structure, and life history parameters. As many as three subspecies have been recognized (Riemer 1958), but most authorities currently recognize only two that are allopatric forms found in the Coast Range and Sierra Nevada (Riemer 1958; Stebbins 1985). A geographic isolate in San Diego Co., California, has exceptionally warty skin and has been treated by some as a separate subspecies (*T. t. klauberi*).

The Sierra newt (*T. t. sierrae*) is reddish to chocolate brown above and burnt orange to yellow below. The eyelids and snout have conspicuous light coloring and the tail fins of breeding males are less developed than those in the nominate subspecies. The Coast Range newt (*T. t. torosa*) is yellowish to dark brown above and pale yellow to orange below. The eyelids and snout are not as conspicuously colored as those of *T. t. sierrae*, and breeding males develop more pronounced tail fins.

The two subspecies are geographically isolated from each other and have diverged moderately in adult morphology, breeding habitats, egg size, and larval coloration. Hedgecock (1976) and Hedgecock and Ayala (1974) found much greater genetic similarities between subspecies of *T. torosa* than between *T. torosa*

and *T. granulosa*. Three genetic subgroups are evident that correspond to *T. t. sierrae* and northern and southern groups of *T. t. torosa*. Genetic distances (*D*) between five populations vary from 0.13 to 0.27. Coates (1967) also found evidence of northern and southern races of coastal forms based on analyses of serum proteins.

Tan and Wake (1995) examined mitochondrial DNA variation in cytochrome *b* and recognize five subgroups within *T. torosa* that include a southern and northern cluster of *T. t. torosa* and northern, central, and southern clusters of *T. t. sierrae*. All of these studies indicate significant genetic substructuring of *T. torosa*. Data presented by Tan and Wake (1995) suggest a rather complex biogeographic history within *Taricha* that led to the current distributions of the genetic subgroups. Collins (1991) recognized *T. t. sierrae* as a separate species using arguments based on evolutionary species concepts. His suggested nomenclatural change was immediately challenged (Frost et al. 1992; Montanucci 1992; Van Devender et al. 1992) and is not used here.

DISTRIBUTION AND ADULT HABITAT. The California newt occurs as two major allopatric groups, one in the Coast Range that extends from northwestern to extreme southwestern California, and the second in the Sierra Nevada. Adults inhabit a variety of terrestrial and aquatic habitats from near sea level to 2000 m.

Populations in northern California are found in mesic forests in rolling or mountainous terrain, whereas populations farther south occupy drier habitats such as oak forests, chaparral, or rolling grasslands. In the Sierra Nevada, California newts inhabit digger pine-blue oak communities and ponderosa pine communities that are drier than forests used by northern coastal populations. Adults breed in both lentic and lotic habitats that include ditches, farm ponds, lakes, and streams. Stream-breeding populations in coastal populations typically breed in sluggish pools in streams, whereas populations in the Sierra Nevada often breed in faster-flowing water. In general, *Taricha t. sierrae* is better adapted to fast-flowing streams than *T. t. torosa*.

BREEDING AND COURTSHIP. The breeding season varies among regional populations and years and may occur anytime between late December and

Range of *Taricha torosa*

- T. t. sierrae
- T. t. torosa

early May depending on elevation, local site conditions, breeding habitats, and seasonal patterns of rainfall. Courting individuals were first observed on 26 April in a stream in southern California, but breeding in a second stream only 10 km away began nearly 2 months earlier (Brame 1968). Amplexed adults have been observed near Berkeley as early as late September, but breeding in this region typically begins in January at the very earliest (Stebbins 1951; D. B. Wake, pers. comm.). At any location, the breeding season typically lasts 6–12 weeks. The breeding season at a central California site lasts from late December to early February (Miller and Robbins 1954).

Large breeding congregations of *T. t. torosa* may form near the mouths of small streams that empty into ponds and lakes (Twitty 1942). In some areas there are two seasonal peaks in breeding: one in ponds in late December and early January that is associated with heavy winter rains and a second in streams around March. The latter is associated with a drop in stream flow as winter flooding subsides. Stream-breeding populations generally curtail breeding during periods of high water.

Adult *T. t. sierrae* normally migrate to breeding streams in January and February (Stebbins 1951). This subspecies shows a stronger tendency to use streams, but occasionally uses ditches and other bodies of water with minimal current (Twitty 1942). Breeding activity has been observed from early March through early May depending on local site conditions (Riemer 1958; Twitty 1942).

Secondary sexual characteristics begin to develop in male *T. t. torosa* about 2 months prior to the initiation of breeding and are lost shortly after males depart from the breeding sites. Development progresses slowly and reaches a peak soon after males enter aquatic sites to mate (Miller and Robbins 1954). Males begin arriving at breeding sites before females and remain for a longer time after breeding (Stebbins 1951; Twitty 1942). Adults monitored by Miller and Robbins (1954) require 6–8 weeks to reach the breeding sites, and adults of both sexes breed every other year.

Courtship of *Taricha t. torosa* is nearly identical to that of *T. rivularis* (Davis and Twitty 1964; Smith 1941a). When courting, a male *T. torosa* first amplexes a female just posterior to her shoulder region, then hooks his chin over her nose and rubs it with lateral head motions while pressing the cloaca against her dorsum. After about 1 hour of intermittent periods of head rubbing and tail fluttering, the male dismounts and deposits a spermatophore. He orients at a right angle to the female's body and makes lateral hip movements. The female then moves forward and picks up the spermatophore. Details of this general sequence are presented in the account of *T. rivularis*.

One difference in courtship behavior between these species is that the male of *Taricha t. torosa* curls the tail forward and periodically flutters it when amplexing the female. In addition, the male does not curve his pelvic region as markedly after spermatophore deposition, and often swims about with the female during the rubbing phase.

Males usually produce fewer than four spermatophores. Each spermatophore consists of a broad, oval, gelatinous base about 16 mm in greatest diameter. The base tapers into a three-pronged stalk containing an oval opening near the middle. The stalk slants posteriorly, whereas the three prongs project anteriorly. The two lateral prongs are relatively short; the median prong is longer and supports the sperm cap at the top (Smith 1941b). The entire structure is 11–12 mm high.

Males typically outnumber females at the breeding sites and compete intensely for mates. During the early part of the breeding season, when males greatly outnumber females, females are sometimes surrounded by tangled masses of males (Smith 1941a).

REPRODUCTIVE STRATEGY. Shortly after mating, females attach spherical masses of eggs to stones, roots, twigs, or branches in ponds or streams. The ova are pale brown to dark brown above, cream colored below, and surrounded by a vitelline membrane and two jelly layers. Freshly laid eggs are 1.9–2.8 mm in diameter on average.

The jelly coats swell within a few hours of egg deposition to form a firm mass. The masses measure 15–25 mm in diameter on average and typically contain 7–30 eggs. The mean number of eggs per mass in 24 *T. t. torosa* masses from a pond in west-central California was 17 (Storer 1925) compared with 19–23 in several populations in southern California (Brame 1956, 1968; Mosher et al. 1964). Female *T. t. sierrae* in streams attach masses of 11–22 eggs to the sides and bottoms of stones in relatively swift water. Females that breed in ditches with little current may attach the egg masses to exposed roots or other support structures (Twitty 1942).

Female *T. t. torosa* often cover the bottoms of pond margins with egg masses and sometimes oviposit en masse where small streams enter ponds and lakes (Twitty 1942). Females often lay eggs in water <15 cm

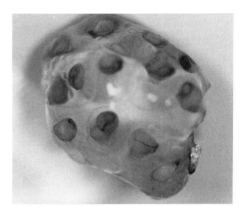

Fig. 300. *Taricha torosa;* egg mass; southern California (L. B. Kats).

deep, and sometimes deposit their eggs directly on pond bottoms unattached to support structures (Mosher et al. 1964; Stebbins 1951). Unlike the Sierra newt, *T. t. torosa* shows little tendency to oviposit in cryptic sites. These differences may reflect the fact that *T. t. sierrae* oviposits in faster-flowing water where exposed egg masses could be washed from rocks.

Taricha t. sierrae produces larger eggs than *T. t. torosa* (Kaplan 1985). Estimates of mean egg diameters are 2.8 mm for *T. t. sierrae* (Twitty 1942) compared with 2.1 mm (Miller and Robbins 1954), 2.06 mm (Storer 1925), and 2.0 mm and 2.2 mm (estimated from Fig. 1 in Kaplan 1985) for *T. t. torosa*.

Mean clutch size presumably differs among subspecies, but few comparative data are available. Estimates of clutch size for *T. t. torosa* are 130–160 mature ova in specimens examined by Miller and Robbins (1954) and 133 eggs laid by a captive female (Brame 1968). Data on clutch size of *T. t. sierrae* are not available.

Duration of the incubation period depends on local water temperatures and varies substantially among populations. Estimates include 52 days for a pond population in west-central California (Storer 1925) and 14–21 days in southern California populations (Mosher et al. 1964).

AQUATIC ECOLOGY. Most aspects of the natural history of the larval stage have not been studied. The larval period lasts for several months, with transformation occurring during the summer or early autumn (Riemer 1958). Larvae of *T. t. torosa* usually transform at 50–60 mm TL beginning in early September and continuing for several months (Bishop

Fig. 301. *Taricha torosa;* larva; Los Angeles Co., California (L. B. Kats).

1943). Specimens of *T. t. sierrae* taken in late August vary from 55 to 62 mm TL. There are no records of overwintering larvae.

Kaplan (1985) conducted laboratory experiments to examine the ecological consequences of variation in egg size. When food is not limiting, *T. t. torosa* larvae from large eggs transform sooner, and at a larger size, than larvae from small eggs. When larvae are food stressed, larger larvae require longer to transform and are larger at metamorphosis. Thus both large and small eggs may be advantageous depending on environmental conditions, such as food level and the seasonal duration of the breeding site.

Unlike adult *T. rivularis,* adult *T. torosa* feed when in breeding streams and regularly cannibalize their own eggs and larvae. Elliott et al. (1993) and Kats et al. (1994) examined chemically mediated avoidance of cannibalistic adults by larvae in a southern California stream. At their study site adult density declines abruptly after breeding ceases, then continues to decline throughout the summer to near zero as the few adults that remain after breeding slowly exit the stream. When exposed to water containing chemical cues from potentially cannibalistic adults, 2-week-old larvae respond by moving under cover, but 5-week-old larvae do not. This behavioral shift corresponds to a time when most adults are leaving the stream and cannibalism risk is declining markedly. In natural streams, larvae that move beneath rocks may have reduced foraging opportunities and be more vulnerable to predators that reside beneath cover. Thus the ontogenetic shift in larval behavior appears to be an adaptive response to conflicting risks associated with finding food and being eaten.

Cannibalism of eggs and larvae may be an important source of mortality in some populations, and a female may sometimes attempt to eat the eggs protruding from the cloaca of a second female (Kaplan and Sherman 1980). As many as 71 instances of oophagy have been observed during a 9-hour period in a population in Contra Costa Co., California (Marshall et al. 1990). Females eat eggs much more often than males, and individuals often eat eggs as they protrude from a female's vent. Individuals in this population primarily eat egg masses that are recently laid and have not swollen into firm masses. In a southern California stream, the adults regularly feed on firm egg masses (L. B. Kats, pers. comm.). Oophagy not only provides a high-quality meal, but also eliminates conspecific competitors during the larval stage.

TERRESTRIAL ECOLOGY. Metamorphs disperse from streams and ponds and live in surrounding habitats until sexually mature. The juveniles appear to spend much of their time beneath ground or cover and are rarely observed above ground in large numbers. The duration of the juvenile stage is not known, but it probably lasts 5–8 years as in other *Taricha* species. The smallest sexually mature males and females collected by Miller and Robbins (1954) measured 70 and 63 mm SVL, respectively.

Most adults leave the water within a few weeks after breeding and remain in subsurface retreats during the dry summer months. The adults emerge with the onset of autumn rains and are active on the ground surface during the autumn and winter months. Stebbins (1951) notes that 14 aestivating adults were dug from a hole next to a boulder about 1.5 m from a stream, but one animal was collected 61 m from water.

Adults make a series of sounds that may serve various purposes (Davis and Brattstrom 1975). Clicking sounds are produced when newts encounter other newts or are placed in an unfamiliar location. In some instances, individuals may click during what appears to be the defense of perch sites in aquaria. When approached by an intruder, a resident will sometimes rise high on its legs, wag its tail, and click. The intruder may withdraw or engage in similar posturing. Squeaking sounds are emitted whenever newts are picked up, whereas a whistling sound is produced when the middle of the back is contacted by other newts or by an experimenter. The latter sound has only been observed during the mating season and may possibly function in species recognition.

The diet of adults consists of earthworms, snails, slugs, sowbugs, numerous types of insects, and the eggs and larvae of conspecifics (Elliott et al. 1993; Kats et al. 1992, 1994; Ritter 1897). A hatchling bird was found in the stomach of one adult, along with invertebrates and larval newts (Hanson et al. 1994).

PREDATORS AND DEFENSE. Like other species of *Taricha*, this species possesses a potent neurotoxin (tetrodotoxin) in its skin, ova, and ovaries that serves as an antipredator defense (Buchwald et al. 1964; Hubbard 1903; Mosher et al. 1964; Twitty 1937; Wakeley et al. 1966). Despite their toxicity, the eggs and larvae are preyed upon by crayfish and mosquitofish (Gamradt and Kats 1996). When molested, adults assume an "unken" (defensive) posture in which the back is depressed, the head is raised vertically, the tail is raised forward over the body, the eyes are closed, and the limbs are extended outward (Johnson and Brodie 1975; Riemer 1958).

COMMUNITY ECOLOGY. *Taricha torosa* is the only species of salamander that breeds in streams in central and southern California, and larvae have a monopoly on invertebrate food resources. The importance of larvae in organizing aquatic invertebrate communities in breeding streams has not been determined.

CONSERVATION BIOLOGY. Gamradt and Kats (1996) provide evidence that the introduction of crayfish (*Procambarus clarkii*) and mosquitofish (*Gambusia affinis*) into streams in southern California is adversely affecting local populations of *T. torosa*. These species prey heavily on the eggs and larvae of California newts and can cause catastrophic mortality. Field surveys indicate that newt populations have declined markedly following the introduction of these exotics into streams in the Santa Monica Mountains.

COMMENTS. Storer (1925) recognized only one *Taricha* species (*T. torosa*). Because his detailed accounts of the geographic variation and natural history of "*T. torosa*" probably represent mixed accounts of two or three species, I have relied primarily on more recent literature for summarizing the natural history of each species.

Family Sirenidae
Sirens

The sirenids are a small group consisting of only two extant genera (*Pseudobranchus* and *Siren*) and four extant species that are restricted to the eastern United States and northeastern Mexico. Fossils of sirenids extend back to the upper Cretaceous. Fossils of *Siren* have been found from the middle Eocene of Wyoming, the middle Miocene of Nebraska and Texas, and the lower Miocene and Pleistocene of Florida, whereas fossils of *Pseudobranchus* occur in Pliocene and Pleistocene deposits in Florida (Estes 1963, 1981; Holman and Voorhies 1985; Martof 1974).

Adults have long, slender, eel-shaped bodies, and the largest species (*Siren lacertina*) can reach nearly 1 m in length. All species are perennibranchs and exhibit larval traits such as the absence of eyelids and the presence of gill slits (one in *Pseudobranchus* and three in *Siren*) and external gills. Sirens lack rear legs and have a horny beak instead of premaxillary teeth. The front limbs are greatly reduced and have only three (*Pseudobranchus*) or four (*Siren*) toes per limb. Females lack spermathecae and males lack cloacal glands. Males do not produce spermatophores and fertilization is presumed to be external. Duellman and Trueb (1986) provide a detailed list of technical features that define the family.

Pseudobranchus axanthus (Netting and Goin)
Southern Dwarf Siren
PLATE 168

Fig. 302. *Pseudobranchus a. belli*; adult; Glades Co., Florida (R. W. Van Devender).

IDENTIFICATION. The southern dwarf siren is a slender, elongated, slimy salamander that lacks hindlimbs and has minute forelimbs, each with three toes (Martof 1972). The adults are permanently gilled and have conspicuous, bushy gills. The body is circular in cross section and there are 29–37 costal grooves. The adults have a brownish black to light gray ground color. Parallel yellow or tan stripes occur on the back and sides from the head to the tip of the tail. Striping pattern and coloration vary among subspecies and are described in detail in the next section. Males and females are similar in external coloration and morphology, but differ in maximum size. The average TL of the five longest females collected by Netting and Goin (1942) was 28% greater than that of the five longest males. Adults measure 10–25 cm TL.

The hatchlings have pond-type morphology with dorsal fins that extend from the head to the tail tip and reach 10.0–11.5 mm SVL and 14.5–16.0 mm TL. Hatchlings of *P. a. axanthus* are brown above with cream-colored stripes on the middorsal and lateral portions of the body and head (Goin 1947). This species is closely allied to *P. striatus*, and specimens of both species are best identified based on the collection locality.

SYSTEMATICS AND GEOGRAPHIC VARIATION. Until recently *Pseudobranchus* was thought to be a monotypic genus consisting of one species with five recognized subspecies (Conant 1975; Martof 1972). Chromosomal studies by Moler and Kezer (1993) indicate that *Pseudobranchus* includes at least two species, one with n = 24 chromosomes (*P. striatus*) and a second with n = 32 chromosomes (*P. axanthus*). *Pseudobranchus striatus* corresponds to three of the five previously recognized subspecies (*P. s. lustricolus*, *P. s. spheniscus*, and *P. s. striatus*), whereas *P. axanthus* corresponds to the remaining two (*P. s. axanthus* and *P. s. belli*). These species have mostly parapatric distributions, but areas of sympatry occur in portions of the upper Florida peninsula. *Pseudobranchus axanthus* and *P. striatus* occur syntopically in at least three locations and show evidence of microspatial segregation; *P. axanthus* prefers open marshes and prairie ponds and *P. striatus* prefers cypress swamps in acid pine flatwoods (Moler and Kezer 1993).

The southern dwarf siren consists of two subspecies. The Everglades dwarf siren (*P. a. belli*) has three narrow light lines within the middorsal stripe and two wider buffy bands along each side of the body. The venter is gray, the dorsal ground color is brownish, and there are usually 29–33 costal grooves. The narrow-striped dwarf siren (*P. a. axanthus*) has a gray ground color that lacks sharply defined light stripes. The lateral stripes are poorly defined and grayish, and there are 34–37 costal grooves. Zones of intergradation are not well documented.

DISTRIBUTION AND ADULT HABITAT. The southern dwarf siren inhabits cypress ponds, swamps, ditches, marshes, limestone sinks, and other permanent and semipermanent aquatic habitats in peninsular Florida (Moler and Kezer 1993). Individuals often congregate in the fibrous roots of water hyacinth, a floating aquatic plant that has been introduced throughout the southern Coastal Plain.

BREEDING AND COURTSHIP. No information is available on the mating season and courtship behavior.

REPRODUCTIVE STRATEGY. The oviposition period lasts from early November through March.

☐ P. a. axanthus
☐ P. a. belli

Range of *Pseudobranchus axanthus*

Eggs of *P. a. axanthus* have been found in November, February, and March attached singly by their adhesive outer capsules to water hyacinth roots 15–30 cm below the water surface (Goin 1947). The eggs are widely scattered throughout the breeding site and the outer egg capsules are 7–9 mm in diameter. Netting and Goin (1942) found a single egg attached to a water hyacinth root on 31 March and collected females with 18–36 large, pigmented eggs in February, March, April, and October (one specimen for the latter month). The maximum diameter of mature ovarian eggs is 3.1 mm.

Freshly laid ova are brownish above, average 3 mm in diameter, and are surrounded by the vitelline membrane and three jelly envelopes. The inner envelope is so thin that the eggs often superficially appear to be surrounded by only two envelopes (Noble and Richards 1932). The embryos have limb buds at hatching (Noble 1927b).

AQUATIC ECOLOGY. Dwarf sirens feed on a variety of aquatic invertebrates. Dietary items include oligochaetes in 12 specimens from southern Florida (Duellman and Schwartz 1958), chironomids and amphipods in 25 specimens from Florida (Carr 1940), and amphipods, chironomids, and ostracods in specimens from northern Florida (Freeman 1967). Captive individuals will eat the eggs of conspecifics (Noble 1930).

Data on growth rates, age and size at first reproduction, and many other aspects of the juvenile and adult stages are not available. Dwarf sirens often inhabit semipermanent habitats and may aestivate in subsurface burrows if the habitats dry. Individuals reduce their metabolic rates by as much as 60–70% when aestivating compared with the rates when they are living in water (Etheridge 1990b). Individuals at one pond site aestivate in S-shaped burrows 10–30 cm below the ground surface (Freeman 1958). Aestivating

Fig. 303. *Pseudobranchus a. axanthus;* hatchling; Florida (K. Haker).

animals have reduced gills and nonslimy bodies. In the laboratory, animals can survive in dried mud for >2 months with little ill effect.

Adult *Pseudobranchus* use their lungs, gills, and integument for gas exchange. Individuals can remain submerged and respire through their gills and skin whenever oxygen levels are relatively high in ponds (Ultsch 1971). However, when oxygen levels drop, individuals must surface periodically to breathe through their lungs. Water hyacinth mats provide cover, abundant food, and a low-oxygen environment that may possibly serve as a physiological refuge from predatory centrarchid fishes that are restricted to open water with higher oxygen levels.

PREDATORS AND DEFENSE. *Pseudobranchus axanthus* is undoubtedly preyed upon by a variety of predators that inhabit swampy habitats in the Coastal Plain, such as wading birds, turtles, alligators, and aquatic snakes. Known predators include the striped crayfish snake (*Regina alleni;* Van Hyning 1932), the mud snake (*Farancia abacura;* Carr 1940), and the southern water snake (*Nerodia fasciata;* P. E. Moler, pers. comm.). Specimens sometimes emit high-pitched, faint yelps when picked up or prodded (Carr 1940; Neill 1952).

COMMUNITY ECOLOGY. Moler and Kezer (1993) found evidence of microspatial segregation between *P. axanthus* and *P. striatus*, which suggests that competitive exclusion may be occurring in some regions.

CONSERVATION BIOLOGY. We currently have no data on long-term changes in populations of *P. axanthus,* although many local populations have undoubtedly been eliminated by the destruction of wetlands in Florida. This species is often locally abundant where suitable habitat remains. A long-term program to monitor these and other amphibians that are at risk of decline is needed for many species in the South.

Pseudobranchus striatus LeConte
Northern Dwarf Siren
PLATE 169

Fig. 304. *Pseudobranchus s. striatus;* adult; Long Co., Georgia (R. W. Van Devender).

IDENTIFICATION. The northern dwarf siren is an eel-like, permanently gilled salamander that lacks hindlimbs and has minute forelimbs that each have three toes (Martof 1972). Adults measure 10–22 cm TL, and females are on average slightly larger than males. The adults have a brownish to black dorsal ground color. Parallel yellow or tan stripes occur along the back and sides, and extend from the head to the tip of the tail. Striping pattern and coloration vary among subspecies and are described in the next section. Hatchlings are of the pond type, with dorsal fins that extend from the head to the tail tip. Color patterns of hatchlings have not been described in detail. This species is closely allied to *P. axanthus* and is most easily identified based on the geographic locality from which specimens are collected.

SYSTEMATICS AND GEOGRAPHIC VARIATION. Until recently *Pseudobranchus* was thought to be a monotypic genus consisting of one species with five recognized subspecies (Conant 1975; Martof 1972). Chromosomal studies by Moler and Kezer (1993) indicate that *Pseudobranchus* includes at least two species. Of the five subspecies of *Pseudobranchus* that are recognized by Conant and Collins (1991) and Martof (1972), only three (*lustricolus, spheniscus,* and *striatus*) correspond to *P. striatus.* Moler and Kezer (1993) ex-

Range of *Pseudobranchus striatus*

amined specimens from six localities within the range of *P. s. lustricolus*. None of the specimens resembles the descriptions of this subspecies and most were more similar to *P. s. spheniscus*. The range of *P. s. lustricolus* as defined by Conant and Collins (1991) contains specimens referable to *P. axanthus, P. s. lustricolus,* and *P. s. spheniscus*. Both the distribution and the taxonomic affinity of *P. s. lustricolus* are poorly understood (P. E. Moler, pers. comm.). Here I continue to treat this form as a subspecies of *P. striatus* (range not shown).

The broad-striped dwarf siren (*P. s. striatus*) is a stocky race that reaches a maximum length of 20.3 cm

Fig. 305. *Pseudobranchus s. spheniscus;* adult; Madison Co., Florida (R. W. Van Devender).

TL (Moler and Mansell 1986). Adults have a dark brown to blackish middorsal stripe with a narrow yellow line down the middle. This contrasts sharply with a broader, yellow lateral stripe that occurs on each side of the body. The venter is slightly lighter than the dorsal ground color and is heavily mottled with yellow. The slender dwarf siren (*P. s. spheniscus*) is a smaller and more slender race that reaches a maximum size of 15 cm TL. The head and snout are narrow and wedge shaped, and two (rarely three) narrow tan or yellow stripes occur along each side of the body. The Gulf Hammock dwarf siren (*P. s. lustricolus*) is a stocky form that reaches 22 cm TL. This subspecies has a dark broad middorsal stripe that contains three narrow yellow stripes. In addition there are two light stripes along each side of the body. The upper one is orangish brown and the lower one is silvery white.

DISTRIBUTION AND ADULT HABITAT. Northern dwarf sirens inhabit cypress swamps, flooded ditches, marshes, and other permanent and semipermanent aquatic habitats from southern South Carolina to the panhandle of Florida (Harper 1935; Martof 1972). In areas of sympatry with *P. axanthus,* the former species is often restricted to cypress ponds in acid pine flatwoods, whereas the latter species prefers open marshes and prairie ponds (Moler and Kezer 1993).

COMMENTS. Most of the published life history data for *Pseudobranchus* species concern *P. axanthus*. Moler and Kezer's (1993) discovery that northern populations of *Pseudobranchus* are a distinct species from more southern populations has resulted in an almost complete lack of information on the natural history of *P. striatus*. The life history of this species is presumably similar to that of *P. axanthus,* but detailed life history studies are needed. Neill (1952) notes that specimens sometimes emit high-pitched yelps when molested.

Harper (1935) reports unearthing two aestivating specimens with greatly reduced gills in southern Georgia while digging a fire trench through a dried portion of a swamp. A female collected by Noble (1930) from southern Georgia on 23 February laid 11 eggs in the laboratory over a 5-week period beginning in early March. The eggs were attached singly without stalks and the outer capsule had an average diameter of about 5 mm.

Siren intermedia Barnes
Lesser Siren
PLATES 170, 171

Fig. 306. *Siren i. intermedia;* adult; Berkeley Co., South Carolina (R. W. Van Devender).

IDENTIFICATION. The lesser siren is a large perennibranch that lacks rear limbs and has four toes on each front limb. The body is elongated and eel shaped. The margins of the jaws are covered with horny sheaths, and a dorsal fin extends from the vent posteriorly to the tail tip. The dorsal ground color of adults varies from olive green to grayish blue or black. Lighter-colored individuals often have scattered brown or black spots on the dorsum. The venter is slightly lighter than the dorsum, and white to yellowish flecks may occur along the sides of the body (Martof 1973). Males have enlarged masseter muscles that cause the temporal region of the head to appear swollen (Gehlbach and Kennedy 1978; Godley 1983; Martof 1973; Sugg et al. 1988). Males in a sample of 358

adults from Arkansas are on average 14% longer in SVL than females (Sugg et al. 1988). Adults measure 18–69 cm TL. The number of costal grooves between the front limbs and the cloaca varies from 31 to 38 and differs significantly among subspecies.

Hatchlings are the pond type and have conspicuous dorsal fins, short tails, elongated bodies, and partially developed limbs (Godley 1983; Noble and Marshall 1932). The young are boldly marked with longitudinal stripes along the body (except *S. i. intermedia*) and yellow to red banding on the head (Martof 1973; Noble and Marshall 1932). A broad, dark band with a light streak in the middle occurs on either side of the body. The venter and middorsal region are light colored. Hatchlings from two nests measured 11.5 and 11.6 mm TL on average (Godley 1983).

Small juveniles have a triangular yellow to red patch on the snout that often extends laterally to the gills (Hanlin and Mount 1978; Neill 1949). In some individuals a red or yellow band leads from each eye toward the base of the gills, and a transverse band may occur across the back of the head. Yellow and red pigmentation may also occur on the feet, near the base of the gills, and on portions of the dorsal fin. Juveniles in some Alabama populations lack the conspicuous head bands (Mount 1975). Juveniles usually have nail-like cornifications on the toe tips.

The dorsal fins, light body stripes, and red bands on the head progressively disappear as juveniles grow. Juvenile *S. i. nettingi* nearing sexual maturity are olive green with tiny black spots, whereas those of *S. i. intermedia* lack spots. A large juvenile (113 mm TL)

Fig. 307. *Siren i. nettingi;* adult; western Kentucky (R. W. Barbour).

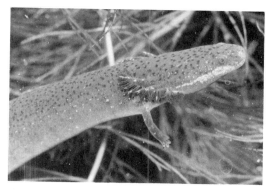

Fig. 308. *Siren i. intermedia;* adult; Charleston Co., South Carolina (R. W. Van Devender).

from an Alabama site had yellow speckling similar to that of *S. lacertina* (Hanlin and Mount 1978).

SYSTEMATICS AND GEOGRAPHIC VARIATION. Preliminary electrophoretic and karyological data indicate that the genus *Siren* is more speciose than currently recognized (P. E. Moler, pers. comm.). The description of several new taxa is anticipated, but detailed studies have not been completed, and only two species are recognized in this treatment.

Three subspecies are currently recognized (Goin 1942, 1957; Martof 1973). The Rio Grande siren (*S. i. texana*) is a relatively large subspecies that reaches 69 cm TL. The dorsum is gray to brownish gray above and is marked with tiny black spots. The modal number of costal grooves is 37 and the venter is light gray. This subspecies is restricted to the lower Rio Grande Valley and adjoining regions in northern Mexico.

The western lesser siren (*S. i. nettingi*) grows to 50 cm TL, has minute black spots on an olive green to gray dorsum, and has a modal number of 35 costal grooves. The venter is dark with numerous light flecks. This subspecies occurs from Alabama westward to central Texas and northward through the Mississippi Valley to Michigan. It is reported to intergrade with *S. i. texana* in southern Texas (Goin 1957) and with *S. i. intermedia* in a broad area extending from western Alabama to southeastern Louisiana (Boyd and Vickers 1963; Conant and Collins 1991; Mount 1975). However, both zones of intergradation are poorly defined and may not exist (McDaniel 1977; P. E. Moler, pers. comm.).

The eastern lesser siren (*S. i. intermedia*) reaches a maximum length of about 38 cm. This form has a black or dark brown dorsum that sometimes has minute black spots, and a modal number of 32–33 costal grooves. The venter is uniformly dark, but lighter than the dorsum. This subspecies occurs from central Alabama to southeastern Virginia.

DISTRIBUTION AND ADULT HABITAT. The lesser siren ranges from the Coastal Plain of Virginia to Florida, then westward to southern Texas and northeastern Mexico. Populations extend northward in the Mississippi Valley to Illinois, Indiana, and southwestern Michigan. Geographic isolates occur in northern Indiana, southwestern Michigan, northeastern North Carolina, and southeastern Virginia.

Lesser sirens inhabit a variety of permanent or semipermanent habitats, including marshes, swamps, farm ponds, ditches, canals, sloughs, and sluggish, vegetation-choked creeks. This species is common in habitats ranging from temporary floodplain pools to shallow, heavily vegetated sections of ponds with deep sediments that provide burrowing sites (Funderburg and Lee 1967; Gehlbach and Kennedy 1978).

BREEDING AND COURTSHIP. Adult *S. i. nettingi* in Louisiana become more active during the fall and winter months with the onset of the breeding season as individuals presumably search for mates (Raymond 1991). Specimens taken during the breeding season are often scarred, and this finding suggests that individuals aggressively defend mates. Courtship behavior has not been observed, and it may also involve a biting phase. Spent or ovulated female *S. i. intermedia* from Florida often have bite marks on their bodies, which are possibly inflicted by courting males (Godley 1983). How-

Range of *Siren intermedia*

ever, bite marks have been found on both sexes in specimens collected from Louisiana in December, January, and March (Raymond 1991). Average sex ratios of *S. intermedia* are about 1:1 (Cagle and Smith 1939; Gehlbach and Kennedy 1978).

REPRODUCTIVE STRATEGY. Time of oviposition is geographically variable, and southern populations tend to breed earlier than northern populations. *Siren i. nettingi* oviposits as early as January in Louisiana (Raymond 1991), whereas *S. i. intermedia* oviposits from late December through March in populations in south-central Florida (Godley 1983). In the Carolinas egg laying occurs primarily in March and April (Collette and Gehlbach 1961; Gibbons and Semlitsch 1991). Breeding records for western populations include a single ovulating female collected on 31 March in Arkansas (Trauth et al. 1990), nests with late-term embryos in Arkansas in early April (Noble and Marshall 1932), nesting females on 12 March in southern Texas, and nesting by captive females in late February and early March in east-central Texas (Gehlbach and Kennedy 1978). Most females collected in late April from central Texas were still gravid, suggesting a peak in breeding in May in some Texas populations (Davis and Knapp 1953).

Females deposit the eggs singly in large masses or aggregates that consist of three to four layers of eggs, although a few single eggs have been found up to 20 cm from the main cluster (Godley 1983). The eggs have sticky outer capsules that adhere to each other as well as to vegetation or plant roots. Egg masses with females or adults of undetermined sex have been found at the base of rooted macrophytes, in fibrous mats of water hyacinth roots, and in muddy depressions or plant debris in pond bottoms (Godley 1983; Hubbs 1962; Noble and Marshall 1932).

The only nesting record for *S. i. intermedia* is for two nests with late-term embryos found in a canal in southern Florida on 8 February (Godley 1983). The nests were 4 m from shore in water 35–40 cm deep and had 206 and 381 eggs. One mass measured 65 × 60 mm. Nesting records for *S. i. nettingi* in northeastern Arkansas include a nest with 555 late-term embryos found on 8 April (Noble and Marshall 1932) and a nest with 260 eggs in similar stages of development found on 10 April. Both nests were in muddy pockets in pond bottoms. Hubbs (1962) reports finding several *S. i. texana* nests with eggs or larvae in depressions in mud and vegetation in ponds in southern Texas on 12 March.

Clutch size estimates based on the weight of eggs in gravid females are 224–706 eggs for 24 southern Illinois specimens (Cagle and Smith 1939). Estimates based on counts of ovarian eggs are 299 ova for a female from northeastern Arkansas (Noble and Marshall 1932); a mean of 851 ova (range = 98–1506) for eight other Arkansas specimens (Trauth et al. 1990); 130 and 269 large ova from North Carolina specimens (Collette and Gehlbach 1961); and 151, 190, and 226 ova from east-central Texas specimens (Gehlbach and Kennedy 1978). In the study by Trauth et al. (1990), number of ova was positively correlated with female SVL.

Freshly laid eggs are dark brown above, 2.5–3.0 mm in diameter, and surrounded by three jelly envelopes (Collette and Gehlbach 1961; Noble and Marshall 1932). The diameter of the outer capsule is 4.4–7.5 mm (Godley 1983; Noble and Marshall 1932). The incubation period lasts 1.5–2.5 months based on the first appearance of small larvae in seasonal samples.

AQUATIC ECOLOGY. Dietary studies of lesser sirens often reveal large amounts of filamentous algae, pieces of vascular plants, and mud and bottom debris in the gastrointestinal tract (Bennett and Taylor 1968; Davis and Knapp 1953; Noble and Marshall 1932; Scroggin and Davis 1956). Sirens obtain little nourishment from algae since it tends to pass through the gastrointestinal tract undigested (Davis and Knapp 1953). The large amount of debris in specimens may reflect a common feeding behavior that involves gulping in bottom debris or plant material, then filtering out food and other items as water is passed through the branchial openings. Laboratory animals filter-feed on tiny crustaceans, but also capture and eat larger prey, such as amphipods, individually (Altig 1967). Young sirens feed mostly on zooplankton but also eat larger prey, such as amphipods, craneflies, and lumbriculid worms (Carr 1940).

Sirens in ponds treated with rotenone typically are not found until after dark, suggesting that individuals remain in their burrows during the day and move into the water column to feed at night. In one study (Davis and Knapp 1953), only 30 of 116 specimens that were collected at night following a rotenone application contained food in their gastrointestinal tracts, and none had food in its stomach.

Food items of *S. i. nettingi* from a farm pond included chironomid larvae, mayflies, water mites, snails, caddisflies, and large amounts of algae, submerged plants, and bottom debris (Davis and Knapp 1953). The diet of 68 *S. i. nettingi* in a southern Illinois swamp included a variety of benthic and pelagic prey (Altig 1967). Cladocerans, ostracods, and sphaeriid clams are the most numerically important prey. Other prey include copepods, isopods, amphipods, odonates, beetles, dipterans, and miscellaneous insects. The prey of three North Carolina specimens of *S. i. intermedia* included odonates, isopods, and siren eggs; filamentous algae and detritus were also found (Collette and Gehlbach 1961). Those of 91 *S. i. nettingi* from central Texas included over 32,000 items (Scroggins and Davis 1956). Snails, ostracods, and mayflies were the most important prey. Other items in the gut included crayfish, earthworms, pill bugs, crickets, bugs, leaf hoppers, beetles, spiders, millipedes, siren eggs, filamentous algae, remains of vascular plants, bottom debris, and pieces of cardboard and paper. Terrestrial prey are eaten when the pond expands and floods land following heavy rains.

Lesser sirens frequently cannibalize their own eggs. Records of oophagy include 10 eggs in three specimens from North Carolina (Collette and Gehlbach 1961) and 126 siren eggs in 91 specimens from Texas (Scroggins and Davis 1956). It is uncertain if oophagy reflects nest predation by other sirens or cases of females eating their own eggs.

Juveniles are most active at night and live in burrows or in thick mats of aquatic vegetation, where they forage on small invertebrates. Data from several studies indicate that most juveniles mature during their second year of growth and reproduce for the first time the following winter or spring. Male and female *S. i. nettingi* in a central Texas farm pond grow on average about 13 and 10 cm SVL, respectively, during their first year of life (Davis and Knapp 1953). Growth during the second year is on average about 9 cm for females and 7.5 cm for males. Females mature sexually during their second year of life when they measure about 15 cm SVL.

A winter aggregate of 138 *S. i. nettingi* collected in southern Illinois consisted of two size groups: one with mostly juveniles measuring 15–22 cm TL and a second with mostly adults measuring 23–32 cm TL (Cagle and Smith 1939). Individuals in this population mature sexually during their second year of life when they reach 22–25 cm TL. Summer samples of *S. i. intermedia* from South Carolina populations contain three distinct size classes. Juveniles in these populations also require about 2 years to reach sexual maturity (Gibbons and Semlitsch 1991), as do those of *S. i. nettingi* in Arkansas (Trauth et al. 1990). The smallest female with yolked ova in the latter study measured 16.5 cm SVL. Four weight classes are evident in samples from Texas *S. i. nettingi* that are thought to correspond to year classes, but juvenile and adult groups are not readily distinguishable (Gehlbach and Kennedy 1978).

Sirens hide in burrows, beneath logs or bark, or in thick vegetation during the day (Asquith and Altig 1987; Bennett and Taylor 1968; Davis and Knapp 1953). At night they move about in vegetation or prowl about on pond bottoms in search of food (Minton 1972; Noble and Marshall 1932). Individuals hide in individual retreats and rarely aggregate, except when shelters are very limiting (Asquith and Altig 1987).

Lesser sirens have a limited repertoire of sounds, including clicking and yelping, that may in part play a role in defending burrows. *Siren i. nettingi* and *S. i. texana* make clicking sounds shortly before or after moving from their burrows toward the water surface to gulp air (Gehlbach and Walker 1970). When resting in burrows, animals in the immediate vicinity of each other frequently click while simultaneously engaging in head-jerking motions. Resident sirens also tend to click more frequently when nonresidents are introduced into aquaria. The clicking sounds appear to function in intraspecific communication because they occur most often when other sirens are present, are emitted at different pulse rates by individuals clicking together, and are associated with head-jerking movements that may visually reinforce an individual's acoustically defended space. A yelping sound that is associated with distress or alarm is made by individuals when butted or bitten by others.

Siren i. nettingi in a Texas beaver pond tend to be sedentary (Gehlbach and Kennedy 1978). In general, larger individuals move farther than smaller ones, and males move farther than females. The maximum distances moved from initial release points are <12 m. The density and biomass in optimal habitats at this site are 1.1 animals/m^2 and 46.2 g/m^2, respectively. These standing biomass estimates are perhaps the highest of any salamander.

Cagle and Smith (1939) found a winter aggregate consisting of 138 *S. i. nettingi* and 30 newts on 1–2 January in a cement culvert connecting two ponds in southern Illinois. The sirens were found in a small area within the culvert, were inactive, and had empty stomachs. Aggregates of this sort have not been observed during the warmer months of the year.

Lesser sirens often inhabit semipermanent aquatic habitats that dry for several months of the year. During periods of drought, the animals burrow into the bottom muck of drying ponds or enlarge existing crayfish burrows. Individuals have often been dug up by excavation crews and by farmers plowing previously flooded lowlands (Blatchley 1900; Cockrum 1941). A work crew excavating a ditch in southern Illinois unearthed 60–70 *S. i. nettingi* in late December and early January (Cockrum 1941). Most of the specimens were 46–102 cm below the ground surface at the bottom of the ditch. Some animals were in J-shaped burrows; others were deep within crayfish burrows.

Animals that become trapped in dried mud on pond beds secrete a mucuslike envelope around themselves. Over a period of several weeks or months, mucoid secretions and cells from the skin accumulate to form a parchmentlike cocoon that retards dehydration (Gehlbach et al. 1973; Reno et al. 1972). It is not uncommon for sirens to aestivate for several months without food before rains refill the ponds (Cockrum 1941; Gehlbach et al. 1973). Aestivating animals have lower heart rates and lower oxygen demands than nonaestivating animals. These adaptations lower energy expenditures and increase the time that animals can remain without food (Gehlbach et al. 1973).

During aestivation the gills become reduced, and in extreme cases sirens may lose up to 50% of their initial body weight. In laboratory studies, small sirens starve to death within 12–14 weeks after initiating aestivation, whereas large sirens are able to survive at least 16 weeks. Smaller sirens have higher metabolic demands and proportionately less fat reserves than larger animals. Susceptibility of small sirens to starvation during recurrent droughts may set limits on the westernmost distribution of this species in Texas (Gehlbach et al. 1973).

PREDATORS AND DEFENSE. Natural predators are poorly documented but undoubtedly include water snakes, fishes, alligators, and wading birds. The tendency for animals to be nocturnally active is presumably an antipredator behavior that minimizes predation risk from diurnal predators such as fish and wading birds.

COMMUNITY ECOLOGY. Fauth and Resetarits (1991) used artificial pond communities to examine interactions between *S. i. intermedia*, eastern newts (*Notophthalmus viridescens*), and five species of anuran tadpoles. When at high densities, sirens reduce the growth and survival of newts; however, this effect enhances newt reproductive success by releasing newt larvae from intraspecific competition and cannibalism. The presence of newts has no effect on siren survival or growth rate. Neither direct nor indirect effects of sirens alter the role of *N. viridescens* as a keystone predator in these experimental communities.

CONSERVATION BIOLOGY. Lesser sirens occur in scattered populations throughout their range, and many local populations have been destroyed by the loss of wetlands (e.g., Bury et al. 1980). Although the dispersal mechanisms of this species are not known, movement between isolated habitats is presumably limited because of the lack of an obvious overland dispersal stage in the life cycle. Flooded bottomlands provide temporary connectors between isolated ponds that allow sirens to disperse between ponds. The extent to which flood control programs have eliminated these connectors and altered gene flow between local populations of sirens must be explored by resource managers and wildlife ecologists.

Siren lacertina Linnaeus
Greater Siren
PLATE 172

Fig. 309. *Siren lacertina;* adult; Florida (R. W. Van Devender).

IDENTIFICATION. The greater siren is a very large, eel-shaped salamander that lacks hindlimbs and has greatly reduced forelimbs, each of which has four toes. Juveniles mature into gilled adults. The modal number of costal grooves between the front limbs and the cloaca is 37–38. Adults vary from olive green to light grayish above and sometimes have dark spots on the head, back, and sides. The sides of the body are lighter colored than the back and are usually colored with inconspicuous flecks or blotches of pale green. The bluish gray venter is often flecked with pale green coloration (Martof 1973).

Siren lacertina is one of the largest salamanders in North America, rivaled in size only by *Cryptobranchus* and *Amphiuma*. Adults measure 50–98 cm TL, but most are <70 cm. Males have enlarged masseter muscles that make the head appear larger than that of females of similar size (Hanlin and Mount 1978). The shovel-like head is used for burrowing in pond bottoms. Small *S. lacertina* are easily confused with the eastern lesser siren (*S. i. intermedia*) in areas of sympatry. The eastern lesser siren lacks pale green flecking on the sides and belly and has a modal number of only

Range of *Siren lacertina*

32–33 costal grooves compared with 37–38 in *S. lacertina*.

Hatchlings are the pond type and have prominent dorsal and caudal fins, a pale gray dorsum, and a stripe from the gills to the base of the tail. A hatchling examined by Goin (1947) measured 13 mm SVL and 16 mm TL. Young juveniles have cornified toe tips that resemble claws, a prominent light yellow stripe down each side of the body, and (in some populations) a narrower, less prominent ventrolateral light stripe (Neill 1949). Striping on the body is reduced or lost during the first year of life, and older juveniles are uniformly dark gray with varying amounts of mottling on the tail fin (Duellman and Schwartz 1958).

SYSTEMATICS AND GEOGRAPHIC VARIATION. Villela and Brandon (1992) report that museum specimens from northern Mexico and southern Texas that were previously identified as *S. i. texana* more closely fit the description of *S. lacertina*. The authors tentatively identify these specimens as *S. lacertina* based primarily on their size. Genetic characterization of these and other *Siren* species that is now in progress will help clarify the status of *Siren* populations in southern Texas (P. E. Moler, pers. comm.). Here I consider *S. lacertina* to be restricted to regions east of the Mississippi River. Geographic variation has not been carefully analyzed, and no subspecies are currently recognized.

DISTRIBUTION AND ADULT HABITAT. Greater sirens inhabit a variety of permanent and semipermanent aquatic habitats, including ditches, canals, marshes, farm ponds, rice fields, lakes, and sluggish streams and rivers that are often choked with aquatic plants (Duellman and Schwartz 1958; Martof 1973). They occur in Coastal Plain habitats from the vicinity of Washington, D.C., southward to southern Florida and westward to southwestern Alabama. In many areas in the northern and western portions of its range, *S.*

Fig. 310. *Siren lacertina*; adult; Pitt Co., North Carolina (R. W. Van Devender).

lacertina is uncommon and occurs in scattered populations. Local populations appear to reach their highest densities in Florida, southeastern Georgia, and eastern South Carolina, where the species is often abundant.

BREEDING AND COURTSHIP. The breeding biology of the greater siren is poorly documented. Fertilization is presumed to be external because females lack spermathecae, males lack cloacal glands, and eggs laid by females isolated from males do not develop (Ultsch 1973). Males undoubtedly fertilize the eggs as they are laid, but this hypothesis has not been substantiated. Breeding occurs in February and March in a south Alabama pond (Hanlin and Mount 1978). During this time, males produce spermatozoa and ovarian eggs reach their maximum size. The sex ratio of 42 males and females did not differ significantly from 1:1.

REPRODUCTIVE STRATEGY. Females in Alabama and Florida populations oviposit in February and March. Eggs have almost never been found in nature, and little information is available on nest site selection. Five eggs were found in northern Florida on 4 February in a shallow ditch. The eggs were dipped from a bed of aquatic plants about 1 m from the shore in water 15 cm deep (Goin 1947). A female from Florida laid eggs in the laboratory between 16 and 21 February (Noble and Richards 1932). The eggs were attached singly or in small groups to the substrate by their adhesive outer egg capsules and lacked stalks. Other Florida females collected on 4–6 February laid eggs within 3 days after capture (Ultsch 1973). One deposited 123 eggs that were laid singly or in small groups of <10 eggs. P. E. Moler (pers. comm.) collected a large, grapelike cluster of eggs near hatching in early May while dredging water hyacinths in Florida.

Freshly laid eggs are dark brown above, have an average diameter of 4 mm, and are surrounded by three jelly envelopes (Hanlin and Mount 1978; Noble and Marshall 1932; Salthe 1963). This species is one of the most fecund salamanders in North America, rivaled only by gilled adult *A. tigrinum* in western Texas. One specimen collected by Hanlin and Mount (1978) contained about 1400 enlarged ovarian eggs. The young of the year first appear in late April and early May, suggesting an incubation period of about 2 months (Ultsch 1973).

Fig. 311. *Siren lacertina;* larva; Alachua Co., Florida (R. W. Van Devender).

AQUATIC ECOLOGY. Greater sirens consume a wide variety of prey, but sometimes feed rather heavily on mollusks. Dietary records include 324 snail shells and the valves of a clam passed by a recently captured Florida specimen (Moler 1994) and filamentous algae and a large number of snails in a Virginia specimen (Burch and Wood 1955). Other Florida specimens contained mostly snails, insects, crayfish, and a fish (Duellman and Schwartz 1958; Hamilton 1950). Food items from 33 specimens from a southeastern Alabama pond consisted of filamentous algae, mud and bottom debris, crayfish, spiders, snails, clams, and representatives of several orders of insects (Hanlin 1978).

The juveniles hide in thick aquatic vegetation or other cover during the day and are most active at night in the water column, where they feed on small invertebrates. At one site adults are found in water 76–91 cm deep, whereas juveniles utilize shallower water only 15–30 cm deep (Duellman and Schwartz 1958). Young of the year in Florida may reach 75 mm SVL by mid-October (Ultsch 1973).

The adults are active at night and frequently hide in burrows during the day. At one Florida site, individuals live in a canal bank in burrows that slant downward at about 45° angles before leveling off into enlarged chambers that are 15 cm in diameter (Duellman and Schwartz 1958). The animals are usually found inside the chambers at the ends of burrows. Individuals in an Alabama population are most active 2 hours after dark and 2 hours before dawn (Hanlin and Mount 1978). A seasonal peak in feeding activity occurs during June and July, when average water temperatures are about 28°C, and a seasonal low in activity occurs from October through December. Greater sirens make clicking or yelping noises similar to those of *S. intermedia* that may function in intraspecific communication (Carr 1940; Gehlbach and Walker 1970).

Like dwarf and lesser sirens, greater sirens often become trapped in drying ponds and must survive in underground burrows without food for long periods of time (Carr 1940; Freeman 1958; Hanlin and Mount 1978). Individuals aestivate in cocoons composed of dried squamous epithelial cells and often rest in an S-shaped configuration with the body in contact with the soil (Etheridge 1990a). This posturing allows water exchange with the soil and minimizes the probability of desiccating. The gills typically atrophy during aestivation (Etheridge 1990b).

Greater sirens maintain relatively large fat reserves in their tails and are able to withstand prolonged periods without food (Martof 1969). One siren maintained without food at about 22°C died 26 months later after losing 45% of its initial body weight. A second died after 5.2 years after losing 86% of its initial weight. Resting metabolic rates are 60–70% lower in aestivating animals than in active animals; this differential allows animals to aestivate for years without eating (Etheridge 1990b). Data on survival and longevity in nature are not available. Adults have been kept in captivity for 25 years (Nigrelli 1954).

PREDATORS AND DEFENSE. Natural predators are poorly documented but probably include aquatic snakes, wading birds, and other large predators. American alligators (*Alligator mississippiensis*) regularly feed on *S. lacertina*. In a Florida study 35 *S. lacertina* were found in a sample of 350 American alligators (Delany and Ambercrombie 1986).

COMMUNITY ECOLOGY. The tendency for this species to feed heavily on mollusks suggests that it could play a role in structuring snail communities in some ponds. At present, very few data are available on the ecological importance of sirens in aquatic communities.

CONSERVATION BIOLOGY. The greater siren is patchily distributed in the northern and western portions of its range, although it is common in Florida, southeastern Georgia, and eastern South Carolina. Very little information is available on the conservation status of many of the peripheral populations. A systematic survey of populations is needed to document long-term population trends of this and other Coastal Plain salamanders.

Glossary

adpressed limbs. Limbs that are pressed parallel to the sides of the body (see Fig. 6b).

agonistic behavior. A behavior, such as aggression or submissive posturing, that is involved in conflicts between individuals.

allele. An alternate form of a gene that arises through mutation.

allopatric. Referring to two species or other groups that have nonoverlapping ranges and are geographically isolated from one another.

allotopic. Referring to two species or other groups that co-occur regionally but do not share the same local habitat.

allozymes. Different electrophoretic forms of the same enzyme that are controlled by alleles at the same locus; measures of allozyme variation provide an indirect way to determine the genetic similarity of two or more populations or species (*see* Isozymes).

amplexus (verb, **amplex**). A process wherein a male clasps a female with his limbs prior to mating.

aposematic coloration. Bright coloration or bold color patterning associated with noxious species, which helps predators to learn to avoid noxious prey.

arboreal. Living in trees.

autotomy (verb, **autotomize**). Self-amputation of the tail.

balancer. A fleshy appendage that extends from each side of the head of hatchling (larval) salamanders; they are typically associated with pond-type larvae.

basibranchials. Bones associated with the hyobranchial apparatus in the floor of the mouth.

Batesian mimicry. An evolutionary phenomenon in which one or more palatable species evolve to resemble a noxious species.

biotype. A group of individuals having the same genomic complement.

butterfly. A courtship behavior of certain *Desmognathus* species in which the male rotates the forelimbs in a circular motion that resembles the butterfly stroke of a swimmer.

canthus rostralis. A raised, dorsolateral, bony ridge that extends from the eye to the tip of the snout.

centimeter (cm). A metric unit of measurement; 0.01 meter or 0.39 inch.

cirrus (plural, **cirri**). A downward, fleshy extension of the upper lip that encompasses the nasolabial groove; it functions to transport waterborne chemicals to the nasal passages.

cloaca (vent). A common chamber through which products of the urinary, intestinal, and reproductive systems exit the body.

clonal reproduction. Reproduction that results in the offspring being genetically identical, or nearly so, to its parent.

commissure. A line where two structures join.

compressed. Flattened from side to side, such as the tails of many dusky salamanders.

congeners. Two or more species that belong to the same genus.

conspecific. Belonging to the same species.

costal grooves (folds). Parallel, vertical grooves (folds) on the sides of a salamander's body.

cusp. A curved or straight projection arising from the surface of the crown of the tooth.

dark zone (of a cave). The interior region of a cave, where sunlight is not visible to humans.

deme. A local interbreeding population of organisms.

depressed. Flattened from top to bottom, such as the body of the hellbender.

digit. A toe or finger.

diurnal. Active during daylight.

dorsum (adjective, **dorsal**). The upper surface of an organism.

ectotherm (adjective, **ectothermic**). An organism

that depends on an external heat source to warm its body.

epistasis. An interaction between the genes at two or more loci by which one gene affects the expression of one or more other genes.

erythristic morph. A bright crimson color morph found in some local salamander populations.

erythrocyte. A red blood cell.

euryphagic predator. A predator that eats prey of widely differing types and sizes.

exploitative competition. A situation in which individuals seek an environmental resource that is in short supply and do not interfere with each other in the process.

facultative trait. A trait that may or may not develop in an organism depending on the particular environment that the organism inhabits.

fimbriae. See gill fimbriae.

first-order stream. A headwater stream that is not fed by other headwater streams.

fossorial. Living underground.

F_{st}. The "genetic fixation index," which provides a measure of the extent of gene flow and local differentiation among populations of a species.

gene flow. The movement of genes (and individuals) from one local population to another.

genetic differentiation. A quantitative measure, based on molecular data, of the extent to which two or more groups have diverged evolutionarily from one another.

genetic distance (D). An index that reflects the difference between two or more populations or species based on differences in gene frequencies or other molecular data.

genial gland. A pitlike gland on the side of the head of a newt that presumably functions to mediate courtship behavior.

genomic (noun, **genome**). Referring to all of the genetic material carried by an individual.

gill fimbriae. Hairlike extensions on the surfaces of gills that facilitate gas exchange.

gill rakers. Bony or cartilaginous extensions near the base of the gill arch or rachis.

gill slit. A space separating two adjoining gill arches that allows water to pass from the throat to the outside of the organism.

gular fold. An external flap or fold of tissue that extends across the lower throat region of salamanders.

gynogenesis. A form of asexual reproduction in which sperm stimulates an egg to divide but does not contribute chromosomes to the offspring.

H. A measure of the mean heterozygosity of a series of gene loci sampled from an individual.

hedonic gland. A gland that secretes chemicals that simulate sexual activity; it is often used by males in courtship.

hemiclonal lineage. An ancestral-descent lineage that contains elements of both clonal and syngametic reproduction.

heterosis (heterozygote advantage). A condition in which individuals heterozygous for a trait have higher fitness than homozygous individuals.

heterospecific. Referring to a member of a different species.

heterozygous (noun, **heterozygosity**). Referring to a condition in which an individual carries two different alleles at a gene locus.

holotype. A specimen upon which the description of a new species is based.

hybrid. An individual that is the result of a cross between two genetically distinct parental types.

inclusive fitness. The relative success of leaving genes in the next or subsequent generations based on an individual's own reproductive success and the reproductive success of relatives that carry the same genes.

iridophore. A cell that contains light pigments and produces brassy, silvery, golden, or whitish coloration on the body.

isozymes. Different electrophoretic forms of the same enzyme that are due to differing subunit configurations rather than allelic differences (see allozymes).

juvenile. A sexually immature individual; in species with biphasic life cycles, the stage between the termination of metamorphosis and the onset of sexual maturity.

keel. A thin, raised edge that runs along the dorsal surface of the tail.

lateral. Toward the sides of the body.

lentic habitats. Standing-water habitats, such as ponds, lakes, swamps, and bogs.

locus (plural, **loci**). The site on a chromosome where a gene is located.

lotic habitats. Running-water habitats such as springs, headwater streams, and rivers.

median. The midline of the body.

melanistic. Referring to an exceptionally dark individual or structure that owes its coloration to the presence of high densities of the pigment melanin.

melanophore. A cell that contains the pigment melanin, which produces black, grayish, or brownish coloration on the body.

mental glands. Secretory glands found on the chin of male salamanders that are used in courtship. These are often organized into rounded or oblong, raised patches that are visible with the naked eye.

mesophytic forest. A moist forest that receives moderate to high levels of rainfall.

metamorph. A term applied to an individual that has recently transformed, typically to the time between the resorption of gills and the acquisition of juvenile color patterning.

metamorphosis. In salamanders, the process of changing from a gilled larva into a juvenile that lacks gills. The most conspicuous external changes include the loss of gills and tail fins, the fusion of the gular flap to the body, and the development of eyelids and juvenile coloration.

microhabitat. A smaller component of an organism's overall habitat that tends to have unique physical, chemical, or biotic characteristics. Examples include decaying logs, leaf litter, or clumps of aquatic vegetation.

microsympatric (noun, **microsympatry**). Referring to two or more species that co-occur locally (as used here, synonymous with microsyntopic).

microsyntopic (noun, **microsyntopy**). Referring to two or more species that co-occur locally (as used here, synonymous with microsympatric).

monocuspid (tooth). A tooth with a single cusp on the crown.

monotypic. Referring to a taxonomic group that contains only one subgroup; for example, a monotypic genus contains only one species.

morph. One of a small number of variants in coloration or morphology that occur in a group, e.g., the striped versus unstriped forms of the red-backed salamander.

Müllerian mimicry. An adaptation in which two or more noxious species evolve similar coloration or morphology.

naris (plural, **nares**). A nostril or external opening to the nasal passages.

nasolabial groove. A slitlike channel that extends from the lateral corner of each naris to the margin of the upper lip; it is characteristic of members of the family Plethodontidae.

negative heterosis. A condition in which the fitness of a heterozygous individual (e.g., Aa) is lower than the fitness of homozygous individuals (e.g., AA; aa).

nocturnal. Active at night.

nomenclature. The system of scientific names applied to taxa.

occipital condyle. A projection from the rear of the skull where the neck articulates.

ontogenetic. Describing developmental changes in the traits or behaviors of an organism associated with aging.

oophagy. Inclusion of eggs in the diet.

oviduct. A tube that transports eggs from the ovaries toward the cloaca.

oviposit. To lay eggs or egg masses.

palatopterygoid teeth. Teeth on the palatine and pterygoid bones of the lower jaw.

parapatric. Referring to two species or groups with adjoining but nonoverlapping geographic ranges.

paraphyletic (group). A taxonomic group that includes a common ancestor and some, but not all, of its descendents.

parotoid gland. A swollen gland at the back of the head that secretes noxious or poisonous compounds and aids in defense against predators.

parthenogenesis. The embryonic development of an unfertilized egg.

pentaploid. Having five sets of chromosomes instead of the usual two.

perennibranch (adjective, **perennibranchiate**). An individual or species that is gilled throughout life.

philopatry (adjective, **philopatric**). The tendency of individuals to remain in home areas, or to return to the same home areas, after migrating to and from breeding sites or winter hibernacula.

phreatic cave systems. Cave systems with streams or other underground water.

pleiotropy (adverb, **pleiotropically**). An instance in which a single gene affects two or more phenotypic traits.

plica (plural, **plicae**). A fold or ridge of tissue.

polymorphic trait. A trait, such as coloration, tooth morphology, or allozyme type, that occurs as a few discrete phenotypes in a local population, e.g., the striped versus unstriped color phases of some salamanders or allozyme variation at a gene locus.

polytypic. Referring to a taxonomic group that contains two or more subgroups; for example, a polytypic genus might contain six species.

pond-type larva. A larva having bushy gills and a prominent fin that extends forward to near the shoulder region.

prootic-squamosal crest. A raised dorsal and dorsolateral area on the posterior region of the skull.

pull. A courtship behavior of plethodontid salamanders in which the male presses his chin against the dorsum of the female by bending his head down, then pulls his head backward in several short strokes. This action provides tactile stimulation and may facilitate the introduction of mental gland secretions into the female's circulatory system.

rachis (plural, **rachises**). The central portion of a gill that supports the finely divided gill fimbriae.

rakers. *See* gill rakers.

relict. Referring to a population or species that has a very restricted distribution and that is presumed to be the remnant of a formerly widespread species.

rheotaxis (adjective, **rheotactic**). Orientation of the body with respect to the direction of water flow.

semispecies. As used here, a group of natural populations that occupy a unique subset of a species' range and that have nearly reached the species level of differentiation. Semispecies typically show more morphological and genetic differentiation than geographic races or subspecies. However, they have not reached complete reproductive closure and freely hybridize (often in narrow hybrid zones) with members of other semispecies.

skeletochronology (adjective, **skeletochronological**). Aging of individuals by using counts of bone rings in microscopic sections of long bones.

snap. A courtship behavior of plethodontid salamanders in which the male presses his chin against the dorsum of the female by bending his head down, then draws his head back in a sudden, rapid movement. This action lacerates the skin with the teeth and introduces mental gland secretions into the female's circulatory system.

snout-vent length (SVL). The distance from the tip of the snout to the posterior margin of the cloaca.

sperm cap. A mass of seminal fluid that rests upon the top of the jelly base of a spermatophore.

spermatheca (plural, **spermathecae**). A chamber leading off the internal lining of the female cloaca that is used to store sperm.

spermatophore. A structure deposited on substrates by courting male salamanders that typically consists of a mass of seminal fluid (sperm cap) resting upon a gelatinous base.

steephead. A permanent, spring-fed headwater tributary in a steep-walled ravine of the Coastal Plain.

stratification. A mostly nocturnal behavior displayed by *Ambystoma* larvae in which individuals float in the water column, or at the water surface, and remain in a relatively stable vertical position while feeding.

stream-type larva. A larva having reduced gills and a fin that extends forward only to the insertion of the hindlimbs.

subspecies. A taxonomic group of interbreeding natural populations of a species that occupy a unique subset of the entire species' range. Subspecies are formally recognized geographic races that have one or more diagnostic traits that distinguish them from other such groups.

sympatric. Referring to two or more species or groups that have overlapping ranges and live in the same geographic region.

syngamy. Sexual reproduction involving the production of offspring that contain gametic contributions from both parents.

syntopic. Referring to two species or other groups that occur together locally.

tail-straddle walk. A phase of courtship in plethodontid salamanders in which the female is led forward while straddling the male's tail.

taxon (plural, **taxa**). The taxonomic name applied to a group of organisms, e.g., Plethodontidae, *Ambystoma*, *Taricha*.

taxonomy. A subdiscipline of evolutionary biology that concerns the naming and classification of organisms.

territoriality. The active defense of a home range.

tetraploid. Having four sets of chromosomes instead of the usual two.

tetrapod. A member of one of the terrestrial vertebrate groups.

thyroxin. A hormone that stimulates amphibian larvae to undergo metamorphosis.

total length (TL). The distance from the tip of the snout to the tip of the tail.

transformation. Metamorphosis; the transition from a gilled larval form to a juvenile that lacks gills.

triploid. Having three sets of chromosomes instead of the usual two.

troglobyte (adjective, **troglobitic**). A cave-dwelling organism.

twilight zone (of a cave). The zone near the cave mouth where light is visible.

type I survivorship. A survivorship pattern in which age-specific mortality rates increase with age.

type II survivorship. A survivorship pattern in which age-specific mortality rates are independent of age.

"unken" posture. A defensive posture characteristic of newts in which the back is strongly arched, the head and tail are elevated and arched, and the belly is exposed to reveal bright coloration.

vas deferens (plural, **vasa deferentia**). A duct in males that transports sperm to the cloaca.

vent. The cloaca.

venter (adjective, **ventral**). The lower surface of an organism.

vernal pond. A seasonally ephemeral depression that typically fills with water during winter or early spring and dries before summer or early fall.

vitelline membrane. The innermost protective membrane that surrounds a developing embryo.

vomerine (prevomerine) teeth. Teeth located on paired bones on the roof of the mouth.

vomeronasal organ. An accessory olfactory device located near the nasal cavities.

Literature Cited

Adams, D. R. 1968. Stomach contents of the salamander *Batrachoseps attenuatus* in California. Herpetologica 24:170–172.

Adams, L. 1942. The natural history and classification of the Mount Lyell salamander, *Hydromantes platycephalus*. University of California Publications in Zoology 46:179–204.

Adler, K., and D. Dennis. 1962. *Plethodon longicrus*, a new salamander (Amphibia: Plethodontidae) from North Carolina. Special Publications of the Ohio Herpetological Society 4:1–4.

Alexander, W. P. 1927. The Allegheny hellbender and its habitat. Buffalo Society of Natural Science 7:13–18.

Alford, R. A. 1989. Competition between larval *Rana palustris* and *Bufo americanus* is not affected by variation in reproductive phenology. Copeia 1989:993–1000.

Allen, E. R., and W. T. Neill. 1949. A new subspecies of salamander (genus *Plethodon*) from Florida and Georgia. Herpetologica 5:112–114.

Allen, M. J. 1932. A survey of the amphibians and reptiles of Harrison County, Mississippi. American Museum Novitates 542:1–20.

Allyn, W. P., and C. Shockley. 1939. A preliminary survey of the surviving species of caudata of Vigo County and vicinity. Indiana Academy of Science 48:238–243.

Altig, R. 1967. Food of *Siren intermedia nettingi* in a spring-fed swamp in Illinois. The American Midland Naturalist 77:239–241.

Altig, R., and E. D. Brodie, Jr. 1971. Foods of *Plethodon larselli, Plethodon dunni,* and *Ensatina eschscholtzi* in the Columbia River Gorge, Multnomah County, Oregon. The American Midland Naturalist 85:226–228.

Altig, R., and P. H. Ireland. 1984. A key to the salamander larvae and larviform adults of the United States and Canada. Herpetologica 40:212–218.

Alvarado, R. H. 1967. The significance of grouping on water conservation in *Ambystoma*. Copeia 1967:667–668.

Anderson, J. D. 1960. Cannibalism in *Dicamptodon ensatus*. Herpetologica 16:260.

Anderson, J. D. 1961. The courtship behavior of *Ambystoma macrodactylum croceum*. Copeia 1961:132–139.

Anderson, J. D. 1963. Reactions of the western mole to skin secretions of *Ambystoma macrodactylum croceum*. Herpetologica 19:282–284.

Anderson, J. D. 1965. *Ambystoma annulatum* Cope. Ringed salamander. Catalogue of American Amphibians and Reptiles, pp. 19.1–19.2.

Anderson, J. D. 1967a. A comparison of the life histories of coastal and montane populations of *Ambystoma macrodactylum* in California. The American Midland Naturalist 77:323–355.

Anderson, J. D. 1967b. *Ambystoma opacum*. Catalogue of American Amphibians and Reptiles, pp. 46.1–46.2.

Anderson, J. D. 1968. A comparison of the food habits of *Ambystoma macrodactylum sigillatum, Ambystoma macrodactylum croceum,* and *Ambystoma tigrinum californiense*. Herpetologica 24:273–284.

Anderson, J. D. 1969. *Dicamptodon* and *Dicamptodon ensatus*. Catalogue of American Amphibians and Reptiles, pp. 76.1–76.2.

Anderson, J. D. 1970. Description of the spermatophore of *Ambystoma tigrinum*. Herpetologica 26:304–308.

Anderson, J. D., and R. V. Giacosie. 1967. *Ambystoma laterale* in New Jersey. Herpetologica 23:108–111.

Anderson, J. D., and R. E. Graham. 1967. Vertical migration and stratification of larval *Ambystoma*. Copeia 2:371-474.

Anderson, J. D., and P. J. Martino. 1966. The life history of *Eurycea l. longicauda* associated with ponds. The American Midland Naturalist 75:257-279.

Anderson, J. D., and P. J. Martino. 1967. Food habits of *Eurycea longicauda longicauda*. Herpetologica 23:105-108.

Anderson, J. D., and G. K. Williamson. 1973. The breeding season of *Ambystoma opacum* in the northern and southern parts of its range. Journal of Herpetology 7:320-321.

Anderson, J. D., and G. K. Williamson. 1974. Nocturnal stratification in larvae of the mole salamander, *Ambystoma talpoideum*. Herpetologica 30:28-29.

Anderson, J. D., and G. K. Williamson. 1976. Terrestrial mode of reproduction in *Ambystoma cingulatum*. Herpetologica 32:214-221.

Anderson, J. D., Hassinger, D. D., and G. H. Dalrymple. 1971. Natural mortality of eggs and larvae of *Ambystoma t. tigrinum*. Ecology 52:1107-1112.

Anderson, P. K. 1958. Induced oviposition in *Batrachoseps attenuatus*, and incubation of eggs. Copeia 1958:221-222.

Anderson, P. K. 1960. Ecology and evolution in island populations of salamanders in the San Francisco Bay region. Ecological Monographs 30:359-384.

Andren, C., Marden, M. and G. Nilson. 1989. Tolerance to low pH in a population of moor frogs, *Rana arvalis*, from an acid and a neutral environment: a possible case of rapid evolutionary response to acidification. Oikos 56:215-223.

Angle, J. P. 1969. The reproductive cycle of the northern ravine salamander, *Plethodon richmondi richmondi*, in the Valley and Ridge Province of Pennsylvania and Maryland. Journal of the Washington Academy of Science 59:192-202.

Angleberger, M. A. P., and J. P. Chinnici. 1975. Dimorphism in the red-backed salamander *Plethodon cinereus* (Green) at Mountain Lake, Virginia. Virginia Journal of Science 26:153-158.

Anthony, C. D. 1993. Recognition of conspecific odors by *Plethodon caddoensis* and *P. ouachitae*. Copeia 1993:1028-1033.

Anthony, C. D., and J. A. Wicknick. 1993. Aggressive interactions and chemical communication between adult and juvenile salamanders. Journal of Herpetology 27:261-264.

Anthony, C. D., Mendelson, J. R., III, and R. R. Simons. 1994. Differential parasitism by sex on plethodontid salamanders and histological evidence for structural damage to the nasolabial groove. The American Midland Naturalist 132:302-307.

Antonelli, A. L., Nussbaum, R. A., and S. D. Smith. 1972. Comparative food habits of four species of stream-dwelling vertebrates (*Dicamptodon ensatus, D. copei, Cottus tenuis, Salmo gairdneri*). Northwest Science 46:277-289.

Arnold, S. J. 1976. Sexual behavior, sexual interference, and sexual defense in the salamanders *Ambystoma maculatum, Ambystoma tigrinum*, and *Plethodon jordani*. Zeitschrift für Tierpsychologie 42:247-300.

Arnold, S. J. 1977. The evolution of courtship behavior in New World salamanders with some comments on Old World salamandrids. In: The Reproductive Biology of Amphibians, D. H. Taylor and S. I. Guttman, eds. Plenum Press, New York, pp. 141-183.

Arnold, S. J. 1982. A quantitative approach to antipredator performance: salamander defense against snake attack. Copeia 1982:247-253.

Arnold, S. J., Reagan, N. L., and P. A. Verrell. 1993. Reproductive isolation and speciation in plethodontid salamanders. Herpetologica 1993:216-228.

Arnold, S. J., Verrell, P. A., and S. G. Tilley. 1996. The evolution of asymmetry in sexual isolation: a model and a test case. Evolution 50:1024-1033.

Ash, A. 1988. Disappearance of salamanders from clearcut plots. Journal of the Elisha Mitchell Scientific Society 104:116-122.

Ashton, R. E., Jr. 1975. A study of movement, home range, and winter behavior of *Desmognathus fuscus* (Rafinesque). Journal of Herpetology 9:85-91.

Ashton, R. E., Jr. 1977. The central newt, *Notophthalmus viridescens louisianensis* (Wolterstorff) in Kansas. Transactions of the Kansas Academy of Science 79:15-19.

Ashton, R. E., Jr. 1985. Field and laboratory observations on microhabitat selection, movements, and home range of *Necturus lewisi* (Brimley). Brimleyana 10:83–106.

Ashton, R. E., Jr. 1992. Rare: flatwoods salamander. In: Rare and Endangered Biota of Florida, Volume 3, P. Moler, ed. University Presses of Florida, Gainesville, Florida, pp. 39–43.

Ashton, R. E., Jr., and P. S. Ashton. 1978. Movements and winter behavior of *Eurycea bislineata* (Amphibia, Urodela, Plethodontidae). Journal of Herpetology 12:295–298.

Ashton, R. E., Jr., and P. S. Ashton. 1988. Handbook of Reptiles and Amphibians of Florida, Part 3: The Amphibians. Windward Publishing, Miami, Florida, 191 pp.

Ashton, R. E., Jr., and A. L. Braswell. 1979. Nest and larvae of the Neuse River waterdog, *Necturus lewisi* (Brimley) (Amphibia: Proteidae). Brimleyana 1:15–22.

Ashton, R. E., Jr., Braswell, A. L., and S. I. Guttman. 1980. Electrophoretic analysis of three species of *Necturus* (Amphibia: Proteidae), and the taxonomic status of *Necturus lewisi* (Brimley). Brimleyana 4:43–46.

Ashton, T. E. 1966. An annotated check list of order Caudata (Amphibia) of Davidson County, Tennessee. Journal of the Tennessee Academy of Science 41:106–111.

Asquith, A., and R. Altig. 1987. Phototaxis and activity patterns of *Siren intermedia*. The Southwestern Naturalist 32:146–148.

Attar, E. N., and E. J. Maly. 1980. A laboratory study of preferential predation by the newt *Notophthalmus v. viridescens*. Canadian Journal of Zoology 58:1712–1717.

Aubry, K. B., and P. A. Hall. 1991. Terrestrial amphibian communities in the southern Washington Cascade Range. In: Wildlife and vegetation of unmanaged Douglas-fir forests, L. F. Ruggiero, K. B. Aubry, A. B. Carey, and M. H. Huff, technical coordinators. USDA Forest Service, Pacific Northwest Research Station, Olympia, Washington, General Technical Report PNW-GTR-285, pp. 326–338.

Aubry, K. B., Jones, L. L. C., and P. A. Hall. 1988. Use of woody debris by plethodontid salamanders in Douglas-fir forests in Washington. In: Management of amphibians, reptiles, and mammals in North America, R. C. Szaro, K. E. Severson, and D. R. Patton, eds. USDA Forest Service, Rocky Mountain Forest and Range Experiment Station, Fort Collins, Colorado, Technical Report RM-166, pp. 11–22.

Aubry, K. B., Senger, C. M., and R. L. Crawford. 1987. Discovery of Larch Mountain salamanders (*Plethodon larselli*) in the central Cascade Range of Washington. Biological Conservation 42:147–152.

Austin, N. E., and J. P. Bogart. 1982. Erythrocyte area and ploidy determination in the salamanders of the *Ambystoma jeffersonianum* complex. Copeia 1982:485–488.

Austin, R. M., Jr., and C. D. Camp. 1992. Larval development of black-bellied salamanders, *Desmognathus quadramaculatus*, in northeastern Georgia. Herpetologica 48:313–317.

Axtell, H. H., and R. C. Axtell. 1948. *Pseudotriton ruber* in central New York state. Copeia 1948:64.

Ayala, F. J. 1975. Genetic differentiation during the speciation process. Evolutionary Biology 8:1–78.

Bachmann, M. D. 1984. Defensive behavior of brooding female red-backed salamanders (*Plethodon cinereus*). Herpetologica 40:436–443.

Bachmann, M. D., Carlton, R. G., Burkholder, J. M., and R. G. Wetzel. 1986. Symbiosis between salamander eggs and green algae: microelectrode measurements inside eggs demonstrate effect of photosynthesis on oxygen concentration. Canadian Journal of Zoology 64:1586–1588.

Bahret, R. 1996. Ecology of lake dwelling *Eurycea bislineata* in the Shawangunk Mountains, New York. Journal of Herpetology 30:399–401.

Bailey, J. E., and T. K. Pauley. 1993. Aspects of the natural history of the Cumberland Plateau salamander, *Plethodon kentucki*, in West Virginia. Association of Southeastern Biologists Bulletin 40:133.

Bailey, R. M. 1943. Four species new to the Iowa herpetofauna, with notes on their natural histories. Proceedings of the Iowa Academy of Science 50:347–352.

Baker, C. L. 1945. The natural history and morphology of amphiumae. Journal of the Tennessee Academy of Science 20:55–91.

Baker, C. L. 1947. The species of amphiumae. Journal of the Tennessee Academy of Science 22:9–21.

Baker, C. L. 1963. Spermatozoa and spermateleosis

in *Cryptobranchus* and *Necturus*. Journal of the Tennessee Academy of Science 38:1–11.

Baker, C. L., Baker, L. C., and M. F. Caldwell. 1947. Observation of copulation in *Amphiuma tridactylum*. Journal of the Tennessee Academy of Science 22:87–88.

Baker, J. K. 1957. *Eurycea troglodytes:* a new blind cave salamander from Texas. Texas Journal of Science 9:328–336.

Baker, J. K. 1961. Distribution of and key to the neotenic *Eurycea* of Texas. The Southwestern Naturalist 6:27–32.

Baker, L. C. 1937. Mating habitats and life history of *Amphiuma tridactylum* Cuvier and effects of pituitary injections. Journal of the Tennessee Academy of Science 12:206–218.

Baldauf, R. J. 1947. *Desmognathus f. fuscus* eating eggs of its own species. Copeia 1947:66.

Baldauf, R. J. 1952. Climatic factors influencing the breeding migration of the spotted salamander, *Ambystoma maculatum* (Shaw). Copeia 1952:178–181.

Baldwin, K. S., and R. A. Stanford. 1987. *Ambystoma tigrinum californiense* (California tiger salamander). Herpetological Review 18:33.

Ball, S. C. 1937. Amphibians of Gaspe County, Quebec. Copeia 1937:230.

Banks, R. C., McDiarmid, R. W., and A. L. Gardner. 1987. Checklist of vertebrates of the United States, the U.S. Territories, and Canada. U.S. Fish and Wildlife Service, Washington, D.C., Resource Publication 166, 79 pp.

Banks, R. C., McDiarmid, R. W., Gardner A. L., and W. C. Starnes. 1998. Checklist of vertebrates of the United States, the U.S. Territories, and Canada, 2nd Edition.

Banta, A. M., and W. L. McAtee. 1906. The life history of the cave salamander, *Sperlepes maculicaudus* (Cope). Proceedings of the United States National Museum 30:67–73.

Barach, J. P. 1951. The value of the skin secretions of the spotted salamander. Herpetologica 7:58.

Barbour, R. W. 1953. The amphibians of Big Black Mountain, Harlan County, Kentucky. Copeia 1953:84–89.

Barbour, R. W. 1971. Amphibians and Reptiles of Kentucky. University of Kentucky Press, Lexington, Kentucky, 334 pp.

Barbour, R. W., and R. M. Hays. 1957. The genus *Desmognathus* in Kentucky. I. *Desmognathus fuscus welteri*. The American Midland Naturalist 58:352–359.

Barbour, R. W., and L. Y. Lancaster. 1946. Food habits of *Desmognathus fuscus* in Kentucky. Copeia 1946:48–49.

Barbour, R. W., Hardin, J. W., Schafer, J. P., and M. J. Harvey. 1969. Home range, movements, and activity of the dusky salamander, *Desmognathus fuscus*. Copeia 1969:293–297.

Barden, R. B., and J. Kezer. 1944. The eggs of certain plethodontid salamanders obtained by pituitary gland implantation. Copeia 1944:115–118.

Barr, T. C., Jr. 1949. A preliminary study of cave ecology, with special reference to Tennessee caverns. Bulletin of the National Speleological Society 11:55–59.

Barry, S. J., and H. B. Shaffer. 1994. The status of the California tiger salamander (*Ambystoma californiense*) at Lagunita: a 50-year update. Copeia 1994:159–164.

Bart, H. L., Jr., and R. W. Holzenthal. 1985. Feeding ecology of *Necturus beyeri* in Louisiana. Journal of Herpetology 19:402–410.

Bart, H. L., Jr., Bailey, M. A., Ashton, R. E., Jr., and P. E. Moler. 1997. Taxonomic and nomenclatural status of the Upper Black Warrior River waterdog. Journal of Herpetology 31:192–201.

Barthalmus, G. T., and E. D. Bellis. 1969. Homing in the northern dusky salamander, *Desmognathus fuscus fuscus* (Rafinesque). Copeia 1969:148–153.

Barthalmus, G. T., and E. D. Bellis. 1972. Home range, homing, and the homing mechanism of the salamander, *Desmognathus fuscus*. Copeia 1972:632–642.

Barthalmus, G. T., and I. R. Savidge. 1974. Time: an index of distance as a barrier to salamander homing. Journal of Herpetology 8:247–251.

Bartley, J. A. 1959. Two records of albinism in *Eurycea b. bislineata*. Herpetologica 15:192.

Baumann, W. L., and M. Huels. 1982. Nests of the two-lined salamander, *Eurycea bislineata*. Journal of Herpetology 16:81–83.

Beachy, C. K. 1991. *Ambystoma maculatum* (spotted salamander). Predation. Herpetological Review 22:128.

Beachy, C. K. 1993a. Guild structure in streamside

salamander communities: a test for interactions among larval plethodontid salamanders. Journal of Herpetology 27:465–468.

Beachy, C. K. 1993b. Differences in variation in egg size for several species of salamanders (Amphibia: Caudata) that use different larval environments. Brimleyana 18:71–81.

Beachy, C. K. 1994. Community ecology in streams: effects of two species of predatory salamanders on a prey species of salamander. Herpetologica 50:129–136.

Beachy, C. K. 1995. Effects of larval growth history on metamorphosis in a stream-dwelling salamander (*Desmognathus ochrophaeus*). Journal of Herpetology 29:375–382.

Beachy, C. K. 1996. Reduced courtship success between parapatric populations of the plethodontid salamander *Gyrinophilus porphyriticus*. Copeia 1996:199–203.

Behler, J. L., and F. W. King. 1979. The Audubon Society Field Guide to North American Reptiles and Amphibians. Alfred A. Knopf, New York, 719 pp.

Behre, E. H. 1953. Food of the salamander *Triturus viridescens viridescens*. Copeia 1953:60.

Belcher, D. L. 1988. Courtship behavior and spermatophore deposition by the subterranean salamander, *Typhlomolge rathbuni* (Caudata, Plethodontidae). The Southwestern Naturalist 33:124–125.

Bell, E. L. 1956. Some aspects of the life history of the red salamander, *Pseudotriton r. ruber*, in Huntington County, Pa. The Mengel Naturalist 1956:10–13.

Bellis, E. D. 1968. Summer movements of red-spotted newts in a small pond. Journal of Herpetology 1:86–91.

Beneski, J. T., Jr. 1989. Adaptive significance of tail autotomy in the salamander, *Ensatina*. Journal of Herpetology 23:322–324.

Beneski, J. T., Jr., Zalisko, E. J., and J. H. Larsen, Jr. 1986. Demography and migratory patterns of the eastern long-toed salamander, *Ambystoma macrodactylum columbianum*. Copeia 1986:398–408.

Bennett, A. F., and L. D. Houck. 1983. The energetic cost of courtship and aggression in a plethodontid salamander. Ecology 64:979–983.

Bennett, C., and R. J. Taylor. 1968. Notes on the lesser siren, *Siren intermedia* (Urodela). The Southwestern Naturalist 13:455–457.

Bennett, S. H., Gibbons, J. W., and J. Glanville. 1980. Terrestrial activity, abundance, and diversity of amphibians in differently managed forest types. The American Midland Naturalist 103:412–616.

Berger-Bishop, L. E., and R. N. Harris. 1996. A study of caudal allometry in the salamander *Hemidactylium scutatum* (Caudata: Plethodontidae). Herpetologica 52:515–525.

Berkhouse, C. S., and J. N. Fries. 1995. Critical thermal maxima of juvenile and adult San Marcos salamanders (*Eurycea nana*). The Southwestern Naturalist 40:430–434.

Bernardo, J. 1994. Experimental analysis of allocation in two divergent, natural salamander populations. The American Naturalist 143:14–38.

Besharse, J. C., and J. R. Holsinger. 1977. *Gyrinophilus subterraneus*, a new troglobitic salamander from southern West Virginia. Copeia 1977:624–634.

Bishop, S. C. 1919. Notes on the habits and development of the four-toed salamander, *Hemidactylium scutatum* (Schlegel). New York State Museum Bulletin 219:251–282.

Bishop, S. C. 1924. Notes on salamanders. New York State Museum Bulletin 253:87–96.

Bishop, S. C. 1925. The life history of the red salamander. Natural History 25:385–389.

Bishop, S. C. 1926. Notes on the habits and development of the mudpuppy, *Necturus maculosus* (Rafinesque). New York State Museum Bulletin 268:5–60.

Bishop, S. C. 1928. Notes on some amphibians and reptiles from the southeastern states with a description of a new salamander from North Carolina. Journal of the Elisha Mitchell Scientific Society 43:153–170.

Bishop, S. C. 1932. The spermatophores of *Necturus maculosus* (Rafinesque). Copeia 1932:1–3.

Bishop, S. C. 1941a. Salamanders of New York. New York State Museum Bulletin 324:1–365.

Bishop, S. C. 1941b. Notes on salamanders with descriptions of several new forms. Occasional Papers of the Museum of Zoology, University of Michigan 451:1–27.

Bishop, S. C. 1943. A Handbook of Salamanders. Comstock Publishing, Ithaca, New York, 555 pp.

Bishop, S. C. 1944. A new neotenic plethodont salamander, with notes on related species. Copeia 1944:1–5.

Bishop, S. C. 1947. Supposed cases of mimicry in salamanders. Herpetologica 3:178.

Bishop, S. C., and H. P. Chrisp. 1933. The nests and young of the Allegheny salamander *Desmognathus ochrophaeus* (Cope). Copeia 1933:194–198.

Bishop, S. C., and B. O. Valentine. 1950. A new species of *Desmognathus* from Alabama. Copeia 1950:39–43.

Bizer, J. R. 1978. Growth rates and size at metamorphosis of high elevation populations of *Ambystoma tigrinum*. Oecologia 34:175–184.

Black, J. H. 1969. A cave dwelling population of *Ambystoma tigrinum malvortium* in Oklahoma. Journal of Herpetology 3:183–184.

Black, J. H. 1974. Notes on *Plethodon ouachitae* in Oklahoma. Proceedings of the Oklahoma Academy of Science 54:88–89.

Blair, A. P. 1951. Note on Oklahoma salamanders. Copeia 1951:178.

Blair, A. P. 1961. Metamorphosis of *Pseudotriton palleucus* with iodine. Copeia 1961:499.

Blair, A. P. 1967a. Tail prehensile in *Phaeognathus hubrichti*. Herpetologica 23:67.

Blair, A. P. 1967b. *Plethodon ouachitae*. Catalogue of American Amphibians and Reptiles, p. 40.

Blair, A. P., and H. L. Lindsay, Jr. 1965. Color pattern variation and distribution of two large *Plethodon* salamanders endemic to the Ouachita Mountains of Oklahoma and Arkansas. Copeia 1965:331–335.

Blair, W. F. 1958. Distributional patterns of vertebrates in the southern United States in relation to past and present environments. In: Zoogeography, C. L. Hubbs, ed. American Advances in Science Publication 51:433–468.

Blanchard, F. N. 1922. Discovery of eggs of the four-toed salamander in Michigan. Occasional Papers of the Museum of Zoology, University of Michigan 126:1–3.

Blanchard, F. N. 1923. The life history of the four-toed salamander. The American Naturalist 57:262–268.

Blanchard, F. N. 1928. Topics from the life history and habits of the red-backed salamander in southern Michigan. The American Naturalist 62:156–164.

Blanchard, F. N. 1930. The stimulus to the breeding migration of the spotted salamander, *Ambystoma maculatum* (Shaw). The American Naturalist 64:154–167.

Blanchard, F. N. 1933a. Spermatophores and the mating season of the salamander *Hemidactylium scutatum* (Schlegel). Copeia 1933:40.

Blanchard, F. N. 1933b. Late autumn collections and hibernating situations of the salamander *Hemidactylium scutatum* (Schlegel) in southern Michigan. Copeia 1933:216.

Blanchard, F. N. 1934a. The relation of the female four-toed salamander to her nest. Copeia 1934:137–138.

Blanchard, F. N. 1934b. The date of egg-laying of the four-toed salamander, *Hemidactylium scutatum* (Schlegel), in southern Michigan. Papers of the Michigan Academy of Sciences, Arts and Letters 19:571–575.

Blanchard, F. N. 1935. The sex ratio in the salamander *Hemidactylium scutatum* (Schlegel). Copeia 1935:103.

Blanchard, F. N. 1936. The number of eggs produced and laid by the four-toed salamander, *Hemidactylium scutatum* (Schlegel), in southern Michigan. Papers of the Michigan Academy of Sciences, Arts and Letters 21:567–573.

Blanchard, F. N., and F. C. Blanchard. 1931. Size groups and their characteristics in the salamander *Hemidactylium scutatum* (Schlegel). The American Naturalist 65:149–164.

Blaney, R. M., and P. K. Blaney. 1978. Significance of extreme variation in a cave population of the salamander *Gyrinophilus porphyriticus*. Proceedings of the West Virginia Academy of Science 50:23.

Blaney, R. M., and K. Relyea. 1967. The zigzag salamander, *Plethodon dorsalis* Cope, in southern Alabama. Herpetologica 23:246–247.

Blatchley, W. S. 1897. Indiana caves and their fauna. Department of Geology and Natural Resources of Indiana, 21st annual report, pp. 121–212.

Blatchley, W. S. 1900. Notes on the batrachians and reptiles of Vigo County, Indiana (II). Annual Report of the Indiana Department of Geologic and Natural Resources 24:537–552.

Blaustein, A. R. 1994. Chicken Little or Nero's fiddle? A perspective on declining amphibian populations. Herpetologica 50:85-97.

Blaustein, A. R., and D. B. Wake. 1990. Declining amphibian populations: a global phenomenon? Trends in Ecology and Evolution 5:203-204.

Blaustein, A. R., and D. B. Wake. 1995. The puzzle of declining amphibian populations. Scientific American 273:52-57.

Blaustein, A. R., Edmond, B., Kiesecker, J. M., Beatty, J. J., and D. G. Hokit. 1995. Ambient ultraviolet radiation causes mortality in salamander eggs. Ecological Applications 5:740-743.

Blaustein, A. R., Hoffman, P. D., Hokit, D. G., Kiesecker, J. M., Walls, S. C., and J. B. Hays. 1994. UV repair and resistance to solar UV-B in amphibian eggs: a link to population declines? Proceedings of the National Academy of Sciences 91:1791-1795.

Blaustein, A. R., Hoffman, P. D., Kiesecker, J. M., and J. B. Hays. 1996. DNA repair activity and resistance to solar UV-B radiation in eggs of the red-legged frog. Conservation Biology 10:1398-1402.

Blaustein, A. R., Sousas, W. P., and D. B. Wake. 1993. Amphibian declines: judging stability, persistence, and susceptibility of populations to local and global extinctions. Conservation Biology 8:60-71.

Bleakney, S. 1952. The amphibians and reptiles of Nova Scotia. The Canadian Field-Naturalist 66:125-129.

Bleakney, S. 1957. The egg-laying habits of the salamander, *Ambystoma jeffersonianum*. Copeia 1957:141-142.

Blem, C. R., and L. B. Blem. 1989. Tolerance of acidity in a Virginia population of the spotted salamander, *Ambystoma maculatum* (Amphibia: Ambystomatidae). Brimleyana 15:37-45.

Blem, C. R., and L. B. Blem. 1991. Cation concentrations and acidity in breeding ponds of the spotted salamander, *Ambystoma maculatum* (Shaw) (Amphibia: Ambystomatidae), in Virginia. Brimleyana 15:67-76.

Blymer, M. J., and McGinnes, B. 1977. Observations on possible detrimental effects of clearcutting on terrestrial amphibians. Bulletin of the Maryland Herpetological Society 13:79-83.

Bobka, M. S., Jaeger, R. G., and D. C. McNaught. 1981. Temperature dependent assimilation efficiencies of two species of terrestrial salamanders. Copeia 1981:417-421.

Bock, S. F., and J. E. Fauth. 1992. *Pseudotriton* (northern red salamander). Diet. Herpetological Review 23:58.

Bogart, J. P. 1982. Ploidy and genetic diversity in Ontario salamanders of the *Ambystoma jeffersonianum* complex revealed through an electrophoretic examination of larvae. Canadian Journal of Zoology 60:848-855.

Bogart, J. P. 1989. A mechanism for interspecific gene exchange via all-female salamander hybrids. In: The evolution and ecology of unisexual vertebrates, R. M. Dawley and J. P. Bogart, eds. New York State Museum Bulletin 466:170-179.

Bogart, J. P., and L. E. Licht. 1986. Reproduction and the origin of polyploids in hybrid salamanders of the genus *Ambystoma*. Canadian Journal of Genetics and Cytology 60:848-855.

Bogart, J. P., Elinson, R. P., and L. E. Licht. 1989. Temperature and sperm incorporation in polyploid salamanders. Science 246:1032-1034.

Bogart, J. P., Licht, L. E., Oldham, M. J., and S. J. Darbyshire. 1985. Electrophoretic identification of *Ambystoma laterale* and *Ambystoma texanum* as well as their diploid and triploid interspecific hybrids (Amphibia: Caudata) on Pelee Island, Ontario. Canadian Journal of Zoology 63:340-347.

Bogart, J. P., Lowcock, L. A., Zeyl, C. W., and B. K. Mable. 1987. Genome constitution and reproductive biology of hybrid salamanders, genus *Ambystoma*, on Kellys Island in Lake Erie. Canadian Journal of Zoology 65:2188-2201.

Bogert, C. M. 1960. The influence of sound on behavior of amphibians and reptiles. In: Animal sounds and communication, W. E. Lanyon and W. N. Tavolga, eds. American Institute of Biological Sciences Publication 7:137-320.

Boundy, J., and T. G. Balgooyen. 1988. Record lengths of some amphibians and reptiles from the western United States. Herpetological Review 19:27.

Boyd, C. E., and D. H. Vickers. 1963. Distribution of some Mississippi amphibians and reptiles. Herpetologica 19:202-205.

Bradford, D. F. 1991. Mass mortality and extinction in a high-elevation population of *Rana muscosa*. Journal of Herpetology 25:174-177.

Bradford, D. F., Swanson, C., and M. S. Gordon. 1994. Effects of low pH and aluminum on amphibians at high elevation in the Sierra Nevada, California. Canadian Journal of Zoology 72:1272-1279.

Bradley, S. G., and L. J. Klika. 1981. A fatal poisoning from the Oregon rough-skinned newt (*Taricha granulosa*). Journal of the American Medical Association 246:247.

Bragg, A. N. 1949. Observations on the narrow-mouthed salamander. Proceedings of the Oklahoma Academy of Science 1949:21-24.

Brame, A. H., Jr. 1956. The number of eggs laid by the California newt. Herpetologica 12:325.

Brame, A. H., Jr. 1964. Distribution of the Oregon slender salamander, *Batrachoseps wrighti* (Bishop). Bulletin of the Southern California Academy of Science 63:165-170.

Brame, A. H., Jr. 1968. The number of egg masses and eggs laid by the California newt, *Taricha torosa*. Journal of Herpetology 2:169-170.

Brame, A. H., Jr. 1970. A new species of *Batrachoseps* (slender salamander) from the desert of southern California. Los Angeles County Museum of Natural History, Contributions in Science 200:1-11.

Brame, A. H., Jr., and K. Murray. 1968. Three new slender salamanders (*Batrachoseps*) with a discussion of relationships and speciation within the genus. Los Angeles County Museum of Natural History Bulletin 4:1-35.

Branch, L. C., and R. Altig. 1981. Nocturnal stratification of three species of *Ambystoma* larvae. Copeia 1981:870-873.

Brandon, R. A. 1961. A comparison of the larvae of five northeastern species of *Ambystoma* (Amphibia, Caudata). Copeia 1961:377-383.

Brandon, R. A. 1965a. A new race of the neotenic salamander *Gyrinophilus palleucus*. Copeia 1965:346-352.

Brandon, R. A. 1965b. Morphological variation and ecology of the salamander *Phaeognathus hubrichti*. Copeia 1965:67-71.

Brandon, R. A. 1965c. *Typhlotriton, T. nereus,* and *T. spelaeus*. Catalogue of American Amphibians and Reptiles, p. 20.

Brandon, R. A. 1966a. Additional localities for *Ambystoma texanum* in Alabama, with comments on site of oviposition. Journal of the Ohio Herpetological Society 5:104-105.

Brandon, R. A. 1966b. Systematics of the salamander genus *Gyrinophilus*. Illinois Biological Monographs 35:1-86.

Brandon, R. A. 1966c. A reevaluation of the status of the salamander, *Typhlotriton nereus* Bishop. Copeia 1966:555-561.

Brandon, R. A. 1967a. *Gyrinophilus palleucus*. Catalogue of American Amphibians and Reptiles, pp. 32.1-32.2.

Brandon, R. A. 1967b. Food and intestinal parasite of the troglobitic salamander *Gyrinophilus palleucus necturoides*. Herpetologica 23:52-53.

Brandon, R. A. 1967c. *Gyrinophilus porphyriticus*. Catalogue of American Amphibians and Reptiles, pp. 33.1-33.3.

Brandon, R. A. 1967d. *Haideotriton* and *H. wallacei*. Catalogue of American Amphibians and Reptiles, p. 39.

Brandon, R. A. 1970. *Typhlotriton* and *Typhlotriton spelaeus*. Catalogue of American Amphibians and Reptiles, p. 84.

Brandon, R. A. 1971a. North American troglobitic salamanders: some aspects of modification in cave habitats, with special reference to *Gyrinophilus palleucus*. National Speleological Society Bulletin 33:1-21.

Brandon, R. A. 1971b. Correlation of seasonal abundance with feeding and reproductive activity in the grotto salamander (*Typhlotriton spelaeus*). The American Midland Naturalist 86:93-100.

Brandon, R. A. 1972. Hybridization between the Mexican salamanders *Ambystoma dumerilli* and *Ambystoma mexicanum* under laboratory conditions. Herpetologica 28:199-207.

Brandon, R. A. 1977. Interspecific hybridization between Mexican and United States salamanders under laboratory conditions. Herpetologica 33:133-152.

Brandon, R. A., and D. J. Bremer. 1966. Neotenic newts, *Notophthalmus viridescens louisianensis,* in southern Illinois. Herpetologica 22:213-217.

Brandon, R. A., and D. J. Bremer. 1967. Overwintering of larval tiger salamanders in southern Illinois. Herpetologica 23:67-68.

Brandon, R. A., and J. E. Huheey. 1971. Movements and interactions of two species of

Brandon, R. A., and J. E. Huheey. 1975. Diurnal activity, avian predation, and the question of warning coloration and cryptic coloration in salamanders. Herpetologica 31:252-255.

Brandon, R. A., and J. E. Huheey. 1981. Toxicity in the plethodontid salamanders *Pseudotriton ruber* and *Pseudotriton montanus* (Amphibia, Caudata). Toxicon 19:25-31.

Brandon, R. A., and J. E. Huheey. 1985. Salamander skin toxins, with special reference to *Necturus lewisi*. Brimleyana 10:75-82.

Brandon, R. A., and E. J. Maruska. 1982. *Phaeognathus hubrichti* (Red Hills salamander). Reproduction. Herpetological Review 13:46.

Brandon, R. A., and J. M. Rutherford. 1967. Albinos in a cavernicolous population of the salamander *Gyrinophilus porphyriticus* in West Virginia. The American Midland Naturalist 78:537-540.

Brandon, R. A., Labanick, G. M., and J. E. Huheey. 1977. Defensive behavior in the plethodontid salamanders *Pseudotriton montanus* and *P. ruber*. Herpetological Review 8 (suppl. 2).

Brandon, R. A., Labanick, G. M., and J. E. Huheey. 1979a. Learned avoidance of brown efts, *Notophthalmus viridescens louisianensis* (Amphibia, Urodela, Salamandridae), by chickens. Journal of Herpetology 13:171-176.

Brandon, R. A., Labanick, G. M., and J. E. Huheey. 1979b. Relative palatability, defensive behavior, and mimetic relationships of red salamanders (*Pseudotriton ruber*), mud salamanders (*Pseudotriton montanus*), and red efts (*Notophthalmus viridescens*). Herpetologica 35:289-303.

Brandt, B. B. 1952. Albino *Ambystoma maculatum*. Copeia 1952:3.

Branin, M. L. 1935. Courtship activities and extra-seasonal ovulation in the four-toed salamander, *Hemidactylium scutatum* (Schlegel). Copeia 1935:172-175.

Braswell, A. L., and R. E. Ashton, Jr. 1985. Distribution, ecology, and feeding habits of *Necturus lewisi* (Brimley). Brimleyana 10:13-35.

Brattstrom, B. H. 1953. Records of Pleistocene reptiles and amphibians from Florida. Journal of the Florida Academy of Sciences 16:243-248.

Breder, R. B. 1927. The courtship of the spotted salamander. Bulletin of the New York Zoological Society 30:51-56.

Breitenbach, G. L. 1982. The frequency of joint nesting and solitary brooding in the salamander, *Hemidactylium scutatum*. Journal of Herpetology 16:241-346.

Brimley, C. S. 1896. Batrachia found at Raleigh, N. C. The American Naturalist 30:500-501.

Brimley, C. S. 1910. Records of some reptiles and batrachians from the southeastern United States. Proceedings of the Biological Society of Washington 23:9-18.

Brimley, C. S. 1920a. Reproduction of the marbled salamander. Copeia 1920:25.

Brimley, C. S. 1920b. Notes on *Amphiuma* and *Necturus*. Copeia 1920:5-7.

Brimley, C. S. 1921. Breeding dates of *Ambystoma maculatum* at Raleigh, N. C. Copeia 1921:26-27.

Brimley, C. S. 1923. The dwarf salamander at Raleigh, N. C. Copeia 120:81-83.

Brimley, C. S. 1924. The water dogs (*Necturus*) of North Carolina. Journal of the Elisha Mitchell Scientific Society 40:166-168.

Brimley, C. S. 1928. Yellow-cheeked *Desmognathus* from Macon Co., North Carolina. Copeia 1928:21-23.

Bristow, C. E. 1991. Interactions between phylogenetically distant predators: *Notophthalmus viridescens* and *Enneacanthus obesus*. Copeia 1991:1-8.

Brock, J., and P. Verrell. 1994. Courtship behavior of the seal salamander, *Desmognathus monticola* (Amphibia: Caudata: Plethodontidae). Journal of Herpetology 28:411-415.

Brode, W. E. 1961. Observations on the development of *Desmognathus* eggs under relatively dry conditions. Herpetologica 17:202-203.

Brode, W. E., and G. Gunter. 1958. Egg clutches and prehensilism in the slimy salamander. Herpetologica 13:279-280.

Brodie, E. D., III. 1989. Individual variation in antipredator response of *Ambystoma jeffersonianum* to snake predators. Journal of Herpetology 23:307-309.

Brodie, E. D., III, and E. D. Brodie, Jr. 1990. Tetrodotoxin resistance in garter snakes: an evolutionary response of predators to dangerous prey. Evolution 44:651-659.

Brodie, E. D., III, and E. D. Brodie, Jr. 1991. Evolutionary response of predators to dangerous prey: reduction of toxicity of newts and resistance of garter snakes in island populations. Evolution 45:221-224.

Brodie, E. D., Jr. 1968a. Investigations on the skin toxin of the red-spotted newt, *Notophthalmus viridescens viridescens*. The American Midland Naturalist 80:276-280.

Brodie, E. D., Jr. 1968b. Observations on the mental hedonic gland-clusters of western salamanders of the genus *Plethodon*. Herpetologica 24:248-250.

Brodie, E. D., Jr. 1968c. Investigations on the skin toxin of the adult rough-skinned newt, *Taricha granulosa*. Copeia 1968:307-313.

Brodie, E. D., Jr. 1970. Western salamanders of the genus *Plethodon*: systematics and geographic variation. Herpetologica 26:468-516.

Brodie, E. D., Jr. 1971a. Two more toxic salamanders: *Ambystoma maculatum* and *Cryptobranchus alleganiensis*. Herpetological Review 3:8.

Brodie, E. D., Jr. 1971b. *Plethodon stormi*. Catalogue of American Amphibians and Reptiles, pp. 103.1-103.2.

Brodie, E. D., Jr. 1976. Additional observations on the Batesian mimicry of *Notophthalmus viridescens* efts by *Pseudotriton ruber*. Herpetologica 32:68-70.

Brodie, E. D., Jr. 1977. Salamander antipredator postures. Copeia 1977:523-535.

Brodie, E. D., Jr. 1978. Biting and vocalization as antipredator mechanisms in terrestrial salamanders. Copeia 1978:127-129.

Brodie, E. D., Jr. 1981. Phenological relationships of model and mimic salamanders. Evolution 35:988-994.

Brodie, E. D., Jr., and R. G. Altig. 1967. Morphological variation of the Jemez Mountains salamander, *Plethodon neomexicanus*. Copeia 1967:670-672.

Brodie, E. D., Jr., and E. D. Brodie, III. 1980. Differential avoidance of mimetic salamanders by free-ranging birds. Science 208:181-182.

Brodie, E. D., Jr., and D. R. Formanowicz, Jr. 1981. Larvae of the predaceous diving beetle *Dytiscus verticalis* acquire an avoidance response to skin secretions of the newt *Notophthalmus viridescens*. Herpetologica 37:172-176.

Brodie, E. D., Jr., and L. S. Gibson. 1969. Defensive behavior and skin glands of the northwestern salamander, *Ambystoma gracile*. Herpetologica 25:187-194.

Brodie, E. D., Jr., and R. R. Howard. 1972. Behavioral mimicry in the defensive displays of the urodele amphibians *Notophthalmus viridescens* and *Pseudotriton ruber*. Bioscience 22:666-667.

Brodie, E. D., Jr., and R. R. Howard. 1973. Experimental study of Batesian mimicry in the salamanders *Plethodon jordani* and *Desmognathus ochrophaeus*. The American Midland Naturalist 90:38-46.

Brodie, E. D., Jr., and R. M. Storm. 1971. *Plethodon elongatus*. Catalogue of American Amphibians and Reptiles, pp. 102.1-102.2.

Brodie, E. D., Jr., Dowdey, T. G., and C. D. Anthony. 1989. Salamander antipredator strategies against snake attack: biting by *Desmognathus*. Herpetologica 45:167-171.

Brodie, E. D., Jr., Hensel, J. L., and J. A. Johnson, 1974a. Toxicity of the urodele amphibians *Taricha, Notophthalmus, Cynops,* and *Paramesotriton* (family Salamandridae). Copeia 1974:506-511.

Brodie, E. D., Jr., Johnson, J. A., and C. K. Dodd, Jr. 1974b. Immobility as a defensive behavior in salamanders. Herpetologica 30:79-85.

Brodie, E. D., Jr., Nowak, R. T., and W. R. Harvey. 1979. The effectiveness of antipredator secretions and behavior of selected salamanders against shrews. Copeia 1979:270-274.

Brodman, R. 1993. The effect of acidity on interactions of *Ambystoma* salamander larvae. Journal of Freshwater Ecology 8:209-214.

Brodman, R. 1995. Annual variation in breeding success of two syntopic species of *Ambystoma* salamanders. Journal of Herpetology 29:111-113.

Brodman, R. 1996. Effects of intraguild interactions on fitness and microhabitat use of larval *Ambystoma* salamanders. Copeia 1996:372-378.

Brooks, M. 1945. Notes on amphibians from Bickle's Knob, West Virginia. Copeia 1945:231.

Brooks, M. 1946. Burrowing of *Plethodon jordani*. Copeia 1946:102.

Brooks, M. 1948a. Clasping in the salamanders *Aneides* and *Desmognathus*. Copeia 1948:65.

Brooks, M. 1948b. Notes on the Cheat Mountain salamander. Copeia 1948:239-244.

Brophy, T. E. 1980. Food habits of sympatric larval *Ambystoma tigrinum* and *Notophthalmus viridescens*. Journal of Herpetology 14:1-6.

Brown, B. C. 1942. Notes on *Eurycea neotenes*. Copeia 1942:176.

Brown, B. C. 1967a. *Eurycea latitans*. Catalogue of American Amphibians and Reptiles, p. 34.

Brown, B. C. 1967b. *Eurycea nana*. Catalogue of American Amphibians and Reptiles, p. 35.

Brown, C. W. 1968. Additional observations on the function of the nasolabial grooves of plethodontid salamanders. Copeia 1968:728–731.

Brown, C. W. 1974. Hybridization among the subspecies of the plethodontid salamander *Ensatina eschscholtzii*. University of California Publications in Zoology 98:1–56.

Brown, E. E. 1979. Some snake food records from the Carolinas. Brimleyana 1:113–124.

Brown, H. A. 1976. The time-temperature relation of embryonic development in the northwestern salamander, *Ambystoma gracile*. Canadian Journal of Zoology 54:552–558.

Brown, J. L. 1965. Stability of color phase ratio in populations of *Plethodon cinereus*. Copeia 1965:95–98.

Brown, W. C., and S. C. Bishop. 1948. Eggs of *Desmognathus aeneus*. Copeia 1948:129.

Brown, W. C., and J. R. Slater. 1939. The amphibians and reptiles of the islands of the state of Washington. Occasional Papers of the Department of Biology, College of Puget Sound 4:6–31.

Bruce, R. C. 1966. Occurrence of the salamander *Plethodon jordani* in the Piedmont of northwestern South Carolina. Copeia 1966:888–889.

Bruce, R. C. 1967. A study of the salamander genus *Plethodon* on the southeastern escarpment of the Blue Ridge Mountains. Journal of the Elisha Mitchell Scientific Society 82:74–82.

Bruce, R. C. 1968. The role of the Blue Ridge Embayment in the zoogeography of the green salamander, *Aneides aeneus*. Herpetologica 24:185–194.

Bruce, R. C. 1969. Fecundity in primitive plethodontid salamanders. Evolution 23:50–54.

Bruce, R. C. 1970. The larval life of the three-lined salamander, *Eurycea longicauda guttolineata*. Copeia 1970:776–779.

Bruce, R. C. 1971. Life cycle and population structure of the salamander *Stereochilus marginatus* in North Carolina. Copeia 1971:234–246.

Bruce, R. C. 1972a. Variation in the life cycle of the salamander *Gyrinophilus porphyriticus*. Herpetologica 28:230–245.

Bruce, R. C. 1972b. The larval life of the red salamander, *Pseudotriton ruber*. Journal of Herpetology 6:43–51.

Bruce, R. C. 1974. Larval development of the salamanders *Pseudotriton montanus* and *P. ruber*. The American Midland Naturalist 92:173–190.

Bruce, R. C. 1975. Reproductive biology of the mud salamander, *Pseudotriton montanus*, in western South Carolina. Copeia 1975:129–137.

Bruce, R. C. 1976. Population structure, life history, and evolution of paedogenesis in the salamander *Eurycea neotenes*. Copeia 1976:242–249.

Bruce, R. C. 1977. The pygmy salamander, *Desmognathus wrighti* (Amphibia, Urodela, Plethodontidae), in the Cowee Mountains, North Carolina. Journal of Herpetology 11:244–246.

Bruce, R. C. 1978a. Life-history patterns of the salamander *Gyrinophilus porphyriticus* in the Cowee Mountains, North Carolina. Herpetologica 34:53–64.

Bruce, R. C. 1978b. A comparison of the larval periods of Blue Ridge and Piedmont mud salamanders (*Pseudotriton montanus*). Herpetologica 34:325–332.

Bruce, R. C. 1978c. Reproductive biology of the salamander *Pseudotriton ruber* in the southern Blue Ridge Mountains. Copeia 1978:417–423.

Bruce, R. C. 1979. Evolution of paedogenesis in salamanders of the genus *Gyrinophilus*. Evolution 33:998–1000.

Bruce, R. C. 1980. A model of the larval period of the spring salamander, *Gyrinophilus porphyriticus*, based on size-frequency distributions. Herpetologica 36:78–86.

Bruce, R. C. 1982a. Larval periods and metamorphosis in two species of salamanders of the genus *Eurycea*. Copeia 1982:117–127.

Bruce, R. C. 1982b. Egg-laying, larval periods, and metamorphosis of *Eurycea bislineata* and *E. junaluska* at Santeetlah Creek, North Carolina. Copeia 1982:755–762.

Bruce, R. C. 1985a. Larval periods, population structure and the effects of stream drift in larvae of the salamanders *Desmognathus quadramaculatus* and *Leurognathus marmoratus* in a southern Appalachian stream. Copeia 1985:847–854.

Bruce, R. C. 1985b. Larval period and metamorphosis in the salamander *Eurycea bislineata*. Herpetologica 41:19-28.

Bruce, R. C. 1986. Upstream and downstream movements of *Eurycea bislineata* and other salamanders in a southern Appalachian stream. Herpetologica 42:149-155.

Bruce, R. C. 1988a. Life history variation in the salamander *Desmognathus quadramaculatus*. Herpetologica 44:218-227.

Bruce, R. C. 1988b. An ecological life table for the salamander *Eurycea wilderae*. Copeia 1988:15-26.

Bruce, R. C. 1989. Life history of the salamander *Desmognathus monticola*, with a comparison of the larval periods of *D. monticola* and *D. ochrophaeus*. Herpetologica 45:144-155.

Bruce, R. C. 1990. An explanation for differences in body size between two desmognathine salamanders. Copeia 1990:1-9.

Bruce, R. C. 1991. Evolution of ecological diversification in desmognathine salamanders. Herpetological Review 22:44-46.

Bruce, R. C. 1993. Sexual size dimorphism in desmognathine salamanders. Copeia 1993:313-318.

Bruce, R. C. 1995. The use of temporary removal sampling in a study of population dynamics of the salamander *Desmognathus monticola*. Australian Journal of Ecology 20:403-412.

Bruce, R. C. 1996. Life-history perspective of adaptive radiation in desmognathine salamanders. Copeia 1996:783-790.

Bruce, R. C., and N. G. Hairston, Sr. 1990. Life history correlates of body-size differences between two populations of the salamander, *Desmognathus monticola*. Journal of Herpetology 24:124-134.

Brunkow, P. E., and J. P. Collins. 1996. Effects of individual variation in size on growth and development of larval salamanders. Ecology 77:1483-1492.

Brussock, P. P. III, and A. V. Brown. 1982. Selection of breeding ponds by the ringed salamander, *Ambystoma annulatum*. Proceedings of the Arkansas Academy of Science 36:82-83.

Buchwald, H. D., Durham, L., Fischer, H. G., Harada, R., Mosher, H. S., Kao, C. Y., and F. A. Fuhrman. 1964. Identity of tarichatoxin and tetrodotoxin. Science 143:474-475.

Buhlmann, K. A., Pague, C. A., Mitchell, J. C., and R. B. Glasgow. 1988. Forestry operations and terrestrial salamanders: techniques in a study of the Cow Knob salamander, *Plethodon punctatus*. In: Management of amphibians, reptiles, and mammals in North America, R. C. Szaro, K. E. Severson, and D. R. Patton, eds. USDA Forest Service, Rocky Mountain Forest and Range Experiment Station, Fort Collins, Colorado, Technical Report RM-166, pp. 38-44.

Burch, P. R., and J. T. Wood. 1955. The salamander *Siren lacertina* feeding on clams and snails. Copeia 1955:255-256.

Burger, J. W. 1935. *Plethodon cinereus* (Green) in eastern Pennsylvania and New Jersey. The American Naturalist 64:578-586.

Burger, W. L., Smith H. M., and F. E. Potter, Jr. 1950. Another neotenic salamander from the Edwards Plateau. Proceedings of the Biological Society of Washington 63:51-58.

Burke, C. V. 1911. Note on *Batrachoseps attenuatus* Esch. The American Naturalist 45:413-414.

Burns, D. M. 1954. A new subspecies of the salamander *Plethodon vandykei*. Herpetologica 10:83-87.

Burns, D. M. 1962. The taxonomic status of the salamander *Plethodon vandykei larselli*. Copeia 1962:177-181.

Burns, D. M. 1964. *Plethodon larselli*. Catalogue of American Amphibians and Reptiles, p. 13.

Burt, C. E. 1938. Contributions to Texas herpetology VII. The salamanders. The American Midland Naturalist 20:374-380.

Burton, T. M. 1976. An analysis of the feeding ecology of the salamanders (Amphibia: Urodela) of the Hubbard Brook Experimental Forest, New Hampshire. Journal of Herpetology 10:187-204.

Burton, T. M. 1977. Population estimates, feeding habits and nutrient and energy relationships of *Notophthalmus v. viridescens* in Mirror Lake, New Hampshire. Copeia 1977:139-143.

Burton, T. M., and G. E. Likens. 1975. Salamander populations and biomass in the Hubbard Brook Experimental Forest, New Hampshire. Copeia 1975:541-546.

Bury, R. B. 1970. Food similarities in the tailed frog, *Ascaphus truei*, and the olympic salamander, *Rhyacotriton olympicus*. Copeia 1970:170-171.

Bury, R. B. 1972. Small mammals and other prey in the diet of the Pacific giant salamander (*Dicamptodon ensatus*). The American Midland Naturalist 87:524-526.

Bury, R. B. 1983. Differences in amphibian populations in logged and old growth redwood forest. Northwest Science 57:167–178.

Bury, R. B. 1988. Habitat relationships and ecological importance of amphibians and reptiles. In: Streamside management: riparian wildlife and forestry interactions, K. J. Raedeke, ed. Institute of Forest Resources, University of Washington, Seattle, Washington, Contribution 59:61–76.

Bury, R. B., and P. S. Corn. 1988a. Douglas-fir forests in the Oregon and Washington Cascades: abundance of terrestrial herpetofauna related to stand age and moisture. In: Management of amphibians, reptiles, and small mammals in North America, R. C. Szaro, K. E. Severson, and D. R. Patton, technical coordinators. USDA Forest Service, Rocky Mountain Forest and Range Experiment Station, Fort Collins, Colorado, Technical Report RM-166, pp. 11–22.

Bury, R. B., and P. S. Corn. 1988b. Responses of aquatic and streamside amphibians to timber harvest: a review. In: Streamside management: riparian wildlife and forestry interactions, K. J. Raedeke, ed. Institute of Forest Resources, University of Washington, Seattle, Washington, Contribution 59:165–181.

Bury, R. B., and C. R. Johnson. 1965. Note on the food of *Plethodon elongatus* in California. Herpetologica 21:68–69.

Bury, R. B., and M. Martin. 1967. The food of the salamander *Rhyacotriton olympicus*. Copeia 1967:487.

Bury, R. B., and M. Martin. 1973. Comparative studies on the distribution and foods of plethodontid salamanders in the redwood region of northern California. Journal of Herpetology 7:331–335.

Bury, R. B., Dodd, C. K., Jr., and G. M. Fellers. 1980. Conservation of the amphibia of the United States: a review. U. S. Department of the Interior, Fish and Wildlife Service, Washington, D.C., Resource Publication 134:1–34.

Bush, F. M. 1959. The herpetofauna of Clemmons Fork, Breathitt County, Kentucky. Transactions of the Kentucky Academy of Science 20:11–18.

Caetano, M. H., and R. LeClair, Jr. 1996. Growth and population structure of red-spotted newts (*Notophthalmus viridescens*) in permanent lakes of the Laurentian Shield, Quebec. Copeia 1996:866–874.

Cagle, F. R. 1942. Herpetological fauna of Jackson and Union counties, Illinois. The American Midland Naturalist 28:164–200.

Cagle, F. R. 1948. Observations on a population of the salamander, *Amphiuma tridactylum* Cuvier. Ecology 29:479–491.

Cagle, F. R. 1954. Observations on the life history of the salamander *Necturus louisianensis*. Copeia 1954:257–260.

Cagle, F. R., and P. E. Smith. 1939. A winter aggregation of *Siren intermedia* and *Triturus viridescens*. Copeia 1939:232–233.

Cahn, A. R., and W. Shumway. 1925. Color variations in larvae of *Necturus maculosus*. Copeia 1925:106–107.

Caldwell, R. S. 1975. Observations on the winter activity of the red-backed salamander, *Plethodon cinereus*, in Indiana. Herpetologica 31:21–22.

Caldwell, R. S. 1980. Lens morphology as an identification tool in the salamander subfamily Desmognathinae. Journal of the Tennessee Academy of Science 55:15–17.

Caldwell, R. S., and W. C. Houtcooper. 1973. Food habits of larval *Eurycea bislineata*. Journal of Herpetology 7:386–388.

Caldwell, R. S., and G. S. Jones. 1973. Winter congregations of *Plethodon cinereus* in ant mounds, with notes on their food habits. The American Midland Naturalist 90:482–485.

Caldwell, R. S., and S. E. Trauth. 1979. Use of toe pad and tooth morphology in differentiating three species of *Desmognathus* (Amphibia, Urodela, Plethodontidae). Journal of Herpetology 13:491–497.

Calef, R. T. 1954. The salamander *Ambystoma tigrinum nebulosum* in southern Arizona. Copeia 1954:223.

Camp, C. D. 1986. Distribution and habitat of the southern red-back salamander, *Plethodon serratus* Grobman (Amphibia: Plethodontidae), in Georgia. Georgia Journal of Science 44:136–146.

Camp, C. D. 1988. Aspects of the life history of the southern red-back salamander *Plethodon serratus* Grobman in the southeastern United States. The American Midland Naturalist 119:93–100.

Camp, C. D. 1989. Fishing for "spring lizards": a technique for collecting blackbelly salamanders. Herpetological Review 20:47.

Camp, C. D. 1996. Bite scar patterns in the black-bellied salamander, *Desmognathus quadramaculatus*. Journal of Herpetology 30:543–546.

Camp, C. D., and L. L. Bozeman. 1981. Foods of two species of *Plethodon* (Caudata: Plethodontidae) from Georgia and Alabama. Brimleyana 6:163–166.

Camp, C. D., and T. P. Lee. 1996. Intraspecific spacing and interaction within a population of *Desmognathus quadramaculatus*. Copeia 1996:78–84.

Camper, J. D. 1986. *Ambystoma tigrinum tigrinum* (eastern tiger salamander). Herpetological Review 17:19.

Camper, J. D. 1990. Mode of reproduction in the small-mouthed salamander, *Ambystoma texanum* (Ambystomatidae), in Iowa. The Southwestern Naturalist 35:99–100.

Canterbury, R. A., and T. K. Pauley. 1990. Gut analysis of the green salamander (*Aneides aeneus*) in West Virginia. Proceedings of the West Virginia Academy of Science 62:47–50.

Canterbury, R. A., and T. K. Pauley. 1991. Intra- and interspecific competition in the green salamander, *Aneides aeneus*. Association of Southeastern Biologists Bulletin 38:114.

Canterbury, R. A., and T. K. Pauley. 1994. Time of mating and egg deposition of West Virginia populations of the salamander *Aneides aeneus*. Journal of Herpetology 28:431–434.

Carey, C. 1993. Hypothesis concerning the causes of the disappearance of boreal toads from the mountains of Colorado. Conservation Biology 7:355–362.

Carey, S. D. 1984. *Amphiuma pholeter*. Herpetological Review 15:77.

Carl, G. C. 1942. The long-toed salamander on Vancouver Island. Copeia 1942:56.

Carl, G. C. 1943. The amphibians of British Columbia. British Columbia Provincial Museum Handbook 2:1–62.

Carl, G. C., and I. M. Cowan. 1945. Notes on the salamanders of British Columbia. Copeia 1945:43–44.

Carlin, J. L. 1997. Genetic and morphological differentiation between *Eurycea longicauda longicauda* and *E. guttolineata* (Caudata: Plethodontidae). Herpetologica 53:206–217.

Carpenter, C. C. 1953. An ecological survey of the herpetofauna of the Grand Teton–Jackson Hole area of Wyoming. Copeia 1953:170–174.

Carpenter, C. C. 1955. Aposematic behavior in the salamander *Ambystoma tigrinum melanostictum*. Copeia 1955:311.

Carr, A. F., Jr. 1939. *Haideotriton wallacei*, a new subterranean salamander from Georgia. Occasional Papers of the Boston Society of Natural History 8:333–336.

Carr, A. F., Jr. 1940. A contribution to the herpetology of Florida. University of Florida Biological Sciences Series 3(1):1–118.

Carr, D. E. 1996. Morphological variation among species and populations of salamanders in the *Plethodon glutinosus* complex. Herpetologica 52:56–65.

Carr, D. E., and D. H. Taylor. 1985. Experimental evaluation of population interactions among three sympatric species of *Desmognathus*. Journal of Herpetology 19:507–514.

Carreno, C. A., Vess, T. J., and R. N. Harris. 1996. An investigation of kin recognition abilities in larval four-toed salamanders *Hemidactylium scutatum* (Caudata: Plethodontidae). Herpetologica 52:293–300.

Castanet, J., Francillon-Vieillot, H., and R. C. Bruce. 1996. Age estimation in desmognathine salamanders assessed by skeletochronology. Herpetologica 52:160–171.

Chadwick, C. S. 1940. Some notes on the burrows of *Plethodon metcalfi*. Copeia 1940:50.

Chadwick, C. S. 1944. Observations on the life cycle of the common newt in western North Carolina. The American Midland Naturalist 32:491–494.

Chadwick, C. S. 1950. Observations on the behavior of the larvae of the common American newt during metamorphosis. The American Midland Naturalist 43:392–398.

Chandler, A. C. 1918. The western newt or waterdog (*Notophthalmus torosus*), a natural enemy of mosquitoes. Oregon Agricultural College Experimental Station Bulletin 152:1–24.

Chaney, A. H. 1951. The food habits of the salamander *Amphiuma tridactylum*. Copeia 1951:45–49.

Channell, L. S., and B. D. Valentine. 1972. A yellow albino *Desmognathus fuscus* from West Virginia. Journal of Herpetology 6:144–146.

Chippindale, P. T. 1995. Evolution, phylogeny, biogeography, and taxonomy of central Texas perennibranchiate salamanders, *Eurycea* and *Typhlomolge* (Plethodontidae: Hemidactyliini). Ph.D. dissertation, University of Texas, Austin, Texas.

Chippindale, P. T., Price, A. H., and D. M. Hillis. 1993. A new species of perennibranchiate salamander (*Eurycea*: Plethodontidae) from Austin, Texas. Herpetologica 49:248-259.

Chivers, D. P., Kiesecker, J. M., Anderson, M. T., Wildy, E. L., and A. R. Blaustein. 1996. Avoidance response of a terrestrial salamander (*Ambystoma macrodactylum*) to chemical alarm cues. Journal of Chemical Ecology 22:1709-1716.

Christman, S. P., and L. R. Franz. 1973. Feeding habits of the striped newt, *Notophthalmus perstriatus*. Journal of Herpetology 7:133-135.

Christman, S. P., and H. I. Kochman. 1975. The southern distribution of the many-lined salamander, *Stereochilus marginatus*. The Florida Scientist 38:139-141.

Christman, S. P., and D. B. Means. 1991. Striped newt. In: Rare and Endangered Biota of Florida, Volume 3, Amphibians and Reptiles, R. W. McDiarmid, ed. University Presses of Florida, Gainesville, Florida, pp. 14-15.

Clanton, W. 1934. An unusual situation in the salamander *Ambystoma jeffersonianum* (Green). Occasional Papers of the Museum of Zoology, University of Michigan 290:1-15.

Clark, J. M. 1985. Fossil plethodontid salamanders from the latest Miocene of California. Journal of Herpetology 19:41-47.

Clark, K. L. 1986. Responses of *Ambystoma maculatum* populations in central Ontario to habitat acidity. The Canadian Field-Naturalist 100:463-469.

Clark, K. L., and R. J. Hall. 1985. Effects of elevated hydrogen ion and aluminum concentrations on the survival of amphibian embryos and larvae. Canadian Journal of Zoology 63:116-123.

Clay, W. M., Case, R. B., and R. Cunningham. 1955. On the taxonomic status of the slimy salamander, *Plethodon glutinosus* (Green), in southeastern Kentucky. Transactions of the Kentucky Academy of Science 16:57-65.

Cliburn, J. W., and A. B. Porter. 1986. Comparative climbing abilities of the salamanders *Aneides aeneus* and *Plethodon glutinosus* (Caudata, Plethodontidae). Journal of the Mississippi Academy of Science 31:91-96.

Cliburn, J. W., and A. B. Porter. 1987. Vertical stratification of the salamanders *Aneides aeneus* and *Plethodon glutinosus* (Caudata: Plethodontidae). Journal of the Alabama Academy of Science 58:18-22.

Clovis, J. F. 1979. Tree importance values in West Virginia red spruce forests inhabited by the Cheat Mountain salamander. Proceedings of the West Virginia Academy of Science 51:58-64.

Coates, M. 1967. A comparative study of the serum proteins of the species *Taricha* and their hybrids. Evolution 21:130-140.

Coates, M., Benedict, E., and C. L. Stephens. 1970. An unusual aggregation of the newt *Taricha granulosa granulosa*. Copeia 1970:176-178.

Cockran, M. E. 1911. The biology of the red-backed salamander (*Plethodon cinereus erythronotus* Green). Biological Bulletin 20:332-349.

Cochran, P. A., and J. D. Lyons. 1985. *Necturus maculosus* (mudpuppy). Juvenile ecology. Herpetological Review 16:53.

Cockrum, L. 1941. Notes on *Siren intermedia*. Copeia 1941:265.

Cody, M. L. 1969. Convergent characteristics in sympatric species: a possible relation to interspecific competition and aggression. Condor 71:222-239.

Cohen, N. W. 1952. Comparative rates of dehydration and hydration in some California salamanders. Ecology 33:462-479.

Coker, C. M. 1931. Hermit thrushes feeding on salamanders. The Auk 48:277.

Cole, C. J. 1990. When is an individual not a species? Herpetologica 46:104-108.

Collazo, A., and S. B. Marks. 1994. Development of *Gyrinophilus porphyriticus*: identification of the ancestral developmental pattern in the salamander family Plethodontidae. Journal of Experimental Zoology 268:239-258.

Collette, B. B., and F. R. Gehlbach. 1961. The salamander *Siren intermedia intermedia* LeConte in North Carolina. Herpetologica 17:203-204.

Colley, S. A., Keen, W. H., and R. W. Reed. 1989.

Effects of adult presence on behavior and microhabitat use of juveniles of a desmognathine salamander. Copeia 1989:1–7.

Collins, J. P. 1981. Distribution, habitats, and life history variation in the tiger salamander, *Ambystoma tigrinum*, in east-central and southeast Arizona. Copeia 1981:666–675.

Collins, J. P., and J. E. Cheek. 1983. Effects of food and density on development of typical and cannibalistic salamander larvae in *Ambystoma tigrinum nebulosum*. American Zoologist 23:77–84.

Collins, J. P., and J. R. Holomuzki. 1984. Intraspecific variation in diet within and between trophic morphs in larval tiger salamanders (*Ambystoma tigrinum nebulosum*). Canadian Journal of Zoology 62:168–174.

Collins, J. P., Jones, T. R., and H. J. Berna. 1988. Conserving genetically distinctive populations: the case of the Huachuca tiger salamander (*Ambystoma tigrinum stebbinsi* Lowe). In: Management of amphibians, reptiles, and small mammals in North America, R. C. Szaro, K. E. Severson, and D. R. Patton, technical coordinators. USDA Forest Service, Rocky Mountain Forest and Range Experiment Station, Fort Collins, Colorado, Technical Report RM-166, pp. 45–53.

Collins, J. P., Minton, J. B., and B. A. Pierce. 1980. *Ambystoma tigrinum*: a multispecies conglomerate? Copeia 1980:938–941.

Collins, J. P., Zerba, K. E., and M. J. Sredl. 1993. Shaping intraspecific variation: development, ecology and the evolution of morphology and life history variation in tiger salamanders. Genetica 89:167–183.

Collins, J. T. 1965. A population study of *Ambystoma jeffersonianum*. Journal of the Ohio Herpetological Society 5:61.

Collins, J. T. 1990. Standard common and current scientific names for North American amphibians and reptiles, 3rd Edition. SSAR Herpetological Circular 19:1–418.

Collins, J. T. 1991. Viewpoint: a new taxonomic arrangement for some North American amphibians and reptiles. Herpetological Review 22:42–43.

Collins, J. T. 1992. The evolutionary species concept: a reply to VanDevender et al. and Montanucci. Herpetological Review 23:43–46.

Collins, J. T. 1993. Amphibians and reptiles in Kansas, 3rd Edition. University of Kansas Museum of Natural History, Lawrence, Kansas, Public Education Series 13.

Collins, J. T., Conant, R., Huheey, J. E., Knight, J. L., Runquist, E. M., and H. M. Smith. 1982. Standard common and current scientific names for North American amphibians and reptiles, 2nd Edition. SSAR Herpetological Circular 12:1–28.

Collins, J. T., Huheey, J. E., Knight, J. L., and H. M. Smith. 1978. Standard common and current scientific names for North American amphibians and reptiles, 1st Edition. SSAR Herpetological Circular 7:1–36.

Conant, R. 1957. The eastern mud salamander, *Pseudotriton montanus montanus*: a new state record for New Jersey. Copeia 1957:152–153.

Conant, R. 1958. A Field Guide to Reptiles and Amphibians of the United States and Canada East of the 100th Meridian. Houghton Mifflin, Boston, 366 pp.

Conant, R. 1975. A Field Guide to Reptiles and Amphibians of Eastern and Central North America, 2nd Edition. Houghton Mifflin, Boston, 429 pp.

Conant, R., and J. T. Collins. 1991. A Field Guide to Reptiles and Amphibians of Eastern and Central North America, 3rd Edition. Houghton Mifflin, New York, 450 pp.

Cook, M. L., and B. C. Brown. 1974. Variation in the genus *Desmognathus* (Amphibia: Plethodontidae) in the western limits of its range. Journal of Herpetology 8:93–105.

Cook, R. P. 1983. Effects of acid precipitation on embryonic mortality of *Ambystoma* salamanders in the Connecticut Valley of Massachusetts. Biological Conservation 27:77–88.

Cooper, J. E. 1955. Notes on the amphibians and reptiles of southern Maryland. The Maryland Naturalist 23:90–100.

Cooper, J. E. 1956. Aquatic hibernation of the redbacked salamander. Herpetologica 1956:165–166.

Cooper, J. E. 1960. The mating antic of the long-tailed salamander. The Maryland Naturalist 30:17–18.

Cooper, J. E. 1961. Cave records for the salamander *Plethodon r. richmondi* Pope, with notes on additional cave-associated species. Herpetologica 17:250–255.

Cooper, J. E. 1968. The salamander *Gyrinophilus*

palleucus in Georgia with notes on Alabama and Tennessee populations. Journal of the Alabama Academy of Science 39:182-185.

Cooper, J. E., and R. E. Ashton, Jr. 1985. The *Necturus lewisi* study: introduction, selected literature review, and comments on the hydrologic units and their faunas. Brimleyana 10:1-12.

Cooper, J. E., and M. R. Cooper. 1968. Cave-associated herpetozoa II: salamanders of the genus *Gyrinophilus* in Alabama caves. Bulletin of the National Speleological Society 30:19-24.

Corn, P. S., and R. B. Bury. 1989. Logging in western Oregon: responses of headwater habitats and stream amphibians. Forest Ecology and Management 29:39-57.

Corn, P. S., and R. B. Bury. 1991. Terrestrial amphibian communities in the Oregon Coast Range. In: Wildlife and vegetation of unmanaged Douglas-fir forests, L. F. Ruggiero, K. B. Aubry, A. B. Carey, and M. H. Huff, technical coordinators. USDA Forest Service, Pacific Northwest Research Station, Olympia, Washington, General Technical Report PNW-GTR-285, pp. 304-317.

Corn, P. S., and F. A. Vertucci. 1992. Descriptive risk assessment of the effects of acid deposition on Rocky Mountain amphibians. Journal of Herpetology 26:361-369.

Cortwright, S. A. 1988. Intraguild predation and competition: an analysis of net growth shifts in larval amphibian prey. Canadian Journal of Zoology 66:1813-1821.

Cory, L., and J. L. Manion. 1953. Predation on eggs of the wood-frog, *Rana sylvatica*, by leeches. Copeia 1953:66.

Couture, M. R., and D. M. Sever. 1979. Developmental mortality of *Ambystoma tigrinum* (Amphibia: Urodela) in northern Indiana. Proceedings of the Indiana Academy of Science 88:173-175.

Craddock, J. E., and W. L. Minckley. 1964. Amphibians and reptiles from Meade Co., Kentucky. The American Midland Naturalist 71:382-391.

Crump, M. L., Hensley, F. R., and K. L. Clark. 1992. Apparent decline of the golden toad: underground or extinct? Copeia 1992:413-420.

Culver, D. C. 1973. Feeding behavior of the salamander *Gyrinophilus porphyriticus* in caves. International Journal of Speleology 5:369-377.

Cunningham, J. D. 1960. Aspects of the ecology of the Pacific slender salamander, *Batrachoseps pacificus*, in southern California. Ecology 41:88-99.

Cupp, P. V., Jr. 1971. Fall courtship of the green salamander, *Aneides aeneus*. Herpetologica 27:308-310.

Cupp, P. V., Jr. 1980. Territoriality in the green salamander, *Aneides aeneus*. Copeia 1980:463-468.

Cupp, P. V., Jr. 1991. Aspects of the life history and ecology of the green salamander, *Aneides aeneus*, in Kentucky. Journal of the Tennessee Academy of Science 66:171-174.

Cupp, P. V., Jr. 1994. Salamanders avoid chemical cues from predators. Animal Behaviour 48:232-235.

Cupp, P. V., Jr., and D. T. Towles. 1983. A new variant of *Plethodon wehrlei* in Kentucky and West Virginia. Transactions of the Kentucky Academy of Science 44:157-158.

Curd, M. R. 1950. The salamander *Amphiuma tridactylum* in Oklahoma. Copeia 1950:324.

Dalrymple, G. H. 1970. Caddisfly larvae feeding upon eggs of *Ambystoma t. tigrinum*. Herpetologica 26:128-129.

Danforth, C. G. 1950. New locality for Mt. Lyell salamander. Yosemite Nature Notes 29:18-19.

Daniel, P. M. 1989. *Hemidactylium scutatum*. In: Salamanders of Ohio, R. A. Pfingsten and F. L. Downs, eds. Ohio Biological Survey Bulletin, New Series 7(2):223-228.

Danstedt, R. T., Jr. 1975. Local geographic variation in demographic parameters and body size of *Desmognathus fuscus* (Amphibia: Plethodontidae). Ecology 56:1054-1067.

Danstedt, R. T., Jr. 1979. A demographic comparison of two populations of the dusky salamander (*Desmognathus fuscus*) in the same physiographic province. Herpetologica 35:164-168.

Daugherty, C. H., Allendorf, F. W., Dunlap, W. W., and K. L. Knudsen. 1983. Systematic implications of geographic patterns of genetic variation in the genus *Dicamptodon*. Copeia 1983:679-691.

Davic, R. D. 1983. Microgeographic body size variation in *Desmognathus fuscus fuscus* salamanders from western Pennsylvania. Copeia 1983: 1101-1104.

Davic, R. D. 1991. Ontogenetic shift in diet of *Desmognathus quadramaculatus*. Journal of Herpetology 25:108-111.

Davic, R. D., and L. P. Orr. 1987. The relationship

between rock density and salamander density in a mountain stream. Herpetologica 43:357–361.

David, R. S., and R. G. Jaeger. 1981. Prey location through chemical cues by a terrestrial salamander. Copeia 1981:435–440.

Davidson, J. A. 1956. Notes on the food habits of the slimy salamander *Plethodon glutinosus glutinosus*. Herpetologica 12:1–88.

Davidson, M., and H. Heatwole. 1960. Late summer oviposition in the salamander, *Plethodon cinereus*. Herpetologica 16:141–142.

Davis, J. 1952. Observations on the eggs and larvae of the salamander *Batrachoseps pacificus major*. Copeia 1952:272–274.

Davis, J. R., and B. H. Brattstrom. 1975. Sounds produced by the California newt, *Taricha torosa*. Herpetologica 31:409–412.

Davis, W. B., and F. T. Knapp. 1953. Notes on the salamander *Siren intermedia*. Copeia 1953:119–121.

Davis, W. C., and V. C. Twitty. 1964. Courtship behavior and reproductive isolation in the species of *Taricha* (Amphibia, Caudata). Copeia 1964:601–610.

Dawley, E. M. 1984a. Identification of sex through odors by male red-spotted newts, *Notophthalmus viridescens*. Herpetologica 40:101–105.

Dawley, E. M. 1984b. Recognition of individual, sex and species odours by salamanders of the *Plethodon glutinosus–P. jordani* complex. Animal Behaviour 32:353–361.

Dawley, E. M. 1986a. Behavioral isolating mechanisms in sympatric terrestrial salamanders. Herpetologica 42:156–164.

Dawley, E. M. 1986b. Evolution of chemical signals as a premating isolating mechanism in a complex of terrestrial salamanders. In: Chemical Signals in Vertebrates IV, D. Duvall, D. Müller-Schwarze, and R. M. Silverstein, eds. Plenum Press, New York, pp. 221–224.

Dawley, E. M. 1987. Species discrimination between hybridizing and non-hybridizing terrestrial salamanders. Copeia 1987:924–931.

Dawley, E. M. 1992a. Correlation of salamander vomeronasal and main olfactory system anatomy with habitat and sex: behavioral interpretations. In: Chemical Signals in Vertebrates VI, R. L. Doty and D. Müller-Schwarze, eds. Plenum Press, New York, pp. 403–409.

Dawley, E. M. 1992b. Sexual dimorphism in a chemosensory system: the role of the vomeronasal system in salamander reproductive behavior. Copeia 1992:113–120.

Dawley, E. M., and A. H. Bass. 1989. Chemical assess to the vomeronasal organ of a plethodontid salamander. Journal of Morphology 200:163–174.

Dawley, E. M., and J. Crowder. 1995. Sexual and seasonal differences in the vomeronasal epithelium of the red-backed salamander (*Plethodon cinereus*). Journal of Comparative Neurology 359:282–390.

Dawley, E. M., and R. M. Dawley. 1986. Species discrimination by chemical cues in a unisexual-bisexual complex of salamanders. Journal of Herpetology 20:114–116.

Deckert, R. F. 1916. Note on *Ambystoma opacum*, Grav. Copeia 1916:23–24.

Deevey, E. S., Jr. 1947. Life tables for natural populations of animals. Quarterly Review of Biology 22:283–314.

Delany, M. F., and C. L. Ambercrombie. 1986. American alligator food habits in northcentral Florida. Journal of Wildlife Management 50:348–353.

deMaynadier, P. G. 1995. *Plethodon cinereus cinereus* (redback salamander). Coloration. Herpetological Review 26:199.

deMaynadier, P. G, and M. L. Hunter, Jr. 1995. The relationship between forest management and amphibian ecology: a review of the North American literature. Environmental Review 3:230–261.

Dempster, W. T. 1930. The growth of larvae of *Ambystoma maculatum* under natural conditions. Biological Bulletin 58:182–192.

Dennis, D. M. 1962. Notes on the nesting habits of *Desmognathus fuscus* (Raf.) in Licking County, Ohio. Journal of the Ohio Herpetological Society 3:28–35.

Dent, J. N., and J. S. Kirby-Smith. 1963. Metamorphic physiology and morphology of the cave salamander *Gyrinophilus palleucus*. Copeia 1963:119–130.

Dethlefsen, E. S. 1948. A subterranean nest of the Pacific giant salamander, *Dicamptodon ensatus* (Eschscholtz). The Wasmann Collector 7:81–84.

Dieckmann, J. M. 1927. The cloaca and spermatheca of *Gyrinophilus porphyriticus*. Biological Bulletin 53:258-280.

DiGiovanni, M., and E. D. Brodie, Jr. 1981. Efficacy of skin glands in protecting the salamander *Ambystoma opacum* from repeated attacks by the shrew *Blarina brevicauda*. Herpetologica 37:234-237.

Diller, J. S. 1907. A salamander-snake fight. Science 26:907-908.

Diller, L. V., and R. L. Wallace. 1994. Distribution and habitat of *Plethodon elongatus* in managed, young growth forests in north coastal California. Journal of Herpetology 28:310-318.

Diller, L. V., and R. L. Wallace. 1996. Distribution and habitat of *Rhyacotriton variegatus* on managed, young growth forests in north coastal California. Journal of Herpetology 30:184-191.

Dodd, C. K., Jr. 1977. Preliminary observations on the reactions of certain salamanders of the genus *Ambystoma* (Amphibia, Urodela, Ambystomatidae) to a small colubrid snake (Reptilia, Serpentes, Colubridae). Journal of Herpetology 11:222-223.

Dodd, C. K., Jr. 1980. Notes on the feeding behavior of the Oklahoma salamander, *Eurycea tynerensis* (Plethodontidae). The Southwestern Naturalist 25:111-113.

Dodd, C. K., Jr. 1989a. Duration of immobility in salamanders, genus *Plethodon* (Caudata: Plethodontidae). Herpetologica 45:467-473.

Dodd, C. K. Jr. 1989b. Status of the Red Hills salamander is reassessed. Endangered Species Technical Bulletin 14:10-11.

Dodd, C. K., Jr. 1990a. Line transect estimation of Red Hills salamander burrow density using a Fourier series. Copeia 1990:555-557.

Dodd, C. K., Jr. 1990b. The influence of temperature and body size on duration of immobility in salamanders of the genus *Desmognathus*. Amphibia-Reptilia 11:401-410.

Dodd, C. K., Jr. 1990c. Postures associated with immobile woodland salamanders, genus *Plethodon*. Florida Scientist 53:43-49.

Dodd, C. K., Jr. 1991. The status of the Red Hills salamander *Phaeognathus hubrichti*, Alabama, USA, 1976-1988. Biological Conservation 55:57-75.

Dodd, C. K., Jr. 1992. Biological diversity of a temporary pond herpetofauna in north Florida sandhills. Biodiversity and Conservation 1:125-142.

Dodd, C. K., Jr. 1993. Cost of living in an unpredictable environment: the ecology of striped newts *Notophthalmus perstriatus* during a prolonged drought. Copeia 1993:605-614.

Dodd, C. K., Jr. 1996. Use of terrestrial habitats by amphibians in the Sandhill Uplands of north-central Florida. Alytes 14:42-52.

Dodd, C. K., Jr., and E. D. Brodie, Jr. 1976. Observations on the mental hedonic gland-cluster of eastern salamanders of the genus *Plethodon*. Chesapeake Science 17:129-131.

Dodd, C. K., Jr., and B. G. Charest. 1988. The herpetofaunal community of temporary ponds in north Florida sandhills: species composition, temporal use, and management implications. In: Management of amphibians, reptiles, and small mammals in North America, R. C. Szaro, K. E. Severson, and D. R. Patton, eds. USDA Forest Service, Rocky Mountain Forest and Range Experimental Station, Fort Collins, Colorado, General Technical Report RM-166, pp. 87-97.

Dodd, C. K., Jr., and L. V. LaClaire. 1995. Biogeography and status of the striped newt (*Notophthalmus perstriatus*) in Georgia, USA. Herpetological Natural History 3:37-46.

Dodd, C. K., Jr., Johnson, J. A., and E. D. Brodie, Jr. 1974. Noxious skin secretions of an eastern small *Plethodon, P. nettingi hubrichti*. Journal of Herpetology 8:89-92.

Dodson, S. I. 1970. Complementary feeding niches sustained by size-selective predation. Limnology and Oceanography 15:131-137.

Dodson, S. I., and V. E. Dodson. 1971. The diet of *Ambystoma tigrinum* larvae from western Colorado. Copeia 1971:614-624.

Donovan, A., and P. A. Verrell. 1991. The effect of partner familiarity on courtship success in the salamander *Desmognathus ochrophaeus*. Journal of Herpetology 25:93-95.

Donovan, L. A., and G. W. Folkerts. 1972. Foods of the seepage salamander *Desmognathus aeneus* Brown and Bishop. Herpetologica 28:35-37.

Doody, J. S. 1996. Larval growth rate of known age *Ambystoma opacum* in Louisiana under natural conditions. Journal of Herpetology 30:294-297.

Douglas, M. E. 1979. Migration and sexual selection in *Ambystoma jeffersonianum*. Canadian Journal of Zoology 57:2303-2310.

Douglas, M. E., and B. L. Monroe, Jr. 1981. A comparative study of topographical orientation in *Ambystoma* (Amphibia: Caudata). Copeia 1981:460–463.

Dowdey, T. G., and E. D. Brodie, Jr. 1989. Antipredator strategies of salamanders: individual and geographical variation in responses of *Eurycea bislineata* to snakes. Animal Behaviour 37:707–711.

Dowling, H. G. 1956. Geographic relations of Ozarkian amphibians and reptiles. The Southwestern Naturalist 1:174–189.

Dowling, H. G. 1993. Viewpoint: a reply to Collins (1991, 1992). Herpetological Review 24:11–13.

Downs, F. L. 1978. Unisexual *Ambystoma* from the Bass Islands of Lake Erie. Occasional Papers of the Museum of Zoology, University of Michigan 685:1–36.

Downs, F. L. 1989. Family Ambystomatidae. In: Salamanders of Ohio, R. A. Pfingsten and F. L. Downs, eds. Ohio Biological Survey Bulletin, New Series 7(2):87–172.

Drost, C. A., and G. M. Fellers. 1996. Collapse of a regional frog fauna in the Yosemite area of the California Sierra Nevada, USA. Conservation Biology 10:414–425.

Ducey, P. K. 1989. Agonistic behavior and biting during intraspecific encounters in *Ambystoma* salamanders. Herpetologica 45:155–160.

Ducey, P. K., and E. D. Brodie, Jr. 1983. Salamanders respond selectively to contacts with snakes: survival advantage of alternative antipredator strategies. Copeia 1983:1036–1041.

Ducey, P. K., and J. Dulkiewicz. 1994. Ontogenetic variation in antipredator behavior of the newt *Notophthalmus viridescens:* comparisons of terrestrial adults and efts in field and laboratory studies. Journal of Herpetology 28:530–533.

Ducey, P. K., and J. Heuer. 1991. Effects of food availability on intraspecific aggression in salamanders of the genus *Ambystoma*. Canadian Journal of Zoology 69:288–290.

Ducey, P. K., and P. Ritsema. 1988. Intraspecific aggression and responses to marked substrates in *Ambystoma maculatum* (Caudata: Ambystomatidae). Copeia 1988:1008–1013.

Ducey, P. K., Schramm, K., and N. Cambry. 1994. Interspecific aggression between the sympatric salamanders, *Ambystoma maculatum* and *Plethodon cinereus*. The American Midland Naturalist 131:320–329.

Duellman, W. E. 1948. An *Ambystoma* eats a snake. Herpetologica 4:164.

Duellman, W. E. 1954a. Observations on autumn movements of the salamander *Ambystoma tigrinum tigrinum* in southeastern Michigan. Copeia 1954:156–157.

Duellman, W. E. 1954b. The salamander *Plethodon richmondi* in southwestern Ohio. Copeia 1954:40–45.

Duellman, W. E. 1955. Notes on reptiles and amphibians from Arizona. Occasional Papers of the Museum of Zoology, University of Michigan 569:1–14.

Duellman, W. E., and A. Schwartz. 1958. Amphibians and reptiles of southern Florida. Bulletin of the Florida State Museum 3:181–324.

Duellman, W. E., and L. Trueb. 1986. Biology of Amphibians. McGraw-Hill, New York, 670 pp.

Duellman, W. E., and J. T. Wood. 1954. Size and growth of the two-lined salamander, *Eurycea bislineata rivicola*. Copeia 1954:92–96.

Dumas, P. C. 1955. Eggs of the salamander *Plethodon dunni* in nature. Copeia 1955:65.

Dumas, P. C. 1956. The ecological relations of sympatry in *Plethodon dunni* and *Plethodon vehiculum*. Ecology 37:484–495.

Duncan, R., and R. Highton. 1979. Genetic relationships of the eastern large *Plethodon* of the Ouachita Mountains. Copeia 1979:95–110.

Dundee, H. A. 1947. Note on salamanders collected in Oklahoma. Copeia 1947:117–120.

Dundee, H. A. 1957. Partial metamorphosis induced in *Typhlomolge rathbuni*. Copeia 1957:52–53.

Dundee, H. A. 1958. Habitat selection by aquatic plethodontid salamanders of the Ozarks, with studies of their life histories (abstract). Dissertation Abstracts International 19:1480–1481.

Dundee, H. A. 1962. Response of the neotenic salamander *Haideotriton wallacei* to a metamorphic agent. Science 135:1060–1061.

Dundee, H. A. 1965a. *Eurycea multiplicata*. Catalogue of American Amphibians and Reptiles, pp. 21.1–21.2.

Dundee, H. A. 1965b. *Eurycea tynerensis*. Catalogue of American Amphibians and Reptiles, p. 22.

Dundee, H. A. 1971. *Cryptobranchus* and

Cryptobranchus alleganiensis. Catalogue of American Amphibians and Reptiles, pp. 101.1–101.4.

Dundee, H. A. 1974. Rediscovery of *A. tigrinum* in eastern Louisiana, with comments on the biology of the species. Journal of Herpetology 8:265–267.

Dundee, H. A., and D. S. Dundee. 1965. Observations on the systematics and ecology of *Cryptobranchus* from the Ozark Plateaus of Missouri and Arkansas. Copeia 1965:369–370.

Dunn, E. R. 1915. The transformation of *Sperlepes ruber* (Daudin). Copeia 1915:28–30.

Dunn, E. R. 1916. Two new salamanders of the genus *Desmognathus*. Proceedings of the Biological Society of Washington 29:73–76.

Dunn, E. R. 1917. The breeding habits of *Ambystoma opacum* (Gravenhorst). Copeia 1917:41–43.

Dunn, E. R. 1920. Some reptiles and amphibians from Virginia, North Carolina, Tennessee, and Alabama. Proceedings of the Biological Society of Washington 33:129–137.

Dunn, E. R. 1926. The Salamanders of the Family Plethodontidae. Smith College 50th Anniversary Publication, Northhampton, Massachusetts, 441 pp.

Dunn, E. R. 1942. An egg cluster of *Aneides ferreus*. Copeia 1942:52.

Dunn, E. R. 1944. Notes on the salamanders of the *Ambystoma gracile* group. Copeia 1944:129–130.

Dunson, W. A., Wyman, R. L., and E. S. Corbett. 1992. A symposium on amphibian declines and habitat acidification. Journal of Herpetology 26:349–352.

DuShane, G. P., and C. Hutchinson. 1944. Differences in size and developmental rate between eastern and midwestern embryos of *Ambystoma maculatum*. Ecology 25:414–423.

Dye, R. L. 1982. Sandhill cranes prey on amphiumas. The Florida Field Naturalist 10:76.

Dyrkacz, S. 1981. Recent instances of albinism in North American amphibians and reptiles. Society for the Study of Amphibians and Reptiles Herpetological Circular 11:1–31.

Eagleson, G. W. 1976. A comparison of the life histories and growth patterns of populations of the salamander *Ambystoma gracile* (Baird) from permanent low-altitude and montane lakes. Canadian Journal of Zoology 54:2098–2111.

Easterla, D. A. 1968. Melanistic spotted salamanders in northeast Arkansas. Herpetologica 24:330–331.

Easterla, D. A. 1971. A breeding concentration of four-toed salamanders, *Hemidactylium scutatum*, in southeastern Missouri. Journal of Herpetology 5:194–195.

Eaton, T. H., and G. T. Eaton. 1956. A new locality for the green salamander and woodfrog in North Carolina. Herpetologica 12:312.

Eaton, T. H., Jr. 1953. Salamanders of Pitt County, North Carolina. Journal of the Elisha Mitchell Scientific Society 69:49–53.

Eaton, T. H., Jr. 1956. Larvae of some Appalachian plethodontid salamanders. Herpetologica 12:303–311.

Echelle, A. A. 1990. In defense of the phylogenetic species concept and the ontological status of hybridogenetic taxa. Herpetologica 46:109–113.

Edgren, R. A. 1949. An autumnal concentration of *Ambystoma jeffersonianum*. Herpetologica 6:137–138.

Efford, I. E., and J. A. Mathias. 1969. A comparison of two salamander populations in Marion Lake, British Columbia. Copeia 1969:723–736.

Efford, I. E., and K. Tsumura. 1973. A comparison of the food of salamanders and fish in Marion Lake, British Columbia. Transactions of the American Fisheries Society 1:33–47.

Eigenmann, C. H., and C. Kennedy. 1903. Variation notes. Biological Bulletin 4:227–229.

Elinson, R. P., Bogart, J. P., Licht, L. E., and L. A. Lowcock. 1992. Gynogenetic mechanisms in polyploid hybrid salamanders. Journal of Experimental Zoology 264:93–99.

Elliott, S. A., Kats, L. B., and J. A. Breeding. 1993. The use of conspecific chemical cues for cannibal avoidance in California newts (*Taricha torosa*). Ethology 95:186–192.

Emerson, E. T. 1905. General anatomy of *Typhlomolge rathbuni*. Proceedings of the Boston Society of Natural History 32:43–76.

Endler, J. 1970. Kinesthetic orientation in the California newt (*Taricha torosa*). Behaviour 37:15–23.

Engelhardt, G. P. 1916a. *Ambystoma tigrinum* on Long Island. Copeia 1916:20–22.

Engelhardt, G. P. 1916b. *Ambystoma tigrinum* on Long Island. II. Records of larvae. Copeia 1916:32–35.

Ernst, C. H., and R. W. Barbour. 1989. Snakes of

Eastern North America. George Mason University Press, Fairfax, VA.

Espinoza, F. A., Jr., Deacon, J. E., and A. Simmin. 1970. An Economic and Biostatistical Analysis of the Bait Fish Industry in the Lower Colorado River. University of Nevada Las Vegas Special Publication 1-87.

Estes, R. 1963. Early Miocene salamanders and lizards from Florida. Quarterly Journal of the Florida Academy of Science 26:234-256.

Estes, R. 1981. Gymnophiona, Caudata. In: Encyclopedia of Paleoherpetology, Part 2, P. Wellnhofer, ed. Gustav-Fischer-Verlag, Stuttgart, pp. 1-115.

Etheridge, K. 1990a. Water balance in estivating sirenid salamanders (*Siren lacertina*). Herpetologica 46:400-406.

Etheridge, K. 1990b. The energetics of estivating sirenid salamanders (*Siren lacertina* and *Pseudobranchus striatus*). Herpetologica 46:407-414.

Evans, A. L., and D. C. Forester. 1996. Conspecific recognition by *Desmognathus ochrophaeus* using substrate-borne odor cues. Journal of Herpetology 30:447-451.

Evans, H. E. 1947. Herpetology of Crystal Lake, Sullivan County, New York. Herpetologica 4:19-21.

Evenden, F. G. 1943. Notes on amphibia of the Cascade Mountains in Oregon. Copeia 1943:251-252.

Evenden, F. G. 1948. Food habits of *Triturus granulosus* in western Oregon. Copeia 1948:219-220.

Eycleshymer, A. C. 1906. The habits of *Necturus maculosus*. The American Naturalist 40:123-136.

Farner, D. S. 1947. Notes on the food habits of the salamanders of Crater Lake, Oregon. Copeia 1947:259-261.

Farner, D. S., and J. Kezer. 1953. Notes on the amphibians and reptiles of Crater Lake National Park. The American Midland Naturalist 50:448-462.

Fauth, J. E. 1990. Interactive effects of predators and early larval dynamics of the treefrog *Hyla chrysoscelis*. Ecology 71:1609-1616.

Fauth, J. E., and W. J. Resetarits, Jr. 1991. Interactions between the salamander *Siren intermedia* and the keystone predator *Notophthalmus viridescens*. Ecology 72:827-838.

Fauth, J. E., Buchanan, B. W., Wise, S. E., and M. J. Komoroski. 1996. *Cryptobranchus alleganiensis alleganiensis* (hellbender). Coloration. Herpetological Review 27:135.

Feder, J. H., Wurst, G. Z., and D. B. Wake. 1978. Genetic variation in western salamanders of the genus *Plethodon*, and the status of *Plethodon gordoni*. Herpetologica 34:64-69.

Feder, M. E., and S. J. Arnold. 1982. Anaerobic metabolism and behavior during predatory encounters between snakes (*Thamnophis elegans*) and salamanders (*Plethodon jordani*). Oecologia 53:93-97.

Feder, M. E., and P. L. Londos. 1984. Hydric constraints upon foraging in a terrestrial salamander, *Desmognathus ochrophaeus* (Amphibia: Plethodontidae). Oecologia 64:413-418.

Feminella, J. W., and C. P. Hawkins. 1994. Tailed frog tadpoles differentially alter their feeding behavior in response to non-visual cues from four predators. Journal of the North American Benthological Society 13:310-320.

Fenster, T. L. D., and C. B. Fenster. 1996. *Plethodon cinereus* (redback salamander). Predation. Herpetological Review 27:194.

Ferguson, D. E. 1954. An annotated list of the amphibians and reptiles of Union County, Oregon. Herpetologica 10:149-152.

Ferguson, D. E. 1961. The geographic variation of *Ambystoma macrodactylum* Baird, with the description of two new subspecies. The American Midland Naturalist 65:311-338.

Ferguson, D. E., and J. R. Rhodes. 1958. A new locality for the zigzag salamander in Mississippi. Herpetologica 14:129.

Fernandez, P. J., and J. P. Collins. 1988. Effect of environment and ontogeny on color pattern variation in Arizona tiger salamanders (*Ambystoma tigrinum nebulosum* Hallowell). Copeia 1988:928-938.

Figiel, C. R., and R. D. Semlitsch. 1990. Population variation in survival and metamorphosis of larval salamanders (*Ambystoma maculatum*) in the presence and absence of fish predation. Copeia 1990:818-826.

Figiel, C. R., Jr., and R. D. Semlitsch. 1995. Experimental determination of oviposition site selection in the marbled salamander, *Ambystoma opacum*. Journal of Herpetology 29:452-454.

Fisher, R. N., and H. B. Shaffer. 1996. The decline of amphibians in California's Great Central Valley. Conservation Biology 10:1387-1397.

Fitch, F. W. 1947. A record *Cryptobranchus alleganiensis*. Copeia 1947:210.

Fitch, H. S. 1936. Amphibians and reptiles of the Rogue River Basin, Oregon. The American Midland Naturalist 17:634–652.

Fitch, H. S. 1940. A biogeographical study of the *ordinoides* artenkreis of garter snakes (genus *Thamnophis*). University of California Publications in Zoology 44:1–150.

Fitch, K. L. 1959. Observations on the nesting habits of the mudpuppy, *Necturus maculosus* Rafinesque. Copeia 1959:339–340.

Fitzpatrick, L. C. 1973. Energy allocation in the Allegheny Mountain salamander, *Desmognathus ochrophaeus*. Ecological Monographs 43:43–58.

Flageole, S., and R. Leclair, Jr. 1992. Etude démographique d'une population de salamandres (*Ambystoma maculatum*) a l'aide de la methode squeletto-chronologique. Canadian Journal of Zoology 70:740–749.

Foard, T., and D. L. Auth. 1990. Food habits and gut parasites of the salamander, *Stereochilus marginatus*. Journal of Herpetology 24:428–431.

Folkerts, G. W. 1971. Notes on South Carolina salamanders. Journal of the Elisha Mitchell Scientific Society 87:206–208.

Fontenot, C. L., Jr., and L. W. Fontenot. 1989. *Amphiuma tridactylum* (three-toed amphiuma). Feeding. Herpetological Review 20:48.

Forester, D. C. 1977. Comments on the female reproductive cycle and philopatry by *Desmognathus ochrophaeus* (Amphibia, Urodela, Plethodontidae). Journal of Herpetology 11:311–316.

Forester, D. C. 1978. Laboratory encounters between attending *Desmognathus ochrophaeus* (Amphibia, Urodela, Plethodontidae) females and potential predators. Journal of Herpetology 12:537–541.

Forester, D. C. 1979a. The adaptiveness of parental care in *Desmognathus ochrophaeus* (Urodela: Plethodontidae). Copeia 1979:332–341.

Forester, D. C. 1979b. Homing to the nest by female mountain dusky salamanders (*Desmognathus ochrophaeus*) with comments on the sensory modalities essential to clutch recognition. Herpetologica 35:330–335.

Forester, D. C. 1981. Parental care in the salamander *Desmognathus ochrophaeus*: female activity pattern and trophic behavior. Journal of Herpetology 15:29–34.

Forester, D. C. 1983. Duration of the brooding period in the mountain dusky salamander (*Desmognathus ochrophaeus*) and its influence on aggression towards conspecifics. Copeia 1983:1098–1101.

Forester, D. C. 1984. Brooding behavior by the mountain dusky salamander: can the female's presence reduce clutch desiccation? Herpetologica 40:105–109.

Forester, D. C. 1986. The recognition and use of chemical signals by a nesting salamander. In: Chemical Signals in Vertebrates IV, D. Duvall, D. Müller-Schwarze, and R. M. Silverstein. Plenum Press, New York, pp. 205–219.

Forester, D. C., and D. V. Lykens. 1991. Age structure in a population of red-spotted newts from the Allegheny Plateau of Maryland. Journal of Herpetology 25:373–376.

Forester, D. C., Harrison, K., and L. McCall. 1983. The effects of isolation, the duration of brooding, and non-egg olfaction cues on clutch recognition by the salamander, *Desmognathus ochrophaeus*. Journal of Herpetology 17:308–314.

Formanowicz, D. R., Jr., and E. D. Brodie, Jr. 1993. Size-mediated predation pressure in a salamander community. Herpetologica 49:265–270.

Fowler, J. A. 1940. A note on the eggs of *Plethodon glutinosus*. Copeia 1940:133.

Fowler, J. A. 1946. The eggs of *Pseudotriton montanus montanus*. Copeia 1946:105.

Fowler, J. A. 1947. Record for *Aneides aeneus* in Virginia. Copeia 1947:144.

Fowler, J. A. 1952. The eggs of *Plethodon dixi*. National Speleological Society Bulletin 14:61.

Fowler, J. A. 1962. Another Virginia record for the eggs of *Pseudotriton r. ruber*. Bulletin of the Virginia Herpetological Society 31:4.

Franz, R. 1964. The eggs of the long-tailed salamander from a Maryland cave. Herpetologica 20:216.

Franz, R. 1967. Notes on the long-tailed salamander, *Eurycea longicauda* (Green), in Maryland caves. Bulletin of the Maryland Herpetological Society 3:1–6.

Franz, R., and H. Harris. 1965. Mass transformation and movement of larval long-tailed salamanders, *Eurycea longicauda longicauda* (Holbrook). Journal of the Ohio Herpetological Society 5:32.

Fraser, D. F. 1976a. Empirical evaluation of the hypothesis of food competition in salamanders of the genus *Plethodon*. Ecology 57:459–471.

Fraser, D. F. 1976b. Coexistence of salamanders in the genus *Plethodon*: a variation of the Santa Rosalia theme. Ecology 57:238–251.

Fraser, D. F. 1980. On the environmental control of oocyte maturation in a plethodontid salamander. Oecologia 46:302–307.

Freda, J. 1983. Diet of larval *Ambystoma maculatum* in New Jersey. Journal of Herpetology 17:177–179.

Freda, J., and W. A. Dunson. 1985. Field and laboratory studies of ion balance and growth rates of ranid tadpoles chronically exposed to low pH. Copeia 1985:415–423.

Freeman, J. R. 1958. Burrowing in the salamanders *Pseudobranchus striatus* and *Siren lacertina*. Herpetologica 14:130.

Freeman, J. R. 1967. Feeding behavior of the narrow-striped siren *Pseudobranchus striatus axanthus*. Herpetologica 23:313–314.

French, T. W., and R. H. Mount. 1978. Current status of the Red Hills salamander, *Phaeognathus hubrichti*, and factors affecting its distribution. Journal of the Alabama Academy of Science 49:172–179.

Friet, S. C. 1995. *Plethodon cinereus* (eastern red-backed salamander). Nest behavior. Herpetological Review 26:198–199.

Frisbie, M. P., and R. L. Wyman. 1991. The effects of soil pH on sodium balance in the red-backed salamander, *Plethodon cinereus*, and three other terrestrial salamanders. Physiological Zoology 64:1050–1068.

Frisbie, M. P., and R. L. Wyman. 1992. The effect of soil chemistry on sodium balance in the red-backed salamander: a comparison of two forest types. Journal of Herpetology 26:434–442.

Frost, D. R., and D. M. Hillis. 1990. Species in concept and practice: herpetological applications. Herpetologica 46:87–104.

Frost, D. R., Kluge, A. G., and D. M. Hillis. 1992. Species in contemporary herpetology: comments on phylogenetic inference and taxonomy. Herpetological Review 23:46–54.

Funderburg, J. B. 1955. The amphibians of New Hanover County, North Carolina. Journal of the Elisha Mitchell Scientific Society 71:19–28.

Funderburg, J. B., and D. S. Lee. 1967. Distribution of the lesser siren, *Siren intermedia*, in central Florida. Herpetologica 23:65.

Gabor, C. R. 1995. Correlational test of Mathis' hypothesis that bigger salamanders have better territories. Copeia 1995:729–735.

Gabor, C. R., and R. G. Jaeger. 1995. Resource quality affects agonistic behaviour of territorial salamanders. Animal Behaviour 49:71–79.

Gage, S. H. 1891. The life history of the vermilion-spotted newt (*Diemyctylus viridescens* Raf.). The American Naturalist 25:1084–1103.

Gamradt, S. C., and L. B. Kats. 1996. The effect of introduced crayfish and mosquitofish on California newts (*Taricha torosa*). Conservation Biology 10:1155–1162.

Garber, D. P., and C. E. Garber. 1978. A variant form of *Taricha granulosa* (Amphibia, Urodela, Salamandridae) from northwestern California. Journal of Herpetology 12:59–64.

Garcia-Paris, M., and S. M. Deban. 1995. A novel antipredator mechanism in salamanders: rolling escape in *Hydromantes platycephalus*. Journal of Herpetology 29:149–151.

Garton, J. S. 1972. Courtship of the small-mouthed salamander, *Ambystoma texanum*, in southern Illinois. Herpetologica 28:41–45.

Gates, J. E., and E. L. Thompson. 1982. Small pool habitat selection by red-spotted newts in western Maryland. Journal of Herpetology 16:7–15.

Gates, J. E., Hocutt, C. H., Stauffer, J. R., Jr., and G. J. Taylor. 1985. The distribution and status of *Cryptobranchus alleganiensis* in Maryland. Herpetological Review 16:17–18.

Gatz, A. J. 1973. Algal entry into the eggs of *Ambystoma maculatum*. Journal of Herpetology 7:137–138.

Gehlbach, F. R. 1965. Herpetology of the Zuni Mountains region, northwestern New Mexico. Proceedings of the U.S. National Museum 116:243–332.

Gehlbach, F. R. 1967a. *Ambystoma tigrinum* (Green). Catalogue of American Amphibians and Reptiles, pp. 52.1–52.4.

Gehlbach, F. R. 1967b. Evolution of tiger salamanders (*Ambystoma tigrinum*) on the Grand Canyon rims, Arizona. Yearbook of the American Philosophical Society 1967:266–269.

Gehlbach, F. R. 1969. Determination of the relationships of tiger salamander larval populations to different stages of pond succession at the Grand Canyon, Arizona. Yearbook of the American Philosophical Society 1969:299-302.

Gehlbach, F. R., and S. E. Kennedy. 1978. Population ecology of a highly productive aquatic salamander (*Siren intermedia*). The Southwestern Naturalist 23:423-430.

Gehlbach, F. R., and B. Walker. 1970. Acoustic behavior of the aquatic salamander, *Siren intermedia*. BioScience 20:1107-1108.

Gehlbach, F. R., Gordon, R., and J. B. Jordan. 1973. Aestivation of the salamander, *Siren intermedia*. The American Midland Naturalist 89:455-463.

Gehlbach, F. R., Kimmel, J. R., and W. A. Weems. 1969. Aggregations and body water relations in tiger salamanders (*Ambystoma tigrinum*) from the Grand Canyon rims, Arizona. Physiological Zoology 42:173-182.

Gentry, G. 1955. An annotated checklist of the amphibians and reptiles of Tennessee. Journal of the Tennessee Academy of Science 30:168-176.

George, C. J., Boylen, C. W., and R. B. Sheldon. 1977. The presence of the red-spotted newt, *Notophthalmus viridescens* Rafinesque (Amphibia, Urodela, Salamandridae), in waters exceeding 12 meters in Lake George, New York. Journal of Herpetology 11:87-90.

Gergits, W. F., and R. G. Jaeger. 1990a. Site attachment by the red-backed salamander, *Plethodon cinereus*. Journal of Herpetology 24:91-93.

Gergits, W. F., and R. G. Jaeger. 1990b. Field observations of the behavior of the red-backed salamander (*Plethodon cinereus*): courtship and agonistic interactions. Journal of Herpetology 24:93-95.

Gibbons, J. W., and S. Nelson, Jr. 1968. Observations on the mudpuppy, *Necturus maculosus*, in a Michigan lake. The American Midland Naturalist 80:562-564.

Gibbons, J. W., and R. D. Semlitsch. 1991. Guide to the Reptiles and Amphibians of the Savannah River Site. University of Georgia Press, Athens, Georgia, 128 pp.

Giguere, L. 1979. An experimental test of Dodson's hypothesis that *Ambystoma* (a salamander) and *Chaoborus* (a phantom midge) have complementary feeding niches. Canadian Journal of Zoology 57:1091-1097.

Gilbert, F. F., and R. Allwine. 1991. Terrestrial amphibian communities in the Oregon Cascade Range. In: Wildlife and vegetation of unmanaged Douglas-fir forests, L. F. Ruggiero, K. B. Aubry, A. B. Carey, and M. H. Huff, technical coordinators. USDA Forest Service, Pacific Northwest Research Station, Olympia, Washington, General Technical Report PNW-GTR-285, pp. 318-324.

Gilbert, P. W. 1941. Eggs and nests of *Hemidactylium scutatum* in the Ithaca region. Copeia 1941:47.

Gilbert, P. W. 1942. Observations on the eggs of *Ambystoma maculatum* with especial reference to the green algae found within the egg envelopes. Ecology 23:215-227.

Gilbert, P. W. 1944. The algae-egg relationship in *Ambystoma maculatum*, a case of symbiosis. Ecology 25:366-369.

Gilhen, J. 1974. Distribution, natural history, and morphology of the blue-spotted salamanders, *Ambystoma laterale* and *A. tremblayi* in Nova Scotia. Nova Scotia Museum Curatorial Report 22:1-38.

Gilhen, J. 1984. Amphibians and Reptiles of Nova Scotia. Nova Scotia Museum, Halifax, Nova Scotia, 162 pp.

Gill, D. E. 1978a. The metapopulation ecology of the red-spotted newt, *Notophthalmus viridescens* (Rafinesque). Ecological Monographs 48:145-166.

Gill, D. E. 1978b. Effective population size and interdemic migration rates in a metapopulation of the red-spotted newt, *Notophthalmus viridescens* (Rafinesque). Evolution 32:839-849.

Gill, D. E. 1979. Density dependence and homing behavior in adult red-spotted newts *Notophthalmus viridescens* (Rafinesque). Ecology 60:800-813.

Gill, D. E. 1985. Interpreting breeding patterns from census data: a solution to the Husting dilemma. Ecology 66:344-354.

Gill, D. E., Berven, K. A., and B. A. Mock. 1983. The environmental component of evolutionary biology. In: Population Biology—Retrospect and Prospect, C. E. King and P. S. Dawson, eds. Columbia University Press, New York. 235 pp.

Glass, B. P. 1951. Age at maturity of neotenic *Ambystoma t. mavortium* Baird. The American Midland Naturalist 46:391-393.

Gloyd, H. K. 1928. The amphibians and reptiles of

Franklin County, Kansas. Transactions of the Kansas Academy of Science 31:115-141.

Gnaedinger, L. M., and C. A. Reed. 1948. Contribution to the natural history of the plethodont salamander *Ensatina eschscholtzii*. Copeia 1948:187-196.

Godley, J. S. 1983. Observations on the courtship, nests and young of *Siren intermedia* in southern Florida. The American Midland Naturalist 110:215-219.

Goin, C. J. 1942. Description of a new race of *Siren intermedia* LeConte. Annals of the Carnegie Museum 29:211-217.

Goin, C. J. 1947. Notes on the eggs and early larvae of three Florida salamanders. Chicago Academy of Science Natural History Miscellanea 10:1-4.

Goin, C. J. 1950. A study of the salamander *Ambystoma cingulatum*, with the description of a new subspecies. Annuals of the Carnegie Museum 31:299-321.

Goin, C. J. 1951. Notes on the eggs and early larvae of three more Florida salamanders. Annals of the Carnegie Museum 32:253-263.

Goin, C. J. 1957. Description of a new salamander of the genus *Siren* from the Rio Grande. Herpetologica 13:37-42.

Good, D. A. 1989. Hybridization and cryptic species in *Dicamptodon* (Caudata: Dicamptodontidae). Evolution 43:728-744.

Good, D. A., and D. B. Wake. 1992. Geographic variation and speciation in the torrent salamanders of the genus *Rhyacotriton* (Caudata: Rhyacotritonidae). University of California Publications in Zoology 126:1-91.

Good, D. A., Wurst, G. Z., and D. B. Wake. 1987. Patterns of geographic variation in allozymes of the Olympic salamander, *Rhyacotriton olympicus* (Caudata: Dicamptodontidae). Fieldiana Zoology 32:1-15.

Goodwin, O. K., and J. T. Wood. 1953. Note on egg-laying of the four-toed salamander, *Hemidactylium scutatum* (Schlegel), in eastern Virginia. Virginia Journal of Science 4:65-66.

Gordon, R. E. 1952. A contribution to the life history and ecology of the plethodontid salamander *Aneides aeneus* (Cope and Packard). The American Midland Naturalist 47:666-701.

Gordon, R. E. 1953. A population of Holbrook's salamander, *Eurycea longicauda guttolineata* (Holbrook). Tulane Studies in Zoology 1:55-60.

Gordon, R. E. 1960. Young of the salamander, *Plethodon jordani melaventris*. Copeia 1960:26-29.

Gordon, R. E. 1961. Movements of displaced green salamanders. Ecology 42:200-202.

Gordon, R. E. 1966. Some observations on the biology of *Pseudotriton ruber schencki*. Journal of the Ohio Herpetological Society 5:163-164.

Gordon, R. E. 1967. *Aneides aeneus*. Catalogue of American Amphibians and Reptiles, p. 30.

Gordon, R. E. 1968. Terrestrial activity of the spotted salamander, *Ambystoma maculatum*. Copeia 1968:879-880.

Gordon, R. E., and R. L. Smith. 1949. Notes on the life history of the salamander *Aneides aeneus*. Copeia 1949:173-175.

Gordon, R. E., MacMahon, J. A., and D. B. Wake. 1962. Relative abundance, microhabitat and behavior of some southern Appalachian salamanders. Zoologica 47:9-14.

Gore, J. A. 1983. The distribution of desmognathine larvae (Amphibia: Plethodontidae) in coal surface impacted streams of the Cumberland Plateau, USA. Journal of Freshwater Ecology 2:13-23.

Gorman, J. 1954. A new species of salamander from central California. Herpetologica 10:153-158.

Gorman, J. 1956. Reproduction in plethodont salamanders of the genus *Hydromantes*. Herpetologica 12:249-259.

Gorman, J. 1964. *Hydromantes brunus, H. platycephalus*, and *H. shastae*. Catalogue of American Amphibians and Reptiles, p. 11.

Gorman, J., and C. L. Camp. 1953. A new cave species of salamander of the genus *Hydromantes* from California, with notes on habits and habitats. Copeia 1953:39-43.

Gosner, K. K., and I. H. Black. 1957. The effects of acidity on the development of New Jersey frogs. Ecology 38:256-262.

Graf, W. 1949. Observations on the salamander *Dicamptodon*. Copeia 1949:79-80.

Graf, W., Jewett, S. G., Jr., and K. L. Gordon. 1939. Records of amphibians and reptiles from Oregon. Copeia 1939:101-104.

Grant, B. W., Brown, K. L., Ferguson, G. W., and J. W. Gibbons. 1994. Changes in amphibian biodiversity associated with 25 years of pine forest regeneration: implications for biodiversity management. In: Biological Diversity: Problems and Challenges, S. K. Majumdar, F. J. Brenner, J. E. Lovich, J. F. Schalles, and E. W. Miller, eds. Pennsylvania Academy of Science, Philadelphia, Pennsylvania, pp. 355–367.

Grant, C. 1958. Irruption of young *Batrachoseps attenuatus*. Copeia 1958:222.

Grant, D., Anderson, O., and V. Twitty. 1968. Homing orientation by olfaction in newts, *Taricha rivularis*. Science 160:1354–1356.

Grant, K. P., and L. E. Licht. 1995. Effects of ultraviolet radiation on life-history stages of anurans from Ontario, Canada. Canadian Journal of Zoology 73:2292–2301.

Grant, W. C., Jr. 1955. Territorialism in two species of salamanders. Science 121:137–138.

Graves, B. M. 1994. The role of nasolabial grooves and the vomeronasal system in recognition of home area by red-backed salamanders. Animal Behaviour 47:1216–1219.

Gray, I. E. 1941. Amphibians and reptiles of the Duke Forest and vicinity. The American Midland Naturalist 26:652–658.

Green, H. T. 1925. The egg-laying of the purple salamander. Copeia 1925:32.

Green, N. B. 1933. *Cryptobranchus alleganiensis* in West Virginia. Proceedings of the West Virginia Academy of Science 7:28–30.

Green, N. B. 1935. Further notes on the food habits of the water dog, *Cryptobranchus alleganiensis* Daudin. Proceedings of the West Virginia Academy of Science 9:36.

Green, N. B. 1939. The pygmy salamander *Desmognathus wrighti* King, on White Top Mountain, Virginia. Copeia 1939:49.

Green, N. B. 1941. The four-toed salamander in Kentucky. Copeia 1941:53.

Green, N. B. 1956. The ambystomatid salamanders of West Virginia. Proceedings of the West Virginia Academy of Science 27:16–18.

Green, N. B., and P. Brant, Jr. 1966. Salamanders found in West Virginia caves. Proceedings of the West Virginia Academy of Science 38:42–45.

Green, N. B., and T. K. Pauley. 1987. Amphibians and Reptiles in West Virginia. University of Pittsburgh Press, Pittsburgh, Pennsylvania, 241 pp.

Green, N. B., and N. D. Richmond. 1944. Courtship of *Plethodon metcalfi*. Copeia 1944:256.

Green, N. B., Brant, P., Jr., and B. Dowler. 1967. *Eurycea lucifuga* in West Virginia: its distribution, ecology, and life history. Proceedings of the West Virginia Academy of Science 39:297–304.

Greer, A. E., Jr. 1973. Adaptive significance of the color phases of the red-backed salamander. Yearbook of the American Philosophical Society 1973:308–309.

Gregory, P. T. 1978. Feeding habits and dietary overlap of three species of garter snake (*Thamnophis*) on Vancouver Island. Canadian Journal of Zoology 56:1967–1974.

Grinnell, J., and T. I. Storer. 1924. Animal Life in the Yosemite. University of California Press, Berkeley, California, 752 pp.

Grizzell, R. A., Jr. 1949. The hibernation site of three snakes and a salamander. Copeia 1949:231–232.

Grobman, A. B. 1943. Notes on salamanders with the description of a new species of *Cryptobranchus*. Occasional Papers of the Museum of Zoology, University of Michigan 470:1–5.

Grobman, A. B. 1944. The distribution of the salamanders of the genus *Plethodon* in the eastern United States and Canada. Annals of the New York Academy of Science 45:261–316.

Groves, C. R., Cassirer, E. F., Genter, D. L., and J. D. Reichel. 1996. Coeur d' Alene salamander (*Plethodon idahoensis*). Elemental Stewardship Abstract. Natural Areas Journal 6:238–247.

Gruberg, E. R., and R. V. Stirling. 1972. Observations on the burrowing habits of the tiger salamander (*Ambystoma tigrinum*). Herpetological Review 4:85–89.

Gunter, G. 1968. Further notes on weight changes of starving *Amphiuma means*. Herpetologica 24:180–181.

Gunter, G., and W. E. Brode. 1964. *Necturus* in the state of Mississippi, with notes on adjacent areas. Herpetologica 20:114–126.

Gustafson, M. P. 1994. Size-specific interactions among larvae of the plethodontid salamanders *Gyrinophilus porphyriticus* and *Eurycea cirrigera*. Journal of Herpetology 28:470–476.

Guttman, S. I. 1989a. *Eurycea bislineata*. In: Salamanders of Ohio, R. A. Pfingsten and F. L. Downs, eds. Ohio Biological Survey Bulletin, New Series 7(2):195–204.

Guttman, S. I. 1989b. *Eurycea longicauda*. In: Salamanders of Ohio, R. A. Pfingsten and F. L. Downs, eds. Ohio Biological Survey Bulletin, New Series 7(2):204-209.

Guttman, S. I., and A. A. Karlin. 1986. Hybridization of cryptic species of two-lined salamanders (*Eurycea bislineata* complex). Copeia 1986:96-108.

Guttman, S. I., Karlin, A. A., and G. M. Labanick. 1978. A biochemical and morphological analysis of the relationship between *Plethodon longicrus* and *Plethodon yonahlossee* (Amphibia, Urodela, Plethodontidae). Journal of Herpetology 12:445-454.

Guttman, S. I., Weigt, L. A., Moler, P. E., Ashton, R. E., Jr., Mansell, B. W., and J. Peavey. 1990. An electrophoretic analysis of *Necturus* from the southeastern United States. Journal of Herpetology 24:163-175.

Hacker, V. R. 1956. Biology and management of lake trout in Green Lake, Wisconsin. Transactions of the American Fisheries Society 86:1-13.

Hairston, N. G. 1949. The local distribution and ecology of the plethodontid salamanders of the southern Appalachians. Ecological Monographs 19:47-73.

Hairston, N. G. 1950. Intergradation in Appalachian salamanders of the genus *Plethodon*. Copeia 1950:262-273.

Hairston, N. G. 1951. Interspecific competition and its probable influence upon the vertical distribution of Appalachian salamanders of the genus *Plethodon*. Ecology 32:266-274.

Hairston, N. G. 1980a. The experimental test of an analysis of field distributions: competition in terrestrial salamanders. Ecology 61:817-826.

Hairston, N. G. 1980b. Evolution under interspecific competition: field experiments of terrestrial salamanders. Evolution 34:409-420.

Hairston, N. G. 1980c. Species packing in the salamander genus *Desmognathus*: what are the interspecific interactions involved? The American Naturalist 115:354-366.

Hairston, N. G. 1981. An experimental test of a guild: salamander competition. Ecology 62:65-72.

Hairston, N. G. 1983a. Growth, survival, and reproduction of *Plethodon jordani*: trade-offs between selective pressures. Copeia 1983:1024-1035.

Hairston, N. G. 1983b. Alpha selection in competing salamanders: experimental verification of an a priori hypothesis. The American Naturalist 122:105-13.

Hairston, N. G. 1986. Species packing in *Desmognathus* salamanders: experimental demonstration of predation and competition. The American Naturalist 127:266-291.

Hairston, N. G. 1987. Community Ecology and Salamander Guilds. Cambridge University Press, Cambridge, England, 230 pp.

Hairston, N. G. 1993. On the validity of the name *teyahalee* as applied to a member of the *Plethodon glutinosus* complex (Caudata: Plethodontidae): a new name. Brimleyana 18:65-69.

Hairston, N. G., and C. H. Pope. 1948. Geographic variation and speciation in Appalachian salamanders (*Plethodon jordani* group). Evolution 2:266-278.

Hairston, N. G., and R. H. Wiley. 1993. No decline in salamander (Amphibia: Caudata) populations: a twenty year study in the southern Appalachians. Brimleyana 18:59-64.

Hairston, N. G., Nishikawa, K. C., and S. L. Stenhouse. 1987. The evolution of competing species of terrestrial salamanders: niche partitioning or interference? Evolutionary Ecology 1:247-262.

Hairston, N. G., Wiley, R. H., and C. K. Smith. 1992. The dynamics of two hybrid zones in Appalachian salamanders of the genus *Plethodon*. Evolution 46:930-938.

Hall, R. J. 1976. Summer foods of the salamander, *Plethodon wehrlei* (Amphibia, Urodela, Plethodontidae). Journal of Herpetology 10:129-131.

Hall, R. J. 1977. A population analysis of two species of streamside salamanders, Genus *Desmognathus*. Herpetologica 33:109-1134.

Hall, R. J., and D. P. Stafford. 1972. Studies in the life history of Wehrle's salamander, *Plethodon wehrlei*. Herpetologica 28:300-309.

Hamilton, R. 1948. The egg-laying process in the tiger salamander. Copeia 1948:212-213.

Hamilton, W. J., Jr. 1932. The food and feeding habits of some eastern salamanders. Copeia 1932:83-86.

Hamilton, W. J., Jr. 1940. The feeding habits of larval newts with reference to availability and predilection of food items. Ecology 21:351-356.

Hamilton, W. J., Jr. 1943. Winter habits of the

dusky salamander, in central New York. Copeia 1943:192.

Hamilton, W. J., Jr. 1946. Summer habitat of the yellow-barred tiger salamander. Copeia 1946:51.

Hamilton, W. J., Jr. 1950. Notes on the food of the congo eel, *Amphiuma*. Natural History Miscellanea 62:1-3.

Hammen, C. S., and V. H. Hutchison. 1962. Carbon dioxide assimilation in the symbiosis of the salamander *Ambystoma maculatum* and the alga *Oophila amblystomatis*. Life Sciences 10:527-532.

Hanlin, H. G. 1978. Food habits of the greater siren, *Siren lacertina*, in an Alabama Coastal Plain pond. Copeia 1978:358-360.

Hanlin, H. G., and R. H. Mount. 1978. Reproduction and activity of the greater siren, *Siren lacertina* (Amphibia: Sirenidae), in Alabama. Journal of the Alabama Academy of Science 49:31-39.

Hanlin, H. G., Beatty, J. J., and S. W. Hanlin. 1979. A nest site of the western red-backed salamander *Plethodon vehiculum* (Cooper). Journal of Herpetology 13:212-214.

Hansen, R. W. 1990. *Hydromantes platycephalus* (Mount Lyell salamander). Toxicity. Herpetological Review 21:91.

Hanson, K., Snyder, J., and L. Kats. 1994. *Taricha torosa* (California newt). Diet. Herpetological Review 25:62.

Hardin, J. W., Schafer, J. P., and R. W. Barbour. 1969. Observations on the activity of a seal salamander, *Desmognathus monticola*. Herpetologica 25:150-151.

Hardy, J. D., and J. Olmon. 1974. Restriction of the range of the frosted salamander, *Ambystoma cingulatum*, based on a comparison of the larvae of *Ambystoma cingulatum* and *Ambystoma mabeei*. Herpetologica 30:156-160.

Hardy, J. D., Jr. 1952. A concentration of juvenile spotted salamanders, *Ambystoma maculatum* (Shaw). Copeia 1952:181-182.

Hardy, J. D., Jr. 1969a. Reproductive activity, growth, and movements of *Ambystoma mabeei* Bishop in North Carolina. Bulletin of the Maryland Herpetological Society 5:65-76.

Hardy, J. D., Jr. 1969b. A summary of recent studies of the salamander, *Ambystoma mabeei*. Chesapeake Biological Laboratory, Solomons, Maryland, Reference 69-20:3.

Hardy, J. D., Jr. and J. D. Anderson. 1970. *Ambystoma mabeei*. Catalogue of American Amphibians and Reptiles, pp. 81.1-81.2.

Hardy, L. M., and M. C. Lucas. 1991. A crystalline protein is responsible for dimorphic egg jellies in the spotted salamander, *Ambystoma maculatum* (Shaw) (Caudata: Ambystomatidae). Comparative Biochemistry and Physiology 100A:653-660.

Hardy, L. M., and L. R. Raymond. 1980. The breeding migration of the mole salamander, *Ambystoma talpoideum*, in Louisiana. Journal of Herpetology 14:327-335.

Hardy, M. P., and J. N. Dent. 1988. Behavioral observations on the transfer of sperm from the male to the female red-spotted newt (*Notophthalmus viridescens*, Salamandridae). Copeia 1988:789-792.

Harestad, A. S., and J. J. Stelmock. 1983. Size of clouded salamanders and their prey. Syesis 16:39-42.

Harper, F. 1935. Records of amphibians in the southeastern states. The American Midland Naturalist 16:275-310.

Harris, J. P., Jr. 1959. The natural history of *Necturus*, III. Food and feeding. Field and Laboratory 27:105-111.

Harris, J. P., Jr. 1961. The natural history of *Necturus*, IV. Reproduction. Journal of the Graduate Research Center (Field and Laboratory) 29:69-81.

Harris, P. M. 1995. Are autecologically similar species also functionally similar? A test in pond communities. Ecology 76:544-552.

Harris, R. N. 1980. The consequences of within-year timing of breeding in *Ambystoma maculatum*. Copeia 1980:719-722.

Harris, R. N. 1981. Intrapond homing behavior in *Notophthalmus viridescens*. Journal of Herpetology 15:355-356.

Harris, R. N. 1984. Transplant experiments with *Ambystoma* larvae. Copeia 1984:161-169.

Harris, R. N. 1987. Density-dependent paedomorphosis in the salamander *Notophthalmus viridescens dorsalis*. Ecology 68:705-712.

Harris, R. N., and D. E. Gill. 1980. Joint nesting, brooding behavior, and embryonic survival of the four-toed salamander *Hemidactylium scutatum*. Herpetologica 36:141-144.

Harris, R. N., Alford, R. A., and H. M. Wilbur.

1988. Density and phenology of *Notophthalmus viridescens dorsalis* in a natural pond. Herpetologica 44:234-242.

Harris, R. N., Hames, W. W., Knight, I. T., Carreno, C. A., and T. J. Vess. 1995. An experimental analysis of joint nesting in the salamander *Hemidactylium scutatum* (Caudata: Plethodontidae): the effects of population density. Animal Behaviour 50:1309-1316.

Harris, R. N., Semlitsch, R. D., Wilbur, H. M., and J. E. Fauth. 1990. Local variation in the genetic basis of paedomorphosis in the salamander *Ambystoma talpoideum*. Evolution 44:1588-1603.

Harrison, J. R. 1967. Observations on the life history, ecology and distribution of *Desmognathus aeneus aeneus* Brown and Bishop. The American Midland Naturalist 77:356-370.

Harrison, J. R. 1973. Observations on the life history and ecology of *Eurycea quadridigitata* (Holbrook). HISS NEWS Journal 1:57-58.

Harrison, J. R. 1992. *Desmognathus aeneus*. Catalogue of American Amphibians and Reptiles, pp. 534.1-534.4.

Harte, J., and E. Hoffman. 1989. Possible effects of acid deposition on a Rocky Mountain population of the tiger salamander *Ambystoma tigrinum*. Conservation Biology 3:149-158.

Hassinger, D. D., Anderson, J. D., and G. H. Dalrymple. 1970. The early life history and ecology of *Ambystoma tigrinum* and *Ambystoma opacum* in New Jersey. The American Midland Naturalist 84:474-495.

Hassler, W. G. 1932. New locality records for two salamanders and a snake in Cattaraugus County, New York. Copeia 1932:94-96.

Hawkins, C. P., Murphy, M. L., Anderson, N. H., and M. A. Wilzbach. 1983. Density of fish and salamanders in relation to riparian canopy and physical habitat in streams of the northwestern United States. Canadian Journal of Fisheries and Aquatic Sciences 40:1173-1185.

Hay, O. P. 1888. Observations on *Amphiuma* and its young. The American Naturalist 22:315-321.

Hay, O. P. 1892. The batrachians and reptiles of the state of Indiana. Annual Report of the Indiana Department of Geology and Natural Resources 17:412-602.

Hayes, M. P., and M. R. Jennings. 1986. Decline of ranid frog species in western North America: are bullfrogs (*Rana catesbeiana*) responsible? Journal of Herpetology 20:490-509.

Healy, W. R. 1970. Reduction in neoteny in Massachusetts populations of *Notophthalmus viridescens*. Copeia 1970:578-581.

Healy, W. R. 1973. Life history variation and the growth of juvenile *Notophthalmus viridescens* from Massachusetts. Copeia 1973:641-647.

Healy, W. R. 1974a. Population consequences of alternative life histories in *Notophthalmus v. viridescens*. Copeia 1974:221-229.

Healy, W. R. 1974b. Sex ratio variation in samples of adult *Notophthalmus viridescens*. The American Midland Naturalist 92:492-495.

Healy, W. R. 1975. Terrestrial activity and home range in efts of *Notophthalmus viridescens*. The American Midland Naturalist 93:131-138.

Heath, A. G. 1975. Behavioral thermoregulation in high altitude tiger salamanders, *Ambystoma tigrinum*. Herpetologica 31:84-93.

Heath, D. R., Saugey, D. A., and G. A. Heidt. 1986. Abandoned mine fauna of the Ouachita Mountains, Arkansas: vertebrate taxa. Proceedings of the Arkansas Academy of Sciences 40:33-36.

Heatwole, H. 1960. Burrowing ability and behavioral responses to desiccation of the salamander, *Plethodon cinereus*. Ecology 41:661-668.

Heatwole, H. 1962. Environmental factors influencing local distribution and activity of the salamander, *Plethodon cinereus*. Ecology 43:460-472.

Heatwole, H., and F. H. Test. 1961. Cannibalism in the salamander, *Plethodon cinereus*. Herpetologica 17:143.

Hecht, M. K. 1958. A synopsis of the mud puppies of eastern North America. Proceedings of the Staten Island Institute of Arts and Sciences 21:5-38.

Hedgecock, D. 1976. Genetic variation in two widespread species of salamander, *Taricha torosa* and *Taricha granulosa*. Biochemical Genetics 14:561-576.

Hedgecock, D., and F. J. Ayala. 1974. Evolutionary divergence in the genus *Taricha* (Salamandridae). Copeia 1974:738-747.

Hedges, S. B., Bogart, J. P., and L. R. Maxson. 1992. Ancestry of unisexual salamanders. Nature 356:708-710.

Heintzel, S. J., and C. R. Rossell, Jr. 1995. A new record of the mole salamander, *Ambystoma talpoideum*, in Buncombe County, North Carolina. Journal of the Elisha Mitchell Scientific Society 111:130–131.

Henderson, B. A. 1973. The specialized feeding behavior of *Ambystoma gracile* in Marion Lake, British Columbia. The Canadian Field-Naturalist 87:151–154.

Hendricks, L. J., and J. Kezer. 1958. An unusual population of a blind cave salamander and its fluctuation during one year. Herpetologica 14:41–43.

Hendrickson, J. R. 1954. Ecology and systematics of salamanders of the genus *Batrachoseps*. University of California Publications in Zoology 54:1–46.

Henry, W. V., and V. C. Twitty. 1940. Contributions to the life histories of *Dicamptodon ensatus* and *Ambystoma gracile*. Copeia 1940:247–250.

Hensel, J. L., Jr., and E. D. Brodie, Jr. 1976. An experimental study of aposematic coloration in the salamander *Plethodon jordani*. Copeia 1976:59–65.

Hensley, M. 1959. Albinism in North American amphibians and reptiles. Publications of the Museum of Michigan State University, Series 1, pp. 135–139.

Hensley, M. 1964. The tiger salamander in northern Michigan. Herpetologica 20:203–204.

Herring, K., and P. Verrell. 1996. Sexual incompatibility and geographical variation in mate recognition systems: tests in the salamander *Desmognathus ochrophaeus*. Animal Behaviour 52:279–287.

Herrington, R. E. 1988. Talus use by amphibians and reptiles in the Pacific Northwest. In: Management of amphibians, reptiles, and small mammals in North America, R. C. Szaro, K. E. Severson, and D. R. Patton, technical coordinators. USDA Forest Service, Ft. Collins, Colorado, General Technical Report RM-166, pp. 216–221.

Herrington, R. E., and J. H. Larsen, Jr. 1985. Current status, habitat requirements and management of the Larch Mountain salamander *Plethodon larselli* Burns. Biological Conservation 34:169–179.

Herrington, R. E., and J. H. Larsen, Jr. 1987. Reproductive biology of the Larch Mountain salamander (*Plethodon larselli*). Journal of Herpetology 21:48–56.

Hershey, J. L., and D. C. Forester. 1980. Sensory orientation in *Notophthalmus v. viridescens* (Amphibia: Salamandridae). Canadian Journal of Zoology 58:266–276.

Hewitt, G. M. 1989. The subdivision of species by hybrid zones. In: Speciation and Its Consequences, D. Otte and J. A. Endler, eds. Sinauer Associates, Sunderland, Massachusetts, pp. 85–110.

Highton, R. 1956. The life history of the slimy salamander, *Plethodon glutinosus*, in Florida. Copeia 1956:75–93.

Highton, R. 1959. The inheritance of the color phases of *Plethodon cinereus*. Copeia 1959:33–37.

Highton, R. 1961. A new genus of lungless salamander from the Coastal Plain of Alabama. Copeia 1961:65–68.

Highton, R. 1962a. Revision of North American salamanders of the genus *Plethodon*. Bulletin of the Florida State Museum 6:235–367.

Highton, R. 1962b. Geographic variation in the life history of the slimy salamander. Copeia 1962:597–613.

Highton, R. 1970. Evolutionary interactions between species of North American salamanders of the genus *Plethodon*. Part 1. Genetic and ecological relationships of *Plethodon jordani* and *P. glutinosus* in the southern Appalachian Mountains. Evolutionary Biology 4:211–241.

Highton, R. 1972. Distributional interactions among eastern North American salamanders of the genus *Plethodon*. In: The distributional history of the biota of the southern Appalachians, P. C. Holt, ed. Research Division Monograph 4. Virginia Polytechnic Institute and State University, Blacksburg, Virginia, pp. 139–188.

Highton, R. 1975. Geographic variation in genetic dominance of the color morphs of the red-backed salamander, *Plethodon cinereus*. Genetics 80:363–374.

Highton, R. 1977. Comparison of microgeographic variation in morphological and electrophoretic traits. Evolutionary Biology 10:397–436.

Highton, R. 1979. A new cryptic species of salamander of the genus *Plethodon* from the southeastern United States (Amphibia: Plethodontidae). Brimleyana 1:31–36.

Highton, R. 1983. A new species of woodland salamander of the *Plethodon glutinosus* group from the southern Appalachian Mountains. Brimleyana 9:1–20.

Highton, R. 1985. The width of the contact zone

between *Plethodon dorsalis* and *P. websteri* in Jefferson County, Alabama. Journal of Herpetology 19:544–546.

Highton, R. 1986a. *Plethodon aureolus*. Catalogue of American Amphibians and Reptiles, p. 381.

Highton, R. 1986b. *Plethodon hoffmani*. Catalogue of American Amphibians and Reptiles, p. 392.1.

Highton, R. 1986c. *Plethodon fourchensis*. Catalogue of American Amphibians and Reptiles, p. 391.

Highton, R. 1986d. *Plethodon kentucki*. Catalogue of American Amphibians and Reptiles, p. 382.

Highton, R. 1986e. *Plethodon websteri*. Catalogue of American Amphibians and Reptiles, pp. 384.1–384.2.

Highton, R. 1986f. *Plethodon serratus*. Catalogue of American Amphibians and Reptiles, pp. 394.1–394.2.

Highton, R. 1987. *Plethodon wehrlei*. Catalogue of American Amphibians and Reptiles, pp. 402.1–402.3.

Highton. R. 1990. Taxonomic treatment of genetically differentiated populations. Herpetologica 46:114–121.

Highton, R. 1991. Molecular phylogeny of plethodonine salamanders and hylid frogs: statistical analysis of protein comparisons. Molecular Biology and Evolution 8:796–818.

Highton, R. 1995. Speciation in eastern North American salamanders of the genus *Plethodon*. Annual Review of Ecology and Systematics 26:579–600.

Highton, R. 1997. Geographic protein variation and speciation in the *Plethodon dorsalis* complex. Herpetologica 53:345–356.

Highton, R., and S. A. Henry. 1970. Evolutionary interactions between species of North American salamanders of the genus *Plethodon*. Evolutionary Biology 4:211–256.

Highton, R., and D. A. Jones. 1965. A striped color phase of *P. richmondi* in Virginia. Copeia 1965:371–372.

Highton, R., and A. Larson. 1979. The genetic relationships of salamanders of the genus *Plethodon*. Systematic Zoology 28:579–599.

Highton, R., and J. R. MacGregor. 1983. *Plethodon kentucki* Mittleman: a valid species of Cumberland Plateau woodland salamander. Herpetologica 39:189–200.

Highton, R., and T. Savage. 1961. Functions of the brooding behavior in the female red-backed salamander, *Plethodon cinereus*. Copeia 1961:95–98.

Highton, R., and T. P. Webster. 1976. Geographic protein variation and divergence in populations of the salamander *Plethodon cinereus*. Evolution 30:33–45.

Highton, R., and R. D. Worthington. 1967. A new salamander of the genus *Plethodon* from Virginia. Copeia 1967:617–626.

Highton, R., Maha, G. C., and L. R. Maxson. 1989. Biochemical evolution in the slimy salamanders of the *Plethodon glutinosus* complex in the eastern United States. University of Illinois Biological Monographs 57:1–153.

Hileman, K. S., and E. D. Brodie, Jr. 1994. Survival strategies of the salamander *Desmognathus ochrophaeus*: interaction of predator-avoidance and anti-predator mechanisms. Animal Behaviour 47:1–6.

Hill, J., Formanowicz, D. R., Jr., and R. G. Jaeger. 1982. Patch foraging by a terrestrial salamander. Journal of Herpetology 16:405–408.

Hillis, D. M. 1977. Sex ratio, mortality rate, and breeding stimulus in a Maryland population of *Ambystoma maculatum*. Bulletin of the Maryland Herpetological Society 13:84–91.

Hillis, R. E., and E. D. Bellis. 1971. Some aspects of the ecology of the hellbender, *Cryptobranchus alleganiensis alleganiensis,* in a Pennsylvania stream. Journal of Herpetology 5:121–126.

Hinderstein, B. 1971. Studies on the salamander genus *Desmognathus*: variation of lactate dehydrogenase. Copeia 1971:636–644.

Hirschfeld, C. J., and J. T. Collins. 1963. Range extensions for three amphibians in north-central Kentucky. Copeia 1963:438–439.

Hoff, J. G. 1977. A Massachusetts hibernation site of the red-backed salamander, *Plethodon cinereus*. Herpetological Review 8:33.

Hoffman, R. L. 1951. A new species of salamander from Virginia. Journal of the Elisha Mitchell Scientific Society 67:249–254.

Hoffman, R. L. 1953. *Plethodon welleri* Walker in Tennessee. Journal of the Tennessee Academy of Science 28:86–87.

Hokit, D. G., Walls, S. C., and A. R. Blaustein. 1996. Context-dependent kin discrimination in larvae of the marbled salamander, *Ambystoma opacum*. Animal Behaviour 52:17–31.

Holland, D. C., Hayes, M. P., and E. McMillan. 1990. Late summer movement and mass mortality in the California tiger salamander (*Ambystoma californiense*). The Southwestern Naturalist 35:217-220.

Holman, J. A. 1955. Fall and winter food of *Plethodon dorsalis* in Johnson County, Indiana. Copeia 1955:143.

Holman, J. A. 1968. Lower Oligocene amphibians from Saskatchewan. Quarterly Journal of the Florida Academy of Science 31:273-289.

Holman, J. A. 1975. Herpetofauna of the WaKeeney local fauna (lower Pliocene: Claredonian) of Trego County, Kansas. Contributions of the Museum of Paleontology, University of Michigan 12:49-66.

Holman, J. A. 1977. The Pleistocene (Kansan) herpetofauna of Cumberland Cave, Maryland. Annals of Carnegie Museum 46:157-172.

Holman, J. A. 1996. The large Pleistocene (Sangamonian) herpetofauna of the Williston IIIA site, north-central Florida. Herpetological Natural History 4:35-47.

Holman, J. A., and M. R. Voorhies. 1985. *Siren* (Caudata: Sirenidae) from the Barstovian Miocene of Nebraska. Copeia 1985:264-266.

Holomuzki, J. R. 1980. Synchronous foraging and dietary overlap of three species of plethodontid salamanders. Herpetologica 36:109-115.

Holomuzki, J. R. 1982. Homing behavior of *Desmognathus ochrophaeus* along a stream. Journal of Herpetology 16:307-309.

Holomuzki, J. R. 1985a. Diet of larval *Dytiscus dauricus* (Coleoptera: Dytiscidae) in east-central Arizona. Pan-Pacific Entomology 61:229.

Holomuzki, J. R. 1985b. Life history aspects of the predaceous diving beetle, *Dytiscus dauricus* (Gebler), in Arizona. The Southwestern Naturalist 30:485-490.

Holomuzki, J. R. 1986a. Predator avoidance and diel patterns of microhabitat use by larval tiger salamanders. Ecology 1986:737-748.

Holomuzki, J. R. 1986b. Intraspecific predation and habitat use by tiger salamanders (*Ambystoma tigrinum nebulosum*). Journal of Herpetology 20:439-441.

Holomuzki, J. R. 1986c. Effect of microhabitat on fitness components of larval tiger salamanders, *Ambystoma tigrinum nebulosum*. Oecologia 71: 142-148.

Holomuzki, J. R. 1989a. Predation risk and macroalga use by the stream-dwelling salamander *Ambystoma texanum*. Copeia 1989:22-28.

Holomuzki, J. R. 1989b. Salamander predation and vertical distributions of zooplankton. Freshwater Biology 21:461-472.

Holomuzki, J. R. 1991. Macrohabitat effects on egg deposition and larval growth, survival, and instream dispersal in *Ambystoma barbouri*. Copeia 1991:687-694.

Holomuzki, J. R., and J. P. Collins. 1983. Diel movement of larvae of the tiger salamander, *Ambystoma tigrinum nebulosum*. Journal of Herpetology 17:276-278.

Holomuzki, J. R., and J. P. Collins. 1987. Trophic dynamics of a top predator, *Ambystoma tigrinum nebulosum* (Caudata: Ambystomatidae), in a lentic community. Copeia 1987:949-957.

Holomuzki, J. R., Collins, J. P., and P. E. Brunkow. 1994. Trophic control of fishless ponds by tiger salamander larvae. Oikos 71:55-64.

Holsinger, J. R. 1982. A preliminary report on the cave fauna of Burnsville Cove, Virginia. Bulletin of the National Speleological Society 44:98-101.

Hom, C. L. 1987. Reproductive ecology of female dusky salamanders, *Desmognathus fuscus* (Plethodontidae), in the southern Appalachians. Copeia 1987:768-777.

Hom, C. L. 1988. Cover object choice by female dusky salamanders, *Desmognathus fuscus.* Journal of Herpetology 22:247-249.

Hood, H. H. 1934. A note on the red-backed salamander at Rochester, New York. Copeia 1934:141-142.

Horne, E. A. 1988. Aggressive behavior of female red-backed salamanders. Herpetologica 44:203-209.

Horne, E. A., and R. G. Jaeger. 1988. Territorial pheromones of female red-backed salamanders. Ethology 78:143-152.

Houck, L. D. 1980. Courtship behavior in the plethodontid salamander, *Desmognathus wrighti* (abstract). The American Zoologist 20:825.

Houck, L. D. 1982. Mail tail loss and courtship success in the plethodontid salamander *Desmognathus ochrophaeus.* Journal of Herpetology 16:335-340.

Houck, L. D. 1986. The evolution of salamander courtship pheromones. In: Chemical Signals in Vertebrates IV, D. Duvall, D. Müller-Schwarze, and R. M. Silverstein, eds. Plenum Press, New York, pp. 173–190.

Houck, L. D. 1988. The effect of body size on male courtship success in a plethodontid salamander. Animal Behaviour 36:837–842.

Houck, L. D., and H. Francillon-Vieillot. 1988. Tests for age and size effects on male mating success in a plethodontid salamander. Amphibia-Reptilia 9:135–144.

Houck, L. D., and N. L. Reagan. 1990. Male courtship pheromones increase female receptivity in a plethodontid salamander. Animal Behaviour 39:729–734.

Houck, L. D., and K. Schwenk. 1984. The potential for long-term sperm competition in a plethodontid salamander. Herpetologica 40:410–415.

Houck, L. D., and P. A. Verrell. 1993. Studies of courtship behavior in plethodontid salamanders: a review. Herpetologica 49:175–184.

Houck, L. D., Arnold, S. J., and A. R. Hickman. 1988. Tests for sexual isolation in plethodontid salamanders (genus *Desmognathus*). Journal of Herpetology 22:186–191.

Houck, L. D., Arnold, S. J., and R. A. Thisted. 1985a. A statistical study of mate choice: sexual selection in a plethodontid salamander (*Desmognathus ochrophaeus*). Evolution 39:370–386.

Houck, L. D., Tilley, S. G., and S. J. Arnold. 1985b. Sperm competition in a plethodontid salamander: preliminary results. Journal of Herpetology 19:420–423.

Houck, W. J. 1969. Albino *Aneides ferreus*. Herpetologica 25:54.

Howard, J. H., and R. L. Wallace. 1981. Microgeographic variation of electrophoretic loci in populations of *Ambystoma macrodactylum columbianum* (Caudata: Ambystomatidae). Copeia 1981:466–471.

Howard, J. H., and R. L. Wallace. 1985. Life history characteristics of populations of the long-toed salamander (*Ambystoma macrodactylum*) from different altitudes. The American Midland Naturalist 113:361–373.

Howard, J. H., Seeb, L. W., and R. L. Wallace. 1993. Genetic variation and population divergence in the *Plethodon vandykei* group (Caudata: Plethodontidae). Herpetologica 49:238–247.

Howard, J. H., Wallace, R. L., and J. H. Larsen, Jr. 1983. Genetic variation and population divergence in the Larch Mountain salamander (*Plethodon larselli*). Herpetologica 39:41–46.

Howard, R. R. 1971. Avoidance learning of spotted salamanders, *Ambystoma maculatum*, by domestic chickens. The American Zoologist 11:637.

Howard, R. R., and E. D. Brodie. 1971. Experimental study of mimicry in salamanders involving *Notophthalmus viridescens viridescens* and *Pseudotriton ruber schencki*. Nature 233:277.

Howard, R. R., and E. D. Brodie. 1973. A Batesian mimetic complex in salamanders: responses to avian predators. Herpetologica 29:33–41.

Howard, W. E. 1950. Eggs of the salamander *Ensatina eschscholtzii platensis*. Copeia 1950:236.

Howell, T., and A. Hawkins. 1954. Variation in topotypes of the salamander *Plethodon jordani melaventris*. Copeia 1954:32–36.

Howell, T., and V. Switzer. 1953. Intergrades of the two-lined salamander, *Eurycea bislineata*, in Georgia. Herpetologica 9:152.

Huang, C., and A. Sih. 1990. Experimental studies on behaviorally mediated, indirect interactions through a shared predator. Ecology 71:1515–1522.

Huang, C., and A. Sih. 1991. Experimental studies of direct and indirect interactions in a three trophic-level stream system. Oecologia 85:530–536.

Hubbard, M. E. 1903. Correlated protective devices in some California salamanders. University of California Publications in Zoology 1:157–170.

Hubbs, C. 1962. Effects of a hurricane on the fish fauna of a coastal pool and drainage ditch. Texas Journal of Science 14:289–296.

Huckabee, J. W., C. P. Goodyear, and R. D. Jones. 1975. Acid rock in the Great Smokies: unanticipated impact on aquatic biota of road construction in regions of sulfide mineralization. Transactions of the American Fisheries Society 4:677–684.

Hudson, R. G. 1954. An annotated list of the reptiles and amphibians of the Unami Valley, Pennsylvania. Herpetologica 10:67–72.

Hudson, R. G. 1955. Observations on the larvae of the salamander *Eurycea bislineata bislineata*. Herpetologica 11:202–204.

Huheey, J. E. 1960. Mimicry in the color patterns of certain Appalachian salamanders. Journal of the Elisha Mitchell Scientific Society 76:246–251.

Huheey, J. E. 1966a. Variation in the Blue Ridge

Mountain dusky salamander in western North Carolina. Journal of the Elisha Mitchell Scientific Society 82:118–126.

Huheey, J. E. 1966b. Studies in warning coloration and mimicry. V. Red-cheeked dusky salamanders in North Carolina. Journal of the Elisha Mitchell Scientific Society 82:126–131.

Huheey, J. E., and R. A. Brandon. 1961. Further notes on mimicry in salamanders. Herpetologica 17:63–64.

Huheey, J. E., and R. A. Brandon. 1973. Rock-face populations of the mountain salamander, *Desmognathus ochrophaeus*, in North Carolina. Ecological Monographs 43:59–77.

Huheey, J. E., and R. A. Brandon. 1974. Studies of warning coloration and mimicry. VI. Comments on the warning coloration of red efts and their presumed mimicry by red salamanders. Herpetologica 30:149–155.

Huheey, J. E., and R. A. Brandon. 1975. Another function of maternal brooding behavior in salamanders of the genus *Desmognathus*. Journal of Herpetology 9:257.

Huheey, J. E., and R. A. Brandon. 1977. Novel toxins and the question of warning coloration and mimicry in salamanders. Herpetological Review 8 (suppl. 10).

Huheey, J. E., and A. Stupka. 1967. Amphibians and Reptiles of the Great Smoky Mountains National Park. University of Tennessee Press, Knoxville, Tennessee, 98 pp.

Humphries, A. A., Jr. 1955. Observations on the mating behavior of normal and pituitary-implanted *Triturus viridescens*. Physiological Zoology 28:73–79.

Humphries, R. L. 1956. An unusual aggregation of *Plethodon glutinosus* and remarks on its subspecific status. Copeia 1956:122–123.

Hunsaker, D., and F. E. Potter, Jr. 1960. "Red leg" in a natural population of amphibians. Herpetologica 16:285–286.

Hurlbert, S. H. 1969. The breeding migrations and interhabitat wandering of the vermilion-spotted newt *Notophthalmus viridescens* (Rafinesque). Ecological Monographs 39:465–488.

Hurlbert, S. H. 1970a. Predator responses to the vermillion-spotted newt (*Notophthalmus viridescens*). Journal of Herpetology 4:47–55.

Hurlbert, S. H. 1970b. The post-larval migration of the red-spotted newt *Notophthalmus viridescens* (Rafinesque). Copeia 1970:515–528.

Husting, E. L. 1965. Survival and breeding structure in a population of *Ambystoma maculatum*. Copeia 1965:352–362.

Hutcherson, J. E., Peterson, C. L., and R. F. Wilkinson. 1989. Reproductive and larval biology of *Ambystoma annulatum*. Journal of Herpetology 23:181–183.

Hutchison, V. C. 1959. Aggressive behavior in *Plethodon jordani*. Copeia 1959:72–73.

Hutchison, V. H. 1956. Notes on the plethodontid salamanders *Eurycea lucifuga* (Rafinesque) and *Eurycea longicauda longicauda* Green. Occasional Papers of the National Speleological Society 3:1–24.

Hutchison, V. H. 1958. The distribution and ecology of the cave salamander, *Eurycea lucifuga*. Ecological Monographs 28:1–20.

Hutchison, V. H. 1966. *Eurycea lucifuga*. Catalogue of American Amphibians and Reptiles, pp. 24.1–24.2.

Hutchison, V. H., and C. S. Hammen. 1958. Oxygen utilization in the symbiosis of embryos of the salamander *Ambystoma maculatum* and the alga *Oophila amblystomatis*. Biological Bulletin 115:483–489.

Ingersol, C. A., Wilkinson, R. F., Peterson, C. L., and R. H. Ingersol. 1991. Histology of the reproductive organs of *Cryptobranchus alleganiensis* (Caudata: Cryptobranchidae) in Missouri. The Southwestern Naturalist 36:60–66.

Ireland, P. H. 1973. Overwintering of larval spotted salamanders, *Ambystoma maculatum* (Caudata) in Arkansas. The Southwestern Naturalist 17:435–437.

Ireland, P. H. 1974. Reproduction and larval development of the dark-sided salamander, *Eurycea longicauda melanopleura* (Green). Herpetologica 30:338–343.

Ireland, P. H. 1976. Reproduction and larval development of the gray-bellied salamander *Eurycea multiplicata griseogaster*. Herpetologica 32:233–238.

Ireland, P. H. 1979. *Eurycea longicauda*. Catalogue of American Amphibians and Reptiles, pp. 221.1–221.4.

Ireland, P. H. 1989. Larval survivorship in two populations of *Ambystoma maculatum*. Copeia 1989:209–215.

Jackman, T. R., and D. B. Wake. 1994. Evolutionary and historical analysis of protein variation in the blotched forms of salamanders of the *En-*

satina complex (Amphibia: Plethodontidae). Evolution 48:876–897.
Jackson, M. E., and R. D. Semlitsch. 1993. Paedomorphosis in the salamander *Ambystoma talpoideum*: effects of a fish predator. Ecology 74:342–350.
Jackson, M. E., Scott, D. E., and R. A. Estes. 1989. Determinants of nest success in the marbled salamander (*Ambystoma opacum*). Canadian Journal of Zoology 67:2277–2281.
Jacobs, A. J., and D. H. Taylor. 1992. Chemical communication between *Desmognathus quadramaculatus* and *Desmognathus monticola*. Journal of Herpetology 26:93–95.
Jacobs, J. F. 1987. A preliminary investigation of geographic variation and systematics of the two-lined salamander, *Eurycea bislineata* (Green). Herpetologica 43:423–446.
Jaeger, R. G. 1970. Potential extinction through competition between two species of terrestrial salamanders. Evolution 24:632–642.
Jaeger, R. G. 1971a. Competitive exclusion as a factor influencing the distributions of two species of terrestrial salamanders. Ecology 52:632–637.
Jaeger, R. G. 1971b. Moisture as a factor influencing the distributions of two species of terrestrial salamanders. Oecologia 6:191–207.
Jaeger, R. G. 1972. Food as a limited resource in competition between two species of terrestrial salamanders. Ecology 53:535–546.
Jaeger, R. G. 1974. Interference or exploitation? A second look at competition between salamanders. Journal of Herpetology 8:191–194.
Jaeger, R. G. 1978. Plant climbing by salamanders: periodic availability of plant-dwelling prey. Copeia 1978:686–691.
Jaeger, R. G. 1979. Seasonal spatial distributions of the terrestrial salamander *Plethodon cinereus*. Herpetologica 35:90–93.
Jaeger, R. G. 1980a. Microhabitats of a terrestrial forest salamander. Copeia 1980:265–268.
Jaeger, R. G. 1980b. Density-dependent and density-independent causes of extinction of a salamander population. Evolution 34:617–621.
Jaeger, R. G. 1980c. Fluctuations in prey availability and food limitation for a terrestrial salamander. Oecologia 44:335–341.
Jaeger, R. G. 1981. Dear enemy recognition and the costs of aggression between salamanders. The American Naturalist 117:962–974.
Jaeger, R. G. 1984. Agonistic behavior of the red-backed salamander. Copeia 1984:309–314.
Jaeger, R. G. 1986. Pheromonal markers as territorial advertisement by terrestrial salamanders. In: Chemical Signals in Vertebrates IV, D. Duvall, D. Müller-Schwarze, and R. M. Silverstein, eds. Plenum Press, New York, pp. 191–203.
Jaeger, R. G. 1988. A comparison of territorial and non-territorial behavior in two species of salamander. Animal Behaviour 36:307–310.
Jaeger, R. G., and D. E. Barnard. 1981. Foraging tactics of a terrestrial salamander: choice of diet in structurally simple environments. The American Naturalist 117:639–664.
Jaeger, R. G., and D. C. Forester. 1993. Social behavior of plethodontid salamanders. Herpetologica 49:163–175.
Jaeger, R. G., and C. R. Gabor. 1993. Intraspecific chemical communication by a terrestrial salamander via the postcloacal gland. Copeia 1993:1171–1174.
Jaeger, R. G., and W. F. Gergits. 1979. Intra- and interspecific communication in salamanders through chemical signals on the substrate. Animal Behaviour 27:150–156.
Jaeger, R. G., and A. M. Rubin. 1982. Foraging tactics of a terrestrial salamander: judging prey profitability. Journal of Animal Ecology 51:167–176.
Jaeger, R. G., and J. K. Schwarz. 1991. Gradational threat postures by the red-backed salamander. Journal of Herpetology 25:112–114.
Jaeger, R. G., and S. E. Wise. 1991. A reexamination of the male salamander "sexy faeces hypothesis." Journal of Herpetology 25:370–373.
Jaeger, R. G., Barnard, D. E., and R. G. Joseph. 1982a. Foraging tactics of a terrestrial salamander: assessing prey density. The American Midland Naturalist 119:885–890.
Jaeger, R. G., Fortune, D., Hill, G., Palen, A., and G. Risher. 1993. Salamander homing behavior and territorial pheromones: alternative hypotheses. Journal of Herpetology 27:236–239.
Jaeger, R. G., Goy, J. M, Tarver, M., and C. E. Marquez. 1986. Salamander territories: pheromonal markers as advertisement by males. Animal Behaviour 34:860–864.

Jaeger, R. G., Joseph, R. G., and D. E. Barnard. 1981. Foraging tactics of a terrestrial salamander: sustained yield in territories. Animal Behaviour 30:490–496.

Jaeger, R. G., Kalvarsky, D., and N. Shimizu. 1982b. Territorial behaviour of the red-backed salamander: expulsion of intruders. Animal Behaviour 30:490–496.

Jaeger, R. G., Nishikawa, K. C., and D. E. Barnard. 1983. Foraging tactics of a terrestrial salamander: costs of territorial defence. Animal Behaviour 31:191–197.

Jaeger, R. G., Schwarz, J. K., and S. E. Wise. 1995a. Territorial male salamanders have foraging tactics attractive to gravid females. Animal Behaviour 49:633–639.

Jaeger, R. G., Wicknick, J. A., Griffis, M. R., and C. D. Anthony. 1995b. Socioecology of a terrestrial salamander: juveniles enter adult territories during stressful foraging periods. Ecology 76:533–543.

Jameson, E. W., Jr. 1944. Food of the red-backed salamander. Copeia 1944:145–147.

Janzen, F. J., and E. D. Brodie, III. 1989. Tall tails and sexy males: sexual behavior of rough-skinned newts (*Taricha granulosa*) in a natural breeding pond. Copeia 1989:1068–1071.

Jennings, M. R. 1996. *Ambystoma californiense* (California tiger salamander). Burrowing ability. Herpetological Review 27:194.

Jewell, R. D., and T. K. Pauley. 1995. Notes on the reproductive biology of the salamander *Plethodon richmondi* (Netting and Mittleman) in West Virginia. Herpetological Natural History 3:91–93.

Jiang, S., and D. L. Claussen. 1993. The effects of temperature on food passage time through the digestive tract in *Notophthalmus viridescens*. Journal of Herpetology 27:414–419.

Jobson, H. G. M. 1940. Reptiles and amphibians from Georgetown County, South Carolina. Herpetologica 2:39–43.

Johnson, C. R., and C. B. Schreck 1969. Food and feeding of larval *Dicamptodon ensatus* from California. The American Midland Naturalist 81:280–281.

Johnson, J. A., and E. D. Brodie, Jr. 1975. The selective advantage of the defensive posture of the newt, *Taricha granulosa*. The American Midland Naturalist 93:139–148.

Johnson, J. E., and A. S. Goldberg. 1975. Movement of larval two lined salamanders (*Eurycea bislineata*) in the Mill River, Massachusetts. Copeia 1975:588–589.

Johnston, R. F., and G. A. Schad. 1959. Natural history of the salamander, *Aneides hardii*. University of Kansas Publications of Museum of Natural History 10:573–585.

Jones, L. L. C. 1984. *Aneides flavipunctatus flavipunctatus* (speckled black salamander). Behavior. Herpetological Review 15:17.

Jones, L. L. C. 1989. *Plethodon vandykei* (Van Dyke's salamander). Reproduction. Herpetological Review 20:48.

Jones, L. L. C., and K. B. Aubry. 1985. *Ensatina eschscholtzii* (Oregon ensatina). Reproduction. Herpetological Review 16:26.

Jones, L. L. C., and R. B. Bury. 1986. *Dicamptodon ensatus* (Pacific giant salamander). Coloration. Herpetological Review 16:26.

Jones, R. G. 1991. *Ambystoma texanum* (small-mouthed salamander). Albinism. Herpetological Review 22:128–129.

Jones, R. L. 1982. Distribution and ecology of the seepage salamander *Desmognathus aeneus* Brown and Bishop (Amphibia: Plethodontidae), in Tennessee. Brimleyana 7:95–100.

Jones, R. L. 1986. Reproductive biology of *Desmognathus fuscus* and *Desmognathus santeetlah* in the Unicoi Mountains. Herpetologica 42:323–334.

Jones, T. R. 1980. A reevaluation of the salamander, *Eurycea aquatica* Rose and Bush (Amphibia: Plethodontidae). M.S. Thesis, Auburn University, Auburn, Alabama.

Jones, T. R., and J. P. Collins. 1992. Analysis of a hybrid zone between subspecies of the tiger salamander (*Ambystoma tigrinum*) in central New Mexico, USA. Journal of Evolutionary Biology 5:375–402.

Jones, T. R., Collins, J. P., Kocher, T. D., and J. B. Mitton. 1988. Systematic status and distribution of *Ambystoma tigrinum stebbinsi* Lowe (Amphibia: Caudata). Copeia 1988:621–635.

Jones, T. R., Kluge, A. G., and A. J. Wolf. 1993a. When theories and methodologies clash: a phylogenetic reanalysis of the North American ambystomatid salamanders (Caudata: Ambystomatidae). Systematic Biology 42:92–102.

Jones, T. R., Routman, E. J., Begun, D. J., and J. P. Collins. 1995. Ancestry of an isolated subspecies of salamander, *Ambystoma tigrinum stebbinsi* Lowe: the evolutionary significance of hybridization. Molecular Phylogenetics and Evolution 4:194-202.

Jones, T. R., Skelly, D. K., and E. E. Werner. 1993b. *Ambystoma tigrinum tigrinum* (eastern tiger salamander). Developmental polymorphism. Herpetological Review 24:147-148.

Jordan, E. O. 1891. The spermatophores of *Diemyctylus*. Journal of Morphology 5:263-270.

Jordan, R., Jr., and R. H. Mount. 1975. The status of the Red Hills salamander, *Phaeognathus hubrichti* Highton. Journal of Herpetology 9:211-215.

Judd, W. W. 1957. The food of Jefferson's salamander, *Ambystoma jeffersonianum*, in Rondeau Park, Ontario. Ecology 38:77-81.

Juterbock, J. E. 1978. Sexual dimorphism and maturity characteristics of three species of *Desmognathus* (Amphibia, Urodela, Plethodontidae). Journal of Herpetology 12:217-230.

Juterbock, J. E. 1984. Evidence for the recognition of specific status for *Desmognathus welteri*. Journal of Herpetology 18:240-255.

Juterbock, J. E. 1986. The nesting behavior of the dusky salamander, *Desmognathus fuscus*. I. Nesting phenology. Herpetologica 42:457-471.

Juterbock, J. E. 1987. The nesting behavior of the dusky salamander, *Desmognathus fuscus*. II. Nest site tenacity and disturbance. Herpetologica 43:361-368.

Juterbock, J. E. 1989. *Aneides aeneus*. In: Salamanders of Ohio, R. A. Pfingsten and F. L. Downs, eds. Ohio Biological Survey Bulletin, New Series 7(2):190-195.

Juterbock, J. E. 1990. Variation in larval growth and metamorphosis in the salamander *Desmognathus fuscus*. Herpetologica 46:291-303.

Kaplan, D. L. 1977. Exploitative competition in salamanders: test of a hypothesis. Copeia 1977:234-238.

Kaplan, R. H. 1979. Ontogenetic variation in "ovum" size in two species of *Ambystoma*. Copeia 1979:348-350.

Kaplan, R. H. 1980. The implications of ovum size variability for offspring fitness and clutch size within several populations of salamanders (*Ambystoma*). Evolution 34:51-64.

Kaplan, R. H. 1985. Maternal influences on offspring development in the California newt, *Taricha torosa*. Copeia 1985:1028-1035.

Kaplan, R. H., and M. L. Crump. 1978. The noncost of brooding in *Ambystoma opacum*. Copeia 1978:99-103.

Kaplan, R. H., and P. W. Sherman. 1980. Intraspecific oophagy in California newts. Journal of Herpetology 14:183-185.

Karlin, A. A., and S. I. Guttman. 1981. Hybridization between *Desmognathus fuscus* and *Desmognathus ochrophaeus* (Amphibia: Urodela: Plethodontidae) in northeastern Ohio and northwestern Pennsylvania. Copeia 1981:371-377.

Karlin, A. A., and S. I. Guttman. 1986. Systematics and geographic isozyme variation in the plethodontid salamander *Desmognathus fuscus*. Herpetologica 42:283-301.

Karlin, A. A., and D. B. Means. 1994. Genetic variation in the aquatic salamander genus *Amphiuma*. The American Midland Naturalist 132:1-9.

Karlin, A. A., and R. A. Pfingsten. 1989. *Desmognathus fuscus*. In: Salamanders of Ohio, R. A. Pfingsten and F. L. Downs, eds. Ohio Biological Survey Bulletin, New Series 7(2):174-180.

Karlin, A. A., Guttman, S. I., and D. B. Means. 1993. Population structure in the Ouachita Mountain dusky salamander, *Desmognathus brimleyorum* (Caudata: Plethodontidae). The Southwestern Naturalist 38:36-42.

Karns, D. R. 1992. Effects of acidic habitats on amphibian reproduction in a northern Minnesota peatland. Journal of Herpetology 26:401-412.

Kats, L. B. 1986. *Nerodia sipedon* (northern water snake). Feeding. Herpetological Review 17:61-62.

Kats, L. B. 1988. The detection of certain predators via olfaction by small-mouthed salamander larvae, *Ambystoma texanum*. Behavioral and Neural Biology 50:126-131.

Kats, L. B., and A. Sih. 1992. Oviposition site selection and avoidance of fish by streamside salamanders (*Ambystoma barbouri*). Copeia 1992:468-473.

Kats, L. B., Breeding, J. A., Hanson, K. M., and P. Smith. 1994. Ontogenetic changes in California newts (*Taricha torosa*) in response to chemical cues from conspecific predators. Journal of the North American Benthological Society 13:321-325.

Kats, L. B., Elliott, S. A., and J. Currens. 1992. Intraspecific oophagy in stream-dwelling California newts (*Taricha torosa*). Herpetological Review 23:7-8.

Kats, L. B., Petranka, J. W., and A. Sih. 1988. Antipredator defenses and the persistence of amphibian larvae with fishes. Ecology 69:1865-1870.

Keen, W. H. 1975. Breeding and larval development of three species of *Ambystoma* in central Kentucky (Amphibia: Urodela). Herpetologica 31:18-21.

Keen, W. H. 1979. Feeding and activity patterns in the salamander *Desmognathus ochrophaeus* (Amphibia, Urodela, Plethodontidae). Journal of Herpetology 13:461-467.

Keen, W. H. 1982. Habitat selection and interspecific competition in two species of plethodontid salamanders. Ecology 63:94-102.

Keen, W. H. 1984. Influence of moisture on the activity of a plethodontid salamander. Copeia 1984:684-688.

Keen, W. H., and L. P. Orr. 1980. Reproductive cycle, growth, and maturation of northern female *Desmognathus ochrophaeus*. Journal of Herpetology 14:7-10.

Keen, W. H., and R. W. Reed. 1985. Territorial defence of space and feeding sites by a plethodontid salamander. Animal Behaviour 32:58-65.

Keen, W. H., and S. Sharp. 1984. Responses of a plethodontid salamander to conspecific and congeneric intruders. Animal Behaviour 32:58-65.

Keen, W. H., McManus, M. G., and M. Wohltman. 1987. Cover site recognition and sex differences in cover site use by the salamander, *Desmognathus fuscus*. Journal of Herpetology 21:363-365.

Keen, W. H., Travis, J., and J. Julianna. 1984. Larval growth in three sympatric *Ambystoma* salamander species: species differences and the effects of temperature. Canadian Journal of Zoology 62:1043-1047.

Keiser, E. D., Jr., and P. J. Conzelmann. 1969. The red-backed salamander, *Plethodon cinereus* (Green) in Louisiana. Journal of Herpetology 3:190-191.

Kern, W. H., Jr. 1986. Reproduction of the hellbender, *Cryptobranchus alleganiensis*, in Indiana. Proceedings of the Indiana Academy of Science 95:521.

Kesler, D. H., and W. R. Munns, Jr. 1991. Diel feeding by adult red-spotted newts in the presence and absence of sunfish. Journal of Freshwater Ecology 6:267-273.

Kessel, E. L., and B. B. Kessel. 1942. An egg cluster of *Aneides lugubris lugubris* (Hallowell). The Wasmann Collector 5:141-142.

Kessel, E. L., and B. B. Kessel. 1943a. The rate of growth of the young larvae of the Pacific giant salamander, *Dicamptodon ensatus* (Eschscholtz). The Wasmann Collector 5:108-111.

Kessel, E. L., and B. B. Kessel. 1943b. Rate of growth of the older larvae of the Pacific giant salamander, *Dicamptodon ensatus* (Eschscholtz).The Wasmann Collector 5:141-142.

Kessel, E. L., and B. B. Kessel. 1944. Metamorphosis of the Pacific giant salamander, *Dicamptodon ensatus* (Eschscholtz). The Wasmann Collector 6:38-48.

Kezer, J. 1952a. The eggs of *Typhlotriton spelaeus* Stejneger, obtained by pituitary gland implantation. National Speleological Society Bulletin 14:58-59.

Kezer, J. 1952b. Thyroxin-induced metamorphosis of the neotenic salamanders *Eurycea tynerensis* and *Eurycea neotenes*. Copeia 1952:234-237.

Kezer, J., and D. S. Farner. 1955. Life history patterns of the salamander *Ambystoma macrodactylum* in the high Cascade Mountains of southern Oregon. Copeia 1955:127-131.

Kezer, J. and S. K. Sessions. 1979. Chromosome variation in the plethodontid salamander, *Aneides ferreus*. Chromosoma 71:65-80.

Kiesecker, J. 1996. pH-mediated predator-prey interactions between *Ambystoma tigrinum* and *Pseudacris triseriata*. Ecological Applications 6:1325-1331.

King, W. 1935. Ecological observations of *Ambystoma opacum*. Ohio Journal of Science 35:4-17.

King, W. 1939. A survey of the herpetology of Great Smoky Mountain National Park. The American Midland Naturalist 21:531-582.

Kirk, J. J. 1991. *Batrachoseps wrighti*. Catalogue of American Amphibians and Reptiles, pp. 506.1-506.3.

Kirk, J. J., and R. B. Forbes. 1991. *Batrachoseps wrighti* (Oregon slender salamander). Herpetological Review 22:22-23.

Kleeberger, S. R. 1984. A test of competition in two sympatric populations of desmognathine salamanders. Ecology 65:1846-1856.

Kleeberger, S. R. 1985. Influence of intraspecific density and cover on the home range of a plethodontid salamander. Oecologia 66:404–410.

Kleeberger, S. R., and J. K. Werner. 1982. Home range and homing behavior of *Plethodon cinereus* in northern Michigan. Copeia 1982:409–415.

Kleeberger, S. R., and J. K. Werner. 1983. Postbreeding migration and summer movement of *Ambystoma maculatum*. Journal of Herpetology 17:176–177.

Klein, H. G. 1960. Population estimate of the red-backed salamander. Herpetologica 16:52–54.

Knepton, J. C., Jr. 1954. A note on the burrowing habits of the salamander *Amphiuma means means*. Copeia 1954:68.

Knopf, G. N. 1962. Paedogenesis and metamorphic variation in *Ambystoma tigrinum malvortium*. The Southwestern Naturalist 7:75–76.

Knudsen, J. W. 1960. The courtship and egg mass of *Ambystoma gracile* and *Ambystoma macrodactylum*. Copeia 1960:44–46.

Kramer, P., Reichenbach, N., Hayslett, M., and P. Sattler. 1993. Population dynamics and conservation of the Peaks of Otter salamander, *Plethodon hubrichti*. Journal of Herpetology 27:431–435.

Kraus, F. 1985a. Unisexual salamander lineages in northwestern Ohio and southeastern Michigan: a study of the consequences of hybridization. Copeia 1985:309–324.

Kraus, F. 1985b. A new unisexual salamander from Ohio. Occasional Papers of the Museum of Zoology, University of Michigan 709:1–24.

Kraus, F. 1988. An empirical evaluation of the use of the ontogeny polarization criterion in phylogenetic inference. Systematic Zoology 37:106–141.

Kraus, F. 1989. Constraints on the evolutionary history of the unisexual salamanders of the *Ambystoma laterale-texanum* complex as revealed by mitochondrial DNA analysis. In: The evolution and ecology of unisexual vertebrates, R. M. Dawley and J. P. Bogart, eds. New York State Museum Bulletin 466:218–227.

Kraus, F. 1991. Intra-individual ploidy consistency among unisexual *Ambystoma*. Copeia 1991:38–43.

Kraus, F. 1995. The conservation status of unisexual vertebrate populations. Conservation Biology 9:956–959.

Kraus, F., and M. M. Miyamoto. 1990. Mitochondrial genotype of a unisexual salamander of hybrid origin is unrelated to either of its nuclear haplotypes. Proceedings of the U.S. Academy of Science 87:2235–2238.

Kraus, F., and J. W. Petranka. 1989. A new sibling species of *Ambystoma* from the Ohio River drainage. Copeia 1989:94–110.

Kraus, F., Ducey, P. K., Moler, P., and M. M. Miyamoto. 1991. Two new triparental unisexual *Ambystoma* from Ohio and Michigan. Herpetologica 47:429–439.

Kreeger, F. B. 1942. The cloaca of the female *Amphiuma tridactylum*. Copeia 1942:240–245.

Krenz, J. D., and D. E. Scott. 1994. Terrestrial courtship affects mating locations in *Ambystoma opacum*. Herpetologica 50:46–50.

Krenz, J. D., and D. M. Sever. 1995. Mating and oviposition in paedomorphic *Ambystoma talpoideum* proceeds the arrival of terrestrial males. Herpetologica 51:387–393.

Krzysik, A. J. 1979. Resource allocation, coexistence, and the niche structure of a streambank salamander community. Ecological Monographs 49:173–194.

Krzysik, A. J. 1980a. Microhabitat selection and brooding phenology of *Desmognathus fuscus fuscus* in western Pennsylvania. Journal of Herpetology 14:291–292.

Krzysik, A. J. 1980b. Trophic aspects of brooding behavior in *Desmognathus fuscus fuscus*. Journal of Herpetology 14:426–428.

Krzysik, A. J., and E. B. Miller. 1979. Substrate selection by three species of desmognathine salamanders from southwestern Pennsylvania: an experimental approach. Annals of the Carnegie Museum 48:111–117.

Kucken, D. J., Davis, J. S., Petranka, J. W., and C. K. Smith. 1994. Anakeesta stream acidification and metal contamination: effects on a salamander community. Journal of Environmental Quality 23:1311–1317.

Kumpf, K. F. 1934. The courtship of *Ambystoma tigrinum*. Copeia 1934:7–10.

Kumpf, K. F., and S. C. Yeaton. 1932. Observations on the courtship behavior of *Ambystoma jeffersonianum*. American Museum Novitates 546:1–7.

Labanick, G. M. 1983. Inheritance of the red-leg and red-cheek traits in the salamander

Desmognathus ochrophaeus. Herpetologica 39:114–120.

Labanick, G. M. 1984. Anti-predator effectiveness of autotomized tails of the salamander *Desmognathus ochrophaeus.* Herpetologica 40:110–118.

Labanick, G. M. 1988. Non-random association of red-leg and red-cheek coloration in the salamander *Desmognathus ochrophaeus.* Herpetologica 44:185–189.

Labanick, G. M., and R. A. Brandon. 1981. An experimental study of Batesian mimicry between the salamanders *Plethodon jordani* and *Desmognathus ochrophaeus.* Journal of Herpetology 15:275–281.

Labanick, G. M., and G. T. Davis. 1978. The spermatophore of the small-mouthed salamander, *Ambystoma texanum* (Amphibia, Urodela, Ambystomatidae). Journal of Herpetology 12:111–114.

Lagler, K. F., and K. E. Goellner. 1941. Notes on *Necturus maculosus* (Rafinesque), from Evans Lake, Michigan. Copeia 1941:96–98.

Lamb, T. 1984. The influence of sex and breeding condition on microhabitat selection and diet in the pig frog *Rana grylio.* The American Midland Naturalist 111:311–318.

Lancaster, D. L., and R. G. Jaeger. 1995. Rules of engagement for adult salamanders in territorial conflicts with heterospecific juveniles. Behavioral Ecology and Sociobiology 37:25–29.

Lancaster, D. L., and S. E. Wise. 1996. Differential response by the ringneck snake, *Diadophis punctatus,* to odors of tail-autotomizing prey. Herpetologica 52:98–108.

Landreth, H. F., and D. E. Ferguson. 1967. Newt orientation by sun-compass. Nature 215:516–518.

Langebartel, D. 1946. A giant tiger salamander. Copeia 1946:51.

Lannoo, M. J., and M. D. Bachmann. 1984. Aspects of cannibalistic morphs in a population of *Ambystoma t. tigrinum* larvae. The American Midland Naturalist 112:103–109.

Lannoo, M. J., Lowcock, L., and J. P. Bogart. 1989. Sibling cannibalism in noncannibal morph *Ambystoma tigrinum* larvae and its correlation with high growth rates and early metamorphosis. Canadian Journal of Zoology 67:1911–1914.

Larson, A. 1980. Paedomorphosis in relation to rates of morphological and molecular evolution in the salamander *Aneides flavipunctatus* (Amphibia, Plethodontidae). Evolution 34:1–17.

Larson, A. 1983a. A molecular phylogenetic perspective on the origins of a lowland tropical salamander fauna. I. Phylogenetic inferences from protein comparisons. Herpetologica 39:85–99.

Larson, A. 1983b. A molecular phylogenetic perspective on the origins of a lowland tropical salamander fauna. II. Patterns of morphological evolution. Evolution 37:1141–1153.

Larson, A. 1984. Neontological inferences of evolutionary pattern and process in the salamander Family Plethodontidae. Evolutionary Biology 17:119–217.

Larson, A. 1991. A molecular perspective on the evolutionary relationships of the salamander families. Evolutionary Biology 25:211–277.

Larson, A., and P. Chippindale. 1993. Molecular approaches to the evolutionary biology of plethodontid salamanders. Herpetologica 49:204–215.

Larson, A., and R. Highton. 1978. Geographic protein variation and divergence in the salamanders of the *Plethodon welleri* group (Amphibia, Plethodontidae). Systematic Zoology 27:431–448.

Larson, A., Wake, D. B., Maxson, L. R., and R. Highton. 1981. A molecular phylogenetic perspective on the origins of the morphological novelties in the salamanders of the tribe Plethodontini. Evolution 35:405–422.

Larson, D. W. 1968. The occurrence of neotenic salamanders, *Ambystoma tigrinum diaboli* Dunn, in Devil's Lake, North Dakota. Copeia 1968:620–621.

Lazell, J. D., Jr., and R. A. Brandon. 1962. A new stygian salamander from the southern Cumberland Plateau. Copeia 1962:300–306.

Lazell, J. D., Jr., and T. Mann. 1991. *Plethodon websteri* (Webster's salamander). Geographic distribution. Herpetological Review 22:62.

Leclair, R., Jr., and J. P. Bourassa. 1981. Observation et analyse de la predation des oeufs d'*Ambystoma maculatum* (Shaw) (Amphibia, Urodela) par des larves de Dipteres chironomides, dans la region de Trois-Rivieres (Quebec). Canadian Journal of Zoology 59:1339–1343.

Lee, D. S. 1969a. A food study of the salamander *Haideotriton wallacei* Carr. Herpetologica 25:175-177.

Lee, D. S. 1969b. Notes on the feeding behavior of cave-dwelling bullfrogs. Herpetologica 25:211-212.

Lee, D. S., and R. Franz. 1974. Comments on the feeding behavior of larval tiger salamanders, *Ambystoma tigrinum*. Bulletin of the Maryland Herpetological Society 10:105-107.

Lee, D. S., and A. W. Norden. 1973. A food study of the green salamander, *Aneides aeneus*. Journal of Herpetology 7:53-54.

Leff, L. G., and M. D. Bachmann. 1986. Ontogenetic changes in predatory behavior of larval tiger salamanders (*Ambystoma tigrinum*). Canadian Journal of Zoology 64:1337-1344.

Leonard, W. P. 1996. *Plethodon vehiculum* (western red-backed salamander). Habitat. Herpetological Review 27:195.

Leonard, W. P., and D. M. Darda. 1995. *Ambystoma tigrinum* (tiger salamander). Reproduction. Herpetological Review 26:29-30.

Leonard, W. P., Brown, H. A., Jones, L. L. C., McAllister, K. R., and R. M. Storm. 1993. Amphibians of Washington and Oregon. Seattle Audubon Society, Seattle, Washington, 168 pp.

Licht, L. E. 1969. Observations on the courtship behavior of *Ambystoma gracile*. Herpetologica 25:49-52.

Licht, L. E. 1973. Behavior and sound production by the northwestern salamander *Ambystoma gracile*. Canadian Journal of Zoology 51:1055-1056.

Licht, L. E. 1975. Growth and food of larval *Ambystoma gracile* from a lowland population in southwestern British Columbia. Canadian Journal of Zoology 53:1716-1722.

Licht, L. E. 1989. Reproductive parameters of unisexual *Ambystoma* on Pelee Island, Ontario. In: The evolution and ecology of unisexual vertebrates, R. M. Dawley and J. P. Bogart, eds. New York State Museum Bulletin 466:209-217.

Licht, L. E. 1992. The effect of food level on growth rate and frequency of metamorphosis and paedomorphosis in *Ambystoma gracile*. Canadian Journal of Zoology 70:87-93.

Licht, L. E., and J. P. Bogart. 1987. Ploidy and developmental rate in the salamander hybrid complex (genus *Ambystoma*). Evolution 41:918-920.

Licht, L. E., and J. P. Bogart. 1989a. Embryonic development and temperature tolerance in diploid and polyploid salamanders (Genus *Ambystoma*). The American Midland Naturalist 122:401-407.

Licht, L. E., and J. P. Bogart. 1989b. Growth and sexual maturity in diploid and polyploid salamanders (genus *Ambystoma*). Canadian Journal of Zoology 67:812-818.

Licht, L. E., and J. P. Bogart. 1990. Courtship behavior of *Ambystoma texanum* on Pelee Island, Ontario. Journal of Herpetology 24:450-452.

Licht, L. E., and D. M. Sever. 1993. Structure and development of the paratoid gland in metamorphosed and neotenic *Ambystoma gracile*. Copeia 1993:116-123.

Licht, P., and A. G. Brown. 1967. Behavioral thermoregulation and its role in the ecology of the red-bellied newt, *Taricha rivularis*. Ecology 48:598-611.

Lindsey, C. C. 1966. Temperature-controlled meristic variation in the salamander *Ambystoma gracile*. Nature 209:1152-1153.

Liner, E. A. 1954. The herpetofauna of Lafayette, Terrebonne, and Vermilion parishes, Louisiana. Louisiana Academy of Science 12:65-85.

Ling, R. L., and J. K. Werner. 1988. Mortality in *Ambystoma maculatum* larvae due to *Tetrahymena* infection. Herpetological Review 19:26.

Ling, R. W., VanAmberg, J. P., and J. K. Werner. 1986. Pond acidity and its relationship to larval development of *Ambystoma maculatum* and *Rana sylvatica* in upper Michigan. Journal of Herpetology 20:230-236.

Little, E. L., Jr., and J. G. Keller. 1937. Amphibians and reptiles of the Jordana Experimental Range, New Mexico. Copeia 1937:216-222.

Livezey, R. L. 1959. The egg mass and larvae of *Plethodon elongatus* Van Denburgh. Herpetologica 15:41-42.

Loafman, P., and L. Jones. 1996. *Dicamptodon copei* (Cope's giant salamander). Metamorphosis and reproduction. Herpetological Review 27:136.

Loeb, M. L. G., Collins, J. P., and T. J. Maret. 1994. The role of prey in controlling expression of a trophic polymorphism in *Ambystoma tigrinum nebulosum*. Functional Ecology 8:151-158.

Lombard, R. E., and D. B. Wake. 1977. Tongue evolution in the lungless salamanders, family

Plethodontidae. II. Functional and evolutionary diversity. Journal of Morphology 153:39–79.

Long, L. E., Saylor, L. S., and M. E. Soule. 1995. A pH/UV-B synergism in amphibians. Conservation Biology 9:1301–1303.

Longley, G. 1978. Status of *Typhlomolge* (= *Eurycea rathbuni*), the Texas blind salamander. Species Report, U.S. Fish and Wildlife Service 2:1–45.

Loomis, R. B., and O. L. Webb. 1951. *Eurycea multiplicata* collected at the restricted type locality. Herpetologica 7:141–142.

Loredo, I., and D. Van Vuren. 1996. Reproductive ecology of a population of the California tiger salamander. Copeia 1996:895–901.

Loredo, I., Van Vuren, D., and M. L. Morrison. 1996. Habitat use and migration behavior of the California tiger salamander. Journal of Herpetology 30:282–285.

Lotter, F. 1978. Reproductive ecology of the salamander *Plethodon cinereus* (Amphibia, Urodela, Plethodontidae) in Connecticut. Journal of Herpetology 12:231–236.

Lotter, F., and N. J. Scott, Jr. 1977. Correlation between climate and distribution of the color morphs of the salamander *Plethodon cinereus*. Copeia 1977:681–690.

Lowcock, L. A. 1989. Biogeography of hybrid complex of *Ambystoma*: interpreting unisexual-bisexual generic data through time and space. In: The evolution and ecology of unisexual vertebrates, R. M. Dawley and J. P. Bogart, eds. New York State Museum Bulletin 466:180–208.

Lowcock, L. A. 1994. Biotype, genomotype, and genotype: variable effects of polyploidy and hybridity on ecological partitioning in a bisexual-unisexual community of salamanders. Canadian Journal of Zoology 72:104–117.

Lowcock, L. A., and J. P. Bogart. 1989. Electrophoretic evidence for multiple origins of triploid forms in the *Ambystoma laterale-jeffersonianum* complex. Canadian Journal of Zoology 67:350–356.

Lowcock, L. A., and R. W. Murphy. 1991. Pentaploidy in hybrid salamanders demonstrates enhanced tolerance of multiple chromosome sets. Experientia 47:490–493.

Lowcock, L. A., Griffith, H., and R. W. Murphy. 1991. The *Ambystoma laterale-jeffersonianum* complex in central Ontario: ploidy structure, sex ratio and breeding dynamics in bisexual-unisexual communities. Copeia 1991:87–105.

Lowcock, L. A., Griffith, H., and R. W. Murphy. 1992. Size in relation to sex, hybridity, ploidy, and breeding dynamics in central Ontario populations of the *Ambystoma laterale-jeffersonianum* complex. Journal of Herpetology 26:46–53.

Lowe, C. H., Jr. 1950. The systematic status of the salamander *Plethodon hardii*, with a discussion of biogeographical problems in *Aneides*. Copeia 1950:92–99.

Lowe, C. H., Jr. 1954. A new salamander (genus *Ambystoma*) from Arizona. Proceedings of the Biological Society of Washington 67:243–246.

Lutterschmidt, W. I., Marvin, G. A., and V. H. Hutchison. 1994. Alarm response by a plethodontid salamander (*Desmognathus ochrophaeus*): conspecific and heterospecific "Schreckstoff." Journal of Chemical Ecology 20:2751–2759.

Lynch, J. E., Jr., and R. L. Wallace. 1987. Field observations of courtship behavior in Rocky Mountain populations of Van Dyke's salamander, *Plethodon vandykei*, with a description of its spermatophore. Journal of Herpetology 21:337–340.

Lynch, J. F. 1974. *Aneides flavipunctatus*. Catalogue of American Amphibians and Reptiles, pp. 158.1–158.2.

Lynch, J. F. 1981. Patterns of ontogenetic and geographic variation in the black salamander, *Aneides flavipunctatus* (Caudata: Plethodontidae). Smithsonian Contributions to Zoology 324:1–53.

Lynch, J. F. 1985. The feeding ecology of *Aneides flavipunctatus* and sympatric plethodontid salamanders in northwestern California. Journal of Herpetology 19:328–352.

Lynn, W. G., and J. N. Dent. 1941. Notes on *Plethodon cinereus* and *Hemidactylium scutatum* on Cape Cod. Copeia 1941:113–114.

McAllister, C. T., Tumlison, C. R., and L. M. Cook. 1981. *Necturus maculosus louisianensis* (Red River mudpuppy). Coloration. Herpetological Review 12:78.

Macartney, J. M., and P. T. Gregory. 1981. Differential susceptibility of sympatric garter snake species to amphibian skin secretions. The American Midland Naturalist 106:271–281.

McAtee, W. L. 1907. A list of the mammals, reptiles, and batrachians of Monroe County, Indiana. Proceedings of the Biological Society of Washington 20:1–16.

McComb, W. C., Chambers, C. L., and M. Newton. 1993a. Small mammal and amphibian abundance in streamside and upslope habitats of mature Douglas-fir stands, western Oregon. Northwest Science 67:181–188.

McComb, W. C., McGarigal, K., and R. G. Anthony. 1993b. Small mammal and amphibian abundance in streamside and upslope habitats of mature Douglas-fir stands, western Oregon. Northwest Science 67:7–15.

McCoy, C. J. 1982. Amphibians and reptiles in Pennsylvania. Special Publications of the Carnegie Museum of Natural History 6:1–91.

McCoy, C. J. 1992. Rediscovery of the mud salamander (*Pseudotriton montanus*, Amphibia, Plethodontidae) in Pennsylvania, with restriction of the type locality. Journal of the Pennsylvania Academy of Science 66:92–93.

McCrady, E. 1954. A new species of *Gyrinophilus* (Plethodontidae) from Tennessee caves. Copeia 1954:200–206.

MacCulloch, R. D., and J. R. Bider. 1975. Phenology, migrations, circadian rhythm and the effect of precipitation on the activity of *Eurycea b. bislineata* in Quebec. Herpetologica 31:433–439.

McDaniel, V. R. 1977. A morphometric analysis of sirens from southern Texas. In: South Texas fauna: a symposium honoring Dr. Allen W. Chaney, B. R. Chapman and J. W. Tunnell, Jr., eds. Caeser Kleberg Wildlife Research Institute, Texas A & I University, Kingsville, Texas, pp. 31–35.

McDaniel, V. R., and D. A. Saugey. 1977. Geographic distribution: *Ambystoma annulatum*. Herpetological Review 8:38.

McDowell, W. T. 1988. Egg hatching season of *Eurycea longicauda longicauda* and *Eurycea lucifuga* (Caudata: Plethodontidae) in southern Illinois. Bulletin of the Chicago Herpetological Society 23:145.

McGavin, M. 1978. Recognition of conspecific odors by the salamander *Plethodon cinereus*. Copeia 1978:356–358.

Macgregor, H. C., and T. M. Uzzell, Jr. 1964. Gynogenesis in salamanders related to *Ambystoma jeffersonianum*. Science 143:1043–1045.

Macgregor, H. C., Horner, H., Owen, C. A., and I. Parker. 1973. Observations on centrometric heterochromatin and satellite DNA in salamanders of the genus *Plethodon*. Chromosoma 43:329–348.

McGregor, J. H., and W. R. Teska. 1989. Olfaction as an orientation mechanism in migrating *Ambystoma maculatum*. Copeia 1989:779–781.

McKenzie, D. S. 1970. Aspects of the autecology of the plethodontid salamander, *Aneides ferreus* (Cope) (abstract). Dissertation Abstracts International 30B:5299.

McKenzie, D. S., and R. M. Storm. 1970. Patterns of habitat selection in the clouded salamander, *Aneides ferreus* (Cope). Herpetologica 26:450–454.

McKnight, M. L., Dodd, C. K., Jr., and C. M. Spolsky. 1991. Protein and mitochondrial DNA variation in the salamander *Phaeognathus hubrichti*. Herpetologica 1991:440–447.

MacMahon, J. A. 1964. Additional observations on the courtship of Metcalf's salamander, *Plethodon jordani* (*metcalfi* phase). Herpetologica 20:67–69.

McMillan, M. A., and R. D. Semlitsch. 1980. Prey of the dwarf salamander, *Eurycea quadridigitata*, in South Carolina. Journal of Herpetology 14:422–424.

McMillian, J. E., and R. F. Wilkinson, Jr. 1972. The effects of pancreatic hormones on blood glucose in *Ambystoma annulatum*. Copeia 1972:664–668.

MacNamara, J. A. 1977. Food habits of terrestrial adult migrants and immature red efts of the red-spotted newt *Notophthalmus viridescens*. Herpetologica 33:127–132.

McWilliams, S. R., and M. Bachmann. 1989a. Foraging ecology and prey preference of pond-form larval small-mouthed salamanders, *Ambystoma texanum*. Copeia 1989:948–961.

McWilliams, S. R., and M. Bachmann. 1989b. Predatory behavior of larval small-mouthed salamanders (*Ambystoma texanum*). Herpetologica 45:459–467.

Madison, D. M. 1969. Homing behaviour of the red-cheeked salamander, *Plethodon jordani*. Animal Behaviour 17:25–39.

Madison, D. M. 1972. A mechanism of homing orientation in salamanders involving chemical cues. In: Animal orientation and navigation, S. R. Galler, K. Schmidt-Koenig, G. J. Jacobs, and R. E. Belleville, eds. NASA Special Publication 262:485–498.

Madison, D. M. 1975. Intraspecific odor preferences between salamanders of the same sex: dependence on season and proximity of residence. Canadian Journal of Zoology 53:1356–1361.

Madison, D. M., and C. R. Shoop. 1970. Homing behavior, orientation, and home range of salamanders tagged with Tantalum-182. Science 168:1484–1487.

Maglia, A. M. 1996. Ontogeny and feeding ecology of the red-backed salamander, *Plethodon cinereus*. Copeia 1996:576–586.

Maha, G. C., Maxson, L. R., and R. Highton. 1983. Immunological evidence for the validity of *Plethodon kentucki*. Journal of Herpetology 17:398–400.

Mahrdt, C. R. 1975. The occurrence of *Ensatina eschscholtzii eschscholtzii* in Baja California, Mexico. Journal of Herpetology 9:240–242.

Maiorana, V. C. 1976. Size and environmental predictability for salamanders. Evolution 30:599–613.

Maiorana, V. C. 1977a. Tail autotomy, functional conflicts and their resolution by a salamander. Nature 265:533–535.

Maiorana, V. C. 1977b. Observations of salamanders (Amphibia, Urodela, Plethodontidae) dying in the field. Journal of Herpetology 11:1–5.

Maiorana, V. C. 1978a. Differences in diet as an epiphenomenon: space regulates salamanders. Canadian Journal of Zoology 56:1017–1025.

Maiorana, V. C. 1978b. Behavior of an unobservable species: diet selection by a salamander. Copeia 1978:664–672.

Maksymovitch, E., and P. Verrell. 1992. The courtship behavior of the santeetlah dusky salamander, *Desmognathus santeetlah* Tilley (Amphibia: Caudata: Plethodontidae). Ethology 90:236–246.

Maksymovitch, E., and P. Verrell. 1993. Divergence of mate recognition systems among conspecific populations of the plethodontid salamander *Desmognathus santeetlah* (Amphibia: Plethodontidae). Biological Journal of the Linnaean Society 49:19–29.

Mansueti, R. 1941. Sounds produced by the slimy salamander. Copeia 1941:266–267.

Marangio, M. S. 1978. The occurrence of neotenic rough-skinned newts (*Taricha granulosa*) in montane lakes of southern Oregon. Northwest Science 52:343–350.

Marangio, M. S., and J. D. Anderson. 1977. Soil moisture preference and water relations of the marbled salamander, *Ambystoma opacum* (Amphibia, Urodela, Ambystomatidae). Journal of Herpetology 11:169–176.

Maret, T. J., and J. P. Collins. 1994. Individual responses to population size structure: the role of size variation in controlling expression of a trophic polyphenism. Oecologia 100:279–285.

Marlow, R. W., Brode, J. M., and D. B. Wake. 1979. A new salamander, genus *Batrachoseps*, from the Inyo Mountains of California, with a discussion of relationships in the genus. Natural History Museum of Los Angeles County, Contributions in Science 308:1–17.

Marshall, C. J., Doyle, L. S., and R. H. Kaplan. 1990. Intraspecific and sex-specific oophagy in a salamander and a frog: reproductive convergence of *Taricha torosa* and *Bombina orientalis*. Herpetologica 46:395–399.

Marshall, J. L. 1996. *Eurycea cirrigera* (southern two-lined salamander). Nest site. Herpetological Review 27:75–76.

Martin, D. L., Jaeger, R. G., and C. P. Labat. 1986. Territoriality in an *Ambystoma* salamander? Support for the null hypothesis. Copeia 1986:725–730.

Martin, J. B., Witherspoon, N. B., and M. H. A. Keenleyside. 1974. Analysis of feeding behavior in the newt *Notophthalmus viridescens*. Canadian Journal of Zoology 52:277–281.

Martof, B. 1953. The "spring lizard" industry, a factor in salamander distribution and genetics. Ecology 34:436–437.

Martof, B. S. 1955. Observations on the life history and ecology of amphibians of the Athens area, Georgia. Copeia 1955:166–170.

Martof, B. S. 1956. Three new subspecies of *Leurognathus marmorata* from the southern Appalachian Mountains. Occasional Papers of the Museum of Zoology, University of Michigan 575:1–14.

Martof, B. S. 1962. Some aspects of the life history and ecology of the salamander *Leurognathus*. The American Midland Naturalist 67:1–35.

Martof, B. S. 1968. *Ambystoma cingulatum*. Catalogue of American Amphibians and Reptiles, pp. 57.1–57.2.

Martof, B. S. 1969. Prolonged inanition in *Siren lacertina*. Copeia 1969:285–289.

Martof, B. S. 1972. *Pseudobranchus, P. striatus*. Catalogue of American Amphibians and Reptiles, pp. 118.1–118.4.

Martof, B. S. 1973. *Siren intermedia*. Catalogue of American Amphibians and Reptiles, pp. 127.1–127.3.

Martof, B. S. 1974. Sirenidae. Catalogue of American Amphibians and Reptiles, pp. 151.1–151.2.

Martof, B. S. 1975a. *Pseudotriton ruber*. Catalogue of American Amphibians and Reptiles, pp. 167.1–167.3.

Martof, B. S. 1975b. *Pseudotriton montanus*. Catalogue of American Amphibians and Reptiles, pp. 166.1–166.2.

Martof, B. S., and H. C. Gerhardt. 1965. Observations on the geographic variation in *Ambystoma cingulatum*. Copeia 1965:342–346.

Martof, B. S., and F. L. Rose. 1963. Geographic variation in southern populations of *Desmognathus ochrophaeus*. The American Midland Naturalist 69:376–425.

Martof, B. S., and D. C. Scott. 1957. The food of the salamander *Leurognathus*. Ecology 38:494–501.

Martof, B. S., Palmer, W. M., Bailey, J. R., and J. R. Harrison III. 1980. Amphibians and Reptiles of the Carolinas and Virginia. University of North Carolina Press, Chapel Hill, North Carolina, 264 pp.

Marvin, G. A. 1996. Life history and population characteristics of the salamander *Plethodon kentucki* with a review of *Plethodon* life histories. The American Midland Naturalist 136:385–400.

Marvin, G. A., and V. H. Hutchison. 1995. Avoidance response by adult newts (*Cynops pyrrhogaster* and *Notophthalmus viridescens*) to chemical alarm cues. Behaviour 132:95–105.

Marvin, G. A., and V. H. Hutchison. 1996. Courtship behavior of the Cumberland Plateau woodland salamander, *Plethodon kentucki* (Amphibia: Plethodontidae), with a review of courtship in the genus *Plethodon*. Ethology 102:285–303.

Maslin, T. P. 1950. The production of sound in caudate Amphibia. University of Colorado Studies in Biology 1:29–45.

Maslin, T. P., Jr. 1939. Egg-laying of the slender salamander (*Batrachoseps attenuatus*). Copeia 1939:209–212.

Massey, A. 1988. Sexual interactions in red-spotted newt populations. Animal Behaviour 36:205–210.

Massey, A. 1990. Notes on the reproductive ecology of red-spotted newts (*Notophthalmus viridescens*). Journal of Herpetology 24:106–107.

Mathews, E. 1952. Erythrism in the salamander *Plethodon cinereus cinereus*. Copeia 1952:277.

Mathews, R. C., Jr. 1982. Predator stoneflies: role in freshwater stream communities. Journal of the Tennessee Academy of Science 57:82–83.

Mathews, R. C., Jr., and E. L. Morgan. 1982. Toxicity of Anakeesta leachates to shovel-nosed salamanders, Great Smoky Mountains National Park. Journal of Environmental Quality 11:102–106.

Mathis, A. 1989. Do seasonal spatial patterns in a terrestrial salamander reflect reproductive behavior or territoriality? Copeia 1989:788–791.

Mathis, A. 1990a. Territoriality in a terrestrial salamander: the influence of resource quality and body size. Behaviour 112:162–175.

Mathis, A. 1990b. Territorial salamanders assess sexual and competitive information via chemical signals. Animal Behaviour 40:953–962.

Mathis, A. 1991a. Large male advantage for access to females: evidence of male-male competition and female discrimination in a territorial salamander. Behavioral Ecology and Sociology 29:133–138.

Mathis, A. 1991b. Territories of male and female terrestrial salamanders: costs, benefits, and intersexual spatial associations. Oecologia 86:433–440.

Mathis, A., and R. R. Simons. 1994. Size-dependent responses of resident male red-backed salamanders to chemical stimuli from conspecifics. Herpetologica 50:335–344.

Mathis, A., Jaeger, R. G., Keen, W. H., Ducey, P. K., and B. W. Buchanan. 1995. Aggression and territoriality by salamanders and a comparison with the territorial behaviour of frogs. In: Amphibian Biology, Volume 2, Social Behavior, H. Heatwole and B. K. Sullivan, eds. Surry Beatty & Sons, Chipping Norton, NSW, Australia, pp. 633–676.

Maughan, O. E., Wickham, M. G., Laumeyer, P., and R. L. Wallace. 1976. Records of the Pacific giant salamander, *Dicamptodon ensatus* (Amphibia, Urodela, Ambystomatidae), from the Rocky Mountains of Idaho. Journal of Herpetology 10:249–251.

Maxson, L. R., and R. D. Maxson. 1979. Comparative albumin and biochemical evolution in plethodontid salamanders. Evolution 33:1057–1062.

Maxson, L. R., Highton, R., and D. Ondrula.

1984. Immunological evidence on genetic relationships of *Plethodon dorsalis*. Journal of Herpetology 18:341-344.

Maxson, L. R., Highton, R., and D. B. Wake. 1979. Albumin evolution and its phylogenetic implications in the plethodontid salamander genera *Plethodon* and *Ensatina*. Copeia 1979:502-508.

Maxson, L. R., Moler, P. E., and B. W. Mansell. 1988. Albumin evolution in salamanders of the genus *Necturus*. Journal of Herpetology 22:231-235.

Meade, G. P. 1934. Feeding *Farancia abacura* in captivity. Copeia 1934:91-92.

Means, D. B. 1972. Notes on the autumn breeding biology of *Ambystoma cingulatum* (Cope) (Amphibia: Urodela: Ambystomatidae). Association of Southeastern Biologist Bulletin 19:84.

Means, D. B. 1974. The status of *Desmognathus brimleyorum* Stejneger and an analysis of the genus *Desmognathus* (Amphibia: Urodela) in Florida. Bulletin of the Florida State Museum 18:1-100.

Means, D. B. 1975. Competitive exclusion along a habitat gradient between two species of salamanders (*Desmognathus*) in western Florida. Journal of Biogeography 2:253-263.

Means, D. B. 1992a. Rare one-toed Amphiuma. In: Rare and Endangered Biota of Florida, Volume 3, Amphibians and Reptiles, P. Moler, ed. University Presses of Florida, Gainesville, Florida, pp. 34-38.

Means, D. B. 1992b. Rare Georgia blind salamander. In: Rare and Endangered Biota of Florida, Volume 3, Amphibians and Reptiles, P. Moler, ed. University Presses of Florida, Gainesville, Florida, pp. 9-11.

Means, D. B. 1993. *Desmognathus apalachicolae*. Catalogue of American Amphibians and Reptiles, pp. 556.1-556.2.

Means, D. B., and A. A. Karlin. 1989. A new species of *Desmognathus* from the eastern gulf coastal plain. Herpetologica 45:37-46.

Means, D. B., and C. J. Longden. 1970. Observations on the occurrence of *Desmognathus monticola* in Florida. Herpetologica 26:396-399.

Means, D. B., Palis, J. G., and M. Baggett. 1996. Effects of slash pine silviculture on a Florida population of flatwoods salamander (*Ambystoma cingulatum*). Conservation Biology 10:426-437.

Mecham, J. S. 1967a. *Notophthalmus perstriatus*. Catalogue of American Amphibians and Reptiles, p. 38.

Mecham, J. S. 1967b. *Notophthalmus viridescens*. Catalogue of American Amphibians and Reptiles, pp. 53.1-53.4.

Mecham, J. S. 1968. On the relationships between *Notophthalmus meridionalis* and *Notophthalmus kallerti*. Journal of Herpetology 2:121-127.

Mecham, J. S., and R. E. Hellman. 1952. Notes on the larvae of two Florida salamanders. Quarterly Journal of the Florida Academy of Science 15:127-133.

Meffe, G. K., and A. L. Sheldon. 1987. Habitat use by dwarf waterdogs (*Necturus punctatus*) in South Carolina streams, with life history notes. Herpetologica 43:490-496.

Merchant, H. C. 1972. Estimated population size and home range of salamanders *Plethodon jordani* and *Plethodon glutinosus*. Journal of the Washington Academy of Science 62:248-257.

Merkle, D. A., and S. I. Guttman. 1977. Geographic variation in the cave salamander *Eurycea lucifuga*. Herpetologica 33:313-321.

Merkle, D. A., Guttman, S. I, and M. A. Nickerson. 1977. Genetic uniformity throughout the range of the hellbender, *Cryptobranchus alleganiensis*. Copeia 1977:549-553.

Merritt, R. B., Kroon, W. H., Wienski, D. A., and K. A. Vincent. 1984. Genetic structure of natural populations of the red-spotted newt, *Notophthalmus viridescens*. Biochemical Genetics 22:669-686.

Meshaka, W. E., Jr., and S. E. Trauth. 1995. Reproductive cycle of the Ozark zigzag salamander, *Plethodon dorsalis angusticlavius* (Caudata, Plethodontidae), in north central Arkansas. Alytes 12:175-182.

Metter, D. 1963. Stomach contents of Idaho larval *Dicamptodon*. Copeia 1963:435-436.

Miller, L. 1944. Notes on the eggs and larvae of *Aneides lugubris*. Copeia 1944:224-230.

Miller, M. R., and M. E. Robbins. 1954. The reproductive cycle in *Taricha torosa* (*Triturus torosus*). Journal of Experimental Zoology 125:415-445.

Minckley, W. L. 1959. An atypical *Eurycea lucifuga* from Kansas. Herpetologica 15:240.

Minton, S. A. 1954. Salamanders of the *Ambystoma jeffersonianum* complex in Indiana. Herpetologica 10:173-179.

Minton, S. A., Jr. 1972. Amphibians and reptiles

of Indiana. Indiana Academy of Science Monographs 3:1-346.

Mitchell, J. C., and S. B. Hedges. 1980. *Ambystoma mabeei* Bishop (Caudata: Ambystomatidae): an addition to the salamander fauna of Virginia. Brimleyana 3:119-121.

Mitchell, J. C., and J. A. Taylor. 1986. Predator-prey size relationships in a North Carolina population of *Plethodon jordani*. Journal of Herpetology 20:562-566.

Mitchell, J. C., and W. S. Woolcott. 1985. Observations of the microdistribution, diet, and predator-prey size relationships in the salamander *Plethodon cinereus* from the Virginia Piedmont. Virginia Journal of Science 36:281-288.

Mitchell, R. W., and J. R. Reddell. 1965. *Eurycea tridentifera*, a new species of troglobitic salamander from Texas and a reclassification of *Typhlomolge rathbuni*. Texas Journal of Science 17:12-27.

Mitchell, R. W., and R. E. Smith. 1972. Some aspects of the osteology and evolution of the neotenic spring and cave salamanders (*Eurycea*, Plethodontidae) of central Texas. Texas Journal of Science 23:343-362.

Mittleman, M. B. 1950. Cavern-dwelling salamanders of the Ozark Plateau. National Speleological Society Bulletin 12:12-15.

Mittleman, M. B. 1951. American caudata. VII. Two new salamanders of the genus *Plethodon*. Herpetologica 7:105-112.

Mittleman, M. B. 1966. *Eurycea bislineata*. Catalogue of American Amphibians and Reptiles, pp. 45.1-45.4.

Mittleman, M. B. 1967. *Manculus* and M. *quadridigitatus*. Catalogue of American Amphibians and Reptiles, p. 44.1-45.2.

Mizuno, S., and H. C. Macgregor. 1974. Chromosomes, DNA sequences, and evolution in salamanders of the genus *Plethodon*. Chromosoma 48:239-296.

Mizuno, S., Andrews, C., and H. C. Macgregor. 1976. Interspecific "common" repetitive DNA sequences in the salamanders of the genus *Plethodon*. Chromosoma 58:1-31.

Mohr, C. E. 1930. The ambystomatid salamanders of Pennsylvania. Proceedings of the Pennsylvania Academy of Science 4:50-55.

Mohr, C. E. 1931. Observations on the early breeding habits of *Ambystoma jeffersonianum* in central Pennsylvania. Copeia 1931:102-104.

Mohr, C. E. 1943. The eggs of the long-tailed salamander, *Eurycea longicauda* (Green). Proceedings of the Pennsylvania Academy of Science 17:86.

Mohr, C. E. 1944. A remarkable salamander migration. Proceedings of the Pennsylvania Academy of Science 18:51-54.

Mohr, C. E. 1950. Ozark cave life. National Speleological Association Bulletin 12:3-11.

Mohr, C. E. 1952. The eggs of the zig-zag salamander, *Plethodon cinereus dorsalis*. National Speleological Society Bulletin 14:59-60.

Moir, W. H., and H. M. Smith. 1970. Occurrence of an American salamander, *Aneides hardyi* (Taylor), in tundra habitat. Arctic and Alpine Research 2:155-156.

Moler, P. E. 1994. *Siren lacertina* (greater siren). Diet. Herpetological Review 25:62.

Moler, P. E., and J. Kezer. 1993. Karyology and systematics of the salamander genus *Pseudobranchus* (Sirenidae). Copeia 1993:39-47.

Moler, P. E., and B. W. Mansell. 1986. *Pseudobranchus striatus striatus* (broad-striped dwarf siren). Maximum size. Herpetological Review 17:45.

Montague, J. R. 1979. Note on the larval feeding behavior in *Desmognathus fuscus fuscus*, the northern dusky salamander. Copeia 1979:354.

Montague, J. R., and J. W. Poinski. 1978. Note on the brooding behavior in *Desmognathus fuscus fuscus* (Raf.) (Amphibia, Urodela, Plethodontidae) in Columbiana County, Ohio. Journal of Herpetology 12:104.

Montanucci, R. R. 1992. Commentary on a proposed taxonomic arrangement for some North American amphibians and reptiles. Herpetological Review 23:9-10.

Moore, G. A., and R. C. Hughes. 1939. A new plethodontid from eastern Oklahoma. The American Midland Naturalist 22:696-699.

Moore, G. A., and R. C. Hughes. 1941. A new plethodont salamander from Oklahoma. Copeia 1941:139-142.

Moore, J. E., and E. H. Strickland. 1955. Further notes on the food of Alberta amphibians. The American Midland Naturalist 54:253-256.

Moore, R. D., Newton, B., and A. Sih. 1996. Delayed hatching as a response of streamside salamander eggs to chemical cues from predatory fish. Oikos 77:331–335.

Morafka, D. J., and B. H. Banta. 1976. Biogeographical implications of pattern variation in the salamander *Aneides lugubris*. Copeia 1976:580–586.

Moreno, G. 1989. Behavioral and physiological differentiation between color morphs of the salamander, *Plethodon cinereus*. Journal of Herpetology 23:335–341.

Morescalchi, A. 1975. Chromosome evolution in the caudate amphibia. Evolutionary Biology 8:339–387.

Morgan, A. H., and M. C. Grierson. 1932. Winter habits and yearly food consumption of adult spotted newts, *Triturus viridescens*. Ecology 13:54–62.

Morin, P. J. 1981. Predatory salamanders reverse the outcome of competition among three species of anuran tadpoles. Science 212:1284–1286.

Morin, P. J. 1983a. Competitive and predatory interactions in natural and experimental populations of *Notophthalmus viridescens dorsalis* and *Ambystoma tigrinum*. Copeia 1983:628–639.

Morin, P. J. 1983b. Predation, competition, and the composition of larval anuran guilds. Ecological Monographs 53:119–138.

Morin, P. J. 1995. Functional redundancy, non-additive interactions, and supply-side dynamics in experimental pond communities. Ecology 76:133–149.

Morin, P. J., Wilbur, H. M., and R. N. Harris. 1983. Salamander predation and the structure of experimental communities: responses of *Notophthalmus* and microcrustacea. Ecology 64:1430–1436.

Moritz, C., Schneider, C. J., and D. B. Wake. 1992. Evolutionary relationships within the *Ensatina eschscholtzii* complex confirm the ring species interpretation. Systematic Zoology 41:273–291.

Morris, M. A. 1985. A hybrid *Ambystoma platineum* × *A. tigrinum* from Indiana. Herpetologica 41:267–271.

Morris, M. A., and R. A. Brandon. 1984. Gynogenesis and hybridization between *Ambystoma platineum* and *Ambystoma texanum* in Illinois. Copeia 1984:324–337.

Morse, M. 1904. Batrachians and reptiles of Ohio. Ohio State University Bulletin, Series 8(18):96–143.

Mosher, H. S., Fuhrman, F. A., Buchwald, H. D., and H. G. Fischer. 1964. Tarichatoxin-tetrodotoxin: a potent neurotoxin. Science 144:1100–1110.

Mosimann, J. E., and G. B. Rabb. 1948. The salamander *Ambystoma mabeei* in South Carolina. Copeia 1948:304.

Mosimann, J. E., and T. M. Uzzell, Jr. 1952. Description of the eggs of the mole salamander, *Ambystoma talpoideum* (Holbrook). Natural History Miscellanea of the Chicago Academy of Science 104:1–3.

Moulton, J. M. 1954. Notes on the natural history, collection, and maintenance of the salamander *Ambystoma maculatum*. Copeia 1954:64–65.

Mount, R. H. 1975. The Reptiles and Amphibians of Alabama. Auburn Printing, Auburn, Alabama, 345 pp.

Mueller, C. F., and M. Himchak. 1983. *Plethodon cinereus* (redback salamander). Coloration. Herpetological Review 14:72–73.

Murphy, M. L., and J. D. Hall. 1981. Varied effects of clear-cut logging on predators and their habitat in small streams of the Cascade Mountains, Oregon. Canadian Journal of Fisheries and Aquatic Science 38:137–145.

Murphy, M. L., Hawkins, C. P., and N. H. Anderson. 1981. Effects of canopy modification and accumulated sediment on stream communities. Transactions of the American Fisheries Society 110:469–478.

Murphy, T. D. 1961. Predation on eggs of the salamander, *Ambystoma maculatum*, by caddis fly larvae. Copeia 1961:495–496.

Myers, C. W. 1958a. Amphibia in Missouri caves. Herpetologica 14:35–36.

Myers, C. W. 1958b. Notes on the eggs and larvae of *Eurycea lucifuga* Rafinesque. Journal of the Florida Academy of Science 21:125–130.

Myers, G. S. 1942. Notes on Pacific Coast *Triturus*. Copeia 1942:77–82.

Myers, G. S. 1943. Notes on *Rhyacotriton olympicus* and *Ascaphus truei* in Humboldt County, California. Copeia 1943:125–126.

Myers, G. S., and T. P. Maslin, Jr. 1948. The California plethodont salamander, *Aneides flavipunctatus* (Strauch), with descriptions of a new subspecies and notes on other western *Aneides*. Proceedings of the Biological Society of Washington 61:127–138.

Nagel, J. W. 1977. Life history of the red-backed

salamander, *Plethodon cinereus*, in northeastern Tennessee. Herpetologica 33:13–18.

Nagel, J. W. 1979. Life history of the ravine salamander (*Plethodon richmondi*) in northeastern Tennessee. Herpetologica 35:38–43.

Neill, W. T. 1947. A collection of amphibians from Georgia. Copeia 1947:271–272.

Neill, W. T. 1948. Salamanders of the genus *Pseudotriton* from Georgia and South Carolina. Copeia 1948:134–136.

Neill, W. T. 1949. Juveniles of *Siren lacertina* and *S. i. intermedia*. Herpetologica 5:19.

Neill, W. T. 1950. A new species of salamander, genus *Desmognathus*, from Georgia. Publications of the Research Division of Ross Allen's Reptile Institute 1:1–6.

Neill, W. T. 1952. Remarks on salamander voices. Copeia 1952:195–196.

Neill, W. T. 1955. Posture of chilled newts (*Diemyctylus viridescens louisianensis*). Copeia 1955:61.

Neill, W. T. 1963. Notes on the Alabama waterdog, *Necturus alabamensis* Viosca. Herpetologica 19:166–174.

Neill, W. T. 1964. A new species of salamander, Genus *Amphiuma*, from Florida. Herpetologica 20:62–66.

Neill, W. T., and F. L. Rose. 1949. Nest and eggs of the southern dusky salamander, *Desmognathus fuscus auriculatus*. Copeia 1949:234.

Neish, I. C. 1971. Comparison of size, structure, and distributional patterns of two salamander populations in Marion Lake, British Columbia. Journal of the Fisheries Research Board of Canada 28:49–58.

Netting, M. G. 1929. The food of the hellbender *Cryptobranchus alleganiensis* (Daudin). Copeia 1929:23–24.

Netting, M. G. 1939. The ravine salamander, *Plethodon richmondi* Netting and Mittleman, in Pennsylvania. Proceedings of the Pennsylvania Academy of Science 13:50–51.

Netting, M. G., and C. J. Goin. 1942. Descriptions of two new salamanders from peninsular Florida. Annals of the Carnegie Museum 29:175–196.

Netting, M. G., and M. B. Mittleman. 1938. Description of *Plethodon richmondi*, a new salamander from West Virginia and Ohio. Annals of the Carnegie Museum 27:287–293.

Netting, M. G., and N. Richmond. 1932. The green salamander, *Aneides aeneus*, in northern West Virginia. Copeia 1932:101–102.

Netting, M. G., Green, N. B., and N. D. Richmond. 1946. The occurrence of Wehrle's salamander, *Plethodon wehrlei* Fowler and Dunn, in Virginia. Proceedings of the Biological Society of Washington 59:157–160.

Newman, W. B. 1954. A new plethodontid salamander from southwestern Virginia. Herpetologica 10:9–14.

Ng, M. Y., and H. M. Wilbur. 1995. The cost of brooding in *Plethodon cinereus*. Herpetologica 51:1–8.

Nicholas, J. S. 1925. A balancer in larvae of *Amblystoma tigrinum*. The American Naturalist 59:191–192.

Nicholls, J. T. 1949. A new salamander of the genus *Desmognathus* from east Tennessee. Journal of the Tennessee Academy of Science 24:127–129.

Nickerson, M. A., and C. E. Mays. 1973a. The hellbenders: North American giant salamanders. Milwaukee Public Museum Publications in Biology and Geology 1. 106 pp.

Nickerson, M. A., and C. E. Mays. 1973b. A study of the Ozark hellbender *Cryptobranchus alleganiensis bishopi*. Ecology 54:1164–1165.

Nickerson, M. A., and M. D. Tohulka. 1986. The nests and nest site selection by Ozark hellbenders, *Cryptobranchus alleganiensis bishopi* Grobman. Transactions of the Kansas Academy of Science 89:66–69.

Nickerson, M. A., Ashton, R. E., and A. L. Braswell. 1983. Lampreys in the diet of hellbender *Cryptobranchus alleganiensis* (Daudin), and the Neuse River waterdog *Necturus lewisi* (Brimley). Herpetological Review 14:10.

Nigrelli, R. F. 1954. Some longevity records for vertebrates. Transactions of the New York Academy of Science 16:296–299.

Nishikawa, K. C. 1985. Competition and the evolution of aggressive behavior in two species of terrestrial salamanders. Evolution 39:1282–1294.

Nishikawa, K. C. 1990. Intraspecific spatial relationships of two species of terrestrial salamanders. Copeia 1990:418–426.

Noble, G. K. 1925. An outline of the relation of ontogeny to phylogeny within the Amphibia. II. American Museum Novitates 166:1–10.

Noble, G. K. 1926. The Long Island newt: a contribution to the life history of *Triturus viridescens*. American Museum Novitates 228:1–11.

Noble, G. K. 1927a. The plethodontid salamanders; some aspects of their evolution. American Museum Novitates 249:1–26.

Noble, G. K. 1927b. The value of life history data in the study of the evolution of the amphibia. Annals of the New York Academy of Science 30:31–128.

Noble, G. K. 1929a. The relationship of courtship to the secondary sexual characteristics of the two-lined salamander, *Eurycea bislineata* (Green). American Museum Novitates 362:1–5.

Noble, G. K. 1929b. Further observations on the life-history of the newt, *Triturus viridescens*. American Museum Novitates 348:1–22.

Noble, G. K. 1930. The eggs of *Pseudobranchus*. Copeia 1930:52.

Noble, G. K. 1931. The Biology of the Amphibia. McGraw-Hill, New York. 557 pp.

Noble, G. K., and M. K. Brady. 1930. The courtship of the plethodontid salamanders. Copeia 1930:52–54.

Noble, G. K., and M. K. Brady. 1933. Observations on the life history of the marbled salamander, *Ambystoma opacum*. Zoologica 11:89–132.

Noble, G. K., and G. Evans. 1932. Observations and experiments on the life history of the salamander, *Desmognathus fuscus fuscus* (Rafinesque). American Museum Novitates 533:1–16.

Noble, G. K., and B. C. Marshall. 1929. The breeding habits of two salamanders. American Museum Novitates 347:1–12.

Noble, G. K., and B. C. Marshall. 1932. The validity of *Siren intermedia* LeConte, with observations on its life history. American Museum Novitates 532:1–17.

Noble, G. K., and L. B. Richards. 1932. Experiments on the egg-laying of salamanders. American Museum Novitates 513:1–25.

Noble, G. K., and J. A. Weber. 1929. The spermatophores of *Desmognathus* and other plethodontid salamanders. American Museum Novitates 351:1–15.

Noeske, T. A., and M. A. Nickerson. 1979. Diel activity rhythms in the hellbender, *Cryptobranchus alleganiensis* (Caudata: Cryptobranchidae). Copeia 1979:92–95.

Norman, B. R. 1986. *Ensatina eschscholtzii oregonensis* (Oregon ensatina). Reproduction. Herpetological Review 17:89.

Norman, B. R., and M. Swartwood. 1991. *Plethodon vehiculum* (western red-backed salamander). Reproduction. Herpetological Review 22:55.

Norman, C. E., and B. R. Norman. 1980. Notes on the egg clusters and hatchlings of *Ensatina eschscholtzii oregonensis*. Bulletin of the Chicago Herpetological Society 15:99–100.

Norman, W. W. 1900. Remarks on the San Marcos salamander, *Typhlomolge rathbuni* Stejneger. The American Naturalist 34:179–183.

Norris, D. O. 1989. Seasonal changes in diet of paedogenetic tiger salamanders (*Ambystoma tigrinum mavortium*). Journal of Herpetology 23:87–89.

Nunes, V. da S. 1988. Feeding asymmetry affects territorial disputes between males of *Plethodon cinereus*. Herpetologica 44:386–391.

Nunes, V. da S., and R. G. Jaeger. 1989. Salamander aggressiveness increases with length of territorial ownership. Copeia 1989:712–718.

Nussbaum, R. A. 1969a. A nest site of the olympic salamander, *Rhyacotriton olympicus* (Gaige). Herpetologica 25:277–278.

Nussbaum, R. A. 1969b. Nests and eggs of the Pacific giant salamander, *Dicamptodon ensatus* (Eschscholtz). Herpetologica 25:257–262.

Nussbaum, R. A. 1970. *Dicamptodon copei*, n. sp., from the Pacific Northwest, U.S.A. (Amphibia: Caudata: Ambystomatidae). Copeia 1970:506–514.

Nussbaum, R. A. 1976. Geographic variation and systematics of salamanders of the genus *Dicamptodon* Strauch (Ambystomatidae). Miscellaneous Publications of the Museum of Zoology, University of Michigan 149:1–94.

Nussbaum, R. A. 1983. *Dicamptodon copei* Nussbaum. Catalogue of American Amphibians and Reptiles, pp. 334.1–334.2.

Nussbaum, R. A., and E. D. Brodie, Jr. 1971. The taxonomic status of the rough-skinned newt, *Taricha granulosa* (Skilton), in the Rocky Mountains. Herpetologica 27:260–270.

Nussbaum, R. A., and G. W. Clothier. 1973. Population structure, growth, and size of larval *Dicamptodon ensatus* (Eschscholtz). Northwest Science 47:218–227.

Nussbaum, R. A., and C. Maser. 1969. Observations of *Sorex palustris* preying on *Dicamptodon ensatus*. The Murrelet 50:23.

Nussbaum, R. A., and C. K. Tait. 1977. Aspects of the life history and ecology of the Olympic salamander, *Rhyacotriton olympicus* (Gaige). The American Midland Naturalist 98:176-199.

Nussbaum, R. A., Brodie, E. D., Jr., and R. M. Storm. 1983. Amphibians and Reptiles of the Pacific Northwest. University Press of Idaho, Moscow, Idaho, 332 pp.

Nyman, S. 1987. *Ambystoma maculatum* (spotted salamander). Reproduction. Herpetological Review 18:15.

Nyman, S. 1991. Ecological aspects of syntopic larvae of *Ambystoma maculatum* and the *A. laterale-jeffersonianum* complex in two New Jersey ponds. Journal of Herpetology 25:505-509.

Nyman, S., Ryan, M. J., and J. D. Anderson. 1988. The distribution of the *Ambystoma jeffersonianum* complex in New Jersey. Journal of Herpetology 22:224-228.

Nyman, S., Wilkinson, R. F., and J. E. Hutcherson. 1993. Cannibalism and size relations in a cohort of larval ringed salamanders (*Ambystoma annulatum*). Journal of Herpetology 27:78-84.

O'Donnell, J. D. 1937. Natural history of the ambystomatid salamanders of Illinois. The American Midland Naturalist 18:1063-1071.

Oliver, G. V., Jr. 1967. Food habits of the white-throated slimy salamander in central Texas. Transactions of the Oklahoma Junior Academy of Science 1967:500-503.

Oliver, M. G., and H. M. McCurdy. 1974. Migration, overwintering, and reproductive patterns of *Taricha granulosa* on southern Vancouver Island. Canadian Journal of Zoology 52:541-545.

Organ, J. A. 1958. Courtship and spermatophore of *Plethodon jordani metcalfi*. Copeia 1958:251-259.

Organ, J. A. 1960a. The courtship and spermatophore of the salamander *Plethodon glutinosus*. Copeia 1960:34-40.

Organ, J. A. 1960b. Studies on the life history of the salamander, *Plethodon welleri*. Copeia 1960:287-297.

Organ, J. A. 1961a. Studies of the local distribution, life history, and population dynamics of the salamander genus *Desmognathus* in Virginia. Ecological Monographs 31:189-220.

Organ, J. A. 1961b. Life history of the pygmy salamander, *Desmognathus wrighti*, in Virginia. The American Midland Naturalist 66:384-390.

Organ, J. A. 1961c. The eggs and young of the spring salamander, *Pseudotriton porphyriticus*. Herpetologica 17:53-56.

Organ, J. A. 1968a. Courtship behavior and spermatophore of the cave salamander, *Eurycea lucifuga* (Rafinesque). Copeia 1968:576-580.

Organ, J. A. 1968b. Time of courtship activity of the slimy salamander, *Plethodon glutinosus*, in New Jersey. Herpetologica 24:84-85.

Organ, J. A., and L. A. Lowenthal. 1963. Comparative studies of macroscopic and microscopic features of spermatophores of some plethodontid salamanders. Copeia 1963:659-669.

Organ, J. A., and D. J. Organ. 1968. Courtship behavior of the red salamander, *Pseudotriton ruber*. Copeia 1968:217-223.

Orr, L. P. 1967. Feeding experiments with a supposed mimetic complex in salamanders. The American Midland Naturalist 77:147-155.

Orr, L. P. 1989. *Desmognathus ochrophaeus*. In: Salamanders of Ohio, R. A. Pfingsten and F. L. Downs, eds. Ohio Biological Survey Bulletin, New Series 7(2):181-189.

Orr, L. P., and W. T. Maple. 1978. Competition avoidance mechanisms in salamander larvae of the genus *Desmognathus*. Copeia 1978:679-685.

Orser, P. N., and D. J. Shure. 1972. Effects of urbanization on the salamander *Desmognathus fuscus fuscus*. Ecology 53:1148-1154.

Orser, P. N., and D. J. Shure. 1975. Population cycles and activity patterns of the dusky salamander, *Desmognathus fuscus fuscus*. The American Midland Naturalist 93:403-410.

Orton, G. 1942. Noted on the larvae of certain species of *Ambystoma*. Copeia 1942:170-172.

Ovaska, K. 1987. Seasonal changes in agonistic behaviour of the western red-backed salamander, *Plethodon vehiculum*. Animal Behaviour 35:67-74.

Ovaska, K. 1988a. Recognition of conspecific odours by the western red-backed salamander, *Plethodon vehiculum*. Canadian Journal of Zoology 66:1293-1296.

Ovaska, K. 1988b. Spacing and movements of the salamander *Plethodon vehiculum*. Herpetologica 44:377-386.

Ovaska, K. 1989. Pheromonal divergence between populations of the salamander *Plethodon vehiculum* in British Columbia. Copeia 1989:770–775.

Ovaska, K., and T. M. Davis. 1992. Faecal pellets as burrow markers: intra- and interspecific odour recognition by western plethodontid salamanders. Animal Behaviour 43:931–939.

Ovaska, K., and P. T. Gregory. 1989. Population structure, growth, and reproduction in a Vancouver Island population of the salamander *Plethodon vehiculum*. Herpetologica 45:133–143.

Ovaska, K., and M. A. Smith. 1988. Predatory behavior of two species of ground beetles (Coleoptera: Carabidae) towards juvenile salamanders (*Plethodon vehiculum*). Canadian Journal of Zoology 66:599–604.

Packer, W. C. 1960. Bioclimatic influences on the breeding migration of *Taricha rivularis*. Ecology 41:509–517.

Packer, W. C. 1961. Feeding behavior in adult *Taricha*. Copeia 1961:351–352.

Packer, W. C. 1962. Aquatic homing behavior in *Taricha rivularis*. Copeia 1962:207–208.

Packer, W. C. 1963. Observations on the breeding migration of *Taricha rivularis*. Copeia 1963:378–382.

Palis, J. G. 1995. Larval growth, development, and metamorphosis of *Ambystoma cingulatum* on the Gulf Coastal Plain of Florida. The Florida Scientist 1995:352–358.

Palis, J. G. 1996a. Flatwoods salamander (*Ambystoma cingulatum* Cope). Element stewardship abstract. Natural Areas Journal 16:49–54.

Palis, J. G. 1996b. *Ambystoma opacum* (marbled salamander). Communal nesting. Herpetological Review 27:134.

Palis, J. G. 1997. Breeding migration of *Ambystoma cingulatum* in Florida. Journal of Herpetology 31:71–78.

Parham, J. F., Dodd, C. K., Jr., and G. R. Zug. 1996. Skeletochronological age estimates for the Red Hills Salamander, *Phaeognathus hubrichti*. Journal of Herpetology 30:401–404.

Parker, M. S. 1991. Relationship between cover availability and larval pacific giant salamander density. Journal of Herpetology 25:355–357.

Parker, M. S. 1993. Size-selective predation on benthic macroinvertebrates by stream-dwelling salamander larvae. Archives Hydrobiologica 128:385–400.

Parker, M. S. 1994. Feeding ecology of stream-dwelling Pacific giant salamander larvae (*Dicamptodon tenebrosus*). Copeia 1994:705–718.

Parker, M. V. 1937. Some amphibians and reptiles from Reelfoot Lake. Journal of the Tennessee Academy of Science 12:60–87.

Parmelee, J. R. 1993. Microhabitat segregation and spatial relationships among four species of mole salamander (genus *Ambystoma*). Occasional Papers of the Museum of Natural History, University of Kansas 160:1–33.

C. L. 1922. Some amphibians and reptiles from British Columbia. Copeia 1922:74–79.

Patterson, K. K. 1978. Life history aspects of paedogenic populations of the mole salamander, *Ambystoma talpoideum*. Copeia 1978:649–655.

Pauley, T. K. 1978a. Moisture as a factor regulating habitat partitioning between two sympatric *Plethodon* (Amphibia, Urodela, Plethodontidae) species. Journal of Herpetology 12:491–493.

Pauley, T. K. 1978b. Food types and distribution as a *Plethodon* habitat partitioning factor. Bulletin of the Maryland Herpetological Society 14:79–82.

Pauley, T. K. 1980. Field notes on the distribution of terrestrial amphibians and reptiles of the West Virginia Mountains above 975 meters. Proceedings of the West Virginia Academy of Science 52:84–92.

Pauley, T. K. 1981. The range and distribution of the Cheat Mountain Salamander, *Plethodon nettingi*. Proceedings of the West Virginia Academy of Science 53:31–35.

Pauley, T. K. 1993. Amphibians and reptiles of the upland forests. In: Upland Forests of West Virginia, S. L. Stephenson, ed. McClain Printing, Parsons, West Virginia, pp. 179–196.

Pauley, T. K., and W. H. England. 1969. Time of mating and egg deposition in the salamander, *Plethodon wehrlei* Fowler and Dunn, in West Virginia. Proceedings of the West Virginia Academy of Science 41:155–160.

Pawling, R. O. 1939. The amphibians and reptiles of Union County, Pennsylvania. Herpetologica 1:165–170.

Peacock, R. L., and R. A. Nussbaum. 1973. Reproductive biology and population structure of the western red-backed salamander, *Plethodon vehiculum* (Cooper). Journal of Herpetology 7:215–224.

Pearse, A. S. 1921. Habits of the mud-puppy, *Necturus,* an enemy of food fishes. Bureau of Fisheries Economic Circular 49:1-8.

Pechmann, J. H. K. 1995. Use of large field enclosures to study the terrestrial ecology of pond-breeding amphibians. Herpetologica 51:434-450.

Pechmann, J. H. K., and H. Wilbur. 1994. Putting declining amphibian populations in perspective: natural fluctuations and human impacts. Herpetologica 50:65-84.

Pechmann, J. H. K., Scott, D. E., Semlitsch, R. D., Caldwell, J. P., Vitt, L. J., and J. W. Gibbons. 1991. Declining amphibian populations: the problem of separating human impacts from natural fluctuations. Science 253:892-895.

Peck, S. B. 1973. Feeding efficiency in the cave salamander *Haideotriton wallacei.* International Journal of Speleology 5:15-19.

Peck, S. B. 1974. The food of the salamanders *Eurycea lucifuga* and *Plethodon glutinosus* in caves. National Speleological Society Bulletin 36:7-10.

Peck, S. B., and B. L. Richardson. 1976. Feeding ecology of the salamander *Eurycea lucifuga* in the entrance, twilight zone, and dark zone of caves. Annales de Speleologie 31:175-182.

Peckham, R. S., and C. F. Dineen. 1954. Spring migrations of salamanders. Proceedings of the Indiana Academy of Science 64:278-280.

Pedersen, S. C. 1993. Skull growth in cannibalistic tiger salamanders, *Ambystoma tigrinum.* The Southwestern Naturalist 38:316-324.

Peterson, C. L. 1987. Movement and catchability of the hellbender, *Cryptobranchus alleganiensis.* Journal of Herpetology 21:197-204.

Peterson, C. L. 1988. Breeding activities of the hellbender in Missouri. Herpetological Review 19:28-29.

Peterson, C. L. and R. F. Wilkinson. 1996. Home range size of the hellbender (*Cryptobranchus alleganiensis*) in Missouri. Herpetological Review 27:126-127.

Peterson, C. L., Ingersol, C. A., and R. F. Wilkinson. 1989a. Winter breeding of *Cryptobranchus alleganiensis bishopi* in Arkansas. Copeia 1989:1031-1035.

Peterson, C. L., Metter, D. E., Miller, B. T., Wilkinson, R. F., and M. S. Topping. 1988. Demography of the hellbender *Cryptobranchus alleganiensis* in the Ozarks. The American Midland Naturalist 119:291-303.

Peterson, C. L., Reed, J. W., and R. F. Wilkinson. 1989b. Seasonal food habits of *Cryptobranchus alleganiensis* (Caudata: Cryptobranchidae). The Southwestern Naturalist 34:438-441.

Peterson, C. L., Wilkinson, R. F., Moll, D., and T. Holder. 1991. Premetamorphic survival of *Ambystoma annulatum.* Herpetologica 47:96-100.

Peterson, C. L., Wilkinson, R. F., Moll, D., and T. Holder. 1992. Estimating the number of female *Ambystoma annulatum* (Caudata: Ambystomatidae) based on oviposition. The Southwestern Naturalist 37:425-426

Peterson, C. L., Wilkinson, R. F., Jr., Topping, M. S., and D. E. Metter. 1983. Age and growth of the Ozark hellbender (*Cryptobranchus alleganiensis bishopi*). Copeia 1983:225-231.

Peterson, J. A., and A. R. Blaustein. 1991. Unpalatability in anuran larvae as a defense against natural salamander predators. Ethology, Ecology and Evolution 3:63-72.

Petranka, J. W. 1979. The effects of severe winter weather on *Plethodon dorsalis* and *Plethodon richmondi* populations in central Kentucky. Journal of Herpetology 13:369-371.

Petranka, J. W. 1982a. Geographic variation in the mode of reproduction and larval characteristics of the small-mouthed salamander in the east-central United States. Herpetologica 38:252-262.

Petranka, J. W. 1982b. Courtship behavior of the small-mouthed salamander (*Ambystoma texanum*) in central Kentucky. Herpetologica 38:333-336.

Petranka, J. W. 1983. Fish predation: a factor affecting the spatial distribution of a stream-dwelling salamander. Copeia 1983:624-628.

Petranka, J. W. 1984a. Breeding migrations, breeding season, clutch size, and oviposition of stream-breeding *Ambystoma texanum.* Journal of Herpetology 18:106-112.

Petranka, J. W. 1984b. Incubation, larval growth, and embryonic and larval survivorship of small-mouthed salamanders (*Ambystoma texanum*) in streams. Copeia 1982:862-868.

Petranka, J. W. 1984c. Sources of interpopulational variation in growth responses of larval salamanders. Ecology 65:1857-1865.

Petranka, J. W. 1984d. Ontogeny of the diet and feeding behavior of *Eurycea bislineata* larvae. Journal of Herpetology 18:48-55.

Petranka, J. W. 1987. *Notophthalmus viridescens dor-*

salis (broken-striped newt). Behavior. Herpetological Review 18:72-73.

Petranka, J. W. 1989. Density-dependent growth and survival of larval *Ambystoma*: evidence from whole-pond manipulations. Ecology 70:1752-1767.

Petranka, J. W. 1990. Observations on nest site selection, nest desertion, and embryonic survival in marbled salamanders. Journal of Herpetology 24:229-234.

Petranka, J. W., and J. G. Petranka. 1980. Selected aspects of the larval ecology of the marbled salamander *Ambystoma opacum* in the southern portion of its range. The American Midland Naturalist 104:352-363.

Petranka, J. W., and J. G. Petranka. 1981a. Notes on the nesting biology of the marbled salamander, *Ambystoma opacum*, in the southern portion of its range. Alabama Academy of Science 52:20-24.

Petranka, J. W., and J. G. Petranka. 1981b. On the evolution of nest site selection in the marbled salamander, *Ambystoma opacum*. Copeia 1981:387-391.

Petranka, J. W., and A. Sih. 1986. Environmental instability, competition, and density-dependent growth and survivorship of a stream-dwelling salamander. Ecology 67:729-736.

Petranka, J. W., and A. Sih. 1987. Habitat duration, length of larval period, and the evolution of a complex life cycle of a salamander, *Ambystoma texanum*. Evolution 41:1347-1356.

Petranka, J. W., Brannon, M. P., Hopey, M. E., and C. K. Smith. 1994. Effects of timber harvesting on low elevation populations of southern Appalachian salamanders. Forest Ecology and Management 67:135-147.

Petranka, J. W., Eldridge, M. E., and K. E. Haley. 1993. Effects of timber harvesting on southern Appalachian salamanders. Conservation Biology 7:363-370.

Petranka, J. W., Just, J. J., and E. C. Crawford, Jr. 1982. Hatching of amphibian embryos: the physiological trigger. Science 217:257-259.

Petranka, J. W., Kats, L. B., and A. Sih. 1987b. Predator-prey interactions among fish and larval amphibians: use of chemical cues to detect predatory fish. Animal Behaviour 35:420-425.

Petranka, J. W., Rushlow, A. W., and M. E. Hopey. 1998. Predation by tadpoles of *Rana sylvatica* on embryos of *Ambystoma maculatum*: implications of ecological role reversals by *Rana* (predator) and *Ambystoma* (prey). Herpetologica 54:1-13.

Petranka, J. W., Sih, A., Kats, L. B., and J. R. Holomuzki. 1987a. Stream drift, size-specific predation, and the evolution of ovum size in an amphibian. Oecologia 71:624-630.

Pfennig, D. W., and J. P. Collins. 1993. Kinship affects morphogenesis in cannibalistic salamanders. Nature 262:836-838.

Pfennig, D. W., Loeb, M. L. G., and J. P. Collins. 1991. Pathogens as a factor limiting the spread of cannibalism in tiger salamanders. Oecologia 88:161-166.

Pfennig, D. W., Sherman, P. W., and J. P. Collins. 1994. Kin recognition and cannibalism in polyphenic salamanders. Behavioral Ecology 5:225-232.

Pfingsten, R. A. 1969. An erythristic population of *Plethodon cinereus* in Ohio. Journal of Herpetology 3:104-105.

Pfingsten, R. A. 1989a. Genus *Plethodon*. In: Salamanders of Ohio, R. A. Pfingsten and F. L. Downs, eds. Ohio Biological Survey Bulletin, New Series 7(2):229-264.

Pfingsten, R. A. 1989b. *Pseudotriton ruber*. In: Salamanders of Ohio, R. A. Pfingsten and F. L. Downs, eds. Ohio Biological Survey Bulletin, New Series 7(2):269-275.

Pfingsten, R. A. 1990. The status and distribution of the hellbender, *Cryptobranchus alleganiensis* in Ohio. Herpetological Review 21:48-51.

Pfingsten, R. A., and F. L. Downs, eds. 1989. Salamanders of Ohio. Ohio Biological Survey Bulletin, New Series 7(2), 315 pp.

Pfingsten, R. A., and C. F. Walker. 1978. Some nearly all black populations of *Plethodon cinereus* (Amphibia, Urodela, Plethodontidae) in northern Ohio. Journal of Herpetology 12:163-167.

Pfingsten, R. A., and A. M. White. 1989. *Necturus maculosus*. In: Salamanders of Ohio, R. A. Pfingsten and F. L. Downs, eds. Ohio Biological Survey Bulletin, New Series 7(2):72-77.

Phillips, C. A. 1992. Variation in metamorphosis in spotted salamanders *Ambystoma maculatum* from eastern Missouri. The American Midland Naturalist 128:276-280.

Phillips, C. A. 1994. Geographic distribution of mitochondrial DNA variants and the historical biogeography of the spotted salamander, *Ambystoma maculatum*. Evolution 48:597-607.

Phillips, C. A., and O. L. Sexton. 1989. Orientation and sexual differences during breeding migrations of the spotted salamander, *Ambystoma maculatum*. Copeia 1989:17–22.

Phillips, J. B. 1985a. Two magnetoreception pathways in a migratory salamander. Science 223:765–767.

Phillips, J. B. 1985b. Magnetic compass orientation in the eastern red-spotted newt (*Notophthalmus viridescens*). Journal of Comparative Physiology 158:103–109.

Phillips, J. B. 1986. Two magnetoreception pathways in a migratory salamander. Science 233:765–767.

Phillips, J. B. 1987. Laboratory studies of homing orientation in the eastern red-spotted newt, *Notophthalmus viridescens*. Journal of Experimental Biology 131:215–229.

Phillips, J. B., and S. C. Borland. 1992. Behavioral evidence for the use of a light-dependent magnetoreception mechanism by a vertebrate. Nature 359:142–144.

Phillips, J. B., and S. C. Borland. 1994. Use of a specialized magnetoreception system for homing by the eastern newt *Notophthalmus viridescens*. Journal of Experimental Biology 188:275–291.

Phillips, J. B., Adler, K., and S. C. Borland. 1995. True navigation by an amphibian. Animal Behaviour 50:855–858.

Piatt, J. 1931. An albino salamander. Copeia 1931:29.

Pierce, B. A. 1985. Acid tolerance in amphibians. BioScience 35:239–243.

Pierce, B. A., and J. M. Harvey. 1987. Geographic variation in acid tolerance of Connecticut wood frogs. Copeia 1987:94–103.

Pierce, B. A., and J. B. Mitton. 1980. Patterns of allozyme variation in *Ambystoma tigrinum malvortium* and *A. t. nebulosum*. Copeia 1980:594–605.

Pierce, B. A., and J. B. Mitton. 1982. Allozyme heterozygosity and growth in the tiger salamander, *Ambystoma tigrinum*. Journal of Heredity 73:250–253.

Pierce, B. A., and H. M. Smith. 1979. Neoteny or paedogenesis? Journal of Herpetology 13:119–121.

Piersol, W. H. 1910a. Spawn and larva of *Ambystoma jeffersonianum*. The American Naturalist 44:732–736.

Piersol, W. H. 1910b. The habits and larval state of *Plethodon cinereus erythronotus*. Transactions of the Royal Canadian Institute 8:469–492.

Piersol, W. H. 1914. The egg-laying habits of *Plethodon cinereus*. Transactions of the Royal Canadian Institute 10:121–126.

Pike, N. 1886. Notes on the life history of *Ambystoma opacum*. Bulletin of the American Museum of Natural History 7:209–212.

Pimentel, R. A. 1958. On the validity of *Taricha granulosa* Skilton. Herpetologica 14:165–168.

Pimentel, R. A. 1960. Inter- and intrahabitat movements of the rough-skinned newt, *Taricha torosa granulosa* (Skilton). The American Midland Naturalist 63:470–496.

Pinder, A. W., and S. C. Friet. 1994. Oxygen transport in egg masses of the amphibians *Rana sylvatica* and *Ambystoma maculatum*: convection, diffusion, and oxygen production by algae. Journal of Experimental Biology 197:17–30.

Pitkin, R. B., and S. G. Tilley. 1982. An unusual aggregate of adult *Notophthalmus viridescens*. Copeia 1982:185–186.

Plummer, M. V. 1977. Observations on breeding migrations of *Ambystoma texanum*. Herpetological Review 8:79–80.

Pope, C. H. 1924. Notes on North Carolina salamanders, with especial reference to the egg-laying habits of *Leurognathus* and *Desmognathus*. American Museum Novitates 153:1–15.

Pope, C. H. 1928. Some plethodontid salamanders from North Carolina and Kentucky with the description of a new race of *Leurognathus*. American Museum Novitates 306:1–19.

Pope, C. H. 1944. Amphibians and Reptiles of the Chicago Area. Chicago Natural History Museum Press, Chicago, Illinois, 275 pp.

Pope, C. H. 1950. A statistical and ecological study of the salamander *Plethodon yonahlossee*. Bulletin of the Chicago Academy of Science 9:79–106.

Pope, C. H. 1964. *Plethodon caddoensis*. Catalogue of American Amphibians and Reptiles, p. 14.

Pope, C. H., and J. A. Fowler. 1949. A new species of salamander (*Plethodon*) from southwestern Virginia. Natural History Miscellanea, Chicago Academy of Science 47:1–4.

Pope, C. H., and N. G. Hairston. 1947. The distribution of *Leurognathus* a southern Appalachian genus of salamanders. Fieldiana Zoology 31:155–162.

Pope, C. H., and S. H. Pope. 1949. Notes on

growth and reproduction of the slimy salamander *Plethodon glutinosus*. Fieldiana Zoology 31:251–261.

Pope, C. H., and S. H. Pope. 1951. A study of the salamander *Plethodon ouachitae* and the description of an allied form. Bulletin of the Chicago Academy of Science 9:129–152.

Pope, M. H., and R. Highton. 1980. Geographic genetic variation in the Sacramento Mountain salamander, *Aneides hardii*. Journal of Herpetology 14:343–346.

Pope, P. H. 1924. The life-history of the common water newt, *Notophthalmus viridescens*, together with observations on the sense of smell. Annals of the Carnegie Museum 15:305–368.

Pope, P. H. 1928. The longevity of *Ambystoma maculatum* in captivity. Copeia 1928:99–100.

Pope, P. H. 1937. Notes on the longevity of an *Ambystoma* in captivity. Copeia 1937:140–141.

Potter, F. E., Jr., and S. S. Sweet. 1981. Generic boundaries in Texas cave salamanders, and a redescription of *Typhlomolge robusta* (Amphibia: Plethodontidae). Copeia 1981:64–75.

Pough, F. H. 1974. Comments on the presumed mimicry of red efts (*Notophthalmus*) by red salamanders (*Pseudotriton*). Herpetologica 30:24–27.

Pough, F. H. 1976. Acid precipitation and embryonic mortality in spotted salamanders, *Ambystoma maculatum*. Science 192:68–70.

Pough, F. H., and R. E. Wilson 1970. Natural daily temperature stress, dehydration, and acclimation in juvenile *Ambystoma maculatum* (Shaw) (Amphibia: Caudata). Physiological Zoology 43:194–205.

Pough, F. H., and R. E. Wilson. 1977. Acid precipitation and reproductive success of *Ambystoma* salamanders. Water Air and Soil Pollution 7:307–316.

Pough, H. F., Smith, E. M., Rhodes, D. H., and A. Collazo. 1987. The abundance of salamanders in forest stands with different histories of disturbance. Forest Ecology and Management 20:1–9.

Pounds, J. A., and M. L. Crump. 1994. Amphibian declines and climate disturbance: the case of the golden toad and the harlequin frog. Conservation Biology 8:72–85.

Powders, V. N. 1973. Cannibalism by the slimy salamander, *Plethodon glutinosus* in eastern Tennessee. Journal of Herpetology 7:139–140.

Powders, V. N., and R. Cate. 1980. Food of the dwarf salamander, *Eurycea quadridigitata* in Georgia. Journal of Herpetology 14:81–82.

Powders, V. N., and W. L. Tietjen. 1974. The comparative food habits of sympatric and allopatric salamanders, *Plethodon glutinosus* and *Plethodon jordani* in eastern Tennessee and adjacent areas. Herpetologica 30:167–175.

Powers, J. H. 1907. Mophological variation and its causes in *Ambystoma tigrinum*. University of Nebraska Studies 7:197–274.

Promislow, D. E. L. 1987. Courtship behavior of a plethodontid salamander, *Desmognathus aeneus*. Journal of Herpetology 21:298–306.

Propper, C. R. 1991. Courtship in the rough-skinned newt *Taricha granulosa*. Animal Behaviour 41:547–557.

Prosser, D. T. 1911. Habits of *Ambystoma tigrinum* at Tolland, Colorado. University of Colorado Studies 8:257–263.

Pylka, J. M., and R. D. Warren. 1958. A population of *Haideotriton* in Florida. Copeia 1958:334–336.

Rabb, G. B. 1956. Some observations on the salamander, *Stereochilus marginatum*. Copeia 1956:119.

Rabb, G. B. 1966. *Stereochilus* and *S. marginatus*. Catalogue of American Amphibians and Reptiles, p. 25.

Ramsey, L. W., and J. W. Forsyth. 1950. Breeding dates for *Ambystoma texanum*. Herpetologica 6:70.

Rand, A. S. 1954. Defensive display in the salamander *Ambystoma jeffersonianum*. Copeia 1954:223–224.

Raphael, M. G. 1988. Long-term trends in abundance of amphibians, reptiles, and mammals in douglas-fir forests of northwestern California. In: Management of amphibians, reptiles, and mammals in North America, R. C. Szaro, K. E. Severson, and D. R. Patton, eds. USDA Forest Service, Rocky Mountain Forest and Range Experiment Station, Fort Collins, Colorado, Technical Report RM-166, pp. 11–22.

Ray, C. 1958. Vital limits and rates of desiccation in salamanders. Ecology 39:75–83.

Raymond, L. R. 1991. Seasonal activity of *Siren intermedia* in northwestern Louisiana (Amphibia: Sirenidae). The Southwestern Naturalist 36:144–147.

Raymond, L. R., and L. M. Hardy. 1990. Demography of a population of *Ambystoma talpoideum* (Caudata: Ambystomatidae) in northwestern Louisiana. Herpetologica 46:371–382.

Raymond, L. R., and L. M. Hardy. 1991. Effects of a

clearcut on a population of mole salamander, *Ambystoma talpoideum,* in an adjacent unaltered forest. Journal of Herpetology 25:513-517.

Reagan, D. P. 1972. Ecology and distribution of the Jemez Mountains salamander, *Plethodon neomexicanus.* Copeia 1972:486-492.

Reagan, N. L. 1984. Courtship behavior of *Eurycea bislineata wilderae* complex. M.S. Thesis, Western Carolina University, Cullowhee, North Carolina.

Redmond, W. H. 1980. Notes on the distribution and ecology of the Black Mountain dusky salamander *Desmognathus welteri* Barbour (Amphibia: Plethodontidae) in Tennessee. Brimleyana 4: 123-131.

Redmond, W. H., and R. L. Jones. 1985. *Plethodon wehrlei.* Herpetological Review 16:31.

Redmond, W. H., and A. F. Scott. 1996. Atlas of amphibians in Tennessee. The Center for Field Biology, Austin Peay State University, Clarksville, Tennessee, Miscellaneous Publication 12, 94 pp.

Reese, A. M. 1933. The fauna of West Virginia caves. Proceedings of the West Virginia Academy of Science 7:39-53.

Reese, R. W., and H. M. Smith. 1951. Pattern neoteny in the salamander *Eurycea lucifuga* Rafinesque. Copeia 1951:243-244.

Reigle, N. J., Jr. 1967. The occurrence of *Necturus* in the deeper waters of Green Bay. Herpetologica 23:232-233.

Reilly, S. M. 1990. Biochemical systematics and evolution of the eastern North American newts, Genus *Notophthalmus* (Caudata: Salamandridae). Herpetologica 46:51-59.

Reilly, S. M., and R. A. Brandon. 1994. Partial paedomorphosis in the Mexican stream ambystomatids and the taxonomic status of the genus *Rhyacosiredon* Dunn. Copeia 1994:656-662.

Reilly, S. M., Lauder, G. V., and J. P. Collins. 1992. Performance consequences of a trophic polymorphism: feeding behavior in typical and cannibal phenotypes of *Ambystoma tigrinum.* Copeia 1992:672-679.

Reinbold, S. L. 1979. Habitat comparisons of two sympatric salamander species of the genus *Plethodon* (Amphibia, Caudata, Plethodontidae). Journal of Herpetology 13:504-506.

Reno, H. W., Gelbach, F. R., and R. A. Turner. 1972. Skin and aestivational cocoon of the aquatic amphibian, *Siren intermedia.* Copeia 1972:625-631.

Resetarits, W. J., Jr. 1991. Ecological interactions among predators in experimental stream communities. Ecology 72:1782-1793.

Resetarits, W. J., Jr. 1995. Competitive asymmetry and coexistence in size-structured populations of brook trout and spring salamanders. Oikos 73:188-198.

Resetarits, W. J., Jr., and H. M. Wilbur. 1989. Choice of oviposition site by *Hyla chrysoscelis*: role of predators and competitors. Ecology 70:220-228.

Richmond, N. D. 1945. Nesting of the two-lined salamander on the Coastal Plain. Copeia 1945:170.

Richmond, N. D. 1952. First record of the green salamander in Pennsylvania, and other range extensions in Pennsylvania, Virginia, and West Virginia. Annals of the Carnegie Museum 32:313-318.

Riemer, W. J. 1958. Variation and systematic relationships within the salamander genus *Taricha.* University of California Publications in Zoology 56:301-390.

Ries, K. M., and E. D. Bellis. 1966. Spring food habits of the red-spotted newt in Pennsylvania. Herpetologica 22:152-155.

Riesecrer, J. M., Anderson, M. T., Chivers, D. P., Wildy, E. L., DeVito, J., Marco, A., Blaustein, A. R., Beatty, J. J., and R. M. Storm. 1996. *Plethodon dunni* (Dunn's salamander). Cannibalism. Herpetological Review 27:194.

Rising, J. D., and F. W. Schueler. 1980. Screech owl eats fish and salamander in winter. Wilson Bulletin 92:250-251.

Ritter, W. E. 1897. The life-history and habits of the Pacific coast newt (*Diemyctylus torosus* Esch.). Proceedings of the California Academy of Science Series 3, 1:73-114.

Ritter, W. E. 1903. Further notes on the habits of *Autodax lugubris.* The American Naturalist 37:883-886.

Ritter, W. E., and L. Miller. 1899. A contribution to the life history of *Autodax lugubris* Hallow., a Californian salamander. The American Naturalist 33:691-704.

Roberts, D. T., Schleser, D. M., and T. L. Jordan. 1995. Notes on the captive husbandry and reproduction of the Texas salamander *Eurycea neotenes* at the Dallas aquarium. Herpetological Review 26:23-25.

Robertson, W. B., and E. L. Tyson. 1950. Herpetological notes from eastern North Carolina. Journal of the Elisha Mitchell Scientific Society 66:130-147.

Robinson, T. S., and K. T. Reichard. 1965. Notes on the breeding biology of the midland mud salamander, *Pseudotriton montanus diastictus*. Journal of the Ohio Herpetological Society 5:29.

Rogers, K. L. 1985. Facultative metamorphosis in a series of high altitude fossil populations of *Ambystoma tigrinum* (Irvingtonian: Alamosa County, Colorado). Copeia 1985:926-932.

Rose, F. L. 1966a. Homing to nests by the salamander *Desmognathus auriculatus*. Copeia 1966:251-253.

Rose, F. L. 1966b. Weight change during starvation in *Amphiuma means*. Herpetologica 22:312-313.

Rose, F. L. 1966c. Reproductive potential of *Amphiuma means*. Copeia 1966:598-599.

Rose, F. L. 1967. Seasonal changes in lipid levels of the salamander *Amphiuma means*. Copeia 1967:662-666.

Rose, F. L., and D. Armentrout. 1976. Adaptive strategies of *Ambystoma tigrinum* Green inhabiting the Llano Estacado of west Texas. Journal of Animal Ecology 45:713-729.

Rose, F. L., and F. M. Bush. 1963. A new species of *Eurycea* (Amphibia: Caudata) from the southeastern United States. Tulane Studies in Zoology 10:121-128.

Rose, F. L., and J. L. Dobie. 1963. *Desmognathus monticola* in the Coastal Plain of Alabama. Copeia 1963:564-565.

Rosen, M. 1971. An erythristic *Plethodon cinereus cinereus* from Saint Foy, Portneuf County, Quebec. The Canadian Field-Naturalist 85:326-327.

Rosenthal, G. M. 1957. The role of moisture and temperature in the local distribution of the plethodontid salamander *Aneides lugubris*. University of California Publications in Zoology 54:371-420.

Rossman, D. A. 1959. Ecosystematic relationships of the salamanders *Desmognathus fuscus auriculatus* Holbrook and *Desmognathus fuscus carri* Neill. Herpetologica 15:149-155.

Rossman, D. A. 1960. Herpetofaunal survey of the Pine Hills area of southern Illinois. Journal of the Florida Academy of Science 22:207-225.

Roudebush, R. E. 1988. A behavioral assay for acid sensitivity in two desmognathine species of salamanders. Herpetologica 44:392-395.

Roudebush, R. E., and D. H. Taylor. 1987a. Chemical communication between two species of desmognathine salamanders. Copeia 1987:744-748.

Roudebush, R. E., and D. H. Taylor. 1987b. Behavioral interactions between two desmognathine salamander species: importance of competition and predation. Ecology 68:1453-1458.

Routman, E. 1993a. Mitochondrial DNA variation in *Cryptobranchus alleganiensis*, a salamander with extremely low allozyme diversity. Copeia 1993:407-416.

Routman, E. 1993b. Population structure and genetic diversity of metamorphic and paedomorphic populations of the tiger salamander, *Ambystoma tigrinum*. Journal of Evolutionary Biology 6:329-357.

Routman, E., Wu, R., and A. R. Templeton. 1994. Parsimony, molecular evolution, and biogeography: the case of the North American giant salamander. Evolution 48:1799-1809.

Rowe, C. L., and W. A. Dunson. 1993. Relationships among biotic parameters and breeding effort by three amphibians in temporary wetlands of central Pennsylvania. Wetlands 13:237-246.

Rowe, C. L., and W. A. Dunson. 1995. Impacts of hydroperiod on growth and survival of larval amphibians in temporary ponds of central Pennsylvania. Oecologia 102:397-403.

Rowe, C. L., Sadinski, W. J., and W. A. Dunson. 1994. Predation on larval and embryonic amphibians by acid-tolerant caddisfly larvae (*Ptilostomis postica*). Journal of Herpetology 28:357-364.

Rubin, D. 1963. An albino two-lined salamander. Herpetologica 19:72.

Rubin, D. 1969. Food habits of *Plethodon longicrus* Adler and Dennis. Herpetologica 25:102-105.

Rubin, D. 1971. *Desmognathus aeneus* and *D. wrighti* on Wayah Bald. Journal of Herpetology 5:66-67.

Rudolph, D. C. 1978. Aspects of the larval ecology of five plethodontid salamanders of the western Ozarks. The American Midland Naturalist 100:141-159.

Ruth, B. C., Dunson, W. A., Rowe, C. L., and S. B. Blair. 1993. A molecular and functional evaluation of the egg mass color polymorphism of the spotted salamander, *Ambystoma maculatum*. Journal of Herpetology 27:306–314.

Saber, P. S., and W. A. Dunson. 1978. Toxicity of bog water to embryonic and larval anuran amphibians. Journal of Experimental Zoology 204:33–42.

Sadinski, W. J., and W. A. Dunson. 1992. A multilevel study of effects of low pH on amphibians of temporary ponds. Journal of Herpetology 26:413–422.

Salthe, S. N. 1963. The egg capsules in the Amphibia. Journal of Morphology 113:161–171.

Salthe, S. N. 1967. Courtship patterns and the phylogeny of the urodeles. Copeia 1967:100–117.

Salthe, S. N. 1969. Reproductive modes and the number and sizes of ova in the urodeles. The American Midland Naturalist 81:467–490.

Salthe, S. N. 1973a. *Amphiuma means*. Catalogue of American Amphibians and Reptiles, p. 148.1.

Salthe, S. N. 1973b. *Amphiuma tridactylum*. Catalogue of American Amphibians and Reptiles, pp. 149.1–149.2.

Salthe, S. N. 1973c. Amphiumidae, *Amphiuma*. Catalogue of American Amphibians and Reptiles, pp. 147.1–147.4.

Salthe, S. N., and J. S. Mecham. 1974. Reproductive and courtship patterns. In: Physiology of the Amphibia, Volume 2, B. Lofts, ed. Academic Press, New York, pp. 307–521.

Saugey, D. A., and S. E. Trauth. 1991. Distribution and habitat utilization of the four-toed salamander, *Hemidactylium scutatum*, in the Ouachita Mountains of Arkansas. Proceedings of the Arkansas Academy of Science 45:88–91.

Saugey, D. A., Heidt, G. A., and D. R. Heath. 1985. Summer use of abandoned mines by the Caddo Mountain salamander, *Plethodon caddoensis* (Plethodontidae), in Arkansas. The Southwestern Naturalist 30:318–319.

Sayler, A. 1966. The reproductive ecology of the red-backed salamander, *Plethodon cinereus*, in Maryland. Copeia 1966:183–193.

Schad, G. A., Stewart, R. H., and F. A. Harrington. 1959. Geographical distribution and variation of the Sacramento Mountains salamander, *Aneides hardii*. Canadian Journal of Zoology 37:299–303.

Schmidt, K. P. 1920. On the common name of *Amphiuma*. Copeia 1920:41–42.

Schmidt, K. P. 1953. A Check List of North American Amphibians and Reptiles, 6th Edition. American Society of Ichthyologists and Herpetologists.

Schonberger, C. F. 1944. Food of salamanders in the northwestern United States. Copeia 1944:257.

Schwaner, T. D., and R. H. Mount. 1970. Notes on the distribution, habits, and ecology of the salamander *Phaeognathus hubrichti*. Copeia 1970:571–573.

Schwartz, A. 1954. The salamander *Aneides aeneus* in South Carolina. Copeia 1954:296–298.

Schwartz, A. 1955. A clutch of eggs of *Aneides hardyi* (Taylor). Herpetologica 11:70.

Schwartz, A. 1957. "Albinism" in the salamander *Amphiuma means*. Herpetologica 13:75–76.

Schwartz, A., and R. Etheridge. 1954. New and additional herpetological records from the North Carolina Coastal Plain. Herpetologica 10:167–171.

Scott, D. E. 1990. Effects of larval density in *Ambystoma opacum*: an experiment in large-scale field enclosures. Ecology 71:296–306.

Scott, D. E. 1993. Timing in reproduction of paedomorphic and metamorphic *Ambystoma talpoideum*. The American Midland Naturalist 129:397–402.

Scott, D. E. 1994. The effect of larval density on adult demographic traits in *Ambystoma opacum*. Ecology 75:1383–1396.

Scott, D. E., and M. R. Fore. 1995. The effect of food limitation on lipid levels, growth, and reproduction in the marbled salamander, *Ambystoma opacum*. Herpetologica 51:462–471.

Scott, N. J., and C. A. Ramotnik. 1992. Does the Sacramento Mountain salamander require old-growth forests? In: Old-growth forests in the southwest and Rocky Mountain regions, M. R. Kaufmann, W. H. Moir, and R. L. Bassett, technical coordinators. USDA Forest Service, General Technical Report RM-213, pp. 170–178.

Scroggin, J. B., and W. B. Davis. 1956. Food habits of the Texas dwarf siren. Herpetologica 12:231–237.

Seale, D. B. 1980. Influence of amphibian larvae on primary production, nutrient flux, and competition in a pond ecosystem. Ecology 61:1531–1550.

Seeliger, L. M. 1945. A leucistic specimen of the black salamander. Copeia 1945:122.

Seibert, H. C., and R. A. Brandon. 1960. The salamanders of southeastern Ohio. Ohio Journal of Science 60:291–303.

Selby, M. F., Winkel, S. C., and J. W. Petranka. 1996. Geographic uniformity in agonistic behaviors of Jordan's salamander. Herpetologica 52:108–115.

Semlitsch, R. D. 1980a. Growth and metamorphosis of larval dwarf salamanders (*Eurycea quadridigitata*). Herpetologica 36:138–140.

Semlitsch, R. D. 1980b. Geographic and local variation in population parameters of the slimy salamander *Plethodon glutinosus*. Herpetologica 36:6–16.

Semlitsch, R. D. 1981. Terrestrial activity and summer home range of the mole salamander (*Ambystoma talpoideum*). Canadian Journal of Zoology 59:315–322.

Semlitsch, R. D. 1983a. Structure and dynamics of two breeding populations of the eastern tiger salamander, *Ambystoma tigrinum*. Copeia 1983:608–616.

Semlitsch, R. D. 1983b. Terrestrial movements of an eastern tiger salamander, *Ambystoma tigrinum*. Herpetological Review 14:112–113.

Semlitsch, R. D. 1983c. Burrowing ability and behavior of salamanders of the genus *Ambystoma*. Canadian Journal of Zoology 61:616–620.

Semlitsch, R. D. 1985a. Analysis of climatic factors influencing migrations of the salamander *Ambystoma talpoideum*. Copeia 1985:477–489.

Semlitsch, R. D. 1985b. Reproductive strategy of a facultatively paedomorphic salamander *Ambystoma talpoideum*. Oecologia 65:305–313.

Semlitsch, R. D. 1987a. Relationship of pond drying to the reproductive success of the salamander *Ambystoma talpoideum*. Copeia 1987:61–69.

Semlitsch, R. D. 1987b. Density-dependent growth and fecundity in the paedomorphic salamander *Ambystoma talpoideum*. Ecology 68:1003–1008.

Semlitsch, R. D. 1987c. Paedomorphosis in *Ambystoma talpoideum*: effects of density, food, and pond drying. Ecology 68:994–1002.

Semlitsch, R. D. 1988. Allotopic distribution of two salamanders: effects of fish predation and competitive interactions. Copeia 1988:290–298.

Semlitsch, R. D., and J. W. Gibbons. 1985. Phenotypic variation in metamorphosis and paedomorphosis in the salamander *Ambystoma talpoideum*. Ecology 66:1123–1130.

Semlitsch, R. D., and J. W. Gibbons. 1990. Effects of egg size on success of larval salamanders in complex aquatic environments. Ecology 71:1789–1795.

Semlitsch, R. D., and M. A. McMillan. 1980. Breeding migrations, population size structure, and reproduction of the dwarf salamander, *Eurycea quadridigitata*, in South Carolina. Brimleyana 3:97–105.

Semlitsch, R. D., and J. H. K. Pechmann. 1985. Diel pattern of migratory activity for several species of pond-breeding salamanders. Copeia 1985:86–91.

Semlitsch, R. D., and S. C. Walls. 1990. Geographic variation in the egg-laying strategy of the mole salamander, *Ambystoma talpoideum*. Herpetological Review 21:14–15.

Semlitsch, R. D., and S. C. Walls. 1993. Competition in two species of larval salamanders: a test of geographic variation in competitive ability. Copeia 1993:587–595.

Semlitsch, R. D., and C. A. West. 1983. Aspects of the life history and ecology of Webster's salamander, *Plethodon websteri*. Copeia 1983:339–346.

Semlitsch, R. D., and H. M. Wilbur. 1988. Effects of pond drying time on metamorphosis and survival in the salamander *Ambystoma talpoideum*. Copeia 1988:978–983.

Semlitsch, R. D., and H. M. Wilbur. 1989. Artificial selection for paedomorphosis in the salamander *Ambystoma talpoideum*. Evolution 43:105–112.

Semlitsch, R. D., Harris, R. N., and H. M. Wilbur. 1990. Paedomorphosis in *Ambystoma talpoideum*: maintenance of population variation and alternative life-history pathways. Evolution 44:1604–1613.

Semlitsch, R. D., Scott, D. E., and J. H. K. Pechmann. 1988. Time and size at metamorphosis related to adult fitness in *Ambystoma talpoideum*. Ecology 69:184–192.

Semlitsch, R. D., Scott, D. E., Pechmann, J. H. K., and J. W. Gibbons. 1993. Phenotypic variation in the arrival time of breeding salamanders: individual repeatability and environmental influence. Journal of Animal Ecology 62:334–340.

Semlitsch, R. D., Scott, D. E., Pechmann, J. H. K., and J. W. Gibbons. 1996. Structure and dynamics of an amphibian community: evidence from a 16-year study of a natural pond. In: Long-Term Studies of Vertebrate Communities,

M. L. Cody and J. Smallwood, eds. Academic Press, San Diego, California, pp. 217-248.

Sessions, S. K. 1982. Cytogenetics of diploid and triploid salamanders of the *Ambystoma jeffersonianum* complex. Chromosoma 84:599-621.

Sessions, S. K., and J. Kezer. 1987. Cytogenetic evolution in the plethodontid genus *Aneides*. Chromosoma 95:17-30.

Sessions, S. K., and A. Larson 1987. Developmental correlates of genome size in plethodontid salamanders and their implications for genome evolution. Evolution 41:1239-1251.

Sessions, S. K., and J. E. Wiley. 1985. Chromosome evolution in salamanders of the genus *Necturus*. Brimleyana 10:37-52.

Sever, D. M. 1972. Geographic variation and taxonomy of *Eurycea bislineata* (Caudata: Plethodontidae) in the upper Ohio River Valley. Herpetologica 28:314-324.

Sever, D. M. 1975. Morphology and seasonal variation of the mental hedonic glands of the dwarf salamander, *Eurycea quadridigitata* (Holbrook). Herpetologica 31:241-251.

Sever, D. M. 1976. Identity of an enigmatic *Eurycea* (Urodela: Plethodontidae) from the Great Smoky Mountains of Tennessee. Herpetological Review 7:98.

Sever, D. M. 1978. Female cloacal anatomy of *Plethodon cinereus* and *Plethodon dorsalis* (Amphibia, Urodela, Plethodontidae). Journal of Herpetology 12:397-406.

Sever, D. M. 1979. Male secondary sexual characteristics of the *Eurycea bislineata* (Amphibia, Urodela, Plethodontidae) complex in the southern Appalachian Mountains. Journal of Herpetology 13:245-253.

Sever, D. M. 1983a. *Eurycea junaluska*. Catalogue of American Amphibians and Reptiles, pp. 321.1-321.2.

Sever, D. M. 1983b. Observations on the distribution and reproduction of the salamander *Eurycea junaluska* in Tennessee. Journal of the Tennessee Academy of Science 58:48-50.

Sever, D. M. 1985. Sexually dimorphic glands of *Eurycea nana*, *Eurycea neotenes* and *Typhlomolge rathbuni* (Amphibia: Plethodontidae). Herpetologica 41:71-84.

Sever, D. M. 1989a. Caudal hedonic glands in salamanders of the *Eurycea bislineata* complex (Amphibia: Plethodontidae). Herpetologica 45:322-329.

Sever, D. M. 1989b. Comments on the taxonomy and morphology of two-lined salamanders of the *Eurycea bislineata* complex. Bulletin of the Chicago Herpetological Society 24:70-74.

Sever, D. M., and H. L. Bart, Jr. 1996. Ultrastructure of the spermathecae of *Necturus beyeri* (Amphibia: Proteidae) in relation to its breeding season. Copeia 1996: 927-937.

Sever, D. M., and C. F. Dineen. 1978. Reproductive ecology of the tiger salamander, *Ambystoma tigrinum*, in northern Indiana. Proceedings of the Indiana Academy of Science 87:189-203.

Sever, D. M., Dundee, H. A., and C. D. Sullivan. 1976. A new *Eurycea* (Amphibia: Plethodontidae) from southwestern North Carolina. Herpetologica 32:26-29.

Sexton, O. J., and J. R. Bizer. 1978. Life history patterns of *Ambystoma tigrinum* in montane Colorado. The American Midland Naturalist 99:101-118.

Sexton, O. J., Bizer, J., Gayou, D. C., Freiling, P., and M. Moutseous. 1986. Field studies of breeding spotted salamanders, *Ambystoma maculatum*, in eastern Missouri, U.S.A. Milwaukee Public Museum, Contributions in Biology and Geology 67:1-19.

Sexton, O. J., Phillips, C., and J. E. Bramble. 1990. The effects of temperature and precipitation on the breeding migration of the spotted salamander (*Ambystoma maculatum*). Copeia 1990:781-787.

Sexton, O. J., Phillips, C. A,. and E. Routman. 1994. The response of naive breeding adults of the spotted salamander to fish. Behaviour 130:113-121.

Seyle, C. W., Jr. 1985. *Amphiuma means* (two-toed amphiuma). Reproduction. Herpetological Review 16:51-52.

Shaffer, H. B. 1983. Biosystematics of *Ambystoma rosaceum* and *A. tigrinum* in northwestern Mexico. Copeia 1983:67-78.

Shaffer, H. B. 1984a. Evolution in a paedomorphic lineage. I. An electrophoretic analysis of the Mexican ambystomatid salamanders. Evolution 38:1194-1206.

Shaffer, H. B. 1984b. Evolution in a paedomorphic lineage. II. Allometry and form in the Mexican ambystomatid salamanders. Evolution 38:1207–1218.

Shaffer, H. B. 1993. Systematics of model organisms: the laboratory axolotl, *Ambystoma mexicanum*. Systematic Biology 42:508–522.

Shaffer, H. B., and F. Breden. 1989. The relationship between allozyme variation and life history: non-transforming salamanders are less variable. Copeia 1989:1016–1023.

Shaffer, H. B., and M. L. McKnight. 1996. The polytypic species revisited: differentiation and molecular phylogenetics of the tiger salamander *Ambystoma tigrinum* (Amphibia: Caudata) complex. Evolution 50:417–433.

Shaffer, H. B., Clark, J. M., and F. Kraus. 1991. When molecules and morphology clash: a phylogenetic analysis of North American ambystomatid salamanders (Caudata: Ambystomatidae). Systematic Zoology 40:284–303.

Sharbel, T. F., and J. Bonin. 1992. Northernmost record of *Desmognathus ochrophaeus*: biochemical identification in the Chateauguay River Drainage Basin, Quebec. Journal of Herpetology 26:505–508.

Sharbel, T. F., Bonin, J., Lowcock, L. A., and D. M. Green. 1995. Partial genetic compatibility and unidirectional hybridization in syntopic populations of the salamanders *Desmognathus fuscus* and *D. ochrophaeus*. Copeia 1995:466–469.

Shealy, R. M. 1975. Factors influencing activity in the salamanders *Desmognathus ochrophaeus* and *D. monticola* (Plethodontidae). Herpetologica 31:94–102.

Sherman, C. K., and M. L. Morton. 1993. Population declines of Yosemite toads in the eastern Sierra Nevada of California. Journal of Herpetology 27:186–198.

Sherwood, W. L. 1895. The salamanders found in the vicinity of New York City, with notes upon extra-limital or allied species. Proceedings of the Linnaean Society of New York 7:21–37.

Shillington, C., and P. Verrell. 1996. Multiple mating by females is not dependent on body size in the salamander *Desmognathus ochrophaeus*. Amphibia-Reptilia 17:33–38.

Shoop, C. R. 1960. The breeding habits of the mole salamander, *Ambystoma talpoideum* (Holbrook), in southeastern Louisiana. Tulane Studies in Zoology 8:65–82.

Shoop, C. R. 1965a. Orientation of *Ambystoma maculatum*: movements to and from breeding pools. Science 149:558–559.

Shoop, C. R. 1965b. Aspects of reproduction in Louisiana *Necturus* populations. The American Midland Naturalist 74:357–367.

Shoop, C. R. 1968. Migratory orientation of *Ambystoma maculatum*: movements near breeding ponds and displacements of migrating individuals. Biological Bulletin 135:230–238.

Shoop, C. R. 1974. Yearly variation in larval survival of *Ambystoma maculatum*. Ecology 55:440–444.

Shoop, C. R., and T. L. Doty. 1972. Migratory orientation by marbled salamanders (*Ambystoma opacum*) near a breeding area. Behavioral Biology 7:131–136.

Shoop, C. R., and G. E. Gunning. 1967. Seasonal activity and movements of *Necturus* in Louisiana. Copeia 1967:732–737.

Shure, D. J., Wilson, L. A., and C. Hochwender. 1989. Predation on aposematic efts of *Notophthalmus viridescens*. Journal of Herpetology 23:437–439.

Sih, A., and L. B. Kats. 1991. Effects of refuge availability on the responses of salamander larvae to chemical cues from predatory green sunfish. Animal Behaviour 42:330–332.

Sih, A., and L. B. Kats. 1994. Age, experience and the response of streamside salamander hatchlings to chemical cues from predatory sunfish. Ethology 96:253–259.

Sih, A., and E. Maurer. 1992. Effects of cryptic oviposition on egg survival for stream-breeding, streamside salamanders. Journal of Herpetology 26:114–116.

Sih, A., and R. D. Moore. 1993. Delayed hatching of salamander eggs in response to enhanced larval predation risk. The American Naturalist 142:947–960.

Sih, A., and J. W. Petranka. 1988. Optimal diets: simultaneous search and handling of multiple prey loads by salamander larvae. Behavioral Ecology and Sociobiology 23:335–339.

Sih, A., Kats, L. B., and R. D. Moore. 1992. Effects of predatory sunfish on the density, drift, and refuge use of stream salamander larvae. Ecology 73:1418–1430.

Sih, A., Petranka, J. W., and L. B. Kats. 1988. The dynamics of prey refuge use: a model and tests with sunfish and salamander larvae. The American Naturalist 132:463-483.

Simmons, D. 1975. The evolutionary ecology of *Gyrinophilus palleucus*. Unpublished master's thesis, University of Florida, Gainesville, Florida.

Simmons, D. 1976. A naturally metamorphosed *Gyrinophilus palleucus* (Amphibia, Urodela, Plethodontidae). Journal of Herpetology 10:255-257.

Simon, G. S., and D. M. Madison. 1984. Individual recognition in salamanders: cloacal odours. Animal Behaviour 32:1017-1020.

Simons, R. R., and B. E. Felgenhauer. 1992. Identifying areas of chemical signal production in the red-backed salamander, *Plethodon cinereus*. Copeia 1992:776-781.

Simons, R. R., Felgenhauer B. E., and R. G. Jaeger. 1994. Salamander scent marks: site of production and their role in territorial defense. Animal Behaviour 1994:97-103.

Simons, R. R., Jaeger, R. G., and B. E. Felgenhauer. 1995. Juvenile terrestrial salamanders have active postcloacal glands. Copeia 1995:481-483.

Simpson, G. G. 1961. Principles of Animal Taxonomy. Columbia University Press, New York, 247 pp.

Sinclair, R. M. 1950. Notes on some salamanders from Tennessee. Herpetologica 6:49-51.

Sinclair, R. M. 1951. Notes on recently transformed larvae of the salamander *Eurycea longicauda guttolineata*. Herpetologica 7:68.

Sites, J. W., Jr. 1978. The foraging strategy of the dusky salamander, *Desmognathus fuscus* (Amphibia: Urodela: Plethodontidae): an empirical approach to predation theory. Journal of Herpetology 12:373-383.

Skelly, D. K. 1992. Field evidence for a cost of behavioral antipredator response in a larval amphibian. Ecology 1992:704-708.

Slater, J. R. 1933. Notes on Washington salamanders. Copeia 1933:44.

Slater, J. R. 1936. Notes on *Ambystoma gracile* Baird and *Ambystoma macrodactylum* Baird. Copeia 1936:234-236.

Slater, J. R. 1939. *Plethodon dunni* in Oregon and Washington. Herpetologica 1:154.

Slater, J. R., and W. C. Brown. 1941. Island records of amphibians and reptiles for Washington. Occasional Papers of the Department of Biology, College of Puget Sound 13:74-77.

Slater, J. R., and J. W. Slipp. 1940. A new species of *Plethodon* from northern Idaho. Occasional Papers of the Department of Biology, College of Puget Sound 8:38-43.

Smallwood, W. M. 1928. Notes on the food of some Onondaga Urodela. Copeia 1928:89-98.

Smith, B. G. 1907. The life history and habitats of *Cryptobranchus alleghenesis*. Biological Bulletin 13:5-39.

Smith, B. G. 1910. The structure of the spermatophores of *Ambystoma punctatum*. Biological Bulletin 18:204-211.

Smith, B. G. 1911a. Notes on the natural history of *Ambystoma jeffersonianum*, *A. punctatum*, and *A. tigrinum*. Bulletin of the Wisconsin Natural History Society 9:14-27.

Smith, B. G. 1911b. The nests and larvae of *Necturus*. Biological Bulletin 20:191-200.

Smith, B. G. 1912a. The embryology of *Cryptobranchus alleghenesis*, including comparisons with some other vertebrates. I. Introduction; the history of the egg before cleavage. Journal of Morphology 23:61-157.

Smith, B. G. 1912b. The embryology of *Cryptobranchus alleghenesis*, including comparisons with some other vertebrates. II. General embryonic and larval development, with special reference to external features. Journal of Morphology 23:455-579.

Smith, C. C. 1960. Notes on the salamanders of Arkansas. 1. Life history of a neotenic stream-dwelling form. Proceedings of the Arkansas Academy of Science 13:66-74.

Smith, C. K. 1983. Notes on breeding period, incubation period, and egg masses of *Ambystoma jeffersonianum* (Green) (Amphibia: Caudata) from the southern limits of its range. Brimleyana 9:135-140.

Smith, C. K. 1990. Effects of variation in body size on intraspecific competition among larval salamanders. Ecology 71:1777-1788.

Smith, C. K., and J. W. Petranka. 1987. Prey size-distributions and size-specific foraging success of *Ambystoma* larvae. Oecologia 71:239-244.

Smith, C. K., Petranka, J. W., and R. L. Barwick. 1996a. *Desmognathus quadramaculatus* (black-bel-

Smith, C. K., Petranka, J. W., and R. L. Barwick. 1996b. *Desmognathus welteri* (Black Mountain dusky salamander). Reproduction. Herpetological Review 27:136.

Smith, D. D. 1985. *Ambystoma tigrinum* (tiger salamander). Behavior. Herpetological Review 16:77.

Smith, E. M., and F. H. Pough. 1994. Intergeneric aggression in salamanders. Journal of Herpetology 28:41–45.

Smith, H. M. 1934. The amphibians of Kansas. The American Midland Naturalist 15:377–528.

Smith, H. M. 1949. Size maxima in terrestrial salamanders. Copeia 1949:71.

Smith, H. M., and F. E. Potter, Jr. 1946. A third neotenic salamander of the genus *Eurycea* from Texas. Herpetologica 3:105–109.

Smith, H. M., and R. W. Reese. 1968. A record tiger salamander. The Southwest Naturalist 13:370–372.

Smith, L. 1920. Some notes on *Notophthalmus viridescens*. Copeia 1920:22–24.

Smith, L. 1921. A note on the eggs of *Ambystoma maculatum*. Copeia 1921:41.

Smith, P. B., and M. C. Michener. 1962. An adult albino *Ambystoma*. Herpetologica 18:67–68.

Smith, P. W. 1948a. A cestode infestation in *Typhlotriton*. Herpetologica 4:152.

Smith, P. W. 1948b. Food habits of cave dwelling amphibians. Herpetologica 4:205–208.

Smith, P. W. 1961. The amphibians and reptiles of Illinois. Bulletin of the Illinois Natural History Survey 28:1–298.

Smith, P. W. 1963. *Plethodon cinereus*. Catalogue of American Amphibians and Reptiles, pp. 5.1–5.3

Smith, P. W., and S. A. Minton, Jr. 1957. A distributional summary of the herpetofauna of Indiana and Illinois. The American Midland Naturalist 58:341–351.

Smith, R. E. 1941a. Mating behavior in *Triturus torosus* and related newts. Copeia 1941:255–262.

Smith, R. E. 1941b. The spermatophores of *Triturus torosus* and *Triturus rivularis*. Proceedings of the National Academy of Science 27:261–264.

Snyder, D. H. 1973. Some adaptive value of brooding behavior in *Aneides aeneus*. HISS News-Journal (Herpetological Review)1:63.

Snyder, D. H. 1991. The green salamander (*Aneides aeneus*) in Tennessee and Kentucky, with comments on the Carolinas' Blue Ridge populations. Journal of the Tennessee Academy of Science 66:165–169.

Snyder, J. O. 1923. Eggs of *Batrachoseps attenuatus*. Copeia 1923:86–88.

Snyder, R. C. 1956. Comparative features of the life histories of *Ambystoma gracile* (Baird) from populations at low and high altitudes. Copeia 1956:41–50.

Snyder, R. C. 1963. *Ambystoma gracile*. Catalogue of American Amphibians and Reptiles, pp. 6.1–6.2.

Southerland, M. T. 1986a. The effects of variation in streamside habitats on the composition of mountain salamander communities. Copeia 1986:731–741.

Southerland, M. T. 1986b. Coexistence of three congeneric salamanders: the importance of habitat and body size. Ecology 67:721–728.

Southerland, M. T. 1986c. Behavioral interactions among four species of the salamander genus *Desmognathus*. Ecology 67:175–181.

Southerland, M. T. 1986d. Behavioral niche expansion in *Desmognathus fuscus* (Amphibia: Caudata: Plethodontidae). Copeia 1986:235–237.

Southerland, M. T. 1986e. *Leurognathus marmoratus* (shovelnose salamander). Herpetological Review 17:45.

Sparkes, T. C. 1996. Effects of predation risk on population variation in adult size in a stream-dwelling isopod. Oecologia 106:85–92.

Spight, T. M. 1967. Population structure and biomass production by a stream salamander. The American Midland Naturalist 78:437–447.

Spolsky, C., Phillips, C. A., and T. Uzzell. 1992a. Antiquity of clonal salamander lineages revealed by mitochondrial DNA. Nature 356:706–708.

Spolsky, C., Phillips, C. A., and T. Uzzell. 1992b. Gynogenetic reproduction in hybrid mole salamanders (genus *Ambystoma*). Evolution 46:1935–1944.

Spotila, J. R. 1972. Role of temperature and water in the ecology of lungless salamanders. Ecological Monographs 42:95–125.

Spotila, J. R. 1976. Courtship behavior of the ringed salamander (*Ambystoma annulatum*): observations in the field. The Southwestern Naturalist 21:412–413.

Spotila, J. R., and R. J. Beumer. 1970. The breeding habits of the ringed salamander, *Ambystoma annulatum* (Cope), in northwestern Arkansas. The American Midland Naturalist 84:77-89.

Spotila, J. R., and P. H. Ireland. 1970. Notes on the eggs of the gray-bellied salamander, *Eurycea multiplicata griseogaster*. The Southwestern Naturalist 14:366-368.

Sprules, W. G. 1972. Effects of size-selective predation and food competition on high altitude zooplankton communities. Ecology 53:375-386.

Sprules, W. G. 1974a. The adaptive significance of paedogenesis in North American species of *Ambystoma* (Amphibia: Caudata): an hypothesis. Canadian Journal of Zoology 52:393-400.

Sprules, W. G. 1974b. Environmental factors and the incidence of neoteny in *Ambystoma gracile* (Baird) (Amphibia: Caudata). Canadian Journal of Zoology 52:393-400.

Stangel, P. W. 1983. Least sandpiper predation on *Bufo americanus* and *Ambystoma maculatum* larvae. Herpetological Review 14:112.

Stangel, P. W. 1988. Premetamorphic survival of the salamander *Ambystoma maculatum* in eastern Massachusetts. Journal of Herpetology 22:345-347.

Stangel, P. W., and R. D. Semlitsch. 1987. Experimental analysis of predation on the diel vertical migrations of a larval salamander. Canadian Journal of Zoology 65:1554-1558.

Stark, M. A. 1986. Overwintering of an ambystomatid salamander in a prairie rattlesnake hibernaculum. Herpetological Review 17:7.

Staub, N. L. 1993. Intraspecific agonistic behavior of the salamander *Aneides flavipunctatus* (Amphibia: Plethodontidae) with comparisons to other plethodontid species. Herpetologica 49:271-282.

Staub, N. L., Brown, C. W., and D. B. Wake. 1995. Patterns of growth and movements in a population of *Ensatina eschscholtzii platensis* (Caudata: Plethodontidae) in the Sierra Nevada, California. Journal of Herpetology 29:593-599.

Stauffer, J. R., Jr., Gates, J. E., and W. L. Goodfellow. 1983. Preferred temperature of two sympatric *Ambystoma* larvae: a proximate factor in niche segregation? Copeia 1996:1001-1005.

Stebbins, R. C. 1945. Water absorption in a terrestrial salamander. Copeia 1945:25-28.

Stebbins, R. C. 1947. Tail and foot action in the locomotion of *Hydromantes platycephalus*. Copeia 1947:1-5.

Stebbins, R. C. 1949a. Speciation in salamanders of the plethodontid genus *Ensatina*. University of California Publications in Zoology 48:377-526.

Stebbins, R. C. 1949b. Courtship of the plethodontid salamander *Ensatina eschscholtzii*. Copeia 1949:274-281.

Stebbins, R. C. 1949c. Observations on laying, development, and hatching of the eggs of *Batrachoseps wrighti*. Copeia 1949:161-168.

Stebbins, R. C. 1951. Amphibians of Western North America. University of California Press, Berkeley, California, 539 pp.

Stebbins, R. C. 1954. Natural history of the salamanders of the plethodontid genus *Ensatina*. University of California Publications in Zoology 54:47-124.

Stebbins, R. C. 1966. A Field Guide to Western Reptiles and Amphibians. Houghton Mifflin, Boston, 279 pp.

Stebbins, R. C. 1985. A Field Guide to Western Reptiles and Amphibians, 2nd Edition. Houghton Mifflin, Boston, 336 pp.

Stebbins, R. C., and N. W. Cohen. 1995. A Natural History of Amphibians. Princeton University Press, Princeton, New Jersey, 316 pp.

Stebbins, R. C., and C. H. Lowe, Jr. 1949. The systematic status of *Plethopsis* with a discussion of speciation in the genus *Batrachoseps*. Copeia 1949:116-129.

Stebbins, R. C., and C. H. Lowe, Jr. 1951. Subspecific differentiation in the Olympic salamander *Rhyacotriton olympicus*. University of California Publications in Zoology 50:465-484.

Stebbins, R. C., and H. C. Reynolds. 1947. Southern extension of the range of the Del Norte salamander in California. Herpetologica 4:41-42.

Stebbins, R. C., and W. J. Riemer. 1950. A new species of plethodontid salamander from the Jemez Mountains of New Mexico. Copeia 1950:73-80.

Stein, K. F. 1938. Migration of *Triturus viridescens*. Copeia 1938:80-83.

Stelmock, J. J., and A. S. Harestad. 1979. Food habits and life history of the clouded salamander (*Aneides ferreus*) on northern Vancouver Island, British Columbia. Syesis 12:71-75.

Stenhouse, S. L. 1985a. Migratory orientation and homing in *Ambystoma maculatum* and *Ambystoma opacum*. Copeia 1985:631–637.

Stenhouse, S. L. 1985b. Interdemic variation in predation on salamander larvae. Ecology 66:1706–1717.

Stenhouse, S. L. 1987. Embryo mortality and recruitment of juveniles of *Ambystoma maculatum* and *Ambystoma opacum* in North Carolina. Herpetologica 43:496–501.

Stenhouse, S. L., Hairston, N. G., and A. E. Cobey. 1983. Predation and competition in *Ambystoma* larvae: field and laboratory experiments. Journal of Herpetology 17:210–220.

Stevenson, H. M. 1967. Additional specimens of *Amphiuma pholeter* from Florida. Herpetologica 23:134.

Stewart, G. D., and E. D. Bellis. 1970. Dispersion of salamanders along a brook. Copeia 1970:86–89.

Stewart, M. M. 1956. The separate effects of food and temperature differences on development of marbled salamander larvae. Journal of the Elisha Mitchell Scientific Society 72:47–56.

Stewart, M. M. 1958. Seasonal variation in the teeth of the two-lined salamander. Copeia 1958:190–196.

Stewart, M. M. 1968. Population dynamics of *Eurycea bislineata* in New York (abstract). Journal of Herpetology 2:176–177.

Stille, W. T. 1954. Eggs of the salamander *Ambystoma jeffersonianum* in the Chicago area. Copeia 1954:300.

Stine, C. J., Jr., Fowler, J. A., and R. S. Simmons. 1954. Occurrence of the eastern tiger salamander, *Ambystoma tigrinum tigrinum* (Green) in Maryland, with notes on its life history. Annals of the Carnegie Museum 33:145–148.

Stone, L. S. 1964. The structure and visual function of the eye of larval and adult cave salamanders *Typhlotriton spelaeus*. Journal of Experimental Zoology 156:201–218.

Stone, R. 1994. Environmental estrogens stir debate. Science 265:298–444.

Stoneburner, D. L. 1978. Salamander drift: observations on the two-lined salamander (*Eurycea bislineata*). Freshwater Biology 8:291–293.

Storer, T. I. 1925. A synopsis of the Amphibia of California. University of California Publications in Zoology 27:1–342.

Storez, R. A. 1969. Observations on the courtship of *Ambystoma laterale*. Journal of Herpetology 3:87–95.

Storm, R. M. 1947. Eggs and young of *Aneides ferreus*. Herpetologica 4:60–62.

Storm, R. M. 1955. Northern and southern range limits of Dunn's salamander, *Plethodon dunnii*. Copeia 1955:64–65.

Storm, R. M., and A. R. Aller. 1947. Food habits of *Aneides ferreus*. Herpetologica 4:59–60.

Storm, R. M., and E. D. Brodie, Jr. 1970a. *Plethodon dunni*. Catalogue of American Amphibians and Reptiles, pp. 82.1–82.2.

Storm, R. M., and E. D. Brodie, Jr. 1970b. *Plethodon vehiculum*. Catalogue of American Amphibians and Reptiles, pp. 83.1–83.2.

Stout, B. M., III, Stout, K. K., and C. W. Stihler. 1992. Predation by the caddisfly *Banksiola dossuaria* on egg masses of the spotted salamander *Ambystoma maculatum*. The American Midland Naturalist 127:368–372.

Strecker, J. K. 1922. An annotated catalogue of the amphibians and reptiles of Bexer County, Texas. Bulletin of the Scientific Society of San Antonio 4:1–31.

Strecker, J. K., and W. J. Williams. 1928. Field notes on the herpetology of Bowie County, Texas. Contributions of the Baylor University Museum 17:1–19.

Sugg, D. W., Karlin, A. A., Preston, C. R., and D. R. Heath. 1988. Morphological variation in a population of the salamander, *Siren intermedia nettingi*. Journal of Herpetology 22:243–247.

Surface, H. A. 1913. First report on the economic features of the amphibians of Pennsylvania. Zoological Bulletin of The Division of Zoology, Pennsylvania Department of Agriculture 3:68–152.

Svihla, A., and R. D. Svihla. 1933. Amphibians and reptiles of Whitman County, Washington. Copeia 1933:125–128.

Swanson, P. L. 1948. Notes on the amphibians of Venango County, Pennsylvania. The American Midland Naturalist 40:362–371.

Sweet, S. S. 1977a. *Eurycea tridentifera*. Catalogue of American Amphibians and Reptiles, pp. 199.1–199.2.

Sweet, S. S. 1977b. Natural metamorphosis in *Eurycea neotenes*, and the generic allocation of the Texas *Eurycea* (Amphibia: Plethodontidae). Herpetologica 33:364–375.

Sweet, S. S. 1978. On the status of *Eurycea pterophila* (Amphibia: Plethodontidae). Herpetologica 34:101–108.

Sweet, S. S. 1982. A distributional analysis of epigean populations of *Eurycea neotenes* in central Texas, with comments on the origin of troglobitic populations. Herpetologica 38:430–444.

Sweet, S. S. 1984. Secondary contact and hybridization in the Texas cave salamanders *Eurycea neotenes* and *E. tridentifera*. Copeia 1984:428–441.

Tabachnick, W. J. 1977. Geographic variation of five biochemical polymorphisms in *Notophthalmus viridescens*. Journal of Heredity 68:117–122.

Taber, C. A., Wilkinson, R. F., Jr., and M. S. Topping. 1975. Age and growth of hellbenders in the Niangua River, Missouri. Copeia 1975:633–639.

Talentino, K. A., and E. Landre. 1991. Comparative development of two species of sympatric *Ambystoma* salamanders. Journal of Freshwater Ecology 6:395–401.

Tan, A., and D. B. Wake. 1995. MtDNA phylogeography of the California newt, *Taricha torosa* (Caudata, Salamandridae). Molecular Phylogenetics and Evolution 4:383–394.

Tanner, W. W. 1953. Notes on the life history of *Plethopsis wrighti* Bishop. Herpetologica 9:139–140.

Tanner, W. W., Fisher, D. L., and T. J. Willis. 1971. Notes on the life history of *Ambystoma tigrinum nebulosum* Hallowell in Utah. The Great Basin Naturalist 31:213–222.

Taub, F. B. 1961. The distribution of red-backed salamanders, *Plethodon c. cinereus*, within the soil. Ecology 42:681–698.

Taylor, A. S. 1992. Reconstitution of diploid *Ambystoma jeffersonianum* (Amphibia: Caudata) in a hybrid, triploid egg mass with lethal consequences. Journal of Heredity 83:361–366.

Taylor, B. E., Estes, R. A., Pechmann, J. H. K., and R. D. Semlitsch. 1988. Trophic relations in a temporary pond: larval salamanders and their microinvertebrate prey. Canadian Journal of Zoology 66:2191–2198.

Taylor, C. L., Wilkinson, R. F., Jr., and C. L. Peterson. 1990. Reproductive patterns of five plethodontid salamanders from the Ouachita Mountains. The Southwestern Naturalist 35:468–472.

Taylor, J. 1983a. Size-specific associations of larval and neotenic northwestern salamanders, *Ambystoma gracile*. Journal of Herpetology 17:203–209.

Taylor, J. 1983b. Orientation and flight behavior of the neotenic salamander (*Ambystoma gracile*) in Oregon. The American Midland Naturalist 109:40–49.

Taylor, J. 1984. Comparative evidence for competition between the salamanders *Ambystoma gracile* and *Taricha granulosa*. Copeia 1984:672–683.

Teberg, E. K. 1963. An extension into Montana of the known range of the salamander *Plethodon vandykei idahoensis*. Herpetologica 19:287.

Telford, S. R., Jr. 1952. A herpetological survey in the vicinity of Lake Shipp, Polk County, Florida. Quarterly Journal of the Florida Academy of Science 15:175–185.

Telford, S. R., Jr. 1954. A description of the larvae of *Ambystoma cingulatum bishopi* Goin, including an extension of the range. Quarterly Journal of the Florida Academy of Science 17:233–236.

Templeton, A. R., Routman, E., and C. A. Phillips. 1995. Separating population structure from population history: a cladistic analysis of the geographical distribution of mitochondrial DNA haplotypes in the tiger salamander, *Ambystoma tigrinum*. Genetics 140:767–782.

Test, F. H. 1952. Spread of the black phase of the red-backed salamander in Michigan. Evolution 6:197–203.

Test, F. H. 1955. Seasonal differences in populations of the red-backed salamander in southeastern Michigan. Papers of the Michigan Academy of Sciences, Arts and Letters 40:137–153.

Test, F. H., and B. A. Bingham. 1948. Censuses of a population of the red-backed salamander (*Plethodon cinereus*). The American Midland Naturalist 39:362–372.

Test, F. H., and H. Heatwole. 1962. Nesting sites of the red-backed salamander, *Plethodon cinereus*, in Michigan. Copeia 1962:206–207.

Thomas, J. S., Jaeger, R. G., and E. A. Horne. 1989. Are all females welcome? Agonistic behavior of male red-backed salamanders. Copeia 1989:915–920.

Thompson, D. B., and T. R. Jones. 1992. The occurrence of paedomorphic cave-dwelling tiger salamanders in central New Mexico. In: GYPKAP Report 2 1988–1991, D. Belski, ed. Southwestern Region National Speleological Society, Adobe Press, Albuquerque, New Mexico, pp. 3–6.

Thompson, E. L., and J. E. Gates. 1982. Breeding pool segregation by the mole salamanders, *Ambystoma jeffersonianum* and *A. maculatum,* in a region of sympatry. Oikos 38:273–279.

Thompson, E. L., Gates, J. E., and G. S. Taylor. 1980. Distribution and breeding habitat selection of the Jefferson salamander, *Ambystoma jeffersonianum,* in Maryland. Journal of Herpetology 14:113–120.

Thurow, G. R. 1955. An albinistic individual of the salamander *Plethodon dorsalis.* Copeia 1955:62–63.

Thurow, G. R. 1956a. Comparisons of two species of salamander, *Plethodon cinereus* and *Plethodon dorsalis.* Herpetologica 12:177–182.

Thurow, G. R. 1956b. A new subspecies of *Plethodon welleri,* with notes on other members of the genus. The American Midland Naturalist 55:343–356.

Thurow, G. R. 1957a. A new *Plethodon* from Virginia. Herpetologica 13:59–66.

Thurow, G. R. 1957b. Relationships of the red-backed and zig-zag plethodons in the west. Herpetologica 13:91–99.

Thurow, G. R. 1961. A salamander color variant associated with glacial boundaries. Evolution 15:281–287.

Thurow, G. R. 1963. Taxonomic and ecological notes on the salamander, *Plethodon welleri.* University of Kansas Science Bulletin 44:87–108.

Thurow, G. R. 1966. *Plethodon dorsalis.* Catalogue of American Amphibians and Reptiles, p. 29.

Thurow, G. R. 1976. Aggression and competition in eastern *Plethodon* (Amphibia, Urodela, Plethodontidae). Journal of Herpetology 10:277–291.

Tihen, J. A. 1942. A colony of fossil neotenic *Ambystoma tigrinum.* University of Kansas Science Bulletin 28:189–199.

Tihen, J. A. 1955. A new Pliocene species of *Ambystoma,* with remarks on other fossil ambystomatids. Contributions of the Museum of Paleontology, University of Michigan 12:229–244.

Tihen, J. A. 1958. Comments on the osteology and phylogeny of ambystomatid salamanders. Bulletin of the Florida State Museum 3:1–50.

Tihen, J. A. 1969. *Ambystoma.* Catalogue of American Amphibians and Reptiles, pp. 75.1–75.4.

Tihen, J. A. 1974. Two new North American Miocene salamandrids. Journal of Herpetology 8:211–218.

Tilley, S. G. 1968. Size-fecundity relationships and their evolutionary implications in five desmognathine salamanders. Evolution 22:806–816.

Tilley, S. G. 1969. Variation in the dorsal pattern of *Desmognathus ochrophaeus* at Mt. Mitchell, North Carolina, and elsewhere in the southern Appalachian Mountains. Copeia 1969:161–175.

Tilley, S. G. 1972. Aspects of parental care and embryonic development in *Desmognathus ochrophaeus.* Copeia 1972:532–540.

Tilley, S. G. 1973a. Life histories and natural selection in populations of the salamander *Desmognathus ochrophaeus.* Ecology 54:3–17.

Tilley, S. G. 1973b. Observations on the larval period and female reproductive ecology of *Desmognathus ochrophaeus* (Amphibia: Plethodontidae) in western North Carolina. The American Midland Naturalist 89:394–407.

Tilley, S. G. 1974. Structures and dynamics of populations of the salamander *Desmognathus ochrophaeus* Cope in different habitats. Ecology 55:808–817.

Tilley, S. G. 1977. Studies of life histories and reproduction in North American plethodontid salamanders. In: The Reproductive Biology of Amphibians, D. H. Taylor and S. I. Guttman, eds. Plenum Press, New York, pp. 1–41.

Tilley, S. G. 1980. Life histories and comparative demography of two salamander populations. Copeia 1980:806–821.

Tilley, S. G. 1981. A new species of *Desmognathus* (Amphibia: Caudata: Plethodontidae) from the southern Appalachian Mountains. Occasional Papers of the Museum of Zoology, University of Michigan 695:1–23.

Tilley, S. G. 1985. *Desmognathus imitator.* Catalogue of American Amphibians and Reptiles, pp. 359.1–359.2.

Tilley, S. G. 1988. Hybridization between two species of *Desmognathus* (Amphibia: Caudata: Plethodontidae) in the Great Smoky Mountains. Herpetological Monographs 2:27–39.

Tilley, S. G., and J. Bernardo. 1993. Life history evolution in plethodontid salamanders. Herpetologica 49:154–163.

Tilley, S. G., and J. R. Harrison. 1969. Notes on the distribution of the pygmy salamander, *Desmognathus wrighti* King. Herpetologica 25:178–180.

Tilley, S. G., and J. S. Hausman. 1976. Allozymic variation and multiple inseminations in populations of the salamander *Desmognathus ochrophaeus*. Copeia 1976:734–741.

Tilley, S. G., and M. J. Mahoney. 1996. Patterns of genetic differentiation in salamanders of the *Desmognathus ochrophaeus* complex (Amphibia: Plethodontidae). Herpetological Monographs 10:1–42.

Tilley, S. G., and P. M. Schwerdtfeger. 1981. Electrophoretic variation in Appalachian populations of the *Desmognathus fuscus* complex (Amphibia: Plethodontidae). Copeia 1981:109–119.

Tilley, S. G., and D. W. Tinkle. 1968. A reinterpretation of the reproductive cycle and demography of the salamander *Desmognathus ochrophaeus*. Copeia 1968:299–303.

Tilley, S. G., Lundrigan, B. L., and L. B. Brower. 1982. Erythrism and mimicry in the salamander *Plethodon cinereus*. Herpetologica 38:409–417.

Tilley, S. G., Merritt, R. B., Wu, B., and R. Highton. 1978. Genetic differentiation in salamanders of the *Desmognathus ochrophaeus* complex (Plethodontidae). Evolution 32:93–111.

Tilley, S. G., Verrell, P. A., and S. J. Arnold. 1990. Correspondence between sexual isolation and allozyme differentiation: a test in the salamander *Desmognathus ochrophaeus*. Proceedings of the National Academy of Science 87:2715–2719.

Tinkle, D. W. 1952. Notes on the salamander, *Eurycea longicauda guttolineata* in Florida. Field and Laboratory 29:105–108.

Titus, T. A. 1990. Genetic variation in two subspecies of *Ambystoma gracile* (Caudata: Ambystomatidae). Journal of Herpetology 24:107–111.

Titus, T. A., and M. S. Gaines. 1991. Genetic variation in coastal and montane populations of *Ambystoma gracile* (Caudata: Ambystomatidae). Occasional Papers of the Museum of Natural History, University of Kansas 141:1–12.

Titus, T. A., and A. Larson. 1996. Molecular phylogenetics of desmognathine salamanders (Caudata: Plethodontidae): a reevaluation of evolution in ecology, life history, and morphology. Systematic Biology 45:451–472.

Tome, M. E., and F. H. Pough. 1982. Responses of amphibians to acid precipitation. In: Acid rain/fisheries, Proceedings of an international symposium on acid precipitation and fisheries impacts in northeastern North America, T. A. Haines and R. E. Johnson, eds. American Fisheries Society, Bethesda, Maryland, pp. 245–254.

Topping, M. S., and C. A. Ingersol. 1981. Fecundity in the hellbender, *Cryptobranchus alleganiensis*. Copeia 1981:873–876.

Trapido, H., and R. T. Clausen. 1940. The larvae of *Eurycea bislineata major*. Copeia 1940:244–246.

Trapp, M. M. 1956. Range and natural history of the ringed salamander *Ambystoma annulatum* Cope (Ambystomatidae). The Southwestern Naturalist 1:78–82.

Trauth, S. E. 1983. Reproductive biology and spermathecal anatomy of the dwarf salamander (*Eurycea quadridigitata*) in Alabama. Herpetologica 39:9–15.

Trauth, S. E. 1984. Spermathecal anatomy and the onset of mating in the slimy salamander (*Plethodon glutinosus*) in Alabama. Herpetologica 40:314–321.

Trauth, S. E. 1988. Egg clutches of the Ouachita dusky salamander, *Desmognathus brimleyorum* (Caudata: Plethodontidae), collected in Arkansas during a summer drought. The Southwestern Naturalist 33:234–236.

Trauth, S. E., and M. E. Cartwright. 1989. An albino larvae in the ringed salamander, *Ambystoma annulatum*, from Arkansas. Bulletin of the Chicago Herpetological Society 24:128.

Trauth, S. E., and B. G. Cochran. 1991. *Hemidactylium scutatum* (four-toed salamander). Predation. Herpetological Review 22:55.

Trauth, S. E., and C. T. McAllister. 1995. Vertebrate prey of selected Arkansas snakes. Proceedings of the Arkansas Academy of Science 49:190–194.

Trauth, S. E., and B. Richards. 1988. An unusual color pattern of the marbled salamander, *Ambystoma opacum* (Caudata: Ambystomatidae) from Arkansas. Bulletin of the Chicago Herpetological Society 23:87.

Trauth, S. E., Cartwright, M. E., and W. E. Meshaka. 1989a. Winter breeding in the ringed salamander, *Ambystoma annulatum* (Caudata: Ambystomatidae), from Arkansas. The Southwestern Naturalist 34:145–146.

Trauth, S. E., Cochran, B. G., Saugey, D. A., Posey, W. R., and W. A. Stone. 1993a. Distribution of the mole salamander, *Ambystoma talpoideum* (Caudata: Ambystomatidae), in Arkansas with notes on paedomorphic populations. Proceedings of the Arkansas Academy of Science 47:154-156.

Trauth, S. E., Cox, R. L., Butterfield, B. P., Saugey, D. A., and W. E. Meshaka. 1990. Reproductive phenophases and clutch characteristics of selected Arkansas amphibians. Proceedings of the Arkansas Academy of Science 44:107-113.

Trauth, S. E., Cox, R. L., Jr., Wilhide, J. D., and H. J. Worley. 1995. Egg mass characteristics of terrestrial morphs of the mole salamander, *Ambystoma talpoideum* (Caudata: Ambystomatidae), from northeastern Arkansas and clutch comparisons with other *Ambystoma* species. Proceedings of the Arkansas Academy of Science 49:193-196.

Trauth, S. E., Meshaka, W. E., and B. P. Butterfield. 1989b. Reproduction and larval development in the marbled salamander, *Ambystoma opacum* (Caudata: Ambystomatidae), from Arkansas. Proceedings of the Arkansas Academy of Science 43:109-111.

Trauth, S. E., Sever, D. M., and R. D. Semlitsch. 1994. Cloacal anatomy of paedomorphic female *Ambystoma talpoideum* (Caudata: Ambystomatidae), with comments on intermorph mating and sperm storage. Canadian Journal of Zoology 72:2147-2157.

Trauth, S. E., Smith, R. D., Cheng, A., and P. Daniel. 1993b. Caudal hedonic glands in the dark-sided salamander, *Eurycea longicauda melanopleura* (Urodela: Plethodontidae). Proceedings of the Arkansas Academy of Science 47:151-153.

Trauth, S. E., Wilhide, J. D., and P. Daniel. 1992. Status of the Ozark hellbender, *Cryptobranchus bishopi* (Urodela: Cryptobranchidae), in the Spring River, Fulton County, Arkansas. Proceedings of the Arkansas Academy of Science 46:83-86.

Tristram, D. A. 1977. Intraspecific olfactory communications in the terrestrial salamander *Plethodon cinereus*. Copeia 1977:597-600.

Tumlison, R., Cline, G. R., and P. Zwank. 1990a. Morphological discrimination between the Oklahoma salamander (*Eurycea tynerensis*) and the graybelly salamander (*Eurycea multiplicata griseogaster*). Copeia 1990:242-246.

Tumlison, R., Cline, G. R., and P. Zwank. 1990b. Prey selection in the Oklahoma salamander (*Eurycea tynerensis*). Journal of Herpetology 24: 222-225.

Tumlison, R., Cline, G. R., and P. Zwank. 1990c. Surface habitat associations of the Oklahoma salamander (*Eurycea tynerensis*). Herpetologica 46:169-175.

Tupa, D. D., and W. K. Davis. 1976. Population dynamics of the San Marcos salamander, *Eurycea nana* Bishop. Texas Journal of Science 27:179-195.

Twitty, V. C. 1935. Two new species of *Triturus* from California. Copeia 1935:73-80.

Twitty, V. C. 1937. Experiments on the phenomenon of paralysis produced by the toxin occurring in *Triturus* embryos. Journal of Experimental Zoology 76:67-104.

Twitty, V. C. 1941. Data on the life history of *Ambystoma tigrinum californiense* Gray. Copeia 1941:1-14.

Twitty, V. C. 1942. The species of California *Triturus*. Copeia 1942:65-76.

Twitty, V. C. 1955. Field experiments on the biology and genetic relationships of the California species of *Triturus*. Journal of Experimental Zoology 129:129-148.

Twitty, V. C. 1959. Migration and speciation in newts. Science 130:1735-1743.

Twitty, V. C. 1961a. Experiments on homing behavior and speciation in *Taricha*. In: Vertebrate Speciation, W. F. Blair, ed. University of Texas Press, Austin, Texas, pp. 415-459.

Twitty, V. C. 1961b. Second-generation hybrids of the species *Taricha*. Proceedings of the National Academy of Science 47:1461-1486.

Twitty, V. C. 1964. Fertility of *Taricha* species-hybrids and viability of their offspring. Proceedings of the National Academy of Science 51:156-161.

Twitty, V. C. 1966. Of Scientists and Salamanders. W. H. Freeman, San Francisco, 178 pp.

Twitty, V. C., and H. H. Johnson. 1934. Motor inhibitions in *Amblystoma* produced by *Triturus* transplants. Science 80:78.

Twitty, V. C., Grant, D., and O. Anderson. 1964. Long distance homing in the newt *Taricha rivularis*. Proceedings of the National Academy of Science 54:51-58.

Twitty, V. C., Grant, D., and O. Anderson. 1967a. Home range in relation to homing in the newt

Taricha rivularis (Amphibia: Caudata). Copeia 1967:649–653.

Twitty, V. C., Grant, D., and O. Anderson. 1967b. Amphibian orientation: an unexpected observation. Science 155:352–353.

Twitty, V. C., Grant, D., and O. Anderson. 1967c. Initial homeward orientation after long distance displacement of the newt *Taricha rivularis*. Proceedings of the National Academy of Science 57:342–348.

Tyler, J. D., and H. N. Buscher. 1980. Notes on a population of larval *Ambystoma tigrinum* (Ambystomatidae) from Cimarron County, Oklahoma. The Southwestern Naturalist 25:391–395.

Uhlenhuth, E. 1921. Observations on the distribution and habitats of the blind Texan cave salamander, *Typhlomolge rathbuni*. Biological Bulletin 15:73–104.

Uhler, F. M., Cottom, C., and T. E. Clarke. 1939. Food of snakes of the George Washington National Forest, Virginia. Transactions of the North American Wildlife Conference 4:605–622.

Ultsch, G. R. 1971. The relationship of dissolved carbon dioxide and oxygen to microhabitat selection in *Pseudobranchus striatus*. Copeia 1971:247–252.

Ultsch, G. R. 1973. Observations on the life history of *Siren lacertina*. Herpetologica 29:304–305.

Ultsch, G. R., and S. J. Arceneaux. 1988. Gill loss in larval *Amphiuma tridactylum*. Journal of Herpetology 22:347–348.

Uzendoski, K., Maksymovitch, E., and P. A. Verrell. 1993. Do the risks of predation and intermale competition affect courtship behavior in the salamander *Desmognathus ochrophaeus*? Behavioral Ecology and Sociobiology 32:421–427.

Uzendoski, U. V., and P. A. Verrell. 1993. Sexual incompatibility and mate-recognition systems: a study of two species of sympatric salamanders (Plethodontidae). Animal Behaviour 46:267–278.

Uzzell, T. M., Jr. 1963. Natural triploidy in salamanders related to *Ambystoma jeffersonianum*. Science 139:113–115.

Uzzell, T. M., Jr. 1964a. Relations of the diploid and triploid species of the *Ambystoma jeffersonianum* complex (Amphibia, Caudata). Copeia 1964:257–300.

Uzzell, T. M., Jr. 1964b. Gynogenesis in salamanders related to *Ambystoma jeffersonianum*. Science 143:1043–1045.

Uzzell, T. M., Jr. 1967a. *Ambystoma jeffersonianum*. Catalogue of American Amphibians and Reptiles, pp. 47.1–47.2.

Uzzell, T. M., Jr. 1967b. *Ambystoma laterale*. Catalogue of American Amphibians and Reptiles, pp. 48.1–48.2.

Uzzell, T. M., Jr. 1967c. *Ambystoma tremblayi*. Catalogue of American Amphibians and Reptiles, pp. 50.1–50.2.

Uzzell, T. M., Jr. 1969. Notes on spermatophore production by salamanders of the *Ambystoma jeffersonianum* complex. Copeia 1969:602–612.

Uzzell, T. M., Jr., and S. M. Goldblatt. 1967. Serum proteins of salamanders of the *Ambystoma jeffersonianum* complex, and the origin of the triploid species of this group. Evolution 21:345–354.

Valentine, B. D. 1962. Intergrading populations and distribution of the salamander *Eurycea longicauda* in the Gulf States. Journal of the Ohio Herpetological Society 3:42–54.

Valentine, B. D. 1963a. The plethodontid salamander *Phaeognathus*: external morphology and zoogeography. Proceedings of the Biological Society of Washington 76:153–158.

Valentine, B. D. 1963b. The plethodontid salamander *Phaeognathus*: collecting techniques and habits. Journal of the Ohio Herpetological Society 4:49–54.

Valentine, B. D. 1963c. Notes on the early life history of the Alabama salamander, *Desmognathus aeneus chermocki* Bishop and Valentine. The American Midland Naturalist 69:182–188.

Valentine, B. D. 1963d. The salamander genus *Desmognathus* in Mississippi. Copeia 1963:130–139.

Valentine, B. D., and D. M. Dennis. 1964. A comparison of the gill-arch system and fins of three genera of larval salamanders, *Rhyacotriton*, *Gyrinophilus*, and *Ambystoma*. Copeia 1964:196–201.

Van Buskirk, J., and D. C. Smith. 1991. Density-dependent population regulation in a salamander. Ecology 72:1747–1756.

Van Denburgh, J. 1895. Notes on the habits and distribution of *Autodax iecanus*. Proceedings of the California Academy of Science 5:776–778.

Van Denburgh, J. 1905. The reptiles and amphibians of the islands of the Pacific Coast of North America from the Farallons to Cape San Lucas and the Revilla Gigedos. Proceedings of the California Academy of Science 4:1–38.

Van Denburgh, J. 1916. Four species of salamander new to the state of California, with a description of *Plethodon elongatus*, a new species, and notes on other salamanders. Proceedings of the California Academy of Science 6:215-221.

Van Devender, T. R., Lowe, C. H., McCrystal, H. K., and H. E. Lawler. 1992. Viewpoint: reconsider suggested systematic arrangements for some North American amphibians and reptiles. Herpetological Review 23:10-14.

Van Frank, R. 1955. *Paleotaricha oligocenica*, new genus and species of Oligocene salamander from Oregon. Breviora 45:1-12.

Van Hyning, O. C. 1932. Food of some Florida snakes. Copeia 1932:37.

Vernberg, F. J. 1953. Hibernation studies of two species of salamanders, *Plethodon cinereus cinereus* and *Eurycea bislineata bislineata*. Ecology 34:55-61.

Verrell, P. A. 1982a. The sexual behavior of the red-spotted newt, *Notophthalmus viridescens* (Amphibia: Urodela: Salamandridae). Animal Behaviour 30:1224-1236.

Verrell, P. A. 1982b. Male newts prefer large females as mates. Animal Behaviour 30:1254-1255.

Verrell, P. A. 1983. The influence of the ambient sex ratio and intermale competition on the sexual behavior of the red-spotted newt, *Notophthalmus viridescens* (Amphibia: Urodela: Salamandridae). Behavioral Ecology and Sociobiology 13:307-313.

Verrell, P. A. 1985a. Male mate choice for large, fecund females in the red-spotted newt, *Notophthalmus viridescens*: how is size assessed? Herpetologica 41:382-386.

Verrell, P. A. 1985b. Female availability and multiple courtship in male red-spotted newts, *Notophthalmus viridescens* (Amphibia): decisions that maximize male mating success. Behaviour 94:244-253.

Verrell, P. A. 1985c. Is there an energetic cost to sex? Activity, courtship mode and breathing in the red-spotted newt, *Notophthalmus viridescens* (Rafinesque). Monitore Zoologico Italiano 19:121-127.

Verrell, P. A. 1986. Wrestling in the red-spotted newt (*Notophthalmus viridescens*): resource value and contestant symmetry determine contest duration and outcome. Animal Behaviour 34:398-402.

Verrell, P. A. 1988a. Intrinsic male mating capacity is limited in the plethodontid salamander, *Desmognathus ochrophaeus*. Journal of Herpetology 22:394-400.

Verrell, P. A. 1988b. Mating and female sexual responsiveness in the salamander *Desmognathus ochrophaeus*. Herpetologica 44:334-337.

Verrell, P. A. 1988c. Tests for the effects of habitat complexity on mating frequency in mountain dusky salamanders (*Desmognathus ochrophaeus*) in a laboratory environment. Biology of Behaviour 13:1-10.

Verrell, P. A. 1989a. An experimental study of the behavioral basis of sexual isolation between two sympatric plethodontid salamanders, *Desmognathus imitator* and *D. ochrophaeus*. Ethology 80:274-282.

Verrell, P. A. 1989b. Male mate choice for fecund females in a plethodontid salamander. Animal Behaviour 38:1086-1088.

Verrell, P. A. 1990a. A note on the courtship of the broken-striped newt, *Notophthalmus viridescens dorsalis* (Harlan). Journal of Herpetology 24:215-217.

Verrell, P. A. 1990b. Tests for sexual isolation among sympatric salamanders of the genus *Desmognathus*. Amphibia-Reptilia 11:147-153.

Verrell, P. A. 1990c. Sexual compatibility among plethodontid salamanders: tests between *Desmognathus apalachicolae*, and *D. ochrophaeus* and *D. fuscus*. Herpetologica 46:415-422.

Verrell, P. A. 1990d. Frequency of interspecific mating in salamanders of the plethodontid genus *Desmognathus*: different experimental designs may yield different results. Journal of Zoology (London) 221:441-451.

Verrell, P. A. 1991a. Male mating success in the mountain dusky salamander, *Desmognathus ochrophaeus*: are small, young, inexperienced males at a disadvantage? Ethology 88:277-286.

Verrell, P. A. 1991b. Insemination temporarily inhibits sexual responsiveness in female salamanders (*Desmognathus ochrophaeus*). Behavior 119:51-64.

Verrell, P. A. 1994a. Males may choose larger females as mates in the salamander *Desmognathus fuscus*. Animal Behaviour 47:1465-1467.

Verrell, P. A. 1994b. Evidence against a role for experience in the maintenance of sexual incompatibility between sympatric salamanders. Herpetologica 50:475-479.

Verrell, P. A. 1994c. Is decreased frequency of mat-

ing among conspecifics a cost of sympatry in salamanders? Evolution 48:921–925.
Verrell, P. A. 1994d. The courtship behavior of the Apalachicola dusky salamander, *Desmognathus apalachicolae* Means and Karlin (Amphibia: Caudata: Plethodontidae). Ethology, Ecology, and Evolution 6:497–506.
Verrell, P. A. 1994e. Courtship of the salamander *Desmognathus imitator* (Amphibia: Caudata: Plethodontidae). Amphibia-Reptilia 15:135–142.
Verrell, P. A. 1995a. Males choose larger females as mates in the salamander *Desmognathus santeetlah*. Ethology 99:162–171.
Verrell, P. A. 1995b. The courtship behaviour of the spotted dusky salamander, *Desmognathus fuscus conanti* (Amphibia: Caudata: Plethodontidae). Journal of Zoology (London) 235:515–523.
Verrell, P. A. 1997. Courtship in desmognathine salamanders: The southern dusky salamander, *Desmognathus auriculatus*. Journal of Herpetology 31:273–277.
Verrell, P. A., and S. J. Arnold. 1989. Behavioral observations of sexual isolation among allopatric populations of the mountain dusky salamander, *Desmognathus ochrophaeus*. Evolution 43:745–755.
Verrell, P. A., and A. Donovan. 1991. Male-male aggression in the plethodontid salamander *Desmognathus ochrophaeus*. Journal of Zoology (London) 223:203–212.
Verrell, P. A., and S. G. Tilley. 1992. Population differentiation in plethodontid salamanders: divergence of allozymes and sexual compatibility among populations of *Desmognathus imitator* and *Desmognathus ochrophaeus* (Caudata: Plethodontidae). Zoology Journal of the Linnaean Society 104:67–80.
Vial, J. L., and F. B. Preib. 1966. Antibiotic assay of dermal secretions from the salamander, *Plethodon cinereus* (Green). Herpetologica 22:284–287.
Vial, J. L., and L. Saylor. 1993. The Status of Amphibian Populations. DAPTF working document 1, 98 pp.
Vickers, C. R., Harris, L. D., and B. F. Swindel. 1985. Changes in herpetofauna resulting from ditching of cypress ponds in coastal plains flatwoods. Forest Ecology and Management 11:17–29.
Villela, O. F., and R. A. Brandon. 1992. *Siren lacertina* (Amphibia: Caudata) in northeastern Mexico and southern Texas. Annals of Carnegie Museum 61:289–291.
Vincent, W. S. 1947. A checklist of amphibians and reptiles of Crater Lake National Park. Crater Lake National Park Nature Notes 13:19–22.
Viosca, P., Jr. 1937. A tentative revision of the genus *Necturus* with descriptions of three new species from the southern gulf drainage area. Copeia 1937:120–138.
Vogt, R. C. 1981. Natural History of Amphibians and Reptiles of Wisconsin. Milwaukee Public Museum, Milwaukee, Wisconsin, 205 pp.
Volpe, E. P., and C. R. Shoop. 1963. Diagnosis of larvae of *Ambystoma talpoideum*. Copeia 1963:444–447.
Voss, S. R. 1993a. Effect of temperature on body size, developmental stage, and timing of hatching in *Ambystoma maculatum*. Journal of Herpetology 27:329–333.
Voss, S. R. 1993b. Relationship between stream order and length of larval period in the salamander *Eurycea wilderae*. Copeia 1993:736–742.
Voss, S. R., Smith, D. G., Beachy, C. K., and D. G. Heckel. 1995. Allozyme variation in neighboring isolated populations of the plethodontid salamander *Leurognathus marmoratus*. Journal of Herpetology 29:493–497.
Wake, D. B. 1963. Comparative osteology of the plethodontid salamander genus *Aneides*. Journal of Morphology 113:77–118.
Wake, D. B. 1965a. *Aneides ferreus*. Catalogue of American Amphibians and Reptiles, p. 16.
Wake, D. B. 1965b. *Aneides hardii*. Catalogue of American Amphibians and Reptiles, p. 17.
Wake, D. B. 1966. Comparative osteology and evolution of the lungless salamanders, family Plethodontidae. Memoirs of the Southern California Academy of Science 4:1–111.
Wake, D. B. 1970. The abundance and diversity of tropical salamanders. The American Naturalist 104:211–213.
Wake, D. B. 1993. Phylogenetic and taxonomic issues relating to salamanders of the family Plethodontidae. Herpetologica 1993:229–237.
Wake, D. B. 1996. A new species of *Batrachoseps* (Amphibia: Plethodontidae) from the San Gabriel Mountains, southern California. Natural History Museum of Los Angeles County Contributions in Science 463:1–12.
Wake, D. B., and A. H. Brame, Jr. 1969. Systematics and evolution of neotropical salamanders of

Wake, D. B., and J. Castanet. 1995. A skeletochronological study of growth and age in relation to adult size in *Batrachoseps attenuatus*. Journal of Herpetology 29:60–65.

Wake, D. B., and I. G. Dresner. 1967. Functional morphology and evolution of tail autotomy in salamanders. Journal of Morphology 122:265–306.

Wake, D. B., and A. Larson. 1987. Multidimensional analysis of an evolving lineage. Science 238:42–48.

Wake, D. B., and J. F. Lynch. 1976. The distribution, ecology, and evolutionary history of plethodontid salamanders in tropical America. Los Angeles County Museum of Natural History Bulletin 25:1–65.

Wake, D. B., and J. F. Lynch. 1982. Evolutionary relationships among Central American salamanders of the *Bolitoglossa franklini* group, with a description of a new species from Guatemala. Herpetologica 38:257–272.

Wake, D. B., and S. B. Marks. 1993. Development and evolution of plethodontid salamanders: a review of prior studies and a prospectus for future research. Herpetologica 49:194–203.

Wake, D. B., and N. Özeti. 1969. Evolutionary relationships in the family Salamandridae. Copeia 1969:124–137.

Wake, D. B., and K. P. Yanev. 1986. Geographic variation in allozymes in a "ring species," the plethodontid salamander *Ensatina eschscholtzii* of western North America. Evolution 40:702–715.

Wake, D. B., Yanev, K. P., and C. W. Brown. 1986. Intraspecific sympatry in a "ring species," the plethodontid salamander *Ensatina eschscholtzii*, in southern California. Evolution 40:866–868.

Wake, D. B., Yanev, K. P., and M. M. Frelow. 1989. Sympatry and hybridization in a ring species: the plethodontid salamander *Ensatina eschscholtzii*. In: Speciation and Its Consequences, D. Copeland and J. A. Endler, eds. Sinauer Associates, Sunderland, Massachusetts, pp. 134–157.

Wakeley, J. F., Fuhrman, G. J., Fuhrman, F. A., Fischer, H. G., and H. S. Mosher. 1966. The occurrence of tetrodotoxin (tarichatoxin) in Amphibia and the distribution of the toxin in the organs of newts (*Taricha*). Toxicon 3:195–203.

Walker, C. F. 1934. *Plethodon welleri* at White Top Mountain, Virginia. Copeia 1934:190.

Walker, C. F., and W. Goodpaster. 1941. The green salamander, *Aneides aeneus*, in Ohio. Copeia 1941:178.

Wallace, J. T. 1969. A study on *Plethodon richmondi* from Mason County, Kentucky, with notes on its distribution within the state. Proceedings of the Kentucky Academy of Science 29:38–44.

Wallace, J. T., and R. W. Barbour. 1957. Observations on the eggs and young of *Plethodon richmondi*. Copeia 1957:48.

Wallace, R. S. 1984. Use of *Sphagnum* moss for nesting by the four-toed salamander, *Hemidactylium scutatum* Schlegl. (Plethodontidae). Proceedings of the Pennsylvania Academy of Science 58:237–238.

Walls, S. C. 1990. Interference competition in postmetamorphic salamanders: interspecific differences in aggression by coexisting species. Ecology 71:307–314.

Walls, S. C. 1991. Ontogenetic shifts in the recognition of siblings and neighbors by juvenile salamanders. Animal Behaviour 42:423–434.

Walls, S. C. 1995. Differential vulnerability to predation and refuge use in competing larval salamanders. Oecologia 101:86–93.

Walls, S. C., and R. Altig. 1986. Female reproductive biology and larval life history of *Ambystoma* salamanders: a comparison of egg size, hatchling size, and larval growth. Herpetologica 42:334–345.

Walls, S. C., and A. R. Blaustein. 1994. Does kinship influence density dependence in a larval salamander? Oikos 71:459–468.

Walls, S. C., and A. R. Blaustein. 1995. Larval marbled salamanders, *Ambystoma opacum*, eat their kin. Animal Behaviour 50:537–545.

Walls, S. C., and R. G. Jaeger. 1987. Aggression and exploitation as mechanisms of competition in larval salamanders. Canadian Journal of Zoology 65:2938–2944.

Walls, S. C., and R. G. Jaeger. 1989. Growth in larval salamanders is not inhibited through chemical interference competition. Copeia 1989:1049–1052.

Walls, S. C., and R. E. Roudebush. 1991. Reduced aggression toward siblings as evidence of kin recognition in cannibalistic salamanders. The American Naturalist 138:1027–1038.

Walls, S. C., and R. D. Semlitsch. 1991. Visual and

movement displays function as agonistic behavior in larval salamanders. Copeia 1991:936–942.

Walls, S. C., Beatty, J. J., Tissot, B. N., Hokit, D. G., and A. R. Blaustein. 1993b. Morphological variation and cannibalism in a larval salamander (*Ambystoma macrodactylum columbianum*). Canadian Journal of Zoology 71:1543–1551.

Walls, S. C., Belanger, S. S., and A. R. Blaustein. 1993a. Morphological variation in a larval salamander: dietary induction of plasticity in head shape. Oecologia 1993:162–168.

Walls, S. C., Conrad, C. S., Murillo, M. L., and A. R. Blaustein. 1996. Agonistic behaviour in larvae of the northwestern salamander (*Ambystoma gracile*): the effects of kinship, familiarity, and population source. Behaviour 133:965–984.

Walls, S. C., Mathis, A., Jaeger, R. G., and W. F. Gergits. 1989. Male salamanders with high-quality diets have faeces attractive to females. Animal Behaviour 38:546–548.

Walters, B. 1975. Studies of interspecific predation within an amphibian community. Journal of Herpetology 9:267–279.

Ward, D., and O. J. Sexton. 1981. Anti-predator role of salamander egg membranes. Copeia 1981:724–726.

Warner, J. W. 1971. The distribution of *Plethodon glutinosus* (Green) in central Louisiana with a taxonomic comparison to neighboring populations. Journal of Herpetology 5:115–119.

Watermolen, D. J. 1996. *Plethodon cinereus* (redback salamander). Brooding behavior. Herpetological Review 27:136–137.

Watney, G. M. S. 1941. Notes on the life history of *Ambystoma gracile* Baird. Copeia 1941:14–17.

Webb, R. G., and W. L. Roueche. 1971. Life history aspects of the tiger salamander (*Ambystoma tigrinum mavortium*) in the Chihuahuan desert. The Great Basin Naturalist 31:193–212.

Weber, J. A. 1928. Herpetological observations in the Adirondack Mountains, New York. Copeia 1928:106–112.

Weber, J. A. 1944. Observations on the life history of *Amphiuma means*. Copeia 1944:61–62.

Webster, D. A. 1960. Toxicity of the spotted newt, *Notophthalmus viridescens*, to trout. Copeia 1960:74–75.

Weichert, C. K. 1945. Seasonal variation in the mental gland and reproductive organs of the male *Eurycea bislineata*. Copeia 1945:78–84.

Weller, W. F., and B. W. Menzel. 1979. Occurrence of the salamander *Ambystoma platineum* (Cope) in southern Ontario. Journal of Herpetology 13:193–197.

Weller, W. F., and W. G. Sprules. 1976. Taxonomic status of male salamanders of the *Ambystoma jeffersonianum* complex from an Ontario population, with the first record of the Jefferson salamander, *Ambystoma jeffersonianum* (Green), from Canada. Canadian Journal of Zoology 54:1270–1276.

Weller, W. F., Sprules, W. G., and T. P. Lamarre. 1978. Distribution of salamanders of the *Ambystoma jeffersonianum* complex in Ontario. The Canadian Field-Naturalist 92:174–181.

Weller, W. H. 1931. A preliminary list of the salamanders of the Great Smoky Mountains of North Carolina and Tennessee. Proceedings of the Junior Society of Natural Science of Cincinnati 2:21–32.

Wells, K. D. 1980b. Spatial associations among individuals in a population of slimy salamanders (*Plethodon glutinosus*). Herpetologica 36:271–275.

Wells, K. D., and R. A. Wells. 1976. Patterns of movement in a population of the slimy salamander, *Plethodon glutinosus*, with observations on aggregations. Herpetologica 32:156–162.

Wells, M. M. 1963. An incidence of albinism in *Taricha torosa*. Herpetologica 19:291.

Wells, P. H., and W. Gordon. 1958. Brooding slimy salamanders, *Plethodon glutinosus glutinosus* (Green). National Speleological Society Bulletin 20:23–24.

Welsh, H. H., Jr. 1986. *Dicamptodon ensatus* (Pacific giant salamander). Behavior. Herpetological Review 17:19.

Welsh, H. H., Jr. 1990. Relictual amphibians and old-growth forests. Conservation Biology 4:309–319.

Welsh, H. H., Jr., and A. J. Lind. 1988. Old growth forests and the distribution of the terrestrial herpetofauna. In: Management of amphibians, reptiles, and mammals in North America, R. C. Szaro, K. E. Severson, and D. R. Patton, eds. USDA Forest Service, Rocky Mountain Forest and Range Experimental Station, Fort Collins, Colorado, General Technical Report RM-166, pp. 439–455.

Welsh, H. H., Jr., and A. J. Lind. 1991. The structure of the herpetofaunal assemblage in the Douglas-fir/hardwood forests of northwestern

California and southwestern Oregon. In: Wildlife and vegetation of unmanaged Douglas-fir forests, L. F. Ruggiero, K. B. Aubry, A. B. Carey, and M. H. Huff, technical coordinators. USDA Forest Service, General Technical Report PNW-GTR-285, pp. 394–413.

Welsh, H. H., Jr., and A. J. Lind. 1992. Population ecology of two relictual salamanders from the Klamath Mountains of northwestern California. In: Wildlife 2001: Populations, D. R. McCullough and R. H. Barrett, eds. Elsevier Applied Science, London, pp. 419–437.

Welsh, H. H., Jr., and A. J. Lind. 1996. Habitat correlates of the southern torrent salamander, *Rhyacotriton variegatus* (Caudata: Rhyacotritonidae), in northwestern California. Journal of Herpetology 30:385–398.

Welsh, H. H., Jr., and R. A. Wilson. 1995. *Aneides ferreus* (clouded salamander). Reproduction. Herpetological Review 26:196–197.

Welter, W. A., and R. W. Barbour. 1940. Additions to the herpetofauna of northeastern Kentucky. Copeia 1940:132.

Welter, W. A., and K. Carr. 1939. Amphibians and reptiles of northeastern Kentucky. Copeia 1939:128–130.

Werner, E. E., and M. A. McPeek. 1994. Direct and indirect effects of predators on two anuran species along an environmental gradient. Ecology 75:1368–1382.

Werner, J. K. 1971. Notes on the reproductive cycle of *Plethodon cinereus* in Michigan. Copeia 1971:161–162.

Westell, P. A., and F. D. Ross. 1974. Erythristic red-backed salamanders *Plethodon cinereus*, from Ontario. The Canadian Field-Naturalist 88:231–232.

Whipple, A. V., and W. A. Dunson. 1993. Amelioration of the toxicity of H+ to larval stoneflies by metals found in coal mine effluent. Archives of Environmental Contamination and Toxicology 24:194–200.

Whitaker, J. O., Jr., and D. C. Rubin. 1971. Food habits of *Plethodon jordani metcalfi* and *Plethodon jordani shermani* from North Carolina. Herpetologica 27:81–86.

Whitaker, J. O., Jr., Cudmore, W. W., and B. A. Brown. 1982. Foods of larval, subadult, and adult small-mouthed salamanders, *Ambystoma texanum*, from Vigo County, Indiana. Proceedings of the Indiana Academy of Science 90:461–464.

Whitaker, J. O., Jr., Maser, C., Storm, R. M., and J. J. Beatty. 1986. Food habits of clouded salamanders (*Aneides ferreus*) in Curry County, Oregon (Amphibia: Caudata: Plethodontidae). The Great Basin Naturalist 46:228–240.

White, R. L., II. 1977. Prey selection by the rough-skinned newt (*Taricha granulosa*) in two ponds. Northwest Science 51:114–118.

Whiteman, H. H., and S. A. Wissinger. 1991. Differences in the antipredator behavior of three plethodontid salamanders to snake attack. Journal of Herpetology 25:352–355.

Whiteman, H. H., Howard, R. D., and K. A. Whitten. 1995. Effects of pH on embryo tolerance and adult behavior in the tiger salamander, *Ambystoma tigrinum tigrinum*. Canadian Journal of Zoology 73:1529–1537.

Whiteman, H. H., Wissinger, S. A., and A. J. Bohonak. 1994. Seasonal movement patterns in a subalpine population of the tiger salamander, *Ambystoma tigrinum nebulosum*. Canadian Journal of Zoology 72:1780–1787.

Whitford, W. G., and M. Massey. 1970. Responses of a population of *Ambystoma tigrinum* to thermal and oxygen gradients. Herpetologica 26:372–376.

Whitford, W. G., and A. Vinegar. 1966. Homing, survivorship, and overwintering larvae in spotted salamanders, *Ambystoma maculatum*. Copeia 1966:515–519.

Wilbur, H. M. 1971. The ecological relationship of the salamander *Ambystoma laterale* to its all-female, gynogenetic associate. Evolution 25:168–179.

Wilbur, H. M. 1972. Competition, predation, and structure of the *Ambystoma–Rana sylvatica* community. Ecology 53:3–21.

Wilbur, H. M. 1977. Propagule size, number, and dispersion pattern in *Ambystoma* and *Asclepias*. The American Naturalist 111:43–68.

Wilbur, H. M., and J. P. Collins. 1973. Ecological aspects of amphibian metamorphosis. Science 182:1305–1314.

Wilbur, H. M., and J. E. Fauth. 1990. Experimental aquatic food webs: interactions between two predators and two prey. The American Naturalist 135:176–204.

Wilbur, H. M., Morin, P. J., and R. N. Harris. 1983. Salamander predation and the structure of

experimental communities: anuran responses. Ecology 64:1423-1429.
Wilder, H. H. 1899. *Desmognathus fusca* (Rafinesque) and *Sperlerpes bislineatus* (Green). The American Naturalist 33:231-246.
Wilder, I. W. 1913. The life history of *Desmognathus fusca*. Biological Bulletin 24:251-342.
Wilder, I. W. 1924. The developmental history of *Eurycea bislineata* in western Mass. Copeia 1924:77-80.
Wiley, E. O. 1978. The evolutionary species concept reconsidered. Systematic Zoology 27:17-26.
Wilkinson, R. F., Peterson, C. L., Moll, D., and T. Holder. 1993. Reproductive biology of *Plethodon dorsalis* in northwestern Arkansas. Journal of Herpetology 27:85-87.
Williams, A. A. 1980. Fluctuations in a population of the cave salamander. National Speleological Society Bulletin 42:49-52.
Williams, E. E., Highton, R., and D. M. Cooper. 1968. Breakdown of polymorphism of the redbacked salamander on Long Island. Evolution 22:76-86.
Williams, K. L., and R. E. Gordon. 1961. Natural dispersal of the salamander *Aneides aeneus*. Copeia 1961:353.
Williams, R. D., Gates, J. E., Hocutt, C. H., and G. T. Taylor. 1981. The hellbender: a nongame species in need of management. Wildlife Society Bulletin 9:94-100.
Williams, S. R. 1973. *Plethodon neomexicanus*. Catalogue of American Amphibians and Reptiles, pp. 131.1-131.2.
Williams, S. R. 1978. Comparative reproduction of the endemic New Mexico plethodontid salamanders, *Plethodon neomexicanus* and *Aneides hardii* (Amphibia, Urodela, Plethodontidae). Journal of Herpetology 12:471-476.
Wilson, A. G., Jr., and J. H. Larsen, Jr. 1988. Activity and diet in seepage-dwelling Coeur d'Alene salamanders (*Plethodon vandykei idahoensis*). Northwest Science 62:211-217.
Wilson, A. G., Jr., and E. M. Simon. 1985. *Plethodon vandykei idahoensis* (Coeur d'Alene salamander). Predation. Herpetological Review 16:111.
Wilson, A. G., Jr., and E. M. Wilson. 1996. *Plethodon idahoensis* (Coeur d'Alene salamander). Snake predation. Herpetological Review 27:138.
Wilson, A. G., Jr., Larsen, J. H., Jr., and K. R. McAllister. 1995. Distribution of Van Dyke's salamander (*Plethodon vandykei* Van Denburgh). The American Midland Naturalist 134:388-393.
Wilson, F. H. 1940. The life cycle of *Amphiuma* in the vicinity of New Orleans based on a study of the gonads and gonaducts (abstract). Anatomical Record 78 (suppl.):104.
Wilson, F. H. 1941a. The cloaca in the male *Amphiuma tridactylum* (abstract). Anatomical Record 81 (suppl.):63.
Wilson, F. H. 1941b. Age, maturity and growth in a population of *Amphiuma tridactylum* (abstract). Anatomical Record 81 (suppl.):63.
Wilson, F. H. 1942. The cycle of egg and sperm production in *Amphiuma tridactylum* Cuvier (abstract). Anatomical Record 84 (suppl.):532.
Wilson, L. W., and S. B. Friddle. 1950. The herpetology of Hardy County, West Virginia. The American Midland Naturalist 43:165-171.
Wilson, R. L. 1970. *Dicamptodon ensatus* feeding on a microtine. Journal of Herpetology 4:93.
Wiltenmuth, E. B. 1996. Agonistic and sensory behaviour of the salamander *Ensatina eschscholtzii* during asymmetric contests. Animal Behaviour 52:841-850.
Wise, S. E., Siex, K. S., Brown, K. M., and R. G. Jaeger. 1993. Recognition influences social interactions in red-spotted newts. Journal of Herpetology 27:149-154.
Wissinger, S. A., and H. H. Whiteman. 1992. Fluctuation in a Rocky Mountain population of salamanders: anthropogenic acidification or natural variation? Journal of Herpetology 26:377-391.
Wood, J. T. 1945. Ovarian eggs in *Plethodon richmondi*. Herpetologica 2:206-207.
Wood, J. T. 1947a. Juveniles of *Plethodon jordani* Blatchley. Herpetologica 3:185-188.
Wood, J. T. 1947b. Description of juvenile *Plethodon glutinosus shermani* Stejneger. Herpetologica 3:188.
Wood, J. T. 1949. *Eurycea bislineata wilderae* Dunn. Herpetologica 5:61-62.
Wood, J. T. 1953a. The nesting of the two-lined salamander, *Eurycea bislineata*, on the Virginia coastal plain. Natural History Miscellanea of the Chicago Academy of Science 122:1-7.
Wood, J. T. 1953b. Observations on the complements of ova and nesting of the four-toed salamander in Virginia. The American Naturalist 87:77-86.

Wood, J. T. 1955. The nesting of the four-toed salamander, *Hemidactylium scutatum* (Schlegel), in Virginia. The American Midland Naturalist 53:381–389.

Wood, J. T., and R. F. Clarke. 1955. The dusky salamander: oophagy in nesting sites. Herpetologica 11:150–151.

Wood, J. T., and W. E. Duellman. 1951. Ovarian egg complements in the salamander *Eurycea bislineata rivicola* Mittleman. Copeia 1951:181.

Wood, J. T., and M. E. Fitzmaurice. 1948. Eggs, larvae, and attending females of *Desmognathus f. fuscus* in southwestern Ohio and southeastern Indiana. The American Midland Naturalist 39:93–95.

Wood, J. T., and O. K. Goodwin. 1954. Observations on the abundance, food, and feeding behaviour of the newt *Notophthalmus viridescens viridescens* in Virginia. Journal of the Elisha Mitchell Scientific Society 70:27–30.

Wood, J. T., and H. N. McCutcheon. 1954. Ovarian egg complements and nests of the two-lined salamander, *Eurycea b. bislineata* × *cirrigera,* from southeastern Virginia. The American Midland Naturalist 52:433–436.

Wood, J. T., and R. H. Rageot. 1955. The eggs of the slimy salamander in Isle of Wight County, Virginia. Virginia Journal of Science 6:85–87.

Wood, J. T., and R. H. Rageot. 1963. The nesting of the many-lined salamander in the Dismal Swamp. Virginia Journal of Science 14:121–125.

Wood, J. T., and F. E. Wood. 1955. Notes on the nests and nesting of the Carolina mountain dusky salamander in Tennessee and Virginia. Journal of the Tennessee Academy of Science 30:36–39.

Wood, W. F. 1934. Notes on the salamander, *Plethodon elongatus.* Copeia 1934:191.

Wood, W. F. 1936. *Aneides flavipunctatus* in burnt-over areas. Copeia 1936:171.

Woodbury, A. M. 1952. Amphibians and reptiles of the Great Salt Lake Valley. Herpetologica 8:42–50.

Woods, J. E. 1969. The ecology and natural history of Mississippi populations of *Aneides aeneus* and associated salamanders (abstract). Dissertation Abstracts International 29B:3554.

Woodward, B. D. 1982. Local interpopulational variation in clutch parameters in the spotted salamander (*Ambystoma maculatum*). Copeia 1982:157–160.

Worthington, R. D. 1968. Observations on the relative sizes of three species of salamander larvae in a Maryland pond. Herpetologica 24:242–246.

Worthington, R. D. 1969. Additional observations on sympatric species of salamander larvae in a Maryland pond. Herpetologica 25:227–229.

Worthington, R. D., and D. B. Wake. 1971. Larval morphology and ontogeny of the Ambystomatid salamander, *Rhyacotriton olympicus.* The American Midland Naturalist 85:349–365.

Worthylake, K. M., and P. Hovingh. 1989. Mass mortality of salamanders (*Ambystoma tigrinum*) by bacteria (*Acinetobacter*) in an oligotrophic seepage mountain lake. The Great Basin Naturalist 49:364–372.

Wright, A. H. 1908. Notes on the breeding habits of *Amblystoma punctatum.* Biological Bulletin 14:286.

Wright, A. H., and A. A. Allen. 1909. The early breeding habits of *Amblystoma punctatum.* The American Naturalist 43:687–692.

Wright, A. H., and J. M. Haber. 1922. The carnivorous habits of the purple salamander. Copeia 1922:31–32.

Wrobel, D. J., Gergits, W. F., and R. G. Jaeger. 1980. An experimental study of interference competition among terrestrial salamanders. Ecology 61:1034–1039.

Wyman, R. L. 1971. The courtship behavior of the small-mouthed salamander, *Ambystoma texanum.* Herpetologica 27:491–498.

Wyman, R. L. 1988. Soil acidity and moisture and the distribution of amphibians in five forests of south-central New York. Copeia 1988:394–399.

Wyman, R. L., and D. Hawksley-Lescault. 1987. Soil acidity affects distribution, behavior, and physiology of the salamander *Plethodon cinereus.* Ecology 68:1819–1827.

Wyman, R. L., and J. Jancola. 1992. Degree and scale of terrestrial acidification and amphibian community structure. Journal of Herpetology 26:392–401.

Wyman, R. L., and J. H. Thrall. 1972. Sound production by the spotted salamander, *Ambystoma maculatum.* Herpetologica 28:210–212.

Wynn, A. H., Highton, R., and J. F. Jacobs. 1988. A new species of rock-crevice dwelling *Plethodon* from Pigeon Mountain, Georgia. Herpetologica 44:135–143.

Yanev, K. P. 1980. Biogeography and distribution of

three parapatric salamander species in coastal and borderland California. In: The California islands: Proceedings of a multidisciplinary symposium, D. M. Power, ed. Santa Barbara Museum of Natural History, Santa Barbara, California, pp. 531–550.

Yanev, K. P., and D. B. Wake. 1981. Genic differentiation in a relict desert salamander, *Batrachoseps campi*. Herpetologica 37:16–28.

Yeatman, H. C., and H. B. Miller. 1985. A naturally metamorphosed *Gyrinophilus palleucus* from the type-locality. Journal of Herpetology 19:306–308.

Zalisko, E. J., and J. H. Larsen, Jr. 1989. Fate of unused sperm in post-breeding male *Ambystoma macrodactylum columbianum*. Journal of Herpetology 23:463–464.

Zerba, K. E., and J. P. Collins. 1992. Spatial heterogeneity and individual variation in diet of an aquatic top predator. Ecology 73:268–279.

Zweifel, R. G. 1949. Comparison of food habits of *Ensatina eschscholtzii* and *Aneides lugubris*. Copeia 1949:285–287.

Collection Localities and Photographic Credits for Color Plates

The information that follows includes the collection locality of specimens—county and state if available—and (in parentheses) the photographer.

PLATE 1. *Ambystoma annulatum*, Stone Co., Missouri (R. W. Van Devender).
PLATE 2. *Ambystoma barbouri*, Franklin Co., Kentucky (J. W. Petranka).
PLATE 3. *Ambystoma californiense*, Monterey Co., California (R. W. Van Devender).
PLATE 4. *Ambystoma cingulatum*, Chatham Co., Georgia (R. W. Van Devender).
PLATE 5. *Ambystoma cingulatum*, Berkeley Co., South Carolina (R. W. Van Devender).
PLATE 6. *Ambystoma g. gracile*, Thurston Co., Washington (W. P. Leonard).
PLATE 7. *Ambystoma g. gracile*, Lane Co., Oregon (R. W. Van Devender).
PLATE 8. *Ambystoma jeffersonianum*, northern Kentucky (R. W. Barbour).
PLATE 9. *Ambystoma laterale*, Schoolcraft Co., Michigan (R. W. Van Devender).
PLATE 10. *Ambystoma mabeei*, Sampson Co., North Carolina (R. W. Van Devender).
PLATE 11. *Ambystoma mabeei*, Scotland Co., North Carolina (R. W. Van Devender).
PLATE 12. *Ambystoma macrodactylum*, Deschutes Co., Oregon (R. W. Van Devender).
PLATE 13. *Ambystoma macrodactylum columbianum*, Grant Co., Washington (W. P. Leonard).
PLATE 14. *Ambystoma maculatum*, Macon Co., Alabama (J. W. Petranka).
PLATE 15. *Ambystoma maculatum*, Watauga Co., North Carolina (R. W. Van Devender).
PLATE 16. *Ambystoma opacum*, locality unknown (Ken Nemuras, courtesy of R. W. Van Devender).
PLATE 17. *Ambystoma talpoideum*, Beaufort Co., South Carolina (R. W. Van Devender).
PLATE 18. *Ambystoma talpoideum*, Henderson Co., North Carolina (R. W. Van Devender).
PLATE 19. *Ambystoma texanum*, Cleveland Co., Oklahoma (R. W. Van Devender).
PLATE 20. *Ambystoma tigrinum mavortium*, Cimmaron Co., Oklahoma (R. W. Van Devender).
PLATE 21. *Ambystoma tigrinum nebulosum*, Yavapai Co., Arizona (R. W. Van Devender).
PLATE 22. *Ambystoma tigrinum melanostictum*, Douglas Co., Washington (W. P. Leonard).
PLATE 23. *Ambystoma t. tigrinum*, Minnesota (R. W. Barbour).
PLATE 24. *Ambystoma tigrinum mavortium*, Pima Co., Arizona (R. W. Van Devender).
PLATE 25. *Ambystoma platineum*, locality unknown (R. W. Van Devender).
PLATE 26. *Ambystoma tremblayi*, locality unknown (R. W. Van Devender).
PLATE 27. *Amphiuma means*, Long Co., Georgia (R. W. Van Devender).
PLATE 28. *Amphiuma pholeter*, Florida (R. W. Van Devender).
PLATE 29. *Amphiuma tridactylum*, Jefferson Co., Texas (R. W. Van Devender).
PLATE 30. *Cryptobranchus a. alleganiensis*, Tazewell Co., Virginia (R. W. Van Devender).
PLATE 31. *Dicamptodon aterrimus*, Idaho (C. R. Peterson).
PLATE 32. *Dicamptodon copei*, Mason Co., Washington (R. W. Van Devender).
PLATE 33. *Dicamptodon copei*, Grays Harbor Co., Washington (W. P. Leonard).
PLATE 34. *Dicamptodon ensatus*, Santa Clara Co., California (M. Garcia-Paris).

PLATE 35. *Dicamptodon tenebrosus*, Humboldt Co., California (R. W. Van Devender).

PLATE 36. *Dicamptodon tenebrosus*, Pacific Co., Washington (W. P. Leonard).

PLATE 37. *Dicamptodon tenebrosus*, Multnomah Co., Oregon (W. P. Leonard).

PLATE 38. *Desmognathus aeneus*, Graham Co., North Carolina (R. W. Van Devender).

PLATE 39. *Desmognathus apalachicolae*, Liberty Co., Florida (R. W. Van Devender).

PLATE 40. *Desmognathus auriculatus*, Carteret Co., North Carolina (R. W. Van Devender).

PLATE 41. *Desmognathus auriculatus*, New Hanover Co., North Carolina (R. W. Van Devender).

PLATE 42. *Desmognathus brimleyorum*, locality unknown (R. W. Van Devender).

PLATE 43. *Desmognathus carolinensis*, Buncombe Co., North Carolina (J. W. Petranka).

PLATE 44. *Desmognathus fuscus*, Caldwell Co., North Carolina (R. W. Van Devender).

PLATE 45. *Desmognathus fuscus*, Caldwell Co., North Carolina (R. W. Van Devender).

PLATE 46. *Desmognathus imitator*, Swain Co., North Carolina (J. W. Petranka).

PLATE 47. *Desmognathus marmoratus*, Haywood Co., North Carolina (R. W. Van Devender).

PLATE 48. *Desmognathus marmoratus*, larva, Caldwell Co., North Carolina (R. W. Van Devender).

PLATE 49. *Desmognathus monticola*, Watauga Co., North Carolina (R. W. Van Devender).

PLATE 50. *Desmognathus monticola*, Henderson Co., North Carolina (R. W. Van Devender).

PLATE 51. *Desmognathus ochrophaeus*, eastern Kentucky (R. W. Barbour).

PLATE 52. *Desmognathus ocoee*, Macon Co., North Carolina (J. W. Petranka).

PLATE 53. *Desmognathus ocoee*, Macon Co., North Carolina (J. W. Petranka).

PLATE 54. *Desmognathus orestes*, Avery Co., North Carolina (J. W. Petranka).

PLATE 55. *Desmognathus quadramaculatus*, Macon Co., North Carolina (J. W. Petranka).

PLATE 56. *Desmognathus welteri*, Bell Co., Kentucky (J. W. Petranka).

PLATE 57. *Desmognathus wrighti*, Graham Co., North Carolina (R. W. Van Devender).

PLATE 58. *Phaeognathus hubrichti*, Alabama (R. W. Van Devender).

PLATE 59. *Batrachoseps aridus*, Riverside Co., California (M. Garcia-Paris).

PLATE 60. *Batrachoseps attenuatus*, Santa Cruz Co., California (J. W. Petranka).

PLATE 61. *Batrachoseps campi*, Inyo Co., California (M. Garcia-Paris).

PLATE 62. *Batrachoseps nigriventris*, Madera Co., California (M. Garcia-Paris).

PLATE 63. *Batrachoseps pacificus major*, Los Angeles Co., California (R. W. Van Devender).

PLATE 64. *Batrachoseps simatus*, Kern Co., California (M. Garcia-Paris).

PLATE 65. *Batrachoseps stebbinsi*, Kern Co., California (J. T. Collins and the Center for North American Amphibians and Reptiles).

PLATE 66. *Batrachoseps wrighti*, Lane Co., Oregon (R. W. Van Devender).

PLATE 67. *Hydromantes brunus*, Mariposa Co., California (M. Garcia-Paris).

PLATE 68. *Hydromantes platycephalus*, Inyo Co., California (M. Garcia-Paris).

PLATE 69. *Hydromantes shastae*, Shasta Co., California (M. Garcia-Paris).

PLATE 70. *Eurycea bislineata cirrigera*, Powell Co., Kentucky (J. W. Petranka).

PLATE 71. *Eurycea bislineata cirrigera*, Wake Co., North Carolina (R. W. Van Devender).

PLATE 72. *Eurycea junaluska*, Graham Co., North Carolina (R. W. Van Devender).

PLATE 73. *Eurycea junaluska*, Graham Co., North Carolina (R. W. Van Devender).

PLATE 74. *Eurycea guttolineata*, Calloway Co., Kentucky (R. W. Barbour).

PLATE 75. *Eurycea longicauda melanopleura*, Cherokee Co., Oklahoma (R. W. Van Devender).

PLATE 76. *Eurycea l. longicauda*, Bell Co., Kentucky (J. W. Petranka).

PLATE 77. *Eurycea longicauda melanopleura*, Cherokee Co., Oklahoma (R. W. Van Devender).

PLATE 78. *Eurycea lucifuga*, Fayette Co., Kentucky (J. W. Petranka).

PLATE 79. *Eurycea m. multiplicata*, Polk Co., Arkansas (R. W. Van Devender).

PLATE 80. *Eurycea nana*, Hays Co., Texas (R. W. Van Devender).

PLATE 81. *Eurycea neotenes*, Travis Co., Texas (R. W. Van Devender).

PLATE 82. *Eurycea quadridigitata*, Scotland Co., North Carolina (R. W. Van Devender).

Collection Localities and Photographic Credits for Color Plates

PLATE 83. *Eurycea rathbuni*, Texas (R. W. Van Devender).

PLATE 84. *Eurycea sosorum*, Travis Co., Texas (W. Meinzer, courtesy of P. T. Chippindale).

PLATE 85. *Eurycea tridentifera*, Comal Co., Texas (P. T. Chippindale).

PLATE 86. *Eurycea tynerensis*, Oklahoma (R. W. Van Devender).

PLATE 87. *Gyrinophilus palleucus necturoides*, Grundy Co., Tennessee (R. W. Van Devender).

PLATE 88. *Gyrinophilus porphyriticus danielsi*, Buncombe Co., North Carolina (J. W. Petranka).

PLATE 89. *Gyrinophilus porphyriticus duryi*, northeastern Kentucky (R. W. Barbour).

PLATE 90. *Gyrinophilus subterraneus*, Greenbrier Co., West Virginia (R. W. Van Devender).

PLATE 91. *Gyrinophilus subterraneus*, Greenbrier Co., West Virginia (R. W. Van Devender).

PLATE 92. *Haideotriton wallacei*, Jackson Co., Florida (R. W. Van Devender).

PLATE 93. *Hemidactylium scutatum*, Orange Co., North Carolina (J. W. Petranka).

PLATE 94. *Pseudotriton montanus diasticus*, Adams Co., Ohio (R. W. Van Devender).

PLATE 95. *Pseudotriton montanus floridanus*, Alachua Co., Florida (R. W. Van Devender).

PLATE 96. *Pseudotriton m. montanus*, Watauga Co., North Carolina (R. W. Van Devender).

PLATE 97. *Pseudotriton r. ruber*, Powell Co., Kentucky (J. W. Petranka).

PLATE 98. *Pseudotriton r. ruber*, Wolfe Co., Kentucky (J. W. Petranka).

PLATE 99. *Pseudotriton ruber nitidus*, Watauga Co., North Carolina (R. W. Van Devender).

PLATE 100. *Stereochilus marginatus*, Bladen Co., North Carolina (R. W. Van Devender).

PLATE 101. *Stereochilus marginatus*, Bladen Co., North Carolina (R. W. Van Devender).

PLATE 102. *Typhlotriton spelaeus*, Christian Co., Missouri (R. W. Van Devender).

PLATE 103. *Typhlotriton spelaeus*, Stone Co., Missouri (R. W. Van Devender).

PLATE 104. *Aneides aeneus*, eastern Kentucky (R. W. Barbour).

PLATE 105. *Aneides ferreus*, Lane Co., Oregon (W. P. Leonard).

PLATE 106. *Aneides ferreus*, Humboldt Co., California (R. W. Van Devender).

PLATE 107. *Aneides flavipunctatus*, Humboldt Co., California (W. P. Leonard).

PLATE 108. *Aneides flavipunctatus*, Sonoma Co., California (R. W. Van Devender).

PLATE 109. *Aneides hardii*, Otero Co., New Mexico (R. W. Van Devender).

PLATE 110. *Aneides lugubris*, Los Angeles Co., California (R. W. Van Devender).

PLATE 111. *Aneides lugubris*, Humboldt Co., California (W. P. Leonard).

PLATE 112. *Ensatina eschscholtzii klauberi*, California (R. W. Van Devender).

PLATE 113. *Ensatina eschscholtzii platensis*, California (R. W. Van Devender).

PLATE 114. *Ensatina eschscholtzii picta*, Curry Co., Oregon (W. P. Leonard).

PLATE 115. *Plethodon aureolus*, Monroe Co., Tennessee (R. W. Van Devender).

PLATE 116. *Plethodon caddoensis*, collection locality unknown (R. W. Van Devender).

PLATE 117. *Plethodon cinereus*, Shenandoah Co., Virginia (R. W. Van Devender).

PLATE 118. *Plethodon dorsalis*, Edmonson Co., Kentucky (J. W. Petranka).

PLATE 119. *Plethodon dunni*, Multnomah Co., Oregon (W. P. Leonard).

PLATE 120. *Plethodon elongatus*, Siskiyou Co., California (R. W. Van Devender).

PLATE 121. *Plethodon glutinosus*, Washington Co., Virginia (R. W. Van Devender).

PLATE 122. *Plethodon hoffmani*, Monroe Co., West Virginia (R. W. Van Devender).

PLATE 123. *Plethodon hubrichti*, Bedford Co., Virginia (R. W. Van Devender).

PLATE 124. *Plethodon idahoensis*, Kootenai Co., Idaho (W. P. Leonard).

PLATE 125. *Plethodon jordani*, Macon Co., North Carolina (J. W. Petranka).

PLATE 126. *Plethodon jordani*, Oconee Co., South Carolina (R. W. Van Devender).

PLATE 127. *Plethodon jordani*, Swain Co., North Carolina (J. W. Petranka).

PLATE 128. *Plethodon kentucki*, Washington Co., Virginia (R. W. Van Devender).

PLATE 129. *Plethodon larselli*, Skamania Co., Washington (W. P. Leonard).

PLATE 130. *Plethodon neomexicanus*, Sandoval Co., New Mexico (R. W. Van Devender).

PLATE 131. *Plethodon nettingi*, Randolph Co., West Virginia (R. W. Van Devender).

PLATE 132. *Plethodon oconaluftee*, Macon Co., North Carolina (J. W. Petranka).

PLATE 133. *Plethodon ouachitae*, LeFlore Co., Oklahoma (R. W. Van Devender).

PLATE 134. *Plethodon petraeus*, Walker Co., Georgia (R. W. Van Devender).

PLATE 135. *Plethodon punctatus*, Pendleton Co., West Virginia (R. W. Van Devender).

PLATE 136. *Plethodon richmondi*, Jessamine Co., Kentucky (J. W. Petranka).

PLATE 137. *Plethodon serratus*, Macon Co., North Carolina (J. W. Petranka).

PLATE 138. *Plethodon shenandoah*, Page Co., Virginia (J. W. Petranka).

PLATE 139. *Plethodon stormi*, Siskiyou Co., California (W. P. Leonard).

PLATE 140. *Plethodon vandykei*, Clallam Co., Washington (W. P. Leonard).

PLATE 141. *Plethodon vehiculum*, Thurston Co., Washington (W. P. Leonard).

PLATE 142. *Plethodon websteri*, Lee Co., Alabama (R. W. Van Devender).

PLATE 143. *Plethodon wehrlei*, Surry Co., North Carolina (R. W. Van Devender).

PLATE 144. *Plethodon welleri*, Avery Co., North Carolina (R. W. Van Devender).

PLATE 145. *Plethodon yonahlossee*, Avery Co., North Carolina (R. W. Barbour).

PLATE 146. *Necturus alabamensis*, Okaloosa Co., Florida (R. W. Van Devender).

PLATE 147. *Necturus beyeri*, Marion Co., Mississippi (R. W. Van Devender).

PLATE 148. *Necturus lewisi*, Wake Co., North Carolina (R. W. Van Devender).

PLATE 149. *Necturus m. maculosus*, Franklin Co., Ohio (R. W. Van Devender).

PLATE 150. *Necturus m. maculosus*, Tazewell Co., Virginia (R. W. Van Devender).

PLATE 151. *Necturus punctatus*, Moore Co., North Carolina (R. W. Van Devender).

PLATE 152. *Rhyacotriton cascadae*, Skamania Co., Washington (W. P. Leonard).

PLATE 153. *Rhyacotriton cascadae*, Skamania Co., Washington (W. P. Leonard).

PLATE 154. *Rhyacotriton kezeri*, , Pacific Co., Washington (W. P. Leonard).

PLATE 155. *Rhyacotriton olympicus*, Mason Co., Washington (W. P. Leonard).

PLATE 156. *Rhyacotriton variegatus*, Lane Co., Oregon (W. P. Leonard).

PLATE 157. *Notophthalmus m. meridionalis*, Tamaulipas, Mexico (R. W. Van Devender).

PLATE 158. *Notophthalmus perstriatus*, Leon Co., Florida (R. W. Van Devender).

PLATE 159. *Notophthalmus viridescens dorsalis*, Scotland Co., North Carolina (R. W. Van Devender).

PLATE 160. *Notophthalmus v. viridescens*, Bell Co., Kentucky (R. W. Barbour).

PLATE 161. *Notophthalmus v. viridescens*, Harlan Co., Kentucky (R. W. Van Devender).

PLATE 162. *Notophthalmus viridescens dorsalis*, Scotland Co., North Carolina (R. W. Van Devender).

PLATE 163. *Taricha g. granulosa*, Thurston Co., Washington (W. P. Leonard).

PLATE 164. *Taricha g. granulosa*, Klickitat Co., Washington (W. P. Leonard).

PLATE 165. *Taricha rivularis*, California (E. D. Brodie, Jr.).

PLATE 166. *Taricha t. torosa*, California (R. W. Van Devender).

PLATE 167. *Taricha t. torosa*, southern California (W. Swalling, courtesy of S. Cooper and L. B. Kats).

PLATE 168. *Pseudobranchus a. axanthus*, Putnam Co., Florida (R. W. Van Devender).

PLATE 169. *Pseudobranchus striatus spheniscus*, Madison Co., Florida (R. W. Van Devender).

PLATE 170. *Siren intermedia texana*, Texas (R. W. Van Devender).

PLATE 171. *Siren i. intermedia*, Richmond Co., North Carolina (R. W. Van Devender).

PLATE 172. *Siren lacertina*, Florida (R. W. Van Devender).

Taxonomic Index

The following index contains references to genera and species of animals in the text. Taxa other than amphibians are listed only to the genus level, with the common name in parentheses. References to amphibians are listed by species. For salamanders, references to primary species accounts are in **boldface** type.

Acinetobacter (bacterium), 119
Acroneuria (stonefly), 187
Agkistrodon (cottonmouth and copperhead), 134, 138, 303, 360
Alabama waterdog. See *Necturus alabamensis*
Allegheny mountain dusky salamander. See *Desmognathus ochrophaeus*
Alligator (American alligator), 492
Ambystoma, 35
    annulatum, 21, 33, 36, **37-40,** plate 1
    barbouri, 22, 33, 36, **40-46,** 103, 246, plate 2
    californiense, 21, 30, 36, **47-50,** 109, plate 3
    cingulatum, 22, 33, 36, **50-53,** 68, plates 4, 5
    gracile, 22, 30, 36, **53-58,** 154, 468, 469, plates 6, 7
    jeffersonianum, 5, 17, 22, 34, **58-63,** 67, 73, 81, 85-87, 92, 122-129, plate 8
    laterale, 17, 22, 34, 61, **63-67,** 81, 107, 121-129, plate 9
    mabeei, 22, 33, 36, **68-70,** plates 10, 11
    macrodactylum, 21, 30, 36, 56, **70-75,** plates 12, 13
    maculatum, 4, 12, 21, 34, 35, 63, **76-87,** 92, 107, 346, plates 14, 15
    nothagenes, 122
    opacum, 21, 33, 35, 52, 63, 82, 85, 86, **88-96,** 107, plate 16
    platineum, 22, 34, 36, 122-126, plate 25
    talpoideum, 22, 33, 36, 50, 84, 86, 87, 90, 95, **96-102,** 107, 272, 462, plates 17, 18
    texanum, 22, 33, 36, 40, 41, 43, 44, 46, 67, 92, **103-107,** 121-129, plate 19
    tigrinum, 13, 17, 21, 30, 33, 35, 36, 47, 48, 70, 90, 99, 107, **108-122,** plates 21-24
    tremblayi, 22, 34, 121-125, **126-129,** plate 26
    unisexual *Ambystoma*, 22, 34, 62, 67, 107, **122-129**
Ambystomatidae, 35, 36
Amphiuma, 131
    means, 19, 29, **132-134,** 135, plate 27
    pholeter, 19, 29, **134-135,** plate 28
    tridactylum, 19, 29, 132, 135, **136-138,** plate 29
Amphiumidae, 131
Aneides, 157
    aeneus, 12, 26, **310-314,** plate 104
    ferreus, 25, **314-318,** plates 105, 106
    flavipunctatus, 25, **318-320,** plates 107, 108
    hardii, 25, **320-322,** plate 109
    lugubris, 25, 223, **322-325,** plates 110, 111
Apalachicola dusky salamander. See *Desmognathus apalachicolae*
Arboreal salamander. See *Aneides lugubris*
Ascaphus truei (tailed frog), 147, 156, 440

Banksiola (caddisfly), 84
Barton Springs salamander. See *Eurycea sosorum*
Batrachoseps, 324
    aridus, 23, **219-220,** plate 59
    attenuatus, 23, **220-223,** 325, plate 60
    campi, 23, **224-225,** plate 61
    gabrieli, 23, **225-226**
    nigriventris, 23, **226-229,** 231, 232, plate 62
    pacificus, 23, 221, 225, **228-231,** plate 63
    simatus, 23, **231-232,** plate 64
    stebbinsi, 23, 225, **232-233,** plate 65
    wrighti, 23, 225, **234-235,** plate 66
Black-bellied salamander. See *Desmognathus quadramaculatus*
Black-bellied slender salamander. See *Batrachoseps nigriventris*
Black Mountain salamander. See *Desmognathus welteri*
Black salamander. See *Aneides flavipunctatus*
Black-spotted newt. See *Notophthalmus meridionalis*
Blanco blind salamander. See *Eurycea robusta*
Blarina (shrew), 85, 210, 248, 372
Blue Ridge dusky salamander. See *Desmognathus orestes*
Blue-spotted salamander. See *Ambystoma laterale*
Branchinecta (fairy shrimp), 115

Taxonomic Index

Bufo
    americanus (American toad), 96, 115, 461
    boreas (western toad), 15
    canorus (Yosemite toad), 15
    hemiophrys (Canadian toad), 15
Buteo (red-tailed hawk), 461

Caddo Mountain salamander. See *Plethodon caddoensis*
Calidris (least sandpiper), 84
California giant salamander. See *Dicamptodon ensatus*
California newt. See *Taricha torosa*
California slender salamander. See *Batrachoseps attenuatus*
California tiger salamander. See *Ambystoma californiense*
Campostoma (stoneroller), 144
Canis (coyote), 120
Carolina dusky salamander. See *Desmognathus carolinensis*
Cascade torrent salamander. See *Rhyacotriton cascadae*
Catostomus (sucker), 144
Cave salamander. See *Eurycea lucifuga*
Chaoborus (phantom midge), 115, 121
Charadrius (killdeer), 461
Chauliognathus (soldier beetle), 107
Cheat Mountain salamander. See *Plethodon nettingi*
Chelydra (snapping turtle), 138, 461
Chrysemys (painted turtle), 461
Cinclus (American dipper), 406
Clouded salamander. See *Aneides ferreus*
Cnemidophorus (lizard), 120
Coeur d'Alene salamander. See *Plethodon idahoensis*
Columbia torrent salamander. See *Rhyacotriton kezeri*
Comal blind salamander. See *Eurycea tridentifera*
Cope's giant salamander. See *Dicamptodon copei*
Cottus (sculpin), 144
Cryptobranchidae, 139
Cryptobranchus alleganiensis, 19, 29, 139, **140-144**, plate 30
Cumberland Plateau salamander. See *Plethodon kentucki*
Cyanocitta (Stellar's jay), 331, 351

Daphnia (water flea), 106, 115, 121
Del Norte salamander. See *Plethodon elongatus*
Desert slender salamander. See *Batrachoseps aridus*
Desmognathus, 157, 428
    aeneus, 23, **159-161**, 183, 214, 215, plate 38
    apalachicolae, 24, 34, **162-164**, 166, plate 39
    auriculatus, 24, 34, **164-166**, plates 40, 41
    brimleyorum, 24, 34, 164, **167-169**, 195, plate 42
    carolinensis, 24, 34, **169-173**, plate 43
    fuscus, 17, 24, 34, 162, 164, 166, **173-181**, 191, 195, 196, 199, 201, 204, 211, 286, plates 44, 45
    imitator, 24, 34, **182-184**, 372-374, plate 46
    marmoratus, 24, 34, **184-187**, plates 47, 48
    monticola, 5, 17, 24, 34, 164, 173, 181, **187-192**, 196, 199, 201, 210, 211, plate 49, 50
    ochrophaeus, 24, 34, 169, 170, 175, 181, **192-196**, 201, 248, 346, 364, plate 51
    ocoee, 24, 34, 162, 169-171, 182, 184, **196-202**, 210, 285, 286, 372-374, plates 52, 53
    orestes, 24, 34, 162, 169, 172, **202-205**, 210, plate 54
    quadramaculatus, 5, 7, 12, 17, 24, 34, 173, 181, 184, 185, 187, 188, 192, 201, 202, **206-211**, 248, 287, plate 55
    santeetlah. See *D. fuscus*
    welteri, 24, 34, **211-213**, plate 56
    wrighti, 23, 159, 161, 173, 183, 210, **213-216**, 286, 372, 373, plate 57
Diadophis (ringneck snake), 195, 199, 223, 247, 248, 339, 345, 349, 395, 411
Diaptomus (copepod), 121
Dicamptodon, 145
    aterrimus, 21, 30, **146-147**, plate 31
    copei, 21, 30, **147-149**, plates 32, 33
    ensatus, 21, 30, 147, **150-151**, 152, 153, plate 34
    tenebrosus, 7, 21, 30, 147-150, **152-156**, 439, plates 35-37
Dicamptodontidae, 145
Dunn's salamander. See *Plethodon dunni*
Dusky salamander. See *Desmognathus fuscus*
Dwarf salamander. See *Eurycea quadridigitata*
Dwarf waterdog. See *Necturus punctatus*
Dytiscus (predaceous diving beetle), 119, 461

Eastern newt. See *Notophthalmus viridescens*
Enneacanthus (banded sunfish), 462
Ensatina. See *Ensatina eschscholtzii*
Ensatina eschscholtzii, 22, 157, **325-331**, plates 112-114
Etheostoma (darter), 266
Eurycea, 157
    aquatica, 241, 242
    bislineata, 4, 5, 28, 32, 181, 187, 195, 205, 210, 211, 217, **241-248**, 251-253, 285-287, 298, 373-374, 428, plates 70, 71
    guttolineata, 28, 32, 217, **249-251**, 254, plate 72
    junaluska, 28, 32, 242, **251-254**, plates 73, 74
    latitans. See *E. neotenes*
    longicauda, 28, 32, **254-257**, 258, 261, plates 75-77
    lucifuga, 28, 32, **258-261**, 360, plate 78
    multiplicata, 7, 28, 32, **262-264**, 279, plate 79
    nana, 25, 32, **264-266**, plate 80
    neotenes, 25, 32, **266-268**, plate 81
    pterophila. See *E. neotenes*
    quadridigitata, 5, 28, 31, 100, **269-272**, plate 82
    rathbuni, 25, 31, **272-274**, plate 83
    robusta, 25, 31, 275

sosorum, 25, 31, **276-277,** plate 84
tridentifera, 25, 31, 267, **277-278,** plate 85
troglodytes. See E. neotenes
tynerensis, 2, 28, 32, 262, **278-279,** plate 86

Farancia (mud and rainbow snake), 134, 138, 482
Flatwoods salamander. *See Ambystoma cingulatum*
Four-toed salamander. *See Hemidactylium scutatum*

Gambusia (mosquitofish), 478
Georgia blind salamander. *See Haideotriton wallacei*
Gordius (horsehair worm), 251
Greater siren. *See Siren lacertina*
Green salamander. *See Aneides aeneus*
Grotto salamander. *See Typhlotriton spelaeus*
Gulf Coast waterdog. *See Necturus heyeri*
Gyrinophilus, 157
　　palleucus, 28, 32, **280-282,** plate 87
　　porphyriticus, 28, 33, 173, 181, 199, 201, 211, 248, **282-287,** 304, plates 88, 89
　　subterraneus, 28, 32, **287-288,** plates 90, 91

Haideotriton wallacei, 24, 32, 157, **289-290,** plate 92
Hellbender. *See Cryptobranchus alleganiensis*
Hemidactylium scutatum, 12, 22, 31, 157, **290-295,** plate 93
Heterodon (hognose snake), 120, 461
Hydromantes, 158
　　brunus, 25, **236-237,** plate 67
　　platycephalus, 25, **237-238,** plate 68
　　shastae, 25, **239-240,** plate 69
Hyla
　　andersonii, 96
　　arenicolor, 115
　　chrysoscelis, 87, 92
　　versicolor, 92, 119

Ichthyomyzon (lamprey), 144
Ictalurus (bullhead), 266
Idaho giant salamander. *See Dicamptodon aterrimus*
Imitator salamander. *See Desmognathus imitator*
Inyo Mountains slender salamander. *See Batrachoseps campi*

Jefferson salamander. *See Ambystoma jeffersonianum*
Jemez Mountains salamander. *See Plethodon neomexicanus*
Jordan's salamander. *See Plethodon jordani*
Junaluska salamander. *See Eurycea junaluska*

Kern Canyon slender salamander. *See Batrachoseps simatus*

Lampropeltis (kingsnake), 168
Larch Mountain salamander. *See Plethodon larselli*
Lepomis (sunfish), 43, 46, 84, 86, 87, 102, 266
Lesser siren. *See Siren intermedia*
Leurognathus marmoratus. See Desmognathus marmoratus
Limestone salamander. *See Hydromantes brunus*
Lirceus (isopod), 44, 45, 308, 309
Long-tailed salamander. *See Eurycea longicauda*
Long-toed salamander. *See Ambystoma macrodactylum*
Lutra (river otter), 156
Lynx (bobcat), 120

Mabee's salamander. *See Ambystoma mabeei*
Many-lined salamander. *See Stereochilus marginatus*
Many-ribbed salamander. *See Eurycea multiplicata*
Marbled salamander. *See Ambystoma opacum*
Micropterus (bass), 266, 461
Microtus (vole), 156
Mole salamander. *See Ambystoma talpoideum*
Mount Lyell salamander. *See Hydromantes platycephalus*
Mudpuppy. *See Necturus maculosus*
Mud salamander. *See Pseudotriton montanus*
Mustela (weasel), 156

Necturus, 417
　　alabamensis, 20, 31, **418-419,** plate 146
　　beyeri, 20, 31, 418, **419-422,** plate 147
　　lewisi, 20, 31, 418, **422-425,** 430, 431, plate 148
　　maculosus, 12, 20, 31, 418, **425-429,** 430, plates 149, 150
　　punctatus, 20, 31, 418, 424, 425, **429-431,** plate 151
Nerodia (water snake), 45, 107, 134, 181, 187, 286, 298, 429, 482 *Neurotrichus* (shrew-mole), 406
Neuse River waterdog. *See Necturus lewisi*
Northern dwarf siren. *See Pseudobranchus striatus*
Northwestern salamander. *See Ambystoma gracile*
Notemigonus (shiner), 144
Notophthalmus, 445
　　meridionalis, 4, 20, 33, **446-447,** plate 157
　　gerstriatus, 20, 33, 446, **448-451,** plate 158
　　viridescens, 4, 5, 12, 20, 33, 96, 100, 121, 195, 217, 272, 303, 304, 428, 446, 448, 450, **451-462,** 489, plates 159-162
Notropis (shiner), 144

Ocoee salamander. *See Desmognathus ocoee*
Oklahoma salamander. *See Eurycea tynerensis*

Olympic torrent salamander. *See Rhyacotriton olympicus*
Oncorhynchus (rainbow trout), 149
One-toed amphiuma. *See Amphiuma pholeter*
Ophiotaenia (tapeworm), 308
Otus (screech owl), 247, 349
Oregon slender salamander. *See Batrachoseps wrighti*
Ouachita dusky salamander. *See Desmognathus brimleyorum*

Pacific giant salamander. *See Dicamptodon tenebrosus*
Pacific slender salamander. *See Batrachoseps pacificus*
Palaemonetes (shrimp), 274
Parachironomus (midge), 84
Peaks of Otter salamander. *See Plethodon hubrichti*
Percina (logperch), 144
Phaeognathus hubrichti, 24, **216-218**, plate 58
Phagocotus (flatworm), 45
Pigeon Mountain salamander. *See Plethodon petraeus*
Plethodon, 157, 158
    albagula. See P. glutinosus
    aureolus, 27, 158, **332-333**, 355, 358, 369, 385, plate 115
    caddoensis, 26, 158, **333-335**, 387, 389, plate 116
    chattahoochee. See P. glutinosus
    chlorobryonis. See P. glutinosus
    cinereus, 12, 28, 83, 158, 195, 196, 205, 302, **335-346**, 349, 363, 374, 392, 395, 398, 399, 410, 411, plate 117
    cylindraceus. See P. glutinosus
    dixi. See P. wehrlei
    dorsalis, 27, 158, **346-349**, 407-409, plate 118
    dunni, 26, **349-352**, 403, 404, 407, plate 119
    elongatus, 25, 158, **352-354**, 399, plate 120
    fourchensis. See P. ouachitae
    glutinosus, 27, 158, 217, 261, 313, 332, 346, **354-361**, 368, 375, 376, 384, 385, 388, 389, 415, 416, plate 121
    gordani, 350
    grobmani. See P. glutinosus
    hoffmani, 27, 158, 345, 346, **361-363**, 392, plate 122
    hubrichti, 27, 158, 195, **363-365**, 374, plate 123
    idahoensis, 26, 158, **365-367**, 401, 402, plate 124
    jacksoni. See P. wehrlei
    jordani, 4, 27, 158, 210, 286, 332, 356, 358, **367-374**, 384-386, 397, 415, 416, plates 125-127
    kentucki, 27, 158, 313, 355, 358, **374-377**, plate 128
    kiamichi. See P. glutinosus
    kisatchie. See P. glutinosus
    larselli, 26, 158, **377-379**, plate 129
    longicrus, **414-415**
    mississippi. See P. glutinosus
    neomexicanus, 25, 158, **380-381**, plate 130
    nettingi, 27, 158, 205, 363, 374, **381-383**, plate 131
    ocmulgee. See P. glutinosus
    oconaluftee, 27, 158, 286, 332, 333, 355, 356, 368, 369, 372, 373, **383-386**, 397, plate 132
    ouachitae, 26, 158, **386-389**, plate 133
    petraeus, 26, 158, **389-390**, plate 134
    punctatus, 27, 158, 363, **390-392**, plate 135
    richmondi, 27, 158, 195, 363, **392-395**, plate 136
    savannah. See P. glutinosus
    sequoyah. See P. glutinosus
    serratus, 28, 158, 286, 336, 372, 373, **395-397**, plate 137
    shenandoah, 27, 158, 344, 345, 363, **397-399**, plate 138
    stormi, 25, 158, **399-401**, plate 139
    teyahalee. See P. oconaluftee
    vandykei, 26, 158, 351, **401-403**, 407, plate 140
    variolatus. See P. glutinosus
    vehiculum, 26, 158, 351, 352, 354, **403-407**, plate 141
    websteri, 27, 158, **407-409**, plate 142
    wehrlei, 27, 158, 346, 390, 391, **409-411**, plate 143
    welleri, 26, 158, 205, 374, **412-414**, plate 144
    yonahlossee, 26, 158, **414-416**, plate 145
Plethodontidae, 157, 158
Procambarus (crayfish), 478
Procyon (raccoon), 62, 461
Proteidae, 417
Pseudacris
    crucifer, 96, 450
    regilla, 48, 74
    triseriata, 92, 115, 118, 121
Pseudobranchus, 479
    axanthas, 19, 29, **480-482**, 483, plate 168
    striatus, 19, 29, 480, **482-484**, plate 169
Pseudotriton, 157
    montanus, 28, 33, **295-298**, 303, 304, plates 94-96
    ruber, 28, 33, 286, 296, **299-304**, plates 97-99
Ptilostomis (caddisfly), 84
Pygmy salamander. *See Desmognathus wrighti*

Rana
    aurora (red-legged frog), 48, 49, 468
    cascadae (Cascades frog), 15
    catesbeiana (bullfrog), 75, 114, 461
    clamitans (green frog), 114
    muscosa (mountain yellow-legged frog), 15
    sylvatica (wood frog), 63, 84, 86, 92, 115, 121, 462
Ravine salamander. *See Plethodon richmondi*
Red-backed salamander. *See Plethodon cinereus*
Red-bellied newt. *See Taricha rivularis*
Red Hills salamander. *See Phaeognathus hubrichti*
Red salamander. *See Pseudotriton ruber*
Regina (crayfish snake), 482
Rhyacotriton, 145, 404, 433
    cascadae, 20, 30, **434-437**, plates 152, 153
    kezeri, 21, 31, 434, 435, **437-439**, 443, plate 154

olympicus, 20, 30, 435, **439–440,** plate 155
variegatus, 20, 31, 434, 435, 437, 438, **441–443,** plate 156
Rhyacotritonidae, 433
Rich Mountain salamander. *See Plethodon ouachitae*
Ringed salamander. *See Ambystoma annulatum*
Rough-skinned newt. *See Taricha granulosa*

Sacramento Mountain salamander. *See Aneides hardii*
Salamandridae, 445
Salmo (trout), 144
Salvelinus (brook trout), 248
San Gabriel Mountain Slender Salamander. *See Batrachoseps gabrieli*
San Marcos salamander. *See Eurycea nana*
Scapanus (mole), 75
Scaphiopus (=*Spea;* spadefoot toad), 115, 120
Seal salamander. *See Desmognathus monticola*
Seepage salamander. *See Desmognathus aeneus*
Semotilus (creek chub), 461
Shasta salamander. *See Hydromantes shastae*
Shenandoah salamander. *See Plethodon shenandoah*
Shovel-nosed salamander. *See Leurognathus marmoratus*
Siren, 479
  intermedia, 19, 29, 462, **484–489,** plates 170, 171
  lacertina, 19, 29, **489–492,** plate 172
Sirenidae, 479
Siskiyou Mountains salamander. *See Plethodon stormi*
Sistrurus (pygmy ratdesnake), 294
Slimy salamander. *See Plethodon glutinosus*
Small-mouthed salamander. *See Ambystoma texanum*
Sorex (shrew), 149, 156
Southern Appalachian salamander. *See Plethodon oconaluftee*
Southern dusky salamander. *See Desmognathus auriculatus*
Southern dwarf siren. *See Pseudobranchus axanthus*
Southern red-backed salamander. *See Plethodon serratus*
Southern torrent salamander. *See Rhyacotriton variegatus*
Southern zig-zag salamander. *See Plethodon websteri*
Spermophilus (ground squirrel), 49
Spotted salamander. *See Ambystoma maculatum*
Spring salamander. *See Gyrinophilus porphyriticus*
Stenopelmatus (Jerusalem cricket), 231
Stereochilus marginatus, 24, 32, **304–306,** plates 100, 101
Streamside salamander. *See Ambystoma barbouri*

Striped newt. *See Notophthalmus perstriatus*

Taricha, 328, 445
  granulosa, 20, 31,154, **462–469,** 471, 472, plates 163, 164
  rivularis, 20, 31, 467, **469–473,** plate 165
  torosa, 20, 31, 463, 469, 471, **473–478,** plates 166, 167
Taxidea (badger), 120
Tehachapi slender salamander. *See Batrachoseps stebbinsi*
Tellico salamander. *See Plethodon aureolus*
Tennessee cave salamander. *See Gyrinophilus palleucus*
Tetrahymena (protozoan), 84
Texas blind salamander. *See Eurycea rathbuni*
Texas salamander. *See Eurycea neotenes*
Thamnophis (garter snake), 63, 75, 95, 107, 120, 149, 156, 181, 210, 223, 247, 248, 286, 298, 320, 325, 331, 345, 351, 360, 367, 372, 461, 468
Thomomys (pocket gopher), 49
Three-lined salamander. *See Eurycea guttolineata*
Three-toed amphiuma. *See Amphiuma tridactylum*
Tiger salamander. *See Ambystoma tigrinum*
Trypanosoma (protozoan), 458
Two-lined salamander. *See Eurycea bislineata*
Two-toed amphiuma. *See Amphiuma means*
Typhlomolge, 273, 277
Typhlotriton spelaeus, 26, 32, 157, **307–309,** plates 102, 103

Valley and Ridge salamander. *See Plethodon hoffmani*
Van Dyke's salamander. *See Plethodon vandykei*
Virginia (earth snake), 63, 85, 102

Wehrle's salamander. *See Plethodon wehrlei*
Weller's salamander. *See Plethodon welleri*
Western red-backed salamander. *See Plethodon vehiculum*
West Virginia spring salamander. *See Gyrinophilus subterraneus*
White spotted salamander. *See Plethodon punctatus*

Yonahlossee salamander. *See Plethodon yonahlossee*

Zig-zag salamander. *See Plethodon dorsalis*